Special Publication No. 4
of the Society for Geology
Applied to Mineral Deposits

Geology and Metallogeny of Copper Deposits

Proceedings of the Copper Symposium
27th International Geological Congress
Moscow, 1984

Edited by
G.H. Friedrich A.D. Genkin A.J. Naldrett
J.D. Ridge R.H. Sillitoe F.M. Vokes

With 258 Figures

Springer-Verlag
Berlin Heidelberg New York
London Paris Tokyo

Professor
GÜNTHER H. FRIEDRICH
Institut für Mineralogie
und Lagerstättenlehre
RWTH Aachen
Wüllnerstraße 2
5100 Aachen, FRG

Professor
ALEXANDR D. GENKIN
CCCP/IGEM
Staromonetny, 35
Moscow, 109017, USSR

Professor
ANTHONY J. NALDRETT
Department of Geology
University of Toronto
Toronto, Ontario
Canada M5S 1A1

Professor
JOHN D. RIDGE
1402 NW 18th Street
Gainesville, FL 32605, USA

Dr. RICHARD H. SILLITOE
8 West Hill Park
Highgate Village
London N6 6ND
England

Professor
FRANK M. VOKES
Geologisk Institutt
Norges Tekniske Høgskole
Universitetet i Trondheim
7034 Trondheim
Norway

ISBN 3-540-16101-5 Springer-Verlag Berlin Heidelberg New York
ISBN 0-387-16101-5 Springer-Verlag New York Heidelberg Berlin

Library of Congress Cataloging in Publication Data. Geology and metallogeny of copper deposits. (Special publication no. 4 of the Society for Geology Applied to Mineral Deposits) Proceedings of a symposium on copper metallogeny held at the 27th International Geological Congress, Moscow, USSR. 1. Copper Ores–Congresses. I. Friedrich, Günther, 1929–. II. International Geological Congress (27th : Moscow, USSR) III. Series: Special publication . . . of the Society for Geology Applied to Mineral Deposits ; no. 4. QE390.2.C6G46 1986 553.4'3 86-6745

Typesetting: K + V Fotosatz GmbH, Beerfelden.
Offsetprinting and bookbinding: Konrad Triltsch, Graphischer Betrieb, Würzburg.
2132/3130-543210

Preface

Copper belongs to those metals whose concentrations in nature arise from a broad diversity of endogeneous and exogeneous processes, which applies to essentially all genetic classes of ore deposits.

This is the first proceedings volume on copper metallogeny to cover the worldwide distribution of the four main groups of copper deposits, including in Part I: copper-nickel deposits with cobalt and platinum group elements; Part II: copper-molybdenum-gold deposits with silver, zinc, and lead; Part III/IV: copper-zinc-lead deposits (with silver etc.).

On the occasion of the 27th International Geological Congress in Moscow, USSR, a symposium on copper metallogeny was held, dealing with metallogenesis and mineral deposits. The symposium was organized and sponsored by three international societies engaged in the field of ore deposits: The Society of Economic Geology (SEG), the Society of Geology Applied to Mineral Deposits (SGA) and the International Association on the Genesis of Ore Deposits (IAGOD). Invited papers were presented in four sessions: (1) Copper deposits in mafic and ultramafic complexes, (2) Porphyry copper deposits, (3) Copper deposits of volcanic-hydrothermal association, and (4) Sediment-hosted copper deposits. The sessions were chaired by A. D. Genkin, A. J. Naldrett, J. D. Ridge and G. I. Gorbunov; V. I. Sotnikov, A. Soregaroli, R. H. Sillitoe and V. A. Evstrakhin; F. M. Vokes, A. I. Krivtsov, M. Solomon and N. I. Eremin; G. H. Friedrich, Yu. V. Bogdanov, A. C. Brown and F. P. Krendelev.

We wish to express our thanks to the publisher for the work involved in maintaining the known Springer quality of composing and printing. Beside the authors' work, the expenditure of effort and time by numerous referees should be mentioned, who reviewed the manuscripts in detail, especially those from non-English speaking countries. They also proposed improvements to the English translations of foreign manuscripts.

The Editors

Contents

List of Contributors

You will find the addresses at the beginning of the respective contribution

Introduction

The symposium on copper deposits and the publication of its Proceedings are a good example of international scientific cooperation. The extensive exchange of opinions between specialists in different types of copper deposits has provided important new data on their geological setting and condition of formation.

A remarkable number of authors writing about Cu – Ni deposits in Part I of this volume reach the conclusion that these are the consequence of interaction of mafic or ultramafic magma and near-surface country rocks. *Foose* and *Weiblen* draw attention to the contrast between the organized nature of the unmineralized rocks forming the main part of the Kawishiwi intrusion of the Duluth Complex, which have crystallized as a consequence of repetitive influxes of new magma, and the heterogeneous nature of the sulfide-bearing rocks, which show indications of having reacted with volatiles to varying degrees in different places. The volatiles are believed to have been derived from country rocks, and a model is presented of rocks intruding a dynamic tectonic environment during rifting. *Ripley*'s *S, O, H* and *C* isotopic data also indicate that devolatilization of country rocks has accompanied the introduction of sulfur to mafic intrusions at Duluth. At the Dunkha Road deposit, this has occurred in situ, but at Babbitt, devolatilization and partial melting of country is believed to have occurred in the feeder to the present deposit. *Lesher* and *Groves* emphasize the association of mineralized komatiites with sulfide-bearing sediments amongst the country rocks. They point to the ability of komatiites to erode thermally sediments and basalts on which they extrude, and propose that erosion and assimilation of sulfide-bearing sediments makes a major contribution to the sulfides of the ore zones, while erosion of basalt produces the structures in which the ores reside. *Gresham* takes a different view, stressing the contrast between ore-bearing and ore-free zones of komatiites, and proposing that the ore-bearing units are related to rift-produced irregularities in the pre-komatiite volcanic topography. Reporting a study of the Sudbury structure and its rocks, *Naldrett* demonstrates that the Sudbury magmas have assimilated in excess of 50% of country rocks. He proposes that this massive contamination has been responsible for "salting out" sulfides from mafic magma, thus accounting for the high incidence of deposits at Sudbury.

In contrast to these views emphasizing a near-surface origin for immiscible sulfides, both *Distler, Genkin,* and *Dyuzhikov* as well as *Godlevsky* and *Likhachev* emphasize the separation of liquid sulfides from associated mafic magma at depth. All of these authors propose that sulfide-rich magma was introduced separately to the main mass of the associated intrusion. *Distler* et al. explain zoning at Talnakh partly as the consequence of immiscibility between two sulfide liquids. *Godlevsky* and *Likhachev* believe that the shape of a mafic intrusion, chonolithic or lacolithic, depends on the relative density of the magma with respect to its country rocks, which in turn is a function of how much sulfide the magma contains.

Yakovlev and *Yakovleva* emphasize the magmatic origins of deposits of the Kola peninsula, pointing out that metamorphism has changed them, but has not generated any deposits on its own. *Hanski* draws parallels between the gabbro-wehrlite rocks hosting the Proterozoic Pechanga deposits of Kola and those of the Archean Kuhmo belt in Finland, but points out that the latter have distinctive chemical compositions.

Kochnev-Pervukhov, Zaskind and *Proskuryakov* compare komatiites of the SE Baltic shield with those elsewhere and relate the presence of mineralization to the composition of the lavas and to their tectonic setting. *Fan* describes for the first time in western literature chromite and nickel deposits of the People's Republic of China.

The section on porphyry copper deposits, Part II, commences with *Guilbert*'s overview of recent progress in the understanding of porphyry base-metal deposits in the western Americas.

Emphasis is placed on the plate-tectonic settings of porphyry deposits, and on the histories, compositions, sources and flow patterns, and physicochemistry of the mineralizing fluids. A Soviet view of selected physicochemical parameters involved in porphyry copper formation is then presented in two contributions by *Kudrin, Varyash, Pashkov*, and *Rekharsky*, and by *Sotnikov* and *Berzina*. Using northern Mexico as an example, *Damon* presents a hypothesis to explain observed spatial and temporal relationships between batholiths, stocks, and porphyry copper deposits. *Sillitoe* explores the metallogenic implications of the space-time distribution, crustal setting and Cu/Mo ratios of porphyry copper deposits in the central Andes. In contrast, *Evstrakhin, Krivtsov*, and *Migachyov* provide a lithotectonic rather than a plate-tectonic analysis of selected porphyry copper provinces in the USSR. *Golovanov, Nikolayeva*, and *Kazikhin* appraise the geologic controls of high-grade core zones in porphyry copper deposits, using the Almalyk deposits in the USSR as an example. The porphyry copper section is concluded with descriptions of the regional and local geology of the Erdenetuin-Obo deposit in Mongolia, and *Zhamsran, Sotnikov, Berzina, Garamzhav, Saryanov*, and the Recsk deposit in Hungary by *Baksa*.

Part III was planned to comprise a series of papers dealing with sulfide ore deposits hosted in volcanic and volcanosedimentary complexes, characterized metallurgically, chiefly by copper, with varying proportions of zinc and lead, with or without precious metal contents. Mineralogically, the deposits are dominated by the iron sulfides, pyrite and/or pyrrhotite.

These ores, as envisaged by the conveners of the Session, are mainly of the stratabound volcanic-exhalative type and comprise both concordant, often stratiform, often massive, components and cross-cutting, disperse to semi-massive, components representing feeder zones to the massive ores.

By mutual agreement, magmatic ores characterized by Ni – (Cu) occurring in mafic and ultramafic volcanic sequences (Kambalda type) were not included in the papers chosen for this Session.

The final selection of papers published here is, through various causes, much less comprehensive than those originally contemplated, and does not represent all the papers delivered during Session III at the 27th IGC in Moscow on August 1984. The coverage of volcanite-hosted stratabound sulfide ores of Cu – (Zn – Pb) as defined above is admittedly very patchy, both as regards geographical distribution and subject matter.

However, the papers now presented cover a fair range of aspects of the ores defined above. There are descriptions and discussions of two important European stratabound sulfide ore districts, as well as a group of papers which discuss central genetical aspects of the ores, touching on such matters as physicochemical conditions of formation and the causes of chemical and mineralogical zoning both on district- and individual ore deposit scales.

Of the ore district descriptions, the chapter by *Strauss* and *Gray* deals with one of the classical copper ore fields of western Europe, where exploitation began in pre-Roman times, viz., the Iberian Pyrite Belt.

The chapter is in the form of a general review of the geology of the pyrite belt and of the main ore deposits, including a description of the various types of sulfide ore encountered. The ores are considered to be closely linked with a single, submarine, acid-alkaline volcanic cycle within a volcanic-sedimentary complex of Upper Visean to Lower Westphalian age.

The chapter by *Kurbanov* analyses the metallogeny of two pairs of ore-bearing zones in the Caucasian segment of the Mediterranean fold belt, each apparently characterized by sulfide ores of different paleotectonic settings and related to different types of volcanism. The relation of the pyritic deposits to porphyry and vein-type copper deposits can be shown.

The chemical and mineralogical zoning which is exhibited by base-metal-bearing stratabound sulfide ores in many parts of the world is dealt with in other chapters by Russian authors. *Eremin* discusses what he refers to as the "hidden mineralogical and geochemical zoning", which he recognizes in ore deposits from USSR and Japanese ore districts. Two types of zoning are distingushed; repeated, exhibited by hydrothermal replacement ores, and nonrepeated, exhibited by ores of volcanic-sedimentary origin. The latter type of zoning is attributed to variations in temperature and sulfur activity during ore deposition.

Skripchenko, Shcheglov, Ryabov, and *Andreyev* also discuss lateral and vertical zoning shown by individual ores hosted in volcanic and sedimentary sequences. Two models to explain this zoning are presented, relating respectively to volcanite-hosted and to sediment-volcanite-hosted ores.

The origin of base-metal zonation of sulfide ores deposited on the sea floor is also touched on by *Tvalchrelidze* in a consideration of the physicochemical conditions of their formation. This author assigns an important role here to the reaction between pore waters and sulfide rich sediments during diagenesis.

Kalogeropoulos and *Scott* in contrast to the preceeding chapters, deal with a specific restricted zone of one of the ores of the famous Hokuroku ore district in the Eocene Green Tuff Region of Japan. These authors interpret the textural and chemical evidence they find in the barite zone at the Fukazawa mine as indicating the mixing of a circulating barium-rich hydrothermal solution with cold sea water within an unconsolidated pyrite tuff. Of especial interest in respect to the genesis of these ores is their evidence, reinforcing earlier observations, of continuing hydrothermal activity after the "post-ore" hanging wall rocks had been deposited.

Part IV contains 14 chapters from 42 authors and co-authors, dealing with two main subjects. First, mineralogical, petrographical and chemical studies (*Brown* and *Chartrand, Cailteux, Rose* et al., *Sood* et al., *Moine* et al., *Haranczyk* and *Lur'ye*); second, a detailed description and synthesis of copper deposits including genetic interpretations (*Bogdanov, Susura* et al., *Krendelev* et al., *Gablina* and *Lambert* et al.). In addition, results of thermodynamical investigations on sulfide zoning by *Merino* et al. and results of recent exploration of Kupferschiefer-type deposits are presented by *Schmidt* et al.

Brown and *Chartrand* interpret the sequence of mineralization in the sediment-hosted copper deposits at White Pine (Michigan), Redstone (North-West Territories), and Kamoto (Zaire). They suppose an influx of dissolved copper into initially pyritic basal units of grey-bed host rocks during early diagenesis. The plausible sources of copper include the red beds themselves, or pene-exhalative solutions which "exhale" into the red beds.

Cailteux points out the stratiform copper-cobalt mineralization in the Shaban Mines Group with its micro- and macrocycles. This mineralization results from diagenetic processes involving a mixing of stagnant interstitial water with brines which were enriched in metals and migrated through still porous sediments. He additionally gives reaction equations which characterize secondary reactions taking place during chalcocitization and carrolitization.

Rose, Smith, Lustwerk, Ohmoto, and *Hoy* point out geochemical aspects of stratiform redbed copper deposits in the Catskill formation, Pennsylvania, and Redstone Area, North-West Territories. They suppose that the copper minerals were emplaced after earliest diagenesis, but before induration and formation of quartz overgrowth. The copper has been redistributed, depleted from nearly all red (oxidized) sediments. After mobilization of copper from the red beds it has been transported by chloride-bearing pore fluids of marine or evaporative origin. This flow was driven by compaction, sea-level changes, and gravity flow down the alluvial plain.

Sood, Wagner, and *Markazi* describe the characteristics and origin of Kennecott-type copper deposits in the McCarthy Quadrangle, Alaska. They suppose a liberating of copper from the metal oxides and the mafic minerals by hydration-dehydration reactions associated with metamorphism. The circulation and transport of metals took place in meteoric water. The activity in such systems has been driven by the heat of Tertiary intrusives.

Moine, Guilloux, and *Audeoud* investigated the major elements of the host rocks in some Kupferschiefer-type deposits at the Zambian Copperbelt. They presume that the high contents of Mg and K-feldspar cannot be contained by sedimentary processes only and that a large part of these is probably of diagenetic/hydrothermal origin.

Haranczyk shows the complexity of processes active in forming stratiform copper deposits. He gives new data and new ideas, e.g., maps of redox state and paleogeography.

Lur'ye presents a chapter summarizing new data of formation conditions of copper sandstone and shale-type deposits. He points out the relation of ore thickness-grade to underlying lithology and the importance of hydrodynamics, combined with some data of flow rate and rock porosity. Additionally, he believes that there are no copper occurrences where flows of high copper concentration killed sulfate-reducing bacteria.

Bogdanov described major types of copper-bearing zones in the Soviet Union, including orogenic copper-bearing zones, copper-bearing areas in platform regions, and copper-bearing zones of geosynclinal areas. Additionally, he gives information on major epochs of copper accumulation and a prognostic evaluation of copper-bearing areas.

The chapter of *Susura, Glybovsky*, and *Kislitsin* shows interesting ideas on the processes forming the Dzhezkazgan ores, and a summary of sedimentary tectonic environments for numerous sedimentary copper deposits in the western USSR.

Krendelev, Narkelyun, Trubachev, Salikhov, Volodin, Kunitsin, Chechetkin, and *Bakun* show the variety of mineral deposits in the cupriferous sandstones and shales of the Siberian Platform and distinguish six major copper provinces. Within these at least 15 major zones of deposits and occurrences are recognized, occurring in Proterozoic and Paleozoic detrital sediments. The sources of copper are considered to be in the uplifted blocks that supplied the sediments and in volcanic effusives.

Gablina gives an overview of different genetic types of copper mineralization in the Igarka area/Soviet Union. She describes very interesting geological, sedimentological and mineralogical relations of significant copper ores and low grade mineralization.

Lambert, Knutson, Donnelly, and *Etminan* point out a synthesis of features and genesis of various types of sediment-hosted copper Deposits in Australia. Besides hydrothermal fluids associated with igneous activity and regional metamorphism (e.g., Mt. Isa) they describe the widespread, bedded-disseminated mineralization, formed at relatively low temperatures by reactions of cupriferous fluids with biogenic H_2S and early diagenetic iron sulfides.

Merino, Moore, Ortoleva, and *Ripley* give a quantitative kinetic approach, based on systems of thermodynamical and differential equations and equilibrium activity diagrams. The zoning copper/chalcocite/bornite/chalcopyrite/pyrite is portrayed by a $\log f_{O_2}$ versus pH diagram at 100°C, 0.1 m Cl^- and 10^{-4}m total sulfur. It is shown that in the chalcocite field most of the sulfur must be sulfate, and the maximum total of copper concentration is only 0.1 mg/l in the chalcopyrite field.

The chapter of *Schmidt, Schumacher, Spieth*, and *Friedrich* combines scientific investigations (geochemical, mineralogical, petrographical, geological) with results of actual exploration for copper-silver deposits in the Kupferschiefer of West Germany. The relationship between paleoenvironment, the metal distribution, and the diagenetic process of the "Rote Fäule" is shown.

The results of the chapters included in Part IV, indicate a new way to consider the problems of sediment-hosted copper deposits. All authors assume underlying red-bed formation as a plausible source of copper. The controlling factors of mobilizing and transporting copper may be compaction, sea-level changes, gravity flow, hydrodynamic activities, the influence of heat of intrusives and metamorphism. The infiltration and precipitation of the copper-bearing flows (subsurface waters) took place during diagnesis and not during sedimentation.

Part I
Deposits in Mafic and Ultramafic Complexes

The Physical and Petrologic Setting and Textural and Compositional Characteristics of Sulfides from the South Kawishiwi Intrusion, Duluth Complex, Minnesota, USA

M. Foose[1] and P. Weiblen[2]

Abstract

The South Kawishiwi intrusion in the northern part of the Duluth Complex, Minn. is divisable into a sulfide-free upper zone and a sulfide-bearing lower zone. The sulfide-free zone is mostly troctolite containing laterally persistent anorthosite layers. Sequences of troctolite and anorthosite form cyclic units in which the crystallization of plagioclase was followed by precipitation of plagioclase plus olivine. Chemical variations associated with cycles show trends either similar to or reversed from those expected under normal conditions of fractional crystallization. These trends, and the absence of any systematic chemical evolution, indicate that crystallization of the sulfide-free rocks occurred in a chamber that was continuously replenished by compositionally similar liquids. In contrast, the sulfide-bearing zone consists of a heterogeneous mixture of troctolite, picrite, dunite, anorthosite, oxide cumulates, and hornfels in which sulfides are ubiquitously disseminated. The sulfide-bearing zone has no discernible internal stratigraphy. The contact with the sulfide-free zone is sharp.

The sulfides are predominantly pyrrhotite, chalcopyrite with cubanite, and pentlandite occurring mostly as interstitial and included grains. The close association of sulfides with hydrous phases or with grains that show reverse compositional zoning indicates that volatiles derived from adjacent metasedimentary inclusions were important in sulfide formation. The wide dispersion of sulfides, heterogeneous mix of rock types, and absence of traceable units indicates that the rocks accumulated in an extremely dynamic environment that is thought to have been associated with rifting in this area. Faults presumed to be rift-related were active throughout the emplacement of this intrusion and, in part, control the distribution of the sulfide-bearing zone and the overlying rocks.

Introduction

The 1.1 Ga Duluth Complex in the north-central United States (Fig. 1) is a large composite mafic intrusion that contains significant amounts of copper-nickel

[1] US Geological Survey, Reston, VA 22092, USA
[2] Minnesota Geological Survey, Minneapolis, MN 55455, USA

Geology and Metallogeny of Copper Deposits
(ed. by G.H. Friedrich et al.)

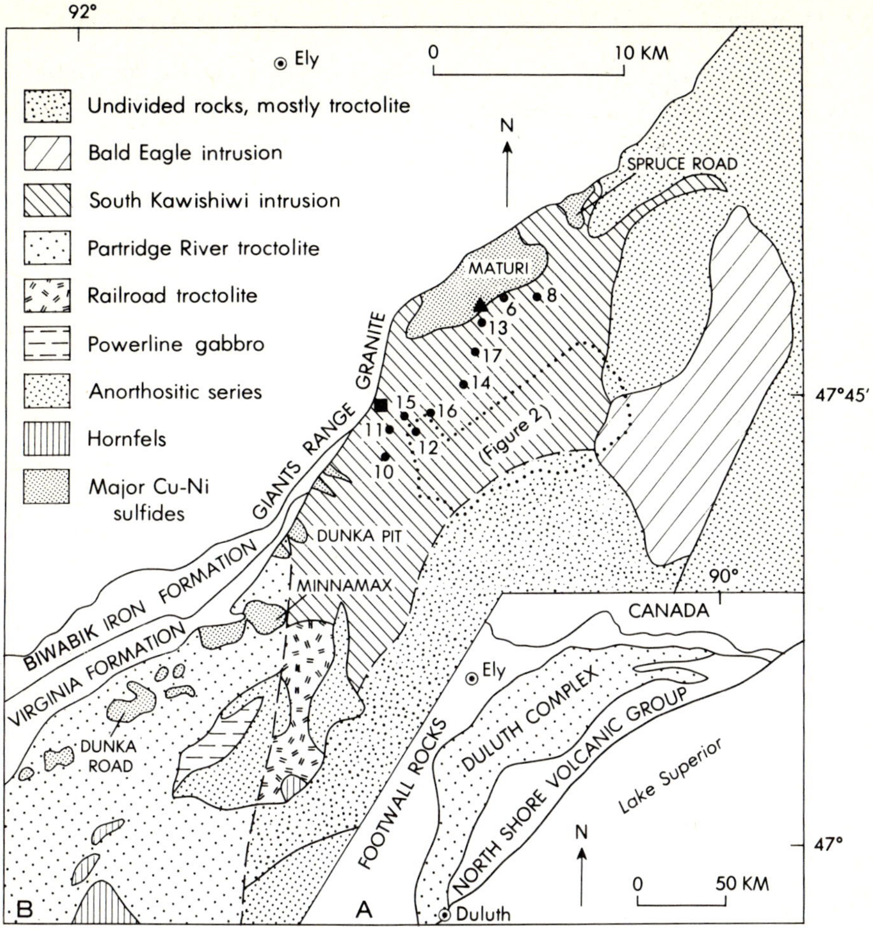

Fig. 1 A, B. Index (**A**) and generalized geologic map (**B**) of the Duluth Complex near Ely, Minnesota, USA. Also shown are major copper-nickel sulfide occurrences (*shaded*), the outline of the area shown in Fig. 2, and location of drill holes shown in Figs. 3 and 4. Location of samples from which sulfur isotopic compositions of +7.5 and +1.7 per mil were collected are shown by *solid square* and *solid triangle*, respectively. (Modified from Morey and Cooper 1977, and Watowich et al. 1981)

sulfides. The full extent of the resource is unknown, but along a 50-km distance at the base of the complex southeast of Ely, Minn., over 4 billion metric tons of mineralized rock containing 0.66% copper and 0.2% nickel has been identified (Listerud and Meineke 1977). In addition to their economic potential, these sulfides are of interest because they are one of the few large concentrations of magmatic sulfides in non-Archean rocks (Naldrett 1981). This contribution describes the physical and petrologic setting of sulfide-bearing rocks and the textures, types, and compositions of sulfide minerals that occur within the South Kawishiwi intrusion which is one part of this composite complex.

Geologic Setting

The early general studies of Grout (1918a, b) and later more specific work of Taylor (1964), Green et al. (1966), Bonnichsen (1970, 1972a), Phinney (1969, 1972), and Weiblen and Morey (1975, 1980) show the Duluth Complex to be a composite intrusion composed predominantly of anorthositic, trocolitic, and gabbroic rocks. These intrusions are divisable into two series. Anorthositic rocks of the older anorthositic series are enclosed in and cut by younger rocks that are predominantly troctolites which contain minor interlayered anorthosite. These younger rocks are known as the troctolitic series (Bonnichsen 1972a). These two series intrude diverse Archean and Proterozoic volcanic, metasedimentary, and igneous rocks that define the northwest margin of the complex. In the northern part, the footwall rocks are either 2.7 Ga granites of the Giants Range Batholith, Early Proterozoic Biwabik Iron Formation, or Early Proterozoic Virginia Formation. To the southeast, the intrusion is overlain by the North Shore Volcanic Group.

Emplacement of the Duluth Complex occurred about 1.1 Ga ago (Goldich et al. 1961, Faure et al. 1969) during an episode of regional crustal rifting which produced what is now referred to as the Midcontinent rift (Wold and Hinze 1982). This rift, as expressed by the Midcontinent gravity high, extends over 1,000 km from Lake Superior to northeast Kansas (King and Zietz 1971, Chase and Gilmer 1973). The Duluth Complex is located on the northwest flank of this gravity anomaly. Magma was injected into fault-bounded voids formed during rifting; crystallization within these separated magma chambers resulted in the multiple intrusions that make up this composite complex (Weiblen and Morey 1975, 1980).

In the part of the Duluth Complex southeast of Ely, Minn. (Fig. 1) rocks of both the older anorthositic series and the younger troctolitic series are exposed. The latter form at least three intrusive bodies which have been given the informal names of the South Kawishiwi intrusion (SKI), the Partridge River troctolite (PRT), and the Bald Eagle intrusion (Green et al. 1966, Bonnichsen and Tyson 1975, Weiblen 1965). The Bald Eagle intrusion is a funnel-shaped body (Weiblen 1965) that clearly cross-cuts anorthositic series rocks. The SKI and the PRT, in contrast, are adjacent and compositionally similar bodies that lack a clearly recognizable mutual contact and which are identified as separate intrusions principally because of slightly different textures.

In this region, significant amounts of copper-nickel sulfide have been located only along the base of the SKI and the PRT. Both general descriptions of these sulfides (Bonnichsen 1972b, 1974, Listerud and Meineke 1977, Watowich et al. 1981, Iwasaki et al. 1983) and specific studies (Boucher 1975, Fukui 1976, Weiblen and Morey 1976, Bonnichsen et al. 1980, Tyson and Chang 1984, and Pasteris 1984) suggest that sulfides in both intrusions have similar mineralogy (dominantly pyrrhotite with less abundant chalcopyrite and cubanite, and minor pentlandite), textures (dominantly interstitial or included), and mode of occurrence (disseminations associated with the footwall or inclusion-rich zones). The composition of the sulfides do, however, change along strike as the copper-nickel ratios increase toward the southwest (Fig. 1) from the Spruce Road deposit

(2.7), to the Maturi deposit (3.2), to the Minnamax deposit (4.1), to the Dunka Road deposit (5.1) (Tyson and Chang 1984).

Petrologic and Structural Setting of Sulfides

The combination of large amounts of surface and subsurface information from the SKI allows the development of an integrated petrologic and structural picture. These relationships are summarized below.

Surface Relations

In outcrop, the SKI is composed mostly of plagioclase-olivine cumulates containing minor (less than 10%) interstitial augite, oxides, or biotite (Bonnichsen and Tyson 1975, Foose and Cooper 1978, 1981). Madal layering is not common and generally is not traceable for more than 100 m. Even less common are phase layers made by plagioclase cumulates. However, many of these phase layers are

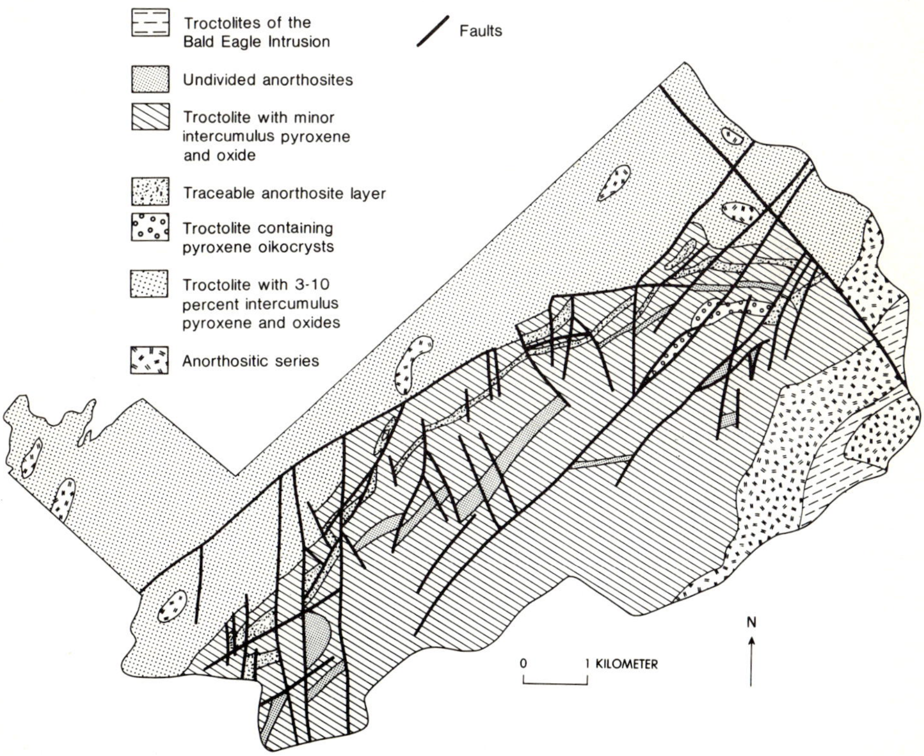

Fig. 2. Generalized geologic map of area outlined in Fig. 1

laterally persistent and can be followed. In areas of good exposure, mapping these layers enables the geologic relationships to be established (Fig. 2).

Most important, the offsetting of these layers shows this intrusion to be highly faulted (Foose and Cooper 1981). Faults strike in three principal directions (N 35°E, N 05°W, and N 40°W), and generally have small offsets. The general lack of offset of one fault set by others indicates that most displacements involved nearly vertical movement of blocks. The faulting parallels regional lineament and joint patterns that have been related to east-west tension (Cooper 1978). The overall displacement, in which movement of blocks is down to the southeast is consistent with the expected motion on the northwest flank of a rift.

Subsurface Relations

Drill holes near the surface exposures allow projection of these relationships into the subsurface. Three major lithologic zones are recognized (Fig. 3). From bottom to top: (a) older, granitic footwall rocks, (b) basal zone sulfide-bearing rocks, and (c) generally sulfide-free (i.e., found in trace amounts only) troctolite with interlayered anorthosite (Bonnichsen et al. 1980, 1984). Contacts between these zones are the only structural and stratigraphic features which have been traced with certainty.

Sulfide-Free Zone. As on the surface, the upper sulfide-free zone is composed predominantly of plagioclase-olivine cumulates with thin plagioclase cumulate interlayers. In addition, some pegmatoids are present. Most pegmatoids are thin (<10 cm) and consists of 50 to 90% plagioclase crystals that are 1 to 4 cm in length with 10 to 40% coarse-grained clinopyroxene, and minor olivine. Typically these pegmatoids form bottoms of 2 to 40 m thick laterally persistent successions which grade upward into medium-grained plagioclase cumulates and then into plagioclase-olivine cumulates. These successions define packages of rocks which can be correlated among drill holes; correlation of individual successions within these packages is less certain (Fig. 3). The most clearly traceable layer is a 30 to 100 m thick plagioclase-rich and commonly pegmatoidal section at the base of the sulfide-free zone which is in sharp contact with the underlying sulfide-bearing zone.

The pegmatoids that form the bottoms of these successions are not always present and, in fact, decrease in abundance to the northeast (Fig. 3). Where pegmatoids are absent, the bottoms are made by medium-grained plagioclase cumulates. The sharp lower contact between the base of these successions and underlying plagioclase-olivine cumulates indicates a break at which the magma changed from crystallizing plagioclase and olivine to crystallizing only plagioclase. The gradation of plagioclase cumulates into overlying plagioclase-olivine cumulates show that olivine eventually rejoined plagioclase on the liquidus.

Sulfide-Bearing Zone. In contrast to overlying rocks, the sulfide-bearing basal zone is a heterogeneous, 70- to 350-m thick sequence of rocks. The most abun-

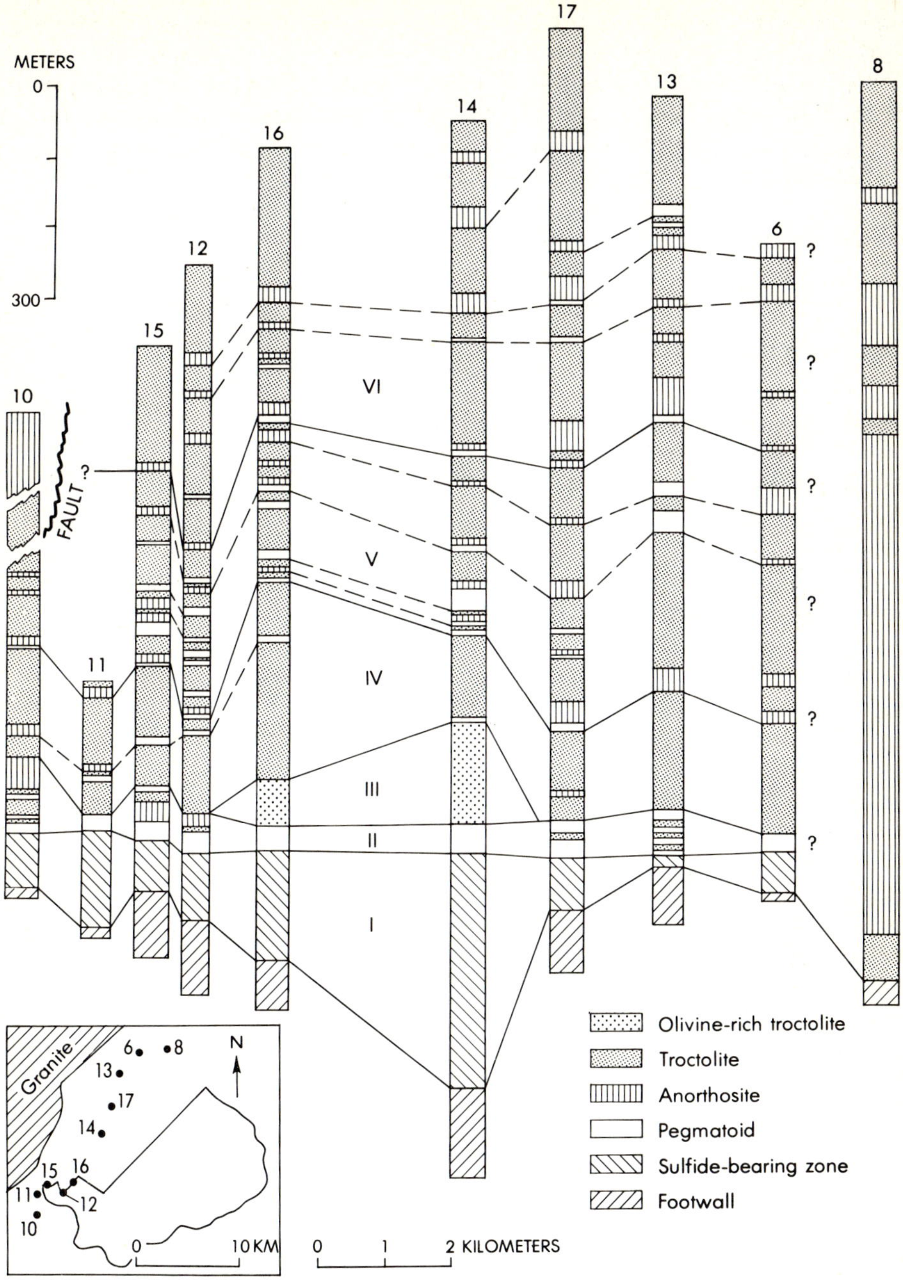

Fig. 3. Generalized subsurface stratigraphy of the South Kawishiwi intrusion. Drill hole locations are shown here in inset and in Fig. 1. Major sequences of rocks are correlated by *solid lines* and are: *I* sulfide-bearing zone; *II* anorthositic and plagioclase-rich pegmatoidal layer; *III* a laterally discontinuous olivine-rich troctolite; *IV* troctolite with few anorthositic layers; *V* troctolite with many anorthositic and pegmatoidal layers; *VI* upper troctolitic sequence having rare pegmatoidal zones. The base of the sulfide-free zone defines the surface to which all holes are referenced. (Modified from Foose 1984)

dant rock type is medium- to fine-grained troctolite; less common are picrites, norites, anorthosites, and thin layers of oxide cumulates. Hornfels inclusions are abundant, and most of the unit contains disseminated sulfides. The sulfide zone lacks a coherent internal stratigraphy and even distinctive lithologic breaks cannot be traced between closely spaced drill holes. The sulfide-bearing zone is distinguished from overlying rocks by its internal heterogeneity, generally greater olivine content, large number of hornfels inclusions, presence of oxide cumulates, and ubiquitous sulfides.

Geochemical Trends

The pegmatoid-anorthosite-troctolite successions in the sulfide-free zone provides important information about the emplacement and evolution of this intrusion. Mineral and whole rock analyses show these sequences to have compositional trends that are either similar to, or reversed from those expected under normal conditions of fractional crystallization (Fig. 4).

Plots of whole-rock nickel show widely variable and relatively high values in the basal zone, reflecting the sporadic distribution of sulfides. In the overlying sulfide-free zone, nickel is partitioned mostly into olivine, and the whole-rock nickel content in these rocks is controlled by the abundance of olivine and by the amount of nickel in olivine. Because nickel content in olivine is, in turn, controlled by the nickel content of the magma from which it crystallized, variations in whole-rock nickel content among rocks having similar amounts of olivine reflect changes in magma composition. Nickel contents from compositionally similar rocks that the have between 7% and 15% olivine are shown separated (dots) from more plagioclase – or olivine-rich rocks (slashes) and define the trends shown by dashed lines (Fig. 4).

Nickel values define two distinct patterns within rocks above the sulfide-bearing zone. Depletion patterns that are expected in normal fractional crystallization in which nickel content decreases upward are more common. Segments in which nickel content increases upward are somewhat less common. These nickel variations parallel trends in mineral compositions as shown by detailed study through a number of these cycles (Foose 1982).

Similar normal and reversed compositional trends have been noted both in the SKI and in the adjacent PRT (Fukui 1976, Bonnichsen et al. 1980, Grant and Molling 1981, Tyson and Chang 1984). Trace element studies in the PRT by Grant and Molling (1981) indicate that some of the reversed trends result from variations in the proportion of intercumulus liquids. Grant and Molling suggest that larger amounts of interstitial liquids are trapped near bottoms of cycles and that subsolidus equilibration with these interstitial liquids causes shifts to compositions that are more sodic and iron-rich than in overlying rocks where less intercumulus liquid was present.

However, subsolidus equilibration with interstitial liquids does not appear to have caused the reversed patterns seen in Fig. 4. Strongly zoned minerals and large amounts of various intercumulus phases are present in the pegmatoids and

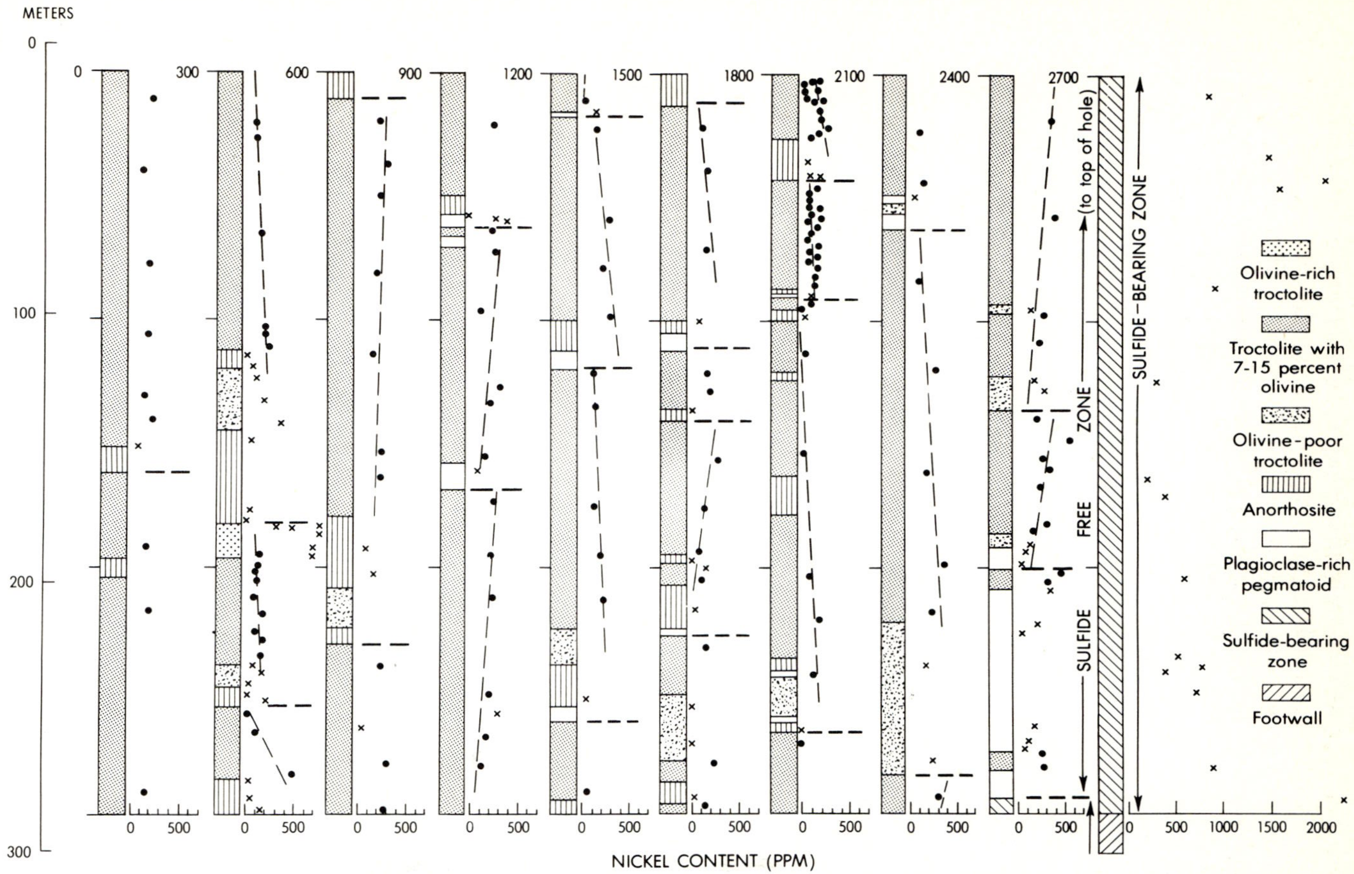

Fig. 4. Whole-rock nickel content and detailed stratigraphy of drill hole 12. Nickel content of rocks having between 7% and 15% olivine (*dots*) are shown separately from more plagioclase- or olivine-rich rocks (*slashes*) and locally define trends shown by *thin dashed lines*. *Dashed horizontal lines* mark distinct breaks in crystallization. (Scale in meters; tic marks and numbers in feet. Analyses on splits from drill core made by atomic absorption methods in USGS laboratory; W.D. Angelo analyst)

clearly show that large amounts of intercumulus liquid were locally present at the bottoms of some cycles. However, the overlying anorthosite and troctolite are virtually unzoned adcumulates, indicating relatively small and uniform amounts of trapped interstitial liquids. Compositional changes within them, therefore, cannot be easily attributable to varying amounts of subsolidus equilibration. Further, in troctolites that have between 7 to 15% olivine, there is good correlation (coefficient = 0.782) between whole-rock nickel content and olivine MgO content, whereas there is no correlation (coefficient = −0.104) between whole-rock nickel and phosphorous contents. Because phosphorous is an incompatible element that is concentrated in the intercumulus phase, a close correlation of Ni and P would be expected if varying amounts of intercumulus liquids had controlled olivine compositions and thereby caused the trends shown in Fig. 4.

It is more likely that formation of the diverse cyclic patterns resulted from the injection and mixing of liquids of slightly different compositions. Episodic injections of liquids more primitive than those within a magma chamber could produce repeated patterns of decreasing nickel content upward. In contrast, increases upward in nickel, Ca/Na, and Mg/Fe could indicate crystallization from injections of more fractionated liquid that were mixed with the larger and compositionally more primitive magma reservoir.

Difficulties in generating, storing, and then emplacing more fractionated liquids into a magma chamber that also would be recharged with batches of more primitive liquids would seem to make injections of batches of more fractionated liquids unlikely. However, because the separate intrusions that constitute the Duluth Complex were emplaced and crystallized in separate fault-bounded chambers created during rifting (Weiblen and Morey 1975, 1980), it is possible that compositionally similar liquids in adjacent magma chambers fractionated to differing degrees. Disruption of these isolated magma chambers by continued rift-related faulting could allow mixing of more evolved and more primitive liquid batches between adjacent chambers.

Support for injection of liquids from adjacent magma chambers is given by the distribution of cyclic sequences (Fig. 3). Cycles are more abundant in the southwestern part of the SKI and become less common to the northeast. Accompanying this lateral change is a decrease in abundance of pegmatoids at the bottoms of cycles. The compositionally similar PRT is present along the southwest boundary of the SKI and may have been a source for injections of diverse batches of liquids. As observed, abundant thin, but not laterally extensive cycles would be expected near this source area, while thicker and more laterally continuous cycles would extend to the northeast. Furthermore, the rapid rates of crystal accumulation expected near areas of magma injection would produce conditions ideal for trapping intercumulus liquids and the formation of the coarse-grained bottoms of cycles. Less rapid accumulation of crystals further from the source of magma injection would produce the more texturally uniform adcumulates that occur to the northeast.

Finally, one should note that mineral compositional changes within individual cycles often exceed the overall average compositional change within the entire exposed portion of SKI. This lack of systematic chemical differentiation provides further evidence that this portion of the intrusion crystallized in an open magmatic system in which there was nearly continuous replenishment of compositionally similar liquids.

Mapping on the surface shows that significant faulting occurred after crystal consolidation. Additionally, some fault-related fractures are filled with pyroxene, indicating that movements occurred prior to complete crystal consolidation. However, the subsurface relationships reveal the importance of syn-intrusion faulting in controlling the emplacement of these rocks. For example, between drill holes 12 and 17, the orientation of the basal contact of the intrusion changes abruptly (Figs. 1 and 3), and the basal zone thickens to over 300 m. Overlying the thicker part of the basal zone is a laterally discontinuous olivine-rich troctolite layer and a thicker sequence of overlying troctolite. These changes indicate that faults were active at the start of the emplacement of the intrusion and formed a fault-bounded trough that accumulated a thicker section of the sulfide-bearing basal zone. The discontinuous lens of olivine-rich troctolite, and thicker sequences of overlying rocks show that faulting continued during the emplacement of the overlying sulfide-free zone.

Characteristics of Sulfides

Three major concentrations of sulfides have been recognized in the troctolitic and anorthositic successions that form the lower parts of the SKI and the adjacent PRT. These are the Spruce Road deposit, the Maturi deposit, and the Minnamax deposit (Fig. 1). All have broadly similar mineralogy, textures, and style of mineralization. Most of the specific features discussed here are in the Spruce Road deposit (Weiblen and Morey 1976), but reference is also made to features that have been observed elsewhere. The Spruce Road deposit has proven resources of 248 million metric tons grading 0.4% Cu and 0.17% Ni (Watowich et al. 1981). The sulfides are mostly in a medium-grained plagioclase-olivine cumulate that contains variable amounts of interstitial augite, orthopyroxene, iron-titanium oxides, and biotite. Particularly important is the relatively high biotite content in the sulfide-bearing rocks as compared with that of the overlying, unmineralized rocks (Weiblen and Morey 1980). Additional rock types are cross-cutting augite troctolite pegmatoids, mafic troctolite, picrite, and abundant hornfels. Because of the great numbers of inclusions, the Spruce Road area may best be described as an igneous breccia.

The sulfides may be grouped into four textural types: (1) interstitial sulfides; (2) included sulfides; (3) sulfide-silicate intergrowths; and (4) fine-grained sulfide veinlets. Representative photomicrographs of these sulfides are shown in Fig. 5.

Interstitial Sulfides. The great majority of sulfide grains occupy roughly triangular shaped primsatic voids that are bounded by a plagioclase framework and less commonly by anhedral olivine. The void fillings may be entirely sulfide or a mixture of silicates, mostly clinopyroxene plus ilmenite, biotite, potassium feldspar, and apatite. Particularly common are void fillings formed by frameworks of plagioclase. Where plagioclase forms the boundary of the voids, it com-

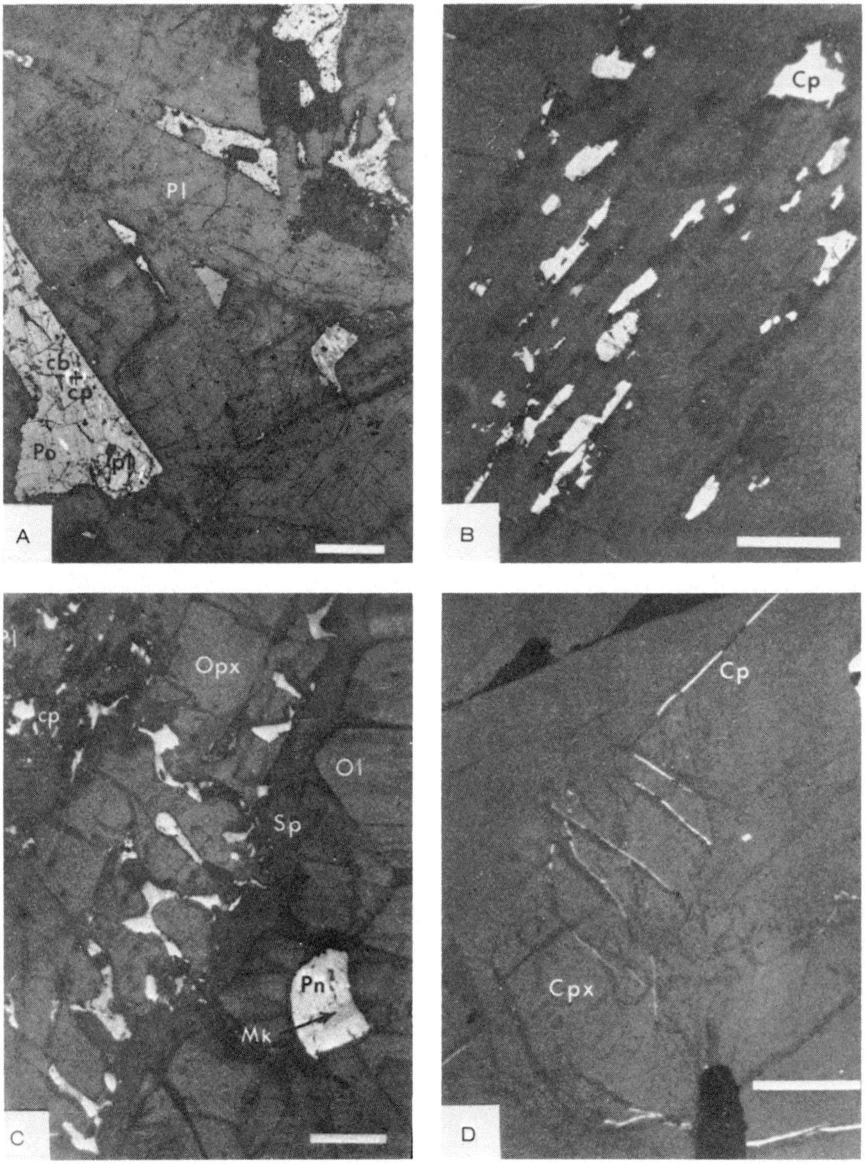

Fig. 5A – D. Sulfide textures. **A** Interstitial sulfides between plagioclase (bar is 1 mm). **B** Included sulfides in plagioclase. Sulfides are mostly chalcopyrite and are concentrated along 001 (bar is 100 μm). **C** Myrmekite-like intergrowths of sulfide with orthopyroxene (bar is 200 μm). **D** Fine-grained veinlets of chalcopyrite cutting interstitial clinopyroxene (bar is 100 μm). *Cb*, cubanite; *Cp*, chalcopyrite; *Po*, pyrrhotite; *Pn*, pentlandite; *Mk*, Mackinawite; *Pl*, plagioclase; *Cpx*, clinopyroxene; *Opx*, orthopyroxene; Ol, olivine; *Sp,* serpentine. (After Weiblen and Morey 1976)

Table 1. Size distribution of interstitial and included sulfides[a]

Phase	Length (mm)			Width (mm)		
	Min	Max	Av	Min	Max	Av
Chalcopyrite						
Interstitial	0.00100	1.29000	0.11979 ± 0.20878	0.00100	0.48000	0.04783 ± 0.07542
Included	0.00067	2.64000	0.06635 ± 0.17783	0.00060	0.83850	0.03034 ± 0.06459
Pyrrhotite						
Interstitial	0.00120	0.72000	0.10151 ± 0.16146	0.00120	0.28800	0.03719 ± 0.05440
Included	0.00100	0.72000	0.01919 ± 0.05358	0.00100	0.09100	0.00825 ± 0.01867
Pentlandite						
Interstitial	0.00174	0.9051	0.01472 ± 0.01457	0.00174	0.04177	0.00811 ± 0.00668
Included	0.02400	0.05280	0.04080 ± 0.01060	0.00480	0.00720	0.00660 ± 0.00104

[a] Measurements of apparent maximum length and width made in thin section on 2,419 sulfide grains. (Modified from Weiblen and Morey 1976).

monly contains included sulfide. Olivine grains are mantled with myrmekite-like overgrowths of orthopyroxene and sulfide. The sizes of sulfides in voids varies widely as a function of the size of the plagioclase framework (Table 1). Sulfide modes vary greatly but averaged about 55% pyrrhotite, 37% chalcopyrite-cubanite, and 8% pentlandite. This average mode is much less copper-rich than the other sulfide types and must indicate a unique sulfide composition. Polyphase assemblages and single phase assemblages of chalcopyrite or pyrrhotite are common; monomineralic occurrences of pentlandite are rare (Pasteris 1984).

Included Sulfides. Included sulfides are the second most important type and occur mostly in plagioclase and clinopyroxene. Some grains also occur in ilmenite, but no inclusions were found in olivine. Inclusions in clinopyroxene are irregular in shape and are scattered throughout the grain, whereas those in plagioclase have rectangular forms and are concentrated near grain margins. The included sulfides average 0.061 mm in length and 0.028 mm in width, which is smaller than the average size of the interstitial sulfides. A striking feature of the included sulfides is that more than 99% of the grains are chalcopyrite. The chalcopyrite is associated with bornite, pyrrhotite, or minor pentlandite.

Sulfide-Silicate Intergrowths. Both orthopyroxene and amphiboles are intergrown with sulfides. Sulfide-orthopyroxene intergrowths typically mantle olivine and, in some cases, are separated from it by serpentine. Interstitial biotite and/or ilmentite may also be present. The sulfides, primarily chalcopyrite, typically have a myrmekitic texture.

The sulfide-amphibole intergrowths occur in interstitial voids and are characterized by an interdigitated texture. The amphiboles are hornblende or actinolite; the sulfides include pyrrhotite, chalcopyrite, cubanite, and pentlandite. Some intergrowths occur either as an independent interstitial phase, or are closely associated with interstitial clinopyroxene and sericititized plagioclase. In this latter occurrence the textural evidence suggests the reaction clinopyroxene + plagioclase + H_2O → hornblende.

Fine-Grained Veinlets. Thin sulfide veins (less than 1 to 50 μm in width) cut all the host silicate phases. Although relatively common, veins constitute only a small fraction of the total sulfide population. The sulfides in the veins are predominantly chalcopyrite and cubanite.

Average sulfide compositions are given in Table 2. Pyrrhotite consists of both monoclinic and hexagonal forms, contains an average of 0.17 wt% nickel and less than 0.05 wt% cobalt. Pentlandite contains between 33 and 37 (average 34.67) wt% nickel and an average of 1.84 wt% cobalt. Pasteris (1984) examined other pentlandites from this deposit and noted a somewhat larger range in nickel contents (29.9 to 42.7 wt%) with the highest values occurring in the most copper-rich assemblages.

Plagioclase and olivine are the two cumulus phases present in these rocks. Olivine associated with sulfide ranges in compositions from Fo_{45} to Fo_{55} and contains between 0.06 to 0.08 wt% nickel. Plagioclase has compositions near An_{55}. Plagioclase is typically overgrown by rims that contain sulfide inclusions. The calcium-sodium ratio of plagioclase overgrowths that contain inclusions is consistently much higher than cores of grains (for example, An_{55} vs An_{70}). Similarly, the magnesium-iron ratios of sulfide-bearing pyroxene overgrowths around olivine are larger than those of framework pyroxene grains that do not include sulfides. The trends are reversed from those expected in a normal differentiation sequence, but are consistent with the depression of the liquidus by introduction of water.

Table 2. Average composition of sulfide phases from the Spruce Road deposit[a]

	Pyrrhotite		Pentlandite	Mackinawite	Chalcopyrite	Cubanite
	(hexagonal)	(monoclinic)				
			wt %			
S	36.67	39.41	33.37	36.06	32.90	36.94
Fe	62.65	60.02	30.55	54.95	31.29	39.46
Co	0.28	0.03	1.84	0.96	0.08	–
Ni	0.01	0.24	34.67	7.68	0.06	0.06
Cu	–	0.01	–	–	35.76	21.97
Total	99.61	99.71	100.43	99.65	100.09	98.43
			Normalized atomic percentages			
S	50.372	53.241	47.102	49.857	47.490	52.238
Fe	49.411	46.553	24.758	43.621	26.041	32.038
Co	0.209	0.022	1.413	0.722	0.063	–
Ni	0.007	0.177	26.727	5.800	0.047	0.046
Cu	–	0.007	–	–	26.157	15.678
M/S	0.985	0.878	1.123	1.005	1.097	0.914

[a] Wavelength dispersive electron microprobe analyses made at 20 kV, 0.03 μA, 3 – 5 μm beam diameter, and 20 s counting interval.

The textural relationships described above are inconsistent with a simple magmatic accumulation of sulfides. Such a process would require significant amounts of gravity accumulation in order to produce the observed ore zone. However, the disseminated, sporadically distributed, and interstitial nature of this mineralization indicates that this accumulation has not happened. Instead, the setting of most sulfides in intercumulus spaces or as inclusions in adcumulus rims shows that sulfide separation occurred well after the accumulation of olivine and plagioclase. Further, the association of sulfides with reversely zoned silicates and hydrous phases (particularly biotite and amphiboles that texturally appear to have formed by hydration reactions) indicates that sulfides formed when volatiles were introduced into the crystallizing magma and that the volatiles were sulfur-bearing. The textural relationships suggest that the volatiles were introduced when the rocks were approximately 75% crystallized.

The relationships are consistent with isotopic studies in the adjacent PRT (Ripley 1981, Rao and Ripley 1983). Sulfides in the PRT show sulfur isotopic compositions between +0.2 to +15.3 per mil, with averages near +7.5 per mil. These values differ significantly from the approximately 0.0 per mil value expected from magmatic sulfides, but are similar to values found in the underlying Virginia Formation. Additionally, $\delta^{18}O$ values in troctolite near footwall rocks have values of up to 9.6 per mil that are similar to $\delta^{18}O$ values of the Viriginia Formation and are much greater than the 5.8 to 7.0 per mil in uncontaminated troctolite. The isotopic similarity between sulfides and country rock clearly shows that assimilation of country rock introduced volatiles into the crystallizing troctolite and that sulfur derived from the breakdown of sedimentary pyrite was added to the magma and was then reprecipitated as copper-nickel sulfides.

Detailed isotopic studies have not been done within the SKI. However, positive values of +7.5 and +1.7 per mil have been determined from sulfides at two different locations (Fig. 1; Ault and Kulp 1959) and are consistent with values observed in the PRT. The footwall of the SKI is mostly granite, however, the abundant metasedimentary inclusions within the sulfide-bearing basal zone indicate that either metasediments capable of contributing sulfur were once present or that metasediments have been transported from another area. Weiblen and Morey (1980) note a consistent, but sharp decrease in biotite content away from the footwall of the SKI, a change attributed to upward diffusion of water into a crystallizing magma. The oxygen isotopic data of Churchill (1978) also provide evidence for introduction of water as they indicate a correlation in anomalous $\delta^{18}O$ values in the troctolites with increased sulfide abundance. These observations support the textural data by indicating that introduction of a water-sulfur rich vapor phase from metasedimentary rocks was an essential part of the genesis of sulfides within the South Kawishiwi intrusion.

The sharp contrast between the relatively pyrrhotite-rich interstitial sulfides and the much more copper-rich included and vein sulfides indicates two episodes of mineralization. The polyphase mineralogy of the interstitial sulfides is consistent with sulfide exsolution from typical magmatic Fe – Cu – Ni sulfide liquids. The chalcopyrite dominated mineralogy of the included and vein sulfides crosscut and replace these polyphase sulfides and are, thus, clearly later. The introduction of the copper-rich solutions that deposited these later sulfides has been attributed to a variety of processes including fractional crystallization of sulfide liquids (Bonnichsen et al. 1980), fractional crystallization of the interstitial silicate liquid (Pasteris 1984), and copper introduction from the adjacent footwall rocks (Tyson and Chang 1984).

Conclusions

The disseminated and sporadically distributed sulfides in the basal zone of the South Kawishiwi intrusion separated from a troctolitic liquid after cumulus olivine and plagioclase had established a crystal framework. Accumulation of sulfide was accompanied by the introduction of water and sulfur-rich volatiles derived from metasedimentary xenoliths and footwall rocks. Reversely-zoned rims on cumulus grains indicate that these volatiles were introduced when the magma was approximately 75% crystallized.

The volatiles could have been introduced into the basal zone by a combination of three mechanisms. These are (1) direct vapor diffusion across the footwall contact, (2) direct vapor diffusion from included metasedimentary rocks, or (3) indirect transfer along faults and fractures. The decrease in biotite away from the footwall contact is consistent with the first process, but the general upward decrease in sulfide abundance which would also be expected has not been documented. Sulfides occur sporadically, but ubiquitously up to the sharp contact with the sulfide-free zone. On the other hand, the apparent lack of concentration of metasedimentary rocks along the footwall of this area and the evidence for dynamic magmatic processes offered by the absence of any traceable silicate layering in the sulfide-bearing zone suggests that conditions were appropriate for the widespread mixing of sulfide-bearing inclusions in the basal zone. Faulting that in part controlled the distribution and thickness of the basal zone also may have provided channels along which volatiles could have concentrated.

The abrupt contact between the sulfide-bearing zone and the overlying sulfide-free zone indicates a radical change in magmatic conditions. It is suggested that the sulfide-bearing zone reflects dynamic magma mixing and assimilation processes in a restricted magma chamber or a restricted part of a larger chamber that was formed during rifting. Surges of magma into the chamber entrained, assimilated, and dispersed pieces of the Virginia Formation and the Biwabik Iron Formation and caused a complete disruption of any internal layering within the chamber. Crystallization of the heterogeneous mix was accompanied by devolatilization of the enclosed metasedimentary xenoliths; vapors derived from the metasediments were important in forming the disseminated and sporadically distributed sulfides. With further rifting, the restricted magmatic conditions yielded to a more quiescent, open magmatic system in which the overlying sulfide-free troctolitic rocks accumulated. The homogeneous sequence of troctolitic rocks that contain thin, but laterally persistent anorthosite layers can be attributed to nearly continuous replenishment of compositionally similar liquids.

References

Ault WU, Kulp JL (1959) Isotopic geochemistry of sulphur. Geochim Cosmochim Acta 16:201 – 235

Bonnichsen B (1970) Geologic map of the Allen, Babbitt, Babbitt NE., Babbitt S.E., and Babbitt S.W. $7\frac{1}{2}$ minute quadranges. Minn Geol Surv Open File Maps

Bonnichsen B (1972a) Southern part of Duluth Complex. In: Sims PK, Morey GB (eds) Geology of Minnesota: A centennial volume. Minnesota Geol Surv, St. Paul, Minn, pp 361 – 387 (see Phinney WC, 1972)

Bonnichsen B (1972b) Sulfide minerals in the Duluth Complex. In: Sims PK, Morey GB (eds) Geology of Minnesota: A centennial volume. Minnesota Geol Surv, St. Paul, Minn, pp 388 – 393

Bonnichsen B (1974) Copper and nickel resources in the Duluth Complex, northeastern Minnesota. Minn Geol Surv Inf Circ 10:24

Bonnichsen B, Tyson RM (1975) Geology of the Ely-Hoyt Lakes region of the Duluth Complex, Minnesota. Geol Soc Am Field Trip Guide, Penrose Conference on Mafic Plutons and Magmatic Sulfides. Ely, Minn, September 14 – 19, p 20

Bonnichsen B, Fukui L, Chang L (1980) Geologic setting, mineralogy, and geochemistry of magmatic sulfides, South Kawishiwi intrusion, Duluth Complex, Minnesota. Proc 5th Q IAGOD Symp, Stuttgart, Germany, pp 545 – 565

Boucher ML (1975) Copper-nickel mineralization in a drill core from the Duluth Complex of northern Minnesota. US Bur Mines Rep Invest 8084:55

Chase CG, Gilmer TH (1973) Precambrian plate tectonics: The midcontinent gravity high. Earth Planet Sci Lett 21:70 – 78

Churchill RK (1978) A geochemical and petrological investigation of the Cu – Ni sulfide genesis in the Duluth Complex, Minnesota. MS Thesis, Univ Minn, Minneapolis, 101 pp (unpublished)

Cooper RW (1978) Lineament and structure analysis of the Duluth Complex, Hoyt Lake-Kawishiwi area, northeastern Minnesota. Ph D Diss, Univ Minn, Minneapolis, 220 pp (unpublished)

Faure G, Chaudhuri S, Fenton MD (1969) Age of the Duluth Complex and the Endion Sill, Duluth, Minnesota. J Geophys Res 74:720 – 725

Foose MP (1982) Structural, stratigraphic, and geochemical features of the South Kawishiwi Intrusion, Duluth Complex. Geol Soc Am (Abstract with Programs) 14:490

Foose MP (1984) Drill logs and correlations within part of the South Kawishiwi Intrusion, Duluth Complex, Minnesota. US Geol Surv Open-File Rep 84-14:230

Foose MP, Cooper RW (1978) Geology of the Harris Lake area, northeastern Minnesota. US Geol Surv Open-File Rep 78-385:34

Foose MP, Cooper RW (1981) Faulting and fracturing in part of the Duluth Complex, northeastern Minnesota. Can J Earth Sci 18:810 – 814

Fukui LM (1976) The mineralogy and petrology of the South Kawishiwi intrusion, Duluth Complex, Minnesota. MS Thesis, Univ Ill, Chicago Circle, 110 pp (unpublished)

Goldich SS, Nier AO, Baadsgaard H, Hoffman JH, Krueger HW (1961) The Precambrian geology and geochronology of Minnesota. Minn Geol Surv Bull 41:193

Grant NK, Molling PA (1981) A strontium isotope and trace element profile through the Partriolge River troctolite, Duluth Complex, Minnesota. Contrib Mineral Petrol 77:296 – 305

Green JC, Phinney WC, Weiblen PW (1966) Gabbro Lake quadrangle, Lake County, Minnesota. Minn Geol Surv Misc Map M-2

Grout FF (1918a) Internal structures of igneous rocks; their significance and origin; with special reference to the Duluth Gabbro. J Geol 26:439 – 458

Grout FF (1918b) The lopolith; an igneous form exemplified by the Duluth Gabbro. Am J Sci 4th Ser 46:516 – 522

Iwasaki I, Malicsi AS, Lipp RJ (1983) By-product recovery from copper-nickel bearing Duluth gabbro flotation tailings. Soc Min Eng AIME, paper presented at SME-AIME meeting. Atlanta Georia, March 6 – 10, preprint No 83-20, p 11

King ER, Zietz I (1971) Aeromagnetic study of the Midcontinent gravity high of central United States. Geol Soc Am Bull 82:2187 – 2208

Listerud WH, Meineke DG (1977) Mineral resources of a portion of the Duluth Complex and adjacent rocks in St. Louis and Lake Counties, northeastern Minnesota. Minn Depart Nat Res Div Miner Rep 93:74

Morey GB, Cooper RW (1977) Bedrock geology of the Hoyt Lakes-Kawishiwi area, St. Louis and Lake Counties, northeastern Minnesota. Minn Geol Surv Open-File Map 1:48,000

Naldrett AJ (1981) Nickel sulfide deposits: classification, composition, and genesis. Econ Geol 75th Anniversary Vol, pp 628 – 685

Pasteris JD (1984) Further interpretation of the Cu – Fe – Ni sulfide mineralization in the Duluth Complex, northeastern Minnesota. Can Miner 22:39 – 53

Phinney WC (1969) The Duluth Complex in the Gabbro Lake quadrangle, Minnesota. Minn Geol Surv Rep Invest 9:20

Phinney WC (1972) Northwestern part of Duluth Complex. In: Sims PK, Morey GB (eds) Geology of Minnesota: A centennial volume. Minnesota Geol Surv, St. Paul, Minn, pp 335 – 345 (see reference by Bonnichsen, for example)

Rao BV, Ripley EM (1983) Petrochemical studies of the Dunka Road Cu – Ni deposit, Duluth Complex, Minnesota. Econ Geol 78:1222 – 1238

Ripley EM (1981) Sulfur isotopic studies of the Dunka Road Cu – Ni deposit, Duluth Complex, Minnesota. Econ Geol 76:610 – 620

Stopler E (1980) A phase diagram for mid-ocean ridge basalts: preliminary results and implications for petrogenesis. Contrib Mineral Petrol 74:13 – 27

Taylor RB (1964) Geology of the Duluth Gabbro Complex near Duluth, Minnesota. Minn Geol Surv Bull 44:63

Tyson RM, Chang LLY (1984) The petrology and sulfide mineralization of the Partridge River troctolite, Duluth Complex, Minnesota. Can Miner 22:23 – 38

Watowich SN, Malcolm JB, Parker PD (1981) A review of the Duluth Gabbro Complex of Minnesota as a domestic source of critical and strategic metals. Soc Min Eng AIME, paper presented at the SME-AIME fall meeting. Denver, Col, Nov. 18 – 20, preprint 81-351, p 9

Weiblen PW (1965) A funnel-shaped, gabbro-troctolite intrusion in the Duluth Complex, Lake County, Minnesota. Ph D Diss, Univ Minn, 161 pp (unpublished)

Weiblen PW, Morey GB (1975) The Duluth Complex – a petrologic and tectonic summary. Proc 48th Annu Meet, Minn Sect Am Inst Min Eng 36th Annu Mining Symp, Univ Minn, Minneapolis, pp 72 – 95

Weiblen PW, Morey GB (1976) Textural and compositional characteristics of sulfide areas from the basal contact zone of the South Kawishiwi intrusion, Duluth Complex, northeastern Minnesota. Proc 49th Annu Meet, Minn Sect Am Inst Min Eng 37th Annu Mining Symp, Univ Minn, Minneapolis, p 24

Weiblen PW, Morey GB (1980) A summary of the stratigraphy, petrology, and structure of the Duluth Complex. Am J Sci 280-A:88 – 133

Wold RJ, Hinze WJ (eds) (1982) Geology and tectonics of the Lake Superior Basin. Geol Soc Am Mem 156:280

Application of Stable Isotopic Studies to Problems of Magmatic Sulfide Ore Genesis With Special Reference to the Duluth Complex, Minnesota

E. M. Ripley[1]

Abstract

Analyses of the isotopic composition of S, C, O, and H in mafic igneous rocks may provide valuable information regarding the extent of contamination by continental crust material. Partial melting and devolatilization of country rocks are important processes in the generation of many Cu – Ni deposits associated with mafic igneous rocks. Sulfur isotopes provide a means of evaluating the possibility of extraneous sulfur addition to a melt. Variability and distribution of $\delta^{34}S$ values may also provide data relating to the timing and mechanisms of sulfur production and incorporation into the melt. Oxygen, hydrogen, and carbon isotopes also can be sensitive indicators of isotopic contamination, whether by partial melting, devolatilization, or solid state exchange. Zones of contamination within an igneous sequence are likely to be areas where factors that control sulfide solubility (T, fS_2, fO_2, melt composition) have been perturbed, and are thus key sites for ore generation.

Stable isotopic studies of two deposits within the Duluth Complex have highlighted the importance of country rock contamination in the generation of Cu – Ni ores. For the Dunka-Road deposit devolatilization of country rocks has resulted in essentially in situ contamination, and the formation of ore that is variable in both its spatial distribution and $\delta^{34}S$ values. Partial melting and major element contamination are restricted to areas near the margins of xenoliths. At the Babbitt deposit, contamination via both partial melting and devolatilization has been significant. Sulfur isotopic distribution between igneous and metasedimentary rocks suggests that sulfur must have been derived prior to or during magma ascent. Partial melting of country rocks and oxygen isotopic exchange may have occurred either before or after magma emplacement. Chemical diffusion, as well as fluid dynamic properties of mixing are thought to control the isotopic inhomogeneity that characterizes sulfide ore zones in the Duluth Complex.

[1] Geology Department, Indiana University, Bloomington, IN 47405, USA

Geology and Metallogeny of Copper Deposits
(ed. by G. H. Friedrich et al.)

Introduction

The chemical interaction between mafic magma and crustal country rocks has received considerable attention in recent years. In particular, isotopic and trace element data has been utilized in evaluating the importance of wall rock assimilation and magma mixing in the generation of compositionally varied igneous suites (Carter et al. 1978, Taylor et al. 1979, Gray et al. 1981, Thompson et al. 1982, Jacquemin et al. 1982). Modelling of trace element and isotopic systematics produced by fractional crystallization and assimilation has been outlined by Taylor (1980) and De Paolo (1981). Most of these studies have focused on large-scale interaction between crust and primitive mafic magma, and its importance in magma petrogenesis.

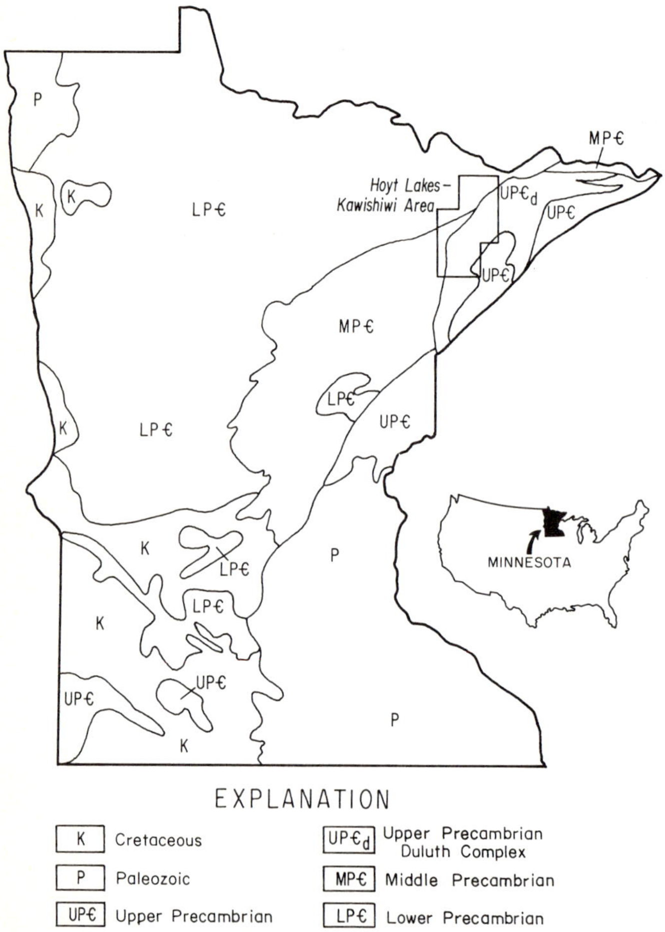

Fig. 1. Generalized bed-rock geology of Minnesota (After Cooper et al. 1981) showing location of the Duluth Complex. The outlined area (Fig. 2) contains most of the Cu – Ni reserves of the Duluth Complex

Contamination of mafic magma by sedimentary country rocks has also been shown to be a potentially important process in the generation of magmatic Cu – Ni and Pt-group element deposits. Mainwaring and Naldrett (1977) have suggested that assimilation of sulfidic country rocks was instrumental in producing ores associated with the Water Hen intrusion of the southern Duluth Complex. Sulfur isotopic studies of sulfide occurrences in the Bushveld Complex (Liebenberg 1970), Muskox intrusion (Sasaki 1969), Norils'k (Vinogradov and Grinenko 1964), and Patridge River troctolite of the Duluth Complex (Ripley 1981) all suggest that most of the sulfur found in the intrusives is of sedimentary derivation. Magma contamination by $CO_2 - H_2O$ fluids has been suggested by de Waal (1977) to explain sulfide mineralization in the base of the layered sequence in the Bushveld Complex. Page (1979), basing his conclusions principally on petrochemical and textural studies, suggests that partial melting of country rocks has occurred in the vicinity of sulfide mineralization in the Basal Zone of the Stillwater Complex. Bonnichsen (1972), Weiblen and Morey (1976), and Rao and Ripley (1983) use petrographic and major element data to illustrate the association between ore zones and areas of contamination in the Duluth Complex.

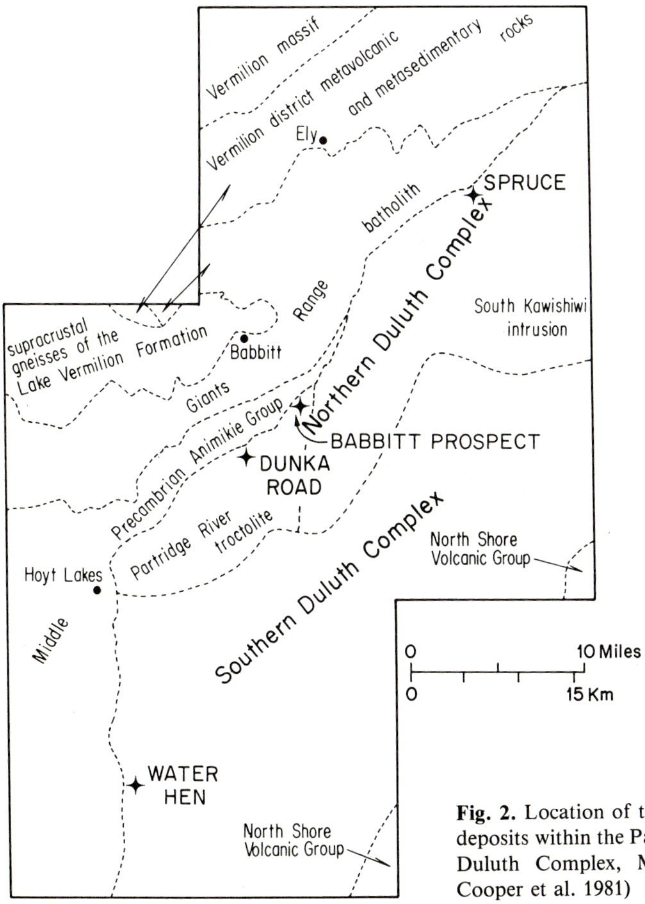

Fig. 2. Location of the Dunka Road and Babbitt deposits within the Patridge River troctolite of the Duluth Complex, Minnesota. (Modified from Cooper et al. 1981)

Groves et al. (1979) outline several models for the generation of Australian Archaean Ni-sulfide deposits, and include sulfur derivation from metasedimentary host rocks for many of the deposits. Addition of sedimentary sulfur and carbon along contact zones has also resulted in the localization of sulfide minerals in the Rana intrusion of Norway (Boyd and Mathiesen 1979). Recent isotopic and chemical studies of the Sudbury ores (Rao et al. 1984) also highlight the role of country rock contamination in the genesis of these important deposits.

The primary purpose of this paper is to review the principles involved in the application of stable isotopic studies to problems of Cu – Ni ore genesis in mafic igneous rocks. Particular emphasis is placed on the origin of sulfide occurrences at two localities in the Duluth Complex (Figs. 1 and 2), where our research group has been involved in stable isotopic and petrochemical studies over the past 6 years.

General Considerations

Evaluation of magma contamination by country rocks is normally based on the premise that one of several possible mixing processes has occurred, and resulted in the production of a melt with an isotopic, trace element, or major element composition ranging between values found in the primary melt and the original contaminant. The process may be a relatively simple one, with mixing of end-members with constant compositions, or extremely complex, characterized by mixing of end-members with variable compositions. A wide spectrum of intermediate results is possible, depending on the degree of partial melting of the country rock and the extent of fractional crystallization of the melt. Devolatilization of country rocks and incorporation of volatiles in the melt may lead to major and trace element, and isotopic systematics that are quite distinct from those that result from a partial melting mechanism.

Haughton et al. (1974), Shima and Naldrett (1975), Buchanan and Nolan (1979), and Danckwerth et al. (1979) have determined that sulfide solubility is dependent on variables such as temperature, sulfur and oxygen fugacities, and melt composition. Clearly all of these variables may be affected by magma mixing or addition of volatiles to a melt. A decrease in magma temperature due to heat transfer associated with contamination may lead to sulfide precipitation or separation of an immiscible sulfide melt. In a dry system sulfur apparently dissolves in a mafic magma by displacing oxygen and bonding to metallic ions, principally Fe^{2+} (Cu^{2+}, Ni^{2+}, etc. may also bond with sulfur, but their abundance is far less than Fe^{2+}) (Fincham and Richardson 1954, Shima and Naldrett 1975). Reactions such as

$$FeO + S^{2-} = FeS + O^{2-}$$

or

$$FeO + \tfrac{1}{2}S_2 = FeS + \tfrac{1}{2}O_2$$

describe the process. Related to fO_2 conditions are iron redox state and overall melt composition. Haughton et al. (1974) suggest that the Fe^{2+} content of mafic magma is proportional to sulfide solubility. A decrease in Fe^{2+} caused either by oxidation ($Fe^{2+} \rightarrow Fe^{3+}$) or magma mixing may lead to a decrease in the sulfide capacity of the melt. Irvine (1977) has also suggested that an increase in SiO_2 content of a melt, accompanied by magma mixing, may lead to sulfide saturation. Such a process is particularly important with respect to partial melting, as SiO_2 constitutes a low melting component in the partial fusion of crustal material. In addition, SiO_2 may be added to a melt in response to volatile production in the country rocks (Walther and Orville 1982). In systems where water or hydrogen is added to a mafic melt, the dissolution mechanism of sulfur is not clear, and may involve reactions such as:

$$H_2S_{(v)} + O^{2-}_{(m)} + Na^{+}_{(m)} = HS^{-}_{(m)} + ONa^{-}_{(m)} + H^{+}_{(m)}$$

$$H_2S_{(v)} + FeO_{(m)} = FeS_{(m)} + H_2O_{(m)}$$

where (v) and (m) refer to vapor and melt, respectively. In such instances, sulfide saturation is reached due to continued incorporation of volatiles from the country rock.

Stable Isotopes as Indicators of Magma Contamination

Stable isotopic studies constitute one type of tool that can, and ideally should, be applied to problems of melt contamination. However, for a comprehensive evaluation such studies should also be combined with detailed petrographic, trace element, bulk chemical, mineral chemical, and radiogenic isotope studies. As an aid in evaluating the types of data that may be obtained from stable isotopic investigations, a brief review of the systematics of S, C, O, and H isotopes in magmatic systems is included below.

Sulfur. For sulfide ores perhaps the key element, and the one that is most commonly isotopically analyzed in evaluating the role of contamination, is sulfur. Primary, or mantle-derived sulfur is normally considered to be characterized by $\delta^{34}S$ values near the meteoritic value of $\simeq 0‰$. Slight fluctuations are possible due to fractionation occurring during magmatic crystallization. Although sulfur isotopic fractionation between crystals, silicate liquid, and sulfide liquid are expected to be small at temperatures in excess of $\simeq 900\,°C$, particularly where conditions are reducing enough that oxidized sulfur species are minimal, evidence for high temperature fractionation in mafic systems does exist. For example, trends in $\delta^{34}S$ values of sulfides found within layered intrusions have been documented (Shima et al. 1963, Ripley 1983), and may be explained by minor fractionation between Ni-, Cu-, and Fe-sulfide, accompanied by a Rayleigh-type process of ^{34}S enrichment during crystallization. Variations caused by such mechanisms rarely exceed 4 to 5‰. In practice, $\delta^{34}S$ values from sulfides in mafic rocks that fall outside a range of $0 \pm 3‰$ have generally been attributed to contamination by sulfur derived from sedimentary country rocks. By simply comparing observed $\delta^{34}S$ values to those that could likely be produced by pristine magmatic processes, external addition of sedimentary sulfur (or other sulfur component characterized by a $\delta^{34}S$ value far removed from 0‰) is relatively easy to document. Evaluation of kinetic isotope effects associated with sulfur liberation in the wall rocks, mechanisms of sulfur transport, and degree of homogeneity within the melt, are impeded if $\delta^{34}S$ values and distribution within the probable contaminant are not well known. Confirmation of external sulfur addition is significantly more complex if $\delta^{34}S$ values of sulfur in the contaminant are not markedly different from $0 \pm 3‰$. Such may be the case for early Archaean deposits, where sulfur in surrounding sedimentary rocks is commonly near 0‰. Variation in the isotopic composition of sedimentary sulfur is linked to the development of sulfate-reducing bacteria, whose origin is not easily documented prior to about 2.6–2.7 b.y. (Ripley and Nicol 1981, Thode and Goodwin 1983).

Oxygen. The successful use of oxygen isotopes in evaluating magma contamination is dependent on the extent of contamination, the difference in $\delta^{18}O$ between primary melt and contaminant, and the ratio of melt to contaminant, at any stage of the process. In a simple two end-member mixing scheme, mass balance expresssions of the form:

$$\delta^{18}O_{mix} = \delta^{18}O_{parent} X_{parent} + \delta^{18}O_{contam.} X_{contam.}$$

describe the resulting $\delta^{18}O$ value of the hybrid magma. In order to properly evaluate the expression the atomic fraction of oxygen (X) contributed from each end-member must be computed. Considering that $\delta^{18}O$ values of most mafic igneous rocks range from 5.5 to 7.0‰ the expression illustrates that a small degree of contamination may not be recognized unless the contaminant $\delta^{18}O$ is far different than that of the melt. For instance, a 20% contribution of oxygen from a country rock characterized by a $\delta^{18}O$ value of 10‰ to a mafic melt with a $\delta^{18}O$ of 6‰, produces a $\delta^{18}O$ mix of only 6.8‰. Careful and detailed isotopic analyses are required to detect such variations.

Oxygen isotopic distribution within the contaminant must be known to evaluate the extent of partial melting or devolatilization, and possible isotopic exchange mechanisms. Analysis of individual minerals are also useful in investigating the timing of contaminant introduction, particularly in cases of small scale or in situ assimilation, or when intrusion of a crystal mush is suspected.

Such a simple end-member mixing process is probably quite rare in geological environments. A more realistic model may be the situation where one or both end-members changes in composition due to oxygen addition or depletion, and to isotopic fractionation. Taylor (1980) and De Paolo (1981) have described the combined effects of fractional crystallization and assimilation on oxygen isotopic systematics. Evaluation assumes a fixed $\delta^{18}O$ of the contaminant, changing proportions of crystallized igneous rock to assimilated contaminant, and may or may not include fractionation between crystals and melt. In a simple situation where one mineral is crystallizing from a melt and crystal-melt isotopic fractionation occurs, oxygen isotopic distributions may be computed by first determining the isotopic effect of removing a specified amount of mineral on the remaining liquid, followed by calculation of the resultant $\delta^{18}O$ value caused by assimilation of a defined amount of country rock. A simple iterative routine allows calculations of this type to be done on most programmable calculators. Necessary expressions are similar to the following:

δ melt $= \delta(o) - \Delta(A/L)$

$\delta(o)$ = Initial isotopic composition of the primary melt (this value is replaced by δ_{melt} (see below) for successive iterations

Δ = Fractionation factor between crystal and melt

A = Grams of mineral crystallized

L = Grams of remaining liquid

$L_{(i)} = L - 4$ (in the case of 5 g of mineral crystallizing to 1 g of country rock dissolving

$X_{\text{ml}} = (L_{(i)} - 1)/L_{(i)}$; mass fraction of melt

$X_{\text{cr}} = 1 - X_{\text{ml}}$; mass fraction of contaminant added to the melt

$\delta_{\text{cr}} = \delta^{18}O$ of contaminant

$\delta_{\text{melt}} = (\delta_{\text{melt}} \cdot X_{\text{ml}}) + (\delta_{\text{cr}} \cdot X_{\text{cr}})$.

Figure 3 illustrates the specific case of assimilation of country rock with a $\delta^{18}O$ value of 14‰ by a magma with an initial $\delta^{18}O$ of 5.7‰, at a ratio of 5 g olivine crystallized to 1 g country rock assimilated. Computation is only slightly more complex for the fractionation of additional minerals. A family of such curves may be generated [see Taylor (1980) and De Paolo (1981)], all with results quite different from the simple mixing model.

The degree of homogeneity in a melt may also be investigated by detailed oxygen isotopic studies. Diffusion plays an important role in the interaction between mafic magma and silicic partial melts (Watson 1982), and may be the limiting process in the case of in situ assimilation. Attainment of isotopic homogeneity is aided by turbulent mixing of liquids, but zones of isotopically normal melt may persist within an otherwise anomalous melt depending on density, velocity, viscosity, etc. which control the fluid dynamics of the system. Oxygen isotopic homogeneity is discussed further below with reference to the Duluth Complex.

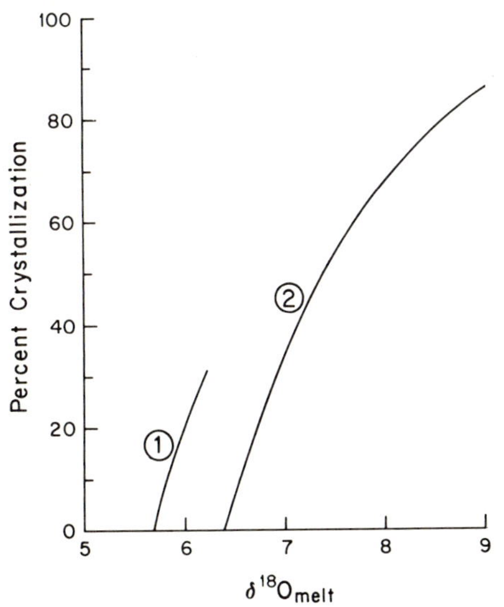

Fig. 3. *Curve 1* shows the $\delta^{18}O_{\text{melt}}$ path for a specific case of assimilation, fractional crystallization of olivine, and fractionation between melt and crystals. Conditions are as follows:

$\delta^{18}O_{\text{initial melt}} = 5.7‰$
$\delta^{18}O_{\text{country rock}} = 14.0‰$
5 g olivine crystallized per 1 g country rock assimilated
$\Delta_{\text{melt-olivine}} = 1.2‰$
$T = 1100\,°C$

Curve 2 is a similar situation for a melt of $\delta^{18}O_{(1)} = 6.4‰$, but without fractionation of a specific mineral and no mineral-melt fractionation

Isotopic variation may also result if the $\delta^{18}O$ value of the contaminant, as well as that of the parent melt, progressively changes. For stable isotopes the magnitude of such an effect may be small, because as temperature increases the fractionation between species decreases. Therefore when partial melting of the contaminant begins, the oxygen isotopic difference between produced melt and residual crystals is expected to be small. Because of possible mineral-fluid fractionations considerable variations in the isotopic composition of the contaminant may result if devolatilization is the primary process responsible for mass transfer. Kinetic effects could in large part control isotopic distributions in such cases.

Hydrogen. Hydrogen isotopes are potentially sensitive indicators of magma contamination because of the small amount of hydrogen present in most basaltic to ultramafic rock types. For isotopic investigation a primary requisite is a significant difference between the δD of the contaminant rocks or released water and that of hydrogen in the primary melt. Often the mere presence of anomalous amounts of hydrous phases in mafic host rocks indicates the introduction of extraneous water. However, δD values of minerals, such as amphibole or biotite (particularly if associated with sulfide ore minerals), may provide direct evidence for melt contamination. If sufficient hydrogen is present in inclusions formed within early crystallizing silicates, such as olivine, pyroxene, or plagioclase, hydrogen isotopic composition of these phases relative to late crystallizing hydrous phases, such as biotite or amphibole, may aid in reconstructing the timing of introduction of country rock hydrogen.

Carbon. Analysis of $\delta^{13}C$ in mafic rocks provides valuable information as to the relative importance of igneous versus sedimentary-derived carbon in the ore-forming process. Many deposits are thought to contain primary graphite associated with sulfides, whereas in deposits, such as those of the Duluth Complex, much of the graphite was derived from country rocks (Hollister 1980, Mainwaring and Douthitt 1981). Primary igneous carbon is normally characterized by $\delta^{13}C$ values near $-5‰$. Sedimentary carbon may be much lighter or heavier, depending on its genesis as biogenic or carbonate carbon of marine or nonmarine origin. Carbon may be incorporated in mafic magmas by partial melting, solid ingestion, or devolatilization of graphite-bearing country rocks. As with the other light stable isotopes, devolatilization provides an opportunity for isotopic variation due to kinetic or equilibrium distributions among produced volatile species (e.g., CH_4, CO_2) and residual carbon. Because of its possible effect on magma oxidation state, and hence on sulfide solubility, knowledge of the mechanism and timing of carbon introduction is desirable in evaluating ore genesis.

Stable Isotopic Studies Applied to Cu – Ni Deposits of the Duluth Complex

Our research group has been involved in detailed mineralogic, chemical, and isotopic investigations of two deposits located in the 1.1 b.y. – old Duluth Complex of Minnesota. Both the Dunka Road and Babbitt deposits occur in the southern portion of the Duluth Complex, within an intrusive sequence known as the Patridge River troctolite (Fig. 2). Rock types present include troctolite, peridotite, norite, olivine gabbro, and gabbro. The sequence constitutes part of what is known as the Troctolitic series (Bonnichsen 1972) and is overlain by a thick succession of predominantly anorthositic rock types (Anorthosite Series). Field evidence suggests that the Troctolitic Series is intrusive into the Anorthositic Series. Mineralogy of the Partridge River troctolite is varied (Bonnichsen 1972, Grant and Molling 1981, Rao and Ripley 1983, Tyson and Chang 1984), with most rock types composed of euhedral to subhedral olivine, plagioclase, or orthopyroxene, and interstitial phases which may include the same minerals, along with clinopyroxene, ilmenite, biotite, and major sulfides. Country rocks in the area include graphitic, and locally sulfidic pelitic metasediments of the Virginia Formation that are underlain by the Biwabik Iron Formation and granitic rocks of the Giant's Range Batholith.

Sulfides are generally localized near the basal contact with the Virginia Formation, but may occur several hundred meters above the lowermost contact. In virtually all cases, however, xenoliths of country rock are found in the general proximity of sulfide accumulations. It is possible that upper ore zones represent previous basal contacts between metasedimentary and igneous rocks. Principal ore minerals are pyrrhotite, chalcopyrite, cubanite, and minor pentlandite. Copper: nickel ratios in this area of the Complex average $\simeq 4-5:1$. Mineralization is dominantly disseminated in form, with sulfides associated with interstitial silicates and oxides, primarily pyroxene, biotite, and ilmenite. Detailed textural descriptions of sulfide occurrences in the Partridge River troctolite are given by Rao and Ripley (1983), Pasteris (1984), and Tyson and Chang (1984). Within most ore zones sulfide volume percentage averages $\simeq 2-3$, although massive occurrences of pyrrhotite or chalcopyrite-cubanite occur locally. Massive ore is more abundant at the Babbitt deposit that at Dunka Road.

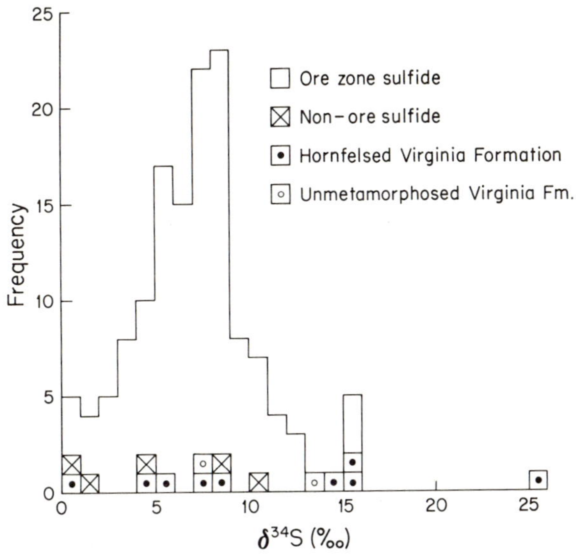

Fig. 4. Histogram of $\delta^{34}S$ values for sulfides in troctolitic rocks of the Dunka Road deposit, Virginia Formation hornfels, and unmetamorphosed Virginia Formation

Sulfur isotopic composition of minerals in the Dunka Road deposit are instructive with regard to sulfur origin and ore formation. The $\delta^{34}S$ values of all sulfide minerals range from 0.2 to 15.3‰, with both mean and mode values near 7.5‰ (Ripley 1981; also see Fig. 4). No trends in sulfur isotopic composition within troctolites are apparent, and values are in general very erratic. Pyrrhotite in the underlying Virginia Formation and pyrite from unmetamorphosed Virginia Formation are characterized by a smilar range in $\delta^{34}S$ values. The high positive $\delta^{34}S$ values and similarity to values found in the underlying Virginia Formation strongly suggest that most of the sulfur in the Dunka Road deposit was derived from the breakdown of pyrite in the Virginia Formation. Sulfur isotopic homogeneity was certainly not attained, and as such is consistent with a relatively late introduction of external sulfur. Nickel contents of olivine (Rao and Ripley 1983) do not indicate that olivine equilibrated with a coexisting sulfide-rich melt. The disseminated form of mineralization, sulfur isotopic variability, and nickel contents of olivine all suggest that sulfide saturation was achieved late in the crystallization sequence of the Dunka intrusive. Such late introduction of sulfur also would have inhibited the development of massive sulfide zones.

Oxygen isotopic results from Dunka (Fig. 5) shed light on the mechanisms responsible for magma contamination. The $\delta^{18}O$ values of troctolites and norites range from 5.8 to 9.6‰. Anomalous $\delta^{1}O$ values (>8‰) are found only near the margins of country rock xenoliths. The $\delta^{18}O$ values of Virginia Formation hornfels in the Dunka Road area range from 8.8 to 11.1‰. High $\delta^{18}O$ values correspond to the presence of orthopyroxene rather than olivine in the igneous rocks. Bulk chemical studies verify the introduction of SiO_2 in these localities, and

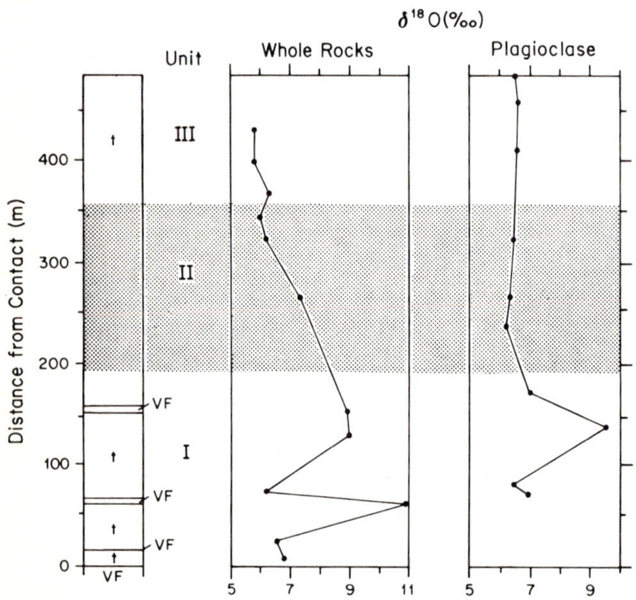

Fig. 5. Representative whole rock and plagioclase $\delta^{18}O$ values for troctolites and Virginia Formation inclusions, Dunka Road deposit

together with isotopic data are suggestive of partial melting along the rims of xenoliths. However, bulk chemical and isotopic anomalies are restricted to within $\simeq 3$ m of inclusion contacts. The $\delta^{18}O$ values of mineral separates confirm the whole rock systematics, with olivine, pyroxene, plagioclase, ilmenite, and biotite values paralleling those of whole rocks.

Sulfur and oxygen isotopic data, together with mineralogical and chemical analyses, strongly suggest that assimilation of nonvolatile elements via partial melting was not a strong control on sulfide genesis at Dunka Road. Isotopic systematics are more consistent with major contamination caused by devolatilization of country rocks. Reactions such as:

$$FeS_2 + 3H_2O + 5/2C = FeS + 3/2CO_2 + CH_4 + H_2S$$

are thought to characterize volatile production in the contact rocks and xenoliths. Experiments done in our lab (Ripley and Snyder, in preparation) indicate that under hydrous conditions, little sulfur isotopic fractionation accompanies the production of H_2S from pyrite. Therefore an in situ model of volatile transfer and melt contamination may best explain the disseminated nature of sulfide mineralization and variable $\delta^{34}S$ values. Sulfur is thought to be derived from local accumulations of pyrite in the Virginia Formation, which are themselves characterized by variable $\delta^{34}S$ values. Because of the late introduction of H_2S into the crystal mush, sulfur mobility and isotopic homogenization were impeded.

Several other mechanisms that may have contributed to sulfur isotopic variability at Dunka Road are discussed by Ripley (1981). Included are mixing of sedimentary and igneous sulfur, isotopic fractionation caused by fluctuations in oxidation states during sulfur volatilization and assimilation, and periodic introduction of isotopically distinct magma batches. Although none of these processes can be entirely eliminated, isotopic, chemical, and mineralogical data are more consistent with the devolatilization model described above.

The Babbitt deposit, located to the northeast of the Dunka Road area (Fig. 2) shares many common characteristics, although isotopic systematics suggest a somewhat different history of ore genesis. Ore at Babbitt is generally disseminated in form, as at Dunka, but massive, high-grade areas occur in greater abundance, and render the deposit economically more attractive. Concentrations of platinum group elements at Babbitt far exceed those at Dunka. Areas of massive to semi-massive (>10% sulfides) ore are locally found in both norites and cordierite-orthopyroxene hornfels. Contacts between igneous and metasedimentary rocks in such areas are difficult to discern, and norite is far more abundant than troctolite, suggesting appreciable introduction of silica. Based on these criteria alone, the degree of probable contamination at Babbitt would appear to be greater than at Dunka.

Oxygen isotopic values of whole rocks and mineral separates from Babbitt confirm mineralogical and chemical evidence for contamination. Igneous rock $\delta^{18}O$ values range from 5.4 to 12.1‰, with a mode at 7 – 7.5‰ (Ripley and Al-Jassar 1983, and in preparation). Metamorphosed Virginia Formation $\delta^{18}O$ values range from 8.0 to 14.5‰, as do representative samples from mildly metamorphosed Virginia Formation graywacke, slate, and argillite. The $\delta^{18}O$ values

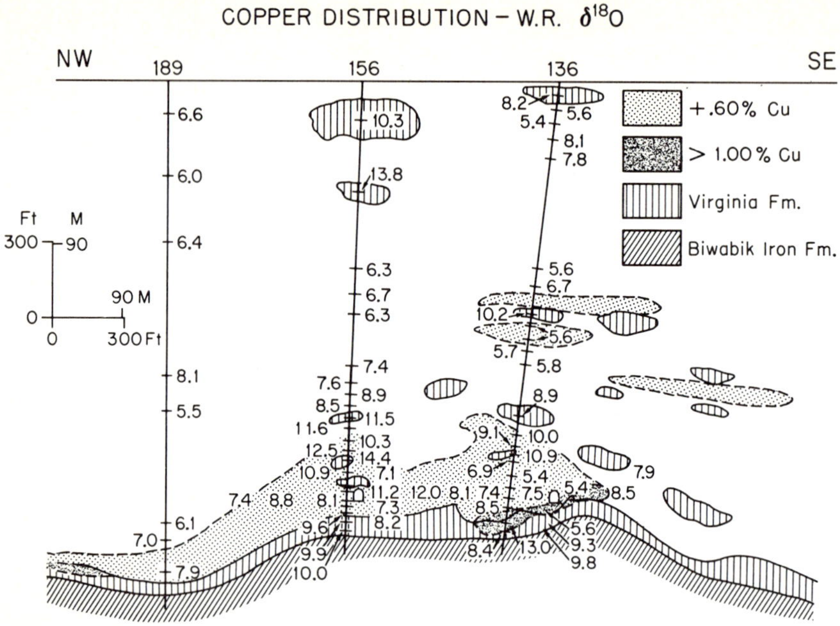

Fig. 6. Representative whole rock $\delta^{18}O$ values for igneous rocks and Virginia Formation hornfels along a drill section, Babbitt deposit

Table 1. Oxygen isotope values through two igneous rock-metasedimentary xenolith profiles, Babbitt deposit

Sample depth (m)	$\delta^{18}O$ (‰ SMOW)	Rock type
Locality 1		
279.8	7.5	Troctolite
280.0	9.1	Troctolite
281.3	11.6	Hornfels
282.8	14.1	Hornfels
284.3	8.8	Hornfels
284.7	8.4	Hornfels
285.9	5.8	Troctolite
Locality 2		
516.0	6.7	Troctolite
517.5	6.5	Troctolite
519.7	11.7	Hornfels
520.9	10.4	Hornfels
523.6	11.1	Hornfels
525.1	12.5	Hornfels
526.7	10.7	Hornfels
528.2	11.0	Hornfels
529.7	10.9	Hornfels
531.2	12.1	Hornfels
532.9	10.9	Hornfels
534.3	8.3	Troctolite
535.8	8.8	Troctolite

show a sporadic distribution within ore zones (Fig. 6). Trends are difficult to distinguish, in part due to the complex mixture of rock types present and in part due to sampling dictated by drill core location. Detailed oxygen isotopic studies around Virginia Formation inclusions show $\delta^{18}O$ variations in both the metasediments and surrounding igneous rocks. Table 1 illustrates the results of two profiles through xenoliths and surrounding igneous rock. Although inclusion hornfels are mineralogically very similar, their $\delta^{18}O$ values may vary by as much as 5‰ over a 2-m interval. Some of this isotopic variability may be attributed to differences in the sedimentary prototypes (graywacke, shale, calcareous units), and some must be due to interaction with the mafic magma. The $\delta^{18}O$ values of igneous rocks in contact with the inclusions decrease away from the contact. This trend is similar to that observed at Dunka Road, but the width of isotopically anomalous zones is generally greater. Within sulfide ore zones over 80% of the igneous rocks analyzed are considered isotopically anomalous. Areas of isotopically normal (5.5 – 7.0‰) igneous rocks occur sporadically within areas having anomalous values.

Locations of isotopically normal troctolites and norites within extensive anomalous zones are difficult to evaluate. Two possibilities seem plausible. One is that assimilation occurred essentially in place, with mass transfer limited by diffusion within the melt. In this manner isotopic "exchange" zones may be produced around xenoliths, some in overlapping fashion. Figure 7 illustrates a case where areas of igneous rock characterized by anomalous $\delta^{18}O$ values surround isotopically normal troctolite.

A second alternative is that contamination occurred during and after intrusion, and zones of turbulent mixing were limited by the presence of xenoliths. The fluid dynamics of partial melting of country rock around an intrusive have been modelled by Kovach and Marsh (1981). Their results suggest that extensive partial melting may occur only if a critical velocity is maintained by the magma. A mafic magma intruding a metasedimentary sequence may remove blocks of wall rock during ascent, but thorough mixing of the partial and primary melts is dependent upon flow patterns as well as diffusion. Local instabilities may be generated along the interface between the two melts (e.g., Huppert and Sparks 1980, Irvine 1980), with the entrainment of contaminants restricted to well-

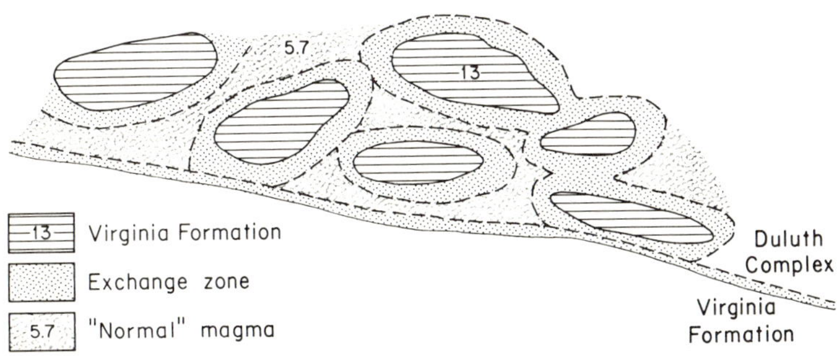

Fig. 7. Oxygen isotopic "exchange" zones developed around the margins of xenoliths, with intervening areas of isotopically "normal" igneous rock

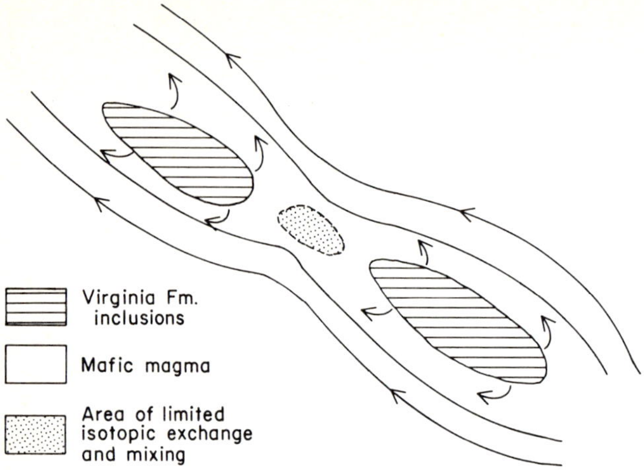

Fig. 8. Hypothetical flow and mixing patterns around xenoliths in a magma of fixed velocity. A small area of essentially unaffected melt occurs between the xenoliths

defined localities. Figure 8 illustrates a situation where flow and mixing patterns around metasedimentary xenoliths produce a small intervening area of melt only moderately affected by the partial melting process. The oxygen isotopic composition obtained in the zone of mixing is dependent on the $\delta^{18}O$ values of xenoliths, diffusion coefficients, fluid dynamic properties of the melts, and time of the mixing process. Variations in any of these parameters could lead to the inhomogeneity observed in isotopically "anomalous" zones.

Sulfur isotopic studies at Babbitt also reveal a pattern of sulfur evolution that differs from that at Dunka. The range and mean of sulfur isotopic values is approximately the same as at Dunka (Fig. 9), and leave little doubt that most of the sulfur in the deposit is characterized by a strong sedimentary component. However, two distinctions are noteworthy. First is the difference between $\delta^{34}S$ values of sulfides in the igneous rocks versus those of pyrrhotite in the Virginia Formation hornfels. Unlike Dunka, where isotopic values were similar to those in troctolites, hornfels from the Babbitt area have $\delta^{34}S$ values that range from 1.5 to 6.2‰, and are lower than over 95% of the values for sulfides in ore zones. A simple explanation of in situ devolatilization or partial melting does not adequately explain the $\delta^{34}S$ distribution. Although the $\delta^{34}S$ values of pyrrhotite from the hornfels are low, they do not fall outside the range for pyrite in weakly metamorphosed Virginia Formation. It is apparent that sulfur has been lost in the conversion to pyrrhotite, but evidence is lacking for the introduction of sufficient sulfur from this source to cause a decrease in ore zone $\delta^{34}S$ values.

Two possible mechanisms are considered for the derivation of sulfur at Babbitt. First is the rather unlikely possibility that partial melting of Virginia Formation has removed all traces of sulfides having $\delta^{34}S$ values in excess of $\simeq$6‰. The fact that $\delta^{34}S$ values of Virginia Formation pyrrhotite from several localities in the Babbitt area show a similar range of values argues against such a hypothesis. A second mechanism involves the derivation of sulfur prior to or during magma ascent. Such a process is similar to that proposed for magma con-

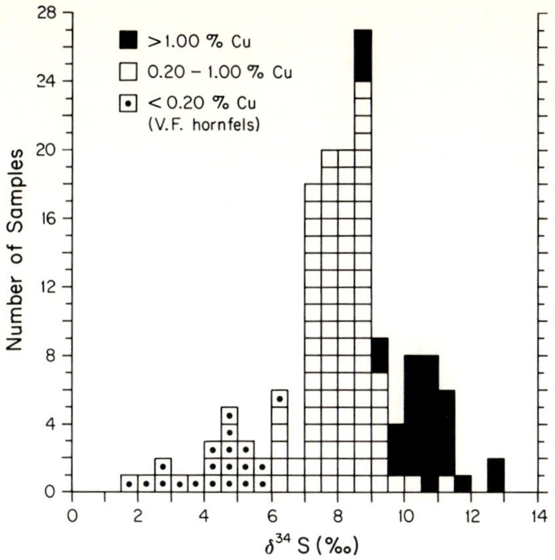

Fig. 9. Histogram of δ^{34}S values for sulfides in igneous rocks and Virginia Formation hornfels from the Babbitt deposit

tamination based on δ^{18}O studies. The δ^{34}S values at Babbitt are consistent with a model of major sulfur derivation from rocks not now represented in the area of the Babbitt deposit. Assimilation of crustal sulfur during ascent, with limited sulfur communication in the melt, could produce the observed δ^{34}S systematics. In addition, a model of sulfur derivation prior to or during magma ascent provides a means for immiscible sulfide liquids to be brought into contact with a greater amount of silicate liquid. Campbell et al. (1983) have presented a model for the generation of Pt-rich sulfide deposits that emphasizes the importance of large silicate/sulfide melt ratios. Analyses of Pt-group elements in the Babbitt and Dunka Road deposits indicate a greater than four times enrichment in the Babbitt deposit versus Dunka Road. Extraction of metal from a larger mass of magma is one mechanism which may account for the compositional differences.

A second unique feature of the δ^{34}S values found at Babbitt is the distribution of isotopes between low grade (Cu-poor) and high grade (>1% Cu) ore (Figs. 9 and 10). Although δ^{34}S values show an overall variability, there is a distinct isotopic fractionation consistent with ore grade. Copper-rich zones are characterized by δ^{34}S values in excess of 9‰, whereas lower grade sulfide occurrences show δ^{34}S values between 6 and 10‰. This isotopic zonation strongly suggests that significant sulfur isotopic fractionation occurred between a separated Cu-enriched sulfide liquid and a Cu-poor residual melt (including any sulfur still dissolved in the silicate melt). A similar ^{34}S enrichment in copper-rich sulfides from the Levack West deposit of the Sudbury areas has been described by Naldrett (1981).

Both sulfur and oxygen isotopic studies suggest that significant contamination via both partial melting and devolatilization, has occurred at Babbitt. Sulfur isotopic values are difficult to explain in terms of in situ sulfur derivation. However, both sulfur and oxygen isotopic compositions are consistent with a model of major contamination prior to or during emplacement, with isotopic homoge-

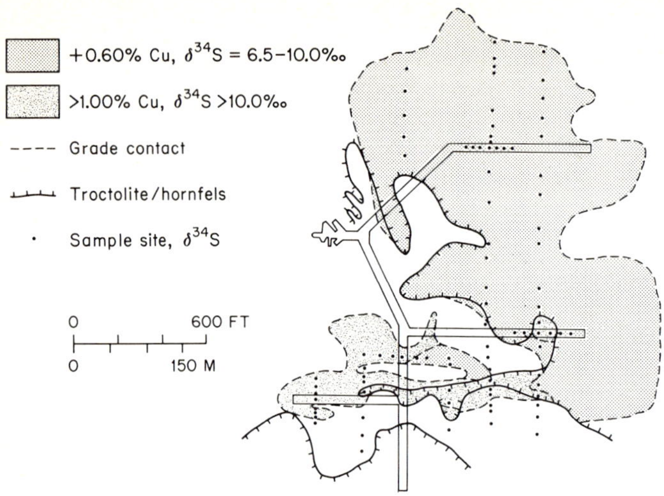

Fig. 10. Copper grade and $\delta^{34}S$ distribution, 1700-ft level, Babbitt deposit

neity prohibited by fluid dynamic properties of the melts. Critical parameters are the size and distribution of crystals and xenoliths, diffusion rates, temperature, density and viscosity of the primary melt and contaminant partial melt, and magma velocity.

Acknowledgments. Appreciation is expressed to the geological staffs of United States Steel Corporation (Dunka Road deposit) and AMAX and Kennecott Company (Babbitt deposit) for cooperation during all phases of our studies in the Duluth Complex. Many of the ideas presented in this paper have benefited by discussions with A.J. Naldrett, B.V. Rao, R. Bonnichsen, P. Weiblen, M. Foose, M. Tyson, N. Grant, T. Al-Jassar, J.D. Pasteris, and D. Cole. Research on the genesis of Cu–Ni ores in the Duluth Complex has been supported through National Science Foundation grants EAR 78-01699 and EAR 81-08536.

References

Bonnichsen BG (1972) Southern part of the Duluth Complex. In: Sims PK, Morey GB (eds) Geology of Minnesota: A centennial volume. Minn Geol Sur, St. Paul, pp 361–388

Boyd R, Mathiesen CO (1979) The nickel mineralization of the Rana mafic intrusion, Nordland, Norway. Can Miner 17, 2:287–299

Buchanan DL, Nolan J (1979) Solubility of sulfur and sulfide immiscibility in synthetic tholeiitic melts and their relevance to Bushveld-complex rocks. Can Miner 17, 2:483–495

Campbell IH, Naldrett AJ, Barnes SJ (1983) A model for the origin of the platinum-rich sulfide horizons in the Bushveld and Stillwater Complexes. J Petrol 24:133–165

Carter SR, Evensen NM, Hamilton PT, O'Nions RK (1978) Nd- and Sr-isotopic evidence for crustal contamination of continental volcanics. Science 202:743–747

Cooper RW, Morey GB, Weiblen PW (1981) Topographic and aeromagnetic lineaments and their relationship to bedrock geology in glaciated Precambrian terrane, northeastern Minnesota. In: O'Leary O (ed) Proc 3rd Int Conf Basement Tectonics. Basement Tecton Commun Inc 3:137–148

Danckwerth PA, Hess PC, Rutherford MJ (1979) The solubility of sulfur in high-TiO_2 mare basalts. Proc Lunar Planet Sci Conf 10:517–530

DePaolo DJ (1981) Trace element and isotopic effects of combined wallrock assimilation and fractional crystallization. Earth Planet Sci Lett 53:189–202

DeWaal SA (1977) Carbon dioxide and water from metamorphic reactions as agents for sulfide and spinel precipitation in mafic magmas. Geol Soc S Afr Trans 80:193–197

Fincham CJB, Richardson FD (1954) The behavior of sulfur in silicate and aluminate melts. Proc R Soc London Ser A 223:40–62

Grant NK, Molling PA (1981) A strontium isotope and trace element profile through the Partridge River troctolite, Duluth Complex, Minnesota. Contrib Mineral Petrol 77:296–305

Gray CM, Cliff RA, Goode ADT (1981) Neodymium-strontium isotope evidence for extreme contamination in a layered basic intrusion. Earth Planet Sci Lett 56:189–198

Groves DI, Barrett RM, McQueen KG (1979) The relative roles of magmatic segregation, volcanic exhalation, and regional metamorphism in the generation of volcanic-associated nickel ores of Western Australia. Can Miner 17, 2:319–337

Haughton D, Roder PL, Skinner BJ (1974) Solubility of sulfur in mafic magmas. Econ Geol 69:451–567

Hollister VF (1980) Origin of graphite in the Duluth Complex. Econ Geol 75:765–766

Huppert HE, Sparks RSJ (1980) The fluid dynamics of a basic magma chamber replenished by hot, dense ultrabasic magma. Contrib Mineral Petrol 75:279–289

Irvine TN (1977) Origin of chromite layer in the Muskox Intrusions and other stratiform intrusions: a new interpretation. Geology 5:273–277

Irvine TN (1980) Magmatic infiltration metasomatism, double diffusive fractional crystallization, and adcumulus growth in the Muskox intrusion, and other layered intrusions. In: Hargraves RB (ed) Physics of magmatic processes. Princeton Univ Press, Princeton, pp 325–385

Jacquemin H, Sheppard SMF, Vidal P (1982) Isotopic geochemistry (O, Sr, Pb) of the Golda Zuelva and Mboutou anorogenic complexes, North Caneroun: mantle origin with evidence for crustal contamination. Earth Planet Sci Lett 61:97–111

Kovach LA, Marsh BD (1981) Magma flow rate and partial fusion of wall rock; Hungtington Lake, CA. Geol Soc Am (Abstracts with Program) 13:490

Liebenberg L (1970) The sulphides in the layered sequence of the Bushveld Igneous Complex. Geol Soc S Afr, Spec Publ 1:108–207

Mainwaring PR, Douthitt CB (1981) Carbon isotope study of graphitic sulfides in the Duluth Complex, Minnesota, USA (Abstracts). Geol Assoc Can 6:A-37

Mainwaring PR, Naldrett AJ (1977) Country rock assimilation and the genesis of Cu–Ni sulfides in the Water Hen intrusion, Duluth Complex, Minnesota. Econ Geol 72:1269–1284

Myers JD, Sinha AK, Marsh BD (1983) Assimilation of crustal material by basaltic magma: Strontium isotopic and trace element data from the Edgecumbe Volcanic Field, S.E. Alaska. J Petrol 25, 1:1–26

Naldrett AJ (1981) Nickel sulfide deposits: classification, composition, and genesis. Econ Geol 75th Anniversary Vol 2:628–685

Page NJ (1979) Stillwater Complex, Montana – structure, mineralogy, and petrology of the Basal zone with emphasis on the occurrence of sulfides. US Geol Surv Prof Pap 1038:1–69

Pasteris JD (1984) Further interpretation of the Cu–Fe–Ni sulfide mineralization in the Duluth Complex, northeastern Minnesota. Can Miner 22:39–53

Rao BV, Ripley EM (1983) Petrochemical studies of the Dunka Road Cu–Ni deposit. Econ Geol 78:1222–1238

Rao BV, Naldrett AJ, Evensen NM (1984) Crustal contamination in the sublayer, Sudbury igneous complex: A combined trace element and strontium isotope study. Geol Assoc Can (Abstrat with Program) 9:98

Ripley EM (1981) Sulfur isotopic studies of the Dunka Road Cu–Ni Deposit, Duluth Complex, Minnesota. Econ Geol 76:610–620

Ripley EM (1983) Sulfide mineralogy and sulfur isotope geochemistry of layered sills in the Deer Lake Complex, Minnesota. Mineral Depos 18:3–15

Ripley EM, Al-Jassar T (1983) Oxygen isotopic studies of the Babbitt Cu–Ni prospect, Duluth Complex, Minnesota (Abstracts). Geol Soc Am (Abstract with Program) 15:671

Ripley EM, Nicol DL (1981) Sulfur isotopic studies of Archean slate and graywacke from Northern Minnesota: Evidence for the existence of sulfate-reducing bacteria. Geochim Cosmochim Acta 45:839–847

Sasaki A (1969) Sulfur isotope study of the Muskox intrusion, District of MacKenzie. Geol Surv Can Pap 68-46:68 pp

Shima H, Naldrett AJ (1975) Solubility of sulfur in an ultramafic melt and the relevance of the system Fe – S – O. Econ Geol 60:960 – 967

Shima M, Gross WH, Thode HG (1963) Sulfur isotope abundances in basic sills, differentiated granites, and meteorites. J Geophys Res 68:2835 – 2847

Taylor HP Jr (1980) The effects of assimilation of country rocks by magmas on $^{18}O/^{16}O$ and $^{87}Sr/^{86}Sr$ systematics in igneous rocks. Earth Planet Sci Lett 47:243 – 254

Taylor HP Jr, Giannetti B, Turi B (1979) Oxygen isotope geochemistry of the potassic igneous rocks from the Roccamonfina volcano, Roman comagmatic region, Italy. Earth Planet Sci Lett 46:81 – 106

Thode HG, Goodwin AM (1983) Further sulfur and carbon isotope studies of late Archaean Iron-Formations of the Canadian Shield and the rise of sulfate-reducing bacteria. Precambrian Res 20:337 – 356

Thompson RN, Dickin AP, Gibson IL, Morrison MA (1982) Elemental fingerprints of isotopic contamination of Hebrideau Paleocene mantle-derived magmas by Archaean sial. Contrib Mineral Petrol 79:159 – 168

Tyson RM, Chang LLY (1984) The petrology and sulfide mineralization of the Patridge River troctolite, Duluth Complex, Minnesota. Can Miner 22:23 – 38

Vinogradov AP, Grinenko LN (1964) On the influence of including rocks on the isotopic composition of ore sulfides. Geokhimiya 9:491 – 499

Walther JV, Orville PM (1982) Volatile production and transport in regional metamorphism. Contrib Mineral Petrol 79:252 – 257

Watson EB (1982) Basalt contamination by continental crust: some experiments and models. Contrib Mineral Petrol 80:73 – 87

Weiblen PW, Morey GB (1976) Textural and compositional characteristics of sulfide ore from the basal contact zone of the South Kawishiwi intrusion, Duluth Complex, northeastern Minnesota. Proc 49th Annu Minn Sect AIME, 24 pp

Controls on the Formation of Komatiite-Associated Nickel-Copper Sulfide Deposits

C. M. LESHER[1] and D. I. GROVES[2]

Abstract

Nickel-copper sulfide deposits associated with rocks of komatiitic affinity may be divided into two associations on the basis of host rock composition and ore distribution: (1) stratiform, massive-matrix-disseminated sulfides hosted by cumulate komatiite lava flows (lava conduits), or strata-bound, coarse disseminated sulfides hosted by cumulate komatiite bodies near volcanic vents, and (2) strata-bound, finely disseminated sulfides and more restricted massive sulfides hosted by subvolcanic komatiitic dunites (high-level magma chambers). These deposits represent a continuum in environments of emplacement from distal volcanic through proximal and central volcanic to subvolcanic.

Field relationships and geochemical data suggest that the sulfides in these deposits are magmatic in origin. Previous models for their formation involve derivation of sulfides directly from the mantle or from low-level magma chambers, but there are several field, physical, and thermodynamic constraints which suggest that this may not be feasible: (1) the spatial concentration of mineralization within certain parts of otherwise virtually barren host units suggests that the latter were not uniformly saturated in sulfide at the time of emplacement; (2) there are physical problems in transporting dense sulfides in low viscosity komatiite magmas, although dispersed sulfide droplets may be supported during rapid ascent; and (3) high temperature komatiitic melts generated by high-degree partial melting, or sequential melting, of mantle source rocks are unlikely to be sulfide-saturated in the source region or to remain sulfide-saturated during adiabatic ascent. These deposits are, however, commonly associated with specific country rocks, particularly sulfidic sediments, and the distribution of the ores appears to be related to the distribution of these rocks. It is most likely that the sulfide ores exsolved from the komatiitic magmas in situ, in response to assimilation of the country rocks during emplacement. The distribution and texture of the mineralization was probably influenced by (1) the mode of emplacement (volcanic vs subvolcanic) and amount of assimilation; (2) the relative volume of magma (lava conduit vs magma chamber) and its capacity to dissolve sulfur; and (3) the timing of sulfide separation and opportunity for segregation.

[1] Department of Geology, University of Alabama, University, AL 35486, USA
[2] Department of Geology, University of Western Australia, Nedlands 6009, Western Australia

Geology and Metallogeny of Copper Deposits
(ed. by G.H. Friedrich et al.)

A fundamental aspect of this model is the coincidence of cumulate komatiitic host units (representing dynamic lava conduits and magma chambers) and siliceous, sulfidic, or ferric iron oxide-rich sediments, the former representing a source of heat for assimilation and of ore metals, and the latter a source of sulfur or of components to induce sulfide separation. Mineralized greenstones appear to have formed in deeper water under conditions of greater extension than unmineralized greenstones; rifting environments represent the most favorable tectonic environment for voluminous komatiite eruption/irruption and for accumulation of sulfidic sediments. Variations in the degree of extension along or within the rift may influence the mode of komatiite emplacement and therefore the type of deposit.

Introduction

Nickel-copper sulfide deposits associated with komatiitic rocks in Archean greenstone belts currently account for over 25% of the world's identified resources of sulfide Ni with greater than 0.8% Ni (Ross and Travis 1981). Field relationships and geochemical data suggest that the sulfides in these deposits are magmatic in origin (Groves et al. 1979, Naldrett 1979). However, several fundamental aspects of ore genesis remain poorly constrained and many details of ore localization are still debated (cf. Lesher et al. 1984, Gresham, this volume).

The most important issue is the timing of separation of the sulfides in these deposits. Most models for the formation of the Ni–Cu ores have stressed their restrictive association with highly magnesian host rocks and have inferred that sulfides were an intrinsic component of the precursor magma (e.g., Nesbitt 1971, Naldrett 1973, Arndt 1976a). For example, Naldrett (1973) suggested that mantle source regions were enriched in sulfides through downward percolation in areas away from major convective overturn. Nesbitt (1971) and Naldrett and Cabri (1976) noted that some deposits contain concentrations of sulfides that are too great to have been dissolved in the overlying host units, and proposed that the sulfides were carried in suspension directly from the mantle. Under the assumption that the olivine in the host units is intratelluric, Ross and Hopkins (1975) and Marston et al. (1981) suggested that immiscible sulfides and olivine crystals accumulated (in a subsurface magma chamber) prior to eruption, and that sulfides were erupted as a segregated phase.

There are several problems with such models. Firstly, the tectonic setting and petrogenesis of komatiites in the Archean have not been established with any degree of certainty. If komatiites in greenstone belts were formed in settings similar to marginal basins (probably the best modern analog: Weaver and Tarney 1979), were generated by fusion of subducted komatiitic ocean crust (e.g., Allegre 1982, Arndt 1983), or were derived from a laterally extensive mantle melt-layer (Nisbet and Walker 1982), then sulfides are unlikely to have become concentrated in the manner suggested by Naldrett (1973). Secondly, iron-nickel sulfide liquids are very dense (4.2–4.3 $g\,cm^{-3}$: Dobrovinski et al. 1969) and should settle to the base of a slowly ascending diapir and be lost in precisely the same manner that they were originally proposed to have concentrated in the source region. Thirdly, the present distribution of sulfides in the ore zones is not necessarily representative of their abundance in the original magma, because the host units are laterally more extensive than the ores (Lesher et al. 1984). Mass balance calculations, therefore, provide only minimum values of magma: sulfide ratios. Finally, dense sulfide liquids that have segregated at the base of a magma chamber (or diapir) are unlikely to be erupted through the less dense lithosphere/crust.

Most models for the formation of these deposits are based on specific features of a few individual deposits and have not considered the more fundamental features of the deposits as a group. There are also several important field, physical, and thermodynamic constraints on the generation and transport of sulfides in komatiitic magmas that require evaluation before models involving eruption of mantle-derived sulfides can be accepted. The aims of this paper are (1) to summarize the fundamental characteristics of komatiite-associated nickel-copper deposits as a group; (2) to discuss some of the field, physical, and thermodynamic constraints on the generation and eruption of sulfide-bearing komatiites; and (3) to propose an alternative, general model for their formation within that context.

Fundamental Features

Komatiite-associated nickel-copper sulfide deposits may be divided into two associations on the basis of host rock composition and ore distribution:

1. Cumulate komatiite-associated deposits:

a) stratiform, massive-matrix-disseminated sulfides hosted by cumulate komatiite lava flows (e.g., Kambalda: Gresham and Loftus-Hills 1981; Wannaway: McQueen 1981b; Windarra: Schmulian 1984; Scotia: Page and Schmulian 1981; Langmuir: Green and Naldrett 1981; Trojan: Chimimba 1984),
b) strata-bound, coarse-disseminated sulfides hosted by cumulate komatiite bodies near volcanic vents (e.g., Shangani: Viljoen et al. 1976; Damba-Silwane: Williams 1979), and

2. Komatiitic dunite-associated deposits:

strata-bound, finely disseminated sulfides and more restricted massive sulfide lenses hosted by subvolcanic komatiitic dunites (e.g., Mt. Keith-Betheno: Groves and Keays 1979; Six Mile Well: Naldrett and Turner 1977; Honeymoon Well: Donaldson and Bromley 1982; Agnew: Billington 1984; Forrestania: Porter and McKay 1981; Dumont: Eckstrand 1975; Katiniq: Barnes et al. 1981).

The first association has been referred to previously as the “volcanic peridotite” association, the second as the “intrusive dunite” association (Marston et al. 1981). The host rocks are here recognized as members of the komatiitic suite (see Arndt and Brooks 1980, Arndt and Nisbet 1982), and are named accordingly.

Table 1. Comparison of the cumulate komatiite and komatiitic dunite hosts to nickel sulfide mineralization in the Norseman-Wiluna Belt, Western Australia

	Cumulate komatiite	Komatiitic dunite
Structure:		
Thickness	Thin (30 – 100 m)	Thick (300 – 1000 m)
Layering	Both: fine-scale cryptic-rhythmic layering	
Textures	Orthocumulate to mesocumulate, crescumulate	Mesocumulate to adcumulate
Grain size	Fine-medium	Coarse
Composition:		
Whole rock	40 – 45% MgO	45 – 50% MgO
Modal olivine	50 – 75%	75 – 95%
Olivine	Fo 90 – 95	Fo 90 – 95
Inferred parent	28 – 32% MgO	28 – 32% MgO
Mineralization:	High-grade, stratiform, massive-disseminated	Low-grade, strata-bound, disseminated, rarely massive
Probable nature:	Lava conduit	Subvolcanic sill

In Western Australia enough detailed work has been done that a comparison (Table 1) can be made between the two types that are found in the Norseman-Wiluna Belt (see Lesher et al. 1982). The cumulate komatiites and komatiitic dunites in Western Australia appear to have formed from parental magmas of similar composition (28 – 32% MgO: Naldrett and Turner 1977, Donaldson and Lesher 1982) and are interpreted to result from flow-through fractional crystallization and in situ accumulation of olivine: the former as linear lava conduits and the latter as subvolcanic magma chambers.

It is important to recognize that these deposits represent a continuum in environments of emplacement from distal volcanic through proximal and central volcanic to subvolcanic. For example, the Shangani, Damba-Silwane, and Langmuir #2 deposits appear to overlie feeders and probably occur in central volcanic settings. The host komatiite sequence at the Scotia deposit contains komatiitic intrusives and komatiitic pyroclastics, and it is considered to occur in a proximal volcanic setting. In contrast, no feeders, komatiitic intrusives, or pyroclastic breccias have been identified at the Kambalda or Widgiemooltha deposits and these most likely occur in distal volcanic settings. The komatiites were probably erupted from discrete volcanic centers within crustal rift zones (see Groves et al. 1984). The absence of sheeted dikes and the presence of multiple horizons of interflow sediments indicates that this was not a midoceanic ridge environment. The setting of most komatiitic dunite-associated deposits is less certain. The better preserved and more extensively studied examples (e.g., Six Mile Well, Honeymoon Well, Forrestania, and Katiniq), exhibit rhythmic or cryptic layering, are conformable with enclosing volcanics, and/or appear to be petrogenetically related to overlying komatiitic volcanics. They are interpreted by most workers as subvolcanic sills that fed the overlying komatiitic volcanics.

The two major deposit types appear to be concentrated in specific tectonic-volcanic provinces and are normally spatially separated from each other. For example, cumulate komatiite-associated deposits are concentrated in the southern part of the Norseman-Wiluna Belt (Marston et al. 1981, Lesher et al. 1982), in the eastern part of the Abitibi Belt (Naldrett and Gasparini 1971, Coad 1979), and in Zimbabwe (Williams 1979, Clutten et al. 1981, Hammerbeck 1984). Komatiitic dunite-associated deposits are concentrated in the northern part of the Norseman-Wiluna Belt (Marston et al. 1981, Lesher et al. 1982), in the southern part of the Southern Cross Province (Porter and McKay 1981), and in the Circum-Superior Belt (Kilburn et al. 1969). Mineralized komatiitic dunites may be overlain by barren komatiite lavas (e.g., Six Mile Well, Honeymoon Well, Forrestania) and barren komatiitic dunites may be overlain by poorly mineralized cumulate komatiites (e.g., Mt. Clifford: Travis 1975), but rarely are both mineralized.

The most important and distinctive characteristic of these deposits, as a group, is their association with *cumulate* rocks of komatiitic affinity. Although the ultramafic portion of the komatiitic suite comprises a complete continuum from aphyric pyroxenitic komatiite (16 – 20% MgO) and ultramafic komatiite (20 – 32% MgO) through porphyritic komatiite (28 – 36% MgO) to orthocumulate-mesocumulate-crescumulate komatiite (36 – 44% MgO) and mesocumulate-adcumulate komatiitic dunite (44 – 50% MgO), the nickel sulfide ores are associated exclusively with the cumulate rocks. This restriction is observed in Western Australia, Canada, and Zimbabwe (references above); greenstone belts that contain komatiites, but only minor cumulate komatiites (e.g., Barberton belt, South Africa; Murchison, Southern Cross, and East Pilbara provinces, Western Australia) generally do not contain significant nickel sulfide deposits. It should be noted that the composition of the liquid from which the cumulates formed does not appear to be very important. Although most Archean deposits are found associated with aphyric lavas ranging between 24 and 32% MgO (Lesher et al. 1981), some Proterozoic deposits are associated with rocks derived from liquids with as low as 19% MgO (Barnes et al. 1981). The petrogenetic affinity of the komatiites (see Nesbitt et al. 1979) also does not appear to be critical (cf. Marston et al. 1981). Although most important deposits are associated with aluminum-undepleted komatiites (e.g., Kambalda: Lesher et al. 1981), some are associated with aluminum-depleted komatiites (e.g., Forrestania: Porter and McKay 1981; Ruth Well: Barley and Bickle 1982; E. G. Nisbet and G. A. Chinner, personal communication, 1984).

As originally noted by Prider (1970) in Western Australia, most of these deposits are also intimately associated with certain country rocks, particularly sulfidic sediments. Most importantly, the distribution of the ores appears to be related to the distribution of these rocks. These relationships are most pronounced for stratiform deposits. For example, the ores at several Kambalda shoots (e.g., parts of Juan, Lunnon, and Jan shoots: Paterson et al. 1984) grade laterally into sulfidic quartz-albite sediments. The ores at Redstone (Robinson and Hutchinson 1982) grade laterally into sulfide facies iron-formation. The ores at Langmuir #2 (Green and Naldrett 1981), Mt. Edwards (INAL Staff 1975), and Foster (Gresham and Loftus-Hills 1981) lap onto sulfidic sediments. The ores at several Kambalda shoots (e.g., parts of Ken, McMahon, and Cruickshank shoots: Gresham and Loftus-Hills 1981), Scotia (Page and Schmulian 1981), Mt. Windarra (Schmulian 1984), Trojan (Chimimba 1984), Texmont and Hart (Coad 1979) directly overlie sulfidic sediments that may be locally thinned beneath the ores. The footwall rocks at the Alexo and Sothman deposits (Coad 1979) are sulfidic volcanics. The ores at Wannaway (McQueen 1981b) overlie altered remnants of sulfidic sediments that are better preserved along strike. The ores at many Kambalda shoots (Gresham and Loftus-Hills 1981), Nepean (Sheppy and Rowe 1975), Widgie 3 (Groves and Lesher 1982), and Miriam (Gemuts 1975) are stratigraphically equivalent to sulfidic sediments. The stratiform ores in several of the Forrestania deposits occur in areas where oxide facies iron-formation is locally thinned or absent (e.g., New Morning, Digger Rocks, Mt. Hope, and Liquid Acrobat: Porter and McKay 1981). The ores at Cosmic Boy (Porter and McKay 1981) apparently occur exclusively on contacts between oxide facies iron-formation and komatiitic dunite; sulfidic sediments are also present in the footwall rocks at Forrestania.

The stratigraphic relationships between ores and specific country rocks are necessarily less specific for strata-bound deposits, but most of these deposits also occur in sequences that contain sulfidic country rocks. For example, the footwall sequence at Damba-Silwane (Williams 1979) contains pyritic shales in the upper part. The Shangani deposit (Viljoen et al. 1976) occurs in a volcanic neck that truncates a thick sulfide facies iron-formation. Mineralization occurs on the upper and lower margins of the komatiitic dunite at Honeymoon Well (Donaldson and Bromley 1982) and both the footwall and hanging wall sequences contain pyritic black shales. The komatiitic dunites at Mt. Keith (Burt and Sheppy 1975) and Six Mile (Naldrett and Turner 1977) are bounded to the west by a thick sulfidic chert.

Massive nickel-copper sulfide ores, irrespective of whether associated with cumulate komatiites (e.g., Kambalda) or with komatiitic dunites (e.g., Agnew, Kataniq), normally occur in footwall embayments at the base of the host units. The geometry of the embayments ranges from shallow, open flexures (e.g., Scotia, Wannaway, Agnew, Katiniq, Forrestania) to reentrant troughs (e.g., most Kambalda shoots). Lesher et al. (1984) argue that the embayments which localize the ores at Kambalda are probably volcanic topographic features, modified by thermal erosion and subsequent deformation. On the basis of theoretical and laboratory experimental studies, Huppert et al. (1984) suggested that the embayments at Kambalda may have been generated entirely by thermal erosion beneath linear lava conduits. This is conceivable, but there are several problems with this interpretation that remain to be resolved (Lesher 1983, Lesher et al. 1984):

1. The original geometry of all embayments has been modified to some degree by superimposed polyphase deformation;
2. Some embayments are elliptical in outline (e.g., parts of the Ken and Juan complexes) and could not have formed beneath linear lava conduits;
3. Although there is evidence of thermal erosion on the lateral margins of some ore horizons, the basal contacts normally appear to be conformable with footwall pillow basalts;

4. The basal contacts of some embayments correlate with cooling-unit boundaries in the adjacent basalt sequence and therefore appear to be stratigraphically conformable within the footwall sequence;
5. The basal host units which define the embayments at Kambalda are derived from the most magnesian liquids (28 – 31% MgO) in the sequence which suggests that thermal erosion and contamination may be less significant than calculated by Huppert et al. (1984);
6. The embayments at the Scotia (Page and Schmulian 1981) and Wannaway (McQueen 1981b) deposits contain preserved or partially preserved sediments that predate the emplacement of the host unit.

The embayments that localize komatiitic dunite-associated deposits are most likely tectonic or thermal-erosional; many (e.g., Agnew) are now so highly tectonized that their original form and origin are virtually impossible to determine. Disseminated ores, including both fine intercumulus grains and coarse blebs, may be marginally or centrally disposed (see Lesher et al. 1982: Table B5). Although many host units, especially komatiitic dunites, contain ubiquitous disseminated sulfides, some host units contain negligible mineralization outside of the ore zones. For example, some host units at Kambalda contain widespread low-grade disseminated mineralization (e.g., Lunnon shoot: Ross and Hopkins 1975, Keays et al. 1981), whereas others are virtually barren of sulfides outside of the basal ore zone (e.g., Juan Main shoot: Marston and Kay 1980). At Honeymoon Well (Donaldson and Bromley 1982) only the margins of the host unit contain disseminated sulfides. Similarly, some parts of the mineralized dunites at Forrestania (Porter and McKay 1981) contain only trace amounts of sulfide. Zones of disseminated mineralization at Six Mile (Naldrett and Turner 1977) alternate with zones containing negligible sulfides.

The majority of komatiite-associated deposits, including most large ones, are of late Archean age (2.7 – 2.9 Ga), but they range from early Archean (3.5 Ga; e.g., Ruth Well: Nisbet and Chinner 1981) to early Proterozoic (1.9 Ga; e.g., Katiniq: Barnes et al. 1982).

Field Constraints

The effects of sulfur mobility during metamorphism are difficult to evaluate (see Groves et al. 1974, Eckstrand 1975, Groves and Keays 1979), but the spatial concentration of nickel sulfide mineralization in certain parts of otherwise virtually barren host units, regardless of the nature of the metamorphism, suggests that those host units were not uniformly saturated in sulfur at the time of emplacement. If sulfur-saturated, the entire host unit should contain ubiquitous disseminated mineralization: (1) sulfides should have been trapped in the chilled margins of the host units; (2) sulfides should have continued to exsolve throughout the host units as the lavas/magmas cooled, oxidized, and crystallized after emplacement; and (3) flow-through fractional crystallization of sulfur-saturated komatiitic lavas/magmas should have resulted in significant accumulation of sulfide as well as olivine. The *absence* of ubiquitous mineralization in many host units, therefore, suggests that the sulfides separated locally, subsequent to emplacement. Note, however, that the *presence* of widespread disseminated mineralization does not necessitate early sulfide saturation, as the sulfides could have separated at a late stage during crystallization.

Geochemical Constraints

Geochemical data have been used to infer that the precursor magmas at some cumulate komatiite-associated deposits were sulfur-saturated prior to eruption (Lesher and Groves 1984):

1. Aphyric and spinifex-textured komatiites from throughout the komatiite sequence at Kambalda, Western Australia are significantly depleted in Ni and Co relative to unmineralized komatiite sequences and to compositions predicted by theoretical sulfide-free fractional crystallization models (Lesher et al. 1981);
2. Chromites in mineralized komatiite sequences in Western Australia are systematically enriched in Zn compared to those from unmineralized sequences (Groves et al. 1977, Groves et al. 1981).

Because the geochemical features are present throughout the host komatiite sequence(s), irrespective of whether the sampled units are mineralized or barren, Lesher et al. (1981) concluded that they probably reflect equilibration of precursor komatiite magmas with sulfides prior to eruption. It has been recently recognized, however, that the host komatiite flows are probably laterally very extensive (see Lesher et al. 1984). As the structure and dimensions of the mineralized volcanic piles are not known, it is quite possible that the barren upper parts of a mineralized sequence in one area of the volcanic belt may be the facies equivalent of mineralized parts of a sequence in another area of the belt. Although the geochemical data may indeed reflect equilibration between the host magmas and sulfides, this does not necessarily restrict that process to the pre-eruptive stage.

Physical Constraints

Despite the fact that sulfides in these deposits are widely assumed to have been derived from the mantle or from low level magma chambers, the feasibility of transporting dense sulfides in low-viscosity komatiite magmas has not been previously assessed. This problem has normally been dismissed under the assumption that the precursor magmas were enriched in intratelluric olivine which would significantly increase their effective viscosity (Marston et al. 1981) and, therefore, their ability to transport sulfides. Relict olivine is rare at most deposits, but the systematic compositional and textural variations exhibited by that which has been studied (Stolz 1981, Lesher 1983, Donaldson 1983) indicates that most is not intratelluric. This, plus the absence of significant amounts of olivine phenocrysts in the chilled margins of most komatiite lava flows, suggests that the precursor komatiitic magmas were phenocryst-poor. The cumulate nature of these units is attributed to flow-through fractional crystallization and in situ accumulation of olivine.

As noted above, dense sulfides that have segregated at the base of a magma chamber are unlikely to be erupted. It is possible, however, that minor amounts of dispersed sulfides may be entrained in a komatiitic magma without significantly increasing its bulk density. Assuming that some concentration of sulfides has occurred, regardless of the precise model for sulfide emplacement (see review by Lesher et al. 1984), the *maximum* amount of sulfides is that presently preserved in the ore zones – which Lesher et al. (1981) calculate to be of the order of 2%. Concentrations of sulfides less than 2% would have negligible influence on the bulk density (and viscosity) of the komatiite magma.

The feasibility of transporting sulfides in an ascending komatiite magma may be evaluated in terms of fluid mechanics. Iron-nickel sulfides of the primary compositions inferred for these deposits will be molten (Kullerud 1963) at all temperatures above the komatiite solidus (Arndt 1976b). Extensive studies of the motion of solid particles, fluid droplets, and gas bubbles in an infinite medium have shown, in general, that the drag coefficient, and consequently the terminal setting velocity, are a function of the dimensionless Reynolds' number:

$$Re = \frac{\varrho d U}{\mu} \tag{1}$$

where ϱ is the density of the continuous phase, d the diameter of the dispersed phase, U the velocity of the dispersed phase, and μ the viscosity of the continuous phase. Analytical solutions are only available when inertial effects are ignored, that is, when $Re \ll 1$, but empirical solutions and theoretical models are available at higher Reynolds' numbers.

Hubbert et al. (1984) have noted that low viscosity komatiites should ascend turbulently. Experimental studies have shown that particles respond to turbulent fluctuations[3] with a scale larger than the particle diameter. The terminal velocity of small particles is not affected by free-stream turbulence, but that of large particles may increase or decrease as a result of changes in the flow regime (Clift et al. 1978). The overall effect of turbulence appears to be to increase the drag coefficient and consequently to decrease the settling velocity. As such, the analytical solutions and empirical correlations discussed below can be considered to represent maximum settling velocities.

The terminal settling velocity of small sulfide droplets at low Reynolds' numbers ($Re < 2$: Stokes flow regime) in a komatiite magma may be estimated from Stokes' law:

[3] Other effects of turbulence that are more difficult to evaluate are (1) increases in the probability of hydrodynamic interference between droplets, leading to coalescence and (2) increases in shear stresses at the surface of the droplet, leading to breakup. The opposing influences of these processes lead to the development of a steady state droplet size distribution.

$$U_{TS} = \frac{gd^2(\varrho_p - \varrho)}{18\mu} \tag{2}$$

where U_{TS} is the terminal settling velocity, g the gravitational acceleration, d the droplet diameter, ϱ_p the density of the sulfide droplet, ϱ the density of the komatiite, and μ the viscosity of the komatiite. There are solutions which contain terms to account for internal circulation in a fluid droplet (i.e., Hadamard-Rybczynski formula: see Clift et al. 1978, Eq. 3–15), but Eq. (2) is sufficiently accurate for our purposes. Note that terminal settling velocity in the Stokes' regime is directly proportional to the difference in density between the two phases and inversely proportional to the viscosity of the continuous phase, but that it is proportional to the surface area of the dispersed phase and, therefore, to the square of the diameter of the droplet. At higher Reynolds' numbers the influences of the viscosity of the continuous phase and the dimension of the dispersed phase become less important.

The terminal settling velocity of moderate-sized droplets at moderate Reynolds' numbers ($2 < Re < 500$: laminar flow regime) may be estimated from the following empirical correlation of the standard drag curve (McCabe and Smith 1976: Eq. 7–51):

$$U_{TL} = \frac{0.153\, g^{0.71} d^{1.14} (\varrho_p - \varrho)^{0.71}}{\varrho^{0.29} \mu^{0.43}} . \tag{3}$$

The terminal settling velocity of larger droplets at higher Reynolds' numbers ($500 < Re < 2 \times 10^5$: turbulent flow regime) may be estimated from the following empirical correlation (McCabe and Smith 1976: Eq. 7–52):

$$U_{TN} = 1.75 \frac{gd(\varrho_p - \varrho)^{1/2}}{\varrho} . \tag{4}$$

Using values of 4.2 g cm^{-3} for ϱ_p of an iron-nickel sulfide melt (Dobrovinski et al. 1969), and 2.75 g cm^{-3} for ϱ and 1 poise for μ of a 32% MgO komatiite (Nisbet 1982, Lesher 1983), terminal settling velocities and Reynolds' numbers for different-sized sulfide droplets are given in Table 2. For a range of plausible fissure widths (0.3–10 m), Huppert et al. (1984) calculated that the ascent velocity of komatiites should be within the range 190–1100 cm s^{-1}. Although this range may be realistic for well-established conduits, most of the ores in

Table 2. Terminal settling velocities and Reynolds' numbers of sulfide droplets in a komatiite magma

d (cm)	U_{TS} (cm s^{-1})	U_{TL} (cm s^{-1})	U_{TN} (cm s^{-1})	Re
0.01	0.0079	–	–	0.00022
0.03	0.071	–	–	0.0059
0.10	0.79	–	–	0.22
0.30	–	5.0	–	4.1
1.0	–	20	–	54
3.0	–	–	69	570

cumulate komatiite-associated nickel sulfide deposits are associated with the initial flow of the komatiite sequence. The first magmas probably expended more energy during intrusion and ascended more slowly than later magmas which passed through previously established conduits. The magnitude of the difference in ascent velocity is difficult to evaluate, but it is clear that at least fine sulfide droplets could be supported and transported in an ascending komatiite magma.

Thermodynamic Constraints

Although it appears physically feasible to transport dispersed sulfide droplets in an ascending komatiite magma, there are several thermodynamic factors that militate against the eruption of sulfide-bearing komatiites:

1. Firstly, komatiitic melts are generated by high-temperature, high-degree partial melting of mantle source rocks. As the dissolution of mantle sulfide will depend primarily on the amount of melt and the temperature of the melt, komatiites are the least likely of all melts to remain sulfur-saturated during formation. For example, the amount of sulfide that can dissolve in a 50% komatiitic melt with a sulfur content of the order of 3000 ppm at sulfide saturation (Shima and Naldrett 1975) is 15 times that of a 10% basaltic melt with a sulfur content of the order of 1000 ppm at sulfide saturation (Haughton et al. 1974). Generation of a 50% partial melt containing 3000 ppm S would require a source with a minimum sulfur content of 1500 ppm. The sulfur content of the mantle is not precisely known, but most estimates are very much lower, of the order of 350 ppm (Ringwood and Kesson 1976, Sun 1981).
2. Secondly, independent experimental studies by Helz (1977), Huang and Williams (1980), and Wentlandt (1982) have shown (a) that there is a strong negative pressure dependence on sulfur solubility in basaltic melts and (b) that sulfur saturation isopleths are parallel to anhydrous silicate liquidi (see Fig. 1). If sulfur-saturated at depth, magmas which ascend along liquidus paths should be sulfur-saturated during ascent and may therefore exsolve an immiscible sulfide liquid during ascent and after eruption. However, as noted above, there is no evidence of significant intratelluric phenocryst contents in komatiites, and magmas of such low viscosity should ascend rapidly and lose little heat. Komatiites should ascend along steep $P-T$ trajectories, nearer to the adiabatic gradient, and should arrive at the surface superheated and sulfur-undersaturated. This is consistent with the field observations.
3. Finally, if komatiites are formed by multistage melting or mixing processes (e.g., Arndt 1977, Bickle et al. 1977), they are even less likely to remain sulfur-saturated. Sulfur may partition into early melts and/or be physically lost as sulfide during melt segregation, and residual melts will be driven off the sulfur saturation surface by continued partial melting/mixing.

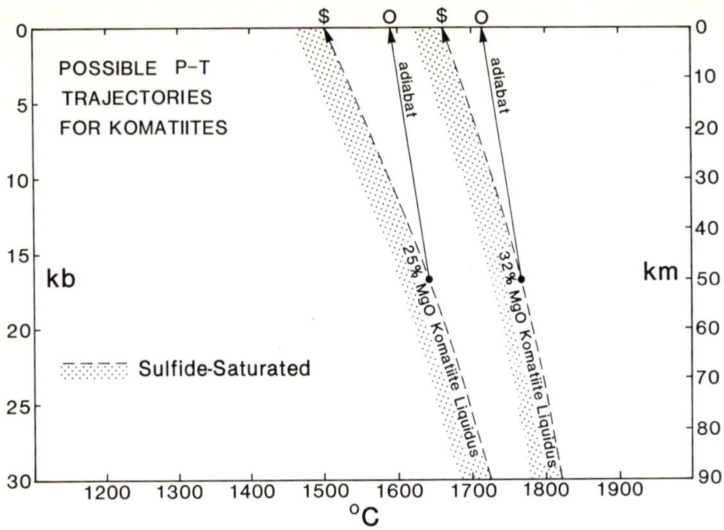

Fig. 1. Sulfide saturation model showing limits on possible pressure-temperature trajectories for ascending komatiite magmas. The *dashed curves* are liquidus ascent paths (Arndt 1976b, Bickle et al. 1977, Bickle 1978). The *solid curves* are adiabatic ascent paths ($dT/dP = 3\,°C/kb$). Sulfide saturation isopleths are assumed to be parallel to the silicate liquidii (see references in text). $ = sulfide-saturated komatiites, 0 = sulfide-undersaturated komatiites

Discussion

Together, the above field, physical, and thermodynamic constraints suggest that most komatiites should not have been sulfide-saturated on emplacement. It may be physically feasible to transport dispersed sulfide droplets in a rapidly ascending komatiite magma, but the field evidence and thermodynamic constraints suggest that this probably did not occur. In the case of finely-disseminated mineralization in komatiitic dunites, sulfide saturation must have occurred late in the crystallization history of the host units in order to produce fine-grained, interstitial textures. The sulfur in these deposits may have been mantle-derived, through dissolution during partial melting and exsolution during crystallization. In contrast, massive sulfides in cumulate komatiites must have formed early in the crystallization history of the host units in order to provide the opportunity for sulfide separation and segregation. If the komatiitic hosts for these deposits were not sulfide-saturated on eruption, as inferred above, then some process must have induced sulfide separation at a later stage.

Sulfide separation occurs in a silicate melt when it becomes saturated in S^{2-}. Sulfur solubility in silicate melts have been shown to decrease with: (1) decreasing temperature; (2) decreasing aFeO or increasing $a SiO_2$; and/or (3) decreasing fS_2 or increasing fO_2 (MacLean 1969, Haughton et al. 1974, Shima and Naldrett 1975, Buchanan and Nolan 1979). As the cumulate komatiite host units are interpreted as dynamic lava conduits and the komatiitic dunite host units as dynamic magma chambers, temperatures were probably maintained at relatively high levels during emplacement. Similarly, variations in fO_2 or fS_2 are unlikely to have been very significant, although sulfur may have been volatilized during eruption (Lesher et al. 1981). Direct contamination of the lava/magma by silica, ferric iron, or sulfide is the most feasible mechanism for inducing rapid, significant sulfide exsolution. The cherty, sulfidic sediments at Kambalda, Windarra, Widgiemooltha, and Nepean, for example, are rich in silica and sulfide (Groves et al. 1976, Bavinton 1981). The oxide facies iron-formations at Forrestania are rich in silica and ferric iron oxide.

On the basis of stratigraphic relationships and physical volcanology, Lesher et al. (1984) proposed a volcanic assimilation model for Kambalda. In light of the intimate spatial and stratigraphic relationships between cumulate komatiite-associated nickel-copper sulfide deposits and certain types of country rocks, assimilation of country rocks is considered to be the most probable mechanism for inducing localized sulfide saturation and sulfide separation in these deposits. The necessity of an external sulfur source or sulfide precipitating mechanism is less pronounced in the case of the disseminated mineralization in komatiitic dunites, but the distribution of sulfides in some of those deposits also appears to be intimately related to the distribution of sulfidic country rocks. Contamination of mafic-ultramafic magmas by either sulfide or siliceous country rocks has been suggested as a mechanism for inducing sulfide saturation for most other magmatic sulfide deposits (see summary by Naldrett 1981), and it is likely that komatiite-associated nickel-copper sulfide deposits are no exception.

In such a model (Fig. 2), the texture and distribution of mineralization will be influenced by (1) the mode of emplacement (volcanic vs subvolcanic) and degree of assimilation; (2) the relative volume of magma (lava flow vs magma chamber) and its capacity to dissolve sulfur; and (3) the timing of sulfide saturation and opportunity for sulfide segregation. For example, local assimilation of siliceous, sulfidic, or ferric iron oxide-rich footwall rocks by thermal erosion beneath

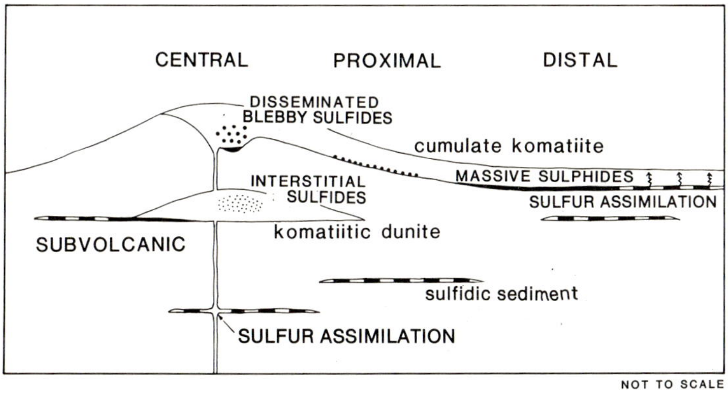

Fig. 2. Generalized model for the formation of komatiite-associated nickel sulfide deposits by assimilation of crustal sulfur. Sulfides are shown in *black*: *solid* for sulfide ores (*dotted* = disseminated; *spotted* = blebby; *filled* = massive), *striped* for sulfidic sediments. Note that sulfide ores may occur in a range of settings from distal volcanic through proximal and central volcanic to subvolcanic

narrow lava conduits, would result in sulfide separation prior to crystallization of the host unit – producing basal stratiform massive ores with overlying disseminated mineralization. Variations in the mode of emplacement of the host unit would influence the degree of assimilation and may account for variations in the intensity of mineralization and for differences in the stratigraphic relationships with the sulfidic sediments that are observed in different cumulate komatiite-associated deposits. In contrast, assimilation of siliceous, sulfidic, or ferric iron oxide-rich country rocks during intrusion of a more voluminous komatiitic dunite body may not induce immediate sulfide saturation and sulfide separation could be delayed until late in the crystallization history – producing disseminated ores. Blebby sulfides are too coarse (av 1 cm diam) to remain suspended in the magma and too abundant to have exsolved in situ (Lesher 1983); they must have formed during flow-through crystallization of the host unit. In all cases, it is the unusual thermal and physical characteristics of komatiitic magmas that influenced their mode of emplacement (Lesher et al. 1984) and promoted thermal erosion and assimilation (Huppert et al. 1984).

At present there are few data available to properly test this model, but the following may be considered to be strong corroborative evidence:

1. *Sulfur Isotopes.* Sulfur isotopic data for ores and possible sulfide sources are available for only a few of the better characterized komatiite-associated deposits. These data (Fig. 3) indicate that sulfides in ores and country rocks at individual deposits generally exhibit indistinguishable sulfur isotopic ratios that are systematically different from those of other deposits of this type. Sulfur/selenium ratios exhibit similar trends. Although the sulfur in the ores and country rocks could have formed independently from the same sulfur source, the sulfur isotopic variations between deposits are not consistent with derivation from a homogeneous mantle source. Fractionation of sulfur isotopes in the hydrothermal-exhalative-sedimentary environment with subsequent incorporation in and exsolution from komatiitic magmas is the most likely explanation for the variation between deposits.

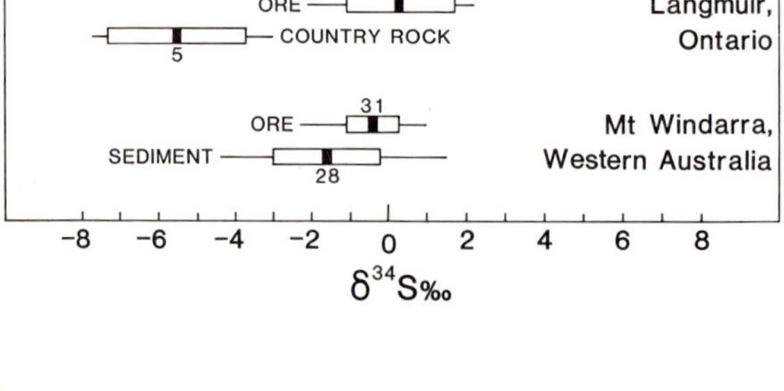

Fig. 3. Sulfur isotopic data for several cumulate komatiite-associated nickel sulfide deposits. Data from Naldrett (1966), Seccombe et al. (1978, 1981), and Green and Naldrett (1981). Data given as mean (*solid bar*), standard deviation (*closed box*), and total range (*horizontal line*). Note that the ores and sulfidic country rocks at most individual deposits exhibit similar and commonly indistinguishable sulfur isotopic ratios that are systematically different from those of the other deposits

2. *Ore Tenor Variations.* The composition of the sulfide fraction of different ore shoots in Western Australia varies between different deposits and between different shoots within a single deposit (Ross and Keays 1979, Marston and Kay 1980, Woolrich et al. 1981). The limited range of parental liquid compositions for the host units (generally 28 – 32% MgO) indicates that this cannot be attributed to variations in the compositions of the precursor magmas. Furthermore, positive correlations of Ni/Cu, Ni/Co, and Pd/Pt with increasing Ni indicates that this cannot be attributed to variations in fO_2 and fS_2 among different magma batches, as major variations in these parameters are unlikely and cannot be expected to change the relative partitioning of the chalcophile elements. The thicker ore shoots in Western Australia are generally of lower grade than the thinner ones (Marston et al. 1981), which indicates a relationship between ore tenor and the amount of sulfide that may be attributed to variations in the degree of sulfide assimilation. Campbell and Naldrett (1979) have shown that the compositions of sulfide melts are markedly influenced by variations in the magma:sulfide ratio.

Lesher et al. (1984) present stratigraphic, volcanologic, and theoretical arguments that specifically support the possibility of sulfidic sediment assimilation at Kambalda. Careful stratigraphic and volcanologic studies at other deposits, in conjunction with systematic sulfur and lead isotopic studies will be required to fully evaluate this as a generalized model.

The metallogenesis of komatiite-associated nickel-copper sulfide deposits is probably tectonically controlled. Groves and Batt (1984) define two general types of Archean greenstone belts that differ in terms of timing and degree of crustal extension. *Platform-phase greenstones* are characterized by: equispaced granitoid batholiths with intervening stellate greenstone belts, a coherent volcanic stratigraphy, abundant komatiitic basalts, and shallow water volcaniclastics or oxide facies iron-formation. They are interpreted to have formed in relatively shallow water under conditions of low crustal extension. *Rift-phase greenstones* are characterized by: elongate granitoid domes, linear tectonic patterns, complex volcanic stratigraphy, abundant komatiites (especially cumulate komatiites), and sulfidic shales and cherts. They are interpreted to have formed in relatively deep water under conditions of high crustal extension. The rift-phase greenstones, typified by the Norseman-Wiluna Belt of Western Australia and the Abitibi Belt of Canada, are distinctly better mineralized. The spatial separation between the two associations, volcanic cumulate komatiite and subvolcanic komatiitic dunite, is probably also related to the degree of tension in the crust (see Roberts 1970). The absence of subvolcanic komatiitic dunites in the southern part of the Norseman-Wiluna Belt, for example, probably indicates a relatively high degree of extension so that all komatiitic magmas were erupted. In contrast, the presence of both subvolcanic komatiitic dunites and komatiite lavas in the northern part of the Norseman-Wiluna Belt, for example, probably indicates a somewhat lower degree of extension or a smaller amount of supracrustal rocks, so that komatiitic magmas were intruded and then erupted. This may result (1) from longitudinal variations within a rift zone; (2) from lateral variations across a rift zone; or (3) from temporal variations during evolution of a rift zone (Groves et al. 1984). In any case, both environments were conducive to the rapid, continued eruption/irruption of komatiitic magma as manifested by dynamic lava conduits/magma chambers.

Conclusions

Previous models for the formation of these deposits inferred derivation of sulfides directly from the mantle or from low-level magma chambers, but there are several field, physical, and thermodynamic constraints which suggest that this may not be feasible: (1) the spatial concentration of mineralization within certain parts of otherwise barren host units suggests that the latter were not uniformly saturated in sulfide at the time of emplacement; (2) there are physical problems in transporting dense sulfides in low viscosity komatiite magmas, although dispersed droplets may be supported during ascent; and (3) high-temperature komatiitic melts generated by high-degree partial melting, or sequential melting, of mantle source rocks are unlikely to be sulfide-saturated in the source region or to remain sulfide-saturated during ascent.

These deposits are intimately associated with specific country rocks, particularly sulfidic sediments, and the distribution of such rocks appears to be related to the distribution of the Fe – Ni – Cu sulfide mineralization. It is most likely that the enhanced sulfur contents of the mineralized komatiitic rocks are due to assimilation of siliceous, sulfidic, or ferric iron-rich country rocks during emplacement. The distribution and texture of the mineralization was probably influenced by (1) the mode of emplacement of the magma (volcanic vs subvolcanic) and degree of assimilation; (2) the relative volume of magma (lava conduit vs magma chamber) and its capacity to dissolve sulfur; and (3) the timing of sulfide separation and opportunity for segregation. For example, local assimilation of sulfidic sediments by thermal erosion beneath narrow lava conduits, such as interpreted for the Kambalda deposits, would result in sulfide separation and segregation prior to crystallization of the host unit – producing stratiform massive ores. In contrast, assimilation of sulfidic country rocks during intrusion of a larger komatiitic dunite body may not induce sulfide saturation immediately and sulfide separation could be delayed until late in the crystallization history – producing strata-bound disseminated ores.

A fundamental aspect of this model is the coincidence of cumulate komatiitic host units and sulfidic country rocks, the former representing a source of heat for assimulation (a consequence of lava/magma replenishment) and of ore metals, and the latter a source of sulfur or of components to induce sulfide separation. Certain rift-phase greenstones appear to have formed in deeper water under conditions of greater extension than other platform-phase greenstones and, therefore, represent the most favorable tectonic environment for both voluminous komatiite eruption/irruption and accumulation of sulfidic sediments. Variations in the degree of extension along or within the rift and/or in the tectonic timing of emplacement may control the mode of emplacement and, therefore, the type of deposit.

Acknowledgments. C.M.L. acknowledges the support of a University of Western Australia research scholarship and a University of Alabama Faculty Research Grant (No. 1254). C.M.L. is very grateful to Kambalda Nickel Operations, Western Mining Corporation, for supporting his research on the Kambalda deposits. Both authors are grateful to the many mining companies in Western Australia who granted access to deposits and extended many courtesies in the field, and to Mmes. S. Simpson and S. Miller for their skillful typing of the manuscript. The paper has been improved as a result of constructive comments by Prof. A. J. Naldrett, Dr. H. E. Huppert, and two anonymous referees.

References

Allegre CJ (1982) Genesis of komatiites in a wet plate. In: Arndt NT, Nisbet EG (eds) Komatiites. Allen & Unwin, London, pp 495 – 500

Arndt NT (1976a) Ultramafic rocks of Munro Township: Economic and tectonic implications. In: Strong DF (ed) Metallogeny and plate tectonics. Geol Assoc Can Spec Pap 14:617 – 657

Arndt NT (1976b) Melting relations of ultramafic lavas (komatiites) at 1 atm and high pressure. Carnegie Inst Washington Yearb 75:555 – 562

Arndt NT (1977) Ultrabasic magmas and high degree melting of the mantle. Contrib Mineral Petrol 64:205 – 221

Arndt NT (1983) The role of a thin, komatiite-rich oceanic crust in the Archean plate-tectonic process. Geology 11:372 – 375

Arndt NT, Brooks C (1980) Komatiites. Geology 8:155 – 156

Arndt NT, Nisbet EG (eds) (1982) Komatiites. Allen & Unwin, London, 526 pp

Barley ME, Bickle MJ (1982) Komatiites in the Pilbara Block, Western Australia. In: Arndt NT, Nisbet EG (eds) Komatiites. Allen & Unwin, London, pp 105 – 116

Barnes SJ, Coats CJA, Naldrett AJ (1981) Petrogenesis of a proterozoic nickel sulfide-komatiite association: The Katiniq sill, Ungava, Quebec. Econ Geol 77:413 – 429

Bavinton OA (1981) The nature of sulfidic sediments at Kambalda and their broad relationships with associated ultramafic rocks and nickel ores. Econ Geol 76:1606 – 1628

Bickle MJ (1978) Melting experiments on peridotitic komatiites. In: Progress in experimental petrology. Natl Energy Res Counc Prog Rep 4 Ser D 11:187 – 195

Bickle MJ, Ford CE, Nisbet EG (1977) The petrogenesis of peridotitic komatiites: Evidence from high-pressure melting experiments. Earth Planet Sci Lett 37:97 – 106

Billington LG (1984) Geological review of the Agnew nickel deposit, Western Australia. In: Buchanan DL, Jones MJ (eds) Sulphide deposits in mafic and ultramafic rocks. Inst Min Metall, London, pp 43 – 54

Buchanan DL, Nolan J (1979) Solubility of sulfur and sulfide immiscibility in synthetic tholeiite melts and their relevance to Bushveld-complex rocks. Can Miner 17:483 – 494

Burt DRL, Sheppy NR (1975) Mount Keith nickel sulphide deposit. In: Knight CL (ed) Economic geology of Australia and Papua-New Guinea, I. Metals. Australas Inst Min Metall Monogr 5:159 – 268

Campbell IH, Naldrett AJ (1979) The influence of silicate:sulfide ratios on the geochemistry of magmatic sulfides. Econ Geol 74:1503 – 1505

Chimimba LR (1984) Geology and mineralogy at Trojan Mine, Zimbabwe. In: Buchanan DL, Jones MJ (eds) Sulphide deposits in mafic and ultramafic rocks. Inst Min Metall, London, pp 147 – 155

Clift R, Grace JR, Weber ME (1978) Bubbles, drops and particles. Academic Press, London New York, 380 pp

Clutten JM, Foster RP, Martin A (1981) Nickel mineralization in Zimbabwe. Episodes 2:10 – 15

Coad PR (1979) Nickel sulphide deposits associated with ultramafic rocks of the Abitibi Belt and economic potential of mafic-ultramafic intrusions. Ontario Geol Surv Stud 20:84 pp

Dobrovinski IE, Esin OA, Barmin LN, Chuchmarev SK (1969) Physicochemical properties of sulphide melts. Russ J Phys Chem 43:1769 – 1771

Donaldson MJ (1983) Progressive alteration of barren and weakly mineralized archaean dunites: A petrological, mineralogical and geochemical study of some intrusive dunites from Western Australia. Ph D Thesis, Univ W Aust, 345 pp (unpublished)

Donaldson MJ, Bromley GJ (1982) The Honeymoon Well nickel sulfide deposits, Western Australia. Econ Geol 76:1550–1564

Donaldson MJ, Lesher CM (1982) Intrusive dunite and extrusive peridotite: Analogous products of komatiitic magmas resulting from different cooling rates (abstract). International Geological Correlation Program Projects 161 and 91, Nickel Suphide Field Conference III, Perth, Western Australia, Abstr Vol, pp 31–32

Eckstrand OR (1975) The Dumont serpentinite: A model for control of nickeliferous opaque assemblages by alternation products in ultramafic rocks. Econ Geol 70:183–201

Gemuts I (1975) Miriam nickel prospect, Coolgardie area. In: Knight CL (ed) Economic geology of Australia and Papus New Guinea. I. Metals. Australas Inst Min Metall Monogr 5:98–99

Green AH, Naldrett AJ (1981) The Langmuir volcanic peridotite-associated nickel deposits: Canadian equivalents of the Western Australian occurrences. Econ Geol 76:1503–1523

Gresham JJ (1986) Depositional environments of volcanic peridotite-associated nickel sulphide deposits with special reference to the Kambalda dome, this Vol.

Gresham JJ, Loftus-Hills GD (1981) The geology of the Kambalda nickel field, Western Australia. Econ Geol 76:1373–1416

Groves DI, Batt WD (1984) Spatial and temporal variations of Archaean metallogenic associations in terms of evolution of granitoid-greenstone terrains with particular emphasis on the Western Australian Shield. In: Kröner A, Hanson GN, Goodwin AM (eds) Archean geochemistry. Springer, Berlin Heidelberg New York, pp 73–98

Groves DI, Keays RR (1979) Mobilization of ore-forming elements during alteration of dunites, Mt. Keith-Betheno, Western Australia. Can Miner 17:373–389

Groves DI, Lesher CM (eds) (1982) Regional geology and nickel deposits of the Norseman-Wiluna Greenstone Belt, Western Australia. Univ W Aust Dep Geol Extension Serv Publ 7:234 pp

Groves DI, Hudson DR, Hack TBC (1974) Modification of iron-nickel sulfides during serpentinization and talc-carbonate alteration at Black Swan, Western Australia. Econ Geol 69:1265–1281

Groves DI, Barrett FM, McQueen KG (1976) Geochemistry and origin of cherty metasediments within ultramafic flow sequences and their relationship to nickel mineralization. In: Glover JE. Groves DI (eds) Archaean cherty metasediments: Their sedimentology, micropaleontology, biogeochemistry, and significance to mineralization. Univ W Aust Dep Geol Extension Serv Publ 2:57–69

Groves DI, Barrett FM, Binns RA, McQueen KG (1977) Spinel phases associated with metamorphosed volcanic-type iron-nickel sulfide ores from Western Australia. Econ Geol 72:1224–1244

Groves DI, Barrett FM, McQueen KG (1979) The relative roles of magmatic segregation, volcanic exhalation and regional metamorphism in the generation of volcanic-associated nickel ores of Western Australia. Can Miner 17:319–336

Groves DI, Barrett FM, Brotherton RH (1981) Exploration significance of chrome-spinels in mineralized ultramafic rocks and Ni–Cu ores. In: Villiers JPR de, Cawthorn GA (eds) ICAM 81: Proc 1st Int Conf Applied Mineralogy, Johannesburg. Geol Soc S Afr Spec Publ 7:21–30

Groves DI, Lesher CM, Gee DG (1984) Tectonic setting of the sulphide nickel deposits of Western Australia. In: Buchanan DL, Jones MJ (eds) Sulfide deposits in mafic and ultramafic rocks. Inst Min Metall, London, pp 1–13

Hammerbeck ECI (1984) Aspects of nickel metallogeny of Southern Africa. In: Buchanan DL, Jones MJ (eds) Sulphide deposits in mafic and ultramafic rocks. Inst Min Metall, London, pp 135–140

Haugthon DR, Roeder PL, Skinner BJ (1974) Solubility of sulfur in mafic magmas. Econ Geol 69:451–462

Helz RT (1977) Determination of the $P-T$ dependence of the first appearance of FeS-rich liquid in natural basalts to 20 kb (abstract). EOS 58:523

Huang W-L, Williams RJ (1980) Melting relations of the system Fe–S–Si–O to 32 kb with implications as to the nature of the mantle-core boundary (abstract). Lunar Planetary Science Conference XI. Lunar Planet Sci Inst, Houston, pp 486–488

Huppert HE, Sparks RSJ, Turner JS, Arndt NT (1984) Emplacement and cooling of komatiite lavas. Nature (London) 309:19–22

INAL Staff (1975) BHP/INAL nickel sulphide occurrences of the Widgiemooltha area. In: Knight CL (ed) Economic geology of Australia and Papua New Guinea. I. Metals. Australas Inst Min Metall Monogr 5:86–89

Keays RR, Ross JR, Woolrich P (1981) Precious metals in volcanic peridotite-associated nickel sulfide deposits in Western Australia, II: Distribution within ores and host rocks at Kambalda. Econ Geol 76:1645 – 1674

Kilburn LC, Wilson HDB, Graham AR, Ogura Y, Coats CJA, Scoates RFJ (1969) Nickel sulfide ores related to ultrabasic intrusions in Canada. In: Wilson HDB (ed) Magmatic ore deposits. Econ Geol Monogr 4:267 – 293

Kullerud G (1963) The Fe – Ni – S system. Carnegie Inst Washington Yearb 62:175 – 189

Lesher CM (1983) Localization and genesis of komatiite-hosted Fe – Ni – Cu sulphide mineralization at Kambalda, Western Australia. Ph D Thesis, Univ W Aust, 318 pp (unpublished)

Lesher CM, Groves DI (1984) Geochemical and mineralogical criteria for the identification of mineralized komatiites in Archaean greenstone belts of Australia. Proc 27th Int Geol Congr. Petrology: Igneous and Metamorphic Rocks 9:283 – 302

Lesher CM, Lee RF, Groves DI, Bickle MJ, Donaldson MJ (1981) Geochemistry of komatiites at Kambalda: I. Chalcophile element depletion – a consequence of sulfide liquid separation from komatiitic magmas. Econ Geol 76:1714 – 1728

Lesher CM, Donaldson MJ, Groves DI (1982) Nickel deposits and their host rocks in the Norseman-Wiluna Belt. In: Groves DI, Lesher CM (eds) Regional geology and nickel deposits of the Norseman-Wiluna Belt, Western Australia. Univ W Aust Dep Geol Extension Serv Publ 7, B:1 – 63

Lesher CM, Arndt NT, Groves DI (1984) Genesis of komatiite-associated nickel sulphide deposits at Kambalda, Western Australia: A distal volcanic model. In: Buchanan DL, Jones MJ (eds) Sulphide deposits in mafic and ultramafic rocks. Inst Min Metall, London, pp 70 – 80

MacLean WH (1969) Liquids phase relations in the $FeS - FeO - Fe_2O_3 - SiO_2$ system, and their application in geology. Econ Geol 64:865 – 884

Marston RJ, Kay BD (1980) The distribution, petrology, and genesis of nickel ores at the Juan complex, Kambalda, Western Australia. Econ Geol 75:546 – 565

Marston RJ, Groves DI, Hudson DR, Ross JR (1981) Nickel sulphide deposits in Western Australia: A review. Econ Geol 76:1330 – 1363

McCabe WL, Smith JC (1976) Unit operations of chemical engineering. McGraw-Hill, New York, 1028 pp

McQueen KG (1981a) Volcanic-associated nickel deposits from around the Widgiemooltha dome, Western Australia. Econ Geol 76:1417 – 1443

McQueen KG (1981b) The nature and metamorphic history of the Wannaway nickel deposit, Western Australia. Econ Geol 76:1444 – 1468

Naldrett AJ (1966) The role of sulphurization in the genesis of iron-nickel sulphide deposits of the Porcupine District, Ontario. Can Inst Min Metall Trans 69:147 – 155

Naldrett AJ (1973) Nickel sulphide deposits – their classification and genesis with special emphasis on deposits of volcanic association. Can Inst Min Metall Bull 66:45 – 63

Naldrett AJ (1979) Partitioning of Fe, Co, Ni, and Cu between sulfide liquid and basaltic melts and the composition of Ni – Cu sulfide deposits – a reply and further discussion. Econ Geol 74:1520 – 1528

Naldrett AJ (1981) Nickel sulfide deposits: Classification, composition and genesis. In: Skinner BJ (ed) Econ Geol 75th Anniversary Vol. Econ Geol Publ Co, El Paso, pp 628 – 685

Naldrett AJ, Cabri LJ (1976) Ultramafic and related rocks: Their classification and genesis with special reference to the concentration of nickel sulfides and platinum-group elements. Econ Geol 71:1131 – 1158

Naldrett AJ, Gasparini EL (1971) Archaean nickel sulphide deposits in Canada: Their classification, geological setting and genesis with some suggestions as to exploration. In: Glover JE (ed) The Archaean Rocks. Geol Soc Aust Spec Publ 3:331 – 347

Naldrett AJ, Turner AR (1977) The geology and petrogenesis of a greenstone belt and related nickel sulfide mineralization at Yakabindie, Western Australia. Precambrian Res 5:43 – 103

Nesbitt RW (1971) The case for liquid immiscibility as a mechanism for nickel sulphide mineralization in the Eastern Goldfields, Western Australia (abstract). In: Glover JE (ed) The Archaean Rocks. Geol Soc Aust Spec Publ 3:253

Nesbitt RW, Sun S-S, Purvis AC (1979) Komatiites: Geochemistry and genesis. Can Miner 17:165 – 186

Nisbet EG (1982) The tectonic setting and petrogenesis of komatiites. In: Arndt NT, Nisbet EG (eds) Komatiites. Allen & Unwin, London, pp 501 – 520

Nisbet EG, Chinner GA (1981) Controls on the eruption of mafic and ultramafic lavas, Ruth Well Ni – Cu prospect, West Pilbara. Econ Geol 76:1719 – 1735

Nisbet EG, Walker D (1982) Komatiites and the structure of the Archaean mantle. Earth Planet Sci Lett 60:105 – 113

Page ML, Schmulian ML (1981) The proximal volcanic environment of the Scotia nickel deposits. Econ Geol 76:1469 – 1479

Paterson HL, Donaldson MJ, Smith RN, Lenard MF, Gresham JJ, Boyack DJ, Keays RR (1984) Nickeliferous sediments and sediment-associated nickel ores at Kambalda, Western Australia. In: Buchanan DL, Jones MJ (eds) Sulphide deposits in mafic and ultramafic rocks. Inst Min Metall, London, pp 81 – 94

Porter DJ, McKay KG (1981) The nickel sulfide mineralization and metamorphic setting of the Forrestania area, Western Australia. Econ Geol 76:1524 – 1549

Prider RT (1970) Nickel in Western Australia. Nature (London) 226:691 – 693

Ringwood AE, Kesson SE (1976) Basaltic magmatism and the composition of the moon, Part II. Siderophile and volatile elements in the moon, earth and chondrites: Implications for lunar origin. Aust Natl Univ, Res School Earth Sci Publ 1222

Roberts JL (1970) The intrusion of magma into brittle rocks. In: Newall G, Rast N (eds) Mechanisms of igneous intrusion. Gallery Press, London, pp 278 – 294

Robinson DJ, Hutchinson RW (1982) Evidence for a volcanogenic-exhalative origin of a massive nickel sulfide deposit at Redstone, Timmins, Ontario. In: Hutchinson RW, Spence CD, Franklin JM (eds) Precambrian sulfide deposits. Geol Assoc Can Spec Pap 25:211 – 254

Ross JR, Hopkins GMF (1975) Kambalda nickel sulfide deposits: In: Knight CL (ed) Economic geology of Australia and Papau New Guinea. I. Metals. Australas Inst Min Metall Monogr 5:100 – 121

Ross JR, Keays RR (1979) Precious metals in volcanic-type nickel sulfide deposits in Western Australia. I. Relationships with the composition of the ores and host rocks. Can Miner 17:417 – 436

Ross JR, Travis GA (1981) The nickel sulfide deposits of Western Australia in global perspective. Econ Geol 76:1291 – 1329

Schmulian ML (1984) Windarra nickel deposits, Western Australia. In: Buchanan DL, Jones MJ (eds) Sulphide deposits in mafic and ultramafic rocks. Inst Min Metall, London, pp 95 – 102

Seccombe PK, Groves DI, Binns RA, Smith JW (1978) A sulfur isotopic study to test a genetic model for Fe – Ni sulphide mineralization at Mt Windarra, Western Australia. In: Robinson BW (ed) Stable isotopes in the earth sciences. N Z Dep Sci Ind Res, Wellingston, DSIR Bull 220:187 – 200

Seccombe PK, Groves DI, Marston RJ, Barrett FM (1981) Sulfide paragenesis and sulfur mobility in Fe – Ni – Cu sulfide ores at Lunnon and Juan Main shoots, Kambalda: Textural and sulfur isotopic evidence. Econ Geol 76:1675 – 1685

Sheppy NR, Rowe J (1975) Nepean nickel deposit. In: Knight CL (ed) Economic geology of Australia and Papau New New Guinea. I. Metals. Australas Inst Min Metall Monogr 5:91 – 98

Shima H, Naldrett AJ (1975) Solubility of sulfur in an ultramafic melt and the relevance of the system Fe – S – O. Econ Geol 70:960 – 967

Stolz GW (1981) A petrographic and geochemical investigation of the archean volcanic succession in the vicinity of the Scotia nickel deposit. Ph D Thesis, Univ Adelaide (unpublished)

Sun SS (1981) Chemical composition and origin of the earth's primitive mantle. Geochim Cosmochim Acta 46:179 – 192

Travis GA (1975) Mount Clifford nickel deposit. In: Knight CL (ed) Economic geology of Australia and Papua-New Guinea. I. Metals. Australas Inst Min Metall Monogr 5:144 – 146

Viljoen MJ, Bernasconi A, VanColler N, Kinloch E, Viljoen RP (1976) The geology of the Shangani nickel deposit, Rhodesia. Econ Geol 71:76 – 95

Weaver BL, Tarney J (1979) Thermal aspects of komatiite generation and greenstone belt models. Nature (London) 279:689 – 692

Wentlandt RF (1982) Sulfide saturation of basalt and andesite melts at high pressures and temperatures. Am Miner 67:877 – 885

Williams DAC (1979) The association of some nickel sulfide deposits with komatiitic volcanism in Rhodesia. Can Miner 17:337 – 349

Woolrich P, Cowden A, Giorgetta NE (1981) The chemical and mineralogical variations in the nickel mineralization associated with the Kambalda dome. Econ Geol 76:1629 – 1644

Depositional Environments of Volcanic Peridotite-Associated Nickel Sulphide Deposits with Special Reference to the Kambalda Dome

J. J. GRESHAM[1]

Abstract

Volcanic peridotite-associated Fe – Ni sulphide deposits in the Archaean Yilgarn Block of Western Australia display a strong volcanological control on the distribution of individual ore shoots. Although considerably modified by tectonism and amphibolite facies metamorphism, there are distinctive stratigraphic relationships within the host ultramafic komatiite sequence that distinguish ore environments from adjacent areas barren of Fe – Ni sulphides.

At the well documented Kambalda Dome deposits, contact ores which occur at the base of the lowermost flow of the ultramafic sequence are commonly confined within narrow (100 – 300 m), elongate embayments or troughs in the underlying tholeiitic footwall metabasalt. Hangingwall ores occur at the base of immediately overlying flows and generally directly overlie the troughs and contact ores.

Compared to adjacent barren areas, ore environments are characterized by thicker more-magnesian basal ultramafic flows, fewer interflow metasediments and better textural development and compositional differentiation in overlying thin flow units. The ultramafic sequence overlying ore environments is generally less ordered than that in adjacent non-ore environments where there is a more consistent trend of diminishing magnesium content upward through the ultramafic sequence.

The major ore-confining troughs are interpreted as original elongate linear depressions within the footwall metabasalt, and some of the local ore-confining structures probably represent original topographic or volcanological features. Ore deposition in an active rift environment may account for the close spatial relationship of contact and hangingwall ores, and the development of the distinctive stratigraphic features of the ore environment. Analogies are drawn with oceanic rift systems and recent submarine volcanism.

[1] Western Mining Corporation, Kambalda Nickel Operations, Kambalda, Western Australia, 6442, Australia

Geology and Metallogeny of Copper Deposits
(ed. by G.H. Friedrich et al.)

Introduction

Volcanic peridotite or komatiite-associated (VPA) nickel deposits occur at the base of komatiitic extrusive sequences (Groves and Hudson 1981, Naldrett 1981). These deposits are usually small (several thousand to 5 million tonnes) and high grade (1 – 5% Ni) and are known to occur in the Archaean mafic and ultramafic sequences (Greenstone belts) of Canada (Green and Naldrett 1981), Zimbabwe (Williams 1979) and Western Australia (Marston et al. 1981). However from the identified metal resource (Ross and Travis 1981) 52% of the nickel metal occurring in this class of deposit occurs in the Australian deposits with 28% and 20% being contained in the Canadian and southern African deposits respectively.

In Western Australia all the major deposits of this type are confined to the Norseman-Wiluna belt (Fig. 1) of the Yilgarn Block (Gee et al. 1980) with almost 90% of the pre-mining reserve occurring between Kalgoorlie and Norseman (Marston et al. 1981). VPA deposits of Western Australia represent a much more significant component of Australia's $\geqslant$0.8% nickel resource and production than similar deposits do in southern Africa and Canada (Ross and Travis 1981). The Western Australian volcanic peridotite-associated deposits constitute close to 50% of the identified nickel metal in ores $\geqslant$0.8% nickel in Australia and to date account for more than 90% of the nickel metal produced by Australia.

The greatest concentration of nickel sulphide mineralisation in Western Australia occurs in the Kambalda nickel field (Woodall and Travis 1969, Ross and Hopkins 1975, Gresham and Loftus-Hills 1981, Lesher 1983) where premined reserves (January 1983) totalled 33.9 million tonnes at 3.78% Ni. Production from ten separate mines from this field to June 1983 since mining commenced from Lunnon shoot (Fig. 2) in March 1967 has totalled 16.6 million tonnes at 3.13% Ni.

Almost all the nickel mineralisation at Kambalda occurs associated with the stratigraphically lowest Kambalda sequence (Gresham and Loftus-Hills 1981) and the most intensely mineralized area of the field is the Kambalda Dome (Fig. 2). Extensive diamond drilling, mine development and geological research in this area has afforded detailed documentation of the nickel ores and their associated host rocks. The features of the environment of the Kambalda Dome will be discussed in detail in this paper and compared to features of other volcanic peridotite-associated deposits of the Norseman-Wiluna belt to evaluate the depositional environment of this important class of deposit.

Geological Features of Western Australian Volcanic Peridotite-Associated Nickel Sulphide Deposits

The Norseman-Wiluna belt (Fig. 1) comprises sequences of mafic and ultramafic rocks, felsic volcanics and sediments separated by granitoid bodies. The greenstone rocks form linear belts, and are commonly dissected by major NNW trending strike faults. Fold styles are dominated by NNW trending axial surfaces but these are thought to post-date an earlier, possibly major phase of recumbent

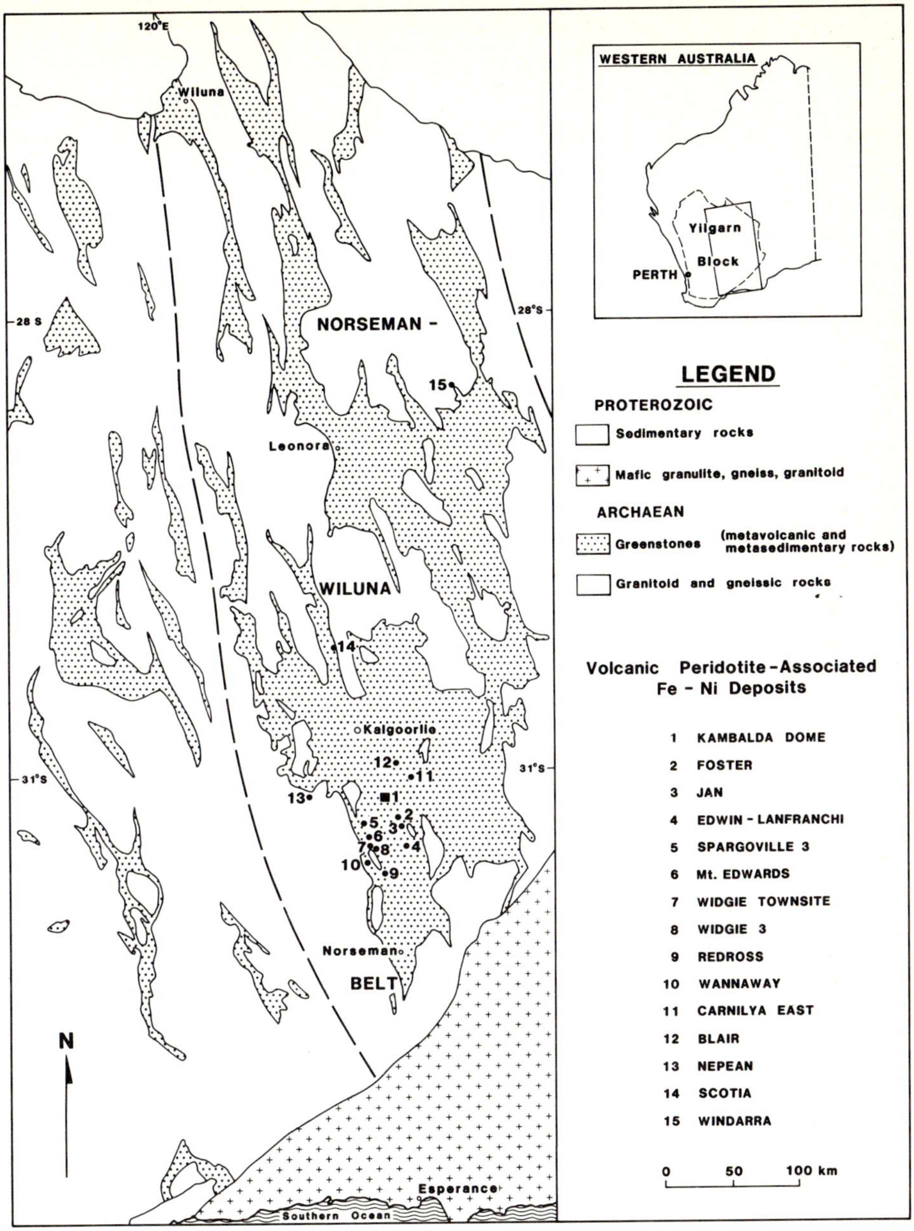

Fig. 1. Geological plan of the Norseman-Wiluna Belt showing location of major volcanic peridotite-associated deposits. (After Marston et al. 1981)

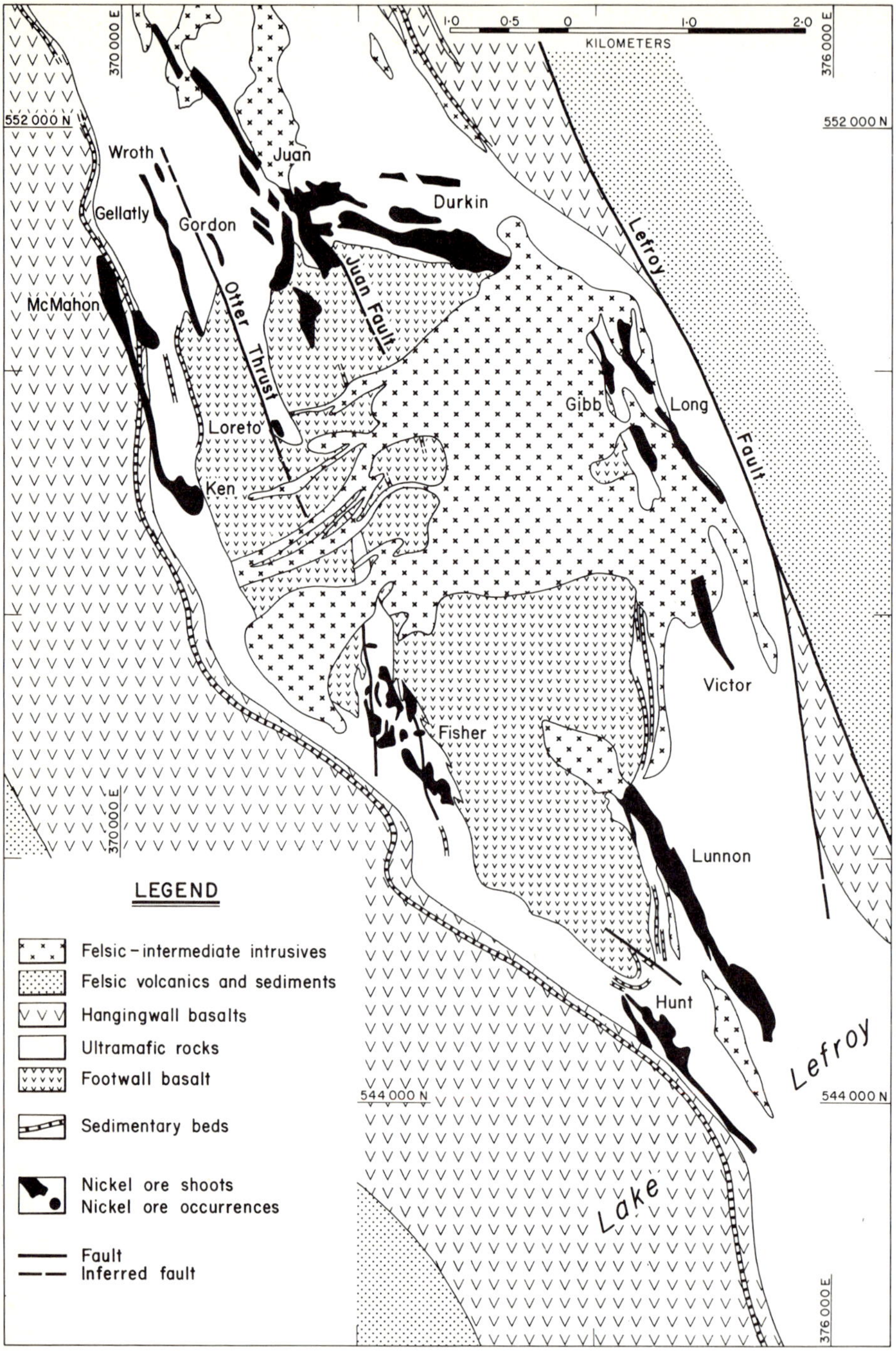

Fig. 2. Geological plan of Kambalda Dome showing major nickel shoots in plan projection. (After Gresham and Loftus-Hills 1981)

folding and nappe development (Archibald et al. 1978, Platt et al. 1978). The volcanic peridotite-associated deposits within the belt dominantly occur in the stratigraphically lowest ultramafic formations (Gemuts and Theron 1975, Gresham and Loftus-Hills 1981), a stratigraphic control also documented for similar deposits in Zimbabwe (Williams 1979) and Canada (Green and Naldrett 1981). Although previous geochronological data had indicated the VPA deposits are concentrated in the 2.8 Ga Archaean sequences (De Laeter et al. 1981, Gee et al. 1980), recent Sm – Nd data (Claoue-Long et al. 1984) on Kambalda samples have indicated ages of 3.2 Ga. This date has yet to be verified.

Volcanic peridotite-associated deposits are invariably located at or near the base of komatiitic volcanic sequences, and footwall rocks usually comprise tholeiitic basalts, although at Windarra (Fig. 1) sulphide and oxide iron formations occur in the footwall (Schmulian 1982). Sequences of tholeiitic and high-magnesium basalts and tholeiitic intrusives overlie the ultramafic rocks. Ore-associated ultramafic sequences can vary in thickness from 40 m at Nepean (Hudson 1973) to over 1000 m at Kambalda and can commonly be divided into a lower sequence of thicker (10 – 100 m) high magnesian (36 – 45% volatile-free MgO) flows and an upper sequence of thinner (<10 m) lower magnesium (20 – 32% MgO) flows. Thin, albite-rich sulphidic interflow sediments occur within the ultramafic sequence and at Kambalda are restricted to the lower part of the sequence.

The nickel sulphide mineralisation generally occurs at, or close to the base of the first high-magnesium ultramafic flow unit and where it occurs directly on the underlying footwall rocks is referred to as *contact* ore. Other ores may occur within the basal part of the ultramafic succession either at the base of immediately overlying flows or as disseminated and blebby sulphides within the host units. These are referred to as *hangingwall* ores which at Kambalda display a close spatial relationship to contact ores. Structurally displaced contact or hangingwall ores are called *offset* ores (Woodall and Travis 1969).

The individual ore bodies or shoots are narrow, elongate bodies which seldom exceed 300 m in width, but may be over 3 km long. Contact ores invariably occur within embayments or depressions of highly variable morphology within the footwall rocks. These embayments vary from open spoon-like depressions, as at Wannaway (McQueen 1981b), Scotia (Page and Schmulian 1981) and Windarra (Schmulian 1982) to strongly re-entrant troughs as at Kambalda (Gresham and Loftus-Hills 1981). Ore thickness is generally less than 5 m although several ore shoots contain thicker low-grade (1 – 2% Ni) disseminated ore zones. A typical ore profile may comprise a thin irregular layer of massive (>80% sulphides) ore, overlain by matrix (40 – 80% sulphides) ore which in turn may be overlain by disseminated (10 – 40% sulphides) ore.

Volcanic peridotite-associated primary nickel ores are typically dominated by pyrrhotite-pentlandite-pyrite assemblages with lesser chalcopyrite, magnetite and ferrochromite. Minor occurrences of millerite-rich ores have been recorded (Keele and Nickel 1974). Nickel grades of massive ore, depending on mineralogy, can vary from 5 – 35% Ni, but generally are in the range 10 – 16% Ni. Ni/Cu varies between 10 – 16 and Ni/Co between 40 – 65 (Marston et al. 1981) and both vary systematically with nickel grade. Platinum group element concentrations generally show a strong positive correlation with nickel grade with typical concentrations in 100% sulphides being, platinum 1000 – 2500 ppb, palladium 1500 – 3500 ppb, iridium 150 – 200 ppb, osmium 200 – 1000 ppb, ruthenium 800 – 1500 ppb and rhodium 250 – 800 ppb (Ross and Keays 1979, Keays et al. 1981, Groves and Hudson 1981, Donaldson 1984a). S/Se values of 5000 – 15000 and δ^{34}S values from 0 to +3.5‰ (Groves and Hudson 1981) for all deposits except Windarra are typical of values for magmatic sulphides.

The nickel ores and their associated host rocks have been subjected to polyphase deformation and greenschist to upper amphibolite facies metamorphism (Binns et al. 1976, Barrett et al. 1977). Metamorphism is extremely heterogeneous and nickel deposits occur in both low- and high-strain environments. Although the metamorphism has resulted in almost complete destruction of original igneous mineralogy, igneous terminology is retained throughout this paper. During metamorphism all ores reverted at least in part to Fe – Ni monosulphide solid solution (Barrett et al. 1977) and present ore distribution and textures also reflect the ductility of the sulphides during deformation.

Geology and Characteristics of Ore Environments at the Kambalda Dome

The stratigraphy of the Kambalda Dome has been well documented (Woodall and Travis 1969, Ross and Hopkins 1975, Gresham and Loftus-Hills 1981) and is summarised in Fig. 3. The footwall rocks are a thick sequence (>2000 m) of low to moderate magnesium basalts (Redman 1982) overlain by a variably thick (200 ≥ 1000 m) series of extrusive ultramafic flows characterised by thick (10 – 100 m) high-magnesium (>40% volatile-free MgO) flows, (lower member), at the base of the sequence with thinner (generally <10 m) lower magnesium (20 – 32% volatile-free MgO) flow units forming the upper part, (upper member), of the ultramafic sequence. The lower member generally constitutes <25% of the total ultramafic sequence. There is a general trend of diminishing Mg content up through the ultramafic sequence, but reversals of this trend do occur. Thin (generally <5 m) sedimentary horizons occur at the footwall basalt-ultramafic contact and at interflow positions within the lower member. These are finely-laminated, albite-rich, sulphidic rocks and cherty, carbonaceous and chloritic varieties occur (Bavinton 1979). The ultramafic rocks are overlain by a complex sequence of basalts with possible major mafic intrusives. A sequence of pillowed, strongly variolitic basalts 60 – 100 m thick immediately overlies the ultramafic rocks. Separating these basalts from a thick sequence of predominantly high magnesium basalts is a thin (<20 m) laterally persistent sulphidic, graphitic sedimentary horizon. The basalts are overlain by a thick, poorly defined sequence of sediments and felsic volcanics. Intruding the rocks of the Kambalda Dome are Archaean sodic-granitoids, rhyolitic and dacitic porphyries, minor mafic dykes and Proterozoic dolerite dykes.

Four deformational events involving fold development with associated faulting occurred in the Kambalda Dome area (Gresham and Loftus-Hills 1981). The early F_1 structures, characterised by tight to isoclinal, recumbent and napped

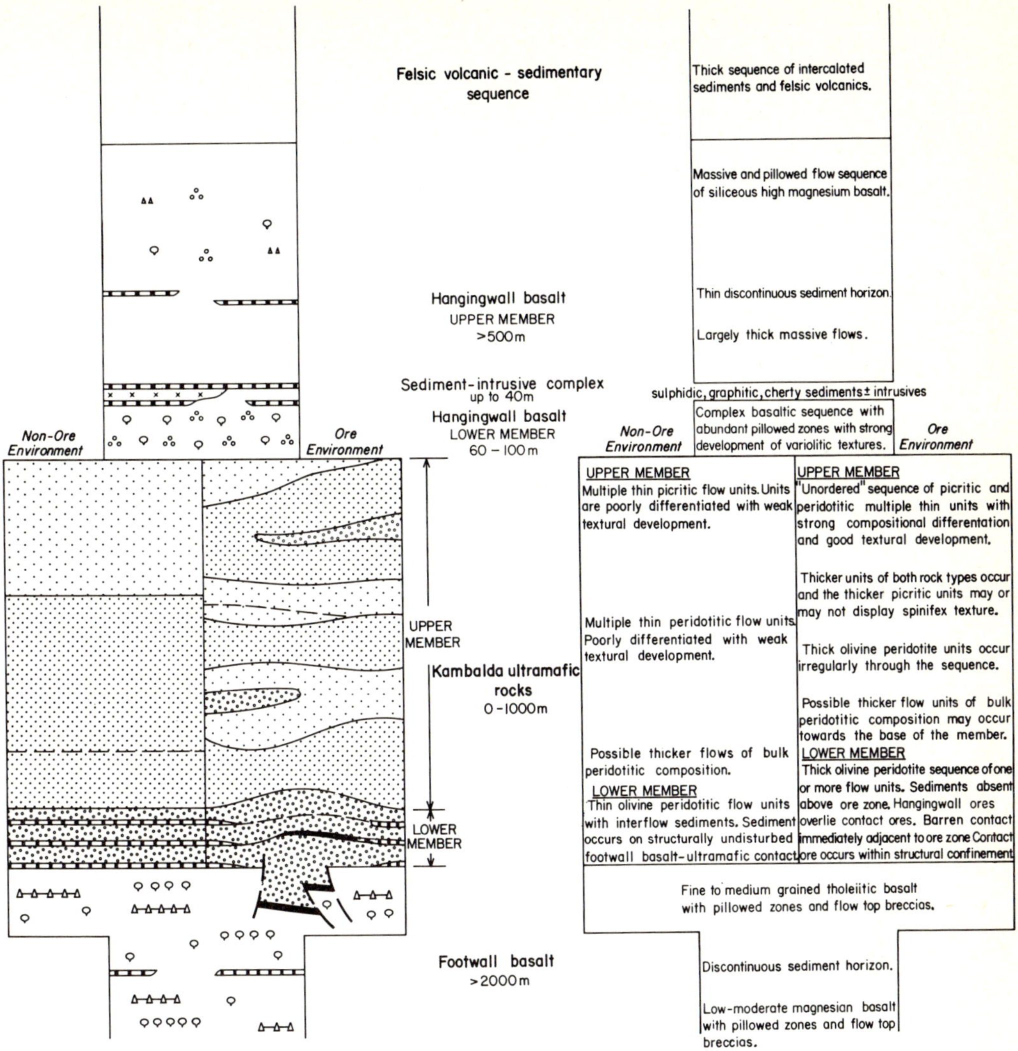

Fig. 3. Stratigraphic column of the Kambalda Dome area with most intrusives removed for clarity. Features of ore and non-ore environments are portrayed and described. (After Gresham and Loftus-Hills 1981)

folds have been refolded by the dominant folding event, F_2, with north-north-west trending axial surfaces. These folds have been refolded by F_3 with east-northeast axial surfaces and the interference of these two folding events is considered to be chiefly responsible for the formation of the Kambalda Dome. F_4 structures are poorly developed upright folds with north trending axial surfaces.

Detailed metamorphic studies (Barrett et al. 1977, Bavinton 1979) indicate metamorphic temperatures slightly in excess of 500 °C and pressures around 2.5 Kb. All rocks have been subjected to varying degrees of hydration, carbonation and potassium metasomatism. Original igneous mineralogy is seldom preserved although rare occurrences of olivine have been recorded at Victor and Durkin shoots (Fig. 2). The Kambalda Dome is an area of generally low strain and original igneous textures are commonly well preserved.

Despite the metamorphic, metasomatic and structural modification of the ores and associated host rocks, distinctive structural, stratigraphic and lithological features at the Kambalda Dome which characterise ore environments can be recognised. These features are displayed within prisms of rock that are elongate along the length of an ore shoot and extend from the immediate footwall upward through the ultramafic sequence stratigraphically above an ore shoot. These features have been discussed in detail by Gresham and Loftus-Hills (1981), are shown schematically in Fig. 3, and can be clearly recognized in the cross-sections in Fig. 5.

Nickel sulphide mineralisation is closely correlated with troughs or embayments within the footwall basalt. Troughs seldom exceed 300 m in width and relief of the trough may vary from a few to several tens of metres. In areas remote from ore the basalt-ultramafic contact, although only defined by relatively widely spaced diamond drill holes, appears to be a relatively undisturbed and planar sediment-covered surface. The cutoff from ore-bearing to non-ore bearing contact is abrupt and generally defined by an irregularity in the basalt-ultramafic contact.

Basal flow units associated with ore generally have thicknesses of 50 – 100 m and are essentially confined to the ore environments and thus form elongate, rather narrow zones coincident with the associated troughs. Thin (10 – 20 m) laterally persistent flow units occur away the ore zones. The relationship between the ore-associated basal flows and the flanking barren flows is not well understood, but in many instances they appear to be stratigraphically continuous. The chemistry of the basal flows has been subject to detailed work by Woolrich and Giorgetta (1978) and by Lesher (1983) and typically they contain >40% volatile-free MgO <45% SiO_2 and 7 – 9% total iron. Lesher indicated considerable olivine enrichment of both flow types although ore-associated units are enriched in MgO and depleted in TiO_2, Al_2O_3, Cr_2O_3 and total FeO compared to barren flanking flows. Woolrich and Giorgetta also indicated that ore-associated flows are depleted in Zn and enriched in Ni compared to barren flanking flows. There are generally two and possibly more, thick high-magnesium flows above the basal unit and, typically these units are thicker directly above ore zones. The total thickness of the lower member may exceed 200 m associated with ore zones, but seldom exceeds 100 m on the barren flanks. Most ore zones are surrounded by a zone of sediment-free contact (Fig. 5), and characterised by an absence of interflow sediments in the overlying sequence although there may be partial overlap of sediments across the ore zone. Remote from ore there may be numerous interflow sediment horizons, but this number diminishes toward the ore zones.

The upper member of the ultramafic sequence is characterised by thin lower magnesium (20 – 32% MgO volatile-free) flow units with rare thicker olivine peridotite units. Remote from ore, particularly on the west flank of the Kambalda Dome, the upper member displays a generally regular trend of diminishing magnesium content upward through the pile. Associated with ore the sequence is much less ordered, but individual flow units associated with ore zones generally display better compositional differentiation and textural development than those remote from ore.

Nickel Deposits of the Kambalda Dome

The Kambalda Dome (Fig. 2), covers and area of approximately 48 km^2 and comprises the richest known concentration of high grade nickel sulphide ore in Western Australia. A pre-mined reserve of 22.4 million tonnes at 4.13% Ni for the Kambalda Dome has been defined (January 1983), representing 72% of the original nickel metal resource of the Kambalda nickel field. To June 1983 production from eight separate mines had amounted to 15.7 million tonnes at 3.15% Ni, 95% of the nickel metal produced from the field. The ore reserve occurs within 15 separate ore bodies or shoots and these shoots contain more than 240 ore surfaces which are defined as geologically and geometrically distinct, essentially continuous zones of ore-grade mineralisation.

During the early stages of exploration and development of these deposits it was recognised that a full range or spectrum of deposit size, quality and nature existed and generalisations about the nature of the deposits were difficult to make. Bavinton (1979), related ore quality to the degree of sediment association and Marston and Kay (1980) and Woolrich et al. (1981) discussed grouping the deposits based on nickel tenor.

In this study of ore bodies have been broadly subdivided into three groups or classes (Fig. 4) based on their ore characteristics and associated stratigraphic features which are summarised in Table 1.

Class I

These ore bodies occur on the eastern and northern flank of the Kambalda Dome: Juan, Otter, Durkin, Durkin North, Gibb, Long, and Victor. They comprise nearly 70% of the pre-mined nickel metal ore reserve of the Kambalda Dome, are high grade and 96% of their contained metal occurs as contact ores. The ore bodies range in size from greater than 5 million tonnes to about 100000 tonnes. No *stratiform* hangingwall ore occurs associated with these ore bodies, the only hangingwall ore occurrences being *strata-bound* blebby ores within the basal host unit. Individual ore surfaces vary from very large elongate bodies containing more than 1 million tonnes of high grade ore to small elliptical surfaces (e.g., at Juan) that may only contain a few tens of thousands of tonnes.

Mineralogy of the ores is dominantly pyrrhotite-pentlandite-pyrite although Gibb and Otter shoots contain millerite-pyrite-magnetite assemblages. The ore tenor of the shoots is variable but Gibb, Durkin, Otter, and Victor are high tenor ores (i.e., Ni $>$ 18% in 100% sulphides), Long a low to moderate tenor ore shoot (Ni $= 14-16$% in 100% sulphides) and Juan contains both high and low tenor ore surfaces that appear, on reconstruction of the original environment (Marston and Kay 1980), to form two sub-parallel belts.

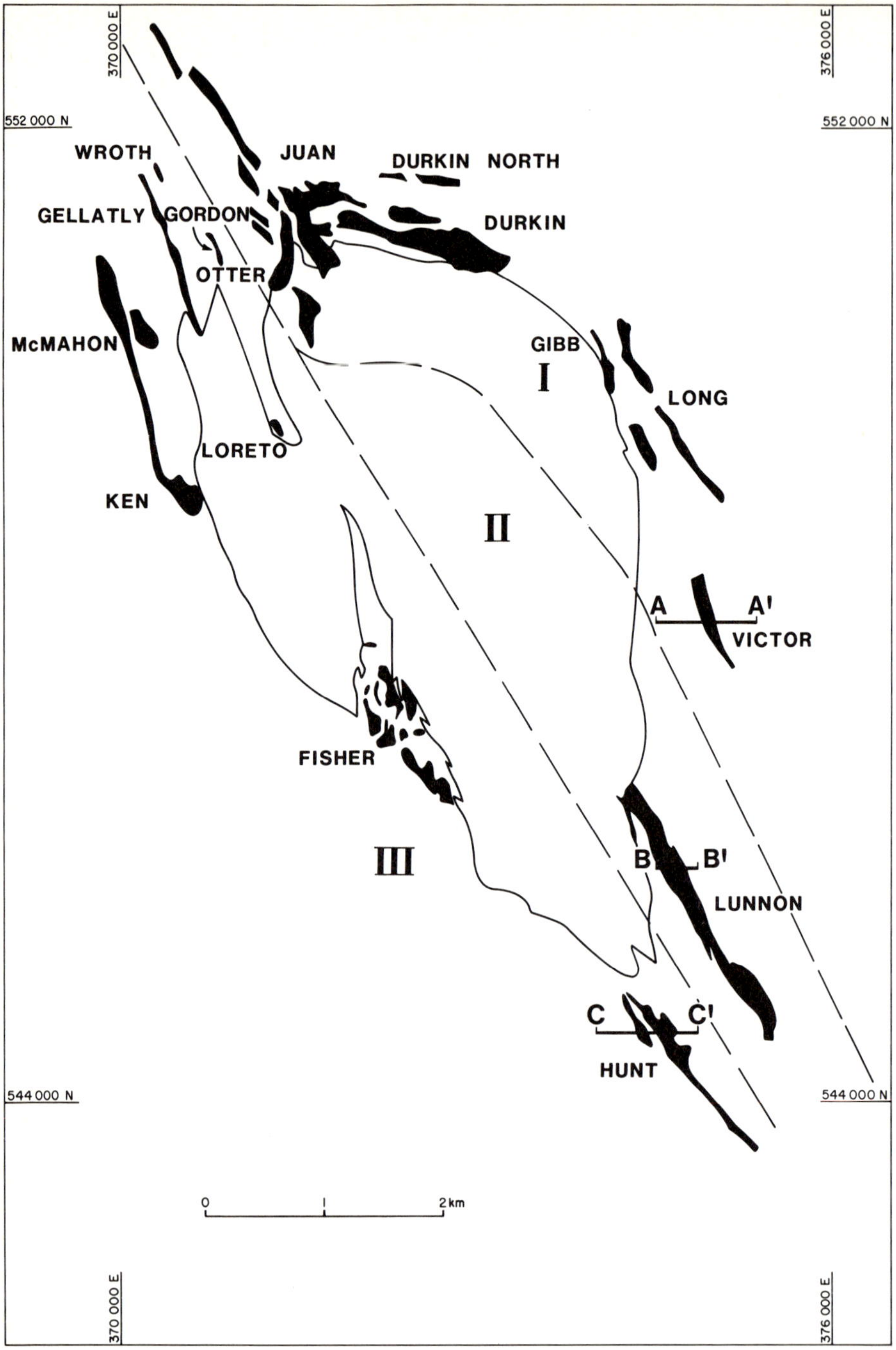

Fig. 4.. Plan of Kambalda Dome showing classification of nickel deposits and location of geological cross-sections in Fig. 5

Table 1. Characteristics of ore Shoots of the Kambalda Dome

Feature Shoots	Class I Juan, Otter, Durkin, Durkin North, Gibb, Long, Victor	Class II Lunnon	Class III Wroth, Gellatly, Gordon, McMahon, Loreto, Ken, Fisher, Hunt
% of total Kambalda Dome ore tonnes	59.4	19.1	21.5
% of total Kambalda Dome Ni metal tonnes	67.8	15.6	16.6
Average Ni grade%	4.71	3.37	3.22
% tonnage as contact ore	92.6	50.4	56.4
% tonnage as hangingwall ore	3.3	42.5	38.2
% tonnage as offset ore	4.1	7.1	5.4
% Ni metal as contact ore	95.6	64.1	62.6
% Ni metal as hangingwall ore	1.3	29.5	31.4
% Ni metal as offset ore	3.1	6.4	6.0
Ore body and ore characteristics	Almost all ore occurs as contact ore with the minor hangingwall occurrences comprising blebby ores within the high-magnesium basal flow units. Juan, Durkin and Long are the major shoots and individual ore surfaces within these ore bodies may contain 1 million tonnes of high-grade ore. Otter, Durkin, Victor, Gibb and parts of Juan comprise high tenor ores whereas Long and the greater part of Juan are moderate tenor (i.e. massive sulphides 10% – 16% Ni)	Ore tonnage almost equally distributed between contact and hangingwall ore and each occurs as extensive, once essentially continuous, ore surfaces. Contact ores are low tenor, but substantially higher grade than the high tenor hangingwall ores. The contact ores at the north end of the ore body are overlain by a thick zone of low-grade (<1% Ni) disseminated sulphides. The intensity of contact mineralisation diminishes from north to south. The hangingwall ores directly overlie the contact ore. Minor hangingwall ore occurs associated with the third flow unit	Contact ore surfaces can be moderate to high grade, but individual ore surfaces seldom exceed 250000 tonnes. Tenor of contact ores is highly variable and may be higher or lower than associated hangingwall ores. Hangingwall ore may be associated with the second or third flow units

Table 1 (continued)

Feature	Class I	Class II	Class III
Shoots	Juan, Otter, Durkin, Durkin North, Gibb, Long, Victor	Lunnon	Wroth, Gellatly, Gordon, McMahon, Loreto, Ken, Fisher, Hunt
Ore-confining features	Most contact ores occur within well-defined troughs or embayments, many with strongly re-entrant structures. Morphology is highly variable, but generally the trough floor is several metres below the barren flanks	Almost all contact ore occurs within the main Lunnon trough although low grade ore does occur on the west flank in places and associated with sediments on the east flank. The trough has been strongly modified by post-depositional faulting and these structures appear to terminate the hangingwall ores	Almost all contact ores occur in troughs and embayments of variable morphology. Definition of structural confinement is generally not as well developed as in the Class I and II deposits and mineralisation may occur as irregular concentrations along a broadly defined structural embayment. Hangingwall ores spatially related to contact ores, but association not as clearly defined as Class II deposits
Stratigraphic features	Ore associated with thick (50 – 100 m) high-magnesium basal flows with thin, fractionated, flow tops. No contact sediments within normal ore environment and interflow sediments generally absent above the troughs. Strongly unordered upper member with individual flow units displaying strong textural development and compositional differentiation	Contact ore associated with thick (30 – 80 m) flow that thins to the south. Basal flow has thick, strongly fractionated flow top. No contact sediments occur on the flanks up to the ore-confining structures. Hangingwall ore associated with 30 – 50 m thick flow also with thick spinifex-textured picritic flow top. Hangingwall ores commonly associated at margins with interflow sediments	Contact ores associated with thick basal flows. Contact sediments generally absent from major troughs, but some sediment-associated contact ore occurs at Ken shoot. Hangingwall ores associated with thick flows and commonly associated with sediments

The ore bodies are associated with thick (50 – 100 m) high-magnesium flow units and there is a total lack of contact sediments within the ore environment. Interflow sediments are generally absent between flow units above the ore-confining troughs although they occur on the flanks away from the ore environment (Fig. 5A). The thin flow units of the upper member above these ore bodies generally display good textural development and strong compositional differentiation, but the sequence of flow units is compositionally unordered.

Class II

Lunnon shoot on the southeast flank of the Kambalda Dome is the only ore body in this class although there are extensive zones of mineralisation south of Lunnon shoot underneath Lake Lefroy which are thought to be plunge extensions of this ore body. This ore body is characterised by extensive fault disrupted contact and hangingwall ore surfaces with the hangingwall ore directly overlying the contact ore. Ore tonnage is almost equally divided between contact and hangingwall surfaces although the higher grade of contact ores results in nearly 65% of the nickel metal occurring as contact ore.

The contact ores are low tenor (massive ores 8% Ni) and may be overlain by a thick (up to 30 m) zone of low-grade disseminated mineralisation (0.5 – 0.8% Ni) within the basal host unit (Ross and Hopkins 1975). The thickness of this zone and the basal host unit, and the intensity of mineralisation diminishes to the south (Ross 1974, Middleton 1980). The majority of the hangingwall ore occurs at the base of the second flow unit occurring on top of the thick spinifex-textured picritic flow-top of the basal flow unit although limited zones of weak mineralisation are recorded at the base of the third flow. Hangingwall ores are high tenor (massive ores up to 25% Ni).

The ores are associated with thick flows, the first flow being 70 – 80 m thick to the north and thinning to 30 – 40 m to the south. Flows thin rapidly to 10 – 15 m on the flanks of the trough (Fig. 5B). Contact sediments are totally lacking from the main Lunnon trough although there is the well-documented ore-sediment relationship on the east flank (Middleton 1980, Paterson et al. 1984). Hangingwall ores commonly terminate against or are associated at their margins with stratigraphically equivalent interflow sediments and occurrences of nickeliferous sediments are common.

Class III

These are the ore bodies of the western flank of the Kambalda Dome: Wroth, Gellatly, Gordon, McMahon, Ken, Loreto, Fisher, and Hunt. They are smaller, less continuous, lower grade, ore positions that comprise both contact and hangingwall ores. Tenor of ores within and between shoots is variable. For example, Ken shoot is composed predominantly of high tenor contact ores. Within Fisher shoot work by Hancock (1978) indicated subparallel belts of high and moderate tenor contact ores and at Hunt shoot, contact ores are moderate to low tenor and the associated hangingwall ore, low tenor. Hangingwall ores are generally associated with the second flow unit although at Fisher there is extensive hangingwall mineralisation associated with the third unit.

Contact ores are associated with thick, high-magnesium flow units and sediments are generally absent from the ore environment apart from the occurrences

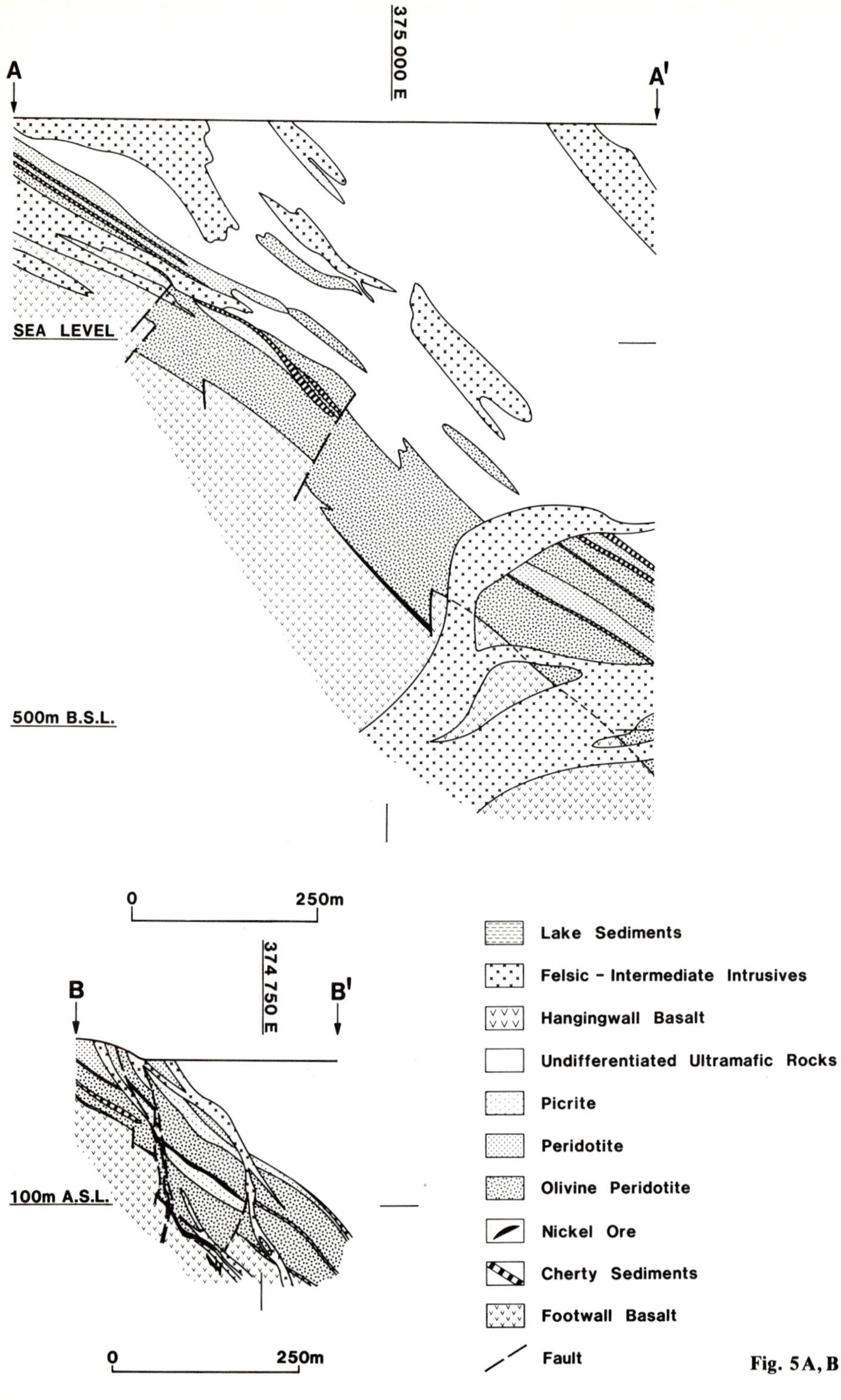
375 000 E
A
A'
SEA LEVEL
500m B.S.L.
0
250m
374 750 E
B
B'
100m A.S.L.
0
250m
Lake Sediments
Felsic - Intermediate Intrusives
Hangingwall Basalt
Undifferentiated Ultramafic Rocks
Picrite
Peridotite
Olivine Peridotite
Nickel Ore
Cherty Sediments
Footwall Basalt
Fault

Fig. 5A, B

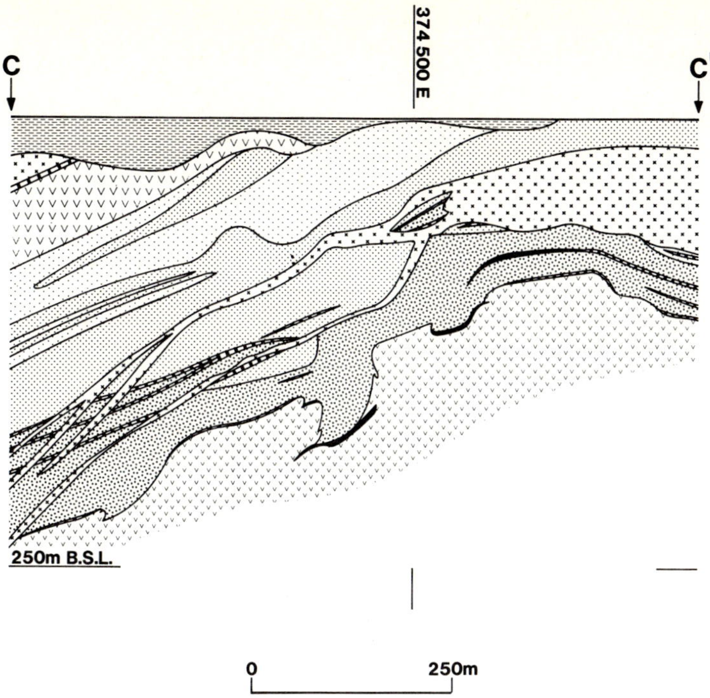

Fig. 5 A – C. Geological cross-sections of the Kambalda Dome. **A** $A-A'$ 547925 N, Victor Shoot; **B** $B-B'$ 545913 N, Lunnon Shoot; **C** $C-C'$ 544572 N, Hunt Shoot

at Ken shoot (Paterson et al. 1984). Hangingwall ores are commonly associated with interflow sediments and at McMahon and Fisher there are extensive zones of nickeliferous sediments.

Reconstruction of the Geological Environment of the Kambalda Dome

Although the rocks and ores of the Kambalda Dome have been strongly deformed and metamorphosed, from the foregoing analysis of the various ore shoots and from a simple structural analysis a generalised reconstruction of the original environment can be made. Reconstructions of individual shoots have been made by Marston and Kay (1980) for Juan shoot, Ross (1974) and Middleton (1980) for Lunnon shoot and Hancock (1978) for Fisher shoot.

The Durkin and Juan shoot areas display a high degree of structural complexity with considerable horizontal shortening involving large-scale thrusting and recumbent folding in the Juan area. These structures are thought to be caused by strong NW-SE compressive forces that may in part be generated by the intrusion of the granite body occupying part of the core of the Kambalda Dome. Considerable movement has been accommodated on the Juan Fault, a major structural feature of the area. Durkin shoot also displays strong development of low angle thrusting (Hayden 1976). Marston and Kay (1980) reconstructed the Juan-Durkin area and defined two sub-parallel belts of ore of differing tenor converging to the northwest. Carrying this reconstruction a stage further, it is suggested that the high tenor Durkin shoot may have once been continuous with the similar tenor ore bodies Gibb and Victor on the east flank of the dome. Long shoot is considered to be a continuation of the low-moderate tenor surfaces of Juan shoot. Although not validated by detailed structural analysis, these ore shoots appear to have once formed essentially linear belts of semi-continuous mineralisation at least 8 – 10 km long. Terminations to the north and south have not been defined and the linear extent of these ore "belts" could be much greater than 10 km (Fig. 8).

The other ore shoots of the Kambalda Dome are extremely structurally complex in detail, but again generalised reconstructions can be made. Lunnon shoot is 2.5 km long and appears to be continuous with mineralisation a further

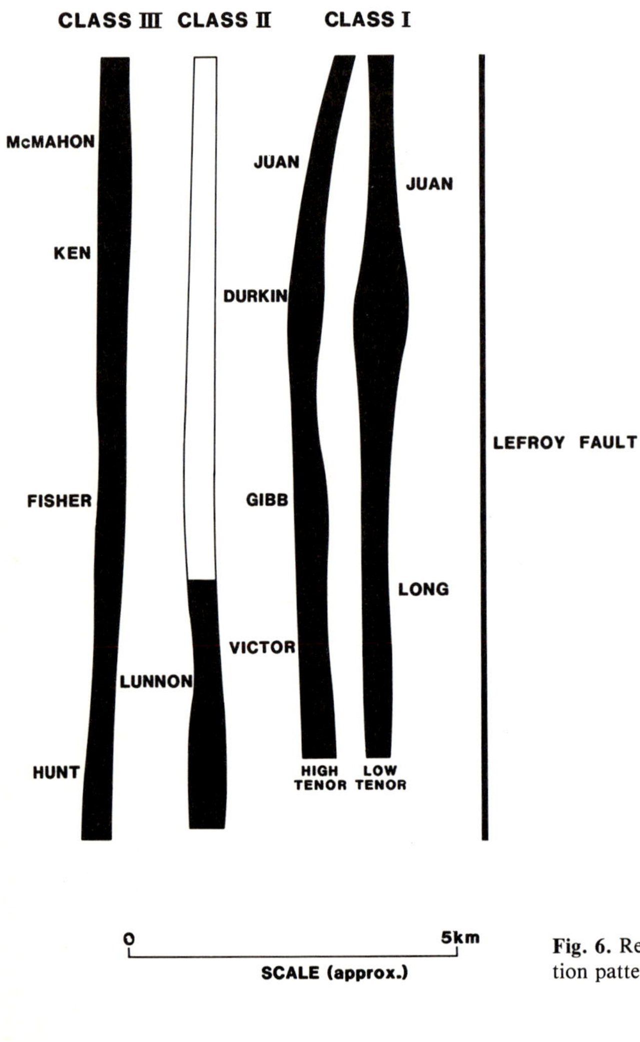

Fig. 6. Reconstruction of ore distribution pattern at the Kambalda Dome

1.5 km to the south. Extensions of Lunnon shoot to the north have been lost through erosion or structural dislocation.

Hunt, Fisher, Ken and McMahon shoots form an essentially linear ore belt, in places comprising sub-parallel high and low tenor belts (e.g., Hancock 1978) that extend for 8 – 10 km and again terminations to this trend have not been recorded. The convergence of Hunt and Lunnon shoot at the south end of the dome is a structural feature.

Therefore a simplistic reconstruction of the Kambalda Dome indicates four major sub-parallel linear belts of ore 600 – 1200 m apart that are at least 10 km in length (Fig. 6). The ore in these belts may occur as large continuous ribbons over 2 km long (i.e., Lunnon, Long) or in discrete pods within the overall linear trend. However, ore zones seldom exceed 300 m in width. The nature and quality of ore is broadly asymmetric with the eastern belt containing essentially all contact ore, the central belt comprising closely-spatially related, and essentially stratigraphically continuous contact and hangingwall ore and the western belt less continuous contact and hangingwall ores.

In addition to the ore-distribution asymmetry across the dome there is a broad asymmetry in the distribution and characteristics of the ultramafic rocks. The features of the ultramafic rocks that characterise an ore environment (see Sect. 3) are common to all types of ore shoots and this "prism" of rock is consistent along the trend of the ore belts. However, the ultramafic sequence thins from east to west (sympathetic with the decreasing ore quality in this direction) and, in general, the ultramafic rocks to the west display a better defined and more consistent trend of diminishing magnesium content upward through the sequence. Combining the cross-sections of Fig. 5 a generalised cross-section

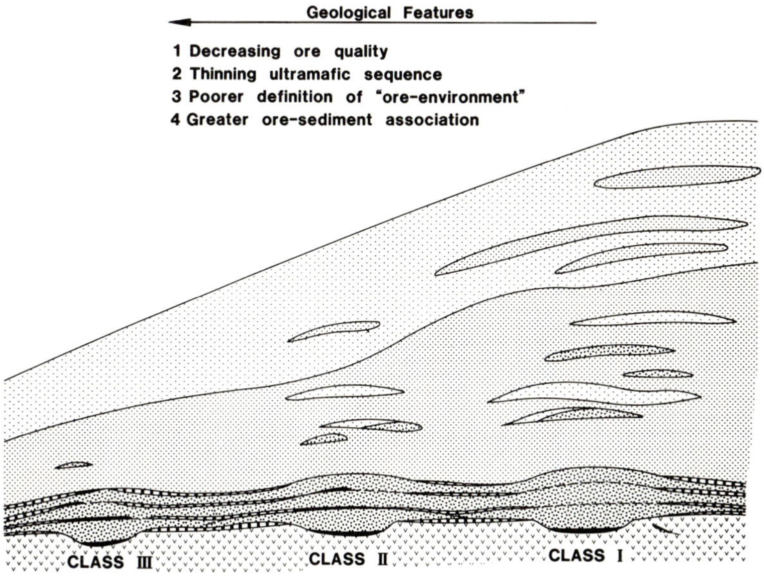

Fig. 7. Cross-sectional reconstruction of the Kambalda Dome. No intrusives or structural detail shown for clarity. (See Fig. 5 for legend)

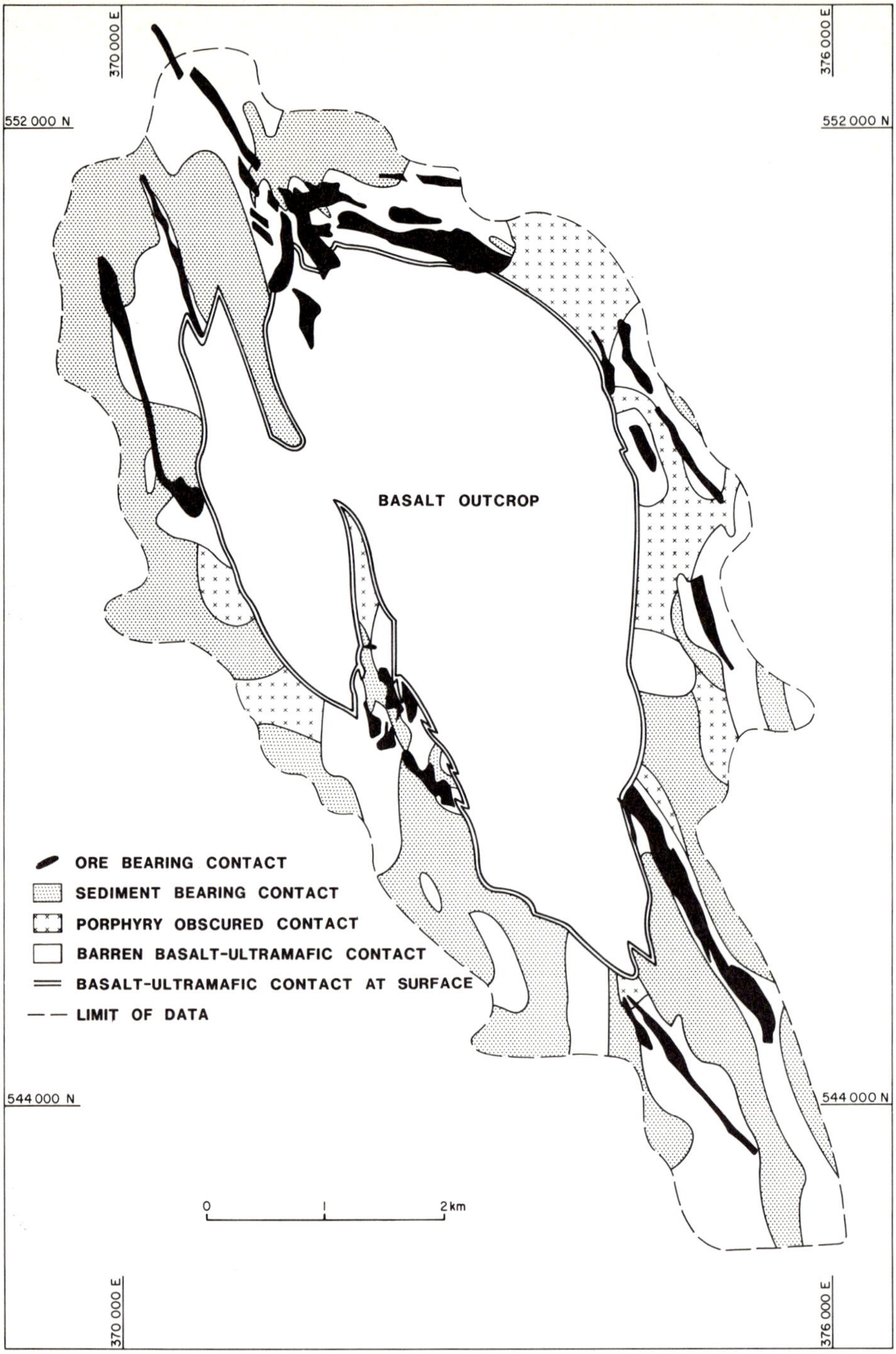

Fig. 8. Plan projection of basalt contact, Kambalda Dome, showing nature of basalt-ultramafic contact

through the Kambalda Dome is shown in Fig. 7. The structural and time relationships between the ore belts are uncertain. It is possible that each belt formed quite independently of the others (i.e., in separate fault-bounded basins). However, Fig. 7 has been drawn without any structural features and intrusives are also omitted.

Sediments occur on the basalt-ultramafic contact throughout the area of the Kambalda Dome. Detailed work by Bavinton (1979) indicated limited chemical and mineralogical variation with the sediments in the Dome area, the most significant variation being the greater of carbonaceous sediments on the west flank of the Dome. A plan of the Kambalda Dome with the ultramafic rocks removed is shown in Fig. 8. Contact sediment distribution is largely taken from Gresham (1978) and is derived from broad-spaced diamond drill information. This plan indicates the sediment-free nature of the contact close to the ore zones.

Comparison with Other Volcanic Peridotite-Associated Deposits

Other volcanic peridotite-associated deposits of the Norseman-Wiluna belt also display features that are observed from the deposits of the Kambalda Dome. Major deposits (Fig. 1) include Nepean (Sanders 1982), Carnilya East, Blair, Spargoville 3 (Andrews 1975), Wannaway (McQueen 1981b) and Redross, Widgie Townsite and Widgie 3 (McQueen 1981a), Mt. Edwards (Turner 1983), Windarra (Schmulian 1982), Scotia (Page and Schmulian 1981) and other deposits at Kambalda including Jan, Foster, Lanfranchi and Edwin (Gresham and Loftus-Hills 1981). Many smaller deposits occur throughout the belt and these are referred to in review articles (Ross and Travis 1981, Marston et al. 1981). The African deposits Shangani (Viljoen et al. 1976) and Damba-Silwane (Williams 1979), and Langmuir (Green and Naldrett 1981) in Canada also display features common with some of those observed at the Kambalda Dome. These include:

1. Ores occur in depressions or embayments in underlying rocks.
2. Ores zones generally form narrow elongate bodies with width: length generally >3:1.
3. Ores are directly associated with thicker zones of high magnesium rocks.
4. Ores and sulphidic sediments are commonly mutually exclusive.
5. The ultramafic sequence is stratigraphically more complex near ore.

Although the deposits of the Norseman-Wiluna belt are highly variable and complex and it is difficult to categorise them, the deposits do display similarities to those of the Kambalda Dome. The Carnilya East, Blair, Nepean, Spargoville 3, Redross, Edwin, Lanfranchi, and Widgie 3 deposits consist almost entirely of contact ore. Although none of the deposits attains the size of the major Class I deposits of the Kambalda Dome they are generally the higher grade deposits of the belt, have no direct association with sediments and generally occur in well-defined footwall embayments. Scotia and Windarra are also primarily contact ore positions, but have features which distinguish them from the other deposits.

Table 2. Classification of volcanic peridotite-associated deposits of the Norseman-Wiluna Belt according to classification of deposits of the Kambalda Dome

Class I	Class II	Class III
Carnilya East	Foster	Jan.
Blair	Mt. Edwards	Many other smaller deposits
Nepean	Wannaway	
Sparqoville 3	Widgie townsite	
Redross		
Edwin-Lanfranchi		
Widgie 3		
(Scotia)		
(Windarra)		

Foster, Wannaway, Widgie Townsite, and Mt. Edwards are ore bodies with extensive contact and hangingwall ores and can be related to the Kambalda Dome Class II deposit. Jan and many of the smaller deposits of Kambalda and Widgiemooltha Domes are characterised by small contact ore positions and associated hangingwall positions and are analogous to the Class III deposits of the Kambalda Dome.

The difficulty in categorising deposits cannot be over emphasised and it is probable that the deposits represent a spectrum of depositional environments. Previous workers, Lesher et al. (1982), Lesher (1983), Lesher et al. (1984) and Groves et al. (1984), have discussed the classification of volcanic peridotite-associated deposits as proximal and distal facies variants.

Aspects of Genesis of Volcanic Peridotite-Associated Deposits

Considerable detailed geological research and documentation of this class of ore deposit have indicated a wide variety of field, geochemical, experimental and theoretical evidence for an original magmatic origin for the nickel sulphide ores. Important evidence for this origin includes:

1. Association of ores with the most magnesian basal units of the komatiite sequence; localisation of ore within footwall embayments or troughs; strong volcanological control as evidenced by original stratigraphic relationships (Ross and Hopkins 1975, Gresham and Loftus-Hills 1981, Lesher 1983).
2. Relationships of massive/matrix/disseminated sulphides at base of host unit and interpillow sulphides within underlying basalts (Woodall and Travis 1969, Ewers and Hudson 1972, Hudson 1972, Naldrett 1973, Gresham and Loftus-Hills 1981).
3. Distinctive ore geochemistry compatible with known or experimentally inferred partitioning behaviour in mafic-ultramafic systems. These include Ni/Cu ratios, precious metal tenors and abundances, S/Se ratios and $\delta^{34}S$

values (Rajamani and Naldrett 1978, Duke and Naldrett 1978, Ross and Keays 1979, Keays et al. 1981, Seccombe et al. 1981, Donaldson 1984a).
4. The presence of distinctive chalcophile ferrochromites (Ewers et al. 1976, Groves et al. 1977, Woolrich et al. 1981, Donaldson 1984b) in the ores.
5. Chalcophile element depletion in the associated host komatiites at Kambalda (Lesher et al. 1981).

Although there is general agreement as to the origin of these ores through magmatic processes, the actual nature and mechanism of the emplacement of the sulphides and their hosts is still uncertain. Early workers (Ross 1974) considered the basal flows to consist of a picritic liquid (23 – 24% MgO) containing varying amounts of olivine phenocrysts. Recent detailed work on the nature of the komatiitic host (Lesher et al. 1981, Arndt 1982, Lesher 1983, Donaldson and Lesher 1982) has indicated liquids containing up to 32% MgO and in situ crystallisation of much or all of the olivine in these flows has been proposed. Combining this work with theoretical modelling of flow characteristics, Lesher et al. (1984) have postulated thermal erosion of the associated sulphidic sediments to induce sulphide segregation. Furthermore Huppert et al. (1984) have postulated that the flows could even thermally erode the underlying footwall basalts hence generating their own trough during flow passage. These concepts are reviewed in the light of critical field evidence and based on the observed field relationships a model for the depositional environment of these deposits is discussed.

A sedimentary origin for the sulphur in the Western Australian nickel deposits was first suggested by Prider (1970) and this concept has been revived recently (Lesher 1983, Lesher et al. 1984, Huppert et al. 1984) largely because of theoretical problems in explaining the capability of low viscosity, phenocryst-poor lavas to transport dense sulphides, and the large volume of sulphides relative to the volume of host ultramafic units. The broadly antithetic relationship between ore and sediments makes the concept of sediment assimilation attractive, but there are many detailed stratigraphic and geochemical relationships at the Kambalda Dome which are inconsistent with this concept.

1. The broad asymmetry of the quality of ore distribution, the variable tenor of sub-parallel ore shoots and of contact and hangingwall ores of the same shoots argue strongly against assimilation of sediments of relatively uniform composition by distal-sourced lava flows.
2. Almost all contact ore occurs in well-defined embayments and there is negligible ore outside these zones. The emplacement conditions and temperatures of ore-associated lava flows were likely to be variable and a range of gradational relationships between ore-bearing and barren contacts would be anticipated. The sharp definition of the ore zones is contrary to the expected relationships if sediment assimilation was important.

   Although the contact ores generally are surrounded by a zone of sediment-free basalt-ultramafic contact (Fig. 8), there are considerable areas of sediment-free contact where there are no defined nickel sulphides. If the sediment-assimilation model assumes that all the basalt was originally covered with sediment, then there should be a more universal distribution of nickel ore than is observed.

Contact sediments are neither thicker nor more sulphur-rich on the east and north flanks of the dome to account for the greater concentration of sulphides in the Class I deposits.

3. Hangingwall ores are invariably lower grade and sulphide abundance is less than associated contact ores (Table 1). However, sulphur content of internal sediments is higher (7.3%) than that of contact sediments (4.0%) (Bavinton 1981) and thicknesses of internal and contact sediments are comparable.
4. At some shoots of the Class I type deposits (e.g., Victor, Fig. 5A) there are 2–3 flow units above the ore zone with no associated interflow sediments. On the flanks of these shoots there are essentially continuous interflow sediments. The complete absence of stratiform hangingwall ores associated with this class of deposit and the absence of interflow sediments directly above the contact ores argue strongly against the concept of sediment assimilation.
5. At some volcanic peridotite-associated deposits, e.g., Mt. Edwards and Wannaway, there are extensive zones of massive sulphides associated with sulphidic sediments with no apparent erosion or attenuation of the sediment in the ore environment.
6. If all ore-bearing and barren contact shown in Fig. 8 was assumed to be sediment covered and subsequently assimilated by the ore-associated flow, there would not be nearly enough sulphur in these sediments (4%) to generate the amount of nickel sulphides observed. A simple calculation assuming a sediment averaging 4 m in thickness and containing 4% sulphur would indicate 28 km^2 of contact sediment would have to be eroded to generate the sulphides observed in the Class I deposits. In actual fact average contact sediment thickness in the Juan area is approximately 1 m (Bavinton 1979).
7. In those shoots that have intimate massive ore-sediment relationships (Paterson et al. 1984) and where some local assimilation of sediment is probable, the ores have significantly higher zinc content (160 ppm) than normal massive ores (40 ppm). This geochemical difference is also reflected at a shoot scale where ores from shoots of the Class II and III deposits generally have higher Zn and As content than those of the Class I deposits.

Although there may be some local assimilation of sediment, the detailed evidence suggests this is an unlikely factor in the generation of the nickel ores and the broadly antithetic ore-sediment relationships must be explained in another way.

Turning now to the origin of the troughs, Huppert et al. (1984) have postulated that the high temperature komatiitic flows could thermally erode the underlying basalt and create their own trough. This largely theoretical work conflicts with some of the detailed observations around the Kambalda Dome.

1. The very strong arguments developed against sediment assimilation.
2. Stratigraphic control of troughs as indicated by Harley (1980) and Lesher (1983). In some areas stratigraphic continuity has been defined between embayment contact (i.e., the one zone) and a unit boundary (e.g., flow-top breccia) in the adjacent basalt. The depth of most embayments is comparable to the thickness of most basalt flow units.
3. The many small elliptical depressions containing nickel ores (i.e., Ken, Juan areas) that are considered primary topographic volcanic features. It is difficult

to conceive of these small scale features being generated through erosional processes.

4. The strongly linear embayments at Kambalda contrast strongly with the highly sinuous lunar rilles (e.g., Hadley rille) thought to have been generated by thermal erosion (Hulme 1973, 1982, Wilson and Head 1983).
5. Complete back of geochemical contamination of basal flows within the trough structures. These have TiO_2 and Al_2O_3 values lower than or equal to the flanking flows (Lesher 1983).

Interpretation of Depositional Environment

The Norseman-Wiluna belt is characterised by thick sequences of volcanic and sedimentary rocks. Accumulation of these sequences is considered to have been rapid (Archibald et al. 1978) and deposition to have taken place in a rapidly subsiding, strongly rifting environment (Groves 1982) with the formation of many discrete, fault-bounded elongate basins of deposition. Stratigraphic correlation between different mafic-ultramafic sequences is uncertain although it is generally thought that there is some stratigraphic control on the mineralisation, with the volcanic peridotite-associated deposits occurring in the lower-most sequences (Gemuts and Theron 1975, Gresham and Loftus-Hills 1981).

Ulramafic sequences hosting volcanic peridotite deposits have in almost all cases been extruded onto tholeiitic basaltic sequences. At the Kambalda Dome these basalts comprise medium to thick flows with, in places, well-developed flow-top breccias, possible intrusive bodies, and abundant development of pillowed sequences. The general lack of interflow sediments or wackes apart from the discontinuous marker horizon 100 – 200 m below the basalt-ultramafic contact indicates rapid accumulation of these basalts which would have resulted in the development of highly irregular topography on the ocean floor. The almost complete lack of vesicles in the pillows of Kambalda indicates water depths possibly exceeding, 5000 m (Jones 1969, Moore 1975) which would indicate extrusion under high confining pressure. Rarely preserved pillows at Carnilya East also lack vesicles whereas at Widgiemooltha there is development of vesicles indicating deposition in shallower water.

The embayments confining the nickel ores are highly variable and obviously no generalised statement can be made regarding their origin. Lesher (1983) has argued that the embayments were largely topographic-volcanological features. The marked continuity and linearity of these features at the Kambalda Dome suggest a tectonic as well as a volcanogenic-topographic control. Abundant small scale faulting is observed in modern ocean ridge environments (Ballard and Moore 1977) and there is evidence at Kambalda for very early faulting (Richards 1983) and specifically at Lunnon Shoot for synvolcanic fault development (Ross 1974, Middleton 1980). Lack of suitable stratigraphic control elsewhere at Kambalda precludes the definition of the universal occurrence of these structures. It is suggested that the broad distribution of the embayments and ore shoots was controlled by the general extensional-rifting tectonics of the area. The

Lefroy Fault, although obviously reactivated during later deformational events may have been a major ore-controlling structure in the area. The troughs have dimensions similar to modern oceanic rifts (Spiess et al. 1980, Francheteau and Ballard 1983, Vinogradov and Udinstsev 1975). Along the trend of these structures a variety of topographic-volcanological embayments would occur.

The asymmetry of the Kambalda environment and ore quality suggests diminishing thermal activity from east to west and it is significant that the major Class I deposits are closest to the Lefroy fault. This relationship of nickel mineralisation to major north-northwest faults has been recorded by Gemuts and Theron (1975) and other workers.

Continued basaltic volcanism and thermal activity along linear zones, now occupied by the ores zones, inhibited sediment accumulation, and eruption of the ultramafic lava and sulphides took place from thin fissures at intervals along these structures. Fissures feeding these eruptions need to be no more than 1 m wide (Wilson and Head 1983) and evidence from Lunnon shoot (Middleton 1980, Lesher 1983) with the thinning basal flow unit and diminishing ore quality indicates individual extrusion sites may be spaced at intervals of 2 – 4 km along the linear trends of the ore belts. The lack of observed feeders in the footwall basalts at Kambalda is not a problem if the fissures were extremely narrow as they may be now unrecognisable or obliterated by the extensive felsic intrusives in the area.

A detailed discussion of the nature of the basal ultramafic flow units is beyond the scope of this paper, but the understanding of the magmatic, physical, mineralogical and chemical characteristics of these flows is critical to the understanding of the origin of volcanic peridotite-associated nickel deposits. Stolz and Nesbitt (1981) at Scotia and Lesher (1983) have documented clear evidence for in situ growth of olivine in flows associated with these deposits. This has led to the development of the concept that these flows originated from almost totally liquid magma of 28 – 32% MgO and the present magnesium rich content (40 – 45% volatile free MgO) was achieved by a process of lava flow-through. The presence of branching crescumulate olivine is evidence of in situ olivine crystallisation, but this type of olivine may in fact represent only a minor component of the basal flows. Flow-through and olivine enrichment as discussed by Lesher (1983) requires the development of lateral facies variants of lower magnesium basal rocks, examples of which are not observed over the 60 km of explored basalt-ultramafic contact of the Kambalda field. The thin, laterally persistent, flanking flows appear to be lateral equivalents of the ore flows, but they are only marginally different in bulk chemical composition. In order to explain the slight chemical differences between ore and non-ore flows, it is suggested that basal flows contained appreciable quantities of intratelluric olivine and some riffling of these olivine phenocrysts did occur as suggested by Naldrett and Campbell (1982). The lateral persistence of the thin flanking flows is indicative of low viscosity lavas and chemical profiles indicate that in situ fractionation occurred in both ore-associated and flanking flows.

The presence of intratelluric olivine would also enhance the sulphide carrying capacity of the flows. As indicated by Lesher et al. (1981), the ultramafic succession at Kambalda was in equilibrium with sulphide liquid at the magmatic stage.

Field evidence such as interpillow sulphides, sulphides in basalt flow-top breccias and the occurrence of sulphide accumulations in topographic lows (i.e., ellipsoidal structures at Juan and Ken) and the along-trough ore relationships at Lunnon are indicative of very early and rapid segregation of the sulphides. Marston and Kay (1980) suggested that this took place during the early stages of horizontal transport although segregation during vertical transport (Ross 1974, Ross and Hopkins 1975) cannot be discounted.

The development of the well-defined ore "prism" (i.e., thicker high-magnesium successive flows, spatially related hangingwall ores, lack of sediment development and textural-compositional development within flows) is strongly suggestive of continued reactivation and extrusion along the trend of the ore shoots. The linearity and consistency of superposition of these features is difficult to reconcile with a single source, distal model.

Deposition of the basal high-magnesium flow units with interflow sediment accumulation was followed by continuous outpourings of thin (generally <10 m) flows of low to moderate magnesium lavas and in turn by thick basaltic, felsic-volcanic and sedimentary sequences.

The ore bodies of Kambalda Dome and other volcanic peridotite-associated deposits display many enigmatic and variable geological features that are difficult to explain by any generalised genetic model. However, the broad stratigraphic and ore relationships at the Kambalda Dome are consistent with a series of sub-parallel extrusion sites that were periodically reactivated.

Sediment assimilation by ultramafic flows was not a major factor in the generation of the nickel ores and thermal erosion of the underlying basalt to generate the ore-confining embayments is not compatible with field relationships.

Acknowledgements. Much of the basic geological information contained in this paper is the result of geological work carried out by company mine and exploration geologists over the past 18 yrs. I would like to thank many of these for stimulating discussions and useful comments. I would particularly like to thank B. Thomson for his help in preparing cross-section 544572N and J. Ross for helpful discussions and suggestions. I would also like to thank N. Archibald, A. Cowden, M. Elias, D. Groves, G. Loftus-Hills, K. Cross, and R. Smith who provided constructive criticism of the manuscript. Thanks are also extended to D. R. Hudson and an anonymous reviewer for their helpful comments.

I am indebted to Debra Woodman and Christine Wood for typing the manuscript and R. Perkin and H. Bush for the preparation of the figures. This paper is published by permission of Western Mining Corporation Limited.

References

Andrews PB (1975) Spargoville nickel deposits. In: Knight CL (ed) Economic geology of Australia and Papua, New Guinea. I. Metals. Australas Inst Min Metall Mon 5:89–91

Archibald NJ, Bettenay LF, Binns RA, Groves DI, Gunthorpe RJ (1978) The evolution of Archean greenstone terrains, Eastern Goldfields province, Western Australia. Precambrian Res 6:103–131

Arndt NT (1982) Post-eruptive olivine and sulphide fractionation and accumulation in komatiite lava flows (abs). IGCP Proj 161, 91: Nickel Sulphide Field Conf III Abs Vol, p 22

Ballard RD, Moore JG (1977) Photographic atlas of the Mid-Atlantic ridge rift valley. Springer, Berlin Heidelberg New York, 114 pp

Barrett FM, Binns RA, Groves DI, Marston RJ, McQueen KJ (1977) Structural history and metamorphic modification of Archean volcanic-type nickel deposits, Yilgarn Block, Western Australia. Econ Geol 72:1195–1223

Bavinton OA (1979) Interflow sedimentary rocks from the Kambalda ultramafic sequence: Their geochemistry, metamorphism and genesis. Ph D Thesis, Aust Natl Univ, Canberra, 196 pp (unpublished)

Bavinton OA (1981) The nature of sulfidic metasediments at Kambalda and their broad relationships with associated ultramafic rocks and nickel ores. Econ Geol 76:1606–1628

Binns RA, Gunthorpe RJ, Groves DI (1976) Metamorphic patterns and development of greenstone belts in the eastern Yilgarn Block, Western Australia. In: Windley BF (ed) Correlation of the Precambrian, vol II. Nauka, Moscow, pp 349–380

Claoue-Long JC, Thirlwell MF, Nesbitt RW (1984) Sm–Nd of Kambalda greenstones – revisited and revised. Nature (London) 307, 5953:697–701

Donaldson MJ (1984a) Geochemistry and mineralogy of the platinum group elements in Kambalda nickel ores and concentrator products. WMC (KNO) Rep K/2781, 33 pp (unpublished)

Donaldson MJ (1984b) Progressive alteration of barren and weakly mineralised Archaean dunites from Western Australia: A petrological, mineralogical and geochemical study of some komatiitic dunites from the Eastern Goldfields Province. Ph D Thesis, Univ W Aust (unpublished)

Donaldson MJ, Lesher CM (1982) Intrusive dunite and extrusive peridotite: Analogous products of komatiitic magmas resulting from different cooling rates (abs). IGCP Proj 91, 161: Nickel Sulphide Field Conf III Abs Vol, pp 31–32

Duke JM, Naldrett AJ (1978) A numerical model of the fractionation of olivine and molten sulfide from komatiite magma. Earth Planet Sci Lett 39:255–266

Ewers WE, Hudson DR (1972) An interpretative study of a nickel-iron sulfide ore intersection, Lunnon Shoot, Kambalda, Western Australia. Econ Geol 67:1075–1092

Ewers WE, Graham J, Hudson DR, Rolls JM (1976) Crystallization of chromite from nickel-iron sulphide melts. Contrib Mineral Petrol 54:61–64

Francheteau J, Ballard RD (1983) The East Pacific rise near 21°N; 13°N, and 20°S: Inferences for along strike variability of axial processes of the mid-ocean ridge. Earth Planet Sci Lett 64:93–116

Gee RD, Williams IR, Wilde SA, Baxter JL (1980) Crustal development of the Yilgarn Block (abs). 2nd. Int Archaean Symp, Perth, Extended Abs, pp 79–80

Gemuts I, Theron A (1975) The Archaean between Coolgardie and Norseman-stratigraphy and mineralization. In: Knight CL (ed) Economic geology of Australia and Papua New Guinea. I. Metals. Australas Inst Min Metall, Melbourne, Mon 5, pp 66–74

Green AH, Naldrett AJ (1981) The Langmuir volcanic peridotite-associated nickel deposits: Canadian equivalents of the Western Australian occurrences. Econ Geol 76:1503–1523

Gresham JJ (1978) The Kambalda dome traverse study final report. WMC (KNO) Rep K/2389, 32 pp (unpublished)

Gresham JJ, Loftus-Hills GD (1981) The geology of the Kambalda nickel field, Western Australia. Econ Geol 76:1373–1416

Groves DI (1982) The Archaean and earliest Proterozoic evolution and metallogeny of Australia. Rev Bras Geocie 12(1–3):135–148

Groves DI, Hudson DR (1981) The nature and origin of Archaean stratabound volcanic-associated nickel-iron copper sulphide deposits. In: Wolf KH (ed) Handbook of stratabound and stratiform ore deposits, vol IX. Elsevier/North Holland Biomedical Press, Amsterdam New York, pp 305–410

Groves DI, Barrett FM, Binns RA, McQueen KG (1977) Spinel phases associated with metamorphosed volcanic-type iron-nickel sulfide ores from Western Australia. Econ Geol 72:1224–1244

Groves DI, Lesher CM, Gee RD (1984) Tectonic setting of the sulphide nickel deposits of the Western Australian Shield. In: Buchanan DL, Jones MJ (eds) Sulphide deposits in mafic and ultramafic rocks. Inst Min Metall, London, pp 1–13

Hancock SL (1978) K.N.O. Progress report – Fisher shoot. Rep K2385, 22 pp

Harley DN (1980) K.N.O. Progress report – Juan shoot. WMC (KNO) Rep K/2497, 24 pp (unpublished)

Hayden P (1976) Fault and contact movements at Durkin shoot and their effects on ore. WMC (KNO) Rep K/2261, 12 pp (unpublished)

Hudson DR (1972) Evaluation of genetic models for Australian sulphide nickel deposits. In: Newcastle Conf, 1972, Papers. Australas Inst Min Metall, Melbourne, pp 59–72

Hudson DR (1973) Genesis of Archaean ultramafic associated nickel-iron sulphides at Nepean, Western Australia. Australas Inst Min Metall, W Aust Conf Perth, Pap, pp 99–100

Hulme G (1973) Turbulent lava fow and the formation of lunar sinuous rilles. Mod Geol 4:107–117

Hulme G (1982) A review of lava flow processes related to the formation of lunar sinuous rilles. Geophys Surv 5:243–279

Huppert HF, Sparks RSJ, Turner SJ, Arndt NT (1984) The emplacement and cooling of komatiite lavas. Nature (London) 309:19–22

Jones JG (1969) Pillow lavas as depth indicators. Am J Sci 267:181–195

Keays RR, Ross JR, Woolrich P (1981) Precious metals in volcanic peridotite-associated nickel sulphide deposits in Western Australia. II. Distribution within the ores and host rocks in Kambalda. Econ Geol 76:1645–1674

Keele RA, Nickel EH (1974) The geology of a primary millerite-bearing sulfide assemblage and supergene alteration at the Otter shoot, Kambalda, Western Australia. Econ Geol 69:1102–1117

Laeter JR, De, Libby WG, Trendall AF (1981) The older Precambrian geochronology of Western Australia. In: Glover JE, Groves DI (eds) Archaean geology. Geol Soc Aust Spec Publ 7:145–157

Lesher CM (1983) Localization and genesis of komatiite-associated Fe–Ni–Cu sulphide mineralization at Kambalda, Western Australia. Ph D Thesis, Univ W Aust, 318 pp (unpublished)

Lesher CM, Lee RF, Groves DI, Bickle MJ, Donaldson MJ (1981) Geochemistry of komatiites from Kambalda, Western Australia. I. Chalcophile element depletion – a consequence of sulfide liquid separation from komatiitic magmas. Econ Geol 76:1714–1728

Lesher CM, Donaldson MJ, Groves DI (1982) Nickel deposits and their host rocks in the Norseman-Wiluna Belt. In: Groves DI, Lesher CM (eds) Regional geology and nickel deposits of the Norseman-Wiluna Belt, Western Australia. Univ West Aust, Dep Geol Ext Serv, Publ 7:B1–B63

Lesher CM, Arndt NT, Groves DI (1984) Genesis of komatiite-associated nickel sulphide deposits at Kambalda, Western Australia: A distal volcanic model. In: Buchanan DL, Jones MJ (eds) Sulphide deposits in mafic and ultramafic rocks. Inst Min Metall, London, pp 70–80

McQueen KG (1981a) Volcanic-associated nickel deposits from around the Widgiemooltha Dome, Western Australia. Econ Geol 76:1417–1443

McQueen KG (1981b) The nature and metamorphic history of the Wannaway Nickel Deposit, Western Australia. Econ Geol 76:1444–1468

Marston RJ, Kay BD (1980) The distribution, petrology and genesis of nickel ores at the Juan complex, Kambalda, Western Australia. Econ Geol 75:546–565

Marston RJ, Groves DI, Hudson DR, Ross JR (1981) Nickel sulphide deposits in Western Australia: A review. Econ Geol 76:1330–1363

Middleton CN (1980) K.N.O. Progress report – Lunnon shoot. WMC (KNO) Rep K/2517, 40 pp (unpublished)

Moore JG (1975) Mechanism of formation of pillow lava. Am Sci 63, 3:269–277

Naldrett AJ (1973) Nickel sulphide deposits – their classification and genesis, with special emphasis on deposits of volcanic association. Can Inst Min Metall Bull 66, 739:45–63

Naldrett AJ (1981) Nickel sulphide deposits: Classification, composition, and genesis. Econ Geol 75th Anniversary Vol, pp 628–685

Naldrett AJ, Campbell IH (1982) Physical and chemical constraints on genetic models for komatiite-related Ni-sulphide deposits. In: Arndt NT, Nisbett EG (eds) Komatiites. Allen & Unwin, London, pp 423–434

Page ML, Schmulian ML (1981) The proximal volcanic environment of the Scotia nickel deposits. Econ Geol 76:1469–1479

Paterson HL, Donaldson MJ, Smith RN, Lenard MF, Gresham JJ, Boyack DJ, Keays RR (1984) Nickeliferous sediments and sediment-associated nickel ores at Kambalda, Western Australia. Buchanan DL, Jones MJ (eds) Inst Min Metall, London, pp 81–94

Platt JP, Allchurch PD, Rutland RWR (1978) Archaean tectonics in the Agnew supracrustal belt, Western Australia. Precambrian Res 7:3–30

Prider RT (1970) Nickel in Western Australia. Nature (London) 226:691–693

Rajamani V, Naldrett AJ (1978) Partitioning of Fe, Co, Ni, and Cu between sulfide liquid and basaltic melts and the composition of Ni–Cu sulfide deposits. Econ Geol 73:82–93

Redman BA (1982) Petrography, petrology and geochemistry of Archaean basic volcanism in the Eastern Goldfields Province, Western Australia – stratigraphic control of gold mineralisation. M Sc Thesis, Univ Melbourne, 304 pp (unpublished)

Richards T (1983) Structural geology of the Hunt shoot environment. WMC Rep (unpublished)

Ross JR (1974) Archean nickel sulphide mineralisation, Lunnon shoot, Kambalda. Ph D Thesis, Univ Calif, Berkeley, 283 pp (unpublished)

Ross JR, Hopkins GMF (1975) Kambalda nickel sulphide deposits. In: Knight CL (ed) Economic geology of Australia and Papua New Guinea. I. Metals. Australas Inst Min Metall, Melbourne, Mon 5:100 – 121

Ross JR, Keays RR (1979) Precious metals in volcanic-type nickel sulfide deposits in Western Australia, I; Relationship with the composition of the ores and their host rocks. Can Miner 17:417 – 435

Ross JR, Travis GA (1981) The nickel sulfide deposits of Western Australia in global perspective. Econ Geol 76:1291 – 1329

Sanders TS (1982) Nepean nickel deposits. In: Groves DI, Lesher CM (eds) Regional geology and nickel deposits of the Norseman-Wiluna Belt, Western Australia. Univ West Geol Aust Dep Ext Serv Publ 7:C35 – C44

Schmulian ML (1982) Windarra nickel deposits. In: Groves DI, Lesher CM (eds) Regional geology and nickel deposits of the Norseman-Wiluna Belt, Western Australia. Univ West Aust Dep Geol Ext Serv Publ 7:C67 – C76

Seccombe PK, Groves DI, Marston RJ, Barrett FM (1981) Sulfide paragenesis and sulfur mobility in Fe – Ni – Cu sulfide ores at Lunnon and Juan Main shoots Kambalda: Textural and sulfur isotopic evidence. Econ Geol 76:1675 – 1685

Spiess FN, McDonald KC, Atwater T, Ballard R, Carranza A, Cordoba D, Cox D, Diaz Garcia VM, Francheteau J, Guerrero J, Hawkins J, Haymon R, Hessler R, Juteau T, Kastner M, Larson R, Luyendyk B, MacDougall JD, Miller S, Normark W, Orcutt J, Rangin C (1980) East Pacific Rise; Hot springs and geophysical experiments. Science 207, 4438:1421 – 1433

Stolz GW, Nesbitt RW (1981) The komatiite nickel sulfide association at Scotia: A petrochemical investigation of the ore environment. Econ Geol 76:1480 – 1502

Turner MStJ (1983) The geology and mining of the Mt. Edwards nickel orebody, Widgiemooltha. WMC (KNO) Rep K/2732, 22 pp (unpublished)

Viljoen MJ, Bernasconi A, Coller N, Van, Kinloch E, Viljoen RP (1976) The geology of the Shangani nickel deposit, Rhodesia. Econ Geol 71:76 – 95

Vinogradov AP, Udinstsev GB (1975) Rift zones of the world oceans. John Wiley & Sons, New York, 503 pp

Williams DAC (1979) The association of some nickel sulfide deposits with komatiitic volcanism in Rhodesia. Can Miner 17:337 – 350

Wilson L, Head JW (1983) A comparison of volcanic eruption processes on Earth, Moon, Mars, Io and Venus. Nature (London) 302:663 – 669

Woodall R, Travis GA (1969) The Kambalda nickel deposits, Western Australia. Proc 9th Commonw Min Metall Congr, London (1969) 2:517 – 533

Woolrich P, Giorgetta N (1978) A study of the geochemistry and mineralogy of the basal ultramafic unit at Kambalda. WMC (KNO) Rep K/2380, 21 pp (unpublished)

Woolrich P, Cowden A, Giorgetta N (1981) The chemical and mineralogical variations in the nickel mineralisation associated with the Kambalda Dome, Western Australia. Econ Geol 76:1629 – 1644

Geochemistry of the Sudbury Igneous Complex: A Model for the Complex and Its Ores

A. J. NALDRETT[1]

Abstract

A recent interpretation of regional gravity and magnetic data (Gupta et al. 1984) has indicated that the Sudbury Igneous Complex is underlain at depths of 5 – 8 km by a 60 × 40 km mass of mafic and ultramafic rock that is not part of the exposed Complex.

The marginal rocks of the Complex are thought to have crystallized in situ. The high SiO_2 and K_2O and low CaO contents, and low Na_2O/K_2O ratio of these rocks in comparison with those of continental flood basalts are suggestive that the magma responsible for the Complex experienced extensive contamination by felsic country rocks. The REE profiles and major elements can be modelled if a 1 : 2 mixture of quartz monzonite and tonalite that form much of the basement at Sudbury, is combined on a 1 : 1 basis with a fairly primitive flood basalt (MgNo = 0.61). The sublayer is more fractionated than the marginal unit of the main mass, but its major element and REE concentrations also suggest significant assimilation of a similar contaminant. The high Sr initial isotope ratios of both main mass and sublayer are consistent with the contamination hypothesis.

Mafic and ultramafic inclusions are restricted to variants of the sublayer that are also mineralized. These inclusions have REE profiles with similar high La/Yb ratios to those of the main mass and sublayer. The Fo content of olivines in the inclusions indicates that they crystallized from liquids spanning the same range of Mg Nos as is spanned by samples of the sublayer. It is proposed that the inclusions have been derived from cumulate layers that formed as the sublayer magmas fractionated.

The contamination that gave rise to the SiO_2-rich composition of the Complex is believed to be the cause of the segregation of large amounts of sulfide. It is suggested that the close association between sulfides, inclusions and sublayer magmas is the consequence of the sulfides and ultramafic and mafic cumulates settling together in hidden sills that are present peripheral to and beneath the Complex. These are responsible in part for the gravity and aeromagnetic anomalies. They were injected into the fractures in the rocks beneath the Complex as offshoots from the main magma conduit. As they cooled and fractionated, residual magma rose to the floor of the crater that now holds the

[1] Department of Geology, University of Toronto, Toronto, M5S 1A1 Canada

Geology and Metallogeny of Copper Deposits
(ed. by G. H. Friedrich et al.)

Complex to form the presently exposed sublayer. Where magma from a deeper sill cut and disrupted an overlying sill, it picked up sulfides and inclusions, and carried them upwards to form the ore deposits.

The magma of the main mass of the Complex, cooling in part within the central conduit, did so more slowly than the sublayer. It thus became contaminated more rapidly, but fractionated less rapidly. It was injected into its present position as a series of pulses at essentially the same time as the sublayer, possibly in response to structural adjustments taking place in the overlying crater.

Introduction

The Ni – Cu ores of the Sudbury district are associated with the Sudbury Igneous Complex (SIC), a layered intrusion ranging from quartz norite at the base, through gabbro to a granophyric cap. The purpose of this paper is to describe the geological setting at Sudbury, discuss the composition of the Complex, show how the composition can account for the profusion of concentrations of magmatic sulfide ore, and propose a model to account for the geology of the Complex and some aspects of its setting.

Regional Geological Setting

The SIC has intruded beneath the Onaping formation (Fig. 1), a breccia composed of fragments of country rocks and recrystallized glassy material set in a matrix of glassy shards, and variably interpreted as an ignimbrite or the "fall-back" breccia resulting from the impact of a meteorite. The SIC and strata overlying it are exposed as a series of concentric, crudely elliptical rings and dip towards the centre, giving rise to the interpretation of the structure as a basin. The basin is located at the contact between gneisses, remnants of greenstone belts and intrusive quartz monzonites all of Archean age to the north, and rocks of the Proterozoic Southern Province, which overlie the Archean basement unconformably and thicken to the south.

In a recent analysis of regional gravity and magnetic data, Gupta et al. (1984) have analyzed the combined residual gravity and magnetic anomaly that marks the Sudbury region itself. After subtracting large-scale "super-regional" gravity trends, they conclude that the broad +20 to 30 mGal anomaly characterizing the Sudbury region cannot be explained by the rocks of the Complex itself. Their modelling indicates that a large (60 × 40 km) mass of mafic rock, with a density similar to gabbro or gabbro-anorthsite ($3.02 \pm / - 0.03$ t/m^3) underlies the Complex at a depth of at least 5 km, extending beyond it, as is indicated in Fig. 2. Their analysis of the regional magnetic data indicates that this can be explained if some of the rocks of this body are partially serpentinized.

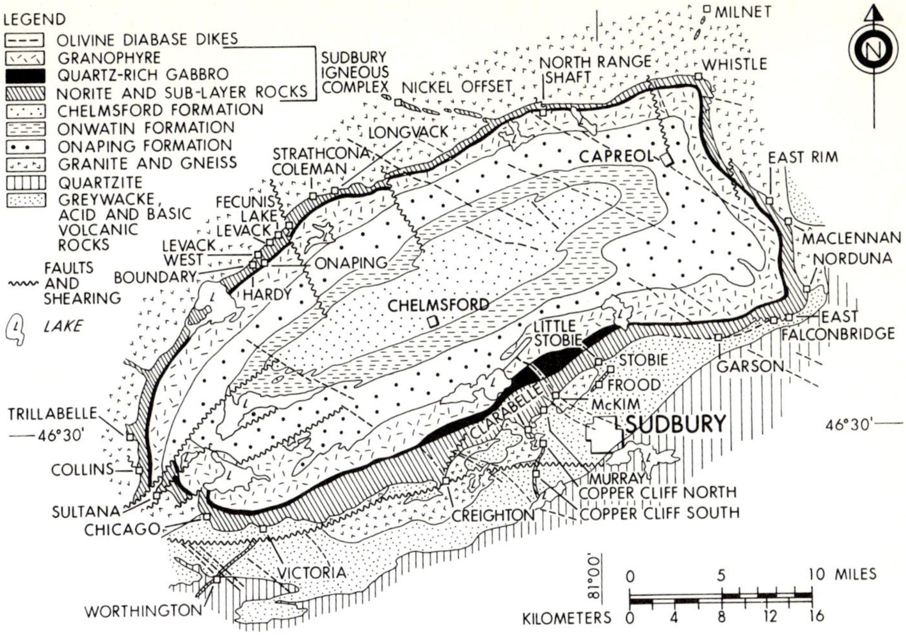

Fig. 1. Geological map of the Sudbury district

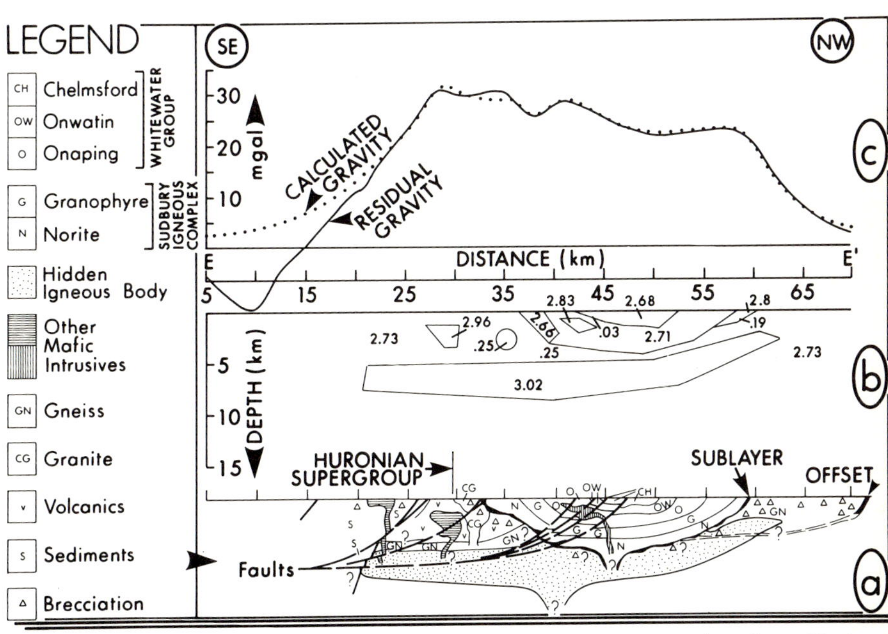

Fig. 2. A northwest–southeast section across the Sudbury district showing the residual and calculated gravity profiles, a two-dimensional reconstruction of the model used to calculate the gravity profile, and a geological interpretation consistent with the reconstruction. (After Gupta et al. 1984)

Local Geological Setting

A detailed description of the local setting is beyond the scope of this paper. It is well covered by a number of authors who have contributed to a recent book on Sudbury geology (Pye et al. 1984). Many aspects of the local setting suggest that an explosion of unusually large intensity gave rise to a crater at Sudbury. These include:

1. The basinal shape of the structure as interpreted from surface and underground mapping and drilling.
2. The presence of an upturned collar around the basin, as seen particularly in the Huronian rocks along the southern margin (Dressler 1984).
3. Evidence of shock metamorphism in the country rocks around the structure (Dressler 1984).
4. The presence of Sudbury breccia (compared with the pseudotachylite of the Vredefort and Ries structures) in the country rocks around the structure and footwall breccia beneath the Complex.
5. Evidence of shock metamorphism in country rock inclusions in the Onaping formation (Muir and Peredery 1984).
6. The 1800 m of Onaping formation itself, the lower part of which is variably interpreted as a meteorite fall-back breccia or an ignimbrite (Peredery and Morrison 1984, Muir 1984).

Opinions are divided between an extra-terrestrial and endogenic origin for the structure. This author believes that meteorite impact is the more likely origin, primarily because so many of the features observed at Sudbury are also found at known impact sites. There are, however, many difficulties with such an origin and these are summarized by Muir (1984).

Petrology of the Sudbury Igneous Complex

The main units of the Complex include (Fig. 3) (1) the sublayer; (2) the marginal quartz-rich norite of the South Range and mafic norite of the North Range;

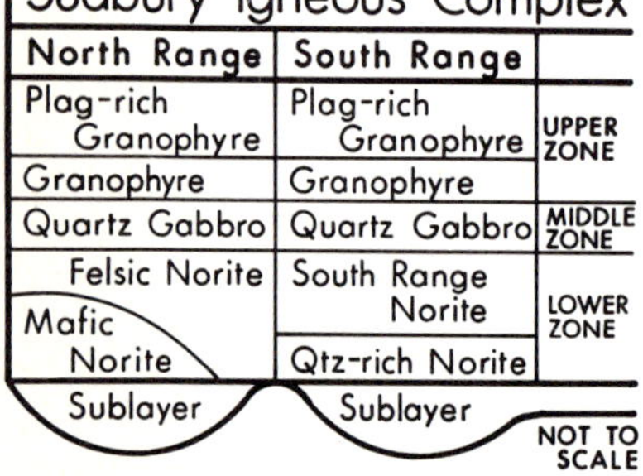

Fig. 3. The main units of the Sudbury Igneous Complex

(3) the South Range norite and Felsic norite; (4) the Quartz Gabbro; and (5) the Granophyre and Plagioclase-rich Granophyre. All except the sublayer are included within the main mass of the Complex.

Main Mass

The marginal unit of the main mass on the South Range is the quartz-rich norite. In this, the quartz content increases very progressively towards the contact over the outer 300 m (Naldrett et al. 1970). This is unlikely to be due to contamination since the increase in quartz occurs as much where the footwall is composed of SiO_2-deficient greenstone as it does where it is composed of granite. This indicates that if contamination is involved, it is not in situ contamination. The increase in quartz is accompanied by an equally progressive decrease in the average Mg/(Mg + Fe) ratio of the pyroxenes (Naldrett et al. 1970). This decrease is due to the pyroxenes becoming progressively more strongly zoned, with Mg/(Mg+Fe) decreasing towards the edges of grains while the cores retain a constant composition. These observations, coupled with a decrease in grain size towards the margins (Naldrett et al.) indicate that the outer part of the quartz-rich norite is a non-cumulate rock that crystallized essentially in situ.

Sublayer

The sublayer occurs discontinuously around the Complex. It can be divided logically into those variants which occur close to the outer contact and those which occur occupying dykes that radiate outwards from the Complex and are known locally as offsets.

The contact sublayer consists of a suite of fine to medium grained norites and gabbros that can be distinguished from the main mass felsic norite and quartz gabbro by their lower quartz content in relation to pyroxene (Naldrett et al. 1972).

Sublayer rocks occurring within the offsets generally have the composition of quartz diorite and are referred to as such. Grant and Bite (1984) recognize a number of variants of quartz diorite which form a continuum grading from hypersthene to biotite quartz diorite, a change that is attributed to varying degrees of contamination.

Geological relationships between the sublayer and the main mass that bear on the relative ages of the two are conflicting. Inclusions of marginal, quartz-rich norite have been observed in sublayer and inclusions of sublayer have been observed in norite of the main mass of the Complex. Contacts are never marked by fine-grained chill zones, suggesting that whichever was the older at any particular location, it was still warm at the time of intrusion of the younger. On the North Range, the distinction between main mass and sublayer is always clear with the sublayer having the finer grain size and lower quartz content. This is not always the case on the South Range, where a number of researchers have commented on gradations that are visible between the two (Slaught 1951, Cochrane

1984). It would seem, therefore, that the introduction of the sublayer and main mass was a complicated process with one preceeding the other and vice versa in different localities.

Inclusions in the Sublayer

In some areas the sublayer is characterized by inclusions. These can be divided into two groups, those of obviously local derivation, and those composed of mafic and ultramafic rocks, many of which do not outcrop in the Sudbury area. The latter group are those of concern here. Scribbins et al. (1984) have described them as ranging from peridotite, through clino- and orthopyroxenites, to olivine gabbro and norite. Most of the inclusions either display cataclastic or cumulate textures. Olivine ranges in composition from Fo_{86} to Fo_{72}, with the Fo content decreasing as the plagioclase content decreases. Scribbins et al. conclude that the inclusions are derived from layered intrusions that have fractionated at moderate depths in the crust.

Geochemistry of the SIC

Major Elements

A selection of average and individual major element analyses of representative material from the Sudbury Igneous Complex are listed in Table 1.

Main Mass. Judging from field and petrographic criteria, the quartz-rich norite is the rock type that is likely to have been closest to the composition of the SIC magma when it was emplaced along the South Range. The Mg/(Mg + Fe) atomic ratio (henceforth referred to as MgNo) of 0.61 indicates that this is a reasonably primitive rock.

Since the SIC is an example of continental tholeiitic magmatism, continental flood basalts provide one of the best yardsticks with which to compare it. In Fig. 4, the quartz-rich norite is compared with Keewanawan and Columbia River flood basalts on the basis of their MgNos. The high SiO_2 and K_2O, low CaO and low Na_2O/K_2O ratio of the quartz-rich norite when compared with flood basalts of similar MgNo is very obvious. The siliceous nature of the Complex has been discussed in the past by Naldrett and Macdonald (1980) who suggested contamination as the cause. Contamination of a relatively unfractionated Columbia River basalt (column 4, Table 2) with 50% of a 1:2 mixture (column 3, Table 2) respectively of Archean quartz monzonite (column 1) and tonalitic gneiss (column 2) gives rise to the composition given in column 6. This is quite similar, considering the arbitrary manner of choosing the end members, to the quartz-rich norite (column 7). As discussed by Rao et al. (1983), any real process of contamination is likely to be more complex than this, with crystallization accompanying assimilation. However, the conclusion on the basis of the major elements is that contamination of the main mass SIC magma by felsic country rocks is permissible, in fact, even likely.

Table 1. Comparison of major element content of rocks of the Sudbury Igneous Complex[a]

	Offset rocks			Main mass						Flodd basalts				Bushveld
	1 Quartz poor Offset (SR)	2 Typical Offset (SR)	3 Offset (NR)	4 Quartz-rich norite (SR)	5	6 South Range norite	7 Felsic norite	8 Quartz gabbro	9 Granophyre	10 Keewan. basalt	11 Columbia basalt	12 Keewan. basalt	13 Columbia basalt	14 norite
SiO_2	50.7	55.3	61.5	56.7	57.0	55.4	60.4	58.9	68.1	49.1	48.7	50.0	48.9	53.3
TiO_2	1.56	0.94	0.73	0.57	1.34	0.49	0.65	1.62	0.65	1.89	2.73	0.99	0.96	0.19
Al_2O_3	16.3	14.3	14.9	16.3	16.4	17.8	15.6	13.5	13.7	16.80	16.4	16.3	17.0	19.2
FeO*	11.4	10.8	6.84	7.44	7.33	7.35	7.19	10.5	5.21	11.91	12.5	9.78	10.0	7.60
MnO	0.17	0.15	0.11	0.12	0.13	0.12	0.13	0.14	0.09	0.17	0.18	0.18	0.20	0.16
MgO	4.91	4.43	3.82	6.10	6.40	6.68	3.81	2.66	1.08	6.61	5.45	8.08	7.90	7.34
CaO	7.92	6.71	4.70	6.66	7.28	7.75	6.05	5.48	2.15	10.03	8.7	11.1	11.1	10.0
Na_2O	2.76	2.61	3.15	–	2.41	2.50	3.12	3.13	3.31	2.76	3.34	2.52	2.49	2.95
K_2O	1.66	1.70	2.60	1.40	1.55	1.10	1.96	2.13	5.00	0.52	0.91	0.43	0.28	0.38
Total	97.28	96.6	98.4	–	99.84	99.2	98.90	98.06	99.29	98.79	98.97	99.38	98.53	101.12
Mg Atomic (Mg + Fe)	0.43	0.42	0.50	0.59	0.61	0.62	0.47	0.31	0.27	0.50	0.44	0.60	0.59	0.63
Na_2O/ K_2O wt.%	1.76	1.54	1.21	–	1.56	2.27	1.59	1.47	0.66	5.31	3.67	5.86	8.89	7.76

[a] Explanation to Table 1: *1* Grant and Bite (t.v.) Table II, No. 4 (Av of 4); *2* Grant and Bite (t.v.) Table II, No. 2 (Av of 14); *3* Grant and Bite (t.v.) Table II, No. 5 (Av of 17); *4* Grant and Bite (t.v.) Table II, No. 6 (Av of 31); *5* Collins (1934) p. 140, No. 92 W; *6* Collins (1934) p. 140, No. 89 W; *7* Grant and Bite (t.v.) Table II, No. 9 (Av of 31); *8* Grant and Bite (t.v.) Table II, No. 10 (Av of 17); *9* Collins (1934) p. 139, No. 56 W; *10* Transitional Basalt, Basaltic Volcanism, Study Project (1981), Table 1.2.2.19, No KEW 4; *11* Basalt, Basaltic Volcanism Study Project (1981), Table 1.2.3.4, No CP-4; *12* Olivine Basalt, Basaltic Volcanism Study Project (1981), Table 1.2.2.19, No KEW-8; *13* Olivine Basalt, Basaltic Volcanism Study Project (1981), Table 1.2.3.4, No CP-3; *14* Naldrett, unpublished data.

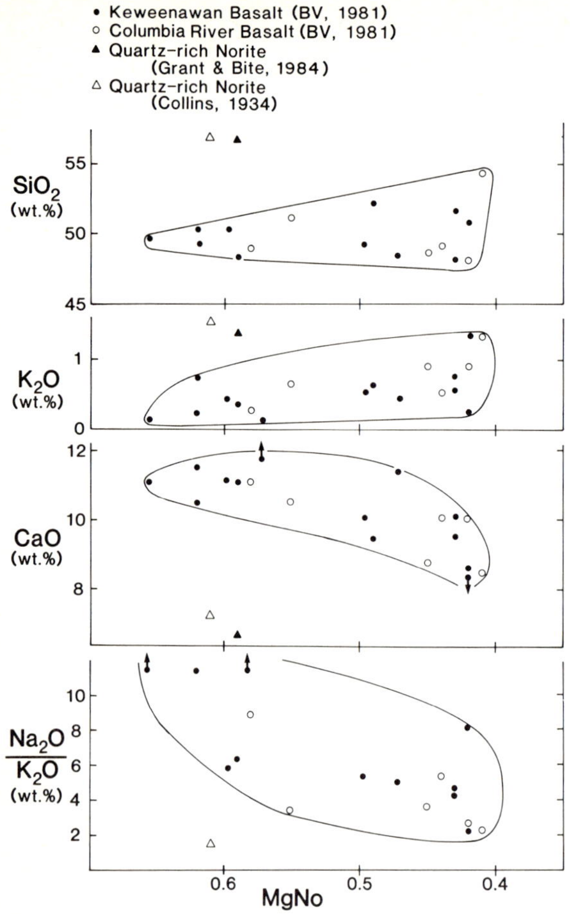

Fig. 4. Comparison of some major element concentrations in the quartz-rich norite with those in typical continental flood basalts

Sublayer. The texture and field relations of the constituent units of the sublayer indicate that these are also not cumulates, but are rocks that have solidified essentially in situ, and thus represent magma compositions. The MgNos are therefore important indices of the degree of fractionation of the rocks and of the magmas. MgNos of North Range samples of contact (as opposed to offset) sublayer given by Rao et al. (1983) range from 0.42 to 0.50, and average 0.47; those from the South Range vary from 0.37 to 0.58 and average 0.51. Thus the magmas are relatively fractionated, considerably more so than the main mass as represented by the quartz-rich norite. At the same time, the MgNos indicate a variable degree of fractionation for the sublayer.

Comparison of the major element compositions (Fig. 5) of sublayer with those of flood basalts indicates that the sublayer is also enriched in SiO_2 and K_2O, and low in CaO and Na_2O/K_2O ratio for a given degree of fractionation as indicated by the MgNo. Contamination by country rocks is therefore indicated by the data.

The data of Rao et al. (1983) on the contact sublayer indicate a wide range of MgNos, and they have called upon fractional crystallization in addition to assim-

Table 2. Compositions relating to the contamination of the Sudbury Igneous Complex magmas[a]

	1	2	3	4	5	6	7	8	9
				Major elements wt.%					
SiO_2	70.4	64.8	66.7	48.9	49.1	57.8	57.0	53.5	51.7
TiO_2	0.46	0.78	0.67	0.96	1.89	0.82	1.34	1.59	1.31
Al_2O_3	15.6	16.2	16.0	17.0	16.80	16.5	16.4	16.6	14.7
FeO[b]	2.2	4.35	3.63	10.0	11.91	6.81	7.83	9.84	12.84
MgO	0.73	2.43	1.86	7.9	6.61	4.88	6.46	5.42	6.21
CaO	1.96	3.48	2.97	11.1	10.03	7.04	7.28	8.27	9.52
Na_2O	3.93	3.89	3.90	2.49	2.76	3.20	2.41	3.05	2.81
K_2O	3.33	1.80	2.31	0.28	0.52	1.30	1.55	0.97	0.81
				Trace elements ppm					
La	–	–	–	6.77	15.2	–	–	–	18.0
Ce	97.9	83.8	89	16.0	33.5	52.3	57.6	47.4	48.2
Nd	31.	37.2	35.1	–	–	–	26.8	–	17.8
Sm	4.55	6.57	5.9	2.72	4.80	4.31	4.52	5.08	4.65
Eu	0.89	1.66	1.37	1.03	1.58	1.20	1.26	1.53	1.96
Gd	2.88	4.79	4.15	3.18[d]	4.8[d]	3.67	4.13	4.64	4.60[d]
Tb	–	–	–	0.59	0.83	–	0.56	–	0.76
Yb	0.696	1.20	1.03	2.20	2.35	1.62	1.88	2.02	1.43
Lu	–	–	–	0.33	0.36	–	–	–	0.27
Ba	807	992	930	180	–	555	–	–	–
Rb	166	75.9	106	2.3	10	54	–	34	21
Sr	290	753	599	338	284	469	–	362	602
MgNo[c]			0.477	0.585	0.497	0.56	0.611	0.495	0.463
K/Rb									
Rb/Sr			0.18	0.007	0.035	0.115		0.094	0.035
Sc/Ba			0.64	1.88	–	0.85			
Ce/Yb									

[a] Explanation to Table 2: *1* Average "Granite" from Table XX.2.1. Trace elements are of Archean quartz monzonite with very similar major element composition from Arth and Hanson (1975), Table 2, analysis 29. *2* Average Gneiss from Table XX.2.1. Trace elements are of tonalitic gneiss of similar major element composition from Arth and Hanson (1975), Table 2, analysis 22. *3* Mixture of 1 and 2 in the proportion 1:2. *4* Relatively primitive continental flood basalt from Columbia River Province, Basaltic Volcanism Study Project (1981), Tables 1.2.3.4 and 1.2.3.5 analysis No. CP-3 (calculated on volatile free basis). *5* Moderately fractionated continental flood basalt from Keewanawan Province, Basaltic Volcanism Study Project (1981), Tables 1.2.2.19, 1.2.2.22 and 1.2.2.23, analysis No. KEW-4 (calculated on volatile-free basis). *6* Calculated composition of South Range initial magma combining columns 3 and 4 in the proportions 1:1. *7* Composition of marginal quartz-rich norite (Table XX.3.1, analysis 5) for comparison with column 6. *8* Calculated composition of a relatively primitive sublayer magma, combining columns 3 and 5 in the proportions 1:3. *9* Composition of one of the least fractionated North Range sublayer samples from Rao et al. (1983), Tables 1 and 3, analysis BVR-11.
All analyses calculated on a volatile-free basis.

[b] FeO^T = Fe calculated as FeO.

[c] MgNo = Mg/(Mg + Fe) atomic ratio.

[d] Chondrite normalized Gd for these two samples assumed to be the weighted average of two parts of Tb and one part of Sm, all on a chondrite normalized basis.

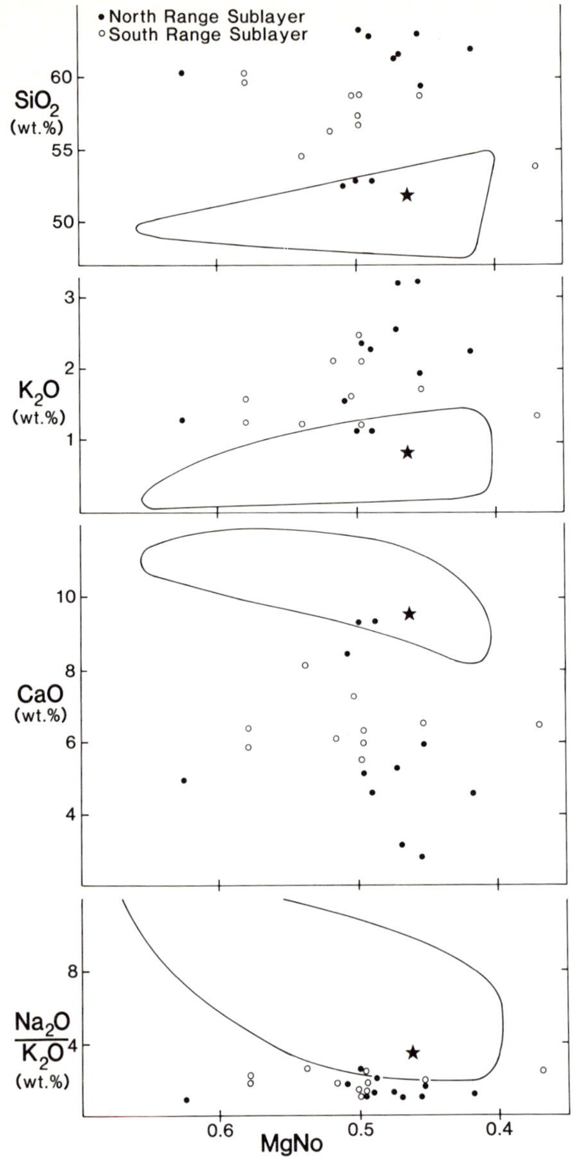

Fig. 5. Comparison of some major element concentrations in the sublayer with those in typical continental flood basalts. The flood basalts fall within the outlined area (see Fig. 4)

ilation to explain the observed compositional variations. One sample (BVR-11, marked by a star in Fig. 5) is one of the more primitive of the North Range samples. In its major element composition and Na_2O/K_2O ratio of 3.47, it is not unlike some of the more fractionated flood basalts. However, its trace element composition is different, as is discussed below.

Trace Elements

Data on trace elements in the SIC are much less abundant in the literature than those on major elements. Data exist for REE elements on most units of the Complex, but that of Rao et al. (1983) on the sublayer are all that are available in the way of modern determinations of other incompatible trace elements.

In Fig. 6, Kuo and Crocket's (1979) REE pattern for the quartz-rich norite is compared with those for the two flood basalts samples of similar MgNo. The much steeper slope of the Sudbury sample (La/Yb = 17) as compared to the flood basalts (La/Yb = 3) is obvious.

The range of REE patterns for the sublayer of the SIC, together with the averages for the North and South Ranges are compared with flood basalts of similar MgNo in Fig. 7. The sublayer patterns, with their La/Yb ratios of 16, are essentially identical in slope and magnitude to the average pattern for the quartz-rich norite and, again, are very much steeper than those for the flood basalts (La/Yb = 6).

The LREE enrichment of the Sudbury rocks could be attributed to contamination by LREE-enriched country rocks. One feature of the Sudbury patterns is the absence of negative Eu anomalies. This would preclude contamination by typical granite or quartz monzonite with their characteristic negative Eu anomalies. Tonalitic gneisses typically are enriched in LREE and show no Eu anomalies (Arth and Hansen 1975), so that they are a possible contaminant. However, they cannot account for the low Na_2O/K_2O ratios of the Sudbury rocks.

A representative set of trace element data for the country rocks of the Sudbury area are not available. However, the major element chemistry of the Sudbury tonalites and quartz monzonites are identical to those studied by Arth and

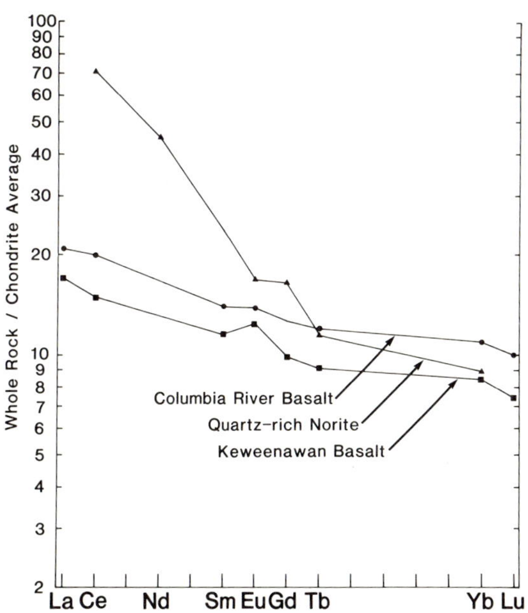

Fig. 6. Comparison of REE data for the quartz-rich norite with that for a Keewanawan and a Columbia River flood basalt of similar MgNo

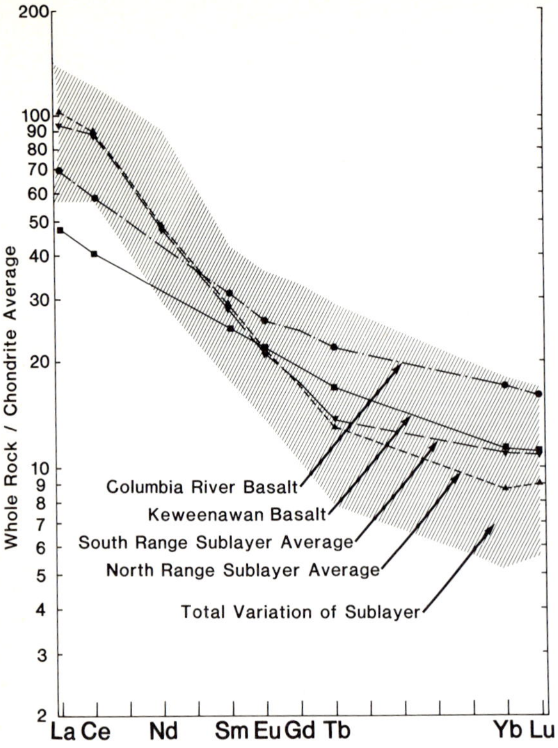

Fig. 7. Comparison of REE data for the sublayer with that for relatively fractioned Keewanawan and Columbia River basalts

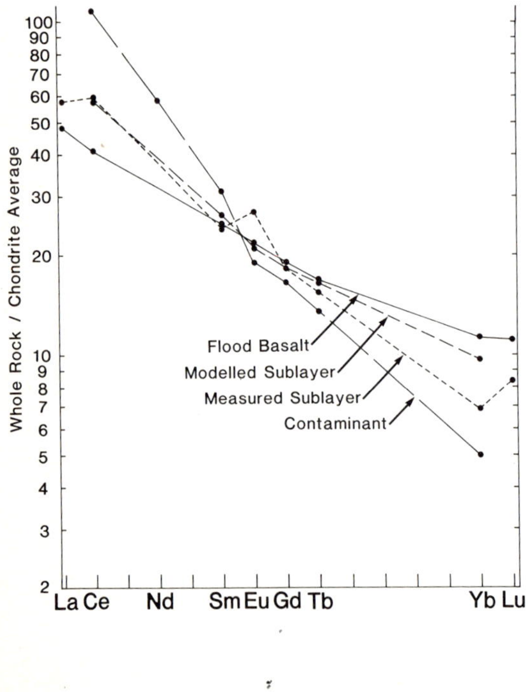

Fig. 8. REE data relating to the mixing of Archean tonalitic gneiss and quartz monzonite with relatively unfractionated Columbia River basalt to produce a composition similar to the quartz-rich norite

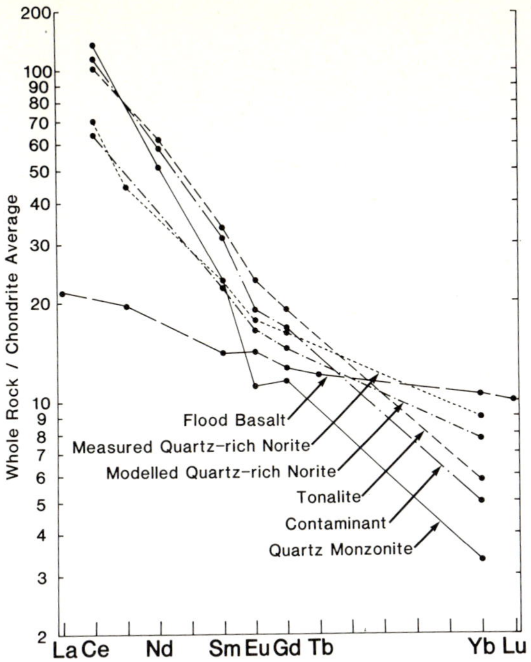

Fig. 9. REE data relating to the mixing of Archean tonalitic gneiss and quartz monzonite with a relatively fractionated Keewanawan basalt to produce a composition similar to one of the less contaminated samples of the sublayer

Hansen in northeastern Minnesota. REE data for a typical tonalitic gneiss and quartz monzonite, and then for a 1:2 mixture respectively of quartz monzonite and tonalite are given in Fig. 8. A linear mixture of this with relatively unfractionated Columbia River basalt is compared with the Sudbury quartz-rich norite pattern in the same figure. It can be seen that the two are similar, consistent with the close match that this mixture also produces for the major elements.

In Fig. 9, one of the least contaminated contact sublayer samples is modelled by mixing 25% of the postulated granite-gneiss contaminant (column 3, Table 2) with 75% of a relatively fractionated Keewanawan basalt (column 5, Table 2). The REE pattern indicates a close match, except that the model composition does not show the positive Eu anomaly of the sublayer sample. However, this positive Eu anomaly is atypical of the sublayer – very few other sublayer samples show it. Rao et al. emphasize that the majority of their sublayer samples are best modelles as the consequence of the fractionation of the more primitive variants, coupled with assimilation of country rocks. From the above discussion, particularly that with reference to the REE profiles, even the more primitive sublayer material seems to have been contaminated to some extent, although this contamination has been less intense than that which affected the initial magma of the main mass before the process described by Rao et al. (1983) started.

Isotopes

Krogh et al. (1984) have shown from Pb – U radiometric dating of zircons and baddeleyite that the SIC has an age of 1.85 Ga. This has been used as a model age

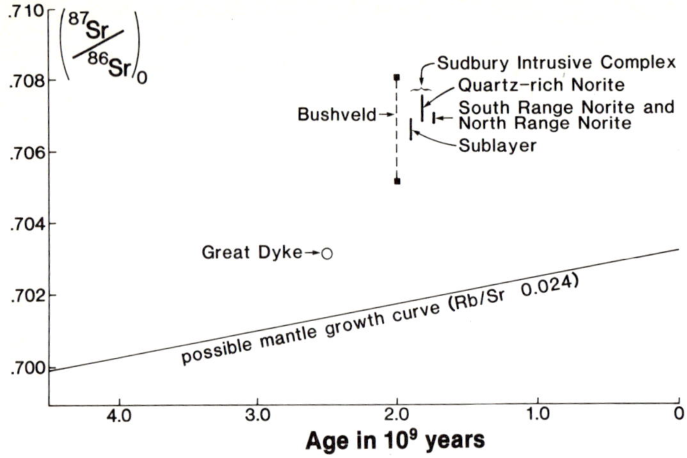

Fig. 10. Sr initial ratios for rocks of the SIC compared with those for other large layered intrusions and a possible mantle growth line

to recalculate the initial Sr^{87}/S^{86} ratios for the data of Gibbins and McNutt (1975). The average and range for a number of units of the SIC are shown in Fig. 10 compared on the basis of age with data for a number of other layered intrusions, and with a hypothetical mantle growth line. While Sudbury is not unique in having high $(Sr^{87}/Sr^{86})_0$ ratios for its age, these values nevertheless suggest either the incorporation of radiogenic strontium into the SIC magma, through interaction with country rocks, or derivation of the magma from a portion of the mantle that contained very radiogenic strontium. Values of the sublayer are more variable, but are somewhat lower than those for the main mass norites. This suggests that if contamination was involved, the main mass was affected more consistently and to a greater degree than the sublayer.

Conclusions

Field relations and radiometric data indicate no substantial age difference between any of the SIC rocks, including the sublayer, main mass norites and granophyre. The unusually quartz-rich composition of the SIC magma relative to its MgNo, and the LREE enrichment and the low Na_2O/K_2O ratios shown by all units are anomalous relative to other layered mafic complexes and to continental flood basalt volcanism, and can be explained as the consequence of the contamination of flood basalt magma by Archean basement rocks.

Relationship of Ore Deposits to the Rocks of the Complex

Several new descriptions and a summary of the general features of the ore deposits are contained within the Sudbury volume (Pye et al. 1984). It is beyond the scope of this paper to repeat them here. Although there is considerable variation in the characteristics of different ore deposits, there are a number of features common to all of them. These include:

1. *Embayments or Other Irregularities at the Base of the SIC.* An increase in sulfide content it usually observed at the lower contact throughout the Complex, but it is where irregularities exist that the zone of sulfide thickens and increases in intensity sufficiently to form ore.
2. *The Presence of Sublayer.* The spatial relationship of ore to sublayer is such that the sulfides constituting the ore bodies appear to have settled out of bodies of sublayer. Sulfides also occur within quartz-rich and mafic norites, but except where they have been moved from their original position by faulting, ore bodies are invariably associated with sublayer.
3. *Ultramafic Inclusions Within Sublayer.* Some sublayer is mineralized, other varieties seem to be devoid of significant sulfide. There is little to distinguish mineralized from unmineralized sublayer in so far as texture, mineralogy and chemical composition of the rocks are concerned, except for the obvious presence of sulfide. All types of sublayer contain inclusions, but both in the contact and the offset environments, the mineralization is associated only with sublayer that carries members of a suite of mafic and ultramafic inclusions that Scribbins et al. (1984) conclude are derived from one or more layered intrusions.

Relationship Between Inclusions, Sublayer, and Main Mass of the SIC

As discussed above, the REE profiles of the Sudbury rocks differ from those of most layered intrusions and suggest that contamination by country rocks has occurred on a large scale. The similarity in the profiles of the sublayer and the main mass rocks is suggestive that the magmas responsible for them have a common heritage. REE profiles for inclusions have the same steep slopes (Fig. 11), although the abundances are lower. Thus there is some evidence that the inclusions were derived from a magma that had suffered similar contamination to the remainder of the SIC rocks.

The MgNos of samples of main mass and sublayer are illustrated in Fig. 12 (contrary to the previous representation of MgNos, in Fig. 12 only 90% of the total Fe has been included in calculating them, to allow for Fe in the magma in the ferric state). Using Roeder and Emslie's (1970) relationship between the Mg/Fe ratio of olivine and of the basaltic liquid that is in equilibrium with it, MgNos of liquids in equilibrium with the olivine of the inclusions have been calculated. It can be seen that they span the range of MgNos exhibited by the sublayer samples. This indicates that ignoring the SiO_2 contents of all of the magmas, the range of olivine compositions observed in the inclusions is precisely that which would be expected in cumulate rocks crystallizing from a magma evolving along a compositional trend such as that recorded within the sublayer. Returning to the question of the SiO_2 contents of the Sudbury magmas, although olivine is not a common constituent of the sublayer, Pattison (1979) has reported it as

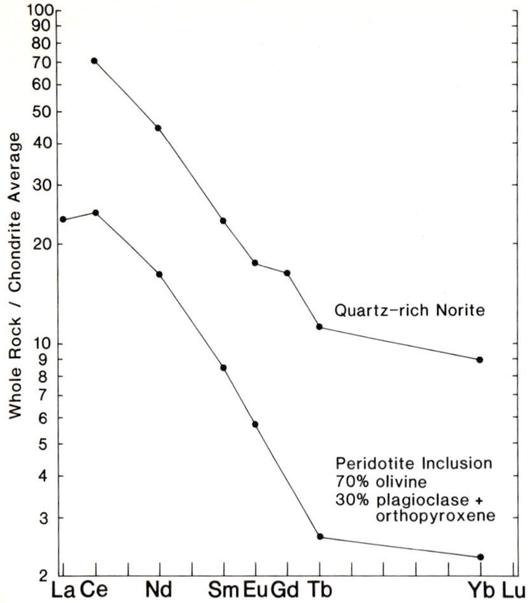

Fig. 11. REE profile for a harzburgite inclusion in the sublayer compared with that for other SIC rocks. (Rao, personal communication, 1984)

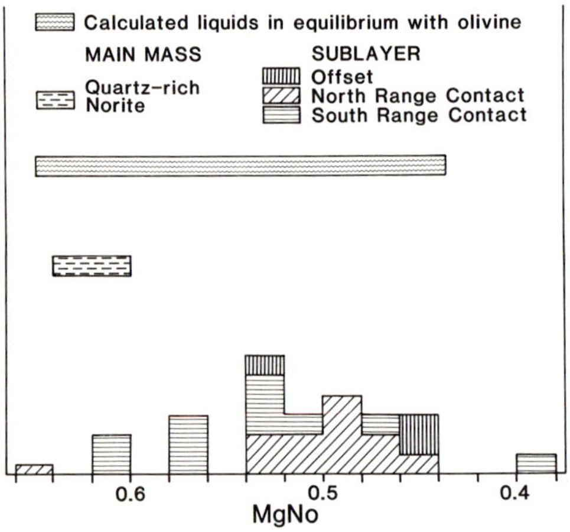

Fig. 12. MgNos of main mass and sublayer samples that are believed to approximate the compositions of liquids, compared with the calculated MgNos of liquids in equilibrium with olivines of the inclusions of the sublayer

present at Whistle. The sublayer here is not obviously different in composition to that elsewhere, which suggests that olivine may have played a role in the fractionation of much of the sublayer, and only has disappeared from the liquidus shortly before it was emplaced.

Thus two lines of evidence, REE profiles and olivine compositions, leads this author to suggest that inclusions, sublayer, and main mass norite are all related, and to propose the model given at the end of this paper.

Segregation of Sulfides

The SIC differs from other layered complexes in a number of significant ways, including:

1. Evidence that the area into which it was intruded had been involved in a catastrophic explosion, in the view of this author probably the consequence of the impact of a meteorite.
2. Rocks of the Complex are very siliceous when compared with other intrusions on the basis of the Mg/(Mg + Fe) ratios of their pyroxenes, implying that the magma responsible was unusually siliceous for its state of fractionation.
3. Other compositional data based on major and trace elements and isotopic analyses that indicate that the high SiO_2 content is the consequence of country rock assimilation.
4. An unusually large number of occurrences of very concentrated sulfide.

Irvine (1975) pointed out that assimilation of SiO_2-rich material by a mafic magma could lower the solubility of sulfur within it, and suggested that this had occurred at Sudbury. Naldrett and Macdonald (1980) used the 1200 °C isothermal section of the FeO – FeS – SiO_2 system, reproduced here as Fig. 13, to illustrate the principle. Composition *A* lies in the field of homogeneous liquid, but addition of SiO_2 to change its composition to *B* moves it into a two-liquid field, and will result in the segregation of two immiscible liquids, one silicate-rich (*Y*) and the other sulfide-rich (*X*). The evidence bearing on the SiO_2-rich nature of the SIC lends support to this hypothesis. The evidence indicating that extensive country rock assimilation has occurred points to how the SiO_2-rich composition has been achieved. The highly fractured ground underlying an explosion site, coupled with the heating and melting resulting from the explosion would have provided the ideal environment in which country rock assimilation could have occurred.

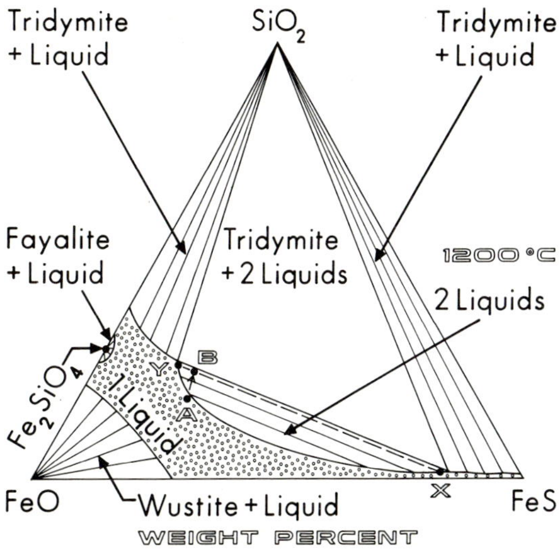

Fig. 13. 1200 °C isotherm of the Fe – S – O system illustrating how the addition of SiO_2 to a homogeneous, sulfide-rich liquid of composition *A* will change its composition to point *B*, where it will consist of two liquids, one silicate-rich (*Y*) and the other sulfide-rich (*X*)

A Model for the Sudbury Igneous Complex

It is suggested that the catastrophic explosion that occurred in the Sudbury region 1.85 Ga ago (this author believes that the weight of the available evidence points to the impact of an extra-terrestrial projectile as the cause, although impact is not essential to the development of this model) gave rise to the Onaping formation, fractured the underlying crust, and triggered ascent of a magma similar in its initial composition to continental flood basalt. Magma probably rose through a centralized plumbing system in the lower crust, but some of it spread out through numerous subsidiary channels as it entered the highly fractured ground beneath the crater (Fig. 14). It is possible that these hidden, flanking intrusions account in part for the buried mass of mafic and ultramafic rock that the analysis of gravity and magnetic data by Gupta et al. (1984) has indicated underlies and extends beyond the present Sudbury structure.

On its initial ascent, the magma fractionated to some degree, achieving an MgNo of 0.65 or slightly less. Batches of magma that infiltrated flanking sub-

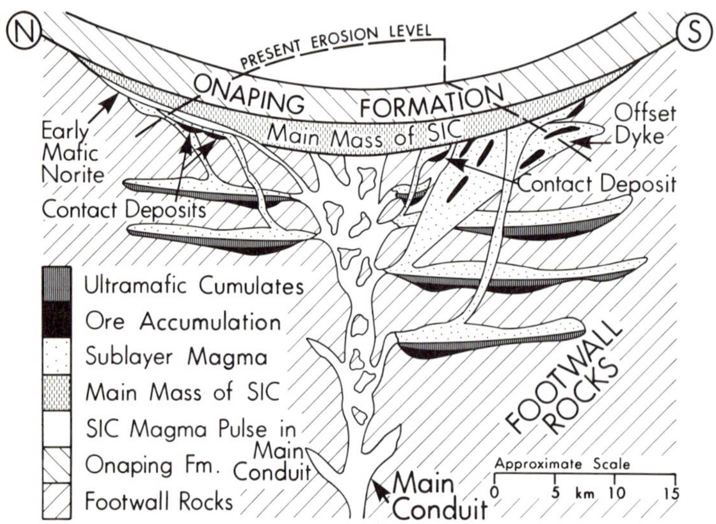

Fig. 14. Cross-section illustrating the emplacement of the SIC. Magma rises up a vertical conduit into a highly fractured zone beneath an explosion (possibly impact) crater. The central body of magma cools somewhat slowly, assimilating much of the fractured country rocks. Subsequently it is emplaced into the crater, beneath the low density Onaping formation, in response to structural readjustments in the area. It is probable that this is a multi-stage process. Magma has also spread out as a series of sills peripheral to the central conduit. Assimilation also occurs within these, although possibly to a lesser extent than in the central conduit due to their cooling more rapidly. The magma of the sills deposits sulfide and fractionates, giving rise to olivine and pyroxene-rich cumulates. Fractionated magma from the flanking sills also works its way up to the crater floor, intruding to form the sublayer of the contact and offsets. Where magma from a lower sill has cut and disrupted an overlying one, it has picked up sulfides and inclusions, intruding with these in suspension within itself. Subsequently both sulfides and inclusions settle out to give rise to the ore deposits with their typical zonation of massive sulfides overlain by inclusion-bearing massive sulfide overlain in turn by sulfide-rich sublayer rock

sidiary channels cooled, became contaminated through reaction with their surroundings, and fractionated further, giving rise to olivine and pyroxene-rich cumulates. The contamination, coupled with the cooling, caused sulfides to precipitate at an early stage: these collected as massive bodies beneath the cumulates in the flanking chambers of the plumbing system. The main body of magma lost heat less quickly than that in the flanking chambers, and thus fractionated less rapidly; because it cooled more slowly, it reacted with country rocks to an even greater extent than was the case in the flanking bodies.

Successive injections of magma occurred along the unconformity between the crater floor and the overlying, relatively unconsolidated, and thus low density, Onaping formation to give rise to the main mass of the Complex.

The faster cooling, and thus more fractionated (consistent with the MgNos), but less contaminated (consistent with the Sr initial ratios) magma of the flanking bodies was also squeezed up to intersect the crater floor, where it spread out, exploiting "troughs" and "embayments" in the floor filled with footwall breccia and also infiltrating fractures in the basement rocks to form the offset dykes. This magma is thus responsible for the sublayer. In some instances (Fig. 14), as it rose up through the basement, it intersected and disrupted overlying bodies, picking up sulfides and ultramafic cumulates in the process, thus accounting for the very important association between the mineralization and the xenoliths. Other batches of magma did not intersect overlying bodies and were emplaced carrying only country rock inclusions and giving rise to the unmineralized sublayer that has been described. The fact that the sublayer intrusions were derived from a number of bodies following parallel, but distinct, evolutionary trends accounts for the greater variability of the sublayer in MgNo and Sr initial ratio in comparison with the main mass.

References

Arth JG, Hanson GN (1975) Geochemistry and origin of the early Precambrian crust of Northeastern Minnesota. Geochim Cosmochim Acta 39:325 – 362

Cochrane LB (1984) Ore deposits of the copper cliff offset. In: Pye EG, Naldrett AJ, Giblin PE (eds) The geology and ore deposits of the Sudbury structure. Ont Geol Surv Spec Vol 1:347 – 360

Dressler BO (1984) The effects of the Sudbury event and the intrusion of the Sudbury Igneous Complex on the footwall of the Sudbury structure. In: Pye EG, Naldrett AJ, Giblin PE (eds) The geology and ore deposits of the Sudbury structure. Ont Geol Surv Spec Vol 1:97 – 138

Gibbins WA, McNutt RH (1975) The age of the Sudbury Nickel Irruptive and the Murray granite. Can J Earth Sci 12:1970 – 1989

Grant RW, Bite A (1984) Sudbury quartz diorite offset dikes. In: Pye EG, Naldrett AJ, Giblin PE (eds) The geology and ore deposits of the Sudbury Structure. Ont Geol Surv Spec Vol 1:275 – 300

Gupta VK, Grant FS, Card KD (1984) Gravity and magnetic characteristics of the Sudbury structure. In: Pye EG, Naldrett AJ, Giblin PE (eds) The geology and the ore deposits of the Sudbury structure. Ont Geol Surv Spec Vol 1:381 – 410

Irvine TN (1975) Crystallization sequence of the Muskox intrusion and other layered intrusions – II. Origin of chromitite layers and similar deposits of other magmatic ores. Geochim Cosmochim Acta 3:991 – 1020

Krogh TE, Davis DW, Corfu F (1984) Precise U – Pb zircon and baddeleyite ages for the Sudbury Area. In: Pye EG, Naldrett AJ, Giblin PE (eds) The geology and ore deposits of the Sudbury structure. Ont Geol Surv Spec Vol 1:431 – 447

Kuo HY, Crocket JH (1979) Rare earth elements in the Sudbury nickel irruptive: Comparison with layered gabbros and implications for nickel irruptive petrogenesis. Econ Geol 79:590–605

Muir L (1984) The Sudbury structure. Consideration for models of an endrogenic origin. In: Pye EG, Naldrett AJ, Giblin PE (eds) The geology and ore deposits of the Sudbury structure. Ont Geol Surv Spec Vol 1:449–490

Muir TL, Peredery WV (1984) The Onaping formation in the geology and ore deposits of the Sudbury structure. In: Pye EG, Naldrett AJ, Giblin PE (eds) Ont Geol Surv Spec Vol 1:139–210

Naldrett AJ, Macdonald AJ (1980) Tectonic settings of some Ni–Cu sulfide ores: importance in genesis and exploration. Geol Assoc Can Spec Pap 20:633–657

Naldrett AJ, Bray JG, Gasparrini EC, Podolsky T, Rucklidge JG (1970) Cryptic variation and the petrology of the Sudbury Nickel Irruptive. Econ Geol 65:122–155

Naldrett AJ, Greenman L, Hewins RH (1972) The main irruptive and the sublayer at Sudbury, Ontario. Proc 24th Int Geol Congr, Montreal, Sect 4:206–214

Pattison EF (1979) The Sudbury sublayers: its characteristics and relationships with the main mass of the Sudbury irruptive. Can Miner 17:257–274

Peredery WV, Morrison GG (1984) Discussion of the origin of the Sudbury structure. In: Pye EG, Naldrett AJ, Giblin PE (eds) The geology and ore deposits of the Sudbury structure. Ont Geol Surv Spec Vol 1:491–512

Pye EG, Naldrett AJ, Giblin PE (eds) (1984) The geology and ore deposits of the Sudbury structure. Geol Surv Spec Vol 1:603 pp

Rao BV, Naldrett AJ, Evensen NM, Dressler BO (1983) Contamination and the genesis of the Sudbury ores. In: Pye EG (ed) Geoscience research grant program, summary of research, 1982–1983. Ont Geol Surv Misc Pap 113:139–151

Roeder PL, Emslie RF (1970) Olivine-liquid equilibrium. Contrib Mineral Petrol 29:275–289

Scribbins B, Rae DR, Naldrett AJ (1984) Mafic and ultramafic inclusions in the sublayer of the Sudbury Igneous Complex. Can Miner 22, 1:67–75

Slaught WH (1951) A petrographic study of the copper cliff offset in the Sudbury district. M Sc Thesis, McGill Univ (unpublished)

Sulfide Petrology and Genesis of Copper-Nickel Ore Deposits

V. V. DISTLER, A. D. GENKIN, and O. A. DYUZHIKOV[1]

Abstract

Some general problems of the formation of copper-nickel ores are discussed on the basis of new data from the Norilsk deposits. These deposits, together with the deposits of greenstone belts and some others, are classified as the deposits connected with ore-bearing volcanic-intrusive complexes of intercontinental rift zones. Massive sulfide ores, sulfide-bearing intrusions, and ultramafic volcanites are shown to have been formed as the result of successive emplacements of magmatic melts formed in the process of deep differentiation of mantle magmas. The factors determining the zonality of distribution of sulfide mineral assemblages in sulfide-bearing intrusions of the bodies of massive sulfides and the formation conditions of platin-bearing horizons of layered intrusions are discussed.

In the recent decade, studies of copper-nickel deposits have been undertaken within the International Program of Geological Correlation. A number of basic works aimed at investigating copper-nickel deposits were carried out in the USSR. These studies have contributed significantly to a better understanding of geological and petrological regularities of sulfide ore formation.

A concept of the magmatic origin of copper-nickel ores as products of the evolution of ultrabasite-basite sulfide-silicate magmatic systems is emerging. Investigations of ancient copper-nickel ore deposits in greenstone belts of the Archean cratons were especially important in this respect. Presently, the results obtained, as well as the achievements in studies of evolution of structures and endogenous ore formation through the Earth's history (Kazanskii 1983, Smirnov 1982), allow us to consider the copper-nickel ores as a specific type of magmatic product formed in the course of definite structural transformations of the continental crust.

Two groups of magmatic assemblages, the formation of which is accompanied by copper-nickel ore formation persist in all metallogenic epochs. The first group includes highly differentiated volcanic-intrusive magmatic assemblages occurring in the zones of near-fault depressions or intercontinental rift structures; the second group comprises magmatic assemblages forming rhythmi-

[1] Institute of the Geology of Ore Deposits, Petrography, Mineralogy and Geochemistry of the USSR Academy of Sciences, Moscow 109017, USSR

Geology and Metallogeny of Copper Deposits
(ed. by G.H. Friedrich et al.)

cally layered stratified plutons occurring in the regional deep faults pertaining to cratonic activity. These two groups show essential differences in ore-genetic processes and the scale of ore matter concentration.

Volcanic-intrusive assemblages give rise to the deposits in the Archean greenstone belts of Western Australia, the Rhodesian and Canadian shields, and also the Proterozoic belt of Pechenga-Imandra-Varzuga and the early Mesozoic deep-seated Norilsk block of the Siberian platform. Here the high degree of deep differentiation of initial magmas that is responsible for the wide diversity of types of melts and for separating of the great volumes of essentially sulfide magma intruding as an independent subphase is characteristic.

Evidence of the separate introduction of sulfide melts could be presented in more detail for the deposits of the Norilsk ore district. The main facts are as follows: (1) The formation of massive sulfide ores is separated from the emplacement of sulfide-bearing intrusives during the time in which magma cools to such a state that the intrusives react during tectonic deformation as solid bodies. The tectonic movements lead to the generation of cavities of different sizes that were filled by sulfide liquid with the formation of veins. The part of such a vein filling, formed by the moving apart of rock blocks and the fabric of the vein contacts are shown on Fig. 1 A, B. (2) The majority of the bodies of massive sulfide ores are separated from the intrusives by unmineralized sedimentary rock seams of different thicknesses. The formation of massive sulfide ores is accompanied by their own aureole of metamorphic transformations that are superimposed on contact-metamorphic rocks formed by the emplacement of intrusive. Recrystallization and remelting of hornfels and metasomatic rocks (Fig. 1 C) are the main types of contact transformations connected with massive sulfide ores. (3) The bodies of massive sulfide ores are localized not only on the lower endocontact of intrusives but, under favorable tectonic conditions, also in the rocks of the roof of the intrusives or intersect the sulfide-bearing intrusives. (4) The absence of correlation of the zonality of sulfide mineralization of layered intrusives and the zonality of massive sulfide ores. (5) The disseminated ore mineralization accompanying the massive ores is superimposed on sedimentary-metamorphic rocks and the rocks of intrusives as well.

It is suggested that magmatic rocks and sulfide ores originate from ultramafic-mafic melts, that are the result of partial melting of garnet peridotites the composition of which is equivalent to that of the integrated differentiated sulfide-rich magmas. Involvement of large masses of sulfide-rich mantle substrate and the fractionation of the parent ultramafic-mafic melt in the conditions of high permeability of the Earth's crust contributed to ore formation. The Norilsk ore basin is a good example of the volcanic-intrusive nickel-rich association activated in the connection with the development of the Yenisei-Khatanga global rift of Early Mesozoic age.

Fig. 1A – C. The relation of sulfide massive ores with enclosing rocks. **A** Vein massive ores in a completely differentiated intrusion. The vein was formed by moving apart the blocks of gabbro-dolerites (shown by *arrows*). Polished hand specimen; natural size. **B** Contact of massive ore with gabbro-dolerite. Thin section. Magnification 20×. *1* Massive ore; *2* reaction rim of magnetite with pyroxene relics; *3* clinopyroxene; *4* plagioclase. **C** Granophyre formed by remelting of sandstones on contact with massive sulfide ores. Thin section; magnification 20×

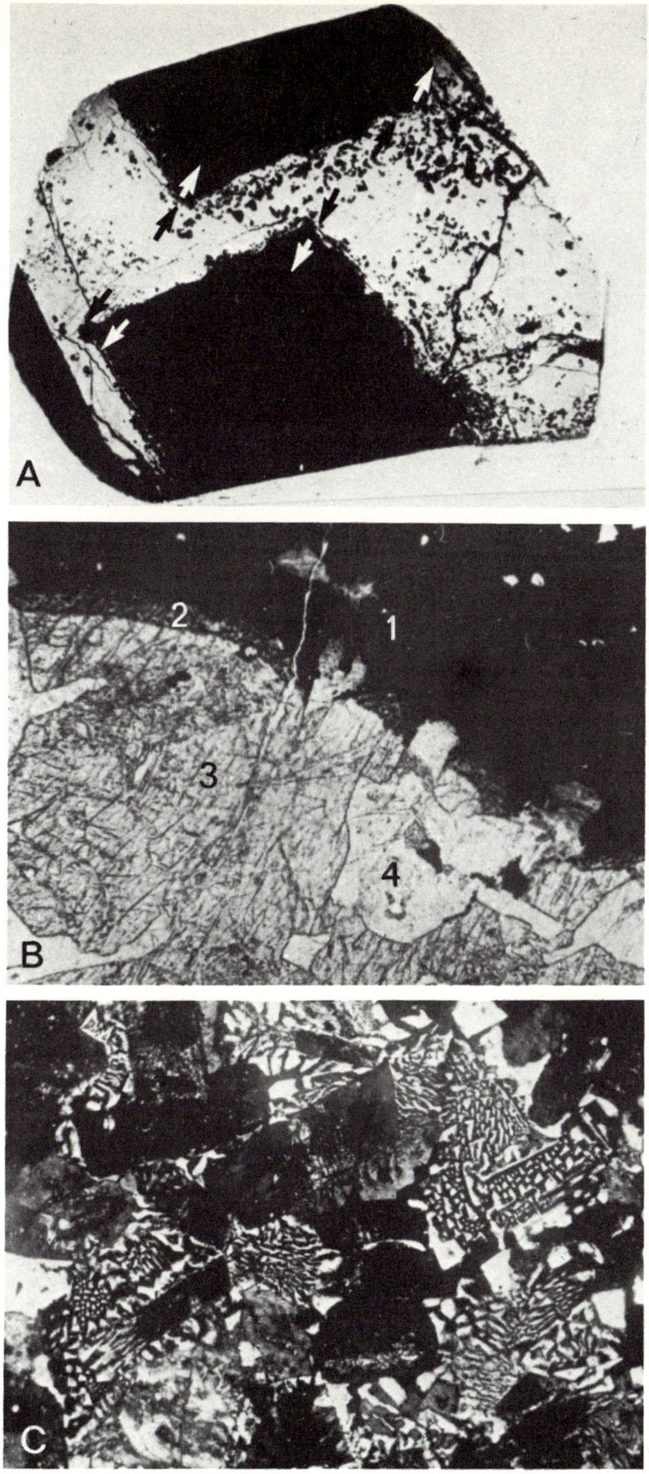

Fig. 1 A – C

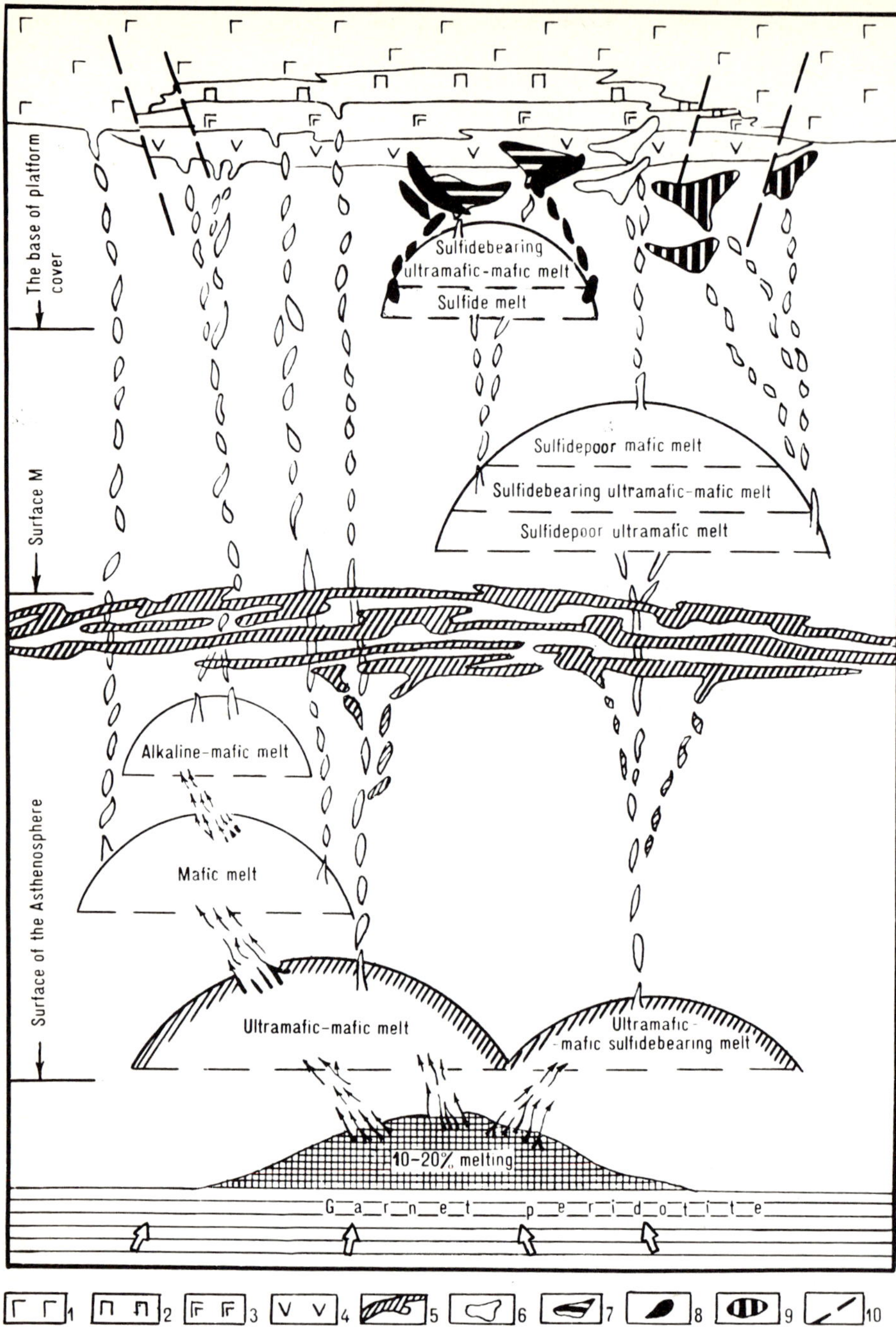

Fig. 2. Genetic model of magmatism and sulfide copper-nickel ore formation in the Norilsk region. *1–4* Late Paleozoic-Early Mesozoic volcanogenic rocks: *1* Tholeiite basalts of later magmatism; *2* hyperbasite-basite basalts (comagmatic to differentiated intrusives); *3* tholeiitic basalts of enhanced alkalinity in the early stage of magmatism; *4* alkaline-basite basalts; *5* Deep sills of hyperbasite-basites; *6–9* sulfide-bearing intrusives and massive ores: *6* leukocratic differentiates (Phase I); *7* differentiates with ore inclusion (Phase II); *8* massive ores; *9* melanocratic differentiates (Phase III); *10* deep faults controlling the alkaline-basite and hyperbasite-basite volcanogenic rock development

The volcanic-intrusive assemblage of the Norilsk region includes the ultramafic-mafic lavas and later emplaced ore-bearing layered intrusives; these are the result of deep differentiation of sulfide-rich juvenile melts (Fig. 2). This process is responsible for the segregation of immiscible sulfide liquids and for the emergence of several groups of intrusives differing in their composition, degree of inter-chamber differentiation, and content of ore. The highest concentration of sulfide liquid results in the formation of intrusions with disseminated ores and of massive ores due to the independent intrusion of sulfide liquid. Also there are grounds to believe that in such structural conditions a much greater part (by volume) of ultrabasite-basite melts does not ascend beyond the upper mantle, or at best reaches only the basement of the consolidated crust. This may be attributed to the density relationship among the lower horizon of the crust, mantle, and ultrabasite-basite melts, and also to the relationship between the lithostatic load and the pressure of the rising ultrabasite-basite astenolith (Cox 1980). The melt spreading in subhorizontal directions and its solidification make the crust heavier and thicker, and lead the border with the mantle to become diffuse. This type of deep structure is, evidently, one of the most essential features defining the areas where volcanic-intrusive nickel-rich associations develop. It also should be stressed that the location of copper-nickel deposits developed under hypabyssal subvolcanic conditions is controled by systems of long, deep-reaching faults and related regional depressions; the latter coincide with the areas of ultrabasite-basite and earlier alkaline-basite magmatism.

It is suggested that prechamber splitting of melts does not occur in magmatic assemblages of rhythmically layered plutons due to the inherent structural conditions of such magma intrusions. As a result, the bulk of such sulfides is brought into the ore-deposition zone with the silicate melt. Ore deposition in areas associated with layered plutons is a process of interrelated consolidation of two immiscible melts – sulfide and silicate.

The formation of copper-nickel ores is controlled by two types of evolution of the sulfide melts: (1) evolution of isolated sulfide liquid without essential interaction with the enclosing environment; (2) evolution of sulfide liquid in close interaction with the silicate melt.

The major purpose of research in sulfide petrology of magmatic ore formation is to ascertain the regularities governing the formation of copper-nickel ores, the stability of their parageneses, factors contributing to the zonality of mineralization in the general process of differentiation of sulfide and sulfide-silicate systems.

Studies of ore formation associated with the evolution of isolated sulfide liquid, interpretation of the paragenetic relationships of the ore-forming minerals – pyrrhotite, chalcopyrite, pentlandite, cubanite, and magnetite groups – are based on fundamental data provided by investigations of sulfide systems undertaken by Kullerud and Donnay (1970), Naldrett et al. (1967), Craig and Kullerud (1968), Yund and Kullerud (1966), Cabri (1973), and others.

The interpretation of the crystallization of ores on the basis of monosulfide solid solutions forming on the liquidus of a sulfide liquid followed by subsolidus transformations at all temperatures from the liquidus downward give an adequate general description of the ore-forming mineral relationships. These concepts are still more convincing since the bulk composition of sulfide aggregates in many deposits indicates the stability of principal ore-forming components. Godlevsky, developing this notion of the formation of copper-nickel ores, came to the conclusion of the inert behavior of sulfur in natural sulfide melts (Godlevskii 1967). The study of the zonality of ore bodies in the Norilsk deposits (Genkin et al. 1981) led us, however, to the conclusion that parageneses that may be ascribed to the inert-sulfur facies are not present at the crystallization of sulfide melts composing them. All ore bodies display cryptic sulfide layering, manifesting through successive changes of parageneses, reflecting the changing fugacity of sulfur in a melt in the process of its crystallization differentiation. Early parageneses are represented by products of low-sulfur monosulfide solid solutions, while the later ones – by high-sulfur. The changing sulfur content of parageneses also can be seen in the nature of fractionation of ore-forming metals in pyrrhotites with various Me/S ratios and pentlandites of varied compositions (Fig. 3).

One more event contributing to the fractionation of components in copper-nickel ore formation should be mentioned. Interpreting the zonal origin of the

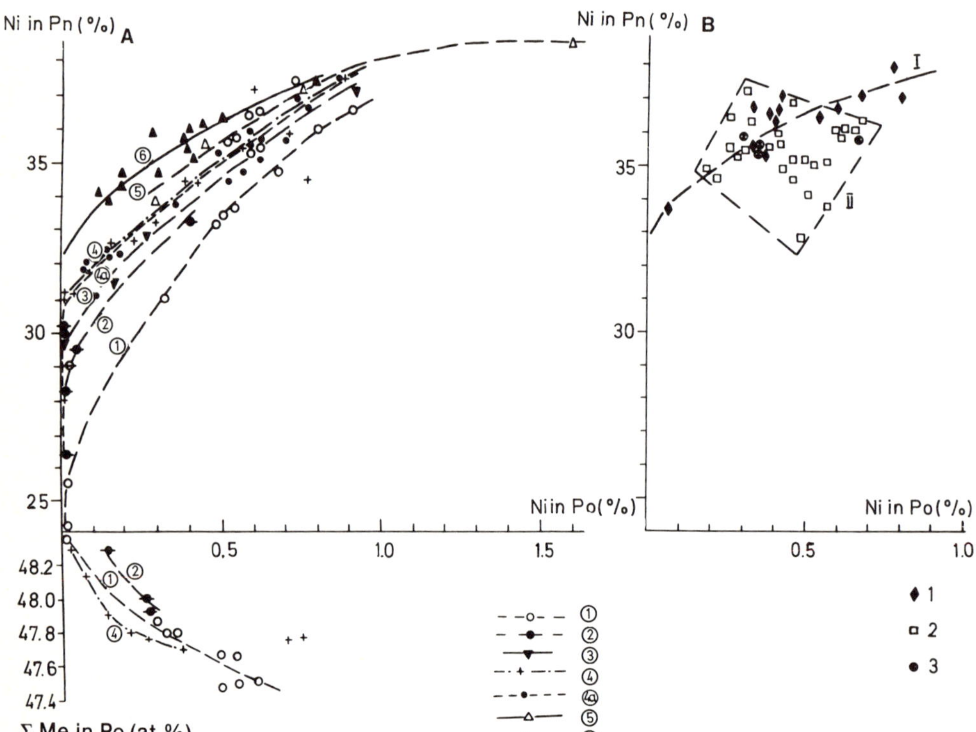

Fig. 3 A, B. Nickel distribution between pyrrhotite (*Po*) and pentlandite (*Pn*), also in pyrrhotites with various iron contents (*ΣMe* in *Po, at.%*) in stratified intrusives of various ore deposits (**A**) and in ores of the Pechenga deposit (**B**), *Circled numbers: 1* Maksut area (Eastern Kazakhstan); *2–5* intrusives of the Norilsk ore region: *2* Lower Norilsk; *3* Zubovsk; *4* Talnakh; *4a* massive ores of the Talnakh intrusive; *5* Manturovsky; *6* Karik-Yarvi area of the East-Pechenga region. *Numbers 1–3* ore types of the Pechenga ore region: *1* disseminated in peridotites; *2* brecciated; *3* massive

ore bodies of the Norilsk deposits that are composed of pyrrhotites, cubanites, and talnakhites, we have revealed that the most convincing proof of the phase relationships of sulfides may be obtained if one assumes the existence of two liquids in a sulfide melt – rich in copper and rich in iron and nickel.

We have obtained direct evidence of the possibility of such liquation events in sulfide liquids. During the study of sulfide mineralization of oceanic basalts, layered sulfide globules, consisting of sulfide solid solutions [iron-nickel (M_{ss}) and iron-copper (I_{ss})] were estabilished (Distler et al. 1983). Phase relations (Fig. 4) witness that both solid solutions are the products of chilled crystallization of two sulfide liquids with a distinct phase boundary between them. Distinct fractionation of components (Table 1) takes place in these liquids with limited solution of copper in M_{ss} and nickel in I_{ss}. The chemical composition of one of the liquids corresponds to nickel-pyrrhotite solid solution with the formula $[Fe_{0.60}Ni_{0.19} \cdot (Co, Cu)_{0.01}]S_{0.80}$; the second with the formula of the solid solution $(Cu_{0.87}Ni_{0.13})_{1.0}Fe_{2.0}S_{3.08}$ is close to the proportions of the components in cubanite. It must be stressed that we are not aware of experimental data reproducing the liquation in a sulfide melt of such composition. On the basis of the analysis of natural sulfide assemblages, it could be assumed that the layering takes place with relatively low concentration (fugacity?) of sulfur in sulfide melts. In the Norilsk intrusions, we have observed, in particular, the layering of troilite-mooihoekite globules. In zonal bodies of massive ores, the separation of great sulfide masses occurs with mooihoekite (talnakhite, putoranite) as the principal ore-forming minerals in one of them and troilite in the other.

The fact that the specific features of evolution resulting in these sulfide parageneses have been disclosed in occurrences of Mesozoic magmatism cannot be overlooked. Variations in the composition of coexisting sulfides in Precambrian deposits are more pronounced if different deposits are compared than is true within any one of them. Moreover, there are cases where the correlation in the distribution of elements between such coexisting minerals such as pyrrhotites of various sulfur content and pentlandites is not observed. One of the reasons here is that ore metamorphism affects more strongly the pyrrhotite composition than those at other coexisting phases. Similar data have been identified for deposits in the Pechenga area (Fig. 3).

Examining the specific nature of ore formation associated with the evolution of a sulfide liquid coexisting with a silicate melt, two stages in the formation of the sulfide melt composition should be outlined. The first pertains to the equilibrium distribution of components between two liquids, which may be attributed to the changed relationship of the sulfur and oxygen fugacity in the course of the differentiation of the melts. The second, includes the consolidation of isolated sulfide liquid coexisting with the solid-phase products of the silicate-melt crystallization.

Comparative analysis made both for the Precambrian and Phanerozoic deposits has shown that the interrelations between sulfide and silicate parageneses are characterized primarily by the stratified distribution of sulfide mineralization being in accord with the general layering of the massifs.

Cryptic layering becomes pronounced in the regular changes of the composition of rock-forming minerals – olivine, clino- and orthopyroxene, and plagio-

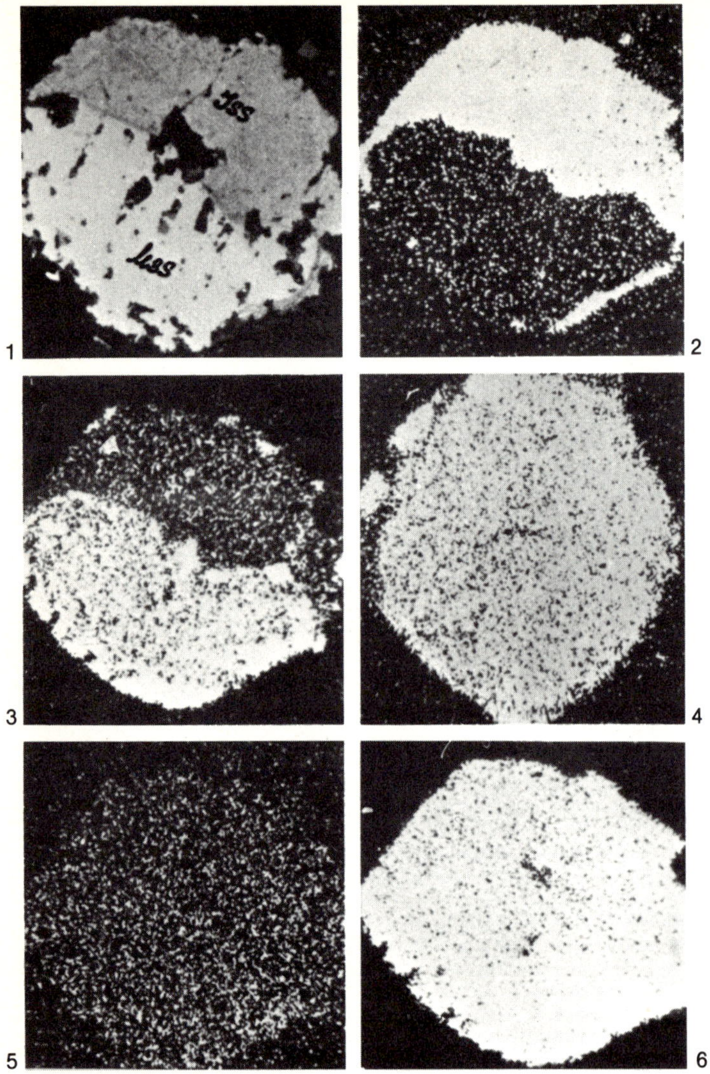

Fig. 4. Characteristic X-ray scans of a layered sulfide globule consisting of quenched copper-iron (I_{ss}) and iron-nickel (M_{ss}) sulfide liquids. The boundary between the two quenched liquids and inflowing of copper-iron liquid along the border with silicate matrix are most pronounced. *1* Reflected light image; *2* copper K_α scanning image; *3* nickel K_α scanning image; *4* iron K_α scanning image; *5* cobalt K_α scanning image; *6* sulfur K_α scanning image

clase – and is a well-known and very important feature of stratified intrusives. Sulfide mineralization has been established also to feature cryptic layering, reflected in the changed composition of coexisting pyrrhotites, pentlandites, and chalcopyrites (Distler et al. 1979). Clearcut types of sulfide parageneses peculiar to basites and ultrabasites have been identified. Parageneses of sulfur-deficient sulfides are mostly developed in ultrabasites, while parageneses of sulfur-rich sul-

Table 1. Chemical composition of sulfide solid solutions of layered globules from oceanic basalts[a] (X-ray microprobe analysis, mass %)

Phase	Fe	Ni	Co	Cu	S	Total
M_{ss}	43.45	14.61	0.28	0.30	41.49	100.13
I_{ss}	41.01	2.68	0.14	20.14	36.27	100.24

[a] Sample 6–1, 112–114, hole 505B (Deep Sea Drilling Project, Costa Rica Rift). Chromite-bearing olivine basalt.

fides – in basites. Thus, the typical minerals of the former are troilite and sulfur-deficient minerals of the chalcopyrite group (such as talnakhite and mooihoekite) or hexagonal pyrrhotite with a low sulfur content, associated with chalcopyrite and cubanite. Pyrrhotites with a high sulfur content and stoichiometric chalcopyrites are common for basite parageneses (Genkin et al. 1981).

In basite and ultrabasite horizons the distribution of ore components is strongly dependent on temperature; for instance, that of nickel between sulfides and olivines or orthopyroxenes as well as coexisting sulfides, corresponding to general temperature parameters of basite and ultrabasite melts. In genetic related series of sulfide-bearing rocks, the value of $K^{\mathrm{Ni}}_{\mathrm{D\,SL-OL}} = (X_{\mathrm{NiS}}/X_{\mathrm{FeS}})_{\mathrm{SL}}$: $(X_{\mathrm{NiSi_{0.5}O_2}}/X_{\mathrm{FeSi_{0.5}O_2}})_{\mathrm{OL}}$ increases from ultrabasites to the gabbroids. In completely differentiated intrusions of the Norilsk ore district (Distler and Genkin 1980) the value in sulfide-bearing picrites is $K^{\mathrm{Ni}}_{\mathrm{D\,SL-Ol}} = 5.4$ and in sulfide-bearing gabbroids accordingly 17.1. The value $K^{\mathrm{Ni}}_{\mathrm{D\,SL-Ol}}$ does not depend on the whole rock concentration of nickel in the melt. In differentiated melanocratic intrusions (Distler and Genkin 1980) for the picrite-troctolite horizon, the value of the distribution coefficient is also 5.5; although the whole rock concentration of nickel in these rocks is nearly one order lower than is its content in picrites of completely differentiated intrusions. The differences in whole rock concentrations are reflected only in the nickel content of olivine. With equal magnesium content of olivines (80–82% Fo), the nickel contents in olivines of completely differentiated intrusions reaches 0.3 mass% and in differentiated melanocratic intrusions 0.10–0.16 mass%. The value $K^{\mathrm{Ni}}_{\mathrm{D\,SL-Ol}}$ depends on the temperature of the equilibrium of the sulfide liquid with coexisting melt. For the sulfide-bearing picrites the value $K_{\mathrm{D}} = 5.4$ corresponds to 1300° – 1400°C and the value 17.1 corresponds to 1150°C. The latter value is very near the K_{D} value, established experimentally for M_{ss}–Fo under 1100°C of 11.3 (Clark and Naldrett 1972). The value increases up to 25.4 under 900°C and up to 39.9 under 700°C. The distribution of nickel in coexisting sulfides and clino- and orthopyroxenes reflects the same tendency established for the sulfides and olivines.

We stress one more aspect of sulfide-silicate interaction in magmatic melts. It concerns the effect of the sulfide liquid on the succession of crystallizing phases peculiar of the silicate melt. Many researches have shown that gravitational events facilitate the parallel concentration of sulfides and olivines with the emergent sulfide-rich peridotites in the bottom part of the intrusives. In the process of sulfide-silicate interaction in magmas, there exist definite influences of the immiscible sulfide liquid on the evolution pecularities of the silicate melt. The presence in the melt of S^{2-} serves as the factor of depolymerization and of increase in the degree of ionization of the melt by means of connecting of part of the petrogenic components, first of all iron, in the Me – S groups. This prevents the formation of Me – O – Si ≡ bonds and, accordingly, results in the isolation of the solid phases and favors more perfect differentiation of silicate melts in the presence of sulfide liquid.

The depolymerizing role of sulfide liquid in sulfide-silicate melts also is felt in the process of formation of mafic pegmatoids rich in sulfides that form isolated discordant bodies that cut the intrusive stratification or individual horizons of the gabbroids of the pegmatoid or taxitic structure, localized nevertheless concordantly with the general stratification of the massifs. Taxitic sulfide-rich rocks of the completely differentiated Norilsk intrusives, ore-bearing pegmatoid gabbroids of the bottom part of the Monchegorsk pluton, and others are examples of such horizons. These are easily melted residues from fractional crystallization of melts. However, the very fact of their emergence and abnormal localization, infringing on the gravitational regularity of layered sequence formation, is indicative of specific interaction of components in magmas. Apart from the sulfide liquid, no less important for the formation of these rocks is the fluid phase of melts coexisting with sulfides. Parallel accumulation of sulfide and fluid phases in products of crystallizing melts emerges as a decisive factor controlling rock localization and inner structure.

The interaction of sulfide and fluid phases with silicate melts is also essential for formation of some other very interesting pecularities of layered plutons, including chromite-rich pegmatoid gabbros belonging to the upper gabbro series of the Norilsk intrusives. Since the development of those rocks is of general significance, let us describe their inner structure, composition, and the specifics of ore mineralization in more detail. Figure 5 shows the location of the pegmatoid gabbro horizon in the intrusive profile and the principal features of its inner structure. The major specific feature of the series is a greatly diverse petrography of the rocks making it up. Three basic horizons may be identified: (1) contact gabbro-dolerites; (2) taxitic chromite-rich anorthositic gabbros; (3) prismatic-granular gabbro-dolerites and diorites. Most interesting are the taxitic chromite-rich gabbros in which the upper level of chromite, sulfide, and platinum mineralization becomes established. The following zones are found in this horizon: Zone I includes predominantly even granular (pegmatoid in places) chromite-rich anorthositic gabbros; Zone II – taxitic chromite-rich anorthositic gabbros with irregular areas of troctolites, picrites, plagioperidotites, and clinopyroxenites. The ultrabasic rock varieties are often rimmed with essentially anorthositic rocks. In this zone, schlierenlike chromitite isolations are found, highest development of sulfide and other ore minerals is observed, and a complex of secondary minerals, including prehnites, biotites, talc, chlorites, carbonates, anhydrites, and others are also widespread. Zone III has the least thickness within the horizon and is composed mainly of even granular gabbros. It acts as a transitional zone to a horizon made up of sulfide-free, chromite-free gabbro-dolerites and quartz-containing gabbro-dolerites (Fig. 5).

Sulfide mineralization is represented by the common assemblage of minerals such as pyrrhotite, pentlandite, chalcopyrite, and also assemblages of later ore

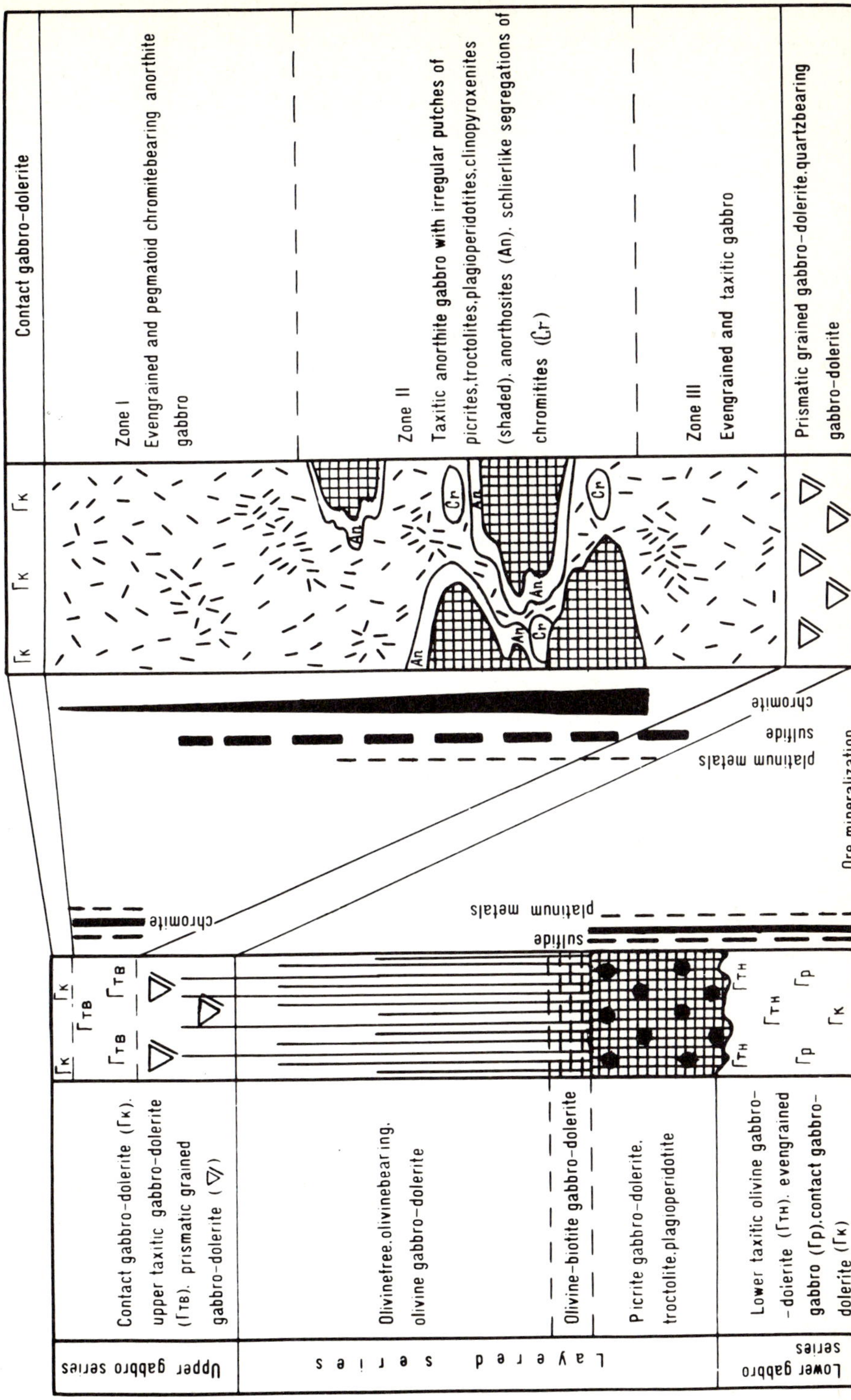

Fig. 5. Profile and inner structure of the Upper gabbro series of the Talnakh intrusive (description see text)

minerals – pyrite, millerite, argentopentlandite, galena, sphalerite, arsenides, and sulfoarsenides of nickel, associated with prehnite, biotite, chlorite, and carbonate. Minerals of platinum metals – sperrylite, stibiopalladinite, merenskyte, mayakite, as well as platinum-palladium-containing nickelite and maucherite, palladium-ruthenium-containing pentlandite, palladium-containing millerite, and others are found in intergrowths with the sulfides and chromian-spinels.

All specific features inherent in the structure and composition of the Upper gabbro series and its ore mineralization emanate from the fluid-magmatic convection that is essential for both rock formation and mineralization. Redistribution of petrogenetic and ore components against the background of normal differentiation in a cooling melt in evidently the major factor responsible for the abnormal location of chromitites and ultrabasites in the upper horizons of a stratified intrusive and the colocation of disequilibrium assemblages of primary and secondary sulfides as well.

The rock types and the ore-bearing nature of the Upper gabbro series in the Norilsk intrusives are very similar to those of the troctolite-anorthosite zone I comprising the $J-M$ Reef with sulfide, chromite, and platinum mineralization in the intrusive Stillwater complex (Todd et al. 1982). The Upper gabbro series differs from the above troctolite-anorthosite zone mainly by the morphology of the rock bodies of definite petrographic composition. In the Norilsk intrusives, these are multiple isolations of irregular shape; in the Stillwater, these are continuous, regularly interbedded horizons. However, if we take into account that the layering in the Stillwater complex develops within a rather large (up to several kilometers thick) magmatic body with the cooling rate much smaller than in the ribbon bodies of the Norilsk intrusives (hundreds of meters thick), then these differences would look quite natural. What is important is that in both cases similar complexes of rocks are developed that crystallized from melts of various compositions, often accompanied by an abnormal distribution of components within them. The latter is manifested, for instance, in the chrome accumulation in the basite melt that favors the crystallization of chromite-rich anorthosites. Later, mineral assemblages, comprising phases with volatile components, are widespread in these horizons too. Such parageneses provide a typomorphic character to these rocks since, in other horizons of intrusives, they are either absent or poorly developed.

Specifics discussed above enable the conclusion that the origin of these rocks may be associated with two interacting processes during the consolidation of magmatic melts. The first is common differentiation of melts by separating solid phases peculiar of the formation of layered intrusives. The second is the fluid-magmatic convection responsible for the changed sequence of melt differentiation and redistribution of components in a melt. The redistribution involves components accumulated in residual liquids and components constituent of early solid phases that dissolve partially while reacting with the residual melt. The effect of such a melt on the olivine + chrome-spinel assemblage causes the migration of its components to the upper horizons of the crystallizing intrusive and the forming of local bodies of ultrabasite melts. The model proposed does not require additional injections of hypothetical ultrabasite melt to explain the origin of rocks and mineralization of the horizons in question (Todd et al. 1982).

In conclusion, it should be stressed that joint analysis of the formation of intrusive rocks and ores of the copper-nickel deposits bears out a close relationship of ore formation with the whole complex of events attending the development of sulfide-silicate magmatic melts from their origin in the zones of magma formation to the final stages of their consolidation under hypabyssal-subvolcanic conditions.

References

Cabri LJ (1973) New data on phase relations in the Cu – Fe – S system. Econ Geol 68:443 – 454

Clark T, Naldrett AJ (1972) The distribution of Fe and Ni between synthetic olivine and sulfide at 900 °C. Econ Geol 67:939 – 952

Cox KG (1980) A model for flood basalt vulcanism. J Petrol 21:629 – 650

Craig JR, Kullerud G (1968) Phase relations in the Cu – Fe – Ni system and their application to magmatic ore deposits. In: Magmatic ore deposits. Econ Geol Monogr 4:344 – 358

Distler VV, Genkin AD (1980) Deposits of sulfide copper-nickel ores of the USSR and their connection with cratonnal volcanism. Proc 6th IAGOD Symp. Schweizerbart, Stuttgart, pp 275 – 295

Distler VV, Smirnov AV, Grokhovskaya TL, Filimonova AA, Muravitskaya GN (1979) Stratification, cryptic layering and formation conditions of sulfide ore mineralization of differentiated trapp intrusions. In: Formation conditions of magmatic ore deposits. Nauka, Moscow, pp 211 – 269 (in Russian)

Distler VV, Pertsev NN, Boronikhin VA (1983) Sulfide petrology of basalts from deep sea drilling project holes 504B and 505B. Initial Rep Deep Sea Drill Proj 69:607 – 617

Genkin AD, Distler VV, Gladishew GD, Filimonova AA et al. (1981) Copper-nickel sulfide ores of the Norilsk deposits. Nauka, Moscow, 234 pp (in Russian)

Godlevskii MN (1967) On differential mobility of the components in the formation of sulfide copper-nickel ores. Geol Rudn Mest 9:17 – 31 (in Russian)

Kazanskii VI (1983) The evolution of the earth's crust and endogenic ore formation. In: Deep structure of ore districts, ore fields, and deposits. Nauka, Moscow, pp 5 – 24 (in Russian)

Kullerud G, Donnay G (1970) Chalcopyrite solid solution. Carnegie Inst Washington Yearb 69

Naldrett AJ, Craig JR, Kullerud G (1967) The central portion of the Fe – Ni – S system and its bearing on pentlandite exsolution in iron-nickel sulfide ores. Econ Geol 62:826 – 847

Smirnov VI (1982) Endogenous ore formation in geological history. Geol Rudn Mest 4:3 – 20 (in Russian)

Todd SG, Keith DW, Le Roy LW et al. (1982) The *J – M* platinum-palladium reef of the Stillwater complex, Montana I. Stratigraphy and petrology. Econ Geol 77:1454 – 1480

Yund RA, Kullerud G (1966) Thermal stability of assemblages in the Cu – Fe – S system. J Petrol 7:454 – 488

Types and Distinctive Features of Ore-Bearing Formations of Copper-Nickel Deposits

M. N. GODLEVSKY† and A. P. LIKHACHEV[1]

Abstract

Ore-bearing formations and copper-nickel deposits are recognized as follows: (1) Duluth-type gabbro-troctolite formation; (2) Norilsk-type gabbro-dolerite formation; (3) Bushveld and Monchegorsk-type gabbro-norite-pyroxenite-peridotite formation; (4) Pechenga-type gabbro-pyroxenite-peridotite formation; (5) Kambalda- and Allarechensk-type pyroxenite-peridotite formation; (6) Mountkeith-type peridotite-dunite; and (7) regenerated Subdury-type diorite-norite formation. Nickel-bearing structures are noted in the greater thicknesses of the Earth's crust and in downwarped Moho boundaries in weakly eroded provinces. There is a clear relationship between magmatic thicknesses of continental depressions and their ore potential: the thickness of volcanogenic units in ore-bearing structures is over 3 km, whereas that of barren structures is less than 2 – 3 km. In undislocated areas, the relation between the morphology of intrusive bodies and their ore potential is observed. Ore-bearing magmatic bodies are typically elongated, band-chonolitelike bodies with flat roofs and downwarped bottoms; this is attributed to the excess of density of ore-bearing magmas over that of environment (at the expense of sulfide load) and gravity field effect during intrusion of the bodies that brought about one-way movement of magmatic masses.

Copper-nickel deposits are associated with mafic-ultramafic magmatic formations that are divided into three series: (1) (Fig. 1 A, Table 1) mafic sulfide-free and slightly sulfide-bearing series (MgO $\leqslant 8$ mass%); (2) mesomafic sulfide-bearing (MgO 8 – 33 mass%); and (3) ultramafic sulfide-free series (MgO >33 mass%). Each of the series is subdivided into magmatic groups that correspond to magma varieties and produce magmatic rock complexes or geological formations that host various copper-nickel deposits.

The following ore-bearing (sulfide-nickel-bearing) formations are recognized for copper-nickel deposits (Fig. 2): (1) Duluth-type gabbro-troctolite formation; (2) Norilsk-type gabbro-dolerite formations; (3) Bushveld- and Monchegorsk-type gabbro-norite-pyroxenite-peridotite formation; (4) Pechenga-type gabbro-pyroxenite-peridotite formation; (5) Kambalda- and Allarechensk-type pyrox-

[1] Central Research Institute of Geological Exploration for Base and Precious Metals, Moscow, USSR

Geology and Metallogeny of Copper Deposits
(ed. by G. H. Friedrich et al.)

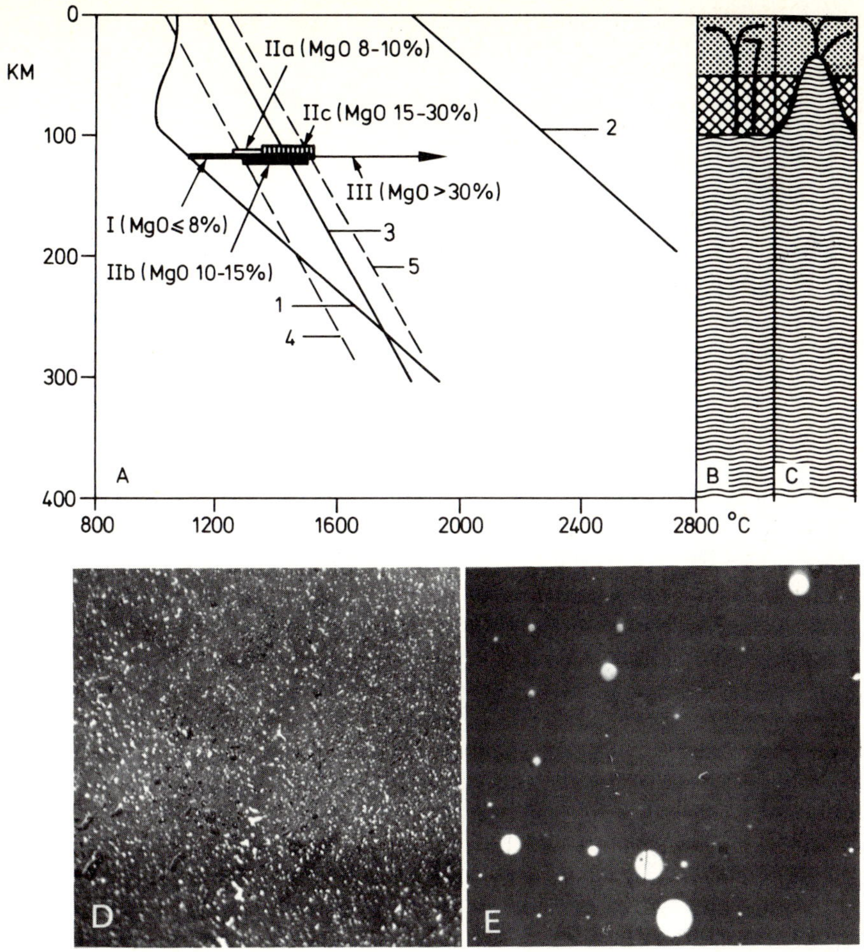

Fig. 1A – E. Types and conditions of formation of mafic-ultramafic magmas. **A** *1, 2* Lines showing solidus and liquidus of pyrolite with 0.1 weight% H_2O (After Wyllie 1971); *3* line showing solidus of pyrrhotite (Sharp 1969, Kennedy and Ryzhenko 1973); *4, 5* inferred lines of mantle sulfide, solidus and liquidus. *I* The area of melting of relatively low-temperature sulfide-free mafic magmas; *II* the area of melting of sulfide-bearing magmas: *IIa* of relatively low-temperature magmas with Cu-rich sulfides (Ni : Cu = 1 : 3 – 4), *IIb* medium-temperature magmas that involve the whole sulfide fraction of initial matter (Ni : Cu = 1 : 1.2 – 2.5), *IIc* high-temperature magmas with Cu-poor sulfides (Ni : Cu = 1 ⩾10 : 1); *III* the area of melting of sulfide-free high-temperature ultramafic magmas. To the *right* are magma generation diagrams: for deep-seated mantle (**B**) and for asthenolith diapir (**C**). *Below* (**D** and **E**) the state of iron-nickel-copper sulfide liquid is shown (*white*) in the oxide-silicate melt of the Norilsk picritic gabbro-dolerite as observed in the experiments: **D** at 1500 °C, **E**) at 1250 °C, with 350 and 150 magnification, respectively

enite-peridotite formation; (6) Mountkeith-type peridotite-dunite, and regenerated Sudbury-type (Likhachev 1978b) diorite-norite formations (Table 1).

All sulfide-nickel-bearing formations and copper-nickel deposits now occur within cratons (Fig. 3). They are located in superimposed continental depressions filled with Mesozoic sedimentary and volcanogenic mafic-mesomafic trap rocks

Table 1. Classification and setting of nickel-bearing complexes and copper-nickel deposits

Magmatic series	Magmatic groups	Ore-bearing formations	Morphology and setting of formations	Types of deposits and ore occurrences, and their examples
Mesomafic sulfide-bearing MgO 8–33 mass%	Weak magnesian group (MgO $\frac{8-10}{1-20}$ mass%)[a] with Cu-rich (relative to nickel) sulfides (Ni:Cu = 1:2.5–10)	Gabbro-troctolite formation with Cu-rich (relative to nickel) platinum-nickel-copper ores	Sill-shaped intrusions in craton cover	Kureika-type; ore occurrences of Kureika distric, Siberian Platform
			Large stratified massifs in craton basements	Duluth-type; Duluth deposits (USA)
	Moderate magnesian group (MgO $\frac{10-15}{1-30}$) with Cu-bearing sulfides (Ni:Cu = 1:1.2–2.5)	Gabbro-dolerite formation with Cu-bearing platinum-nickel-copper ores	Craton-associated elongated chonolith- and bandlike gently dipping bodies of craton cover	Norilsk-type; Norilsk deposits, Insizwa (SAR)
	Magnesian group (MgO $\frac{15-20}{1-40}$) with Cu-poor sulfides (Ni:Cu = 1–2:1)	Gabbro-norite-pyroxenite-peridotite formation with Cu-poor platinum-copper-nickel ores	Craton basement (infrastructures) associated large stratified intrusive complexes	Monchegorsk-Bushveld types; Bushveld and Monchegorsk, Great Dyke, Stillwater, and other deposits
	Higher magnesian group (MgO $\frac{20-25}{1-35}$) with moderate Cu-poor sulfides (Ni:Cu = 2:1)	Gabbro-pyroxenite-peridotite formation with moderate Cu-poor platinum-copper-nickel ores	Sags (suprastructures) associated small interformational bodies	Pechenga-type; Pechenga deposits; Cape Smith-Weikhem Bay deposits in Canada
	High magnesian group (MgO $\frac{25-28}{20-45}$) with Cu-poor sulfides (Ni:Cu = 2–20:1)	Pyroxenite-peridotite formation with Cu-poor platinum-copper nickel ores	Greenstone belts (suprastructures) associated conformably dipping minor bodies	Kambalda-type; greenstone belt associated deposits in Western Australia, Canada, Africa
			Craton basement (infrastructures) associated cross-minor bodies	Allarechensk-type; Allarechensk deposits

[a] The numerator indicates variation limits of MgO content weighted mean; the denominator indicates variation limits of MgO content in magmatic differentiates.

Table 1 (continued)

Magmatic series	Magmatic groups	Ore-bearing formations	Morphology and setting of formations	Types of deposits and ore occurrences, and their examples
Mesomafic sulfide-bearing MgO 8 – 33 mass%	Ultramagnesian group (MgO $\frac{28-33}{20-54}$) with sulfides very poor in Cu (Ni : Cu = 20 + 60 : 1)	Peridotite-dunite formation with platinum-copper-nickel ores very poor in Cu	Greenstone belts (suprastructures) associated cross-cutting (dyke-shaped) bodies	Mount Keith-type; Mount Keith, Six-Mile deposits (Western Australia)
Mafic initially sulfide-free (MgO ≤8 mass%)	Hybrid low magnesian (MgO $\frac{4-8}{0.1-12}$) with sulfides poor and very poor in Cu (Ni : Cu = 1 – 10 : 1)	Diorite-norite formation with ores poor and very poor in Cu	Craton basement associated large and small bodies	Sudbery type; Sudbery lopolith deposits, Elkinsk ore occurrence in the Voronezh crystalline massif

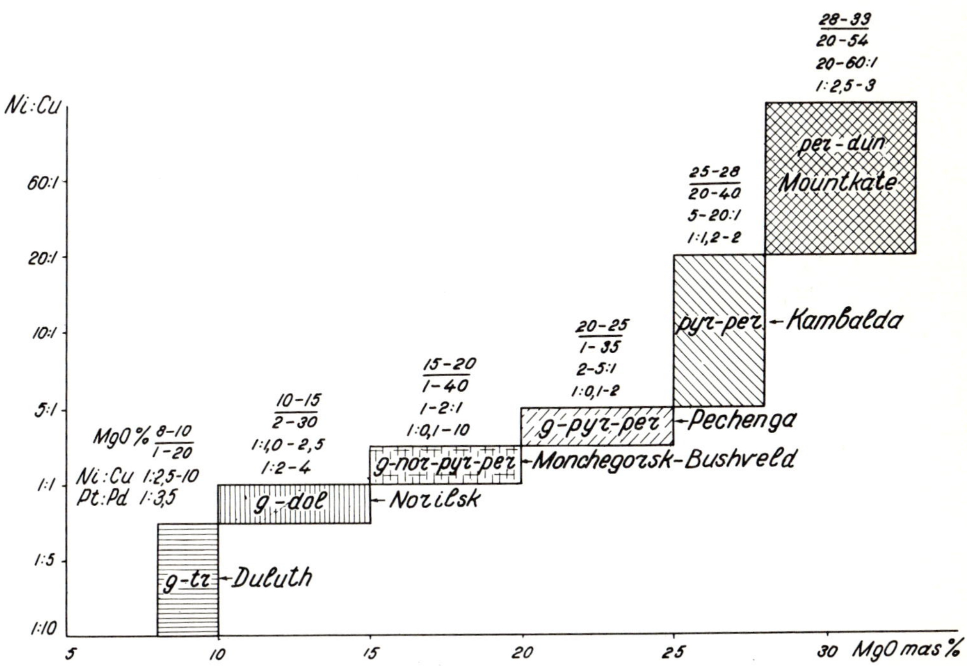

Fig. 2. The relationship of Ni – Cu ratio and magnesian content of the ore-bearing formations. Diagram symbols: The *numerator* indicates variation of MgO limits of the mean weighted contents. The *denominator* indicates alteration of the contents during magmatic differentiates; Ni : Cu and Pt : Pd in sulfides

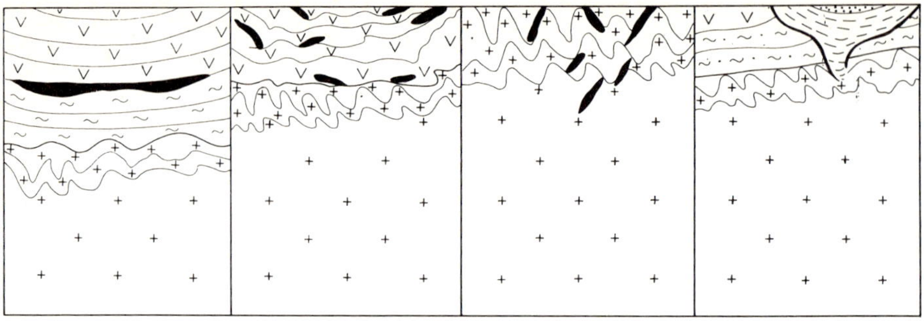

Fig. 3. Environments of nickel-bearing massifs and related copper-nickel deposits: *I* in superimposed continental depressions of Mesozoic craton cover (Norilsk-type); *II* in Archean and Proterozoic supracrustal terrains (deposits related to greenstone belts, the Pechanga structure); *III* in infrastructures: Archean granite-gneiss complexes of shield basement (Allarechensk-type); *IV* Archean and Proterozoic large intrusive massifs (Bushveld-, Monchegorsk-types)

(Norilsk and Insizwa deposits); in supracrustal formations developed in Archean and Proterozoic sedimentary-volcanogenic troughs and depressions superimposed upon the shield basement (deposits related to greenstone belts of Western Australia, Canada, Africa, and the Pechenga structure of the Kola Peninsula); in infrastructures, viz. Archean granite-gneiss complexes of the shield base with sulfide-bearing ultramafic bodies (such as the Allarechensk deposit); in large Archean- and Proterozoic-layered intrusive massifs related to the craton basement, and in old cover (such as the Bushveld and Monchegorsk).

The fact that sulfide-nickel complexes and copper-nickel deposits are invariably associated with continental cratons is attributed to peculiar features of nickel-bearing magmatism: sulfide-bearing magmas generated at deep-seated mantle horizons, with the thick continental crust playing the role of a heat screen to account for the accumulation of heat energy and melting of sulfide-bearing mantle substance at the expense of temperature increase (Fig. 1 B). This results in generation of high-temperature melts were heavy sulfide liquid is present in a dispersed state (Fig. 1 D); this provides the manner of sulfide transport from the deep-seated mantle into the upper-horizons of the Earth's crust in the less dense oxide-silicate melt (thermal magma formation inherent in continental magmatism). In other cases, e.g., in decompressional magma formation (Fig. 1 C) typical of oceanic, island-arc, and continental-rift type magmatism, sulfides of initial matter, being incorporated into the melt, occur as relatively large segregations (Fig. 1 E). The main bulk of the segregations plunges into the deep-seated mantle without reaching the upper horizons of the Earth's crust and its surface (Likhachev 1978a, Godlevsky and Likhachev 1981). Temperature-dependent variation of the mode of occurrence of sulfide liquid in oxide-silicate melt has been demonstrated experimentally (Fig. 1 D) (Godlevsky and Likhachev 1981).

Most copper-nickel deposits are linked to effusive-intrusive complexes developed in superimposed continental depressions filled with sedimentary-volcanogenic rock units. Some of the units are practically undislocated (Norilsk area) others are slightly dislocated (Pechenga structure), still others are intensely dislocated (Archean greenstone belts). Nickel-bearing magmatism produces bodies of various shapes, dimensions, and ore contents depending on the erosional level (Likhachev 1983).

In relatively weakly eroded structures, there is a relation between the magmatic thicknesses and their ore potential. In cases in which effusive magmatic complexes are less than 2 – 3 km-thick, there are practically no ore-bearing magmatic bodies or copper-nickel deposits (traps of India, South America, the southern Tunguska basin), whereas 3.5 – 10 km-thick magmatic complexes are ore-bearing sags in the Norilsk area and the Pechenga sag.

The relation between the thickness and ore potential observed in magmatic formations is attributed to various melting degrees and transport to the surface of initial matter that was incorporated into the melt as easily fusible to infusible fractions (in all cases stage-by-stage). Sulfide-free and weakly sulfide-bearing mafic magmas (MgO $\leqslant$8 mass%) that slightly dissolve sulfur (<0.1 mass%) are formed first and transported toward the Earth's surface, then sulfide-bearing mesomafic magmas (MgO 8 – 33 mass%) producing copper-nickel deposits are formed (Fig. 1 A) (Likhachev 1973, 1978a). The content of sulfides in the initial mantle material is estimated to be <1 mass% (Likhachev, Krivtsov 1975). Therefore, the production of sulfide ore-bearing magma (>1 mass% sulfides) (Godlevsky and Likhachev 1983) requires a preliminary sulfide concentration that can be performed through melting and transport to sulfide-free and slightly sulfide-bearing mafic fractions from the initial material. As has been demonstrated by experiments (Naldrett and Richardson 1967, Shamaraki and Clark 1973, Sharp 1969, Kennedy and Ryzhenko 1973) solidus temperature for silicates should be lower than for sulfides (Naldrett 1973, Naldrett and Turner 1977, Likhachev 1973) (Fig. 1 A). The accumulation of sulfide matter and ascent of sulfide-bearing magma to the surface depends on the volumes of transported mafic material (no matter what their fractionation is: whether direct melting or differentiation of more magnesian magma). This has a bearing on the thickness of magmatic rock sequences in nickel-bearing sags. With the mafic magmas producing less than 2 – 3 km-thick magmatic rock sequences, conditions required for generation, accumulation, and ascent of sulfide ore-bearing products to the upper crust are not achieved. In this case, magmatism is restricted by intrusions and outflow onto the surface mainly of sulfide-free and slightly sulfide-bearing mafic magmas.

Nickel-bearing magmatic areas are noted for the higher thickness of the Earth's crust (30 – 45 km), the downwarping of the Moho boundary, and a flat bottom of sags. In contrast, nickel-free rift troughs are characterized by the uplifted Moho boundary and steplike, en bloc structures.

The contrast is very conspicuous in the Norilsk area as well as in adjacent territories with their typical sags, swell-like uplifts, and extended deep faults.

These sags are filled with basaltic effusive-intrusive complexes in thicknesses of over 3 km. The downwarped structures in geophysical fields are outlined at the levels of the craton cover boundary, granite-gneiss basement, and upper mantle (Fig. 4). This suggests that they formed in the deep-seated mantle and that by their nickel-bearing troughs are actually compensating structures that evolved as

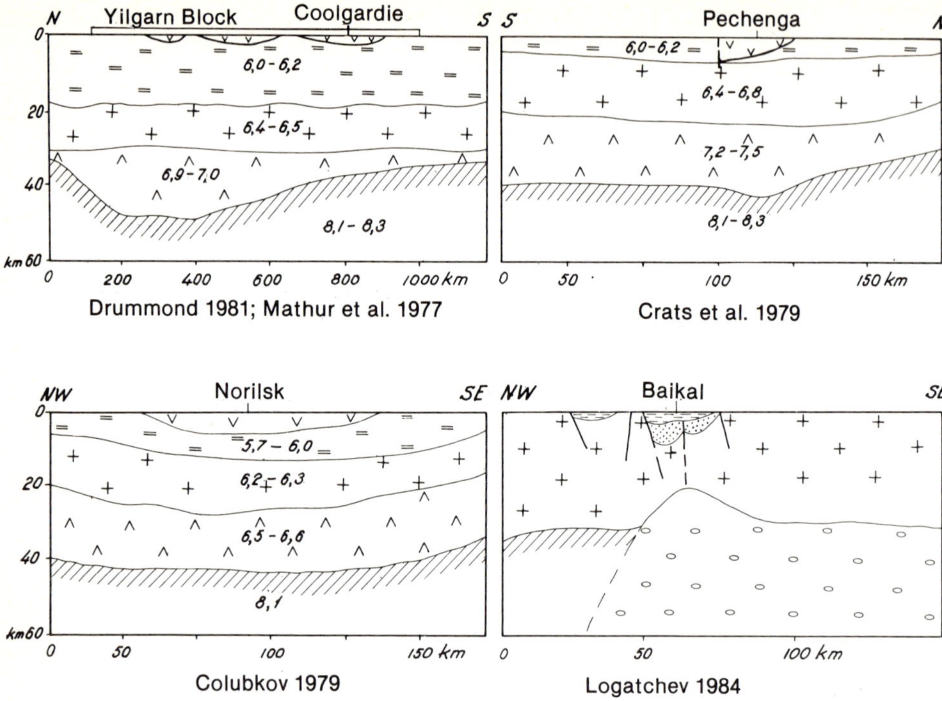

Fig. 4. Seismic sections of nickel-bearing areas

a result of mantle material subtraction and subsequent formation of a depression in the mantle horizon of magma generation. On the other hand, nickel-free structures with a slightly uplifted Moho boundary formed as a result of crustal tension in the period of asthenolithic diapir rise acompanied by sulfide-free decompressional magmatism.

There is a close relation between the morphology and ore contents of magmatic bodies. Thus, intrusive bodies of the Norilsk area are grouped according to the morphology and ore contents into three categories (Fig. 5): barren, sill-shaped bodies; weak ore-bearing, platelike bodies with swells and pinches; ore-bearing, elongated band- and chonolithlike bodies with flat roofs and downwarped bases.

Barren intrusions consist of undifferentiated and slightly differentiated diabase and gabbro-diabase magmatic massifs that are comagmatites of the associated basaltic cover that is composed of sulfide-free mafic magmas (MgO $\leqslant$ 8 mass%) and their magnesian differentiates (MgO 15 – 20 mass%). The intrusions are spread throughout the craton-cover section, giving rise to interlayered subhorizontal bodies in thicknesses of several centimeters to 250 m. The bodies extend for hundreds and thousands of square kilometers. Leukocratic gabbro and taxitic-poikilophytic gabbro-dolerite sill-shaped intrusive bodies also are barren. These bodies making up the frontal parts of ore-bearing intrusions are differentiates of magmatic columns that lost their magnesian and sulfide components

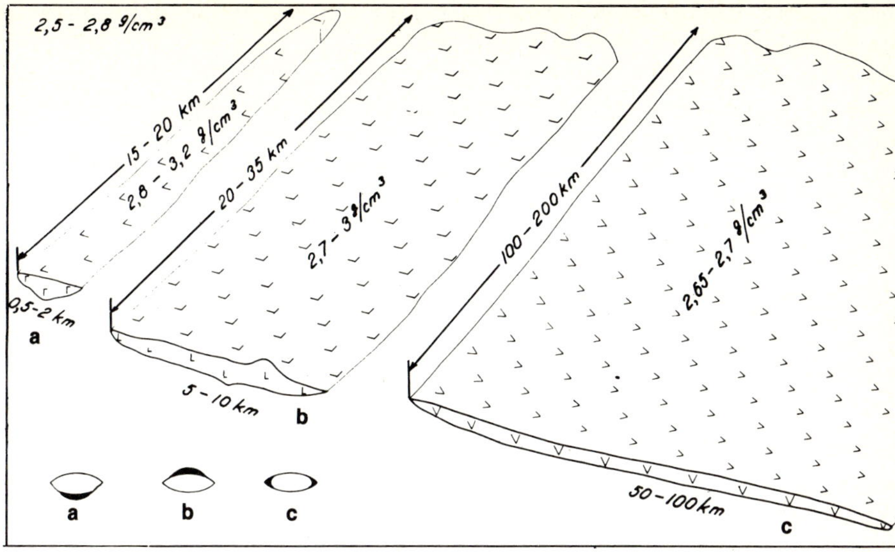

Fig. 5a – c. Morphological varieties of magmatic bodies in the Norilsk area: **a** ore-bearing; **b** weak ore-bearing; **c** barren. *Black patches* at the base show areas of gravity tension in magmatic chambers: **a** with magma density exceeding the density of the environment; **b** with magma density less than that of the environment; **c** with the density identical for magma and environment

in the process of crystallization – gravity differentiation as the magmatic masses rose to the upper crust (Likhachev 1965, Godlevsky and Likhachev 1979, 1983).

Slightly ore-bearing intrusions are of the Lower Talhakn type, being differentiated gabbro-dolerite magmatic massifs localized in sediments, ranging in age from the Permian Tunguss series downward to the Lower Devonian, and, in thickness, from several meters to 150 – 350 m. Gradual transitions between differentiates (from diorite and olivine-bearing gabbro-dolerite at the top to picritic and taxitic gabbro-dolerite at the bottom) are typical. The total higher-magnesia content (MgO weighted mean up to 18 mass%), low-chromium content ($Cr_2O_3 \leqslant 0.01$ mass%), and rather poor disseminated sulfide chalcopyrite-pentlandite-pyrrhotite mineralization (1 – 3 mass%, Ni : Cu = 1 : 1) can be traced throughout the massif cross-section, the lower differentiates being somewhat enriched.

Ore-bearing intrusions of the Norilsk type are strongly differentiated (ranging from olivine-free at the top to picritic and taxitic gabbro-dolerite at the bottom) massifs with copper-nickel sulfides (3 to 10 mass%) concentrated at the bottom of magmatic bodies as well as at exocontacts as disseminated or massive mineralization. They occur at the same horizons as intrusive bodies of two other varieties (barren and weak ore-bearing) displaying echelon structures; magnesium-, chromium-, and sulfide-poor intervals alternate with intervals enriched with these components. The latter make up copper-nickel deposits (Godlevsky and Likhachev 1979, 1983).

The previously mentioned differences in morphology of the intrusive bodies are not controlled by tectonics. This is indicated by the fact that weak ore-bearing and ore-bearing intrusions identical in time and emplacement differ in morphology.

The reasons explaining the link between the morphologies of the intrusive bodies and their ore potential are revealed by analysis of the generation and intrusion of the magmas.

The pattern of magma transport into the Earth's crust and its surface observed in the Norilsk area (from less to more magnesian varieties) suggests fractionated melting of the initial matter and stage-by-stage generation of magmas. This process starts from sulfide-free mafic (MgO $\leqslant$ 8 mass%) magma that gave rise to basaltic sequenes and comagmatic barren intrusions to sulfide-bearing mesomafic varieties (MgO 8 – 15 mass%) which account for weak sulfide-bearing (1% – 3% sulfides) and ore-bearing (3% – 10% sulfides) intrusive bodies and related copper-nickel deposits. The main bulk of the sulfides (over 90%) throughout the existence of magma (from the place of its generation within the mantle to crystallization in the Earth's crust) occurred in the form of liquid with ~4 g cm^{-3} density, thinly dispersed in the oxide-silicate melt with about 2.7 g cm^{-3} density (Godlevsky and Likhachev 1983).

The basic physical factor determining the form of the intruding magmatic masses consisted in the differences of density values that largely depended on the amount of sulfide liquid in the magma.

As is shown by experimental data and direct measurements taken in present-day volcano lavas, the density of sulfide-free mafic magmas varied from 2.65 to 2.70 g cm^{-3}. Weak sulfide-bearing magmas in the oxide-silicate portion possessed a somewhat higher density, and the presence of 1% – 3% sulfide liquid accounted for about a 2.75 – 2.8 g cm^{-3} density increase. Sulfide-bearing mesomafic magma that produced magmatic bodies approximated the density of the oxide-silicate portion of weak sulfide-bearing magma. However, owing to the large volume of 4 g cm^{-3} – density sulfide liquid (3% – 10%), its total density was high: 2.8 – 3.2 g cm^{-3}. The amount if sulfide liquid in the intruding magmas was determined in terms of weighted mean content of observed sulfide matter in the magmatic mass of ore-bearing intrusions and dividing it into liquation (dissolving in silicate melt – ~0.5 mass%) and segregation (surplus, unsolving fractions). The accumulation of up to 10 mass% sulfides during the magma ascent can probably be attributed to the descent of the sulfides, in the form of large drops, from the cooled upper part of the magma column to its deeper and high-temperature levels where sulfides became suspensed because of dispersion effects (Godlevky and Likhachev 1983).

The influence of the density of magmas on the morphology of magmatic bodies is attributed to the effect of gravity.

Mindlin (Roberts 1972) showed that tangential stress in the cyclindrical cavity that occurs below the Earth's surface has minimum, though positive, values at the uppermost point $\phi = \pi/2$, and maximum values at the lowermost point $\phi = 3\pi/2$. The same stress is maintained while the cavity is filled with fluid having a lower density than the environment. On the other hand, the relation is reversed if the density of the fluid filling the cavity is greater than that of the environment.

Since fracturing tends to occur at points $\phi = \pi/2$ and $\phi = 3\pi/2$, the gravity effect is found to limit such ruptures at the point $\phi = \pi/2$ if the magma is lighter than the environment, and at the point $\phi = 3\pi/2$ if the magma is more dense than the host rocks. Similarly, as fracturing related to shearing also tends to occur at the same points, the chamber with a relatively lighter magma will be fractured at the point where $\phi = 3\pi/2$, and the chamber with a heavier magma, at the point where $\phi = \pi/2$.

So the upper portion of the chamber with the relatively dense magma (higher density than the environment) will be compressed, while the lower part will be extended. This latter phenomenon accounts for the increase of the magmatic volume downward and formation of magmatic bodies with flat roofs and downwarped bottoms. On the other hand, in the chamber where density is less than that of the host rocks, the extension of the environment and expansion of the chamber will occur in its upper portion, thus, giving rise to convex bodies. The density of the magma and host rocks being identical, the stress at the roof and at the base of the chamber is balanced. In such cases, at the flanks of the chamber, at points of minimum tangential stress, the negative difference between the density of magma and host rocks and positive hydrostatic pressure are bound to occur, thereby providing lateral increase of the chamber volume. Sill-shaped bodies owe their formation to this phenomenon.

The revealed pattern of relations between the densities of the fluid and of its environment, as well as gravity field affecting the stress and shape of forming chambers, explain fairly well the differences in the morphology of intrusive bodies in the Norilsk area (Fig. 5).

Sulfide-free magmas by their density (2.65 – 2.7 g cm^{-3}) were equivalent in density to their host rocks (2.5 – 2.8 g cm^{-3}), thereby giving rise to environments most in the energy needed to produce sill-shaped bodies. Ore-bearing magmas exceeded the density of the environment (2.8 – 3.2 g cm^{-3}) through effect of the sulfide load, so their extension was determined by the influence of gravity that provided "a channel" in the lower part of the chamber and one-way movement of magmatic masses, including the formation of elongated chonoliths with downwarped bottoms and flat roofs. On the other hand, weak ore-bearing magmas, characterized by intermediate density, accordingly gave rise to magmatic bodies of intermediate morphology.

Roberts (1972) demonstrated that the movement of magma was provoked by minimum tangential stress and positive hydrostatic pressure at the frontal part of the magmatic column. The latter acted as a wedge, the transverse stress of which affected host rocks five times as much as the longitudinal, stress produced by magmatic pressure.

The aforesaid suggests the following conclusions:

1. Sulfide-nickel-bearing formations are associated with continental thermal magmatism that provided for the ascent of heavy sulfide liquid out of the deep-seated mantle to the upper parts of the Earth's crust and formation of copper-nickel deposits.

2. Nickel-bearing structures are characterized by higher thicknesses of the Earth's crust and downwarped Moho boundaries in contrast to nickel-free structures that are generated as a result of the tension in the Earth's crust caused by mantle diapirs and decompressional magmatism.

3. There is a clear relation between the thickness of magmatic complexes associated with continental depressions, and ore potential, viz., in ore-bearing struc-

tures the thickness of volcanogenic units is over 3 km, whereas in barren structures, it is under 2 or 3 km, this being attributed to the variable degree of melting of the initial mantle matter: in barren structures magmatism produces mainly sulfide-free mafic magmas (MgO $\leqslant$ 8 mass%), whereas in ore-bearing structures it produces mafic and sulfide-bearing mesomafic magmas that give rise to copper-nickel deposits.

4. Ore-bearing magmatic bodies in undislocated areas are elongated, band-, chonolith-like bodies with flat roofs and a downwarped bottom, which is attributed to the excess of ore-bearing magma density greater than that of environment, and gravity fields affecting their intrusion and providing one-way movement of magmatic masses.

References

Drummond BJ (1981) Crustal structure of the Precambrian terrains of North-West Australia from seismic refraction data. BMR J Aust Geol Geophys 6:123–135

Godlevsky MN, Likhachev AP (1979) Generation and crystallization of ore-bearing magmas producing copper-nickel deposits. In: Basic parameters of endogenetic ore formation, vol I. Nauka, Novosibirsk, pp 109–118 (in Russian)

Godlesvky MN, Likhachev AP (1981) Formative conditions and evolution of ore-bearing ultrabasic magmas. West. All-Union Miner Soc Iss 6:646–655 (in Russian)

Godlevsky MN, Likhachev AP (1983) Copper-nickel ore formation in the Norilsk area. In: Genetic model for endogenetic ore formations, vol I. Nauka, Novosibirsk, pp 47–54 (in Russian)

Kennedy JR, Ryzhenko BN (1973) The influence of the pressure on eutetics in Fe–FeS systems. Geochimia 9 (in Russian)

Kratz KO et al. (1978) The Earth crust of the eastern Baltic Shield. Nauka, 232 pp (in Russian)

Likhachev AP (1965) The role of leucocratic gabbro in the formation of Norilsk differentiated intrusions. Izv Akad Nauk SSSR Geol Ser 10:75–88 (in Russian)

Likhachev AP (1973) On the nature of magmatic deposits. Sov Geol 5:33–47 (in Russian)

Likhachev AP (1977) Magmatism and nickel potential in North-Central Siberia. Sov Geol 2:30–45 (in Russian)

Likhachev AP (1978a) Formative conditions of ore-bearing and barren mafic and ultramafic magmas. Rep Akad Nauk 238, 2:447–450 (in Russian)

Likhachev AP (1978b) On the genesis of the Sudbury copper-nickel deposits. Sov Geol 6:60–71 (in Russian)

Likhachev AP (1983) Geology and classification of copper-nickel deposits. West. All-Union Miner Soc Iss 1:14–27 (in Russian)

Likhachev AP, Krivtsov AI (1975) Sources and possible ways of ore matter concentration. Sov Geol 5:69–79 (in Russian)

Logatchev NA (1984) The Baikal rift system. Episodes 7, 1:38–42

Naldrett AG (1973) Nickel sulfide deposits. Their classification and genesis, with special emphasis on deposits of volcanic association. Trans Can Inst Min Metall 76:183–201

Naldrett AG, Richardson SW (1967) Effect of water on the melting of pyrrhotite-magnetite assemblages. Carnegie Inst Washington Yearb 66:429–431

Naldrett AG, Turner AR (1977) The geology and petrogenesis of a greenstone belt and related nickel sulfide mineralization of Yakabindie, Western Australia. Precambrian Res 5:43–104

Roberts J (1972) Magma intrusion in loose rocks. In: Mechanism of magma intrusion. Mir, Moscow, pp 230–283

Shamazaki H, Clark LA (1973) Liquids relations in the $FeS-FeO-SiO_2-H_2O$ system and geological implications. Econ Geol 68, 1:79–96

Sharp WE (1969) Melting curves of sphalerite, galena, and pyrrhotite and the decomposition curve of pyrite between 30–65 kilobars. J Geophys Res 74, 6

Wyllie PJ (1971) Experimental limits for melting in the Earth's crust and upper mantle. Geophys Monogr Ser 14:279–301

On the Role of Metamorphism in the Formation of Nickel-Copper Sulfide Deposits in the Kola Peninsula

Yu. N. Yakovlev and A. K. Yakovleva[1]

Abstract

Three stages are apparent in the formation of the nickel-copper sulfide deposits, viz. magmatic, regional metamorphic, and late metamorphic. In the magmatic stage, syngenetic ores are simple and constant in the composition formed. In the regional metamorphic stage, epigenetic ores appear that are distinguished by diversities in mineral composition, typical of the specific tectonic zones formed during such metamorphism. Late metamorphic alteration is confined to local tectonic zones and is characterized by the inconstancy of the mineral assemblages. In the process of metamorphism, the chemical composition of the mineralized rocks experiences less alteration that results in a reduction of the geochemical contrast of their separate varieties. All metamorphic alteration events took place within the primary ore bodies and did not result in the formation of new ore shoots. Thus, the Ni – Cu sulfide deposits of the region can be classed as metamorphosed magmatic deposits rather than metamorphic deposits.

Introduction

Nickel-copper sulfide mineralization of the Kola Peninsula occurs in the massifs of mafites and ultramafites belonging to five magmatic groups (Gorbunov et al. 1982): websterite-gabbro-norite, gabbro-lherzolite-pyroxenite, ultrabasite, peridotite-pyroxenite-gabbro-norite and gabbro-wehrlite. The first two groups (Smirnov 1982) belong to the Kola upper Archean metallogenic stage and the remainder to the Karelian (Lower Proterozoic) stage. Several nickel provinces are distinguished (Fig. 1) and in each, the massifs of one group are well-developed. Thus, ore-bearing intrusions of the Kola stage are concentrated in the Lovnoozero province. The early Karelian is represented by massifs of ultrabasite in the Allarechka province and peridotite-pyroxenite-gabbro-norite in the East-Pechenga (Luostari massif), Monchegorsk (Monchegorsk pluton), and Imandra-Varzuga (Fyodorova and Panskie tundra massifs) provinces. Late Karelian

[1] Institute of Geology, Kola Branch of the USSR Academy of Sciences, Apatity, Murmansk Region, 184200, USSR

Geology and Metallogeny of Copper Deposits
(ed. by G. H. Friedrich et al.)

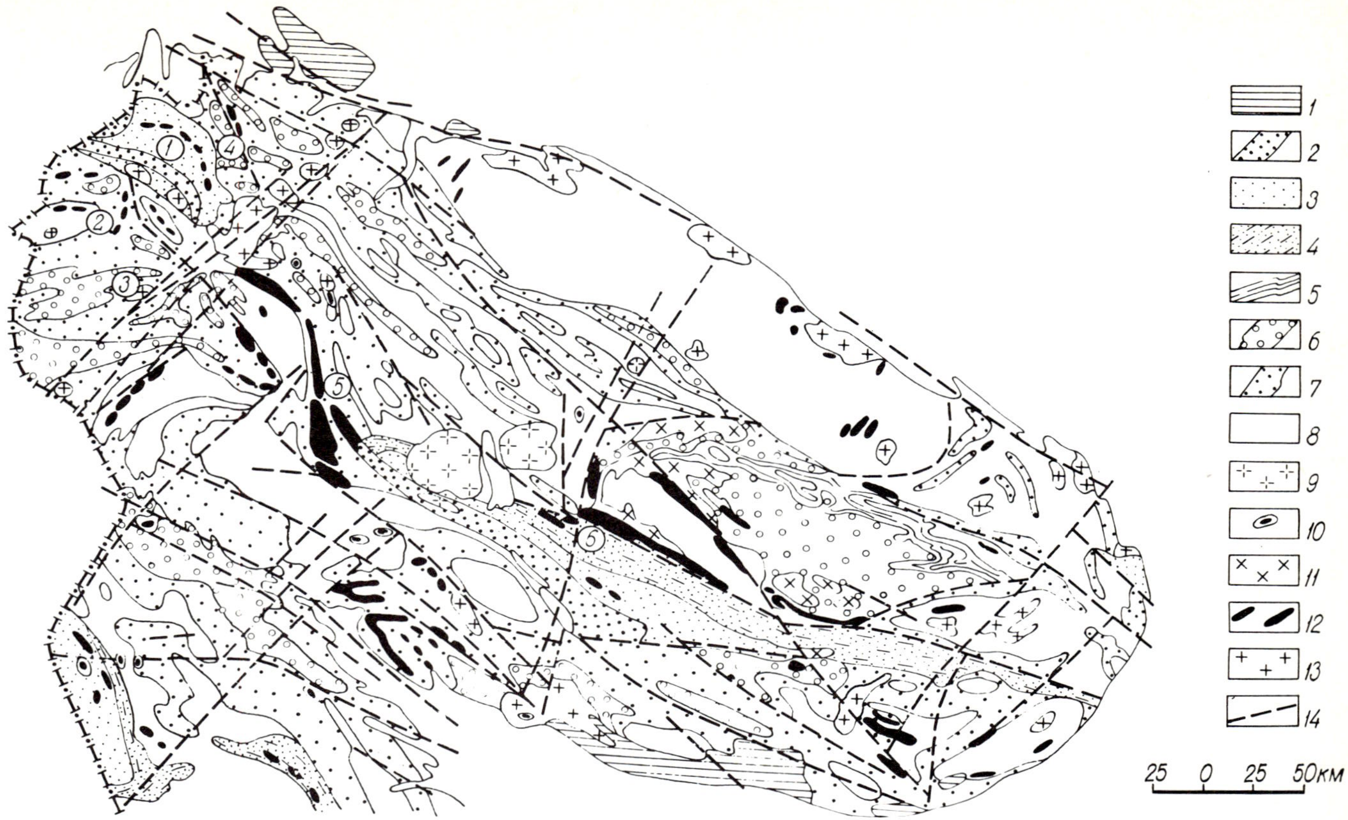

Fig. 1. Geological map of the Kola Peninsula. (After Yakovlev et al. 1981) *1* Rifes terrigenous complex (sandstones, conglomerates, schists); *2 – 4* Karelian volcanogenic-sedimentary complex (*2* South-Pechenga and Tominga series; *3* Pechenga and Varzuga series; *4* Strelna series); *5 – 7* Kola-Belomorian amphibolite-gneiss-crystalline schist complex (*5* Keivy schist series; *6* Lebyazhka, Voronyetundra series and their analogs; *7* Volchelambino, Olenegorsk, Annam series, and their analogs); *8* oldest complex of primary sialic crust (plagiogranites, tonalites, plagiogneisses); *9 – 13* intrusive formations (*9* nepheline alkaline syenites; *10* alkaline gabbroids; *11* alkaline granitoids; *12* mafic and ultramafic rocks; *13* granitoids); *14* most important fractures. *Circled numbers:* nickel areas (*1* Pechenga; *2* Allarechka; *3* Lovnoozero; *4* East-Pechenga; *5* Monchegorsk; *6* Imandra-Varzuga)

examples include intrusions of gabbro-wehrlite in the Pechenga and Imandra-Varzuga provinces.

Detailed geological characteristics of the provinces, the structure of the deposits, the petrology of the mafite-ultramafites, and the ore mineralogy are available in a number of general references (Gorbunov 1968, Gorbunov et al. 1978, Kozlov 1973, Yakovlev et al. 1981). Thus, only brief information on these questions is given in the present paper.

The Pechenga province lies in the NW part of the Pechenga-Varzuga tectonic and metallogenic zone which extends throughout the Kola Peninsula (Fig. 1). The province is made up of volcanogenic-sedimentary rocks of Pechenga strata including four volcanogenic sheets divided by masses of tuffogenic-sedimentary rocks. Nickel-bearing mafic-ultramafic intrusions are concentrated in the fourth tuffogenic-sedimentary mass that is designated as the productive mass and frequently contains pyrite-pyrrhotite mineralization.

The massifs have blanket-, lenticular, and, occasionally, phacolithlike shapes outlining the contours of large folds in the enclosing rocks. They dip SE – SW at 30° – 60°. The massifs extend for 100 – 6000 m along the strike, have a 100 – 1500 m downward dip, and are from 10 – 15 m to 200 – 300 m thick. Major deposits are associated with differentiated massifs in which the following rock types occur from bottom to top: amphibole-chlorite (near contact), serpentinite (after peridotite and olivinite), serpentinized peridotite (wehrlite), pyroxenite, and gabbro. At the western flank of the area, serpentinite and serpentinized peridotite compose up to 70% of the massif section, at the eastern flank gabbro is predominant. The mineralization is confined throughout to the lower parts of the massifs and is represented by low-grade disseminated and high-grade impregnated ores in the serpentinite, by breccia and massive ores in the interstratal tectonic zones extending along the lower contacts of the massifs with the tuffogenic-sedimentary rocks, and by stringer-type impregnated ores in the exocontact phyllite (Fig. 2a, b).

The Allarechka province is made up of Archean amphibolite-gneiss masses. Block-faulted structures and a wide development of domelike structures are common. Numerous small massifs are known in the area. They occur roughly concordantly with the enclosing gneisses and amphibolites, frequently containing disseminated pyrrhotite mineralization. The massifs are located in the marginal parts of the domelike structures and in the interstices between them, frequently in groups of 5 – 10 massifs and in several stages. They are 100 – 2000 m along the strike and 5 – 200 m thick. Most of these were traced for 100 – 200 m and individual massifs for up to 1000 m. The massifs are mainly of blanket, lenticular, and occasionally bandlike shapes. Many massifs are not markedly differentiated: levels of olivinites, peridotites (harzburgites), and pyroxenites are distinguished in them. As a rule, they are cut by the veins of granite pegmatites up to 20 – 25 m thick. Economic mineralization was contained in only several massifs (Allarechka and Vostock deposits, Severnoe and Runnijoki ore showings). Disseminated ores are predominant in the ultramafites of lower and medium parts of the sections; massive and breccia ores are concentrated in the tectonic zones. At the Allarechka deposit, such a zone extends along the western flank and contains mainly massive ores and at the Vostock deposit, the ore zone ex-

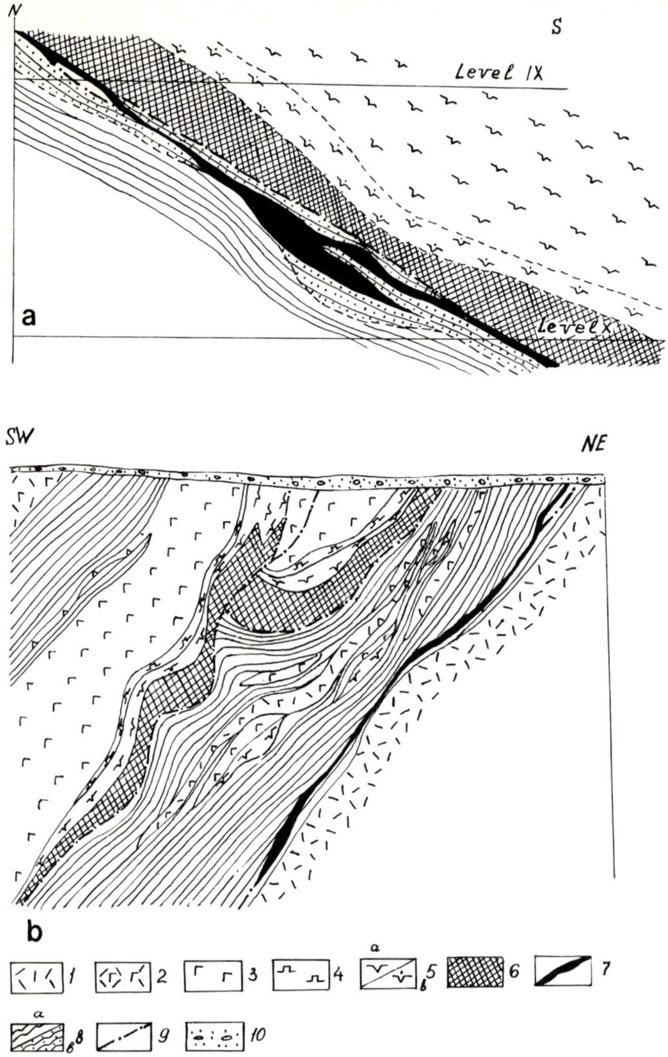

Fig. 2 a, b. Transversal geological sections of the Flangovoye deposit (**a**) and the Eastern ore cluster (**b**) (Gorbunov et al. 1978). *1* Effusive diabase; *2* gabbro-diabase; *3* rocks and ores of differentiated intrusives; *3* gabbro; *4* pyroxenite; *5* serpentinite (*a* ore-free; *b* with low-grade mineralization); *6* high-grade impregnated ore; *7* breccia and massive ores; *8* tuffogenic-sedimentary rocks (*a* ore-free; *b* mineralized); *9* tectonic dislocations; *10* quarternary sediments

tends along the lower contacts of the massifs and contains principally breccia ores (Fig. 3 a, b); stringer-type mineralization is frequently marked in zones of the exo-contacts of the high-grade ores.

The Lovnoozero province lies in the SW part of the belt of Lapland granulites. Two complexes of rocks took part in its formation: (1) granulite (it is predominant in the southern part of the area) and (2) gneiss (it is predominant in

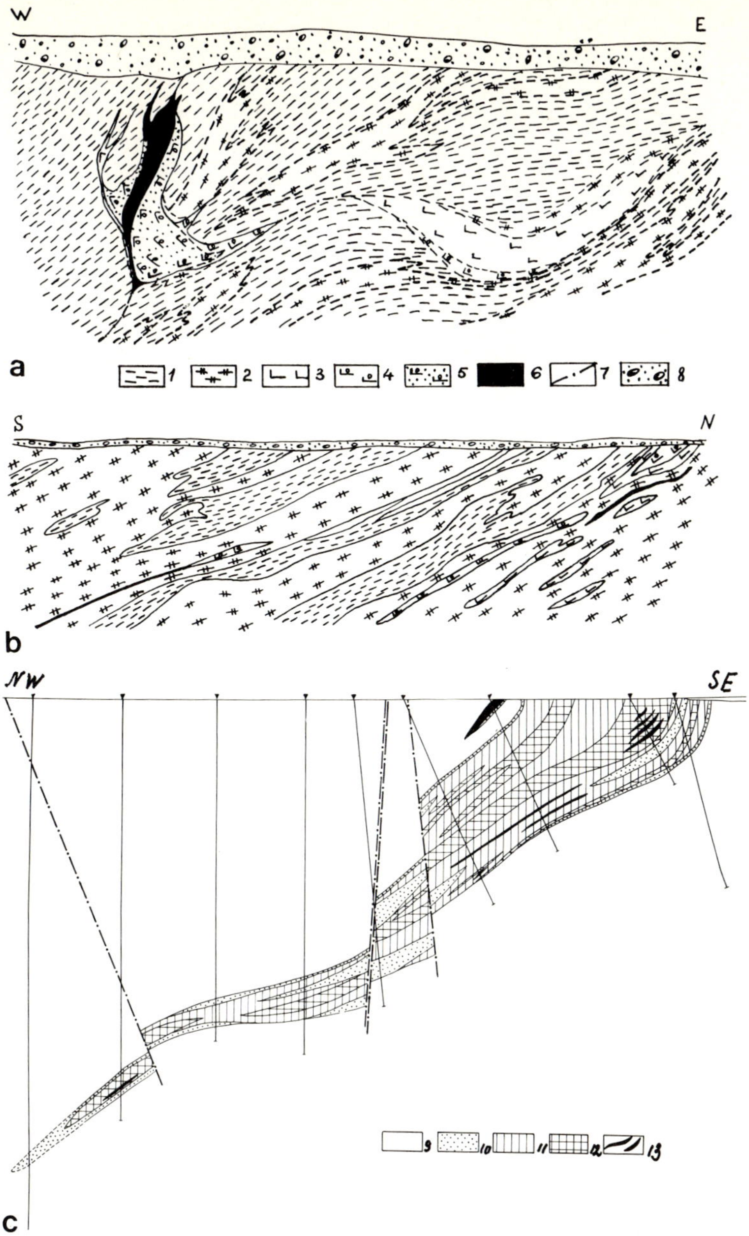

Fig. 3a – c. Transversal geological sections: **a** Allarechka deposit; **b** Vostock deposit; **c** Lovnoozero deposit (Yakovlev and Yakovleva 1974, Zak et al. 1972, Gorbunov et al. 1978, respectively). *1* Gneiss and granite gneiss; *2* feld-spathic amphibolite; *3* ore-free peridotite; *4* peridotite with low-grade mineralization (ore type I); *5* peridotite with high-grade mineralization (ore type II); *6* breccia and massive ores (ore types III and IV); *7* tectonic dislocations; *8* quaternary sediments; *9* hypersthenic plagiogneiss and granulitelike rocks; *10* norite and gabbro-norite with low-grade sulfide mineralization; *11 – 13* mineralized norite and gabbro-norite with nickel and copper content (total): *11* 0.3 – 1.0 wt%; *12* 1.0 – 2.5 wt%; *13* more than 2.5 wt%

the northern part). A great number of small massifs of mafites and about a dozen ultramafites were defined there. All massifs are of lenticular-, cigar-, and band-like shapes. They occur concordantly and rarely discordantly with enclosing rocks. The massifs are 100 – 1000 m (sometimes up to 3000 m) in strike length; and 10 – 100 m (rarely 300 – 600 m) thick. They were traced for 100 – 200 m (seldom for 500 – 600 m) in a downward dip. The massifs of mafites and ultramafites are frequently cut by veins of granite pegmatites with a thickness of 0.5 – 100 m. Mafic rocks are represented by norites and gabbro-norites; ultramafic – by pyroxenites (websterites) and by peridotites (lherzolites). The signs of differentiation are marked both in the mafic massifs and the ultramafic ones.

Sulfide mineralization is mainly associated with mafic massifs (Lovnoozero deposit and a number of small ore showings) and rarely with ultramafic (Sueinlagash ore showing). Disseminated ores develop everywhere. They are irregularly distributed in the mafites (Fig. 3c) and tend to reach the levels of olivine-free rocks close to the flat wall of the ultramafic massifs.

The Monchegorsk province is in the central part of the Kola Peninsula (Fig. 1) at the site of the sharp turn of the Pechenga-Varzuga structural zone. Nickel-copper sulfide mineralization developed in the large Monchegorsk pluton (about 55 km^2 in area) that consists of a peridotite-pyroxenite-gabbro-norite formation. The pluton is interformational and occurs between Archean gneisses and Proterozoic volcanogenic-sedimentary rocks. In plan, the pluton is arched; its sublatitudinal branch includes the Sopcha and Nyud-Poaz massifs, and the submeridional branch includes the Nittis-Kumuzhiya-Travyanaya massifs (NKT). The NKT and the Sopcha massifs are composed mainly of ultramafites forming a layered series, and the Nyud-Poaz massifs are also composed of layered mafites.

Nickel-copper sulfide ores are represented by several morphological types; the major types are disseminated, nestlike impregnated, and veined. Disseminated ores are confined to the levels of the olivine-bearing rocks in the upper parts of the massifs ("beds" of Sopcha and Nyud) or of feldspathic rocks in the lower parts (bottom deposits of Sopcha and NKT massifs). Nestlike impregnated ores are localized in areas of nonuniform structure (the so-called critical level of Nyud). Veins of massive sulfide ores (often with magnetite) developed in the median parts of the NKT and Sopcha massifs; they are spatially associated with diorite-pegmatite.

The knowledge available on the East-Pechenga and Imandra-Varzuga provinces is very slight and is not dealt with, but it is noteworthy that in the East Pechenga province well-defined massifs of mafites-ultramafites are divided into three groups, viz. peridotite-pyroxenite-gabbro-norite (Luostari), ultrabasite (a number of small massifs), and gabbro-wehrlite (Nyasyukki dikes), whereas in the second province the massifs of mafites-ultramafites of two groups were established, viz. peridotite-pyroxenite-gabbro-norite (Fyodorova and Panskie tundras) and gabbro-wehrlite (small massifs in the tuffogenic-sedimentary rocks).

Nickel-bearing massifs of the first three provinces have been metamorphosed together with the enclosing rocks over a wide range of $P-T$ conditions corresponding to greenschist (Pechenga), amphibolite (Allarechka), and granulite (Lovnoozero) facies of regional metamorphism. This corresponds to a range of temperature from 300° – 350 °C to 700° – 715 °C and of pressure from 2 – 3 to 5 – 7 kb. The rocks of large massifs of peridotite-pyroxenite-gabbro-norite formation (Monchegorsk and other provinces) are the least altered, although autometamorphic alteration and other alterations related to contacts, veins, and dislocations are also present. Thus, magmatic, metamorphic, and hydrothermal processes are reflected in the formation of Ni – Cu deposits with the relative importance ranging widely through the different tectonic zones (Fig. 4). The relationship between syngenetic and epigenetic mineralization types also differs in the same manner. The majority of workers correlate the development of the epigenetic ores with the regressive metamorphism or with the transition from the progressive to the regressive substages of metamorphism.

Concepts of the formation of Ni – Cu epigenetic ores from Pechenga, Allarechka, and Lovnoozero provinces in the process of regional metamorphism have developed over a long period. Some workers believe that under metamorphism, primary magmatic sulfides were altered and redeposited in tectonic zones as breccia and massive ores (Kozlov 1960, Likhachov 1974 and other authors). The others think that primary magmatic sulfides were absent or were very small in amount and the major part of the ores formed in the process of metamorphism at the expense of Ni and Fe liberation from primary silicates with the sulfur and copper supplied from the enclosing rocks (Predovsky et al. 1968, Zak et al. 1972). Some researchers, basing their ideas mainly on experimental data, connect the formation of epigenetic ores with sulfurization processes under hydrothermal conditions (Arutyunyan et al. 1978).

On the whole, the majority of supporters of the metamorphogenetic formation of the Ni – Cu ores believe that their positional connection with mafites-ultramafites is paragenetic. On the one hand, these authors believe that these rocks were the metal source and that sulfur came in from the enclosing rocks or was supplied by fluids from deep sources. On the other hand, these rocks provided a favorable environment for ore deposition.

Some of the distinctive features of the metamorphic ores are as follows: higher Cu and S content, variability of mineral composition, and poor correlation between ore components. We shall consider several of these features, designating the following ore types: low-grade disseminated and high-grade impregnations in mafites and ultramafites (I and II); breccia and massive ores in tectonic zones (III and IV); stringer-type ores impregnated in exo-contact metasomatic and enclosing rocks (V and VI). Ore types I and II are always predominant; ore types III and IV are significantly less abundant, and ore types V and VI are rare. The low-grade disseminated ores and parts of the high-grade impregnated and massive ores are purely syngenetic, whereas the remainder are epigenetic (Gorbunov et al. 1973 and others). Besides, data on mineral and chemical ore composition are represented by an average for a specific ore type or deposit and are calculated on the basis of a great number of ordinary determinations, i.e., quantitative mineralogical calculations of thin and polished sections (tens of samples) and the results of sampling (many tens and low hundreds of ordinary samples).

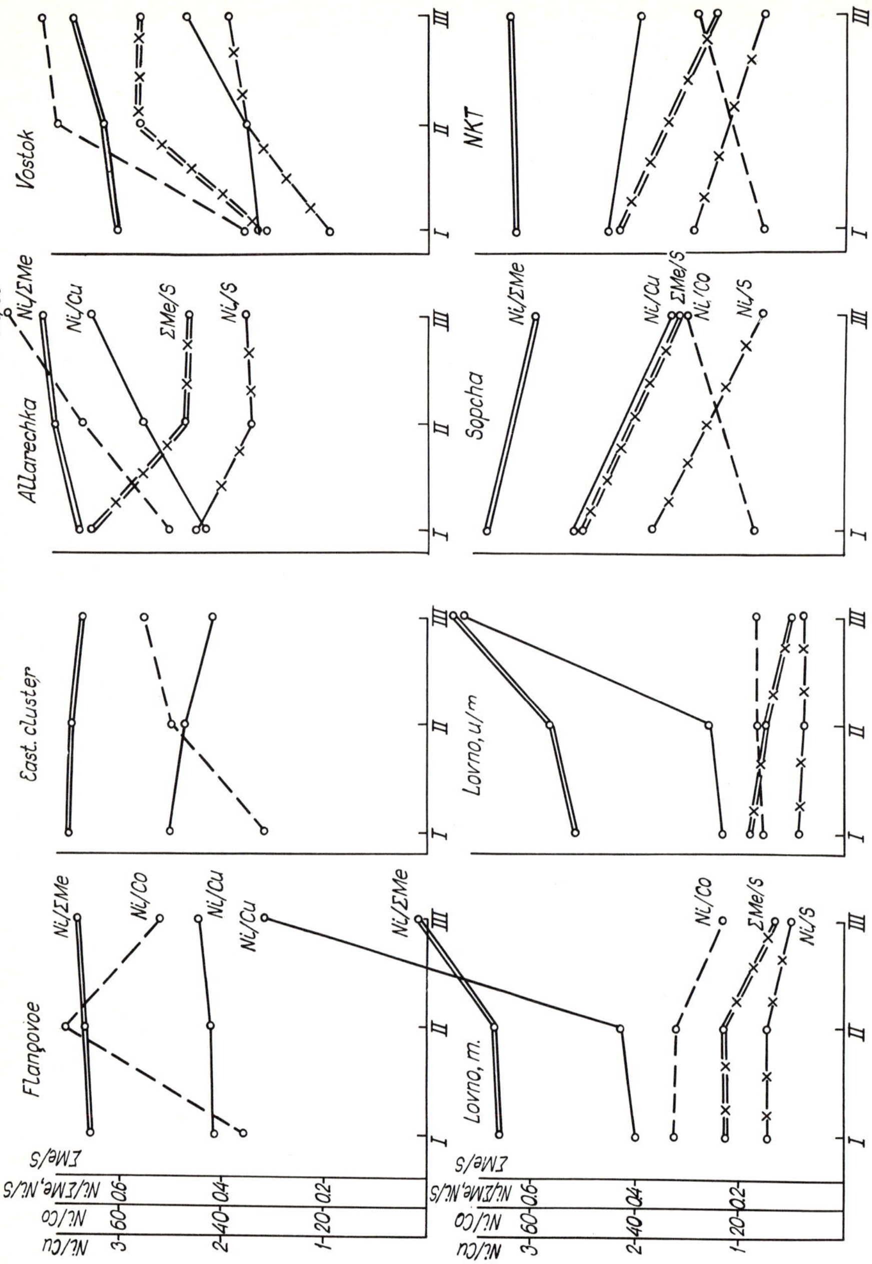

Fig. 4. The relationship between major components in Ni – Cu ores (*I* low-grade disseminated; *II* high-grade impregnated; *III* breccia and massive)

Mineral Composition of the Ores

The paragenesis of the mineral assemblages during deposit formation shows a common sequence (trend) everywhere. Qualitatively each assemblage is present in each area, but there are conspicuous differences in the importance of each assemblage in specific areas. Three stages are defined, viz. magmatic, regional metamorphic, and late metamorphic (hydrothermal), with each having its substages (Yakovlev and Yakovleva 1974, Yakovlev et al. 1981).

The magmatic stage involves: (1) the crystallization of primary silicates and oxides (olivine, pyroxenes, plagioclase, titanomagnetite, chrome-spinel – early magmatic substages); (2) early amphiboles and micas, monosulfides of syngenetic ores (troilite, hexagonal pyrrhotite, pentlandite, chalcopyrite, cubanite – late magmatic substages); and (3) then a partial transformation of both (serpentine, talc, early magnetite – autometamorphic substage).

The regional metamorphic stage is characterized by an intensive replacement of the primary silicates by serpentine, Ca-amphibole, Fe–Mg micas and chlorites, and by the appearance of late generations of major sulfides and numerous subordinate ore minerals (regional metamorphic substages proper). Subsequently, Fe–Mg amphiboles (anthophyllite, cummingtonite), biotite, talc, scattered disulfides, and rare ore minerals (substage of dislocation metamorphism) developed at the contacts with enclosing rocks, in veins of granitic pegmatites near mineralized fractures, and in zones of foliation.

The assemblages of the late metamorphic stage are confined to tectonic breccia zones; these are carbonates, late serpentines, chlorites, and talc, secondary sulfides, rare sulfates, hydroxides, and oxides. They are characterized by intensive alteration of all rocks and ores in specific localities.

This scheme is typical of the Kola Peninsula as a whole, but in specific areas the mineral assemblages of separate metamorphic events are noticeably different (Table 1). All metamorphic transformations are more pronounced in the ultramafites than in mafites.

The extent of metamorphic alteration decreases from Pechenga through Allarechka to Lovnoozero. Alteration affects the entire bulk of the massifs, but neocrystallization is generally restricted, affecting up to 10–15 volume % of the ultramafites and 5–7 volume % of the mafites. Mineralized rocks preserve both primary structures and textures and the main features of the mineral composition, which are characterized by constant assemblages and properties of the main minerals determined by their parent intrusives and the total chemical composition of the ore.

With increasing metamorphic grade, the following change in main assemblages takes place: actinolite-chlorite-serpentine is typical of the Pechenga area, phlogopite-magnesio-hornblende is typical of the Allarechka area, and biotite-hornblende (pargasitic, tschermakitic) of the Lovnoozero area.

The metamorphic transformations affect the massifs, as a whole, both ore-free and ore-bearing, but these are more intensively developed (especially amphibolization) in their endo-contact zones (up to 1–2 m thick). All stages of successive transformation of primary sulfide disseminated ore up to the formation of "negative" (Gorbunov 1968) and subgraphic textures are markedly developed in

Table 1. The sequence of metamorphogenic mineral formation in ore impregnations, NW Kola Peninsula

Areas	Major silicates in the ultramafites	Major silicates in the mafites
Autometamorphism		
All areas	Serpentine, talc, hastingsitic hornblende	Talc, hastingsitic, hornblende, serpentine
Regional metamorphism		
Regional-metamorphic substage		
Pechenga	Serpentine, chlorite, actinolite	Chlorite, actinolite
Allarechka	Magnesiohornblende and actinolitic hornblende, phlogopite, chlorite	–
Lovnoozero	Pargasitic and edenitic hornblende, phlogopite	Tschermakitic hornblende, biotite
Substage of dislocational metamorphism		
All areas	Anthophyllite, cummingtonite, phlogopite, actinolite, talc	
Late metamorphism		
All areas	Carbonates, chlorites, talc, serpentines, micas, actinolite	

the ore-bearing massifs. As a result, in many ultramafic intrusives, the magmatic structures and textures are virtually replaced by metamorphic ones and by recrystallization and partial redeposition of the syngenetic sulfides. However, there are relics of the primary mineralization as "shadow textures", droplike and emulsion inclusions of early ore minerals even in the highly altered rocks. As a rule, sulfide redistribution is virtually confined to the volume of the original ore bodies and, on the whole, the mineralization grade of individual sites in the massifs does not depend on the intensity of metamorphism of the rocks composing them.

The total change of mineral composition during metamorphism is considered, taking as an example the impregnated ores from the northwestern part of the Kola Peninsula (Table 2). These ores are the most abundant of all deposits and are most affected by metamorphic processes. Only the most abundant primary and secondary minerals that define the composition of the impregnated ores are considered. For comparison, Table 2 shows the composition of the least altered impregnated ores from the Monchegorsk pluton (Sopcha and NKT deposits). The mineral compositions of the ores are compared in accordance with increasing $P-T$ conditions of metamorphism, that is from the Pechenga area to Lovnoozero. These data indicate the following features.

The number of individual primary silicates in the ores defined by each group of intrusives differs widely and irregularly, but their relative volume increases considerably with increasing metamorphic grade. The number of primary ore minerals (oxides, sulfides) and their relative abundance ranges only slightly and irregularly. Note that Fe and Ti oxides are equally abundant in all areas, but Ti oxides are predominant in the Lovnoozero area. The Allarechka ores are dis-

Table 2. Mineral composition of the disseminated and impregnated ores[a]

Minerals	Number of mineral species						Mineral content (vol%)					
	P	*A*	*L, u*	*L, m*	*S*	*NKT*	*P*	*A*	*L, u*	*L, m*	*S*	*NKT*
Primary silicates	6	2	4	3	4	4	12	20	55	65	95	95
Primary oxides	4	4	4	4	4	4	2	1	2	2	1	1
Early sulfides	5	7	6	6	5	5	4	7	5	5	3	3
Total	15	13	14	13	13	13	18	28	62	72	99	99
Secondary silicates	20	17	10	8	9	9	70	60	30	20	1	1
Secondary oxides	4	3	3	3	2	2	6	4	1	1	1	1
Late and secondary sulfides	10	10	6	6	6	6	6	8	5	7	1	1
Total	34	30	19	17	17	17	82	72	36	28	1	1
Sum total	49	43	33	30	30	30						

[a] The areas are: *P* Pechenga; *A* Allarechka; *L, u* Lovnoozero – ultramafites; *L, m* Lovnoozero – mafites; *S* Sopcha, peridotite and pyroxenite layers; *NKT* Nittis-Kumuzhia-Travyanaya massif, bentic deposit.

tinguished by the volume of primary ore minerals, since they contain less oxides and more sulfides. On the whole, because of the insignificant role of primary sulfides, the total volume % of primary minerals increases regularly from Pechenga to Lovnoozero which latter ores are similar to the ores of the Monchegorsk pluton.

The number of secondary silicates in the ores decreases systematically and their amount decreases more abruptly from low- to high-metamorphic grade. Principal changes involve layered hydrated silicates (serpentines, chlorites, talc), whereas the volume of amphiboles and, to a lesser extent, micas, increases slightly. The number of secondary oxides is virtually constant, whereas their volume essentially decreases mainly at the expense of secondary magnetite with increasing metamorphic grade. The number of species of late and secondary sulfides gradually decreases, and their content differs irregularly with increasing metamorphic grade.

A gradual reduction occurs in the total number of mineral species with increasing metamorphic grade up to the level of practically nonmetamorphosed ore of the Monchegorsk pluton. The important changes are due to variations in the number and volume of silicates and, in the first place, to the intensity of development of secondary hydrated minerals, viz. serpentines, amphiboles, micas, talc. The ratio of primary to secondary silicates defines the degree of metamorphic ore transformation, whereas the number of mineral species of sulfides and oxides and their volume differ insignificantly and, on the whole, irregularly.

Thus, maximum changes in mineral composition are typical of ore metamorphosed at the greenschist facies and minimum changes take place at the granulite facies.

Chemical Composition of the Ores

The chemical composition of the ores (Table 3) is based on the concentration of major economic elements (Kozlov 1973, Yakovlev and Yakovleva 1974, Yakovlev et al. 1980). Massive and breccia ores are grouped together for all deposits, but it should be noted that breccia ores are dominant (do not constitute less than 80% by volume) in the Pechenga and Lovnoozero areas as well as at the Vostok deposit, but massive ores are predominant at Allarechka and in the Monchegorsk pluton. There are no data on the sulfur contents of the Pechenga ores.

These data indicate the following features. The Ni and Cu contents of monotype-impregnated ores (I and II) from the Pechenga, Allarechka, and Lovnoozero areas differ dramatically (2.3 – 4.4 times), whereas the Ni/Cu ratio has a narrower range (2.1 – 2.2 times). In contrast, the Co content is more constant, but the Ni/Co ratio is quite varied (3.3 – 4.5 times) because of the differences in Ni content. A successive increase in the Ni/Cu ratio from low-grade disseminated ores to high-grade impregnated and further to breccia (massive) ores is apparent at the majority of deposits. The exceptions are the Eastern ore cluster of Pechenga and the deposits of the Monchegorsk pluton in which this parameter

Table 3. The chemical composition of the ores

Ore types	Content (wt%)				Ratio				
	Ni	Cu	Co	S	Ni/Cu	Ni/Co	Ni/ΣMe	Ni/S	ΣMe/S
The Eastern ore cluster (the Pechenga area)									
I	0.51	0.20	0.016		2.55	31.9	0.70		
II	1.78	0.75	0.036		2.37	49.4	0.69		
III	1.96	0.93	0.035		2.11	56.0	0.67		
The Flangovoye deposit (the Pechenga area)									
I	0.65	0.31	0.018		2.10	36.1	0.66		
II	2.39	1.13	0.034		2.11	70.3	0.67		
III	2.25	1.01	0.043		2.23	52.3	0.68		
The Allarechka deposit									
I	0.67	0.30	0.013	1.48	2.23	51.5	0.68	0.45	0.66
II	4.80	1.72	0.043	13.85	2.79	67.6	0.73	0.35	0.48
III	13.17	3.95	0.16	36.95	3.33	82.3	0.76	0.36	0.47
The Vostock deposit									
I	0.83	0.50	0.023	4.13	1.66	36.1	0.61	0.20	0.33
II	3.13	1.75	0.043	8.68	1.79	72.8	0.64	0.36	0.57
III	7.37	3.10	0.097	18.60	2.38	76.0	0.70	0.40	0.57
The Lovnoozero deposit (mafites)									
I	0.62	0.30	0.019	4.06	2.04	32.4	0.66	0.15	0.23
II	1.71	0.79	0.054	11.04	2.15	31.6	0.67	0.15	0.23
III	2.68	0.48	0.114	25.86	5.58	23.5	0.82	0.10	0.13
The Sueinlagash ore showing (ultramafites)									
I	0.33	0.28	0.021	3.62	1.18	15.7	0.52	0.09	0.18
II	1.09	0.82	0.067	13.53	1.33	16.3	0.55	0.08	0.15
III	1.58	0.43	0.093	20.50	3.67	17.0	0.75	0.08	0.10
The Sopcha deposit									
I	0.53	0.21	0.028	1.43	2.52	18.9	0.69	0.37	0.54
III	3.19	2.75	0.127	19.79	1.59	31.0	0.60	0.16	0.31
The NKT deposit									
I	0.32	0.14	0.020	1.10	2.28	16.0	0.64	0.29	0.44
III	5.1	2.6	0.180	31.92	1.96	28.3	0.65	0.16	0.25

diminishes. It shows a maximum range in the ores from the Lovnoozero area (2.7 – 3.1 times), less in the Allarechka ores (1.4 – 1.5 times), slightly less in the ores from the Monchegorsk pluton (1.2 – 1.6 times), and least in the Pechenga ores (1.1 – 1.2 times). The Ni/Co ratio generally differs similarly, that is, it increases toward the high-grade ores, except for the Lovnoozero area and the Flangovoye deposit; it decreases in the same direction in the ore of the former and ranges irregularly in the ore of the latter.

The Ni/S and ΣMe/S ratios are important geochemical characteristics. At all deposits, except for Vostock, the Ni/S value decreases regularly from low-grade disseminated ores to breccia ones. Maximum differences in the ratios are characteristic of the ores of the Monchegorsk pluton and the Allarechka area (1.8 – 2.3 and 1.3 – 2.0 times, respectively) and minimum differences of the ores of Lovnoozero (1.1 – 1.5 times). The ΣMe/S parameter ranges similarly. However, in the ores of the Monchegorsk pluton and the Allarechka area, it lies within a more narrow range than the Ni/S ratios; a wider range in the ratio occurs in the Lovnoozero area. Based on lump ore (about 400 lumps) analyses performed at the Institute of Geology, both parameters gradually diminish (1.6 – 1.7 times) toward the breccia and massive ores at Pechenga (Yakovlev et al. 1980). The absolute value of both parameters for ores from all the areas fluctuates over a wide range: from 0.08 to 0.45 for Ni/S and from 0.10 to 0.66 for ΣMe/S. The sulfide phase of the Allarechka ore is the richest in Ni, one of the Pechenga and Monchegorsk ores is somewhat poorer, and one of the Lovnoozero ores is the poorest, especially in the ultramafites.

The parameter characterizing the Ni fraction of total metal (Ni/ΣMe) is notable for its stability. Its value ranges from 0.61 to 0.70 in low-grade disseminated ores of all the areas, and from 0.64 to 0.73 in high-grade impregnated ores. A similar range is shown in the ores of the Monchegorsk pluton. In all studied ores its range is 0.60 – 0.82, but the maximum value only characterizes localized breccia ore from the Lovnoozero deposit. The minimum value of the parameter (0.52) is established at the Sueinlagash ore showing in the Lovnoozero area. Thus, the Ni fraction of the total metals is persistent everywhere independent of the grade of metamorphism.

Judging from changes in all geochemical characteristics (component contents and ratios), the Allarechka ore is the most unusual of the impregnated ores. If consideration is given to all ores (I – IV types), the Allarechka and Monchegorsk ores show the greatest contrast, the Lovnoozero ores are close to these, and the Pechenga ores prove to be the least changeable. Thus, the most intensively metamorphosed ores (the Allarechka and Lovnoozero areas) and the virtually unmetamorphosed ores (the Monchegorsk pluton) in mafites-ultramafites differ is roughly similar manner.

Discussion

Comparisons of mineral and chemical compositions of disseminated and impregnated Ni – Cu ores from the various areas of the Kola Peninsula indicate that the degree of metamorphic transformation of the nickel-bearing mafites and ultramafites is determined by (1) their original composition; (2) their position within specific tectonic zones; and (3) the structural position of the specific intrusives. As a rule, the ultramafites are more altered than the mafites within one zone, and smaller massifs are more altered than larger ones of similar composition.

The greatest ranges in mineral composition are typical of disseminated and impregnated ores of the Pechenga area (greenschist facies) and the least differences are characteristic of the ores from the Lovnoozero area (granulite facies). The chemical composition of the ores is distinguished by a limited range in most areas except for great changes in the Allarechka area (amphibolite facies). A comparison of geochemical characteristics of syngenetic (low-grade disseminated) ores and epigenetic (breccia) ones indicates a similar trend of variation in all areas. The Ni/Cu and Ni/Co ratios are greater and the Ni/S and ΣMe/S ratios are lower in epigenetic ores. These data, as well as the constant Ni/ΣMe parameter, provide evidence on the persistence of major chemical composition independent of metamorphic grade developed in the original nickel-bearing rocks.

The ore grade of the mafite-ultramafite massifs does not depend on the presence or absence of pyrite-pyrrhotite mineralization in the enclosing series. Ore-bearing and ore-free massifs occur in the enclosing or surrounding rocks with the pyrite-pyrrhotite mineralization occurring in all areas of the northwestern part of the Kola Peninsula. All geochemical parameters of the pyrite-pyrrhotite mineralization are 5 – 10 times less than these of the Ni – Cu mineralization. The evidence presented above does not support a metamorphic hypothesis for the genesis of Ni – Cu sulfide ores, but points to the key role of magmatic processes in their formation. Metamorphism results in significant changes in the composition of primary ores, but is probably insignificant in ore generation.

Major Conclusions

1. The most intensive changes in mineral compositions of disseminated ores occur under low-temperature conditions of regional metamorphism, whereas minimal changes occur under high-temperature conditions. At greenschist facies, deviations from the original magmatic conditions were at a maximum, whereas at granulite facies they were minimal.
2. The more the mineral composition of an ore changes, the less its geochemical characteristics differ from others and vice versa. This fact provides evidence on the primary nature of the contrast in chemical composition of the various ore types and on the gradual decreasing of this contrast by the process of metamorphism.
3. All metamorphic processes involve changes in preexisting ore and are not ore-generating. Thus, the Ni – Cu deposits of the Kola Peninsula should be classed as metamorphosed magmatic deposits.

References

Arutyunyan LA, Petrenko GV, Mityunin YuK (1978) On sulfurization of olivinites under hydrothermal conditions and Ni mobilization. In: Depositional conditions of Ni – Cu ores (natural parageneses and experiment). Kola Branch USSR Acad Sci, Apatity, pp 73 – 85 (in Russian)

Balabonin NL, Yakovlev YuN, Astafyeva VV, Vorobets AA (1974) On geochemical specialization of nickel-copper and kies mineralizations in the Allarechka area. In: Aspects of geology and metallogeny of the Kola Peninsula, vol V, pt 2. Kola Branch USSR Acad Sci, Apatity, pp 185 – 194 (in Russian)

Gorbunov GI (1968) Geology and genesis of Ni – Cu sulfide deposits. Nedra, Moscow, 352 pp (in Russian)

Gorbunov GI, Yakovlev YuN, Astafyev YuA et al. (1973) Atlas of structures and textures of Ni – Cu sulfide ores in the Kola Peninsula. Nauka, Leningrad, 283 pp (in Russian)

Gorbunov GI, Astafyev YuA, Goncharov YuV et al. (1978) Structures of Ni – Cu ore fields and deposits of the Kola Peninsula. Nauka, Leningrad, 160 pp (in Russian)

Gorbunov GI, Vinogradov LA, Makievsky SI, Yakovlev YuN (1982) Basite-ultrabasite magmatism of the Kola Peninsula and its metallogeny. In: Geology, petrology and correlation of crystalline complexes in the European part of the USSR. Proc 3rd Region Petrogr Meet, Dnepropetrovsk 1979. Nedra, Leningrad, pp 82 – 92 (in Russian)

Kozlov EK (1960) Regularities of Ni – Cu sulfide deposit emplacement in the Kola Peninsula. In: Metallogeny of precambrian shields and ancient mobile belts. UkrSSR Acad Sci, Kiev, pp 142 – 149 (in Russian)

Kozlov EK (1973) Natural series of nickel-bearing intrusive rocks and their metallogeny in the Kola Peninsula. Nauka, Leningrad, 288 pp (in Russian)

Likhachov AI (1974) On Genesis of Ni – Cu deposits of the Pechenga and Allarechka Areas. Sov Geol 7:69 – 84 (in Russian)

Predovsky AA, Zhangurov AA, Ilin YuI, Akhmedov AM (1968) New data on regularities of emplacement and conditions of formation of Pechenga Ni – Cu sulfide deposits. In: Geological structure and orebearing in the Kola Peninsula. Kola Branch USSR Acad Sci, Apatity, pp 61 – 67 (in Russian)

Smirnov VI (1982) Endogenetic ore formation in geological history. Geol Ore Depos 4:3 – 20 (in Russian)

Yakovlev YuN, Yakovleva AK (1974) Mineralogy and geochemistry of metamorphosed Ni – Cu ores. Nauka, Leningrad, 330 pp (in Russian)

Yakovlev YuN, Yakovleva AK, Neradovsky YuN et al. (1980) Principal regularities of changes in composition of metamorphosed Ni – Cu disseminated ores: In: Mineral complexes and minerals of the Kola Peninsula. Kola Branch USSR Acad Sci, Apatity, pp 55 – 63 (in Russian)

Yakovlev YuN, Yakovleva AK, Neradovsky YuN et al. (1981) Mineralogy of Ni – Cu sulfide deposits in Kola Peninsula. Nauka, Leningrad, 352 pp (in Russian)

Zak SI, Kochnev-Pervukhov VI, Proskuryakov VV (1972) Ultrabasic rocks of the Allarechka Area, their metamorphism and mineralization. Kareliya, Petrozavodsk, 129 pp (in Russian)

The Gabbro-Wehrlite Association in the Eastern Part of the Baltic Shield

E. HANSKI[1]

Abstract

Proterozoic mafic and ultramafic intrusive rocks are widespread in the Archaean Kuhmo greenstone belt, including three minor Ni – Cu deposits. Similar intrusions have also been found in the Koli area cutting Jatulian metasediments and the underlying granitoid basement. They are incorporated in the same magmatic association "the gabbro-wehrlite association" as the Ni-bearing intrusions of Pechenga. The cumulate sequences in the layered bodies include olivine, olivine-clinopyroxene, clinopyroxene, clinopyroxene-magnetite, and plagioclase-clinopyroxene-magnetite cumulates. Ti-rich amphibole and mica occur abundantly as postcumulus phases in ultramafic rocks. The chemical compositions of the minerals and rocks differ from those of the Pechenga intrusions, reflecting differences in the parent magmas and their fractionation stages. The parent magmas in eastern Finland and Pechenga are similar to picrites of the Suisaarian and Pechenga types, respectively.

Introduction

Large numbers of both intrusive and extrusive ultramafic rocks have been discovered in the course of the long period over which the Kuhmo Archaean greenstone belt has been studied. The close spatial relationship between these and the highly magnesian character of the ultramafic cumulates have led some authors to take for granted the genetic relationship between the intrusions and the komatiitic volcanic rocks (Auvray et al. 1982). In 1981, however, a similarity was detected between the ultramafic and mafic intrusions and those occurring in the Koli area, where they are known to cut Proterozoic Jatulian metasediments (Hanski 1982). Age determinations made by the Geological Survey have later confirmed the Proterozoic age (2.1 – 2.2 Ga) of the intrusions in Kuhmo (Hyppönen 1983).

Further study revealed that the intrusions in Kuhmo and Koli have many features in common with the host rocks of the Ni – Cu deposits of Pechenga in

[1] Department of Geology, University of Oulu, Oulu, Finland.
Present address: Geological Survey of Finland, PL 77, 96101 Rovaniemi, Finland

Geology and Metallogeny of Copper Deposits
(ed. by G. H. Friedrich et al.)

the Kola Peninsula, USSR. Russian geologists have used "formational analysis" to assign the rocks of Pechenga to the gabbro-wehrlite formation (Dagelayskaya 1979). Since the notion of a magmatic formation in the sense defined by the Russian geologists (Masaytis et al. 1979) is not used in the English terminology, the term is replaced here by "association", as is done in Suslova et al. (1981).

This work concentrates on the major geological, mineralogical and geochemical features of the gabbro-wehrlite association in Kuhmo and Koli, comparing the intrusions with the rocks of Pechenga in the light of data collected from the literature and whole rock and mineral analyses carried out on some museum samples.

Regional Geology

The Kuhmo greenstone belt in eastern Finland forms a narrow belt some 200 km in length running from S to N (Fig. 1). The supracrustal rocks include mainly tholeiitic, komatiitic and acid metavolcanics, quartzites and mica schists. The general geology and petrology of the belt and surrounding granitoids are dealt with in many works by Finnish and French geologists (e.g., Taipale 1983, Blais et al. 1978).

The rocks of the gabbro-wehrlite association (henceforth GWA) are concentrated in the greenstone belt, being virtually absent from the surroundings. They form an almost uninterrupted chain of ultramafic rocks, over 100 km in length, in the middle and northern part of the belt. The best-known examples are the Näätäniemi serpentinite and the dykes of Petäjäniemi. Piquet (1982) noted a slight difference in metamorphic grade between ultramafic metacumulates and Archaean metavolcanics, but a more pronounced difference is seen in the deformation grade, for the metacumulates of the GWA have in most cases not suffered any pervasive deformation.

The structure and stratigraphy of the Koli area have been described by Piirainen et al. (1974). Two main stratigraphic units can be distinguished: the Archaean basement complex and the overlying Proterozoic sedimentary formations. The latter rocks are divided into the Jatulian quartzites and the overlying Kalevian graywacke schists.

Both the basement and the Jatulian quartzites are cut by numerous basic dykes and sills, but nowhere are the Kalevian metasediments found to be cut in this way. A differentiated sill at Savilahti, near the shore of Lake Pielinen, has been studied in greater detail. This occurs within the Archaean granitoids near the overlying unconformity and follows the dip and strike of the Jatulian metasediments.

General Features of the Gabbro-Wehrlite Association

The intrusions of the GWA have two principal modes of occurrence. They occur as layered sills and minor intrusions and dykes affected by flowage differentia-

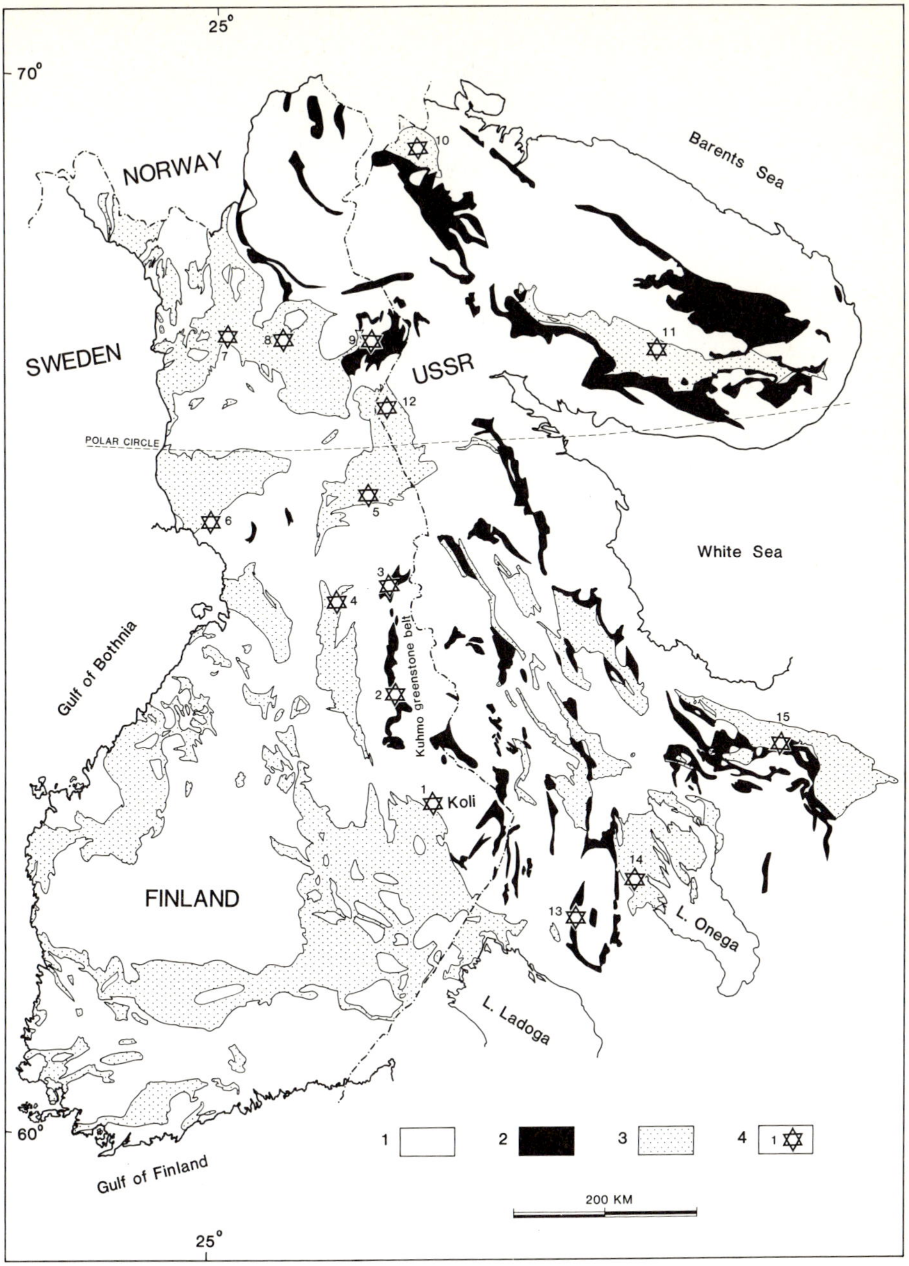

Fig. 1. The occurrences of the gabbro-wehrlite association in the Baltic Shield. *1* Archaean and Proterozoic granitoids; *2* Archaean greenstone belts; *3* Proterozoic schists; *4* Gabbro-wehrlite association. Occurrences of the GWA: *1* Koli; *2* Kuhmo; *3* Suomussalmi; *4* Puolanka; *5* Kuusamo; *6* Kemi; *7* Kittilä; *8* Sodankylä; *9* Savukoski; *10* Pechenga; *11* Imandra-Varzuga; *12* Kuolajärvi; *13* Hautavaara; *14* Onega; *15* Vetrenyi Poyas. Base map simplified mainly after Simonen (1980) and Bilibina (1980)

tion. The steeply dipping dykes, 20 – 50 m in thickness, are composed almost totally of ultramafic olivine or olivine-clinopyroxene cumulate.

The exposed parts of the Näätäniemi intrusion are composed mainly of serpentinized metaperidotite, while in a few places the ultramafic cumulates are found to be overlain by metapyroxenites and metagabbros. The uppermost parts are not exposed at all. The measured dips of the layering vary in most cases between 45° – 70° and the estimated thickness of the layered series is about 400 m.

In contrast with the Näätäniemi intrusions, the last differentiates and the upper marginal zone of the Savilahti sill at Koli are well-exposed. The estimated thickness of the sill is about 300 m and it can be followed along the strike for over 15 km. By combining the available data from both areas, a generalized stratigraphic section with cumulus minerals can be constructed (Fig. 2).

The layered sequence includes the following cumulates: olivine (-chromite), "porphyritic" olivine, olivine-clinopyroxene, clinopyroxene, clinopyroxene-magnetite and plagioclase-clinopyroxene-magnetite cumulates. The latter rock type grades upwards into a plagioclase and quartz-rich rock crystallized from the residual liquid and further into an upper marginal zone. This zone in composed of coarse-grained gabbro and clinopyroxenite. The cpx-magn cumulate is characterized by a small amount of sulphide.

The contacts between each zone represented in Fig. 2 are relatively sharp, except for that between the gabbro and the upper marginal zone. Magmatic

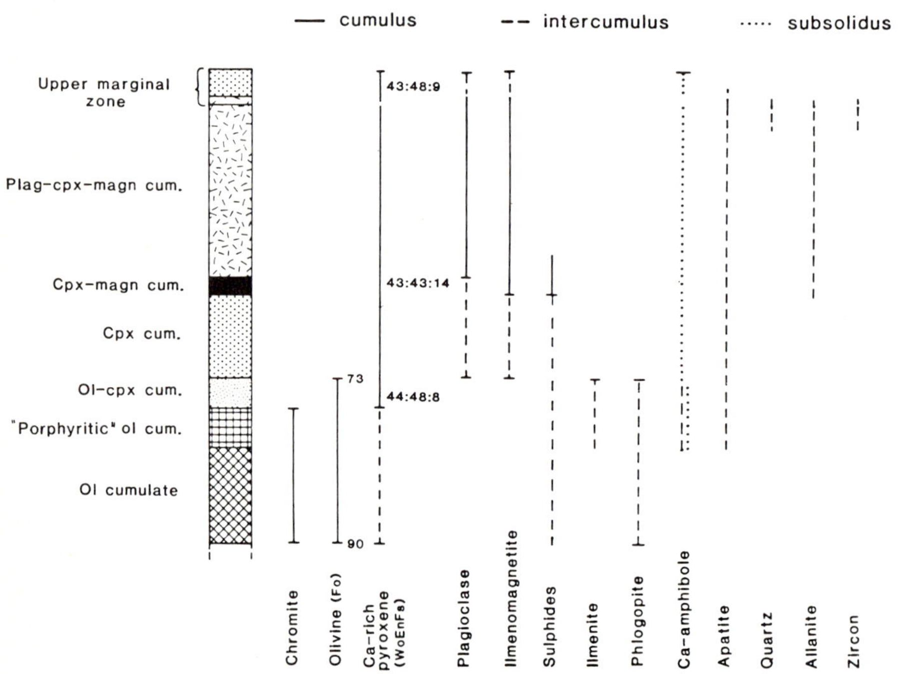

Fig. 2. Generalized stratigraphy of the layered bodies in eastern Finland. Section thickness about 400 m

rhythmic layering is seldom found in any zone. The regular, simple stratigraphy, coupled with geochemical data, suggests that the layered bodies were formed by fractional crystallization in a magma chamber formed by a single magma pulse.

In the light of the discovery of a new magmatic event in the Kuhmo greenstone belt it is interesting to evaluate the possible connection between the known Ni – Cu deposits and this magmatism. Of the four known deposits, that of Arola, which has a high average Ni/Cu ratio (28), is probably related to the Archaean volcanism (Lehtinen 1983). The other three, the Ni – Cu deposits of Peura-aho and Hietaharju in Suomussalmi, described by Kojonen (1981) and a so far little studied mineralization at Siivikkovaara can be interpreted on petrographic grounds (see below) as belonging to the GWA.

The Cu/(Cu + Ni) ratios (0.30, 0.33) of the deposits in Suomussalmi (Kojonen 1981) are similar to the ratios of the syngenetic ores at Pechenga (Zak et al. 1982), but the Pt/(Pt + Pd) ratio at Pechenga (0.55) is much higher than at Suomussalmi (0.176 – 0.333). It should be kept in mind, however, that the published PGE data for Pechenga is so far very limited. Only one analysis of an ore containing 4.6% Ni and 2.45% Cu has been reported by Gorbunov (1968), who also mentions that the ratio of Pt to Pd increases in a regular manner between high-grade and low-grade ores.

Petrography of the Ultramafic Rocks

Although the degree of alteration of the metacumulates, particularly in the Kuhmo area, is high, the magmatic mineralogy can be inferred from the well-preserved pseudomorphous textures.

The cumulus olivine and intercumulus clinopyroxene and phlogopite in the lowermost of cumulates are now replaced by serpentine, tremolite and chlorite, respectively. The quantity of intercumulus minerals in the overlying "porphyritic" ol cumulate increases and sometimes reaches one-half of the volume. The "porphyritic" appearance of this zone is due to poikilitic clinopyroxene crystals 2 – 5 mm in size, which give a mottled texture to the eroded surface. The poikilitic clinopyroxene crystals enclose rounded olivine crystals, and the clinopyroxene is in turn surrounded by poikilitic brown amphibole and phlogopite. Chlorite pseudomorphing phlogopite still sometimes inherits the brownish colour of its precursor. Minor intercumulus minerals are ilmenite and apatite. Apart from the crystaline form of the clinopyroxene and the absence of chromite, the mineralogy and texture of the ol-cpx cumulates are the same as in the zone described above.

In the severely altered host rocks of the Ni – Cu deposits of Peura-aho poikilitic textures can still be distinguished where olivine crystals have been enclosed in clinopyroxene. In addition, there exist relicts of brown intercumulus amphibole and many samples are seen to have contained a large amount of intercumulus phlogopite, now replaced by brownish chlorite.

The olivines poikilitically enclosed by clinopyroxene in the ultramafic cumulates of the Pechenga intrusions often have rounded forms (see Väyrynen

1938, Fig. 18). Other major intercumulus minerals include brown amphibole and brown mica (Yakovleva 1981, Yakovleva and Osokin 1981). The texture of the ultramafic rocks is often said to be "porphyritic" (e.g. Smol'kin 1974a).

The total absence of olivine from the clinopyroxene cumulates and the rounded forms of the olivine in the underlying cumulates in the intrusions of Kuhmo and Koli suggest a peritectic relation between olivine and clinopyroxene.

Mineralogy

Olivines. Two generations of olivine are distinguished on textural and chemical grounds. The first, which is encountered only in the Kuhmo belt, is magnesian (Fo_{86-96}) and relatively high in MnO and very low in CaO (Fig. 3, Table 1). Its composition is compatible with a metamorphic origin (Piquet 1982). In the second, magmatic generation the Fo content varies between 90 – 73%. The range of olivine compositions in Pechenga is more limited, varying in the range Fo_{83-72} (Smol'kin 1978, Smol'kin and Pahomovskiy 1982). It is significant that olivine ceases to crystallize at the same level of Fe-enrichment as in eastern Finland. Olivines from Pechenga have a relatively high CaO content (Fig. 3). As the crystallization depth was not essentially different, the difference in CaO content could be correlated with the lower silica activity of the magma at Pechenga (cf. Stormer 1973).

Both the Fo and the NiO (0.34 – 0.40%) content of most magnesian olivines from Kuhmo are high enough for these to have crystallized from a very primitive magma (Fig. 4). The NiO content decreases below 0.2% with fractionation.

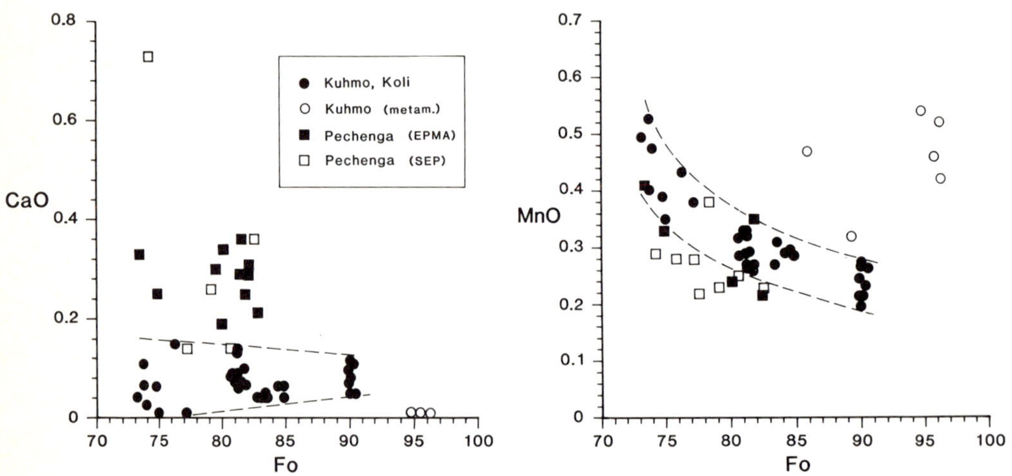

Fig. 3. CaO, MnO, and Fo concentrations in olivines from eastern Finland and Pechenga. The Pechenga analyses are taken mainly from Smol'kin (1978) and Smol'kin and Pahomovskiy (1982)

Table 1. Representative mineral analyses from eastern Finland and Pechenga

	1	*2*	*3*	*4*	*5*	*6*	*7*	*8*	*9*	*10*	*11*	*12*	*13*
SiO_2	40.99	38.24	42.75	54.29	50.99	50.10	49.97	46.67	47.14	41.56	39.72	36.22	35.85
TiO_2	0.01	0.00	0.01	0.40	0.76	1.03	1.38	4.08	4.14	6.72	5.74	7.49	7.20
Al_2O_3	0.02	0.00	0.01	1.57	2.50	2.94	3.42	8.41	7.95	9.60	13.20	11.70	12.46
FeO	9.81	24.01	3.72	4.91	8.63	8.55	10.28	7.88	7.74	9.23	8.26	11.56	12.83
MnO	0.19	0.40	0.52	0.16	0.20	0.18	0.17	0.10	0.09	0.08	0.08	0.40	0.42
MgO	48.18	37.83	52.97	16.41	14.93	15.73	13.53	16.11	16.06	14.09	14.24	20.98	19.23
CaO	0.10	0.06	0.01	21.02	20.64	20.04	19.81	9.80	10.26	11.00	11.47	0.06	0.15
Na_2O	0.01	0.00	n.d.	0.44	0.50	0.28	0.89	3.25	3.12	3.05	2.36	n.d.	0.07
K_2O	0.00	0.00	0.01	0.04	0.00	0.03	0.02	0.40	0.44	0.80	0.55	7.41	7.29
Cr_2O_3	0.02	0.03	0.02	0.48	0.01	0.22	n.d.	0.85	0.72	0.08	0.16	0.76	0.62
NiO	0.37	0.15	0.24	0.04	0.04	0.02	n.d.	0.05	0.05	n.d.	0.06	0.16	n.d.
Total	99.70	100.72	100.26	99.76	99.20	99.12	99.47	97.60	97.71	96.21	95.84	96.74	96.12
Si	1.008	0.996	1.016	1.983	1.914	1.882	1.885	6.709	6.767	6.194	5.893	5.260	5.261
Ti	0.000	0.000	0.000	0.011	0.022	0.029	0.039	0.441	0.447	0.753	0.641	0.818	0.795
Al	0.001	0.000	0.000	0.068	0.111	0.130	0.152	1.452	1.344	1.686	2.308	2.003	2.155
Fe	0.202	0.523	0.074	0.150	0.271	0.269	0.324	0.948	0.929	1.150	1.025	1.404	1.574
Mn	0.004	0.009	0.010	0.005	0.006	0.006	0.005	0.012	0.010	0.010	0.010	0.049	0.052
Mg	1.766	1.469	1.877	0.894	0.835	0.881	0.761	3.452	3.437	3.130	3.148	4.541	4.206
Ca	0.003	0.002	0.000	0.822	0.830	0.807	0.801	1.510	1.577	1.756	1.824	0.009	0.024
Na	0.000	0.000	n.d.	0.031	0.036	0.020	0.065	0.906	0.867	0.881	0.679	n.d.	0.020
K	0.000	0.000	0.000	0.002	0.000	0.001	0.001	0.073	0.080	0.152	0.104	1.373	1.365
Cr	0.000	0.001	0.000	0.014	0.000	0.007	n.d.	0.096	0.082	0.009	0.018	0.087	0.072
Ni	0.007	0.003	0.005	0.001	0.001	0.001	n.d.	0.006	0.006	n.d.	0.007	0.019	n.d.
Fo	89.7	73.7	96.2	–	–	–	–	–	–	–	–	–	–
Wo	–	–	–	44.1	42.9	41.2	42.5	–	–	–	–	–	–
En	–	–	–	47.9	43.1	45.0	40.3	–	–	–	–	–	–
Fs	–	–	–	8.0	14.0	13.7	17.2	–	–	–	–	–	–

1 Olivine, ol cumulate, Kuhmo; *2* Olivine, "porphyritic" ol cumulate, Koli; *3* Metam. oliv., ol cumulate, Kuhmo, (Piquet 1982); *4* Clinopyroxene, ol-cpx cumulate, Kuhmo; *5* Clinopyroxene, cpx-magn cumulate, Koli; *6* Clinopyroxene, ol pyroxenite, Pechenga, (Zhangurov and Predovskiy 1974); *7* Clinopyroxene, plagiopyroxenite, Pechenga, (Smol'kin 1978); *8* Edenitic hornblende, "porphyritic" ol cumulate, Koli; *9* Edenite, "porphyritic" ol cumulate, Koli; *10–11* Kaersutite, ol cumulate, Pechenga; *12–13* Phlogopite, ol cumulate, Pechenga.

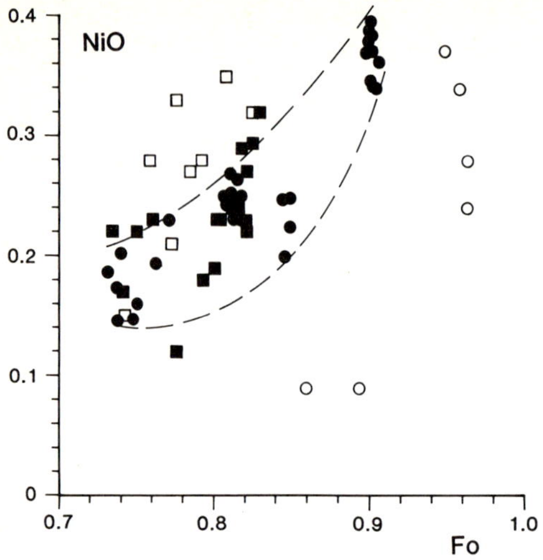

Fig. 4. NiO vs Fo for olivines from eastern Finland and Pechenga. Symbols and data sources as in Fig. 3

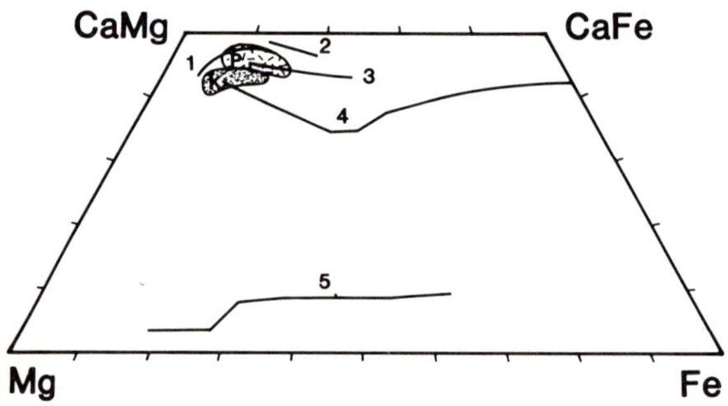

Fig. 5. Comparison of pyroxene composition trends at Koli (field *K*) and Pechenga (*P*) (analyzed at the Univ. of Oulu) with others on a Ca – Mg – Fe diagram. *Lines: 1* Alaskan type (Irvine 1974); *2* Black Jack (Murray 1954); *3* Garbh Eilean (Wilkinson 1957); *4, 5* Skaergaard (Nwe 1976)

Olivine grains from Pechenga separated and analyzed by chemical methods tend to have a higher NiO content than those analyzed by microprobe. Compositions of the latter coincide in most cases within the range of olivines from Kuhmo and Koli, but more microprobe data would be needed to reach any conclusions with certainty. It should be added that Smol'kin and Pahomovskiy (1982) did not notice any differences in Ni content between olivines from ore-bearing and barren intrusions.

Clinopyroxenes. Magmatic and metamorphic varieties can be also distinguished among the pyroxenes. The latter type consists of diopside with very low Al_2O_3, TiO_2, and Na_2O. The magmatic type is preserved from alteration only in cumulates underlying gabbroic rocks. It forms a short trend in a ternary plot of Mg, Ca and total Fe subparallel to the diopside-hedenbergite intersection (Fig. 5). It has lower Wo, Al_2O_3, and TiO_2 than the clinopyroxenes from Pechenga (Table 1). The higher Ti and Al content of the Pechenga pyroxenes could be ascribed to a lower concentration of silica in the magma (cf. Le Bas

1962). Limited Fe-enrichment is also typical of pyroxenes at Pechenga, the most Fe-rich pyroxene reported has an Fs content of about 23% (Smol'kin 1978).

Amphiboles and Micas. Intercumulus amphibole is still well-preserved in many places in the ultramafic cumulates of the Savilahti sill, consisting of Na-rich titanian edenite or edenitic hornblende according to the classification of Leake (1978) (Table 1). The corresponding amphibole in Pechenga is sufficiently rich in TiO_2 and poor in SiO_2 to be classified as kaersutite as the Russian geologists have already called it (e.g. Yakovleva 1981). The Pechenga amphibole also has a higher K/Na ratio, probably reflecting a higher K/Na ratio in the parent magma.

Intercumulus mica has been replaced by penninitic chlorite in the ultramafic rocks of Kuhmo and Koli, but it still has a poikilitic habit. The intercumulus mica found at Pechenga has been referred to as biotite (Väyrynen 1938, Yakovleva and Osokin 1981). According to a few microprobe analyses this mica has an Mg/Fe ratio higher than 2:1 (Table 1) and is rich in TiO_2 (3.4 – 7.5%) and should properly be called titanian phlogopite.

Geochemistry

Major Elements

Representative whole rock analyses from eastern Finland and Pechenga are presented in Table 2. The analyses of cumulates from the layered bodies from eastern Finland define a continuous, strongly curved trend from the MgO corner to the Al_2O_3 corner in the CMA diagram (Fig. 6). The analyses from Pechenga, which have mainly been taken from the compilation of Tkachenko (1982), form a similar but more dispersive pattern.

The trend of the Finnish analyses in the FeO/(FeO + MgO) vs SiO_2 diagram (Fig. 7) begins at a considerably lower FeO/(FeO + MgO) ratio than in the case of Pechenga. On the other hand, the most differentiated rocks in Pechenga have a lower SiO_2 content. The difference in FeO/(FeO + MgO) ratio is consistent with the higher Fa content of the most magnesian olivines at Pechenga. The FeO/(FeO + MgO) ratio of the most magnesian rocks in the Savilahti sill, Petäjäniemi dykes and Ni-bearing intrusions in Suomussalmi is higher than the ratio of layered bodies in Kuhmo, implying fractionation before emplacement. The difference in silica enrichment may be due to the lack of analyses, but it is nevertheless compatible with the probable lower silica content of the parental magma at Pechenga as indicated by the mineralogical data.

Since the Al_2O_3 and TiO_2 concentrations in olivines are very low, the ratio between these elements in olivine cumulates reflects this ratio in the intercumulus liquid and the parental magma. A plot of Al_2O_3 vs TiO_2 is shown in Fig. 8. The lowermost olivine cumulates from Kuhmo have a trend of their own which is probably caused by the removal of TiO_2 during secondary alteration. The concentrations of certain other components have also diminished in these rocks, including CaO and K_2O. The "porphyritic" ol cumulates and ol-cpx cumulates

Table 2. Whole rock analyses of various rock types from eastern Finland and Pechenga

	1	*2*	*3*	*4*	*5*	*6*	*7*	*8*	*9*	*10*	*11*	*12*	*13*
SiO_2	38.57	42.75	45.63	50.93	40.03	50.22	36.29	35.85	36.74	47.97	36.47	46.99	49.59
TiO_2	0.10	0.76	0.77	1.01	2.80	2.07	0.79	0.66	0.86	2.01	4.21	2.70	3.04
Al_2O_3	1.95	4.43	4.24	6.53	6.92	14.22	3.05	2.24	2.69	6.30	4.26	10.38	12.84
Fe_2O_3TOT	10.86	14.55	12.39	11.49	26.19	15.59	14.65	n.d.	n.d.	n.d.	n.d.	n.d.	n.d.
Fe_2O_3	n.d.	n.d.	n.d.	n.d.	n.d.	n.d.	n.d.	9.54	7.45	2.21	9.95	4.45	1.57
FeO	n.d.	n.d.	n.d.	n.d.	n.d.	n.d.	n.d.	8.35	11.51	9.81	16.76	11.35	11.61
MnO	0.09	0.18	0.22	0.29	0.26	0.44	0.20	0.21	0.23	0.15	0.18	0.20	0.18
MgO	36.44	25.52	20.93	12.06	8.40	3.98	29.48	33.21	28.86	11.64	10.62	5.47	3.76
CaO	0.00	5.44	11.17	14.45	12.25	8.02	2.85	2.00	2.60	15.48	13.66	10.53	8.07
Na_2O	0.18	0.09	0.19	1.58	0.89	3.78	0.14	0.04	0.17	1.10	0.21	2.10	4.45
K_2O	0.00	0.06	0.03	0.24	0.25	0.97	0.22	0.18	0.29	0.49	0.11	0.76	1.62
P_2O_5	0.04	0.08	0.08	0.08	0.05	0.14	n.d.	n.d.	0.06	0.19	0.14	n.d.	0.42
Cr_2O_3	0.35	0.47	0.36	0.09	0.01	0.01	n.d.	0.48	0.56	0.10	0.01	0.02	0.01
NiO	0.30	0.13	0.10	0.03	0.05	0.00	n.d.	0.43	0.24	0.01	n.d.	0.01	0.00
Loss	11.28	5.28	3.50	1.10	1.14	0.55	n.d.	6.89	7.53	2.36	1.97	4.19	2.74
Total	100.16	100.28	99.61	99.88	99.24	99.99		100.08	99.79	99.82	98.55	99.15	99.90

Eastern Finland: *1* Ol cumulate, Kuhmo; *2* "Porphyritic" ol cumulate, Koli; *3* Ol-cpx cumulate, Kuhmo; *4* Cpx cumulate, Koli; *5* Cpx-magn cumulate, Koli; *6* Plag-cpx-magn cumulate, Koli. Pechenga: *7* Average ultramafite of ore-bearing intrusions (Zak et al. 1982); *8* Olivinite, (anal. 876, Tkachenko 1982); *9* Wehrlite, (911, op. cit.); *10* Pyroxenite, (1108, op. cit.); *11* Magnetite-rich pyroxenite, (903, op. cit.); *12* Gabbro, (1096, op. cit.); *13* Essexite, (1109, op. cit.).

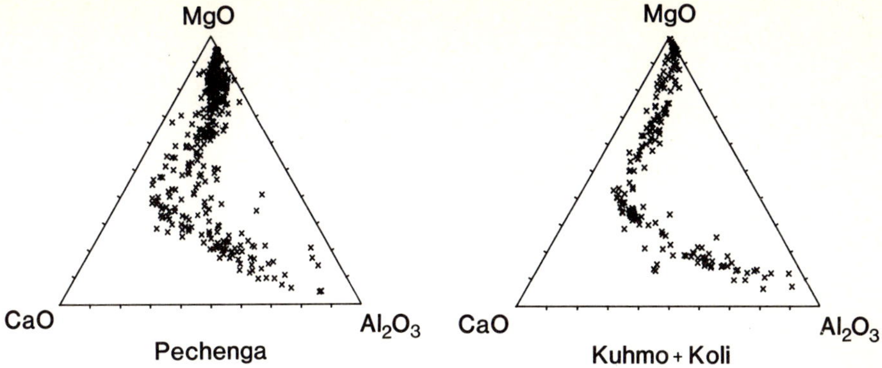

Fig. 6. Chemical compositions of cumulates from Pechenga and eastern Finland on a CMA diagram. The Pechenga analyses are mainly from Tkachenko (1982)

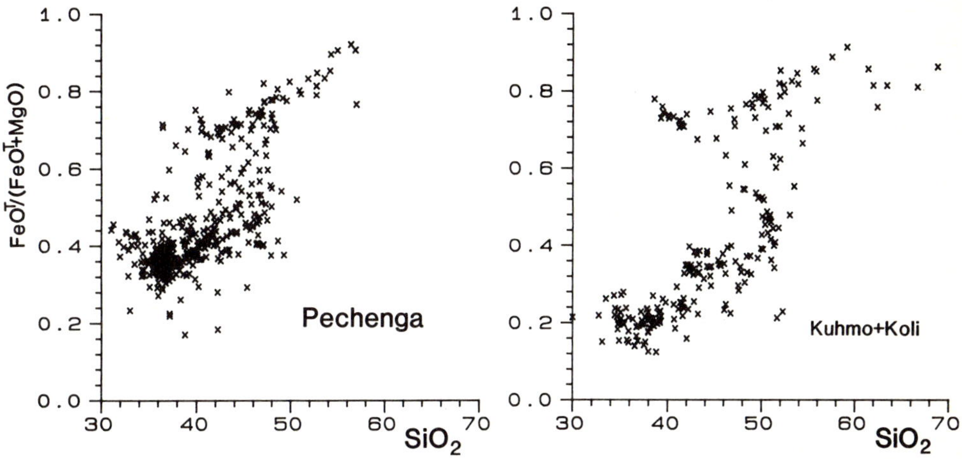

Fig. 7. $FeO^T/(FeO^T + MgO)$ vs SiO_2. The Pechenga analyses are mainly from Tkachenko (1982)

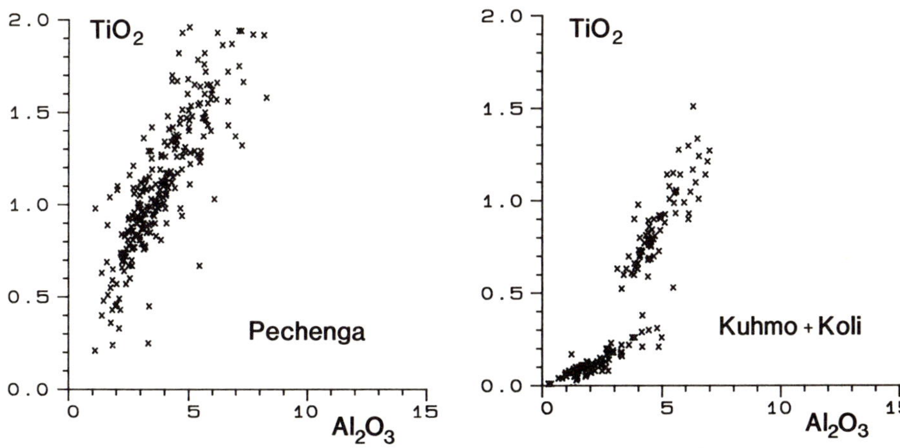

Fig. 8. Composition of ultramafic cumulates from Pechenga and eastern Finland on a plot of TiO_2 vs Al_2O_3. The data for Pechenga are mainly from Tkachenko (1982)

Table 3. REE analyses of various cumulates from Koli and Pechenga

	1	*2*	*3*	*4*	*5*	*6*	*7*	*8*	*9*	*10*	*11*	*12*	*13*	*14*
La	5.8	4.8	4.4	4.0	7.6	9.9	19.1	27.0	54.1	5.8	6.5	13.4	10.4	20.7
Ce	13.7	11.2	11.9	11.5	17.9	24.0	44.0	62.0	119.0	12.3	12.1	30.0	25.0	43.0
Nd	8.1	6.9	10.0	10.5	14.0	17.1	28.0	37.0	72.0	6.1	8.7	20.0	19.6	29.0
Sm	2.1	1.8	2.9	3.2	4.0	5.2	7.1	9.0	14.5	1.6	1.8	4.4	6.4	5.9
Eu	0.75	0.62	0.97	0.92	1.30	1.56	2.10	2.90	3.00	0.60	0.65	1.86	2.20	2.20
Tb	0.39	0.28	0.44	0.48	0.51	0.61	1.05	1.44	2.30	0.26	0.25	0.62	0.75	0.93
Yb	0.84	0.74	1.25	1.16	1.35	1.66	2.70	3.40	6.30	0.42	0.49	1.21	1.28	1.67

Koli: *1–2* "Porphyritic" ol cumulate; *3* Cpx cumulate; *4* Cpx-magn cumulate; *5–8* Plag-cpx-magn cumulate; *9* Plag(-cpx-magn) cumulate. Pechenga: *10–11* Ol cumulate; *12* "Porphyritic" ol cumulate; *13* Cpx cumulate; *14* Plag-cpx-magn cumulate.

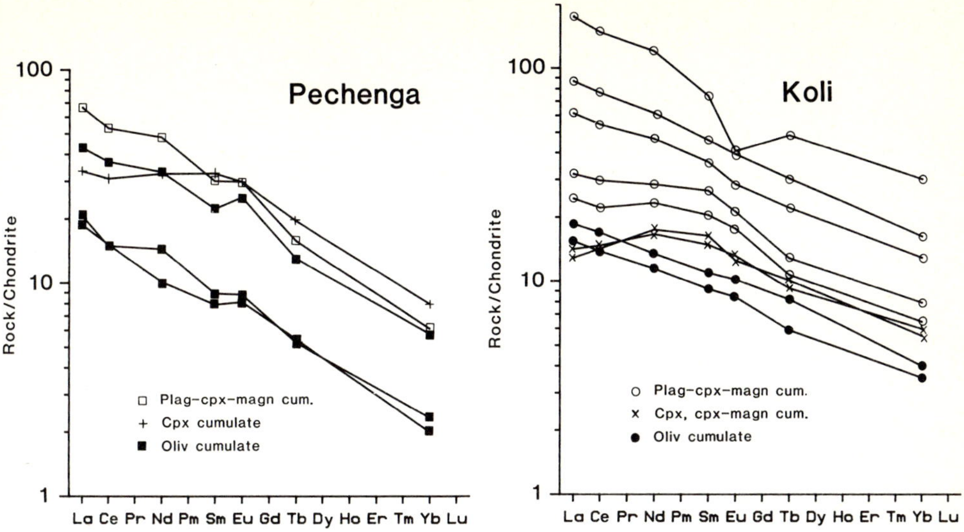

Fig. 9. Chondrite-normalized REE patterns for various cumulates from Pechenga and Koli

define a steeper trend having an Al_2O_3/TiO_2 ratio of about 5.5. The analyses from Pechenga lie on a line which gives a still lower Al_2O_3/TiO_2 ratio. The average ratio calculated from the results of analyses of the well-preserved cumulates from barren, promising and ore-bearing intrusions, as given by Zak et al. (1982), is close to 3.6. This is near the value obtained from analyses of chilled margins or as weighted averages for intrusions (see Table 4).

Rare Earth Elements

The rare earth element concentrations in the various cumulates are seen in Table 3 and are presented in chondrite-normalized form in Fig. 9. The REE patterns for the olivine cumulates are straight and clearly LREE-enriched. The pyroxene cumulates have a sloping curve for HREE but an almost flat or rising one for LREE. The LREE pattern in the gabbroic rocks gradually changes from flat to sloping with the increase in REE content (with height in the layered series). The negative Eu anomally in the most differentiated sample from Koli is probably caused either by an analytical error or by secondary alteration.

In the light of the known crystal/liquid distribution coefficients for REE, the overall REE concentrations in the cumulates from Koli are consistent with the LREE-enriched character of the parental magma and an origin for the various cumulates by fractional crystallization from the same parental magma.

The Parental Magma

A close spatial and genetic relation to basic and ultrabasic volcanic rocks is regarded as a characteristic feature of the GWA (Dagelayskaya 1979). The chem-

Table 4. Analyses of picrites from Onega and Pechenga and of chilled margins and weighted averages of the Pechenga intrusions

	1	*2*	*3*	*4*	*5*	*6*	*7*	*8*
SiO_2	44.18	45.59	46.56	46.25	42.08	43.54	42.66	40.92
TiO_2	1.54	1.58	1.60	2.15	2.67	1.97	1.83	1.88
Al_2O_3	8.80	9.39	9.52	7.46	9.01	7.38	6.37	6.34
Fe_2O_3	1.89	2.11	2.46	1.98	4.45	5.29	3.48	6.05
FeO	9.68	9.10	8.83	12.92	11.60	12.23	14.13	10.34
MnO	0.18	0.15	0.15	0.20	0.19	0.18	0.22	0.18
MgO	17.77	14.56	13.32	17.16	13.76	18.91	20.54	19.02
CaO	8.99	9.10	8.90	6.87	9.56	8.51	6.67	6.38
Na_2O	0.84	1.08	1.43	0.47	0.40	0.62	0.94	0.85
K_2O	0.39	0.27	0.30	0.12	0.16	0.28	0.44	0.36
P_2O_5	0.19	n.d.	n.d.	0.36	0.27	n.d.	n.d.	n.d.
Cr_2O_3	0.14	n.d.	n.d.	0.17	0.11	0.25	0.26	0.20
NiO	0.11	n.d.	n.d.	0.09	0.08	n.d.	0.17	0.14
Loss	5.43	7.32	4.95	3.57	6.08	n.d.	1.72	6.93
Total	100.13	100.29	99.02	99.77	100.42	99.16	99.43	99.59

1 Picrite (no. of anal. = 10), Onega, (Svetov and Sokolov 1978); *2* Picrite ($n = 8$), Onega, (Svetov and Golubev 1978); *3* Picrite ($n = 17$), Onega, (Kulikov 1971); *4* Picrite, Pechenga, (Predovskiy et al. 1974); *5* Picrite, Pechenga, (Predovskiy et al. 1974); *6* Chilled margin ($n = 10$), Pechenga, (Smol'kin 1974b); *7* Weighted average of Pechenga intrusions ($n = 88$), (Zhangurov and Predovskiy 1974); *8* Weighted average of Pechenga intrusions ($n = 88$), (Zhangurov and Predovskiy 1974); *8* Weighted average of Pechenga intrusions ($n = 246$), (Rabinovich 1978).

ical similarity between the average compositions of the intrusions in the Pechenga area or the composition of the chilled margins, and the picrites of the fourth volcanic suite (see Table 4) has led many authors to suggest that the intrusions and picrites represent hypabyssal and effusive facies of the same magmatism (Predovskiy et al. 1974, Smol'kin 1974b). In the Onega trough in southern Karelia minor intrusions of the GWA are very closely associated with Suisaarian picrites (Kulikov et al. 1976).

The molar MgO/(MgO + FeO) ratio (Mg′ value) of the parental magma can be calculated using the composition of the most magnesian olivines and the coefficient determining the distribution of Mg and Fe between the olivine and the liquid (Roeder and Emslie 1970). Using $K_D = 0.3$, Mg′ values 0.59 and 0.73 are obtained for the parental magmas of Pechenga and Kuhmo, respectively, reflecting the more differentiated nature of the magma at Pechenga.

The Al_2O_3/TiO_2 ratios and Mg′ values of the parental magmas of the GWA in Kuhmo and Pechenga are compared in Fig. 10 with corresponding data for picrites from the Pechenga and South Pechenga Complexes, Suisaarian picrites and komatiites from Kuhmo and Vetrenyi Poyas. The Kuhmo parental magma comes very close to the Suisaarian picrites, while the picrites from Pechenga are more primitive than the parental magma of the associated intrusions, but both have very low Al_2O_3/TiO_2 ratios. The picrites from South Pechenga tend to have higher Al_2O_3/TiO_2 ratios.

Many authors have referred to the Pechenga picrites as komatiites (e.g. Suslova 1982). In spite of the descriptions of spinifex textures (Skufin 1980,

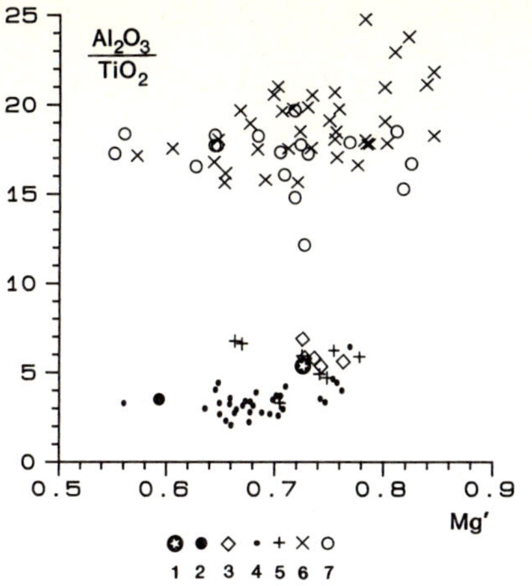

Fig. 10. Al_2O_3/TiO_2 vs Mg′. *Symbols: 1* parental magma of the GWA in Kuhmo; *2* parental magma of the GWA in Pechenga; *3* Suisaarian picrite; *4* Pechenga picrite; *5* picrite from South Pechenga; *6* komatiite from Kuhmo; *7* komatiite from Vetrenyi Poyas. References for the data sources are available from the author

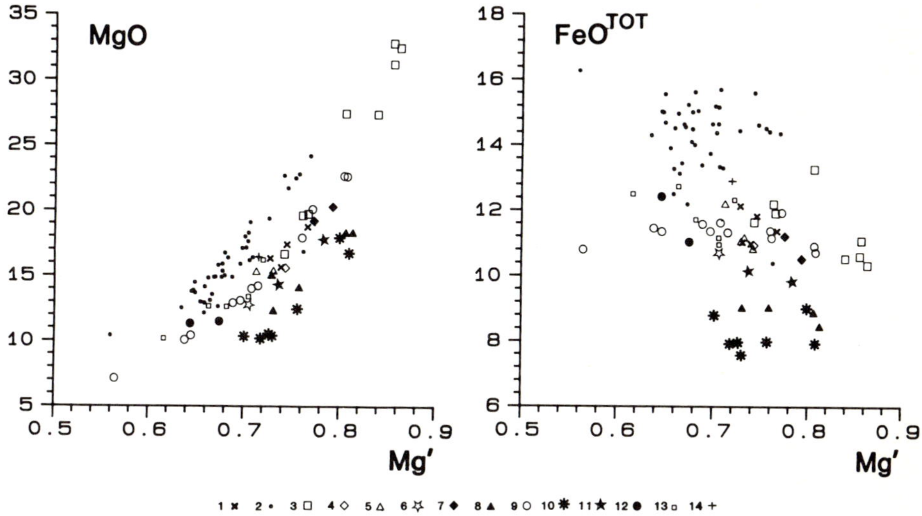

Fig. 11. MgO and total FeO in the Pechenga and Suisaarian picrites compared with other primary magmas. *Symbols: 1* Suisaarian picrite; *2* Pechenga picrite; *3* komatiite; *4* Deccan picrite; *5* Karroo picrite; *6* Skye picrite; *7* Baffin Bay picrite; *8* picrite from the Reykjanes Peninsula; *9* Hawaiian tholeiite; *10* MORB; *11* ophiolite basalt; *12* alkali basalt; *13* basanite; *14* olivine melilite. *3* – *14* taken from Basaltic Volcanism Study Project (1981, Tables 1.4.2.1 and 1.4.2.2; for references see footnotes to these Tables)

Rusanov 1981) they clearly differ chemically from komatiites, and not only in their higher TiO_2 content and lower Al_2O_3/TiO_2 ratio (Fig. 10). In fact, the picrites of Pechenga differ from all other primary magma types in having very low Al_2O_3 and high MgO and total FeO concentrations at the same Mg′ value (Figs. 11 and 12). These features could be ascribed to a very great depth of origin in the mantle. The primary phase field of garnet expands and that of olivine contracts with increasing pressure, resulting in lower Al_2O_3 and higher MgO and FeO in the first melt produced (O'Hara 1968).

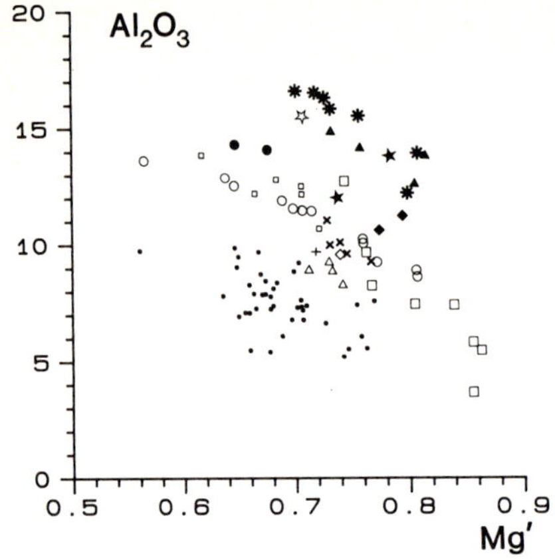

Fig. 12. Al_2O_3 content of the Pechenga and Suisaarian picrites compared with other primary magmas. Symbols and data sources as in Fig. 11

The $(La/Yb)_N$ ratio for the ol cumulates, which reflects the ratio in the parental magma, is clearly greater in the rocks of Pechenga (7.5 – 9.5) than at Kuhmo and Koli (about 4.5). This fact can be interpreted as indicating a lower degree of melting at Pechenga, provided that parental material was the same for both areas. Assuming twice the chondritic REE content and garnet Iherzolite mineralogy ($ol_{60}opx_{20}cpx_{10}gt_{10}$) of the parental material, and using the melting proportions, distribution coefficients and equilibrium melting equation given by Leeman (1977), Terakado (1980) and Shaw (1970), respectively, the melting percentages can be calculated to have been about 5% at Pechenga and 10% in eastern Finland.

Areal and Temporal Correlation

In addition to Pechenga, intrusions of the GWA are known to occur in the Imandra-Varzuga zone (Zhangurov and Smol'kin 1982), in the Kuolajärvi (Lavrov 1979) and possibly Gaikolskaya structures (Sidorenko 1983), in the Hautavaara greenstone belt (Popov 1971), in the Onega trough (Kulikov et al. 1976) and in Vetrenyi Poyas (Slyusarev and Kulikov 1973) within the Soviet part of the Baltic Shield. In Finland rocks petrographically typical of the GWA can be found at least in Puolanka, Kuusamo (Piispanen 1972), Kemi, Kittilä, Sodankylä and Savukoski in addition to the areas discussed earlier in this report (see Fig. 1). In the light of the available geochemical data it seems probable that all the above mentioned occurrences of the GWA, except that at Pechenga and some intrusions in the Imandra-Varzuga zone, are related to a picritic magma of the Suisaarian type rather of the Pechenga type. In contrast to Pechenga, only minor Ni – Cu deposits have been discovered in the other occurrences (Kayryak et al. 1976, Kojonen 1981).

The absolute age of the Pechenga intrusions is as yet somewhat uncertain. The available age determinations for the intrusions and their host rocks, obtained by the Pb isotope, K – Ar and Rb – Sr methods, give ages mostly between

1700 – 1900 Ma (e.g. Gerling et al. 1972). This range is interpreted as reflecting the time of metamorphism (e.g. Gorokhov et al. 1982).

The intrusions of the GWA and related volcanic rocks of Pechenga and other areas in the Soviet part of the Baltic Shield have been correlated by many authors with the Suisaarian volcanic-sedimentary formations (e.g. Sidorenko 1983). Suisaarian formations, including picrites, have been found to be deposited conformably on the Jatulian supracrustal rocks (e.g. Sokolov and Stenar 1980). In Finland intrusions of the GWA are encountered in the Archaean basement or within the Jatulian metasediments but not within the Kalevian metasediments unconformably overlying Jatulian rocks. Age determinations of 2.1 – 2.2 Ga (Sakko 1971, Hyppönen 1983) support a Jatulian age for these, since the upper boundary of the Jatulian is interpreted as being about 2000 Ma (Meriläinen and Sokolov 1981). The Suisaarian rocks have not been dated, but their age is thought to be not far removed from Jatulian (Kratts et al. 1981). Clearly more work is needed to resolve the time-stratigraphic correlation for the GWA in the Baltic Shield.

Conclusions

Many conclusions can be drawn from the data presented above, the most important being:

1. The area of occurrence of the GWA is extensive, being of the order of 500 × 800 km.
2. A reaction pair of olivine and Ca-rich pyroxene seems to be a characteristic feature of the GWA.
3. The lack of orthopyroxene and the presence of primary hydrous minerals indicate a low a_{SiO_2} and a high water content for the parental magma.
4. The parental magma of the layered bodies in Kuhmo was close to the primary magma, but that of the Savilathi sill, Petäjäniemi dykes and the Ni-bearing intrusions of Suomussalmi and Pechenga suffered olivine fractionation before emplacement.
5. In comparison with that in eastern Finland the parental magma in Pechenga may be deduced from the chemical composition of the minerals and rocks to have had a lower SiO_2 concentration and Al_2O_3/TiO_2 ratio and higher K/Na and $(La/Yb)_N$ values and is interpreted as having been produced by a lower degree of melting at a deeper level in the mantle.
6. The parental magmas in eastern Finland and Pechenga correspond to picrites of the Suisaarian and Pechenga types, respectively. The latter type differs from other primary magmas in having a lower Al_2O_3 and a higher MgO and total FeO content at the same Mg′ value.
7. Outside Pechenga only minor Ni – Cu deposits are known to be related to the GWA.

Acknowledgements. I am deeply grateful to Prof. T. Piirainen for many stimulating discussions and continuous interest in my work. I would also like to thank Mr. S. Sivonen, Dipl. Eng., for providing the microprobe analyses and Mr. M. Hicks, M.A., for carefully reviewing the English translation.

References

Auvray B, Blais S, Jahn BM, Piquet D (1982) Komatiites and the komatiitic series of the Finnish greenstone belts. In: Arndt NT, Nisbet EG (eds) Komatiites. Allen & Unwin, London, pp 131 – 146

Basaltic Volcanism Study Project (1981) Basaltic volcanism on the terrestrial planets. Pergamon Press, New York

Bilibina TV (ed) (1980) Map of the geological formations in the eastern part of the Baltic Shield, Scale 1:150000. In: Sidorenko AV, Bilibina TV (eds) Metallogeny of the eastern part of the Baltic Shield. Nedra, Leningrad (in Russian)

Blais S, Capdevila R, Jahn BM, Hameurt J (1978) The Archean greenstone belts of Karelia (Eastern Finland) and their komatiitic and tholeiitic series. In: Windley BF, Naqvi SM (eds) Archean geochemistry. Elsevier North Holland Biomedical Press, Amsterdam New York, pp 87 – 107

Dagelayskaya IN (1979) The gabbro-wehrlite formation. In: Masaytis VL (ed) Magmatic formations of the USSR, vol I. Nedra, Leningrad, pp 257 – 265 (in Russian)

Gerling EK, Lobach-Zhuchenko SB, Gorokhov I, Koltsova TB (1972) Isotopic ages of rocks of the Precambrian in the Baltic Shield. In: Kratts KO (ed) Geochronological boundaries and geological evolution of the Baltic Shield. Nauka, Leningrad, pp 76 – 159 (in Russian)

Gorbunov GI (1968) Geology and origin of the copper-nickel sulphide deposits of Pechenga. Nedra, Moscow (in Russian)

Gorokhov IM, Varshavskaya ES, Kutyavin EP, Melnikov NN (1982) Rb – Sr dating of low-grade metamorphics of the USSR. Precambrian Res 18:145 – 156

Hanski E (1982) A comparative study of albite diabases and related ultramafic rocks in the Kuhmo and Koli areas, eastern Finland. Ark Alueiden Malmiproj Rap 6, Oulun Yliopisto (in Finnish with English summary)

Hyppönen V (1983) Pre-Quaternary rocks of the Ontojoki, Hiisijärvi, and Kuhmo map-sheet areas. Explanation to the maps of Pre-Quaternary rocks, Sheets 4411, 4412 and 4413, 1:100000. Geol Surv Finl (in Finnish with English summary)

Irvine TN (1974) Petrology of the Duke Island Ultramafic Complex, southeastern Alaska. Mem Geol Soc Am 138

Kayryak AI, Morozov SA, Limonova LP, Popov MG (1976) Copper-nickel and polymetallic pyrite ores in eastern Karelia. Razved Okhr Nedr 8:2 – 7 (in Russian)

Kojonen KK (1981) Geology, geochemistry and mineralogy of two Archean nickel-copper deposits in Suomussalmi, eastern Finland. Geol Surv Finl Bull 315

Kratts KO, Khil'tova VYa, Vrevskiy AB, Zapolnov AK, Krylov IN, Lobach-Zhuchenko SB, Manuilova MM, Platunova AP (1981) Stages and types of evolution of the Precambrian crusts in the Ancient Shields. Nauka, Leningrad (in Russian)

Kulikov VS (1971) The basaltic complex of Vetrenyi Poyas and some causes of variation in the Middle Proterozoic volcanism in the southeastern part of the Baltic Shield. In: Bel'kov IV, Shurkin KA (eds) Problems of the magmatism of the Baltic Shield. Nauka, Leningrad, pp 196 – 202 (in Russian)

Kulikov VS, Slyusarev VD, Kochnev-Pervukhov VI, Kravchenko AN (1976) The Suisaarian basic-ultrabasic complex in the Onega synclinorium. In: Kratts KO (ed) Intrusive basic-ultrabasic complexes of the Precambrian of Karelia. Nauka, Leningrad, pp 98 – 109 (in Russian)

Lavrov MM (1979) Hyperbasites and layered peridotite-gabbro-norite intrusions of the Precambrian in Northern Karelia. Nauka, Leningrad (in Russian)

Leake BE (1978) Nomenclature of amphiboles. Can Miner 16:501 – 520

Le Bas MJ (1962) The role of aluminium in igneous clinopyroxenes with relation to their parentage. Am J Sci 260:267 – 288

Leeman WP (1977) Mineral constitution of mantle source regions for Hawaiian basalts – rare earth element evidence for mantle heterogeneity. In: Dick HJB (ed) Magma genesis. Or Dep Geol Miner Ind Bull 96:169 – 183

Lehtinen M (1983) On the geology and Ni deposit of the Arola area, Kuhmo. M Sc Thesis, Univ Turku (in Finnish)

Masaytis VL, Moskaleva VN, Rumyantseva NA (1979) General problems in the study of magmatic formations. In: Masaytis VL (ed) Magmatic formations of USSR, vol I. Nedra, Leningrad, pp 5 – 25 (in Russian)

Meriläinen K, Sokolov VA (1981) Some attainments and problems relating to the geology of the Jatulian formations in Finland and Soviet Karelia. In: Puustinen K (ed) Geological, geochemical and geophysical investigations in the eastern part of the Baltic Shield. Committee for scientific and technical co-operation between Finland and the Soviet Union, Helsinki, pp 9–18

Murray RJ (1954) The clinopyroxenes of the Garbh Eilean sill, Shiant Isles. Geol Mag 91:17–31

Nwe YY (1976) Electron-probe studies of the early pyroxenes and olivines from the Skaergaard intrusion, east Greenland. Contrib Mineral Petrol 55:105–126

O'Hara MJ (1968) The bearing of phase equilibria studies in synthetic and natural systems on the origin and evolution of basic and ultrabasic rocks. Earth Sci Rev 4:69–133

Piirainen T, Honkamo M, Rossi S (1974) A preliminary report on the geology of the Koli area. Bull Geol Soc Finl 46:161–166

Piispanen R (1972) On the spilitic rocks of the Karelidic belt in western Kuusamo, Northeastern Finland. Acta Univ Ouluensis Ser A 4

Piquet D (1982) Mécanismes de recristallisations métamorphiques dans les ultrabasites: exemple des roches vertes archéennes de Finlande orientale (Ceintures de Suomussalmi-Kuhmo). Thèse 3. cycle, Univ Rennes

Popov MG (1971) Proterozoic magmatic complexes of Khjursjulya-Hautavaara (southern Karelia). In: Bel'kov IV, Shurkin KA (eds) Problems of the magmatism of the Baltic Shield. Nauka, Leningrad, pp 211–217 (in Russian)

Predovskiy AA, Fedotov ZhA, Akhmedov AM (1974) Geochemistry of the Pechenga Complex. Nauka, Leningrad (in Russian)

Rabinovich YuI (1978) Petrochemistry and nickel content of basic and ultrabasic rocks of some intrusive complexes in the Pechenga-Allarechensk area. In: Pokrovskiy AD (ed) Additional criteria for prospecting ore deposits in Karelia and the Kola Peninsula. Min Geol RSFSR, Moscow (in Russian)

Roeder PL, Emslie RF (1970) Olivine-liquid equilibrium. Contrib Mineral Petrol 29:275–289

Rusanov MS (1981) Tholeiite-komatiite formations of the Pechenga Complex. Sov Geol 2:98–112 (in Russian)

Sakko M (1971) Radiometric zircon ages of the Early-Karelian metadiabases. Geologi 23:117–118 (in Finnish with English summary)

Shaw DM (1970) Trace element fractionation during anatexis. Geochim Cosmochim Acta 34:237–243

Sidorenko VV (1983) Major geochemical characteristics of basic-ultrabasic magmatism in the Karelia-Kola region. Sov Geol 3:99–106 (in Russian)

Simonen A (1980) Prequaternary rocks of Finland, 1:1000000. Geol Surv Finl

Skufin PK (1980) Characteristics of the volcanism in the Proterozoic Pechenga structure. Pyul. Mosk. O-va Ispyt. Prirody. Otd Geol 55:120–131 (in Russian)

Slyusarev VD, Kulikov VS (1973) Geochemical evolution of the Proterozoic basic-ultrabasic magmatism (SE Baltic Shield). Nauka, Leningrad (in Russian)

Smol'kin VF (1974a) Layering in basic and ultrabasic rocks of the ore-bearing Pil'guyarvi intrusion (Pechenga). In: Bel'kov IV (ed) Petrology, mineralogy and geochemistry. Kol'skiy Filial AN SSSR, Apatity, pp 24–47 (in Russian)

Smol'kin VF (1974b) Composition of the layered intrusions of Pechenga reconstructed according to mineral relicts. In: Bel'kov IV (ed) Petrology, mineralogy and geochemistry. Kol'skiy Filial AN SSSR, Apatity, pp 48–54 (in Russian)

Smol'kin VF (1978) Olivine and clinopyroxene of the Pil'guyarvi gabbro-wehrlite intrusion (Pechenga). In: Tamarinov PM (ed) Minerals and mineral parageneses. Nauka, Leningrad, pp 55–71 (in Russian)

Smol'kin VF, Pahomovskiy YaA (1982) Nickel content of olivines and chrome spinels from ultramafic rocks of the Pechenga Ni-bearing complex. In: Minerals and mineral parageneses of copper-nickel and pyrite ores in the Kola Peninsula. Akademiya Nauk SSSR, Apatity, pp 54–63 (in Russian)

Sokolov VA, Stenar MM (1980) Principles of subdivision and correlation in the Precambrian of the south-eastern part of the Baltic Shield. In: Mitrofanov FP (ed) Principles and criteria of subdivision of the Precambrian in mobile zones. Nauka, Leningrad, pp 140–173 (in Russian with English abstract)

Stormer JC (1973) Calcium zoning in olivine and its relationship to silica activity and pressure. Geochim Cosmochim Acta 37:1815–1821

Suslova SM (1982) Petrochemical characteristics of the early Precambrian basaltic volcanism in the Kola Peninsula. Sov Geol 6:85 – 98 (in Russian)

Suslova SN, Staritsyna GN, Dagelayskaya IN (1981) Nickeliferous ultramafic-mafic volcanic-intrusive complexes of the Karelia-Kola and North Siberian regions. Int Geol Rev 23:1135 – 1147

Svetov AP, Golubev AI (1978) The volcanic complex of Suisaari. In: Sokolov VA (ed) Volcanic complexes of the Proterozoic in Karelia. Nauka, Leningrad, pp 119 – 128 (in Russian)

Svetov AP, Sokolov VA (1978) The volcanic complex of Shidguda. In: Sokolov VA (ed) Volcanic complexes of the Proterozoic in Karelia. Nauka, Leningrad, pp 111 – 119 (in Russian)

Taipale K (1983) The geology and geochemistry of the Archean Kuhmo greenstone-granite terrain in the Tipasjärvi area, eastern Finland. Acta Univ Ouluensis Ser A 151

Terakado Y (1980) Fine structures of rare earth element patterns of Tahitian rocks. Geochem J 14:155 – 166

Tkachenko KN (1982) The Pechenga Complex. In: Gorbunov GI (ed) Chemical analyses of basic-ultrabasic rocks of the Precambrian complexes in the Kola Peninsula. Kol'skiy Filial AN SSSR, Apatity, pp 167 – 206 (in Russian)

Väyrynen H (1938) Petrologie des Nickelerzfeldes Kaulatunturi-Kammikivitunturi in Petsamo. Bull Comm Geol Finl 116

Wilkinson JFG (1957) Clinopyroxenes of a differentiated teschenite sill near Gunnedah, New South Wales. Geol Mag 94:123 – 134

Yakovleva AK (1981) Amphiboles. In: Gorbunov GI (ed) Mineralogy of the copper-nickel ores of the Kola Peninsula. Nauka, Leningrad, pp 96 – 127 (in Russian)

Yakovleva AK, Osokin AS (1981) Micas. In: Gorbunov GI (ed) Mineralogy of the copper-nickel ores of the Kola Peninsula. Nauka, Leningrad, pp 127 – 142 (in Russian)

Zak SI, Makarov VN, Kochnev-Pervukhov VI, Proskuryakov VV, Zaskind ES, Batashev EV, Kolesnikov GP (1982) Geology, magmatism and ore formation in the Pechenga ore belt. Nedra, Leningrad (M-vo Geol RSFSR) (in Russian)

Zhangurov AA, Predovskiy AA (1974) Geochemistry of the rocks and the character of the parental magma of the Ni-bearing intrusives in Pechenga. In: Bel'kov IV (ed) Petrology, mineralogy and geochemistry. Kol'skiy Filial AN SSSR, Apatity, pp 72 – 83 (in Russian)

Zhangurov AA, Smol'kin VF (1982) Basite-hyperbasites of the middle and late Karelian and Caledonian stages. In: Gorbunov GI (ed) The Karelian Imandra-Varzuga Zone (Geology, geochemistry and history of evolution). Nauka, Leningrad, pp 145 – 165 (in Russian)

Zhangurov AA, Petrov VP, Smol'kin VF, Voloshina ZM, Smirnova TN (1974) Pyroxenes of the differentiated intrusives of Pechenga. Mat. Po Miner. Kol'sk. Po-va 10. 52 – 61. Nauka, Leningrad, pp 52 – 61 (in Russian)

Peridotitic Komatiites and the Origin of Ores (Southeastern Part of the Baltic Shield)

V. I. KOCHNEV-PERVUKHOV[1], E. S. ZASKIND[1], and V. V. PROSKURYAKOV[2]

Abstract

Komatiite magmatism of the Archean greenstone belts of the southeastern part of the Baltic shield is considered here. The conformable bedding and interbedding of bodies of komatiites with pillow basalts and horizons of tuffs, with breccias at their tops, pillows, amygdules, glass and spinifex textures permit the determination of these rocks as peridotite komatiite lavas. Plutonic analogs of these rocks are found in ultramafic effusion feeders. Nickel-copper ore mineralization is orthomagmatic and is associated with komatiitic intrusives. The nickel ore potential of units associated with komatiite magmatism seems to be correlated with the appearance of island-arc basalts: the larger these are, the smaller the ore potential of the structures. And, in this sense, the nearest analogs of the belts in the Baltic shield are the Canadian greenstone belts, not the Western Australian. These may be considered geotectonically as nickeliferous provinces.

The southeastern part of the Baltic shield is a region of wide spreading greenstone belts; these are filled with Archean series and form a complicated network of the narrow elongated structures within gneissose granite blocks (Fig. 1). The extension of the belts ranges from 1 – 2 km to a few tens of kilometers. In a number of such structures, komatiitic series are found (Fig. 1). The authors studied the volcanic-plutonic komatiite complex in the Kenozero-Kozhozero greenstone belt.

Komatiitic bodies have various lengths and thicknesses and conform with basalt sheets. Deformational and metamorphic changes in such basalts and komatiites are much the same. The generalized section of the researched bodies consists of two parts (Fig. 2): the upper, less thick part, is composed of breccias and vitreous and spinifex rocks; the lower part of serpentinite after ordinary peridotite and dunite. The ratio of thicknesses of the upper and lower parts fluctuates from 1 : 10 to 1 : 50.

The glass, the skeletal forms of magmatic minerals, and the spinifex textures are doubtless evidences of the crystallization of these rocks from a melt under high temperature-gradient conditions. The composition of the melt is determined from the composition of altered glass and rocks with spinifex textures (Table 1), and corresponds to that of peridotite komatiites (Arndt et al. 1979, Brooks and

[1] The Central Research Exploration Geology Institute of Basic and Precious Metals, Moscow, USSR
[2] The Industrial Geological Trust "Sevzapgeologiya", Leningrad, USSR

Geology and Metallogeny of Copper Deposits
(ed. by G. H. Friedrich et al.)

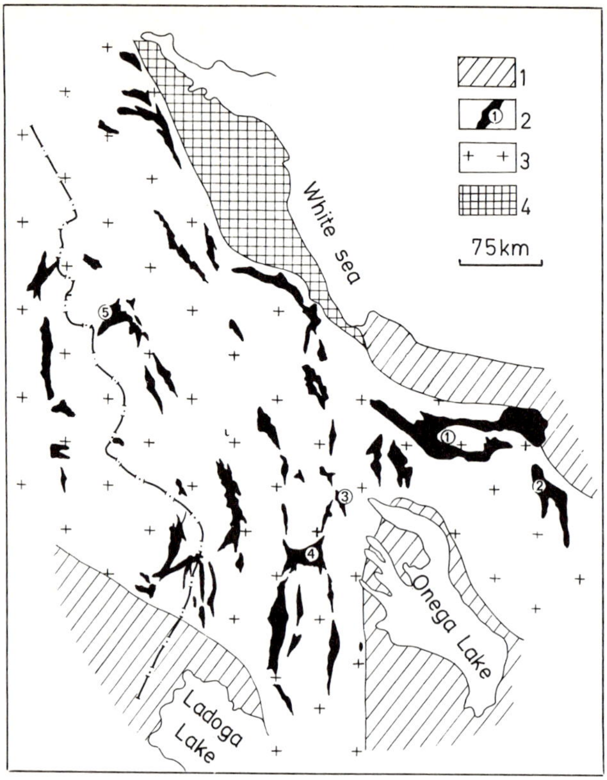

Fig. 1. Extending greenstone belts in the southeastern part of the Baltic shield. *1* proterozoic structures; *2* archaean greenstone belts; *3* granite-gneissic block; *4* belomorides. Greenstone belts designated by the *circles numbers: 1* Kenozero-Kozhozero; *2* Toksha; *3* Koikara; *4* Palaya Lamba; *5* Kostomuksa

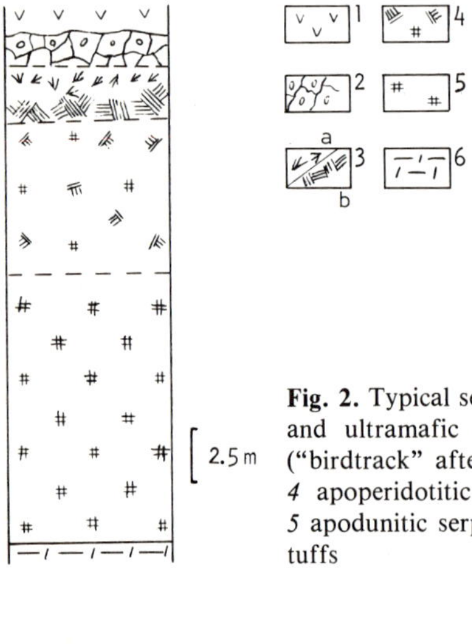

Fig. 2. Typical section of peridotitic komatiite flow. *1* Basalts; *2* breccias and ultramafic glass with amygdules; *3* spinifex texture: *a* random ("birdtrack" after Naldrett and Mason 1968) spinifex, *b* sheaf spinifex; *4* apoperidotitic serpentinites with a small block of sheaf spinifex; *5* apodunitic serpentinites; *6* chlorite-quartz-carbonate shale after acidic tuffs

Table 1. The chemical composition of ultramafic rocks in the Kenozero-Kozhozero greenstone belt[a]

N[b]	SiO_2	TiO_2	Al_2O_3	FeO	MnO	MgO	CaO	NiO	Cr_2O_3	Total
1	44.9	0.34	6.0	11.3	0.22	30.5	5.8	0.11	–	99.2
2	44.8	0.39	6.5	11.3	0.19	29.6	6.3	0.12	0.33	99.5
3	46.2	0.33	4.1	12.2	0.18	31.7	4.5	0.26	0.34	99.8
4	45.1	0.20	2.8	9.4	0.15	39.6	2.1	0.15	0.35	99.8
5	43.7	0.10	0.8	10.1	0.12	42.1	1.5	0.29	0.23	98.9

[a] Analysis was done by AA Kuleschova. The analyses were recalculated as water free to 100%.
[b] *1* Altered glass from peridotitic komatiites; *2* rocks with spinifex texture; *3* average composition of the intrusive bodies (without the olivine cumulate); *4* dunite from peridotitic komatiite flows; *5* dunite in komatiitic intrusives.

Hart 1974), described in Africa (Viljoen and Viljoen 1969), Canada (Pyke et al. 1973), Australia (Ross and Hopkins 1975), and the Russian Platform (Krestin 1978).

The conformable bedding and interbedding of bodies of peridotitic komatiites include flows of pillow basalts and horizons of tuffs, the tops of which contain breccias, pillows, amygdules, and glasses, permit the determination of the rocks studied as effusives.

The actual discovery of ultramafic effusive rocks among the Archean series of the greenstone belt is important because it gives additional possibilities for the discussion of two questions: (1) geologic age and formation type of the ultramafic intrusions of this region, and (2) the general features of the succession formed, and their connection with, the ore potential of the komatiite series that are formed in different geotectonic settings. The former has importance only for the region studied, but the latter is important for any structures containing komatiites.

1. In the terrain where the Kenozero-Kozhozero greenstone belt is situated, ultramafic intrusions are widespread. These intrusions are considered as Proterozoic and are compared with ones of the Vetrenii Poyas structure, which is the younger of the greenstone belts. Within this intrusive complex, it is possible to designate a group of bodies, the composition of which (Table 1) is analogous to the peridotitic komatiites. There intrusives are contained within or near the greenstone belt and are located in horizons that underlie rock masses that include komatiites. Usually these are steep dipping isometric or extensive bodies with considerable predominance of dunite and wehrlite. Peculiarities of the zoning and the position of these bodies in the belt of rock sequences show symmetrical zoning relative to the central part of the intrusive body in which the acid differentiates are confined to the contacts. These are localized in horizons that underlie the komatiite-bearing parts of the column that occur inside the domain of the komatiitic effusive rocks and transitions of steep-dipping bodies to bedding-conformable komatiitic flows with spinifex textures. These relations permit then to be considered as feeders of ultramafic effusions, and to be included in the Archean komatiite series; they are excluded, however, from the magmatism associated with Proterozoic structures. Their communality of origin is confirmed by

the metalliferous specialization of the rocks: the accessory ore mineralization in effusive and plutonic facies of the komatiites is qualitatively the same – in such sulfide material as is present, pentlandite predominates, and the Ni : Cu ratio ranges widely, although essential segregations of the ore minerals occur only in intrusions, and the ores have features of an orthomagmatic origin. The metamorphic process does not lead to accumulation of the sulfide substance, it results only in a modification of textural-structural peculiarities of the mineralization, in the decrease of the Ni : Cu ratio and in the appearance of a series of impurities that are nontypical for magmatic segregations. The metamorphism also leads to the appearance of scattered heazlewoodite mineralization.

2. The rock sequence of the greenstone belt, to which the studied region belongs consists of three formations: (1) the nondifferentiated basaltic, (2) the sequentially differentiated volcanic-sedimentary dacite-liparite-andesite-basaltic, and (3) the komatiite-basaltic. The base of the rock sequence is unknown, and the age of the granitoids, among with the greenstone belt is located, has not reliably been determined. In the interpretation of the rock sequence, the positions of the amygdaloidal zones and of the pillows are confined to basaltic flows, and those of the spinifex are located in the peridotitic komatiites. On these bases, the komatiite-basaltic formation is above and the basaltic nondifferentiated formation is below the sequentially differentiated volcanic-sedimentary formation. It is not excepted, however, that the nondifferentiated basaltic formation would have been a facies of the komatiite-basaltic formation with the transition from one to the other being along the strike. But even if this is true confining of the komatiite-basaltic formation to the upper part of the column and its position above the sequentially differentiated formation would not change.

The lower formation consists of pillow and massive tholeiite basalts, among which the few flows of ferriferous varieties and the subvolcanic bodies of the quartz-porphyry were found. These basaltic flows are separated by thin (less than 1 m) intercalations of the slates. The rock sequence of the formation is wonderfully uniform.

The middle formation consists (from base to top) of basalts Al-basalts, andesite-basalts, andesites, and dacite-liparites that are interbedded with horizons of the intermediate and silicic tuffs and rhythmic tuffaceous flysch (the psammitic parts of the rhythms consist of silicic material, in which there are plagioclase and fragments of acidic effusive rocks, but grains of quartz are never found. The changes in the succession take place along the strike: in the northern part of the belt the succession is essentially volcanogenic and includes intermediate and acidic tuffs; the central part is the tuffogenic flysch with a single dacite-liparitic flow. The southern part of the belt is composed of ferro-cherty rocks of the Algoman type (Dimroth 1975) with rare persistent beds of black shales and acidic tuffs. Flows of andesites occurred only in the northern part of the belt.

The massive sulfide mineralization of three types (after Smirnov 1968) has been widely developed:

1. In the northern part of the structure, this mineralization is hydrothermal copper-zinciferous (Cu > Zn) ores in quartz-sericite and quartz-chlorite metasomatites, following the silicic effusives and tuffs and the exhalative-sedimentary

pyritic mineralization with the copper and zinc in silicic tuffs that are part of the quartz-sericite metasomatites in the footwall.

2. In the middle part of the belt is a sedimentary pyritic mineralization (in the pelitic parts of the tuffogenic flysch) with abundant pyritic concretions and nodules.

3. In the southern part of the belt is a sulfide-oxide mineralization contained in the cherty rocks.

The zoning of the mineralization of the various types and the association of pyrites with one and the same silicic volcanites suggest consanguinity and association in time.

The higher formation consists of pillow and massive tholeiitic basalts with a horizon of the peridotitic komatiites at its base. These basaltic flows interbed with basic tuffs and the black shales, whereby the rare pyritic mineralization, has undergone redeposition and pyrrhotinization. The composition of the basalts is the same as in the lower formation and differs only in the absence of the ferriferous varieties.

It is obvious that the characteristics of the rock sequence and of the ore mineralization of the middle formation are analogous of the ones of the island arcs.

The analysis of the composition of the basalts of the belt by Pearce's method (Pearce et al. 1977) (Fig. 3), has shown that two petrochemical types can be distinguished, which are the oceaniclike basalts (in the lower and higher formations) and the basalts of the island arcs (only in the middle formation).

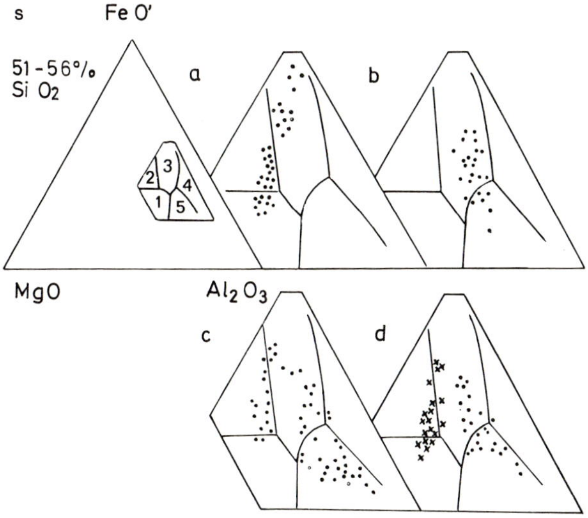

Fig. 3. MgO – FeO – Al_2O_3 for basalts on the Archean greenstone belts: Western Australia, Yilgarn block (*a*); Canadian shield – Yellowknife (*b*), and Abitibi (*c*); Baltic shield, the Eastern Karelia (*d*). In *d x* represents basalts of the upper and lower parts of the rock sequence; indicates the same from the middle part. These are designated by numbers: *1* basalts of middle-oceanic ridges and oceanic floor; *2* basalts of oceanic islands, which are associated with interplate volcanism and the series of oceanic rises; *3* continental platobasalts and traps; *4* basalts of oceanic island, which are located along the axis of middle-oceanic ridges; *5* basalts of island-arcs and active continental margins

Table 2. Nickel-bearing provinces

Intercontinental systems of epicratonic zones of rifting		Zones of rifting in margin continental island-arc systems	
Autonomic intercontinental rift and riftlike zones and regions	Rift zones connected with initial stage of passive continental margins forming	Volcanic island-arcs	Marginal seas, microcontinents or regions of early consolidation
A. Epiarchean			
Siberian platform (Mz_1) Norilsk South Africa (Mz_1) – Insizwa Canadian shield (R_2) Keweenawan district – Duluth Baltic shield (Pt_1) – Pechenga Canadian shield (Pt_1) – Ungawa, Sudbury, Thompson belt Transvaal Craton (Pt_1) – Bushveld	Greenland (Kz) – Skaergaard Canadian shield coppermine district (R_2) – Muskox Baltic shield, Finland (Pt_1) – Kemi-Suhanko	Northwest Pamirs (Pz_3) Baltic shield (Pt_1) south Finland, central Sweden	Kamchatka (Mz_2) Central Kazakhstan (Pz_1) – Zlatogorskii complex Aldan shield (R_3) – North Baikal region Baltic shield (Pt_1) – the main sulfide belt of Finland Voronezh Shield (Pt_1)
B. Archean			
West Australia – Yilgarn block Limpopo belt – Selebi-Pikwe	Transvaal shield – greenstone belt Barberton	Canadian shield – Abitibi; Yellowknife; Rhodesian shield; West Australia – Pilbara block; Baltic shield – Granulitic belt of Lapland, East Karelian belts	

An analogous study was undertaken previously for the Archean greenstone belts of the world according to literature data (Zaskind and Ivanjuta 1981) and this has allowed us to distinguish three types of sequences: without island-arc basalts (Western Australia, Barberton), with island-arc basalts only (Yellowknife in Canada), and sequences of mixed rock type (Abitibi in Canada, Rhodesian belts). If the Barberton belt is not taken into account, the scale of ultramafic magmatism and its nickel ore potential depend on the sequence type. In fact, sequences showing an island-arc rock type does not contain nickel-copper ores at all. Sequences of the mixed succession type contain small deposits and large or prominent deposits of nickel-copper ores are connected with sequences lacking island-arc associations. Therefore, the presence in the sequence of island-arc associations is a negative indication of the lack of nickel-copper ores in them. In this sense, the nearest analogs to the belts of the region studied are the Canadian greenstone belts, not those of Western Australia. However, judging by the example of Western Australia and Barberton, the nickel-copper ore potential of Archean sequences with komatiites belongs only to one of enumerated types; any rock sequence with indications of the island-arc development has low ore potential, and no sequence without komatiites has a high ore potential.

Literature data on the tectonic environments of geologic complexes (Peive et al. 1972, Piirainen et al. 1974, Hain 1977, Archibald et al. 1978, Condie 1980, Baragar and Scoates 1981, Barton and Key 1981, Jackson and Iannelli 1981, Gaal 1982, Sims et al. 1982, Milanovskii 1983 and other) and their comparison with the ones in nickeliferous epiarchean provinces, for which the geotectonic conditions can be determined with more confidence, confirm the above. The initial point of the paleotectonic analysis, the results of which are in Table 2, is the Sawkins' model (Sawkins 1972). This model was enhanced by the ideas of a definite system of the rifting zones (triple junctions discussed by Burke and Dewey 1973). The main conclusion following from this scheme is that all the most important provinces throughout the world are confined to the regions of triple junctions that are intercontinental in the paleotectonic sense and evidently correspond to zones of epicratonic rifting. These regions can have a spatial connection with margin-continental geostructures, and they can develop out of this connection. In this sense, the authors consider them as autonomous. Probably, in Archean time, the belts of the Yilgarn block (Western Australia) corresponded to structures of this kind. Therefore, it is possible to conclude that the tendencies for a connection between magmatism and geotectonic position, which are determined in younger regions, can be traced (though it cannot be so clearly seen) in the older ones and can be used for the classification and evaluating of the latter.

References

Archibald NS, Bettenay LF, Binns RA, Groves DI, Gunthorpe RS (1978) The evolution of Archean greenstone terrains, Eastern Goldfields Province, Western Australia. Precambrian Res 6:103 – 131

Arndt NT, Francis J, Hynes AJ (1979) The field characteristics and petrology of Archean and Proterozoic komatiites. Can Miner 17, 2:147 – 186

Baragar WRA, Scoates RFJ (1981) The Circum-Superior belt: a proterozoic plate margin. In: Kröner A (ed) Precambrian plate tectonics. Elsevier, Amsterdam Oxford New York, pp 297 – 330
Barton SM, Key RM (1981) The tectonic development of the Limpopo mobile belt and the evolution of the Archean cratons of southern Africa. In: Kröner A (ed) Precambrian plate tectonics. Elsevier, Amsterdam Oxford New York, pp 185 – 211
Brooks CK, Hart SR (1974) On the significance of komatiites. Geology 2:107 – 110
Burke KP, Dewey JF (1973) Plume-generated triple junctions: Key indicator in applying plate tectonics to old rocks. J Geol 81, 4:406 – 432
Condie KC (1980) Origin and early development of the earth's crust. Precambrian Res 11:183 – 197
Dimroth E (1975) Paleo-environ of iron-rich sedimentary rocks. Geol Rundsch 64, 3:751 – 766
Eriksson KA (1982) Sedimentation patterns in the Barberton mountain land, South Africa, and the Pilbara block, Australia: Evidence for Archean rifted continental margins. Tectonophysics 81, 3/4:179 – 193
Gaal G (1982) Proterozoic tectonic evolution and late Svekokarelian plate deformation of the Central Baltic Shield. Geol Rundsch 71, 1:120 – 129
Hain VE (1977) Problems of the Precambrian Geology. In: Dagelaiskiy VB (ed) Problems of the geology of Early Pre-Cambrian. Nauka, Leningrad, pp 5 – 12 (in Russian)
Jackson GD, Iannelli TR (1981) Rift-related cyclic sedimentation in the neohelikian borden basin, northern Baffin Island. Geol Surv Can Pap 81-10:269 – 302
Krestin EM (1978) The first find of komatiites in USSR. Rep AN USSR 242-2:412 – 415 (in Russian)
Milanovskii EE (1983) Rifting in the earth's history. Nedra, Moscow (in Russian)
Naldrett AJ, Mason GD (1968) Contrasting Archean ultramafic igneous bodies in Dundonald and Clergue townships, Ontario. Can J Earth Sci 5:111 – 143
Pearce TH, Gorman BE, Birkett TC (1977) The relationship between major element chemistry and tectonic environment of basic and intermediate volcanic rocks. Earth Planet Sci Lett 36:121 – 132
Peive AV, Schtreiss NA, Mossakovskii AA (1972) Paleozoic of the Eurasia and some questiones of evolution of the geosyncline process. Sov Geol 12:7 – 25 (in Russian)
Piirainen T, Hugg R, Isohanni M, Juopperi A (1974) On the geotectonics and ore forming processes in the basic intrusive belts of Kemi-Suhanko, and Syöte-Näränkävaara, northern Finland. Bull Geol Soc Finl 46, 2:93 – 104
Pyke DR, Naldrett AJ, Ekstrand OR (1973) Archean ultramafic flows in Munrotownship, Ontario. Geol Soc Am Bull 84:955 – 978
Ross IR, Hopkins GMF (1975) Kambalda nickel sulfide deposits. In: Knight CL (ed) Economic geology of Australia and Papua New Guinea. I. Metals. Aust Inst Min Met, Mon, pp 100 – 121
Sawkins FJ (1972) Sulfide ore deposits in relation to plate tectonic. J Geol 80, 4:377 – 397
Sims PK, Card KD, Lumbers SB (1982) Evolution of early proterozoic basins of the Great Lakes region. Pap Geol Surv Can 81-10:379 – 397
Smirnov VI (1968) Pyrite deposits. In: Smirnov VI (ed) Genesis of the endogenic ores deposits. Nedra, Moscow, pp 586 – 647 (in Russian)
Viljoen MJ, Viljoen RP (1969) Evidence for the existence of a mobile extrusive peridotitic magma from the Komati Formation of the Onvervaht Group. Geol Soc S Afr Spec Publ 2:87 – 113
Zaskind ES, Ivanjuta ZF (1981) The some peculiarities of the petrochemistry of nickeliferous region's basalts. In: Stepanov VK (ed) The genesis and localization of the nickel-copper ores. Moscow, p 96 (in Russian)

Geologic Setting of Selected Chromium and Nickel Deposits of China

P.-F. FAN[1]

Abstract

Five selected chromium deposits of China are described: Luobusa of Xizang (Tibet), northwestern and northern Inner Mongolia, western Henan, and Panzihua of Sichuan. The first four are podiform deposits and the fifth is stratiform. Six selected nickel deposits of China are described: Jinchuan of Gansu, Dali and Ailaoshan of Yunnan, weathering products near Ailaoshan, and sedimentary nickel deposits of Yangzi craton. These nickel deposits can be classified into four types according to their geologic settings:

1. Rift valley deposits – Jinchuan and Dali
2. Obducted ophiolites – Ailaoshan
3. Weathering products – near Ailaoshan
4. Sedimentary rocks – Lower Cambrian black shales of Yangzi craton

Introduction

The chromium and nickel deposits of China described here were chosen for discussion because of the availability of published information. They represent various types of deposits and most of them are considered major deposits. The chromium deposits are: the Luobusa of Tibet, the northwestern and northern Inner Mongolia, the western Henan, and the Panzihua of Sichuan. The nickel deposits are: the Jinchuan of Gansu, the Dali of western Yunnan, and the Ailaoshan of southern Yunnan (Fig. 1).

A number of chromium occurrences have been reported in China, but most of them are small. The largest occurrences reported are in the Zhayier Mountains, northwest of the Junggar basin, 400 km west of Uramchi, the capital of Xinjiang. The reserves are estimated at 1 million tons of 35% Cr_2O_3 ore. Production of chromium in Xinjiang began in 1967 (Brady 1981, Chin 1983a, Mikami 1983), but no geological reports are available. A number of high-grade chromium deposits have been found in Tibet, and chromium has also been found in Panzihua titanium iron ore.

[1] Hawaii Institute of Geophysics, University of Hawaii, Honolulu, HI 96822, USA

Geology and Metallogeny of Copper Deposits
(ed. by G. H. Friedrich et al.)

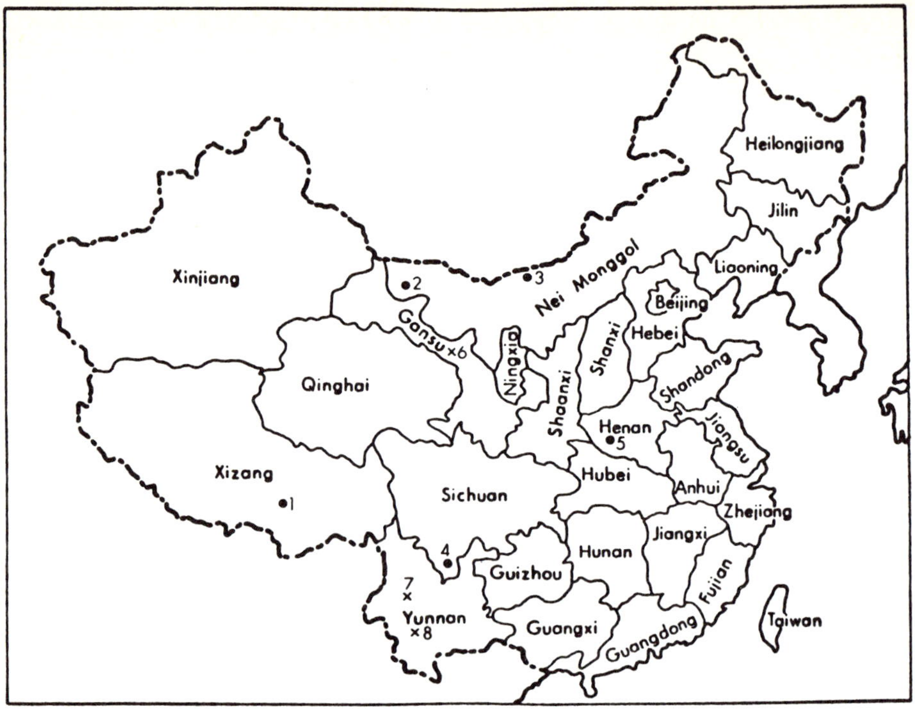

Fig. 1. Location of chromium and nickel deposits of China that are discussed here. Symbols: ● Cr; X Nickel; *1* Luobusa; *2* N.W. Nei Monggol (Inner Mongolia); *3* N. Nei Monggol (Inner Mongolia); *4* W. Henan; *5* Panzihua; *6* Jinchuan; *7* Dali; *8* Ailaoshan

The nickel reserves of China have been estimated at 7 million tons of contained nickel. The largest reserve, estimated at 5 million tons of contained nickel, is in the Jinchuan deposit of Gansu. Jinchuan, the second largest nickel sulfide deposit in the world (after Sudbury of Canada), produced 15000 tons of nickel in 1981. Nickel sulfide deposits are also found in Sichuan, Yunnan, Xinjiang, Hebei, Hunnan, Jiangsu, and Shandong. There are no estimates of the laterite deposits (Chin 1980, 1983a, b).

Tectonic Setting

China is part of the Eurasian plate and is separated from the Siberia plate by the Junggar-Solon Suture, from the India plate by the Yarlungzangbu-Indus Suture, and from the Philippine plate by the Longitudinal Valley of Taiwan (Fig. 2). The nucleus of China comprises the Sino-Korean, Tarim, and Yangzi cratons. Mongolia and northern China were consolidated with Siberia during the Precambrian and Early Paleozoic Caledonian orogenies. The collision of the Silurian and the Sino-Korea cratons formed the Solon-Hegen mountain suture (No. 3 of

Fig. 2; Li 1980). The basement of the Yanzi craton (II of Fig. 2) consolidated about 700 Ma. The Yanzi craton coalesced with the South China fold belt during Silurian time and continued moving northwestward. During Permian time these two cratons were located at 2°N (McElhinny et al. 1981). During Late Triassic time, the Yangzi-South China cratons collided with the Qinling fold belt, and the maritime fold belt of southwestern China consolidated with the Yanzi-South China cratons. The Yarlungzangbu-Indus Cenozoic Suture (No. 13, of Fig. 2) is the result of the subduction of the Tethys Ocean's crust under the Eurasian continent and subsequent collision with the Eurasian and Indian continents (Fan 1984).

Chromium Deposits

There are two types of chromium deposits in China: podiform, exemplified by the Luobusa deposits of Tibet, northwestern and northern Inner Mongolia, and western Henan deposits; and stratiform, represented by the Panzihua deposits of Sichuan.

Luobusa. The Luobusa deposits are located in the eastern segment of the southern Xizang ophiolite belt, south of the Yarlungzangbu rift. The belt, which trends east-west and plunges southward, is 43 km in length and 3.7 km in width. The clastic sediments are sandstone and conglomerate with ultramafic boulders. The belt has a fault contact to the north with Miocene nonmarine clastic sediments, and a southern contact with upper Triassic flysch deposits consisting of metamorphosed sandstone, slate, and phyllite. The ophiolite belt was emplaced during Late Yenshanian to Early Himalayan time (Yen et al. 1979) during the collision of Eurasian and Indian continents.

The ophiolite belt comprises three layers (from north to south or from bottom to top). (1) dunite (bottom) – 150 – 800 m in width and containing veinlets of chromite; (2) dunite and harzburgite – 200 – 1 400 m in width; dunite occurs near the bottom and chromite occurs near the middle of the lower section with harzburgite; (3) harzburgite and dunite (top) – small chromite bodies only (Fig. 3). In the ophiolite belts there are 181 chromite deposits and numerous mineralized regions. Thirteen of the ore bodies are greater than 100 m in length. Most of them are located in the second belt (dunite-harzburgite) and most of the veins are clustered together. Chromite and Cr-spinel are the major ore minerals. Other minerals present are magnetite, millerite, uvarovite, Cr-chlorite, Cr-mica, bornite, olivine, serpentine, and pyroxene. Os, Ru, Rh, and Ir are also reported to be associated with the chromite ore bodies (Yen et al. 1979).

The Luobusa deposits are podiform. The magma while intruded into the deep-seated magma chamber is considered still molten, in a state of crystal mush. The Luobusa deposits were formed in an unstable tectonic environment and underwent structural deformation during the process of crystallization. Chi and Shi (1981) recognized three genetic types ofd chromite in the Luobusa deposits. (1) Deposits formed by segregation of crystals during a late magmatic stage. They

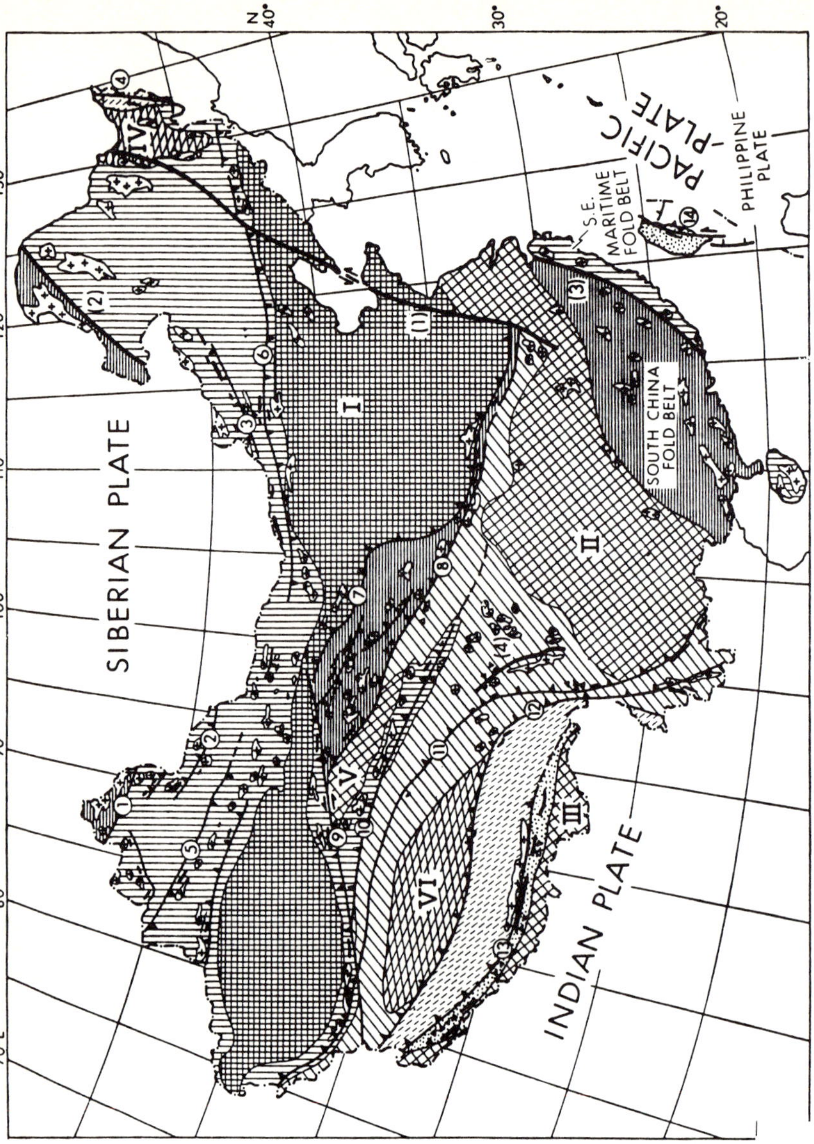
SIBERIAN PLATE
INDIAN PLATE
PACIFIC PLATE
PHILIPPINE PLATE
S.E. MARITIME FOLD BELT
SOUTH CHINA FOLD BELT
I
II
III
IV
V
VI
70°E
80°
90°
100°
110°
120°
130°
N 40°
30°
20°

Fig. 2

Fig. 2. The plate tectonics of China (Li 1980)

Precambrian Massif	Early Mesozoic Fold Belt	Melange
Late Precambrian Massif	Late Mesozoic Fold Belt	Glaucophane Schist
Median Massif	Cenozoic Fold Belt	Subduction Zone
Early Paleozoic Fold Belt	Ophiolite Belt	Deep-Seated Fracture
Late Paleozoic Fold Belt	Granite of the Same Tectonic Stage	

Subduction Zone and Suture

1 Altay Early Paleozoic Subduction Zone
2 Junggar Late Paleozoic Subduction Zone
3 Solon-Hegen Late Paleozoic Subduction Zone
4 Nadanhada Early Mesozoic Subduction Zone
5 Tianshan Late Paleozoic Subduction Zone
6 Yinshan-Tumen Late Paleozoic Subduction Zone
7 Qilian-Qinling Early Paleozoic Subduction Zone
8 Qinghai-Qinling Early Paleozoic Subduction Zone
9 Altun Late Paleozoic Subduction Zone (or Transform Fault)
10 Kunlun Late Paleozoic Subduction Zone
11 Hoh Xil-Jinsha-Ailao Early Mesozoic Subduction Zone
12 North Xizang-West Yunnan Late Mesozoic Subduction Zone
13 Yarlungzangbu-Indus Cenozoic Suture
14 Taiwan Cenozoic Suture (or Transform Fault)

Massif and Median Massif

I. Tarim-Sino-Korean Massif
II. Yangzi Massif
III. Northern margin of Indian Massif
IV. Jiamusi Median Massif
V. Qaidam Median Massif
VI. Qiangtang Median Massif

Depth Fracture

(1) Tan-Lu Depth Fracture
(2) Deierbugan Depth Fracture
(3) Lishui-Haifeng Depth Fracture
(4) Garzê-Litang Depth Fracture

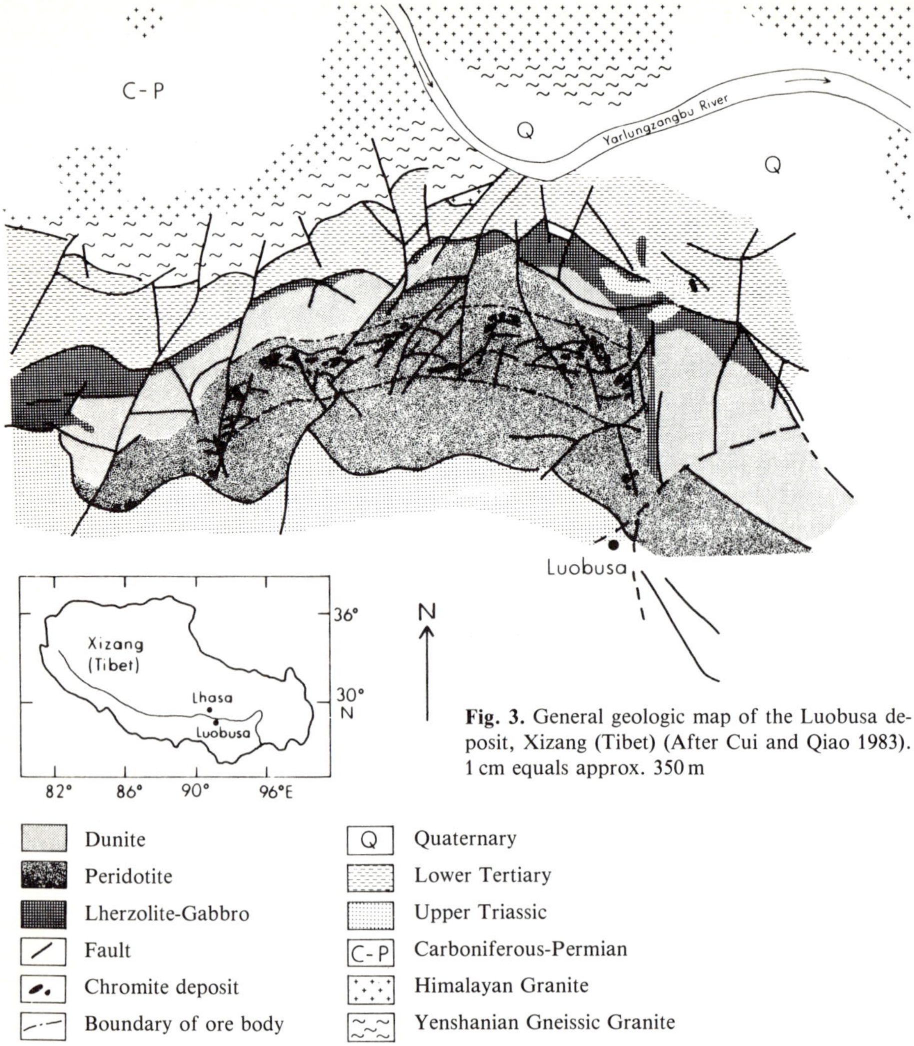

Fig. 3. General geologic map of the Luobusa deposit, Xizang (Tibet) (After Cui and Qiao 1983). 1 cm equals approx. 350 m

could be segregated in situ or by pressure filtering. (2) Primordial cumulative deposits of a late magmatic stage. The ore-forming Cr-spinel has pisolitic, nodular, and ellipsoidal shapes. The pisolites, oriented in graded beds, were formed by gravitational settling and by current differentiation. (3) Deposits formed by injection during a late magmatic stage. The ore bodies form a sharp boundary with the country rocks. The three genetic types occur both separately and in combination. Most of the Luobusa deposits show a zigzag form. Seven types of zigzag forms were recognized by Hu et al. (1981); they were shear fissures developed during the time of ore formation.

Podiform chromite deposits are now interpreted as remnants of on-land oceanic lithosphere. They occur as lenses, pods, slabs, and irregular shapes. The chromite found in ophiolitic harzburgites is suggested to have originated beneath an oceanic spreading ridge. The dunite wall rocks probably resulted from the partial melting of harzburgite during ore body emplacement. The basic texture patterns, seen to be a disruption of the primordial cumulative chromite structure, resulted from deep-seated mantle subsolidus deformation (Thayer 1969). Late Mesozoic and Cenozoic examples include deposits in Greece (Moores 1969), Cyprus (Greenbaum 1977), Turkey (Thayer 1964), Oman (Peters and Kramers 1974), and New Caledonia (Nicolas et al. 1981). Nicolas et al. (1981), in their study of southern New Caledonia nodular chromite, indicated that nodular chromite deposits resulted in the concentration of chromite by a process of elutriation in parts of the magmatic conduits. In the conduits, chromite grains have an opportunity to agglomerate into pisolitic, pelletlike, ellipsoidal, or massive shapes and eventually settle.

Northwestern Inner Mongolia. The northwestern Inner Mongolia chromite deposits, located along the east-west trending rift zone, constitute the eastern segment of the Hercynian Tienshan fold belt. The country rocks consist of hornblende-chlorite schist and mica-quartz schist and are metamorphosed Ordovician-carboniferous marine volcanics. The ultramafic rocks were emplaced during the middle Hercynian. The gabbro intrusion occurred before the ultramafic intrusion and the Permian biotite granite intrusion followed the ultramafic intrusion (Fig. 4). The ultramafic body consists of harzburgite and dunite, with

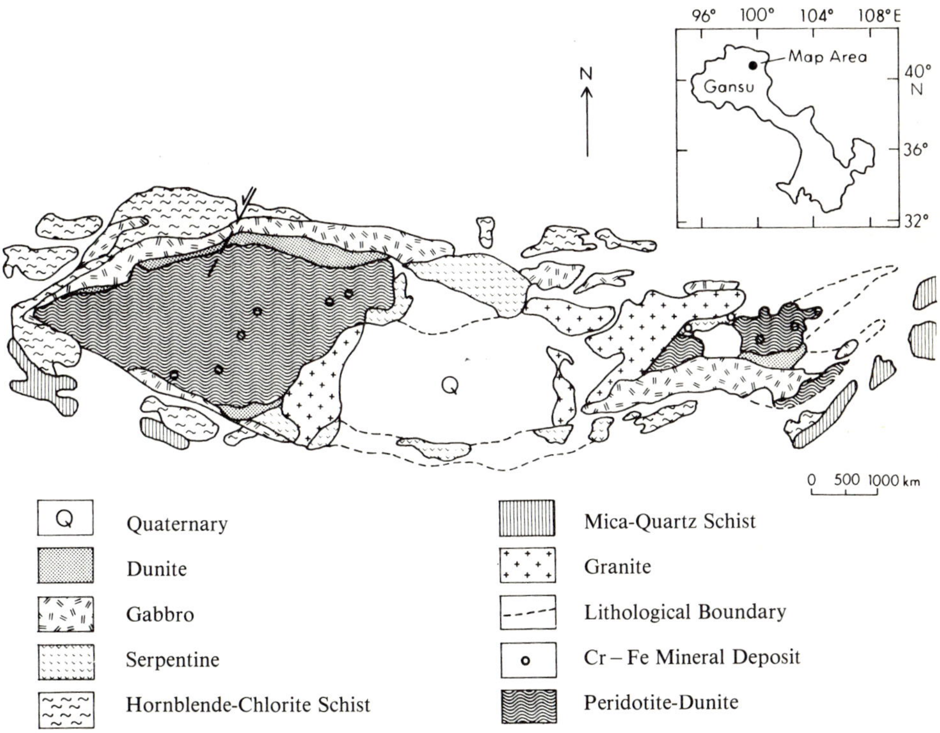

Fig. 4. General geologic map of northwestern Inner Mongolia (After Wei 1978)

chromite occurring in the dunite schlieren, usually in the middle segment of the ultramafic body or in the upper-middle part of the harzburgite zone. The ores occur as magmatic segregations in the form of stringers, lenses, veins, pods, and pillars. Three genetic types of chromite deposits are recognized by Wei (1978): those formed by (1) in situ magmatic segregation, (2) late magmatic injection, and (3) multiple phases of late magmatic segregation.

Northern Inner Mongolia. A chromium deposit is reported by Cao (1976) in the western part of Inner Mongolia (Fig. 1). The chromite-bearing ultramafic body is located in the Inner Mongolia continental margin and western segment of the Hercynian fold belt. The country rocks consist of over 10000 m Carboniferous flysch deposits metamorphosed to arkosic sandstone, slate, phyllite, sericite-quartz schist, and garnet-mica schist. An east-west trending rift south of the deposit is related to the formation of the ultramafic body. The ultramafic body is 32 km in length and 6 km in width and consists of mostly harzburgite with veins of dunite. Ores, which occur as veins, pods, lenses, and stringers, form sharp contacts with the host rocks. The enstatite in harzburgite is characterized by secondary banded structures, banded crystal faces, and undulating extinction under microscope. Cao (1976) suggested that this ultramafic complex was emplaced as a solid intrusion during tectonism.

Western Henan. The chromium deposits in Henan (Fig. 1), described by Wan (1977) and Zhang (1977), are located at the southern flank of an anticline of the eastern segment of the east-west trending Qinling range. A rift cuts through the region and separates two metamorphic facies. Precambrian migmatite occurs north of the rift and Precambrian marble and quartzite to the south. The chrome-bearing ultramafic rocks intrude into the migmatite. Chromite and Cr-spinel are the major ore minerals, although 45 other minerals were identified and their genesis and paragenesis discussed. Forty-seven chromium ore bodies have been discovered in this area. The preexisting chromite deposit, which is magmatic in origin, was altered during migmatization.

Panzihua. The Panzihua deposits are vanadium-bearing titanomagnetite deposits that are also rich in chromium. Panzihua is located in southwestern Sichuan (Fig. 1) at the western edge of Yangzi craton. The country rocks are Sinian limestone. A layered ultramafic-mafic intrusion was emplaced during Middle-Late Paleozoic time (Yen et al. 1979). The intrusive body is 35 km in length and 2 km in width, trending northeast 45° and dipping northwest at 50° – 60° (Fig. 5). Five rhythmic layers are present (from top to bottom). (1) Layered gabbro – the dark-colored gabbro has a thickness from 500 – 1500 m with stockworks. It is poor in ore content. (2) Upper light-colored layered gabbro – iron-bearing gabbro with a thickness from 10 – 120 m, with stockworks consisting of 5% apatite and two ore-bearing zones. (3) Middle dark-colored layered gabbro – iron-bearing gabbro layer with a thickness of 160 – 600 m and stringers or ores near the base. (4) Lower dark-colored medium to coarse-grained ore-bearing gabbro layers. There is a 60 – 570 m thick gabbro layer with a few meters

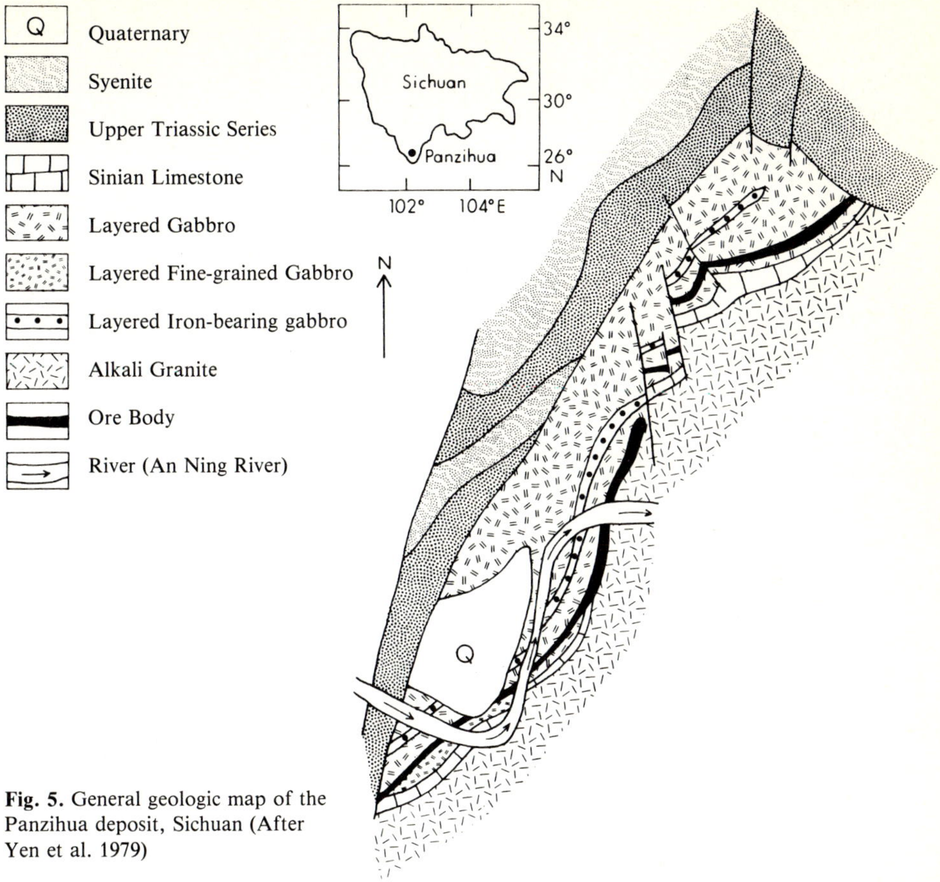

Fig. 5. General geologic map of the Panzihua deposit, Sichuan (After Yen et al. 1979)

of dunite and peridotite, and six ore-bearing zones. (5) Bottom boundary layer – dark-colored, fine-grained gabbro 10 – 300 m thick and poor in ore.

The ore minerals of the Panzihua deposit consist of titanomagnetite, ilmenite, and Mg – Al spinel. Other minerals are magnetite, perovskite, anatase, pyrrhotite, nickel-pyrite, and linnaeite (Yen et al. 1979). Vertically, chrome concentrates in olivine pyroxenite-dunite layers near the bottom of the rhythmic layers. Locally, the ores contain from 1.0 to 3.3% Cr_2O_3, which is mainly concentrated in titanomagnetite (Lu et al. 1980, Luo 1981).

Nickel Deposits

The nickel deposits of China occur in four settings: (1) in ultramafic intrusions in rift valleys – Jinchuan deposits of Gansu and Dali deposit of Yunnan; (2) as obducted ophiolite – Ailaoshan deposit of Yunnan; (3) in deposits of weathering products – area surrounding the Ailaoshan ultramafics and southern China; and (4) in sedimentary deposits in Yangzi craton.

Jinchuan

Jinchuan, also known as the Baijiajuzi deposit, is by far the largest nickel deposit in China. The deposit was discovered in 1958; operations began in the open pit in 1964 and in the underground mine in 1969. Reserves are estimated to be 5 million tons of nickel, 3.5 million tons of copper, and 100000 tons of cobalt. Pt, Os, Rh, Ir, Te, and Se were also reported to occur in the nickel deposit (Chin 1983a).

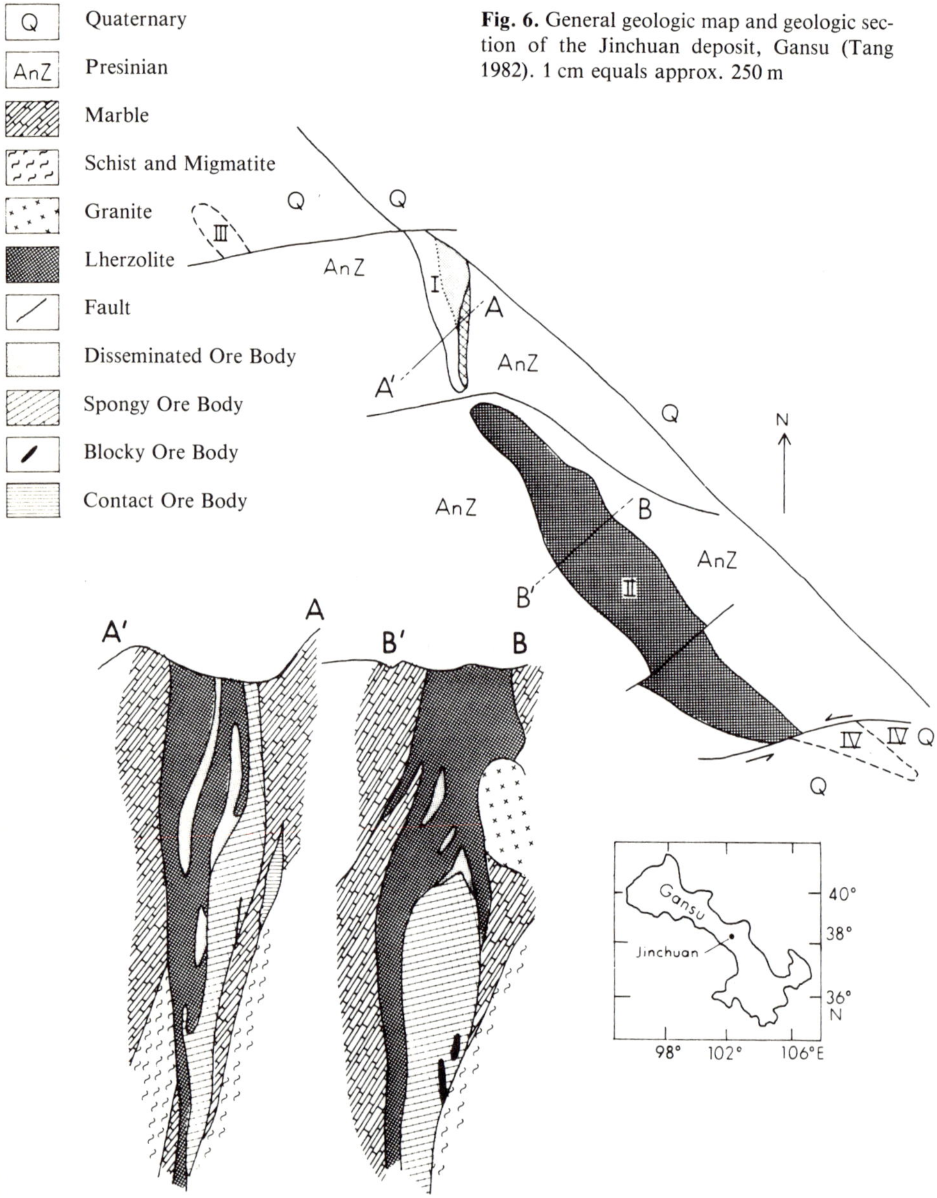

Fig. 6. General geologic map and geologic section of the Jinchuan deposit, Gansu (Tang 1982). 1 cm equals approx. 250 m

The Jinchuan deposits are located at the southeastern uplifted margin of the Sinokorean craton (Fig. 2), and its ore body is in the eastern part of the northern Longshan Mountain ultramafic rocks. The ultramafic rocks occur between the Presinian migmatite and marble of the Baijiajuzi Formation of Changcheng Group, which is 1.5 billion years old (Shi 1980, Tang 1982). The ultramafic body is bounded on both the north and south sides by deep faults trending from northwest 45° – 70° to near-northeast (Shi 1980). The ultramafic body is 6500 m in length, 20 – 500 m in width, strikes in a northwest direction, and dips at 70°. It is dissected by a fault into four sections. The two central sections (I and II of Fig. 6) are exposed, whereas the eastern (IV) and western (III) sections are covered by Quaternary deposits. The ultramafic body consists of lherzolite (64% olivine and 31% bronzite and diopside) and sulfide-bearing dunite. Dunite is concentrated at the middle and upper part of the ultramafic body, and sulfide-bearing dunite is concentrated near the bottom. Diopside-chlorite schists and tremolite-chlorite schists, sometimes up to 1 m thick, occur along the contact. The ore minerals consist of nickel-pyrite, violarite, and chalcopyrite. Other minerals present are Cr-spinel, Fe-spinel with ilmenite inclusions, carbonite, vallerite, marcasite, pyrite, and magnetic pyrite. Elements such as Pt, Rh, Ir, Se, Te, Ru, and Os are also present (Tang 1982).

Four types of ore body are present. (1) The lower magmatic segregated ore body comprises more than 85% of the total reserve. The elliptical ore bodies occur near the ultramafic rocks – sulfide-bearing dunite layers. Ni – Cu sulfides, which occur between olivine crystals, have a spongy, meteoritic iron structure. (2) The upper magmatic segregated ore body is seen in layers, veinlets, spots, and spongy form. The ore content of this layer is less than that of the first type. (3) Intruded ore bodies. Veinlets occur in the lower part of the ultramafic body. (4) Contact ore bodies. They are usually a few meters to several hundred meters in length. Thickness ranges from a few meters to tens of meters (Tang 1982). The tectonic setting of the Jinchuan deposits, an ultramafic intrusion in a rift valley, is similar to that of the Skaergaard Complex, Greenland, with its relationship to the North Atlantic rifting (Naldrett 1981, Chase and Gilmer 1973).

Dali

The Dali deposits are platinum-group mineral deposits with Ni – Cu sulfide as a by-product. Dali is located at the southwestern edge of the Yangzi craton (Fig. 1). The country rock consists of biotite-plagioclase gneiss of the Archean Yuanmou Group, with an age of 4.06 billion years (Tang 1982). The funnel-shaped intrusive body is 600 m long and 430 m wide and consists of gabbro, dunite, lherzolite, and pyroxenite. The ore-bearing dunite is usually located near the base along the contact with the country rocks (Fig. 7) (Yen et al. 1979). Non information is available on relative amounts of nickel and copper.

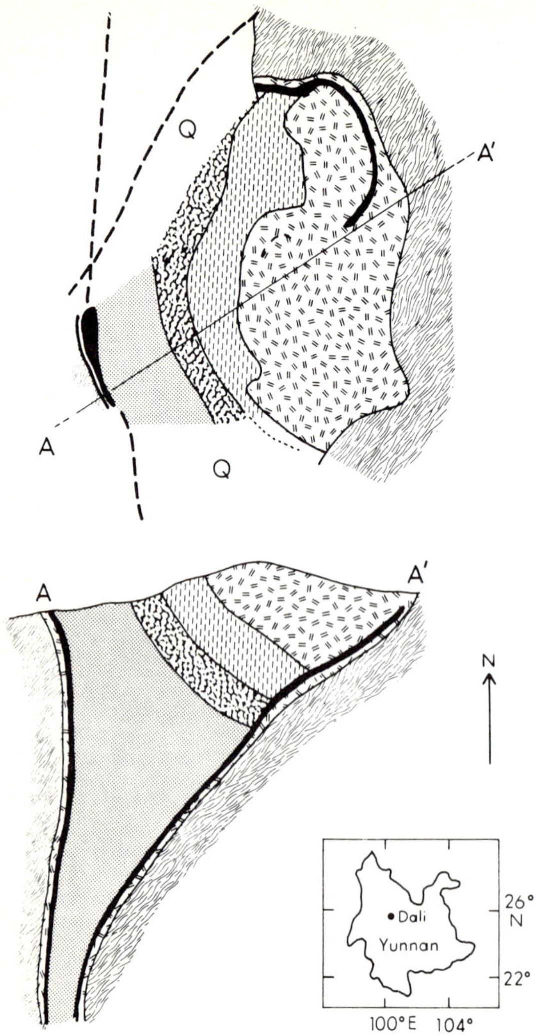

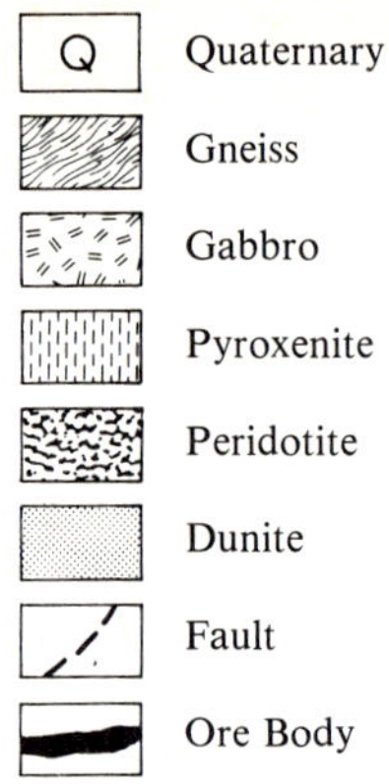

Fig. 7. Geologic map and geologic section of the Dali deposit, Yunnan (After Yen et al. 1979). 2.5 cm equals approx. 150 m

Ailaoshan

The Ailaoshan deposits are located in Yunnan province (Fig. 1). The ultramafic body is 300 – 600 m in length, 50 – 100 m in width, and consists of gabbro, pyroxenite, and lherzolite. Ni – Cu sulfides and some platinum-group minerals occur in the lherzolite (Tang 1982). The nickel-bearing ultramafic body is an obducted ophiolite that resulted from uplifting of tectonic emplacements at the southern portion of the North Tibet-West Yunnan Late Mesozoic subduction zone.

Weathering Products

Deposits of weathering products – nickel silicate clay minerals formed southwest of the Ailaoshan ultramafics – were described by Tang (1982). The following eight weathered surfaces average 80 – 100 m: (1) red residual clay, serpentine detritus present locally; (2) iron crust, loose, joints are recognizable; (3) iron crust, serpentine structure recognizable; (4) greenish kaolinized serpentine structure recognizable; (5) greenish kaolinized serpentine structure recognizable, loose, granular; (6) fracture serpentine belt, kaolization present locally; (7) siliceous serpentine belt, chalcedony and opal present; (8) serpentine belt, carbonation present, secondary calcite, and dolomite filling fissures of serpentine. Eighty percent of the weathered crust are ore bodies. Of these eight layers, numbers 2, 3, and 4 are rich in iron ore; numbers 5, 6, and 7 are rich in magnesium ore; and numbers 4 and 5 are rich in nickel. The nickel content decreases above and below layers 4 and 5. This deposit is similar to the nickel laterites of Soroako, Indonesia (Golightly 1979) and New Caledonia (Troly et al. 1979). They consist of an iron oxide (goethite and hematite) zone overlying saprolite (altered to unaltered serpentine) zone.

Sedimentary Deposits

Sedimentary deposits in the Yangzi craton, such as Lower Cambrian black shales, are good sources for nickel; they are usually associated with pyrite, phosphate, and coal. The ore-breading shale is usually 1 m thick, thin-bedded, and of lens to lenticular shape with a few millimeters of nickel sulfides – millerite, violarite, and nickel-pyrite. The Ni content ranges from 60 – 215 ppm. The sedimentary deposits are also rich in Mo, V, Cu, Pt, Pa, Au, and Ag (Chen et al. 1982). As this is a preliminary report, no data for deposit grade and tonnage are mentioned.

Concluding Remarks

Podiform chromite deposits occurring in Luobusa, northwestern and northern Inner Mongolia, and western Henan are associated with harzburgite and dunite. They are the result of magmatic differentiation and gravity settling and are also associated with a subduction zone. The stratiform chromite in the Panzihua deposit is related to rifting.

The nickel deposits are usually associated with ultramafic bodies, in particular with lherzolite and dunite. The Jinchuan and Dali deposits are related to continental rifting, and the Ailaoshan deposits to a subduction zone. Nickel ores are also found in weathering products around Ailaoshan and in Lower Cambrian black shales in the Yangzi craton.

Acknowledgments. This project is partly supported by the Resource Systems Institute of the East-West Center, Hawaii. I am grateful to Charles Johnson and Allen Clark of the Resource Systems Institute for their support. Sincere thanks to Rita Pujalet of Hawaii Institute of Geophysics for editorial help. Hawaii Institute of Geophysics Contribution No. 1733.

References

Brady ES (1981) China's strategic minerals and metals. Chin Business Rev 8(5):55 – 73

Cao R (1976) On the genesis of the alpine-type ultramafic complex and related chromite deposits in the western part of the Inner Mongolia orogenic belt. Geochimica 2:113 – 125 (in Chinese)

Chase GC, Gilmer TH (1973) Precambrian plate tectonics: The midcontinent gravity high. Earth Planet Sci Lett 21:70 – 78

Chen NS, Yang XZ, Liu DH, Xiao SJ, Fan DL, Wang LF (1982) Lower Cambrian black argillaceous and arenaceous rock series in South China and its associated stratiform deposits. Miner Depos 1:49 – 55 (in Chinese)

Chi S, Shi Q (1981) Genesis of *X* chromite deposits, Xizang autonomous region. Sci Geol Sin 343 – 359 (in Chinese)

Chin E (1980) The mineral industry of China. In: Mineral yearbook, vol III. US Bur Min, pp 237 – 259

Chin E (1983a) China. In: Mineral yearbook, vol III. US Bur Min, pp 225 – 250

Chin E (1983b) Far East-China. Min Annu Rev 349 – 357

Cui JW, Qiao ZJ (1983) A study of the tectonic environment of Lobusha chromite deposit. Min Depos 2(1):48 – 57 (in Chinese)

Fan PF (1984) Geological setting of selected copper deposits of China. Econ Geol 79:9 – 19

Golightly JP (1979) Geology of Soroako nickeliferous laterite deposits. Int Later Symp, New Orleans. Soc Min Eng AIME, pp 38 – 56

Greenbaum D (1977) The chromitiferous rocks of the Troodos ophiolite complex, Cyprus. Econ Geol 72:1175 – 1194

Hu J, Zhang S, Xiang D (1981) The zigzag chromite ore bodies in Luobusa, Xizang (Tibet) and their stress field features. Earth Sci 2:73 – 84 (in Chinese)

Li C (1980) A preliminary study of plate tectonics of China. Bull Chin Acad Geol Sci Ser I-2:11 – 22 (in Chinese)

Lu J, Zhang C, Gu G, Gao X, Fang Y, Zhou J (1980) The distribution and state of chromite in the Hongge layered vanadium-bearing titanomagnetite deposit in Sichuan province. Chin Acad Geol Sci Bull Ser III:1 – 20 (in Chinese)

Luo Y (1981) The characteristics of Ti-chromite mineralization in Xin Jie layered ultramafic-mafic intrusion in Panzihua area, China. Geochimica 66 – 73 (in Chinese)

McElhinny NM, Embleton BJJ, Ma XH, Zhong ZK (1981) Fragmentation of Asia in the Permian. Nature (London) 293:212 – 216

Mikami HM (1983) Chromite. In: Industrial Minerals and rocks (Nonmetallics other than fuels), vol I. Soc Min Eng AIME, New York, pp 567 – 584

Moores EM (1969) Petrology and structure of the Vourinos ophiolite complex, northern Greece. Spec Pap Geol Soc Am 118:1 – 74

Naldrett AJ (1981) Nickel sulfide deposits: Classification, composition, and genesis. Econ Geol 75th Anniversary Vol 628 – 285

Nicolas DC, Rabinovitch M, Moutte J, Leblanc M, Prinzhofer A (1981) Structural classification of chromite pods in southern New Caledonia. Econ Geol 76:805 – 831

Peters TJ, Kramers JD (1974) Chromite deposits in the ophiolite complex of northern Oman. Miner Depos 9:253 – 259

Shi Z (1980) Mineralogy and petrology of Jinchuan ultrabasic rock body with copper-nickel deposits. Chin Acad Geol Sci Bull X 1-1:78 – 91 (in Chinese)

Tang Z (1982) A discussion on the main types of nickel deposits in China with an approach to their relations with the paleoplate tectonics. Miner Depos 1-2:29 – 38 (in Chinese)

Thayer TP (1964) Principal features and origin of podiform chromite deposits and some observations on the Guleman-Soridag District, Turkey. Econ Geol 59:1497 – 1524

Thayer TP (1969) Gravity differentiation and magmatic re-emplacement of podiform chromite deposits. In: Wilson HDB (ed) Magmatic ore deposits. Econ Geol Mon 4:132 – 146

Troly G, Esterle M, Pelletier B, Reibell W (1979) Nickel deposits in New Caledonia, some factors influencing their formation. Int Later Symp, New Orleans. Soc Min Eng AIME 85 – 119

Wan X (1977) A geochemical study of chromite spinel alteration in a chromite deposit, Henan province. Geochimica 4:250 – 261 (in Chinese)

Wei W (1978) On the genesis and genetic types of a chromite deposit in Kansu. Acta Geol Sin 269 – 280 (in Chinese)

Yen JC, Chu SQ, Huo YS (eds) (1979) Ore deposits. Geological Press, Beijing, China, 443 pp (in Chinese)

Zhang B (1977) A migmatization-altered chromite deposit. Geochimica 147 – 155

Part II
Porphyry Deposits

Recent Advances in Porphyry Base Metal Deposit Research

J. M. GUILBERT[1]

Abstract

As an introduction to this portion of the Symposium on Copper Deposits, this paper reviews recent progress and findings in research on deposits associated with hypabyssal epizonal plutons. The *I*-type affinity porphyry copper-molybdenum series is first established as petrogenically distinct from deposits associated with *S*-type, generally 2-mica granites, the porphyry tin – tin granite – uranium granite suite. Then five specific areas of advance in the 1970s and 1980s are described. The first deals with plate tectonic settings and the global geologic-lithologic distribution of porphyry base metal deposits (PBMDs). Second, the significance of the recognition of changes in the ratio of magmatic to meteoric fluids during alteration-mineralization and deposit emplacement is discussed, along with the concept of "phyllic overprint" and its applications. Techniques for observing, measuring, and describing the propagation of fluids and fracturing in time and space are given. Third, experimental studies, thermodynamic modeling, stable and radiogenic isotope analyses, and fluid inclusion chemical analysis and microthermometric studies and results that have advanced our understanding of the geochemical and thermal nature of PBMD fluids are cited. Fourth, geomathematical modeling studies that describe probable fluid source areas, flow paths, kinetics, and the volume hydrothermally influenced by the emplacement of PBMD systems are considered, and finally the present status of our understanding of PBMD fluids in terms of T, P, $f(O_2)$, $f(S_2)$, salinity, and ion complexing are briefly stated. It is pointed out that the promise of the next decade is enormous if field and laboratory studies continue to loop back to support and extend the whole, and if research continues at its present pace.

Porphyry base metal deposits have been the object of global geologic research since 1970, especially in the Cordilleran American countries that are effectively the "type area". The results of that research have been so explicit about pressure-temperature-composition-tectonic aspects of porphyry base metal deposit environments of formation that we have collectively made tremendous advances in understanding both specific ore deposit genesis problems and the more general

[1] Professor of Economic Geology, University of Arizona, Tucson, AZ 85721, USA

Geology and Metallogeny of Copper Deposits
(ed. by G.H. Friedrich et al.)

conditions of granite formation and metallogenesis. "*S*-type"[2] granites are so little known in North and South America that the word "granite" has come to be taken by most American geologists as essentially synonymous with "*I*-type" systems. We are only just now – in large part through the efforts and scientific disclosures of scholars from Australia, Europe, Asia, and the USSR – coming to realize how important the *S*-type occurrences are economically, geochemically, and metallogenically and how varied both *I*- and *S*-types can be. This paper reports the results of research that is heavily directed toward porphyry copper, porphyry copper-molybdenum, and porphyry molybdenum *I*-type metallogeny; it must be extrapolated to porphyry tin and disseminated tin-tungsten-uranium *S*-type-affinity deposits in the Americas and elsewhere, and to *A*-type affinity Climax-type molybdenum deposits, only with great caution. Many of the methods of research may serve all these types of disseminated base metal deposits.

This paper will review recent progress in understanding (1) plate tectonic settings and PBMD distribution; (2) PBMD mineralization-alteration fluid histories; (3) PBMD fluid compositions; (4) fluid sources and flow patterns; and (5) fluid chemistry and physics. It will also be shown how each contributes to improved geologic models, and finally how field, petrographic, and geochemical-geophysical studies loop back to improve overall understanding of these systems.

Plate Tectonics and PBMDs

The impact of plate tectonics in the 1970s was felt as much in ore deposit geology as anywhere else, if not more. Tectonic settings of major ore deposit types have been defined (Guilbert 1981, Mitchell and Garson 1981, Sawkins 1984), and the roles of magma types and ore deposition styles in many lithotectonic terranes have been established. Porphyry copper and porphyry "moly" deposits – "PCDs" and "PMDs" – are associated with compressive, consumptive margins (Sillitoe 1972, 1981, Sawkins 1972), commonly with variations in major and minor elements that appear to be part of an overall metallogenic zoning pattern first established in plate tectonic terms by Mitchell and Garson (1972). Keith (1978) and Westra and Keith (1981) have related the geographic distribution of variations in Cu – Mo – Zn – Pb – Ag – Au in North American porphyry deposits to changes in the angle of plate subduction and to distance from the trench to the arc-orogen hearth of magmatic activity. Recent discoveries of porphyry copper-gold deposits in Argentina east of the porphyry copper-moly deposits of Chile tend to corroborate those findings. As R. L. Nielsen (personal communication, 1984) notes, exploration activities and discoveries in the late 1970s reveal that porphyry copper deposits are much more diverse than was previously

---

[2] For brevity, "*S*-type" will be used except where noted to stand for transformation-type, leucogranite, and ilmenite-series, and "*I*-type" will be used for syntectic-type, granodiorite, and magnetite-series granitoids as used by Chinese, European, and Japanese colleagues respectively, although the terms are not fully equivalent.

thought, especially in terms of associated and genetically related host rocks. We find deposits associated with calcic-diorite clan rocks (Panguna, El Arco); calc-alkalic grandiorite clan rocks (mostly in western USA); alkali shoshonitic clan rocks (Bingham Canyon and OkTedi); and even nepheline-normative alkalic syenite clan rocks (Stikine, B.C.). Some suggestions exist that each of these broad groups may be characteristic of a different tectonic setting. The calcic-diorite rocks seem to be characteristic of island-arc magma systems developed on oceanic crust. Calc-alkalic and alkali-calcic granodiorite to quartz monzonite systems are developed in conventional magmatic arcs over subduction zones in continental margin settings. The syenite host rocks may be related to continental rifts. Some major advances in understanding can be expected as we get more hard data on deposits in the southwest Pacific and southeast Asian region. A topic that has potentially significant economic implications is the variation of gold content and Mo – Cu – Au ratios in porphyry copper systems in various host rocks and tectonic settings. This topic also has some genetic implications regarding the sources of metals, in general, and the sources and evolution of porphyry systems, in particular.

Heidrick and Titley (1982) related the tectonic fabric of southwestern North America to plate tectonic elements and the PCDs within them. Sillitoe et al. (1982) have recently shown that the PCDs of Colombia occur in three subparallel subbelts of different ages (Jurassic, Miocene, and Eocene), each with distinguishable Cu – Mo – Au ratios. Westra and Keith (1981) presented a chemical classification of "Climax-type moly" deposits. They show, as do Mutschler et al. (1981), that those of the Climax-Urad-Henderson-Mount Emmons type are almost certainly atectonic or rift related, the extensional to neutral tectonic result of the annihilation of compressive forces when the North American plate overran the East Pacific Rise in mid-Cenozoic times. Their absence from the South American and Canadian scene is thus understandable. Porphyry tin deposits (Sillitoe et al. 1975) are probably in *S*-type hypabyssal rocks analogous to the tin granites; less is known about them and and their distribution in the Americas. Mutschler et al. (1981) describe "granite molybdenite systems" – the Climax type – as occurring in rocks that are petrographic-geochemical *S*-types; they may actually be closer to *A*-type granitoids. Further progress in understanding granite origin and metallogeny can be anticipated from Nd – Sm and Rb – Sr studies (Farmer and DePaolo 1983) and Lu – Hf systematics (Patchett 1983). Powerful insights to the lithotectonics of granitoids provided by radioisotopic dating methods, especially potassium argon techniques (Damon et al. 1981), are crucial to progress.

Studies are continuing on a broad scale, and we can expect important results in early years from Keith and others on application of magma chemistry and ore deposits to plate tectonic settings and metallogeny, from Sillitoe on continued refinement of PBMD relations to detailed reconstructions of tectonic elements, from Ren and Guilbert on Fe – Ti – O species as determinants in southwestern North American PBMD systems, from several workers in North America and especially Europe on tectonic settings analysis through compilation of geochemical-geochronological-geophysical-tectonic aspects of major districts, and from continued major and minor element whole rock and petrographic-mineralogic analysis of granitoids, as for example at the CNRS-CRPG near Nancy, France. Results from many countries and continents need to be compiled and refined, the potential contribution of the USSR, with several ages of PBMDs, and of China, with four recognized PCD metallotects, representing outstanding examples.

Mineralization-Alteration Fluid History

One of the dramatic discoveries of the 1970s was that although the bulbous to shield-shaped models of alteration-mineralization zoning in PBMDs proposed by Lowell and Guilbert (1970), Hollister et al. (1975), and others are, in general, descriptively correct, they represent a summation of many continuous, time-staged processes rather than a single event. Nielsen (1968) described and Sheppard et al. (1971) documented the phenomenon at Santa Rita that Gustafson and Hunt (1975) were able fully to describe as "phyllic overprint" in the sense of early, prograde magmatic-water mineralization-alteration followed by later, lower temperature alteration as part of a massive retrograde influx of meteoric water at El Salvador, Chile. Sutherland Brown (1976) has further pointed out that PCDs and PMDs can be classified according to pressure, temperature, and time variables. We now recognize that however useful the Lowell and Guilbert model has been in exploration, it is oversimplified in its description of processes. It appears that early alteration is largely alkalic, or alkali metasomatic, and weakly hydrolytic, with the early establishment of a central zone of potassic alteration yielding to an outer zone of propylitic effects, with disseminated and microveinlet copper-moly mineralization peaking in outer potassic volumes. Later phyllic and argillic alteration – with substantial redistribution of sulfides – is then typically "overprinted" onto the mineralized zone, the result of an influx of meteoric waters convecting through the higher permeabilities of that mineralized zone. In many porphyry deposits, veinlets flanked by phyllic alteration do indeed cut potassic assemblages on their outer side and propylitic ones on their inner side; but it is too simplistic to state categorically that all sericite, or phyllic assemblage material, is late in origin. It is noteworthy that we are aware of several deposits in which the age separation is especially clear – Silver Bell and Morenci, Arizona, and Butte, Montana, for example – and several at which the phyllic overprint was weakly developed at best, such as at Ajo and Sierrita, Arizona; El Arco, Baja California del Norte, Mexico; and Bingham Canyon, Utah. Further, as later paragraphs will show, solutions clearly evolved in temperature-pressure-salinity-isotope "space" such that complex fluid histories are more the rule than the exception.

It was assumed as we entered the 1970s that most fracturing was tectonic in nature and, except for breccia pipes and obvious internal effects, that fracturing was an internal strain response to external stress (Mayo 1958, Rehrig and Heidrick 1972, 1976). Just as alteration-mineralization space-time histories have proven resolvable, but more complicated than we thought, so also is the propagation of fractures. A few workers – among them Nielsen and Burnham – suggested in the early 1970s that the stockwork fracture systems in some porphyry deposits were intrinsic and related to stock cooling by a process called "hydrofracting". Several authors, most notably Knapp and Knight (1977) and Norton (1982), have since shown that fluids driven outward from cooling stocks can propagate the fractures along which they then travel by a combination of processes including differential thermal expansion of pore fluids, changes in pore fluid pressure, hydrolysis, and vapor-pressure "wedging". Haynes and Titley (1980) measured these veinlet sets by documenting the progression of fractures and fluids in space and time around the porphyry system at Sierrita, Arizona. They drew squares 1 m on a side on representative outcrops at increasing distances from the center of the porphyry system, and measured the aggregate length and crosscutting relationships of the several types of alteration-mineralization-aperture style veinlet sets that occur in each square meter. By studying $P-T$ conditions indicated by mineralogy and by fluid inclusion filling temperatures and salinities in each of those veinlet types, they were able to describe the evolution of fracturing, of ore and gangue mineralogy, and of geochemical patterns in time and space (Fig. 1). A pattern (Fig. 2) of first advancing, then retreating propylitization, of outward-generated fracturing, and of decreasing temperature and salinity through time is indicated, with the late development of a scanty phyllic overprint event. The first detailed study of structural-mineralizational interdependence and evolution will certainly be followed by other studies elsewhere, since the fracture density-fluid evolution relations are so vital to understanding the processes that are basic to "porphyry" mineralization. Preece

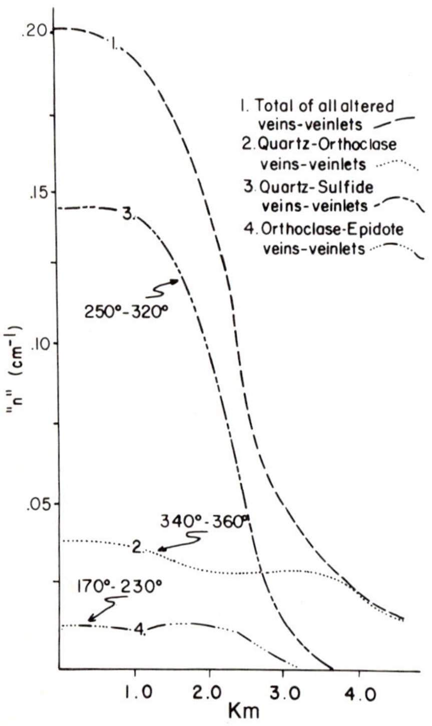

Fig. 1. A plot of veinlet mineralization-alteration evolution in space and time around the Sierrita porphyry copper deposit, Pima County, Arizona. "n" is the aggregate length in meters per square meter – or centimeters per $10^4\ cm^2 = cm^{-1}$) – of each veinlet type. *2* is the earliest, *4* the latest, with emplacement temperatures indicated. Km refers to distance in kilometers from the center of the system at the present moderately deep erosional exposure level. (From Titley 1982)

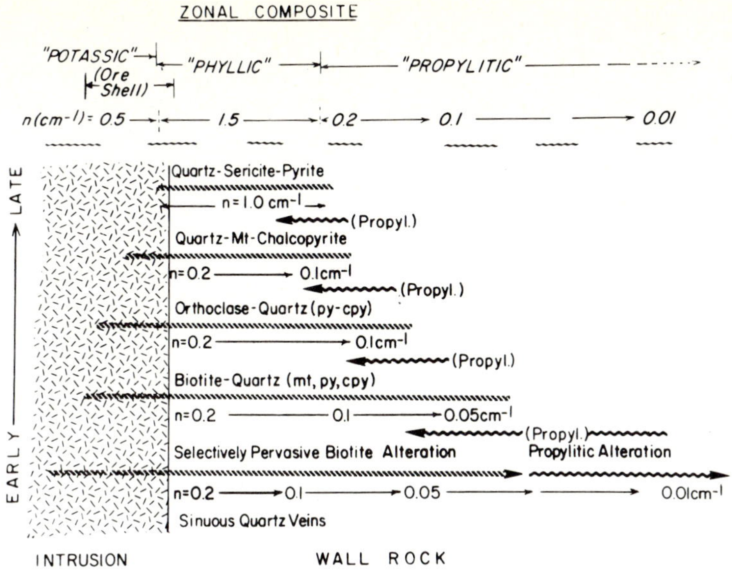

Fig. 2. A summary diagram of alteration-mineralization progress in time (*upward*) and space (*lateral*) at Sierrita. See text and caption to Fig. 1 for explanation. (From Titley 1982)

and Beane (1982) and Ahmad and Rose (1980) reported detailed analyses of the evolution of fluids through the mineralization process at Sierrita, Arizona, and Santa Rita, New Mexico, respectively. In both deposits, high temperature-high salinity fluids, possibly the result of demonstrated boiling, were followed by lower temperature, lower salinity late fluids, a general pattern which seems to be emerging. For example, potassic alteration at Sierrita occurred in association with 35 – 40 wt% NaCl equivalent fluid at 300° – 370°. Slightly later fluids were less saline at 10 – 15 wt% NaCl and hotter at 400° – 430°, at which temperature boiling occurred. Mineralization from these latter solutions ranged from 370°C down to 200°C. At Santa Rita, deposition occurred at 600°C or more down to 200°C, also from two main types of fluid, one of high 30 – 60% NaCl equivalent, the other less saline at 1% – 20%. Boiling and condensation were common phenomena, especially in the high salinity type. In general, fluids responsible for hypogene metallization range from 800° down to 200°C with an apparent zenith of metal sulfide precipitation importance at 375° – 325°C. Complex fluid histories, including boiling, condensation, and resultant variations in oxidation, salinity, and volatile content are common. Indicated pressures range from 200 to 2000 bars, or from depths of 0.5 to 5 km, with emphasis on the shallower depths. The high salinity fluids are distinctive and repetitive enough that their presence in fluid inclusions localized over mineralized volumes is explorationally significant in distinguishing potentially productive from barren occurrences.

Fluid Compositions

A third area of great recent progress is that of the geochemical and thermal nature of hydrothermal fluids. That progress comes from at least four subdisciplines, including synthetic (experimental) studies, thermodynamics, isotope studies, and fluid inclusion analytic techniques. Synthetic studies have culminated in such papers as the Meyer and Hemley (1967) and Rose and Burt (1979) papers on the $K_2O-Al_2O_3-SiO_2-H_2O-HCl$ system and its alteration-mineralization implications and a series by Burnham, most recently Burnham and Ohmoto (1980), on hydrothermal fluid generation in intermediate to felsic magmas and the compositions of those fluids. Thermodynamic approaches, essentially starting with Helgeson's 1964 treatise on complexing and hydrothermal ore deposition, have been dramatically advanced through improved basic thermodynamic data in pertinent systems such as $H_2O-NaCl-CO_2$ through the use of computers for their speed, enormous data bases, and graphic capabilities. The use of activity-activity diagrams – in which the compositions of fluids in equilibrium with minerals and mineral assemblages in terms of ratios of ions in solution can be examined – have greatly clarified hydrothermal fluid evolution patterns. It is now convenient, for example, to relate alteration and mineralization (Fig. 3) by analyzing K^+/H^+ and Cu^+/H^+ behavior with respect to $Fe^{+2}/(H^+)^2$ at appropriate temperatures. Uncertainties are still important at high temperatures where even basic measured thermodynamic data are scant, but each year sees more progress. Computer management of mass transfer, thermal transfer, reaction progress, and solid solution-fluid equilibration is increasingly valuable.

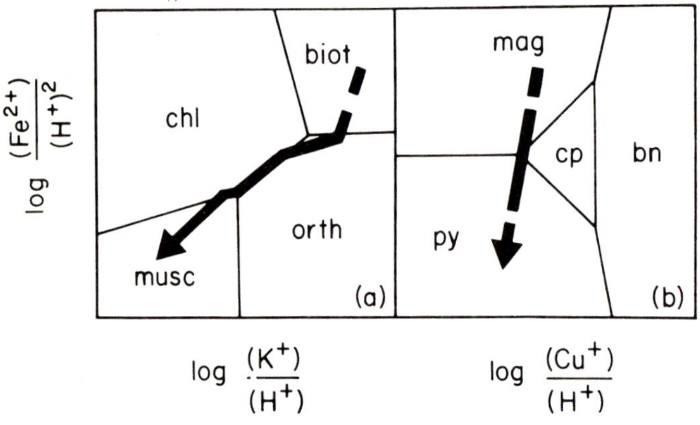

Key to Minerals:

biot = biotite
bn = bornite
chl = chlorite (14Å)
cp = chalcopyrite
mag = magnetite
musc = muscovite
orth = orthoclase
py = pyrite

Fig. 3a, b. Two activity-activity diagrams showing silicate and oxide-sulfide phases in equilibrium during PCD mineralization-alteration as functions of ion ratios and time (*along the arrows*). The ion ratios describe the fluid in equilibrium with potassic biotite-magnetite, biotite-orthoclase-magnetite, orthoclase-chalcopyrite, etc., to muscovite-pyrite phyllic assemblages. (From Beane and Titley 1981)

Isotopic studies have had such a far-reaching and profound impact that their results are difficult even to summarize. Taylor (1979) has reviewed oxygen-hydrogen isotopic systematics that have verified meteoric water involvement in porphyry systems described in the next section, and Ohmoto and Rye (1979) have done much the same for sulfur and carbon isotopes. Sulfur isotopic analysis is now useful for establishing the geochemical origins of sulfur in sulfides, for geothermometry, and for analysis of geochemical conditions and mechanisms of deposition. Ohmoto and Rye report two maxima of $\delta^{34}S$ values in PCDs at -3 to $+1$ and $+8$ to $+15$, suggesting that the complexed sulfur was derived largely from igneous sources either directly or by resolution of earlier magmatic sulfides. Some deposits (Galore Greek, Canada, and Morococha, Peru) may involve some sedimentary sulfur. PBMD sulfate-sulfide isotopic temperatures typically fall in the 450° – 650°C interval. Early stable isotopic studies by Garlick and Epstein (1966) at Butte, Montana, and by Sheppard et al. (1971) at Santa Rita, New Mexico, both in the USA, first established the importance of meteoric water invasion of PBMD systems. Subsequent oxygen-hydrogen isotopic analyses have confirmed that importance.

Fluid inclusion analytic techniques have yielded results as profoundly important as those of isotopic ones. They have indicated – as stated above – that fluids to evolve through time, that they are commonly hot and strongly saline early in mineralization history, and that boiling is common in the environment of deposition. The inclusion petrographer sees daughter mineral species in the vacuoles that include halite (NaCl), sylvite (KCl), anhydrite ($CaSO_4$), gypsum ($CaSO_4 \cdot 2H_2O$), several mixed Ca – Mg – Na – K hydrous sulfate salts, pyrite, chalcopyrite, hematite, magnetite, and several others that reveal the chemical complexity of hydrothermal fluids. Carbon dioxide, methane, carbon disulfide, and several organic and organo-metallic compounds have also been identified. Study of mineral paragenesis and paragenetic sequence in skarn systems has also greatly clarified fluid behavior in porphyry systems and their wall rocks (Einaudi et al. 1981).

Fluid Sources and Flow Paths

Speculation over the last hundred years about the degree to which meteoric fluids mingle with hypogenic ones have been replaced in the last decade by factual information and deduction stemming largely from isotopic studies and geochemical-geophysical-geomathematical modeling. Stable isotopic studies – particularly of hydrogen-oxygen and sulfur – and radiogenic ones – especially strontium-rubidium – have assured us not only that meteoric waters can enter hydrothermal systems, but also that they do, commonly overwhelmingly. Early discussions of the sort by White (e.g., White 1957, 1968) lead to formulation of models of vapor-dominated shallow hydrothermal systems (e.g., White et al. 1971, White 1981), thereby providing conceptual support to the isotopic findings of Taylor and his colleagues and others. These studies documented substantial involvement in PBMD of water with incontrovertible surficial deuterium-hydrogen and

$^{18/16}$oxygen signature. Important single early studies include Sheppard et al. (1971), and many isotopic studies of individual deposits have followed (Field and Gustafson 1976, Eastoe 1982) which require involvement of meteoric waters. How that water enters has been studied by Whitney (1975), Cathles (1977), Henley and McNabb (1978), and especially Norton (1978, 1982). Computer management of geophysical, geochemical, thermal, and kinetic data permit construction of two-dimensional profiles of the consequences of shallow intrusion of a pluton into the crust. "Average" geologic compositing of PBMD characteristics with assigned permeabilities, solution densities and viscosities, and absolute and relative temperatures and heat fluxes permits computation of thermal-hydrothermal dynamics, including kinetics. Norton (1978, et seq.) defined "sourcelines" (the "starting front" of convectively driven fluids), "pathlines" (the travel paths of aliquots of fluid), and "source regions" (the total volume from which fluids convectively move). Norton indicates that the source region might extend as far as 5 km from a 2.7-km-wide 4.5-km-tall stock intruded to within 2 km of the surface. Pathlines are probably not the simple arcuate inward and upward bending arrow shapes so commonly sketched, but rather are complicated by quantitative and qualitative permeability variations, local thermal perturbations, and variations in density-specific gravity-viscosity rela-

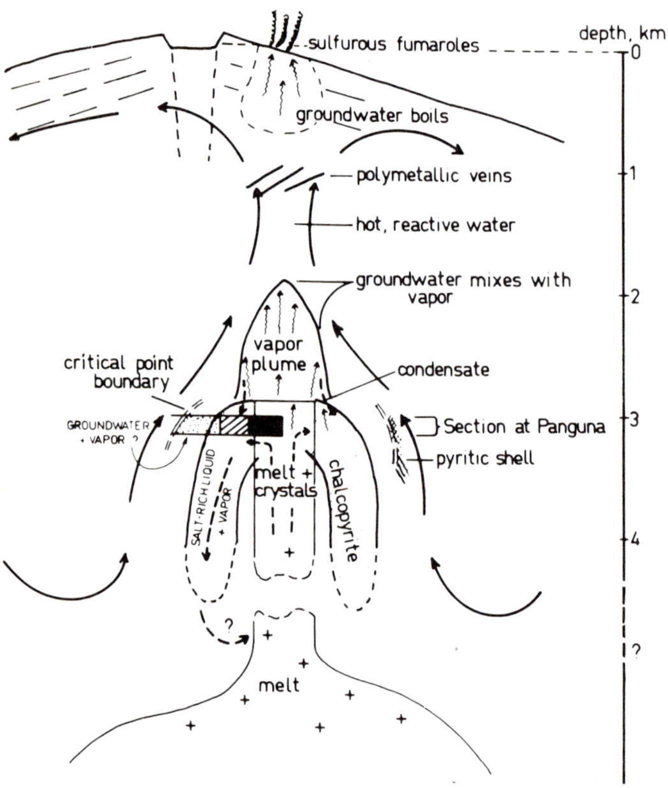

Fig. 4. Interrelationships between magmatic and meteoric waters, salinity, temperature, boiling, and alteration-mineralization in porphyry systems as deduced at Panguna by Eastoe (1982)

tions in the fluid, variations in turn affected by $T-P-X$, alteration progress, and other variables. Figure 4 shows some of the complications of fluid flow predicted by theoretical and observational studies of the sort described in the preceding paragraphs.

Fluid Chemistry and Physics

A clearer understanding of the physicochemical nature of hydrothermal fluids and the nature of metal complexing in them has resulted from the studies briefly described above. Fluid inclusion studies have demonstrated that mineralizing aqueous fluids in PCD systems are NaCl-saturated and on or near the two-phase boiling surface in the $NaCl-H_2O$ system (Ahmad and Rose 1980). Eastoe (1982) concluded that at Panguna "the copper was deposited mainly by salt-rich liquid expelled directly from the magma. Significant salt-rich liquid may also have condensed from a vapor plume" above a boiling zone (Fig. 4). Variations in the fugacities of volatile components are vital; volatile constituents depress the critical point of a hydrothermal fluid, while nonvolatile solutes, in general, increase it. Oxygen fugacity is relatively high (10^{-30} to 10^{-35}) near the hematite-magnetite buffer to sustain the sulfate so commonly recorded in porphyry systems as anhydrite and also the hematite-magnetite paragenesis, although magnetite predominates (Fig. 3). Sulfur fugacities of 10^{-10} to 10^{-12} atm buffer chalcopyrite-pyrite-tennantite assemblages at appropriate temperatures; higher $f(S_2)$ requires enargite-bornite assemblages. Complex ions of the metals are dominated by mixed metal halides, with lesser involvement of complex ions involving $(HS)^-$ and SO_2. Eastoe (1982) determined that vapor transport of complexes of the transition elements is variable. Vapor transport of Fe in the vapor plume of Henley and McNabb (1978) and Fig. 4 is insignificant; it can be important for Cu, Zn, and Mo. Space does not permit further review; the scholar is referred to Eastoe (1982), Beane and Titley (1981), the Barnes volumes (1967 and 1979), and the many articles already referred to for further detail.

Most importantly, the physics and chemistry of the hydrothermal fluid are yielding to modern study. The greatest needs are almost certainly in the area of physical, chemical, and thermodynamic properties of the basic fluids in the system oxygen-hydrogen, ultimately with consideration of important amounts of Na, Cl, K, and so on to include the metals themselves.

Conclusions

The results of each of the bodies of information above – plate tectonics, mineralization-alteration fluid histories, fluid compositions, fluid sources and flow patterns, and fluid physical geochemistry – are continuing to loop back into each other. The flow of field, chemical analytic, isotopic, and petrographic data continues to permit refinement and extension of theoretical analyses. Field studies are of a whole new generation in the 1980s, including such analyses of veinlet histories as that by Haynes and Titley (1980). Careful, quantitative alteration mapping (Guilbert and Park 1986), the determination of mineralogic-chemical equilibrium by phase rule and petrographic analysis, and the measurement of alteration mineral solid solution characteristics and cryptic zoning in alteration assemblages are ongoing and will feed back – with the fracture density, fluid inclusion, and isotopic studies described – to provide increasingly convincing and refined mass balance-mass transfer $P-T-X$ relationships. Quantitative alteration mapping permits computerization of alteration data, and for the first time allows regression analysis to identify academic and explorationally important variables in that context. The data base growth in tectonic, structural, and lithotectonic information improves exploration for these globally important ore-forming systems. Many of the techniques of analysis that are described here for *I*-type systems will apply to *S*-type ones, although the results will of course differ. Computerized *X*-ray diffractometric data, mineral polytype studies, and electron microprobe studies are all generating mineralogic directional variation data that is valuable to both science and exploration. Progress in field, quantitative laboratory, and digital-numerical analytic studies will provide a productive decade of the 1980s for both *I*-type and *S*-type mineralized-altered PBMD-endowed terranes.

References

Ahmad SN, Rose AW (1980) Fluid inclusions in porphyry and skarn ore at Santa Rita, New Mexico. Econ Geol 75:229–250

Beane RE, Titley SR (1981) Porphyry copper deposits, Part II, Hydrothermal alteration and mineralization. Econ Geol 75th Anniversary Vol, pp 235–269

Burnham CW, Ohmoto H (1980) Late-stage processes of felsic magmatism. In: Ishihara S, Takenouchi S (eds) Granitic magmatism and related mineralization. Min Geol Spec Iss 8, Soc Min Geol, Tokyo, Jpn, pp 1–12

Cathles LM (1977) An analysis of the cooling of intrusives by groundwater convection which includes boiling. Econ Geol 72:804–826

Damon PE (1970) Correlation and chronology of ore deposits and volcanic rocks. US Atom Energ Comm Annu Rep C00-689-130

Damon PE, Shafiqullah M, Clark KF (1981) Age trends of igneous activity in relation to metallogenesis in the southern Cordillera. In: Dickinson WR, Payne WD (eds) Ariz Geol Soc Digest 14:137–154

Eastoe CJ (1982) Physics and chemistry of the hydrothermal system at the Panguna porphyry copper deposit, Bougainville, Papua New Guinea. Econ Geol 77:127–153

Einaudi MT, Meinert LD, Newberry RJ (1981) Skarn deposits. Econ Geol 75th Anniversary Vol, pp 317–392

Farmer G, DePaolo DJ (1983) Origin of Mesozoic granite in the western United States and implications for pre-Mesozoic crustal structure: II. Nd and Sr isotopic studies of barren and Cu–Mo mineralized granite in Precambrian craton. J Geophys Res 88:3379–3401

Field CW, Gustafson LB (1976) Sulfur isotopes in the porphyry copper deposit at El Salvador, Chile. Econ Geol 71:1533–1548

Garlick DG, Epstein S (1966) The isotopic composition of oxygen and carbon in hydrothermal minerals at Butte, Montana. Econ Geol 61:1325–1335

Guilbert JM (1981) A plate tectonic-lithotectonic classification of ore deposits. In: Dickinson WR, Payne WD (eds) Ariz Geol Soc Digest 14:1–10. (Available Ariz Bur Geol Univ Ariz, Tucson, USA)

Guilbert JM, Park CF, Jr (1986) The geology of ore deposits. Freeman, New York, 985 pp
Gustafson LB, Hunt JP (1975) The porphyry copper deposit at El Salvador, Chile. Econ Geol 70:857 – 912
Haynes FM, Titley SR (1980) The evolution of fracture-related permeability within the Ruby Star Granodiorite, Sierrita porphyry copper deposit, Pima County, Arizona. Econ Geol 75:673 – 683
Heidrick TL, Titley SR (1982) Fracture and dike patterns in Laramide plutons and their structural and tectonic implications. In: Titley SR (ed) Advances in geology of the porphyry copper deposits. Univ Ariz Press, Tucson, Ariz, pp 73 – 92
Helgeson HC (1964) Complexing and hydrothermal ore deposition. Pergamon Press/MacMillan, New York, 128 pp
Henley RW, McNabb A (1978) Magmatic vapor plumes and ground water interaction in porphyry copper emplacement. Econ Geol 73:1 – 20
Hollister VF, Anzalone SA, Richter DH (1975) Porphyry copper deposits of southern Alaska and contiguous Yukon Territory: CIM Bull 68, 756:104 – 112
Keith SB (1978) Paleosubduction geometries inferred from Cretaceous and Tertiary magmatic patterns in southwestern North America. Geology 6:516 – 521
Knapp RB, Knight JE (1977) Differential thermal expansion of pore fluids: Fracture propagation and microearthquake production in hot pluton environments. J Geophys Res 82:2515 – 2522
Lowell JD, Guilbert JM (1970) Lateral and vertical alteration-mineralization zoning in porphyry ore deposits. Econ Geol 65:373 – 408
Mayo EB (1958) Lineament tectonics and some ore districts of the southwest Min Eng 10:1169 – 1175
Meyer C, Hemley JJ (1967) Wall rock alteration. In: Barnes HL (ed) Geochemistry of hydrothermal ore deposits, Chap 6. Reinhart, New York, pp 167 – 235
Mitchell AHG, Garson MS (1972) Relationship of porphyry copper and circum-Pacific tin deposits to palaeo-Benioff zones. Trans Inst Min Metall 81:B10 – 25
Mitchell AHG, Garson MS (1981) Mineral deposits and global tectonic settings. Academic Press (Geol Ser), London New York, 405 pp
Mutschler FE, Wright EG, Luddington S, Abbott JT (1981) Granite molybdenite systems. Econ Geol 76:874 – 897
Nielson RL (1968) Hypogene texture and mineral zoning in a copper-bearing granodiorite stock, Santa Rita, New Mexico. Econ Geol 63:37 – 50
Norton D (1978) Sourcelines, sourceregions, and pathlines for fluids in hydrothermal systems related to cooling plutons. Econ Geol 73:21 – 28
Norton DL (1982) Fluid and heat transport phenomena typical of copper-bearing plutons environments: Southeastern Arizona. In: Titley SR (ed) Advances in geology of the porphyry copper deposits. Univ Ariz Press, Tucson, Ariz, pp 59 – 72
Norton DL (1984) Theory of hydrothermal systems. Annu Rev Earth Planet Sci 12:155 – 177
Ohmoto H, Rye RO (1979) Isotopes of sulfur and carbon, Chap 10. In: Barnes HL (ed) Geochemistry of hydrothermal ore deposits, 2nd edn. John Wiley & Sons, New York, pp 509 – 567
Patchett PJ (1983) Importance of the Lu – Hf isotopic system in studies of planetary chronology and chemical evolution. Geochim Cosmochim Acta 47:81 – 91
Preece RK III, Beane RE (1982) Contrasting evolutions of hydrothermal alteration in quartz monzonite and quartz diorite wall rocks at the Sierrita porphyry copper deposit, Arizona. Econ Geol 77:1621 – 1641
Rehrig WA, Heidrick TL (1972) Regional fracturing in Laramide stocks of Arizona and its relationship to porphyry copper mineralization. Econ Geol 67:198 – 213
Rehrig WA, Heidrick TL (1976) Regional stress during the Laramide and late Tertiary intrusive periods, Basin and Range Province, Arizona. Ariz Geol Soc Digest 10:205 – 228
Rose AW (1970) Zonal relations of wallrock alteration and sulfide distribution at porphyry copper deposits. Econ Geol 63:920 – 936
Rose AW, Burt DM (1979) Hydrothermal alteration. Chap 5. In: Barnes HL (ed) Geochemistry of hydrothermal ore deposits, 2nd edn. John Wiley & Sons, New York, pp 173 – 235
Sawkins FJ (1972) Sulfide ore deposits in relation to plate tectonics: J Geol 80:377 – 397
Sawkins FJ (1984) Metal deposits in relation to plate tectonics. Springer, Berlin Heidelberg New York Tokyo, 325 pp
Sheppard SMF, Nielsen RL, Taylor HP, Jr (1971) Hydrogen and oxygen isotope ratios in minerals from porphyry copper deposits. Econ Geol 66:515 – 542

Sillitoe RH (1972) Relation of metal provinces in western America to subduction of oceanic lithosphere. Geol Soc Am Bull 83:813 – 818

Sillitoe RH (1981) Ore deposits in Cordilleran and island-arc settings. In: Dickinson WR, Payne WD (eds) Ariz Geol Soc Digest 14:49 – 69 (Available Ariz Bur Geol, Univ Ariz, Tucson, Ariz, USA)

Sillitoe RH, Halls C, Grant JN (1975) Porphyry tin deposits in Bolivia. Econ Geol 70:913 – 927

Sillitoe RH, Jaramillo L, Damon PE, Shafiqullah M, Escovar R (1982) Setting, characteristics, and age of the Andean Porphyry Copper Belt in Colombia. Econ Geol 77:1837 – 1850

Sutherland Brown A (1976) Morphology and classification. In: Sutherland Brown A (ed) Porphyry copper deposits of the Canadian Cordillera. CIM Spec Vol 15, pp 51 – 55

Taylor HP, Jr (1979) Oxygen and hydrogen isotope relationships in hydrothermal mineral deposits, Chap. 6. In: Barnes HL (ed) Geochemistry of hydrothermal ore deposits, 2nd ed. John Wiley & Sons, New York, 798 pp

Titley SR (1982) The style and progress of mineralization and alteration in porphyry copper systems. In: Titley SR (ed) Advances in geology of the porphyry copper deposits. Univ Ariz Press, Tucson, Ariz, USA, pp 93 – 116

Westra G, Keith SB (1981) Classification and genesis of stockwork molybdenum deposits. Econ Geol 76:844 – 873

White DE (1957) Magmatic, connate, and metamorphic waters. Geol Soc Am Bull 68:1659 – 1682

White DE (1968) Environments of deposition of base-metal ore deposits: Econ Geol 63:301 – 335

White DE (1981) Active geothermal systems and hydrothermal ore deposits. Econ Geol 75th Anniversary Vol, pp 392 – 423

White DE, Muffler LJP, Truesdell AH (1971) Vapor-dominated hydrothermal systems compared with hot-water systems. Econ Geol 66:95 – 97

Whitney JA (1975) Vapor generation in a quartz monzonite magma: A synthetic model with application to porphyry copper deposits. Econ Geol 70:346 – 358

The Geochemical Behaviour of Copper and Molybdenum in Ore-Forming Processes

A. V. KUDRIN, L. N. VARYASH, Yu. N. PASHKOV, and V. I. REKHARSKY[1]

Abstract

The geochemical behaviour of molybdenum and copper during the formation of copper-molybdenum ore deposits is considered. Molybdenum and copper mineralization is closely associated with the hydrothermal-metasomatic associations. The probable values of the physicochemical parameters of copper- and molybdenum-bearing hydrothermal solutions are estimated on the basis of calculations and experimental data. The problem of the types of copper and molybdenum species in ore fluids is considered. Hydroxyl alkali chloride or alkali molybdate complexes are suggested as the molybdenum-bearing species responsible for ore transport. Copper chloride complexes are suggested as copper-bearing transport forms. Cooling of the hydrothermal solutions appears to be one of the main ore deposition factors.

Geological and geochemical investigations indicate that the scale of copper and molybdenum mineralization in copper-molybdenum deposits is directly associated with the hydrothermal-metasomatic rock associations (Rekharsky 1980, Shipulin et al. 1975).

The feldspar-quartz hydrothermal-metasomatic association is considered as the main productive ore formation in respect to molybdenum mineralization. It is characterized by feldspar and quartz-feldspar metasomatites with potassium feldspar-biotite-quartz assemblages in their inner zones. The metasomatic rocks are accompanied by quartz and quartz-feldspar veins with molybdenite as the main ore mineral. Molybdenum mineralization in feldspar-quartz metasomatic rocks is present usually in the form of disseminations in the inner zones of the metasomatites.

Copper mineralization is associated mainly with the quartz-sericite metasomatic association and partly with propylites, berezites (quartz-sericite-carbonate metasomatites), etc. The quartz-sericite association is represented by quartz-sericite metasomatites as well as by veins and veinlets of a similar mineral composition. The main copper ore mineral, chalcopyrite, is localized in veins and marginal zones in the quartz-sericite metasomatites. Chalcopyrite is predominantly associated with pyrite, whereas molybdenite, bornite, tetrahedrite and other ore minerals are found only in subordinate quantities.

[1] The Institute of Geology of Ore Deposits, Petrography, Mineralogy, and Geochemistry, USSR Academy of Sciences, Moscow, USSR

Geology and Metallogeny of Copper Deposits
(ed. by G. H. Friedrich et al.)

According to fluid inclusion investigations the temperature of mineralization of the ore-bearing hydrothermal-metasomatic rocks is estimated in the range of 500° – 200°C. The molybdenum ores were formed mainly from 450° to 300°C and copper ores from 320° to 200°C (Pavlova 1978, Popov 1977).

Alkaline metal chlorides and carbon dioxide are considered as the dominant components of the ore-forming hydrothermal solutions on the basis of the fluid-inclusion chemical composition. The latter was determined in samples of quartz coprecipitated with molybdenite and chalcopyrite. In several cases high fluorine concentrations were determined. The published data on fluid-inclusion chemical composition and its dependence on temperature and the character of the altered rocks are sometimes controversial and few in number. The highest temperature inclusions as usual are characterized by high salinity, up to 40 – 50% eq. wt% of NaCl in several cases. The fluid inclusions from early stages of mineralization contain not only "brines", but a low-density water-carbon dioxide fluid with carbon dioxide concentration up to 8 – 10 mol%. The temperature decrease as usual resulted in a decrease of total salinity notwithstanding the reverse relations (Eastoe 1982, Popov 1978, Rekharsky et al. 1980).

Sulphide concentration and redox state of the ore-forming solutions are of great importance for the interpretation of copper and molybdenum geochemical behaviour. The probable range of these parameters may be evaluated using the geochemical interpretation of the mineral assemblages of the ore deposits in question (Distler et al. 1982). Hydrogen sulphide concentrations during ore deposition were estimated as 10^{-2} to 10^{-3} m, gradually decreasing with a decrease in temperature. The oxygen fugacity is evaluated as corresponding to that of the $Fe_3O_4 - Fe_2O_3$ and Ni – NiO buffers.

The data on the physicochemical parameters of the ore-forming solutions based on experimental data are used in examining the role of different molybdenum and copper compounds in ore transport and the causes of ore deposition.

Among the probable forms of molybdenum migration, complex compounds of molybdenum with sulphur, fluoride and chloride complexes as well as molybdation and products of its hydrolization and some other compounds were proposed by a number of investigators (Korenbaum 1970, Rekharsky 1973). Yet according to current experimental data on the stability of these compounds at elevated temperatures and the results of thermodynamic calculations, only the hexavalent molybdenum hydroxyl complexes are considered as important for molybdenum transport (Kolonin and Kosals 1979, Kolonin and Laptev 1975, Turgarinov et al. 1973). It should be emphasized that the conclusions presented are based on extrapolation of the experimental values obtained in the low-temperature range up to 200° – 250°C and published elsewhere.

To test these conclusions as well as to evaluate the possible role of trivalent and tetravalent molybdenum in providing the reduced character of hydrothermal solutions we carried out an experimental determination of the MoO_2 solubility in aqueous solutions. The experiments were undertaken for different Eh and pH values at the temperatures of molybdenite formation. The hexavalent molybdenum hydroxyl complexes, $HMoO^-_{4(sol)}$ and $H_2MoO^0_{4(sol)}$, were found to be the main molybdenum species over a broad interval of redox conditions up to 10^{-5} f_{O_2} for Ni – NiO buffer and acidity from 0.1 m HCl up to 0.001 m NaOH over-

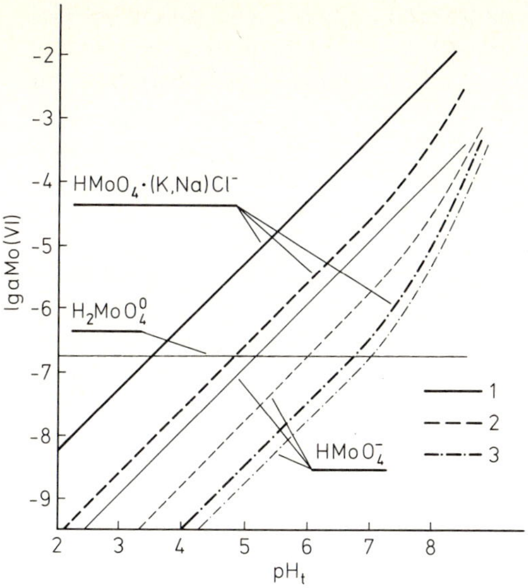

Fig. 1. The solubility of molybdenite in hydrothermal solutions. $\Sigma m(\text{Na, K})\text{Cl} = 1$ mol kg^{-1} H_2O; $\Sigma a_{H_2S\,(sol)} = 0.003$ mol kg^{-1} H_2O; $f_{H_2S} = 0.03$ ($t = 450°C$); f_{O_2} values correspond to Ni – NiO buffer. *1* 450°C; *2* 350°C; *3* 300°C

lapping the maximum probable variations in the hydrothermal solutions. The existence of chlorine-bearing (?) low-valence molybdenum cation forms is probable only in extremely acid solutions with HCl $\geqslant 1$ m.

The solubility of molybdenite was calculated using the experimental solubility data on MoO_2. The latter was found to form hydroxyl complexes in the temperature interval of molybdenite formation (300° – 450°C) (Fig. 1). It was assumed that the total activity of sulphur (S^{-2}) is equal to the average value within the estimated range (0.003 m; $f_{H_2S} = 0.03$ at 450°C) and the redox parameters are equivalent to the equilibrium values of f_{O_2} for the Ni – NiO buffer.

According to a number of workers, the minimum concentration of the metal-bearing component in ore-forming solutions producing economic deposits is estimated as not lower than 10^{-4} m (Crerar and Barnes 1976). These conditions are attained only in fairly alkaline environments at 300°C in ore solutions equilibrated with respect to the molybdenite (Fig. 1). The solubility of molybdenite is moderately increased with increase in temperature up to 450°C, but "ore-forming" molybdenum concentrations in the form of the $HMoO^-_{4(sol)}$ complex are predicted only in highly alkaline solutions. The molybdenum mineralization at porphyry copper deposits is often associated with feldspar-quartz-muscovite mineral assemblages. Thus we assume that the ore-forming solutions at the stage of ore deposition were near-neutral or weakly acidic. Evidently the molybdenum hydroxyl complexes could not support molybdenum transport even in the high-temperature region. The dominant role in molybdenum transport is attributed to some other species.

We assumed a substantial role was played by the alkaline metal chlorides during the formation of molybdenum-bearing complexes, as the chlorides are important ingredients of the hydrothermal solutions. Our experiments on the solubility of MoO_2 in alkaline chloride solutions at 300° – 450°C and under a pressure of 500 bar indicated that the ore-component content in the ore solutions increases with an increase of temperature and salinity. The dependence of solubility of MoO_2 on the solution chemistry is an argument for the existence of two types of Mo^{+6} compounds, such as $HMoO_4 \cdot (K,Na)Cl^-_{(sol)}$ or $(K,Na)HMoO^0_{4(sol)}$.

The formation of the first type of molybdenum-bearing complex is presumably governed by a substantial decrease in the NaCl and KCl dissociation constants with an increase in temperature in the range 300° – 450°C, resulting in an increase in the activities of the neutral molecules $NaCl^0_{(sol)}$ and $KCl^0_{(sol)}$. This prediction is in agreement with the marked increase in $MoO_{2(cryst)}$ solubility within the experimentally investigated temperature interval. On the other hand, the formation of $(K,Na)HMoO^0_{4(sol)}$ could be considered as the result of a general tendency for the formation of neutral complex compounds in the high-temperature region. For an unambiguous solution of the problem, additional experiments are needed. Yet the available data can be used for approximate estimates of MoS_2 solubility resulting in the formation of the assumed species (Fig. 1). The MoS_2 solubility in 1 m (Na, K)Cl solutions at 300°C is similar to its solubility in nonchloride solutions, whereas at 450°C the difference is about 1.5 orders of magnitude. Thus high-temperature neutral and even weakly acidic hydrothermal solutions are to be considered as productive in respect of molybdenum ore deposition.

Copper, like molybdenum, is known to exist in different valence states in hydrothermal solutions. The available data indicate that the predominent transport of copper is performed in the monovalent form, whereas molybdenum is transported in the highest valence state Mo(VI). According to data on the chemical compositions of ore-forming solutions, we predict chloride and sulphide complexes as the most probable copper-transporting species.

The formation of complex compounds of Cu(I) with chlorine was investigated by a number of authors (Ahrland and Kawsthorne 1970, Crerar and Barnes 1976, Nikolaeva et al. 1974, Varajash and Rekharsky 1981). According to these data, copper-bearing chloride complexes are observed to be dominant even at low chlorine concentrations ($\sim 10^{-3}$ m). The formation of $CuCl^0_{(sol)}$, $CuCl^-_{2(sol)}$, $CuOHCl^-_{(sol)}$ and other species is a function of acidity and chlorine concentration in solution. Yet estimates of the stability of these species, especially in the high-temperature region, are controversial. Thus, according to Nikolaeva et al. (1974), the dominant copper-bearing species in dilute ($10^{-2} - 10^{-3}$ m) chloride solutions at temperatures up to 150°C is represented by $CuCl^-_{2(sol)}$, while Crerar and Barnes (1976) reported that $CuCl^0_{(sol)}$ is formed even in highly concentrated solutions (up to 10.4 m NaCl) in the 200° – 350°C temperature range. Evidently the disagreement could lead to erroneous conclusions concerning the ore-forming properties of the complexes.

A number of experiments on metallic copper solubility at 300° – 350°C in chlorine-bearing solutions under specified redox conditions were carried out. It was found that $CuCl^-_{2(sol)}$ complex compounds are the dominant species in acidic and near-neutral solutions with 0.003 – 1.0 m NaCl concentration. In more dilute solutions the monovalent ion, $Cu^+_{(sol)}$, is the dominant copper-bearing species. Thus, our experimental data indicate that $CuCl^0_{(sol)}$ is unstable not only at 25 °C (Ahrland and Kawsthorne 1970), but also in high-temperature solutions.

These experimental data, as well as thermodynamic data on $CuCl^-_{2(sol)}$ and $Cu^+_{(sol)}$ for the low temperature region (Ahrland and Kawsthorne 1970, Naumov et al. 1971, Nikolaeva et al. 1974), were used in the solubility calculations. Chalcopyrite was taken to be in equilibrium with magnetite (350 °C) and pyrite (300° and 200 °C), and the solubility data for a 1.0 m NaCl solution with $CuCl^-_{2(sol)}$ and $Cu^+_{(sol)}$ as the products are plotted on Fig. 2. On the same figure the chalcopyrite solubility with the $Cu(HS)^-_{2(sol)}$ complex as the final product is shown according to Crerar and Barnes' (1976) data. The hydrogen sulphide activity was taken as 0.003 m, similar to that in the molybdenum solubility calculations. From Fig. 2 it is evident that the copper concentration in the solution abruptly increases with increase in temperature from 200° up to 350 °C. At 200 °C the ore-productive copper concentrations are predicted in solutions with pH < 3, while at 350 °C the region is extended up to near-neutral conditions. The dominant copper-bearing forms are represented by the chloride compounds, yet the hydrogen sulphide complexes play a subordinate role even at high H_2S concentrations. It is evident that the cation $Cu^+_{(sol)}$ cannot be considered as one of the transport forms of copper in the hydrothermal solutions.

Thus, the copper-bearing chloride complexes and molybdenum-bearing hydroxyl alkaline chloride complexes (or alkaline molybdate complexes) are the most probable species for the hydrothermal transport of copper and molybdenum.

In this section we discuss the factors involved in copper and molybdenum deposition in the form of chalcopyrite and molybdenite, respectively. On Fig. 3 the evolution of a hypothetical hydrothermal solution containing ore-productive copper and molybdenum concentrations within the 500° – 200 °C temperature interval in terms of calculated and experimental data is shown. The total alkali

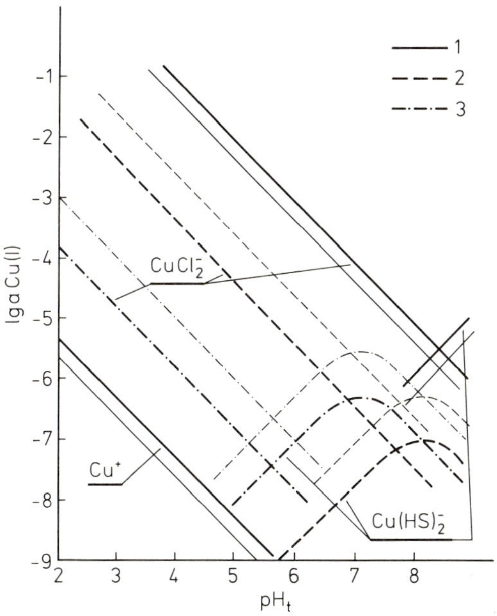

Fig. 2. The solubility of chalcopyrite in equilibrium with magnetite (350 °C) and pyrite (300° and 200 °C) in hydrothermal solutions. $a_{Cl^-} = 1$ mol kg^{-1} H_2O; $\Sigma a_{H_2S\,(sol)} = 0.003$ mol kg^{-1} H_2O. *1* 350 °C; *2* 300 °C; *3* 200 °C; *solid lines* correspond to f_{O_2} Ni – NiO; *thin lines* = f_{O_2} $Fe_3O_4 - Fe_2O_3$

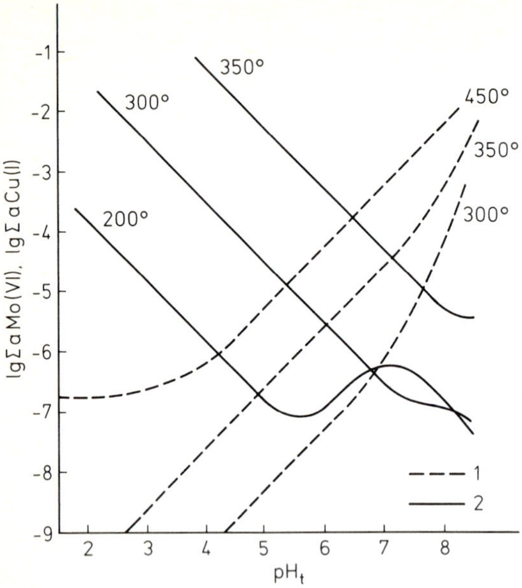

Fig. 3. The solubility of molybdenite (*1*) and chalcopyrite (*2*) in hydrothermal solutions. Σm(Na, K)Cl $= 1$ mol kg^{-1} H_2O; $\Sigma a_{H_2S(sol)} = 0.003$ mol kg^{-1} H_2O; f_{O_2} values correspond to the Ni – NiO buffer

chloride concentration was taken as 1.0 m, $\Sigma a_{H_2S(sol)} = 0.003$ m, redox conditions correspond to the Ni – NiO buffer, copper and molybdenum concentration starting values were taken as 10^{-3} m and 10^{-4} m, respectively. Evidently the acidity of the solution at 500 °C is close to the neutral region, as the opposite case would correspond to ore oversaturation with respect to molybdenum in the acidic region or to copper in the alkaline region at extremely high temperatures exceeding the known interval of ore deposition.

A temperature decrease from 500° to 350 °C would result in molybdenite deposition as its solubility abruptly decreases. At the same time the hydrothermal solution remains undersaturated in relation to chalcopyrite. An increase in acidity as a result of the temperature decrease is considered as an additional factor favouring molybdenite crystallization and preventing copper deposition. Chalcopyrite could be deposited in equilibrium with molybdenite during progressive cooling of the hydrothermal solution (350° to 300 °C) as the chalcopyrite solubility in this temperature range is lower in relation to the initial copper concentration even within the moderately acid region. At 300 °C, practically all molybdenum would be deposited in the solid form, whereas copper deposition is prolonged to 200 °C. The neutralization of the solution is considered as an additional factor favouring a similar evolutionary path.

The stability of molybdenum- and copper-bearing compounds in hydrothermal solutions is governed not only by temperature and acidity variations, but also by changes in redox conditions and variations in S^{-2} and alkali metal chlorides contents in solution. Unfortunately the lack of reliable data on numerical values of these parameters and their dependence on the stages of evolution of the hydrothermal process prevents the estimation of the specific role of each parameter in the ore-forming process. We suggest that the dominant independent factor in the deposition of the molybdenum sulphides and copper sulphides is the temperature decrease of the hydrothermal solutions. Our model of the successive formation of molybdenite and chalcopyrite in terms of gradual cooling of the hypothetical hydrothermal solution is considered to some extent to agree with this conclusion. The model is consistent with the oberserved prevalent association of molybdenum ore mineralization with high-temperature feldspar-quartz metasomatites, whereas the copper ores are observe in assemblage with low-temperature quartz-sericite metasomatites.

References

Ahrland S, Kawsthorne J (1970) The stability of metal halide complexes in aqueous solution. VII. The chloride complexes of copper (I). Acta Chem Scand 24:157 – 172

Crerar DA, Barnes HL (1976) Ore solution chemistry. 5. Solubilities of chalcopyrite and chalcocite assemblages in hydrothermal solution at 200° to 350°C. Econ Geol 71:772 – 794

Distler VV, Rekharsky VI, Pashkov YuN et al. (1982) The geochemistry of ore forming processes. Nauka, Moscow, 270 pp (in Russian)

Eastoe CJ (1982) Physics and chemistry of the hydrothermal system at the Panguna Porphyry copper deposits, Bougainville, Papua New Guinea. Econ Geol 77:127 – 153

Kolonin GR, Kosals YaA (1979) Physico-chemical parameters of molybdenum-tungsten ore formation in ore deposits of quartz-greisen rock association. In: Main parameters of the natural processes of the endogeneous ore formation, vol I. Nauka, Novosibirsk, pp 236 – 251 (in Russian)

Kolonin GR, Laptev YuV (1975) The spectrophotometric investigation of the stability of sulfide molybdenum complexes versus temperature. In: Experimental mineralogical investigations (1974 – 1975). Inst Geol Geophys, Novosibirsk, pp 38 – 43 (in Russian)

Korenbaum SA (1970) Physico-chemical conditions of tungsten- and molybdenum-bearing mineral crystallization in hydrothermal solutions. Nauka, Moscow, 268 pp (in Russian)

Naumov GB, Ryzhenko BN, Khodakovsky IL (1971) Handbook of thermodynamic constants for geologists. Atomizdat, Moscow, 240 pp (in Russian)

Nikolaeva NM, Ehrenburg AM, Skorokhod LS (1974) The influence of temperature on the equilibrium constant values of the reactions of substitution in the halide complexes of copper (I). Izv Sib Otd Akad Nauk SSSR Ser Chim 3:44 – 48 (in Russian)

Pavlova IG (1978) The copper-porphyry ore deposits. Nedra, Leningrad, 274 pp (in Russian)

Popov VS (1977) The geology and genesis of the copper- and molybdenum-porphyry ore deposits. Nauka, Moscow, 201 pp (in Russian)

Rekharsky VI (1973) The geochemistry of molybdenum in the endogeneous processes. Nauka, Moscow, 268 pp (in Russian)

Rekharsky VI (1980) Geochemistry of copper-molybdenum deposits of the Lesser Caucasus. In: European copper deposits. Dep Econ Geol Fac Min Geol, Belgrade, Yug, pp 104 – 107

Rekharsky VI, Pashkov YuN, Ignatjeva IB, Nosik LP (1980) The formation of ore-bearing hydrothermal-metasomatic rock formations in molybdenum ore deposits. In: Thermobarogeochemistry and ore genesis. Vladivstok, pp 83 – 86 (in Russian)

Shipulin FK, Rekharsky VI, Rozbianskaya AA et al. (1975) The intrusions, hydrothermal-metasomatic rocks and copper-molybdenum ore mineralization. Nauka, Moscow, 231 pp (in Russian)

Tugarinov AI, Khodakovsky IL, Zhidikova AP (1973) Physicochemical conditions of molybdenite formation in hydrothermal uranium-molybdenum ore deposits. Geokhimiya 7:975 – 984 (in Russian)

Varjash LN, Rekharsky VI (1981) On monovalent copper behaviour in hydrothermal chloride solutions. Geokhimiya 7:1003 – 1008 (in Russian)

Batholith-Volcano Coupling in the Metallogeny of Porphyry Copper Deposits

P. E. DAMON[1]

Abstract

In the course of our study of the porphyry copper deposits of Mexico, the author became aware of the vast extent of the composite batholiths of Sonora and Sinaloa. We observed that the mineralized porphyritic plutons intrude the pene-contemporaneous batholith and are associated with dominantly andesitic volcanism. An explanation of this phenomenon is provided by the phenomenon of dome-in-dome structures observed by Ramberg (1981) in centrifuge model experiments where less dense material is overlain by material of greater density. I have applied this concept to the case of a cooling batholith. The interior of a batholith is inherently unstable gravitationally with respect to the more rapidly cooling upper part which is in contact with the country rock. As a consequence, the upper reaches cool faster, begin crystallizing sooner, and eventually become denser than the lower reaches. Thus, the interior of the batholith becomes buoyant and intrudes the upper part. The early batholith forms a broad, gentle dome intruded by later domes. The process is repeated with buoyant upwelling of the interior of the intrusive domes forming cupolas, some of which breach the surface and form cones or stratovolcanos.

As a result of the severe temperature regimen and dome-in-dome structures, large-scale hydrothermal fluid transfer occurs dominated by volcanic orifices constituting hydrothermal "pumps" or "artesian vents". Blind porphyries, those that do not breach the surface, have hydrothermal systems that are subsidiary to the master volcanic artesian vents. Such master hydrothermal systems have isotopically dated lifetimes equal to the isotopically dated long-lived hydrothermal activity of stratovolcanos, i.e., up to at least 2.5 m. y. The fluids exported to the volcanic pumps carry ore components extracted from the batholith and country rock. The ore components are distributed and precipitated in the manifold forms of occurrence observed within porphyry systems.

[1] Laboratory of Isotope Geochemistry, Department of Geosciences, University of Arizona, Tucson, AZ 85721, USA

Geology and Metallogeny of Copper Deposits
(ed. by G.H. Friedrich et al.)

Introduction

During the past decade, the author has had the opportunity to work with Ing. Guillermo P. Salas, former Director of the Consejo de Recursos Minerales, and other colleagues (see acknowledgments) on a project entitled "Dating the ore deposits and defining the metallogenetic provinces of the Republic of Mexico". This project had as the initial impulse the construction and modification of the metallogenetic map of Mexico (Salas 1975) and developed the following concepts: (1) Most of the metallic ore deposits of the Southern Cordillera of the U.S.A. and Mexico are related to volcanism; (2) volcanos and related plutons are not randomly spaced and usually occur in volcanic arcs; (3) volcanic arcs of different ages can be distinguished and their paleogeography reconstructed through field work and isotopic dating; (4) volcanic arcs migrate with time; (5) as they migrate away from the associated trench, the silicic magma composition becomes more alkalic (Dickinson and Hatherton 1967, Dickinson 1975, Keith 1978, Clark et al. 1982, for Mexico see Damon et al. 1981b, Fig. 8); (6) the compositions of ore deposits change as magma compositions change (ibid.); (7) this results in metallogenetic provinces that are long, relatively narrow belts, more or less parallel to the paleo-subduction zone (Fig. 1, see caption for explanation); (8) erosion is accelerated by volcanic arc tumescence and by continental uplift proceeding inward from the convergent margin (Damon 1979, 1983); (9) as a result, secondary enrichment takes place, and if erosion is not interrupted, the ore deposit will be removed (Livingston et al. 1968, Damon et al. 1983a); (10) ore preservation depends upon tectonic events, such as faulting and sedimentation, or the ore being buried under a volcanic blanket, such as within the Sierra Madre Occidental and southwestern U.S.A. (Livingston et al. 1968, Damon et al. 1983a).

In recent years, there has been much interest in the Mesozoic-Cenozoic Cordilleran volcanic arcs that are related to the convergent margin of western North America associated with the present cycle of seafloor spreading. It is now generally recognized that these arcs have migrated over long distances. By earlier Jurassic time a volcanic arc appears to have been established throughout the length of the Southern Cordillera from northern Nevada through Chiapas, Mexico (Damon et al. 1981b). By early Cretaceous time, the volcanic arc had regressed westward to the ancestral Pacific margin from which it transgressed slowly eastward again in early Cretaceous time (Lindgren 1915, Henry 1975, Coney and Reynolds 1977, Anderson and Silver 1977, Clark et al. 1979, Damon et al. 1981b, Chen and Moore 1982). During the Laramide orogeny, the volcanic arc moved rapidly inland to the east, toward the Gulf of Mexico, and then following the Laramide orogeny, it rapidly regressed to the Pacific margin (Coney and Reynolds 1977, Clark et al. 1979, Damon et al. 1981b).

In this paper I will restrict my attention to the Mexico states of Sonora and Sinaloa (Fig. 3). Dated igneous rocks from Sonora and Sinaloa are plotted on Fig. 2 relative to the envelope of dates for igneous rocks of all of northern Mexico. It is generally accepted that all of these igneous rocks, both plutonic and volcanic, with the exception of the two-mica granites and pegmatites and post-subduction extrusive rocks, are related to the migrating Cordilleran volcanic arcs

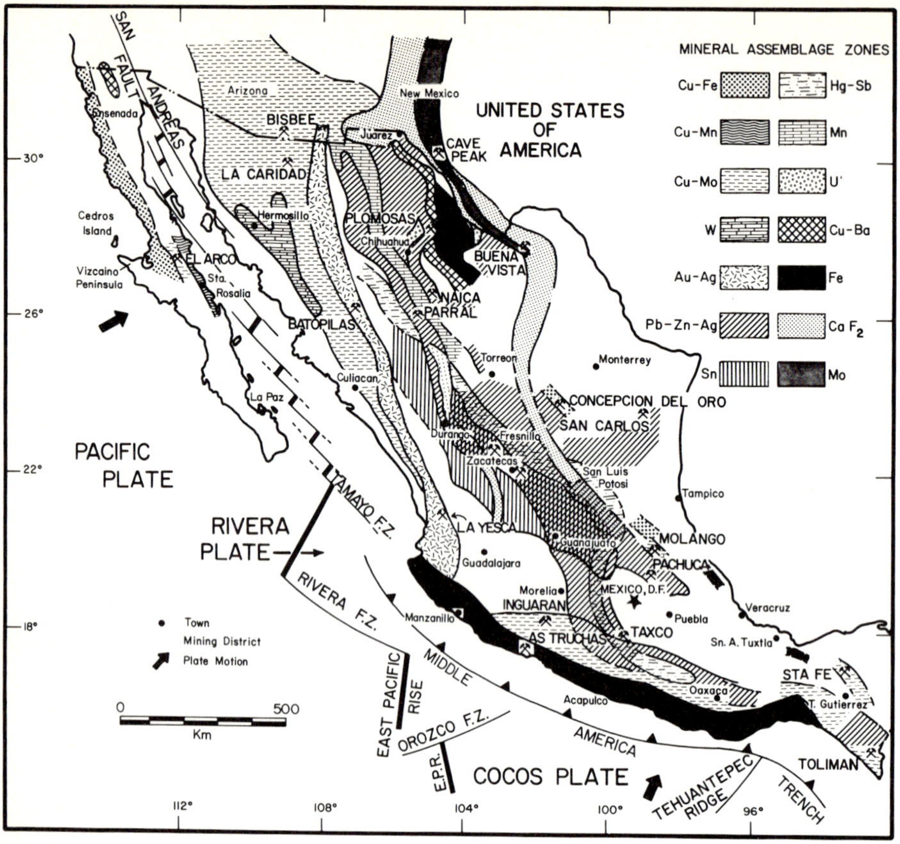

Fig. 1. The metallogenetic provinces of Mexico (modified from Clark et al. 1979). The *patterned areas* enclose provinces in which significant numbers of important deposits are known to occur (see also Salas 1975). Note succession of north-northwest-trending metallogenetic provinces resulting from different stands of the Cordilleran Volcanic Arc which becomes increasingly alkalic with distance from the trench (Damon et al. 1981b). The Cu – Mo province is the focus of interest in this paper. Note the concentration of deposits in that province as shown in the inset of Fig. 3

of North America. The two-mica granites and pegmatites are related to the thermal event accompanying the volcanic arc, but have no known volcanic equivalents. The postsubduction basalts and andesites are related to a taphrogenic event, the Basin and Range Disturbance, that occurred throughout the Basin and Range Province and southward in western Mexico to Cabo Corrientes at the northern end of the Middle American Trench. Figure 3 shows that the Cordilleran Jurassic volcanic arc that terminated in late Jurassic time (145 Ma), was followed by a long magma gap in Sonora and Sinaloa lasting 40 m. y. to the west and 90 m. y. to the east. The arc then rapidly regressed westward through Sonora and Sinaloa during Oligocene and early Miocene time. This was followed by dominantly basaltic and andesitic associated with taphrogenesis.

During the eastward advance of the volcanic arc in the Cretaceous period, porphyry copper deposits were emplaced, forming the metallogenetic province

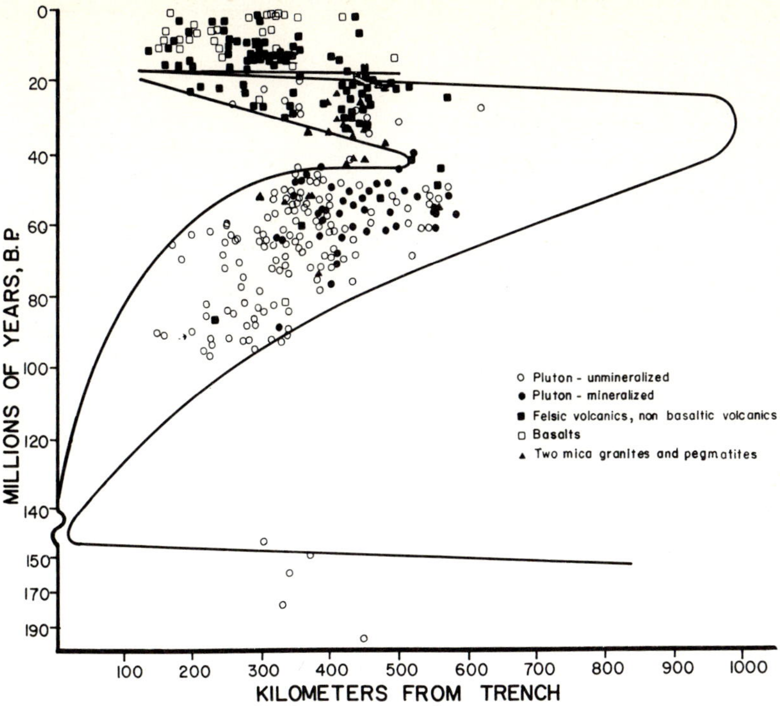

Fig. 2. Migration of the Cordilleran volcanic arcs within northern Mexico with specific reference to Sonora and Sinaloa (modified from Clark et al. 1979). The outer envelope includes over 95% of all of the dated rocks of northern Mexico. The dates plotted are for Sonora and Sinaloa only. Note the magma gap of varying duration between Jurassic and Cretaceous-Cenozoic magmatism. In late Jurassic-early Cretaceous time, the volcanic arc had returned to the Pacific margin, but then migrated slowly eastward until the beginning of rapid convergence during the Laramide orogeny, at 80 Ma. During the Laramide orogeny, 80 to 40 Ma, the Cordilleran Volcanic Arc migrated rapidly eastward. Following extensive changes in plate motions at the end of the Laramide (40 Ma), the Cordilleran Volcanic Arc migrated rapidly back to the Pacific margin. Taphrogeny followed with extrusion of dominantly basaltic and andesitic magma. Data are from Damon and Mauger (1966), Henry (1975), Gastil et al. (1976, 1978), Anderson and Silver (1977, 1979), Gastil and Krummenacher (1977), Anderson and Roldán (1979), Clark et al. (1979), Anderson et al. (1980), Damon et al. (1983a, b)

shown in Fig. 1. Twenty-nine deposits form a long, NNW-trending linear belt (inset, Fig. 3). This was followed by emplacement of tungsten deposits associated with two-mica peraluminous granites. The Au – Ag metallogenetic province shown in Fig. 1 was formed during the regression of the volcanic arc to the Pacific coast.

In this paper, as previously mentioned, I will use the porphyry copper metallogenetic province of Sonora and Sinaloa as an example, in order to demonstrate the importance of batholith-stratovolcano coupling in the formation of porphyry copper deposits.

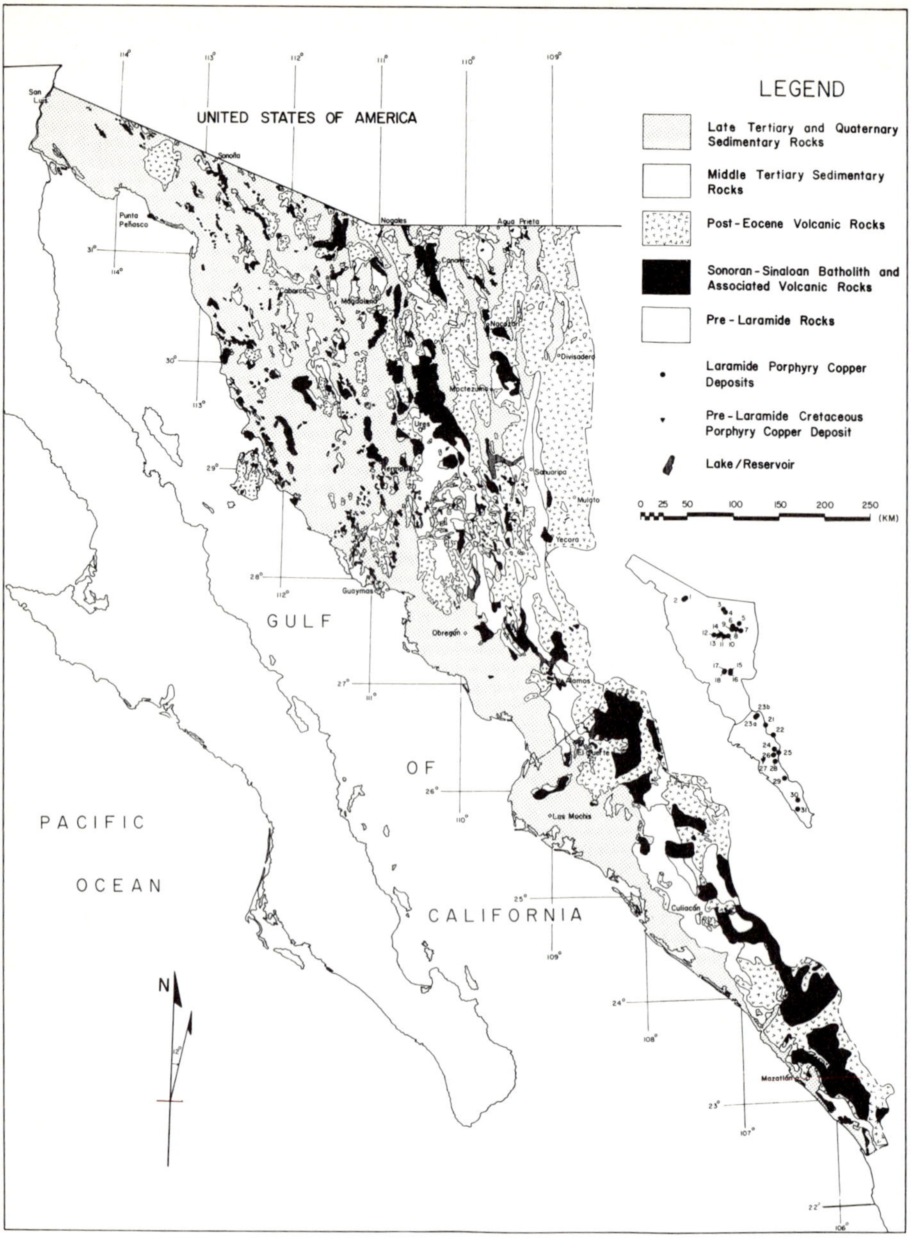

Fig. 3. Simplified geologic map of Sonora and Sinaloa emphasizing the Laramide batholiths. Note that the actual extent of the batholiths must be much greater because voluminous Mid-Tertiary volcanic extrusions, dominantly ignimbritic, and sedimentation in basins have buried much of the batholith. Data are from the Carta Geologica de la Republica Mexicana compiled by López Ramos (1976)

Field and Age Relationships

The late Cretaceous-early Cenozoic (Laramide) batholiths of Sonora and Sinaloa cover much of the two states. The batholiths and their associated volcanic rocks crop out over an area of 13600 km^2 in Sinaloa and 12800 km^2 in Sonora, comprising a total area of 26400 km^2 (Fig. 3). This area may be compared with the area of the Sierra Nevada batholith, which covers some 60000 km^2. The Sonoran-Sinaloan batholiths are covered by Mid-Tertiary volcanic rocks of the Sierra Madre Occidental (King 1939) and of the Basin and Range Province (Shafiqullah et al. 1980), as well as the sedimentary rocks of the Oligocene-Miocene Baucarit Formation (King 1939) and late Tertiary–Quaternary sediments (Fig. 3). Thus their actual extent must rival some of the greatest batholiths of the world. They intrude rocks of varying age ranging from older Precambrian (Livingston and Damon 1968) through marine limestone and clastic rocks of early Cretaceous age (King 1939). The batholiths are composite intrusions consisting primarily of calc-alkaline granodiorite and granite intruded by penecontemporaneous late-stage porphyries, many of which contain economic copper deposits (Fig. 3, inset).

The bulk of the premineralization batholith cooled at a rate which seems, at first, to be surprisingly fast, as demonstrated by the concordance of many hornblende and biotite K – Ar dates for samples of unmineralized portions of the Sonora-Sinaloa batholiths (Table 1). K – Ar mineral dates may be related to the time at which the temperature is too low to activate significant diffusion (the "blocking temperature").

Despite the widely different approximate blocking temperatures for hornblende (500 °C) and biotite (300 °C), 67% of the hornblende-biotite pairs showed no significant difference in age, demonstrating very rapid cooling such as shown for the fast cooling curve in Fig. 4. There is a significant difference in age between the pairs constituting the remaining 33% of the samples. This indicates a slow cooling rate as shown also in Fig. 4. The rate of cooling of a batholith is very much dependent upon its depth of emplacement. For example, Larsen (1945) has calculated that the top of a batholith with a cover of 5 km of sedi-

Table 1. Statistical analysis of 46 biotite-hornblende pairs from unmineralized portions of the Sonora-Sinaloa batholiths

Difference between hornblende and biotite age (Ma)	Number of samples in class interval	%
~0, i.e., number falling between $\pm 2\sigma$ assuming normal distribution	31	67.4
>0 to ≦5 Ma	9	19.6
>5 Ma to ≦10 Ma	5	10.9
>10 Ma	1	2.1
	$\Sigma = 46$	100.0

Note: Data for Sonora are from Damon et al. (1983a), Gastil and Krummenacher (1977), Gastil et al. (1977). Data from Sinaloa are from Henry (1975) and unpublished data from this laboratory.

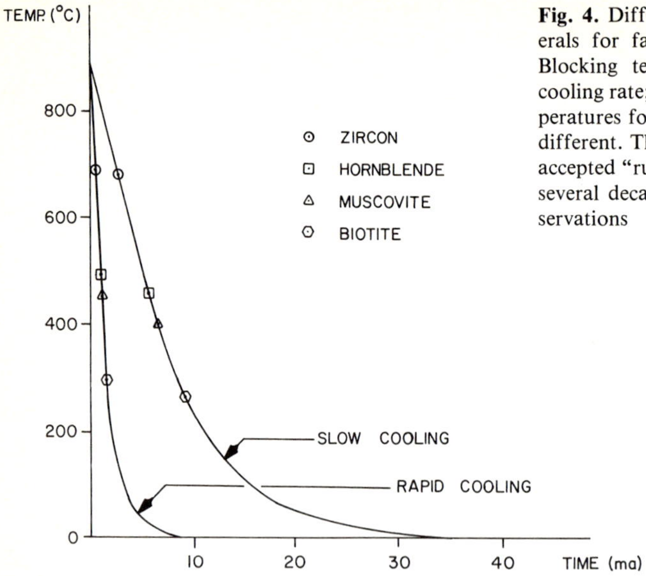

Fig. 4. Difference in age for different minerals for fast and slow cooling regimens. Blocking temperatures are a function of cooling rate; consequently, the blocking temperatures for fast and slow cooling rates are different. The values used here are generally accepted "rule of thumb" values based upon several decades of field and laboratory observations

mentary rock would cool from its emplacement temperature of 820 to 300 °C in about 2 Ma. With a cover of 1 km, it would take only about 0.1 Ma to cool. Seven Ma would be required for cooling to 300 °C at a depth of 3 km within the batholith having a cover of 5 km. With reference to Larsen's graphs, the differences in hornblende-biotite ages (Table 1) are consistent with the upper part of the batholiths having intruded to within 5 km of the surface, and erosion having cut deep enough to expose, for the most part, the upper reaches of the batholith; but in other parts erosion has extended to a depth of as much as 5 km below the upper surface, exposing the slowly cooled depths of the batholith. This would require erosion rates of only about 5 to 20 cm per 1000 yrs. since emplacement of the batholiths. Such rates would be quite low for a mountainous area (Gregory and Walling 1973). The Rocky Mountains of the U.S.A. have been estimated to be eroding at a rate of 50 cm per 1000 yrs. (ibid.). However, the parallel ranges and Basin and Range Province on the west flank of the Sierra Madre Occidental in which the batholiths crop out are not as rugged as the Rocky Mountains and, so, the rate of 5 to 20 cm per 1000 yrs. is reasonable.

During field reconnaissance associated with our study of the porphyry copper deposits of Mexico, we observed that the most common wall rocks of the porphyries are penecontemporaneous early-stage batholith rocks and associated Laramide volcanic rocks (Damon et al. 1983a). Table 2 shows that 69% of the wall rocks for Mexican porphyries are comprised of penecontemporaneous batholithic rocks and associated volcanic rocks.

Epicontinental limestones that are barren in copper comprise 17%, and all other sedimentary rock types comprise only 14% of the wall rocks. Partly as a consequence of this relationship, Damon et al. (1983a) concluded that copper was being transported from the calc-alkalic batholithic wall rocks to the hypabyssal roots of the overlying volcano, "where the metals are distributed and

Table 2. Wall rocks intruded by Mexican porphyry copper deposits

Type of wall rock	Frequency	
	Number	%
Penecontemporaneous composite batholith	15	37
Laramide volcanics	13	32
Epicontinental limestones	7	17
Shales, sandstones, and metasediments	5	12
Pre-Laramide submarine basalts	1	2
	$\Sigma = 41$	100

precipitated in the manifold forms of occurrence observed within the porphyry systems" (Damon et al. 1983a, p. 1069).

Table 3 includes K–Ar dates from six locations (see inset, Fig. 3) for both mineralized and unmineralized facies of the batholiths. If the anomalous datum for Mina Washington is excluded (possibly reset by post-volcanic arc two-mica granite), the average difference is 3.1 ± 0.8 ($\bar{\sigma}$) Ma with the mineralized facies always younger. Thus, about 3 m. y. after emplacement of the early stages of the batholiths, the late-stage porphyries had been intruded and were in the process of accumulating copper mineralization.

Following our early reconnaissance work (Damon et al. 1981a) and K–Ar dating of the batholiths and their associated porphyries and volcanic rocks, Jean Jacques Cochemé, of the Northwestern Regional Station of the Institute of Geology, Autonomous National University of Mexico, mapped an area in Sonora on the edge of the Sierra Madre Occidental (Damon et al. 1983b). A slightly modified version of Cochemé's map is shown in Fig. 5. The map shows that in areas where erosion has not cut too deep, hypabyssal cupolas are still surrounded by Laramide volcanic rocks (NW area). On the other hand, where erosion has exposed the batholith, the hydrothermally altered porphyries clearly intrude the batholith (central and south areas on the map).

Summarizing, I have shown that the vast batholiths of Sonora and Sinaloa rose close enough to the surface for a rapid cooling of their upper surface to have taken place. Deeper areas cooled at a measurably slower rate. Plutons and cupolas intruded the early-stage batholiths several million years after their emplacement. Lastly, the porphyry copper deposits appear to be associated with stratovolcanos (Sillitoe 1973), giving rise to the associated, dominantly andesitic volcanic rocks that are associated with the porphyries (see Fig. 5).

Dome-in-Dome Structures

The relationship of hypabyssal porphyries intruding the penecontemporaneous batholith was incontrovertible, but at first puzzled me. Sillitoe's model (1973, his Fig. 1) has small (<2 km diam) hypabyssal porphyries passing into unminer-

Table 3. Age difference between unmineralized batholith and mineralized porphyry copper

Map No., name, state (Fig. 3) reference	Coordinates	Mineral and rock dated	K – Ar date (Ma)	Difference (Ma)
(11) Mina Washington, Sonora Damon et al. (1983a)	29°54′29″N 110°05′26″W	Biotite, premineralized granodiorite	56.4 ± 1.2	10.7 ± 1.6
	29°53′47″N 110°03′59″W	Sericite, mineralized breccia pipe	45.7 ± 1.0	
(18) Suaqui La Verde, Sonora Damon et al. (1983a)	28°25′12″N 109°49′01″W	Hornblende, premineralized granodiorite	58.8 ± 1.3	2.1 ± 1.7
	28°24′41″N 109°48′11″W	Sericite, mineralized quartz diorite (dissemination)	56.7 ± 1.1	
(22) Cerro Colorado, Chihuahua Damon et al. (1983a)	26°17′10″N 107°20′36″W	Biotite, premineralized granodiorite	48.0 ± 1.2	1.7 ± 1.6
	26°15′08″N 107°19′46″W	Sericite, mineralized breccia	46.3 ± 1.0	
(24) Tameapa, Sinaloa Damon et al. (1983a)	25°39′ N 107°23′ W	Hornblende, premineralized granodiorite	56.9 ± 1.2	2.7 ± 1.6
	25°40′ N 107°21′ W	Secondary biorite, quartz monzonite mineralized stockward	54.1 ± 1.1	
(28) Minas Las Higueras, Sinaloa Damon et al. (1983a)	25°00′39″N 107°13′42″W	Biotite, unmineralized granodiorite	54.9 ± 1.2	5.9 ± 1.6
	25°00′39″N 107°13′45″W	Sericite from dissemination in mineralized granodiorite	49.0 ± 1.0	
(31) Malpica, Sinaloa Henry (1975)	23°14′ N 106°08′ W	Biotite, premineralized granodiorite	57.3 ± 0.6	3.3 ± 1.7
	23°16′ N 106°07′ W	Biotite, breccia filling in granodiorite porphyry	53.8 ± 0.6	
		Average difference		4.4 ± 1.4 $\bar{\sigma}$
		Average difference (excluding anomalous location 11)		3.1 ± 0.7 $\bar{\sigma}$

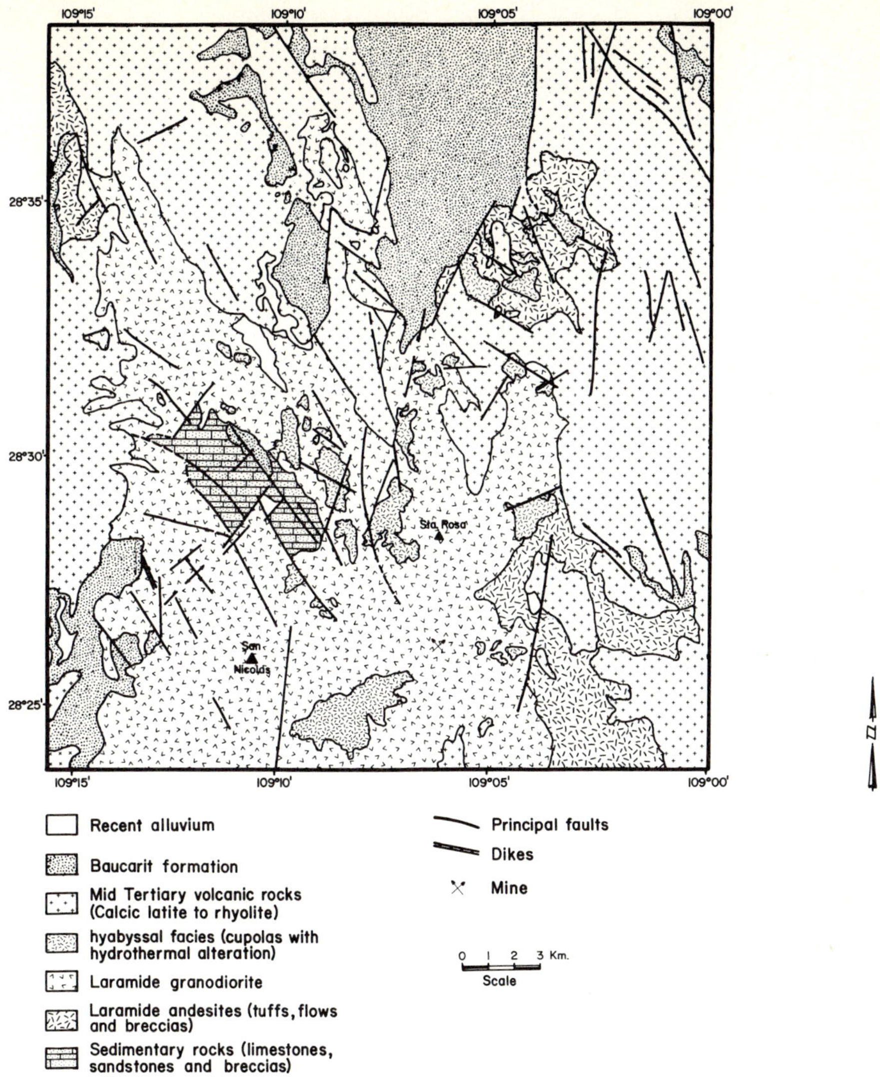

Fig. 5. Geologic map of the region of Santa Rosa de Yecora, Sonora (modified from map by Jean Jaques Cochemé in Damon et al. 1983b). Note that in the southern and central areas of the map, the batholith is intruded by hypabyssal cupolas with hydrothermal alteration. In the northeast, the Laramide volcanic cover has not yet been removed. Erosion has removed all but a small remanent of the Paleozoic-Mesozoic pre-batholith sedimentary cover. This remanent crops out in the west-central area of the map

alized phaneritic granodiorite at a depth of from 7 to 8 km. Branch (1976) suggests that small, high-level stocks (1 – 2 km diam), within and below much larger stratovolcanos, are fed by a main volcanic conduit directly from the mantle. In the Basin and Range Province of the southwestern United States, where taphrogeny has more severely disrupted the basement than in the area of this study on the flanks of the Sierra Madre Occidental in Sonora and Sinaloa, Titley (1982) and Norton (1982) model the porphyry copper deposits as high-level diapirs intruding basement host rocks that are at "slightly higher than normal" temperatures (Norton 1982, p. 61). Norton suggests that a 30 °C km^{-1} gradient seems reasonable. However, where batholithic intrusions precede the hypabyssal, porphyritic plutons, as in Sonora and Sinaloa, much higher temperature gradients must prevail.

The author's perplexity was relieved after a discussion with Dr. Hans Ramberg of the University of Uppsala. After I described the field relationships to Dr. Ramberg, he assured me that his scale models predict this sort of relationship, and referred me to his discussion of dome-in-dome structures in Chap. 11 of his book (Ramberg 1981). I then proceeded to model these structures using his equations and a simple numercial computer model.

The model is shown in Fig. 6. For initial conditions, I chose 8 km of country rock (unit 1) underlain by a granodioritic magma 8 km thick (unit 2 + 3), that in turn is underlain by a more basic magma (unit 4) of greater density than the overlying magma ($\varrho_{(4)} > \varrho_{(2+3)}$), which, consequently, is gravitationally stable. The combined layer ($i = 2+3$) has a density less than the country rock ($\varrho_{(2+3)} < \varrho_{(1)}$) and hence is gravitationally unstable. The viscosity ratio critically determines the wavelength, λ, of the doming of the batholith's upper surface. I chose a viscosity ratio, $\mu_{(1)}/\mu_{(2+3)}$, of 10^4, yielding a λ of 470 km. A larger viscosity ratio would produce an even larger λ. Consequently, the upper surface has the form of a broad, gentle dome. However, as the upper part of the batholith cools more rapidly than the lower part, it will eventually have a higher density than the lower part and the lower part will become gravitationally unstable. To model this, I separated layer ($i = 2+3$) into two layers, $i = 2$ and $i = 3$, with $\varrho_{(1)} > \varrho_{(2)}$. I chose a viscosity ratio of $\mu_2/\mu_3 = 100$, which yielded $\lambda = 50$ km. Upon discussing this with my colleague, Professor Spencer Titley, he pointed out that Pitcher (1972) had mapped similar spacings for centered complexes of arcuate intrusions that form an essential component of the coastal batholith of Peru. Ramberg (1981) gives several other examples with similar spacings.

Layer 3, as it rises into the gently domed upper batholith and country rock, is provided with space by the thinning of the overlying batholith and country rock. The overlying material moves away from the rising domes and then down into the space provided between the domes as layer 3 is "sucked" up into the rising domes. The downward-moving material is compressed and deformed as the dome rises. Ramberg (1981) has modeled a similar situation with two competent modeling-clay sheets (his Fig. 11–13, p. 256). After erosion (Fig. 6f), the resultant geologic terrain would look very much like the map of the Santa Rosa area (Fig. 5).

Summarizing, the lower part of a rising and cooling batholith is inherently unstable with respect to the upper part, because eventually its density will be less than the cooled upper reaches of the batholith. As a consequence, the lower part will rise forming dome-in-dome structures. Ramberg's equations (1981) predict spacings of the rising domes that are in accord with spacings of mapped plutonic complexes intruding into batholiths. This explains the field and age relationships between early-stage batholiths and late-stage porphyries that we have observed in Mexico.

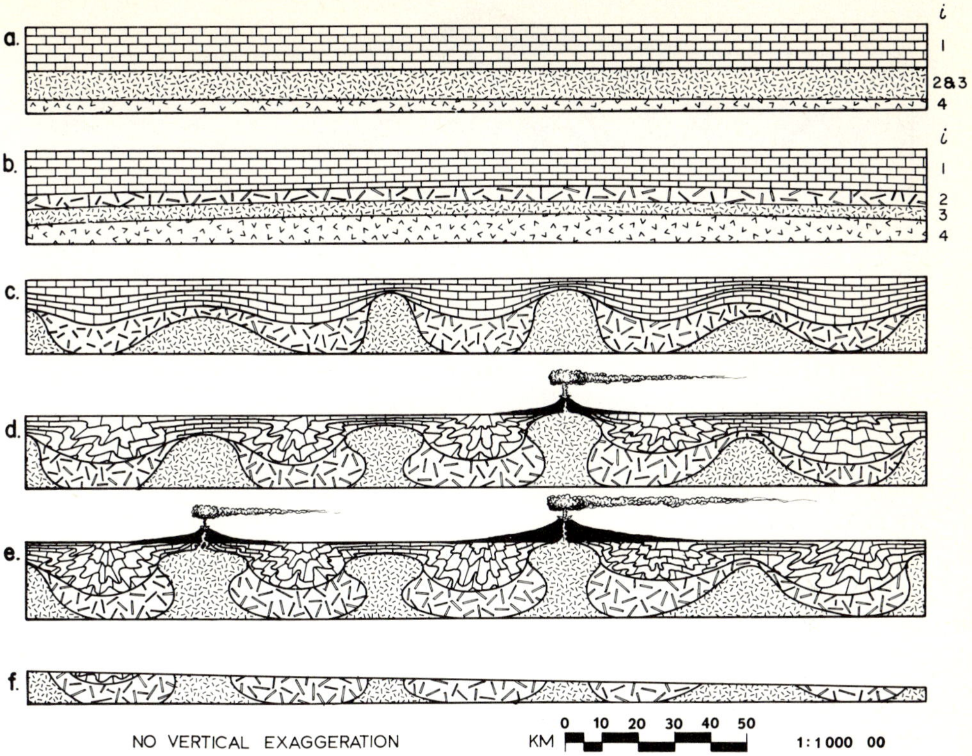

Fig. 6a – f. Dome-in-dome structures: **a** Batholith forms. *Layer 1* represents the covering country rock. *Layers 2+3* represents the internally stable granodioritic batholith underlain by diorite (*layer 4*) which is denser than the granodiorite, hence gravitationally stable. **b** Gentle doming is accompanied by cooling and partial crystallization of the upper part of the granodiorite batholith. The upper part gradually becomes denser than the lower part and gravitationally unstable. *Layers 2+3* becomes *layer 2* and *layer 3*, with densities ϱ_2 and ϱ_3, where $\varrho_2 > \varrho_3$. **c** Buoyant lower granodiorite (*layer 3*) forms rising domes intruding the country rock, and granodiorite *layer 2* sinks to occupy the space left by *layer 3*. **d** One dome breaches the surface to form a juvenile stratovolcano. The sinking part of the batholith and cover rocks continue to be deformed. **e** A second stratovolcano is formed. The previously formed volcano matures and deformation continues. Because the upper part of the domes will also cool and crystallize faster, the lower part will become buoyant and form cupolas, which are not shown here. Surface tumescence over the rising dome is also not shown. **f** Erosion occurs, leaving a small remnant of country rock and exposing the domes that intrude the batholith. The resulting outcrop pattern resembles the geologic outcrops within the southern part of the region of Santa Rosa de Yecora, Sonora (Fig. 5)

Batholith-Volcano Coupling via Dome-in-Dome Structures

Figure 7b is an expanded view of the section around the first stratovolcano to erupt (see Fig. 6d). It is shown as it appears in Fig. 6e. The hypabyssal dome is idealized. Actual domes would usually have a cupola or cupolas because, as in the batholith itself, the exterior of the dome would cool against country rock and batholith at lower temperatures. Consequently, the interior of the dome would become buoyant with respect to the exterior with formation of cupolas. This phe-

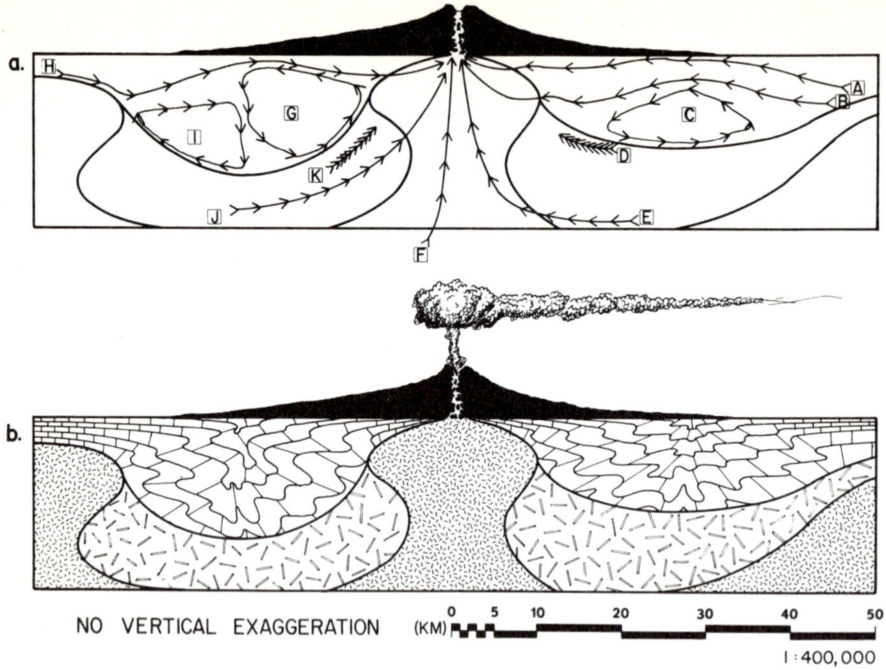

Fig. 7. a Illustration of an extensive hydrothermal system set up by the dome-in-dome structure. *Arrow* show the direction of fluid flow with distance between arrows representing the distance traveled by a fluid packet during 2×10^4 yrs. transit time (total transit time is 1.8×10^5 yrs.). Permeabilities and transit times were derived from Norton (1979, 1982). The permeability of the dome beneath the stratovolcano was assumed to be very high, after Norton, because of extensive fracturing (10 millidarcys, 1 md = 10^{-11} cm^2). The permeability decreases to 2 md in the central part of the dome, then increases to 4 md in less crystallized parts at depth. The sunken batholith was given a permeability of 0.5 md at its contact with the country rock and 2 md at its contact with the diorite below. The permeability of the country rock is high, 4 md, due to deformation with fracturing. The large hydrothermal system with long transport distances is the result of high geothermal gradients and the long lifetime of the stratovolcano (up to 2.5 m. y.). The stratovolcano is an orifice to the surface that acts as a hydrothermal "pump". "Blind" domes, those that do not breach the surface, have hydrothermal systems that are subordinate to the master stratovolcano. Each stratovolcano vents enormous quantities of water and acid gas to the surface. The water transports metals to the hypabyssal roots of the volcano where they are distributed and precipitated in the manifold forms of occurrence observed in porphyry ore deposits. (*Letters A* to *K* designate different fluid pathways during the 1.8×10^5 yrs. transit time). **b** Region of large stratovolcano in Fig. 6e (cupolas and tumescence not modeled)

nomenon can be observed in the mapped area of Fig. 5 where the hydrothermally altered hypabyssal intrusions occur as clusters of small area. These clusters may have fed eruptive centers of a large stratovolcano or clusters of small cones. I intend to numerically model this phenomenon in a future paper.

Norton has modeled fluid and heat transport phenomena around cooling magmas with specific reference to the porphyry copper environment (Norton and Knapp 1977, Norton 1978, 1979, 1982). His work has been based on a model in which the pluton is a high-level diapir intruding country rock with a moderately high geothermal gradient (30 °C km^{-1}). My model differs in that the dome intrudes batholith and country rock that are under a more severe temperature regimen ($\sim 90\,^{\circ}\mathrm{C\ km^{-1}}$, see also Larsen 1945). The dome is also larger, although cupolas, not modeled, would probably be the size of his small plutons. As a result of their small size and less severe temperature gradient, the plutons modeled by Norton cool rapidly, and "by 2×10^5 years elapsed time the thermal anomaly consists only of a slight (3 km) upward perturbation in the 200 °C isotherm, which is centered on the pluton-granite contact, and a broad (3 km) upward perturbation in the 100 °C isotherm" (Norton 1982, p. 65).

Stratovolcanos may have a long, eruptive history. For example, the eruptive history of the silicic eruptive center constituting the San Francisco Peaks stratovolcano in northern Arizona and its satellitic cones began at 2.7 Ma and lasted to 0.2 Ma, indicating a lifetime of 2.5 m. y. (Damon et al. 1974). Page and McDougall (1972) have demonstrated that pluton emplacement and subsequent Au – Cu mineralization in the New Guinea Highlands occurred over a time interval of from 1 to 2 m. y. The model presented here is consistent with such long duration of an eruptive volcano-hydrothermal system. Figure 7a, which is based on parameters from Norton's papers, shows that given the dome-in-dome structure and consequently more severe thermal regime, hydrothermal transport will continue much longer than predicted for high-level diapiric cupolas, and will result in long-distance fluid transport. Neighboring "blind" domes that do not have eruptive centers above them will contribute to the hydrothermal transport system which will be governed by eruptive centers that vent to atmospheric pressure. Fluids at depth will be under a high lithostatic-hydrostatic head, and so the eruptive centers will act as fluid pumps or "artesian vents".

The model in Fig. 7a assumes, after Norton's analysis, high permeability within the hypabyssal roots of the stratovolcano, a somewhat higher permeability than assumed by Norton for the country rock resulting from deformation and consequent fracturing, and increased permeability within crystallizing parts of the batholith and dome. Figure 7a shows the transport after 1.8×10^5 yrs. By that time, fluid packets derived from neighboring domes will have reached the central vent system. Interstitial fluids from a large part of the batholith will also have arrived at the vent system. As pointed out by Norton, convection systems will have been set up in the surrounding country rock. The convection cells will probably be guided by fissuring at the axes of large-scale folds producing higher permeabilities.

Consequence for Ore Deposits of Batholith-Volcano Coupling via Dome-in-Dome Structures

Emmons (1940) emphasized the importance of batholiths in ore formation. Most of his experience was within the Canadian Shield, where batholiths of vast extent dominate the landscape and the imagination. He was less familiar with Neogene volcanism and, as a consequence, volcanism does not appear as an important controlling factor in his models. On the other hand, Sillitoe (1973), with much experience within the Andes, where Neogene stratovolcanos dominate the landscape, produced a brilliant model of the tops and bottoms of porphyry copper deposits in which stratovolcanos played a dominant role. His model used Andean volcanos and porphyry copper deposits to illustrate cross-sections of strato-volcanos and their hypabyssal roots at different levels of dissection. However, Sillitoe's porphyry copper bottoms disappear into unspecified depths. The dome-in-dome structures, as previously pointed out, provide effective batholith-volcano coupling. Thus, the dome-in-dome structures serve to link Emmon's batholiths with Sillitoe's stratovolcanos.

The stratovolcanic "pump" is capable of dominating a very large hydro-thermal system, including interstitial batholith fluids as well as large volumes of water from a more direct meteoric source. Table 4 presents the estimates of Zies (1938) for the amounts of acid gas and water emitted by fumaroles in the Valley of Ten Thousand Smokes, on the flanks of the Katmai andesitic cone, during the year 1919. These amounts may be compared with Norton's (1979) estimates of the fluid mass flux through the central parts of a porphyry copper deposit. He estimates that about 1000 kg cm^{-2} of fluid mass would pass through the central parts of a porphyry copper deposit in 2×10^5 yrs. Thus, a porphyry of 3×10^{10} cm^2 area (3 km^2) would have had 3×10^{10} metric tons of fluid mass pass through it. This amount is equivalent to the flux of water to the surface in 63 yrs. at the 1919 rate of the Valley of the Ten Thousand Smokes (Table 4). When one considers the long duration of the hydrothermal activity of stratovolcanos, their dominant role in fluid transport becomes clearly evident.

Table 4. Amount of water and acid gases emitted by fumaroles in the Valley of Ten Thousand Smokes in 1919

Gas	Metric tons per year
H_2O	480.0×10^6
HCl	1.3×10^6
HF	0.2×10^6
H_2S	0.3×10^6

Source: Zies 1938.

Summarizing, dome-in-dome structures provide effective batholith-strato-volcano coupling. The long lives of stratovolcanos (as much as 2.5 Ma) allow for long-distance transport of fluids containing ore components derived from a large volume of batholith and country rock. Each volcano forms a surface orifice at which there is pressure release and export of enormous quantities of fluid to the surface. Thus, hydrothermal solutions transport ore components from the batholith as well as from the country rock to the hypabyssal roots of the volcano where the metals are distributed and precipitated in the manifold forms of occur-rence observed within the porphyry systems.

Summary and Conclusions

As a result of our study of the Sonoran-Sinaloan batholiths and their hypabyssal plutonic intrusions, I have arrived at the following conclusions:

1. The vast batholiths of Sonora and Sinaloa rose close enough to the Earth's surface for a rapid cooling of their upper surfaces to have taken place. This is shown by 67% of hornblende-biotite K – Ar dates being concordant. The remaining 33% are discordant, having been derived from deeper reaches of the batholiths which cooled at a measurably slower rate. The distribution of concordant and discordant K – Ar dates suggests that erosion occurred on the western flanks of the Sierra Madre Occidental at the modest rate of from 5 to 20 cm per 1000 yrs.
2. Several million years after emplacement, the early-stage batholiths were intruded by hypabyssal domes and porphyritic cupolas, some of which vented to the surface as stratovolcanos of varying size and complexity.
3. The lower part of a rising and cooling batholith is inherently, gravitationally unstable with respect to the upper part because eventually its density will be less than the cooled upper reaches of the crystallizing batholith. As a consequence, the lower part will become buoyant and rise, forming dome-in-dome structures. Similarly, the interior of a rising dome will become buoyant and rise, forming porphyritic cupolas, some of which will be at the roots of stratovolcano vents. This explains the field and age relations between early-stage batholiths and late-stage porphyries that we have observed in Mexico.
4. Modeling, using Ramberg's equations (1981) and assuming reasonable viscosity ratios, predicts spacings of the rising domes that are in accord with mapped spacings between intrusive complexes within batholiths.
5. Dome-in-dome structures from large-scale hydrothermal systems that last for periods of time consistent with lifetimes of stratovolcanos, as much as 2.5 m. y. As a consequence, there will be long-distance transport of fluids, guided by eruptive centers, and intimate hydrothermal coupling, via the dome-in-dome structures, between the batholithic system and the stratovolcano with its hypabyssal porphyritic plutonic roots.
6. Each stratovolcano forms a surface orifice at which there is pressure release and export of enormous quantities of fluid to the surface. Thus, hydrothermal solutions transport ore components from the batholith as well as from the country rock to the hyabyssal roots of the volcano where the metals are distributed and precipitated in the manifold forms of occurrence observed within the porphyry systems.

Lastly, it should be possible to evaluate the model in Figs. 6 and 7 through geophysical and field geologic studies. The model predicts extension and consequent thinning of the cover above the domes with transport to the margins where deformed cover would fill in the space vacated by the rising domes. This magmatotectonic deformation would be superimposed on the back-arc deformation that precedes the advancing volcanic arc. Under favorable conditions in areas where the dome has intruded a relatively thick and layered sedimentary sequence, geologic, gravity, and seismic studies could be used to define the batholith-cover interface and structural analysis could ascertain whether the deformational characteristics are appropriate. The geophysical studies could also be used to detect blind domes that had not breached the surface to form stratovolcanos. Also, it would be interesting to know what, if any, ore deposits are related to the blind domes whose hydrothermal convection cells would be subordinate to the domes coupled to stratovolcanos.

Acknowledgments. This work would not have been possible without the consistent support and encouragement of Ing. Guillermo P. Salas, former Director of the Consejo de Recursos Minerales (C.R.M.). The author is particularly grateful to Prof. Kenneth F. Clark of the Department of Geological Sciences, University of Texas at El Paso, for many stimulating conversations and for his companionship in the field aspects of this work; to Dr. Muhammad Shafiqullah for his constant and able help both in the field and in the laboratory; and to Prof. John M. Guilbert for magnanimously encouraging a geochronologist to enter into the "sacred sanctuary" of economic geology and for giving much helpful advice. The author also benefited from stimulating conversations with his colleagues: Profs. Christopher Eastoe, Jibamitra Ganguly, Denis L. Norton, Spencer R. Titley, and Joaquin Ruiz.

Many Maxican geologists have contributed logistical support and encouraged our effort. In particular, I would like to thank Ings. Jaime Islas-López, Fernando de la Fuente-Lavalle, Antonio Bustamante-Yánez, and Miguel Carrasco Centeno of the C.R.M.; Drs. Ariel Echavarri-P., Director of the Direccion de Minería, Geología y Energeticos of the State of Sonora; Dr. Guillermo A. Salas, formerly head of the Department of Geology at the University of Sonora; and Ing. E. Nestor Silva-Mejia of the Comisión Federal de Electricidad.

I am grateful to my collaborators in the field aspects of this work: Ing. Jaime Roldán-Quintana and Dr. Jean Jaques Cochemé, for stimulating conversation leading to further insights and for many pleasant days in the field together. Dr. Cochemé's map of the Santa Rosa de Yecora area was particularly helpful in my understanding of the relationship between the hypabyssal cupolas and the batholith. Ings. Islas, Roldán, and Silva accompanied the author on several field trips. Ings. Juan Carlos Garcia and Claudio Bartolini-Navarro assisted the author during field work in the Sierra Madre Occidental.

Thomas C. Kirschner, Robert Butcher, and Claudio Bartolini drafted the final figures. The manuscript was reviewed and edited by Austin Long, Joaquin Ruiz, Daniel Lynch, and Yemane Asmerom. Alice Jeanne Woodward typed the manuscript and contributed many editorial comments and corrections.

References

Anderson TH, Roldán JQ (eds) Field trip log. Geology of northern Sonora, guidebook, field trip #27. Annu Meet Geol Soc Am, pp 74–91

Anderson TH, Silver LT (1977) U–Pb isotopic ages of granitic plutons near Cananea, Sonora, Mexico. Econ Geol 72:827–836

Anderson TH, Silver LT (1979) The role of the Mojave-Sonora megashear in the tectonic evolution of northern Sonora. In: Anderson TH, Roldán JQ (eds) Geology of northern Sonora, Guidebook, field trip #27. Annu Meet Geol Soc Am, pp 59–68

Anderson TH, Silver LT, Salas GA (1980) Distribution and U–Pb isotopic ages of some lineated plutons, northwestern Mexico. In: Crittenden MD Jr, Coney PJ, Davis GH (eds) Cordilleran metamorphic core complexes. Geol Soc Am Mem 153:269–283

Branch CU (1976) Development of porphyry copper and stratiform volcanogenic ore bodies during the life cycle of andesitic stratovolcanos. In: Johnson RW (ed) Volcanism in Australia. Elsevier, New York, pp 337–342

Chen JH, Moore JG (1982) Uranium-lead isotopic ages from the Sierra Nevada batholith, California. J Geophys Res 87:4761–4784

Clark KF, Damon PE, Schutter SR, Shafiqullah M (1979) Magmatismo en el Norte de Mexico en relacion a los yacimientos metaliferos, Associacion de Ingenieros de Minas, Metalurgista y Geologos de Mexico. Mem Tec XIII:57

Clark KF, Foster CT, Damon PE (1982) Cenozoic mineral deposits and subduction-related magmatic arcs in Mexico. Geol Soc Am Bull 93:533 – 544

Coney PJ, Reynolds SJ (1977) Cordilleran Benioff Zones. Nature (London) 270:403 – 406

Damon PE (1979) Continental uplift at convergent boundaries. Tectonophysics 61:307 – 319

Damon PE (1983) Continental uplift, compensation and shunting during trench-spreading center collision. Tectonophysics 99:T1 – T8

Damon PE, Mauger RL (1966) Epeirogeny-orogeny viewed from the Basin and Range Province. Trans Soc Min Eng 235:99 – 112

Damon PE, Shafiqullah M, Leventhal JS (1974) K – Ar chronology for the San Francisco volcanic field and the rate of erosion of the Little Colorado River. In: Karlstrom TNV, Swann GA, Eastwood RL (eds) Geology of northern Arizona, pt 1. Regional studies guidebook. Geol Soc Am Rocky Mountain Sect Meet, pp 221 – 235

Damon PE, Clark KF, Shafiqullah M, Roldán JQ, Islas LJ (1981a) Geology and mineral deposits of southern Sonora and the Sonoran Sierra Madre Occidental (excursion 11). In: Ortlieb L, Roldán JQ (eds) Geology of northwestern Mexico and southern Arizona, field guides and papers. Guidebook, 1981. Cordilleran Sect Geol Soc Am, pp 369 – 428

Damon PE, Shafiqullah M, Clark KF (1981b) Age trends of igneous activity in relation to metallogenesis in the southern Cordillera. In: Dickinson WR, Payne WD (eds) Relations of tectonics to ore deposits in the Southern Cordillera. Tucson, Ariz, Geol Soc Digest 4:137 – 154

Damon PE, Shafiqullah M, Clark KF (1983a) Geochronology of the porphyry copper deposits and related mineralization of Mexico. Can J Earth Sci 20:1052 – 1071

Damon PE, Shafiqullah M, Roldán JQ, Cochemé JJ (1983b) El batolito Laramide (90 – 40 Ma) de Sonora. Pap 15th Nat Convent Asoc Ing Min Metal Geol Mex, Guadalajara, Jalisco, October 1983

Dickinson WR (1975) Potash-depth (K – h) in continental margin and intra-oceanic magmatic acrs. Geology 3:43 – 56

Dickinson WR, Hatherton T (1967) Andesitic volcanism and seismicity around the Pacific. Science 157:801 – 803

Emmons WH (1940) Principles of economic geology. McGraw-Hill, New York

Gastil RG, Krummenacher D (1977) Reconnaissance geology of coastal Sonora between Puerto Lobos and Bahia Kino. Geol Soc Am Bull 88:189 – 198

Gastil RG, Krummenacher D, Bushee J, Barthelmy D (1976) La zona batolitica del sur de California y el occidente de Mexico. Bol Soc Geol Mex 37:84 – 90

Gastil RG, Morgan JG, Krummenacher D (1977) Mesozoic history of peninsular California and related areas east of the Gulf of California. In: Mesozoic paleogeography of western United States. Pacific Coast Paleogr Symp

Gastil RG, Krummenacher D, Minch J (1978) The record of Cenozoic volcanism around the Gulf of California. Geol Soc Am Bull 90:839 – 857

Gregory KJ, Walling DE (1973) Drainage basing form and process: A geomorphological approach. Wiley & Sons, New York

Henry CD (1975) Geology and geochronology of the granitic batholithic complex, Sinaloa, Mexico. Ph D Diss, Univ Texas, Austin

Keith SB (1978) Paleo-subduction geometrics inferred from Cretaceous and Tertiary magmatic patterns in southwestern North America. Geology 6:516 – 521

King RE (1939) Geological reconnaissance in northern Sierra Madre Occidental of Mexico. Geol Soc Am Bull 50:1625 – 1722

Larsen ES (1945) Time required for the crystallization of the great batholith of southern and lower California. Am J Sci 243A:399 – 416

Lindgren W (1915) The igneous geology of the Cordilleras and its problems. In: Problems of American geology. Yale Univ Press, New Haven, Conn, pp 234 – 286

Livingston DE, Damon PE (1968) The ages of stratified Precambrian rock sequences in central Arizona and northern Sonora. Can J Earth Sci 5:763 – 772

Livingston DE, Mauger RL, Damon PE (1968) Geochronology of the emplacement, enrichment and preservation of Arizona porphyry copper deposits. Econ Geol 63:30 – 36

López Ramos E (1976) Carta geologica de la Republica Mexicana. Com Carta Geol Mex, 4th edn, scale 1 : 2,000,000

Norton DL (1978) Sourcelines, source regions, and pathlines for fluids in hydrothermal systems related to cooling plutons. Econ Geol 73:21 – 28

Norton DL (1979) Transport phenomena in hydrothermal systems: The redistribution of chemical components around cooling magmas. Bull Miner 102:471 – 486

Norton DL (1982) Fluid and heat transport phenomena typical of copper-bearing pluton environments. In: Titley SR (ed) Advances in geology of the porphyry copper deposits: Southwestern North America. Univ Ariz Press, Tuscon, Ariz, USA, pp 59 – 72

Norton DL, Knapp RB (1977) Transport phenomena in hydrothermal systems: Cooling plutons. Am J Sci 277:937 – 981

Page RW, McDougall I (1972) Ages of mineralization of gold and porphyry copper deposits in the New Guinea Highlands. Econ Geol 67:1034 – 1048

Pitcher WS (1972) The coastal batholith of Peru: Some structural aspects. 24th Int Geol Congr Sect 2, pp 156 – 163

Ramberg H (1981) Gravity, deformation and the Earth's crust, 2nd edn. Academic Press, London New York

Salas GP (1975) Metallogenetic chart of Mexico. Geol Soc Am Map Chart Ser, Map MC13, scale 1 : 2,000,000

Shafiqullah M, Damon PE, Lynch DJ, Reynolds SJ, Rehrig WA, Raymond RH (1980) K – Ar chronology and geologic history of southwestern Arizona and adjacent areas. Ariz Geol Soc Digest 12:201 – 260

Sillitoe RH (1973) The tops and bottoms of porphyry copper deposits. Econ Geol 68:799 – 815

Titley SR (1982) The style and progress of mineralization and alteration in porphyry copper systems. In: Titley SR (ed) Advances in geology of the porphyry copper deposits. Southwestern North America. Univ Ariz Press, Tucson, Ariz, USA, pp 93 – 116

Zies EG (1938) The concentration of the less familiar elements through igneous and related activity. Am J Sci 35A:385 – 404

Space-Time Distribution, Crustal Setting and Cu/Mo Ratios of Central Andean Porphyry Copper Deposits: Metallogenic Implications

R. H. SILLITOE[1]

Abstract

At least 80 porphyry copper centres, 18 of them major deposits, are known from the central Andes of Chile, Argentina and southern Peru. They are assignable to four longitudinal sub-belts of Meso-Cenozoic age, which become progressively younger from Late Cretaceous in the west, through Paleocene and Late Eocene-Early Oligocene, to Middle-Late Miocene in the east. A fifth, Late Carboniferous-Early Permian sub-belt is overprinted by the Middle-Late Miocene sub-belt.

The porphyry copper sub-belts are underlain by sialic crust ranging from about 35 to 60 km in present thickness. Outcropping pre-Mesozoic basement rocks range from 2,000-m.y. granulites to Paleozoic magmatic and sedimentary units.

Hypogene Cu/Mo ratios of major deposits range from 22 to 88. No systematic variation in Cu/Mo ratios is discernable within or between the three Cenozoic sub-belts. Neither are ratios dependent on the lithology or age of known pre-Mesozoic basement rocks, nor on crustal thickness.

The lack of correlation between Cu/Mo ratios and spatial or upper-crustal parameters is consistent with derivation of the principal metals in the porphyry copper deposits from subcrustal levels. The fact that the regionally extensive series of temporally and spatially discrete sub-belts parallels the axis of the Peru-Chile trench, and migrated systematically away from it, strongly supports involvement of fundamental subduction-related processes in the liberation of metals incorporated into central Andean porphyry copper deposits.

Introduction

During the last decade or so, the source of metals (Cu, Mo, Au, Ag) and other components (S, K) in porphyry copper deposits has remained a subject of debate. Opinions have been polarized between subcrustal and crustal sources.

Burnham (1981), Oyarzún and Frutos (1980), Sillitoe (1972, 1981) and Westra and Keith (1981) support a subcrustal source, with subducted oceanic crust, both

[1] 8 West Hill Park, Highgate Village, London N6 6ND, England

Geology and Metallogeny of Copper Deposits
(ed. by G. H. Friedrich et al.)

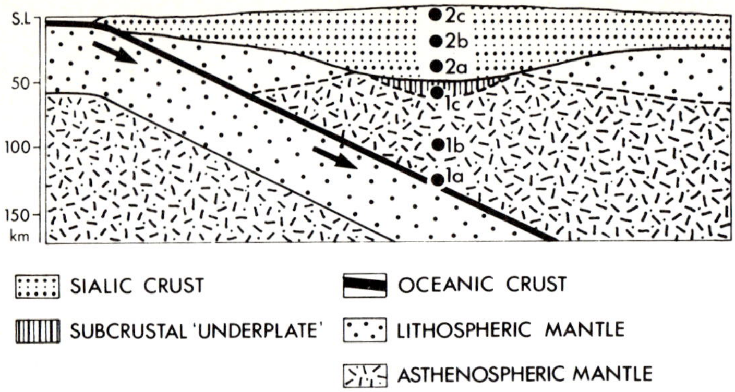

Fig. 1. Possible metal sources for porphyry copper deposits. Subcrustal sources include: subducted oceanic crust (*1a*), the overlying mantle wedge (*1b*), and an "underplate" of magmatic material (derived from sources *1a* and/or *1b*) at the base of the sialic crust (*1c*). Crustal sources include: lower-crustal sites of partial melting (*2a*), sites of selective contamination (*2b*) of subcrustal or lower-crustal magma, and upper-crustal sites of hydrothermal scavenging (*2c*)

layer-2 basalts and layer-1 pelagic sediments, considered to play a preeminent role (Fig. 1). In this model, the continental crust does not contribute a significant proportion of porphyry copper components, although it does provide a site for fractional crystallization of magmas to take place.

In contrast, Griffiths and Godwin (1983), Lowell (1980) and Titley and Beane (1981) favor extraction of metals and other elements from the continental crust, either at the site of partial melting, or by magma contamination resulting from assimilation, or as a consequence of scavenging by circulating hydrothermal fluids (Fig. 1). A number of materials, including shales (Jensen 1971), volcanic rocks (Laznicka 1976) and preexisting mineralized bodies (Lowell 1974), have all been proposed as specific crustal sources of metals in porphyry copper deposits. Moreover, molybdenum, in particular, has been cited as a metal derived as a contaminant from thick sialic crust (e.g., Hollister 1975).

If metals are extracted from sialic crust, then it seems to be a reasonable premise that Cu/Mo ratios of porphyry copper deposits should vary with changes in crustal constitution or thickness. This supposition is inherent in Hollister's (1975) explanation for high Cu/Mo ratios (deficiency of molybdenum) in porphyry copper deposits emplaced into the thinner, more mafic crust of island arcs. Conversely, if Cu/Mo ratios fail to correlate with crustal parameters, then a subcrustal metal source appears the more likely alternative.

The central Andes, comprising northern and central Chile, southern Peru and northwestern Argentina, provide an ideal location to test the validity of the two alternative metal sources. At least 80 porphyry copper deposits and prospects (including groups of tourmaline breccia pipes) in the central Andes (Fig. 2; Table 1) possess an orderly distribution in both space and time, as well as being emplaced into continental crust of different types and thicknesses. In this report, the spatial and temporal distribution of central Andean porphyry copper mineralization and the characteristics of the region's crust are reviewed, preparatory to a consideration of Cu/Mo ratios and their implications.

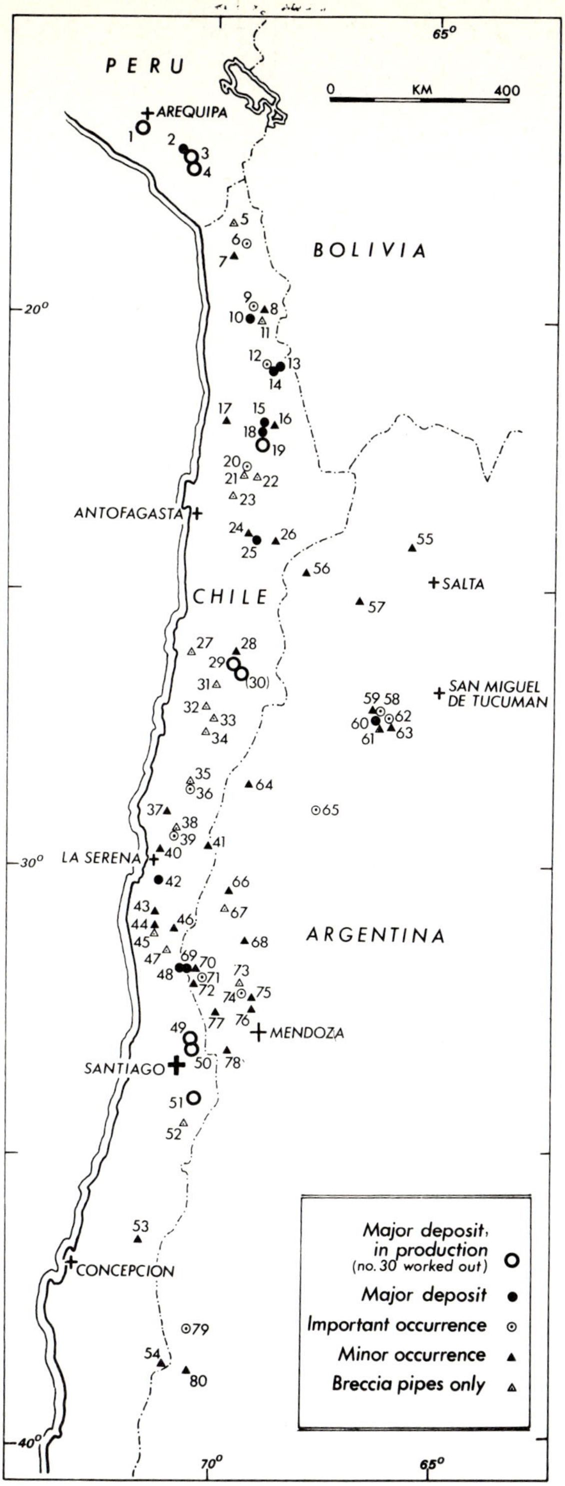

Fig. 2. Distribution of porphyry copper-type mineralization in the central Andes. *Numbers* keyed to names in Table 1

Table 1. Porphyry copper-type mineralization in Chile, Argentina and southernmost Peru (keyed to Fig. 1)

Peru
1. Cerro Verde-Santa Rosa
2. Quellaveco
3. Cuajone
4. Toquepala

Chile
5. Lluta (Campanani)
6. Ticnamar
7. Chilpe (Q. Camarones)
8. Queen Elizabeth
9. Mocha
10. Cerro Colorado
11. Yabricoya (La Planada)
12. Copaquire
13. Collahuasi
14. Quebrada Blanca
15. El Abra
16. Conchi
17. Puntillas
18. Pampa Norte
19. Chuquicamata
20. Sierra Gorda
21. Bella Esperanza
22. Centinela
23. Lomas Bayas
24. Chimberazo
25. La Escondida
26. Imilac
27. Carrizalillo de Las Bombas
28. El Jardín
29. El Salvador
30. Potrerillos
31. San Pedro de Cachiyuyo
32. Cachiyuyo de Llampos
33. Los Azules
34. Cabeza de Vaca-Zapallar
35. El Bolsico
36. Los Morteros (Porvenir)
37. Domeyko
38. Chacritas
39. Pablo y Pablo
40. Los Loros
41. El Indio
42. Andacollo
43. El Altar
44. El Sauce
45. La Lana
46. La Loica
47. Llamuco
48. Los Pelambres
49. Río Blanco
50. Disputada
51. El Teniente
52. Rosario de Rengo
53. Estero San José
54. Galletué

Argentina
55. Pancho Arias
56. Taca Taca
57. Inca Viejo
58. Bajo del Durazno
59. Bajo de Pampitas
60. Bajo de La Alumbrera
61. Bajo de San Lucas
62. Mi Vida (Cerro Rico)
63. Filo Colorado
64. Cordillera de La Brea
65. Los Bayos (Famatina)
66. Arroyo Chita
67. San Francisco de Los Andes
68. Alcaparrosa
69. El Pachón
70. Yunque
71. Cerro Mercedario
72. La Honda
73. La Colorada, Yalguaraz
74. San Jorge, Yalguaraz
75. Paramillos Norte
76. Paramillos Sur
77. Río de Las Vacas
78. Santa Clara
79. Campana Mahuida
80. La Voluntad

Space-Time Distribution

In a previous compilation (Sillitoe 1981), porphyry copper-type mineralization in Chile, Argentina and southernmost Peru was shown to be assignable to five sub-belts (Fig. 3). Each sub-belt was emplaced during a discrete time interval: Late Carboniferous-Early Permian (281 – 267 m.y.), Late Cretaceous (90 – 73 m.y.), Paleocene (67 – 54 m.y.), Late Eocene-Early Oligocene (41 – 29 m.y.) and

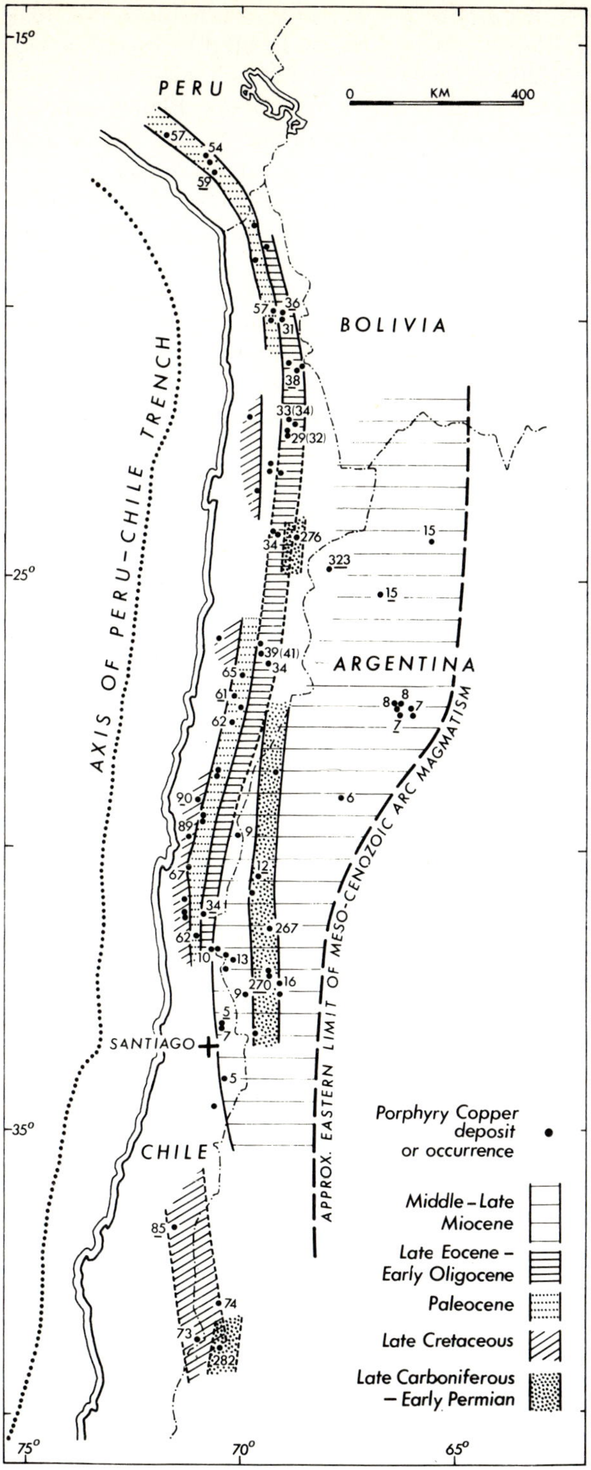

Fig. 3. Porphyry copper subbelts in the central Andes. All porphyry copper centres and breccia pipes from Fig. 2 are represented by *dots*. *Numbers* refer to K – Ar ages determined on hydrothermal or, where *underlined*, magmatic minerals. *Numbers in parenthesis* are alternative ages. K – Ar dates given to nearest m.y.; where more than one apparently reliable date is available, an average is given. Data sources are detailed in Sillitoe (1981), except for new determinations for Quebrada Blanca (Hunt et al. 1983), La Escondida (Minera Utah de Chile, Inc., unpublished data 1983), Domeyko (Shell Chile, unpublished data 1983), El Indio (Araneda 1982), Disputada (Warnaars 1983), and Estero San José and Galletué (Alfaro 1980)

Middle-Late Miocene (16–5 m.y.) (Fig. 3). Recently, seven more porphyry copper centres, at Quebrada Blanca, La Escondida, Domeyko, El Indio, Disputada, Estero San José and Galletué, have been dated radiometrically by the K–Ar method. (See Fig. 3 for data sources). Quebrada Blanca and La Escondida are major deposits and have been confirmed as part of the Late Eocene-Early Oligocene sub-belt. The major Disputada deposit, along with low-grade porphyry-type mineralization related genetically to the recently discovered El Indio Au–Ag–Cu vein deposits (Araneda 1982), lie in the Middle-Late Miocene sub-belt. Domeyko, Estero San José and Galletué are minor occurrences and make up part of the Late Cretaceous sub-belt. Addition of these new data to the previous compilation requires no significant modification in the positions of sub-belts (Fig. 3), although the two separate portions of the Late Eocene-Early Oligocene sub-belt may be united now with reasonable confidence on the basis of the La Escondida determination, and the southern portion of the Late Cretaceous sub-belt may be extended considerably. The Paleocene K–Ar ages of porphyry copper deposits in southern Peru (Sillitoe 1981, Fig. 3) have recently been confirmed by a Rb–Sr whole-rock isochron age (Beckinsale et al. 1985) and by additional K–Ar dating (Zimmermann and Kihien Collado 1983).

Figure 3 documents Paleozoic porphyry copper emplacement, mainly in Argentina, during the Late Carboniferous-Early Permian and, on the basis of a single dated occurrence (Taca Taca, 323 m.y.), the Early Carboniferous. A well-defined Late Carboniferous-Early Permian sub-belt is depicted on the basis of two dated occurrences and four additional occurrences (Cordillera de La Brea, San Francisco de Los Andes, La Colorada at Yalguaraz, Santa Clara; Figs. 2 and 3) inferred to be of similar age. The two other radiometrically dated occurrences of Late Carboniferous-Early Permian age, including Imilac which yielded two imprecise Rb–Sr ages (Halpern 1978) not included in earlier compilations, plot as isolated points north and south of the main sub-belt (Fig. 3).

At the beginning of the Mesozoic, the locus of magmatism stepped abruptly westwards to the vicinity of the present Pacific coastline. Meso-Cenozoic porphyry copper emplacement occurred in at least three narrow sub-belts, which migrated progressively eastwards with time, and appear not to have overlapped one another (Fig. 3). In the Early Miocene, however, the locus of magmatism expanded farther eastwards to form a wide band of dispersed volcanism and stock emplacement, to which the Middle-Late Miocene porphyry copper sub-belt is confined (Fig. 3). The Middle-Late Miocene sub-belt therefore overlaps much of the Late Carboniferous-Early Permian sub-belt.

The well-documented Meso-Cenozoic sub-belts contain all the major porphyry copper deposits in the central Andes. Preeminent is the Late Eocene-Early Oligocene sub-belt, which hosts Quebrada Blanca, Chuquicamata, La Escondida, El Salvador and Potrerillos. Major deposits farther north, at Cerro Verde (Santa Rosa), Quellaveco, Cuajone and Toquepala in Peru, and Cerro Colorado in Chile make up the Paleocene sub-belt, which southwards, from latitudes 26 to 32°S, contains several important groups of tourmaline breccia pipes (Figs. 2 and 3). In the southern part of the central Andes, all major deposits, at Los Pelambres, El Pachón, Río Blanco, Disputada and El Teniente, as well as the El Indio Au – Ag – Cu deposit, constitute the western edge of the Middle-Late Miocene sub-belt. The concentration of major deposits of younger age in the southern part of the central Andes is attributed in large measure to rapid unroofing by glacial erosion south of latitude 30°S (Sillitoe 1981).

The porphyry copper sub-belts, including the Late Carboniferous-Early Permian one, possess a marked parallelism with the axis of the Peru-Chile trench (Fig. 3), a relationship which strongly implies the involvement of subduction-related processes in porphyry copper genesis. It would appear, however, that the Late Cretaceous sub-belt and the present trench axis diverge appreciably in the southern part of the region under study.

Crustal Setting

The Meso-Cenozoic central Andes were formed on a continental, sialic basement, which is known to include both Paleozoic magmatic and sedimentary rocks generated at earlier subducting margins (e.g., Hervé et al. 1981), and at least two Precambrian complexes (Fig. 4). The Paleozoic basement includes the Early Carboniferous and Late Carboniferous-Early Permian porphyry copper mineralization described above.

Although Paleozoic rocks are widespread, especially in western Argentina, the Precambrian is relatively restricted in outcrop. The oldest known Precambrian complex, which includes abundant granulite-facies metamorphic rocks, has been dated by both Rb – Sr and U – Pb methods at nearly 2,000 m.y. (Cobbing et al. 1977, Dalmayrac et al. 1977, Shackleton et al. 1979). Its outcrop is restricted to the Arequipa massif of southwestern Peru, which may possibly extend southwards to include the Belén Schist Formation in northernmost Chile (Fig. 4), for which Pacci et al. (1980) obtained a 1,000-m.y. Rb – Sr isochron. The eastward and southward subsurface extensions of the Arequipa massif are poorly constrained, although it should be pointed out that a late Precambrian Rb – Sr age was obtained for granitic rock encountered during deep drilling at San Andrés in northwestern Bolivia (Fig. 4, Lehmann 1978). Although most investigators have concluded that the mid-Proterozoic rocks of the Arequipa massif are autochthonous and constitute the western leading edge of the Brazilian Shield (Cobbing et al. 1977, Shackleton et al. 1979), a slight possibility exists that they are allochthonous in origin and constitute an exotic suspect terrane in the sense of Coney et al. (1980). However, accretion could not have occurred any later than the Silurian since Devonian strata overlap the outcropping edge of the Arequipa massif.

Precambrian rocks are also known to underlie parts of the Sierras Pampeanas massif in western Argentina. Rocks range in metamorphic grade from greenschist to amphibolite facies, and normally yield late Precambrian radiometric ages, although remnant nuclei of granulite-facies rocks as old as 1,800 m.y. have been reported (Caminos et al. 1982).

Crustal thicknesses in the central Andes (Fig. 4) range from >70 km beneath the Altiplano of southern Peru and western Bolivia to 30 km along the Chilean

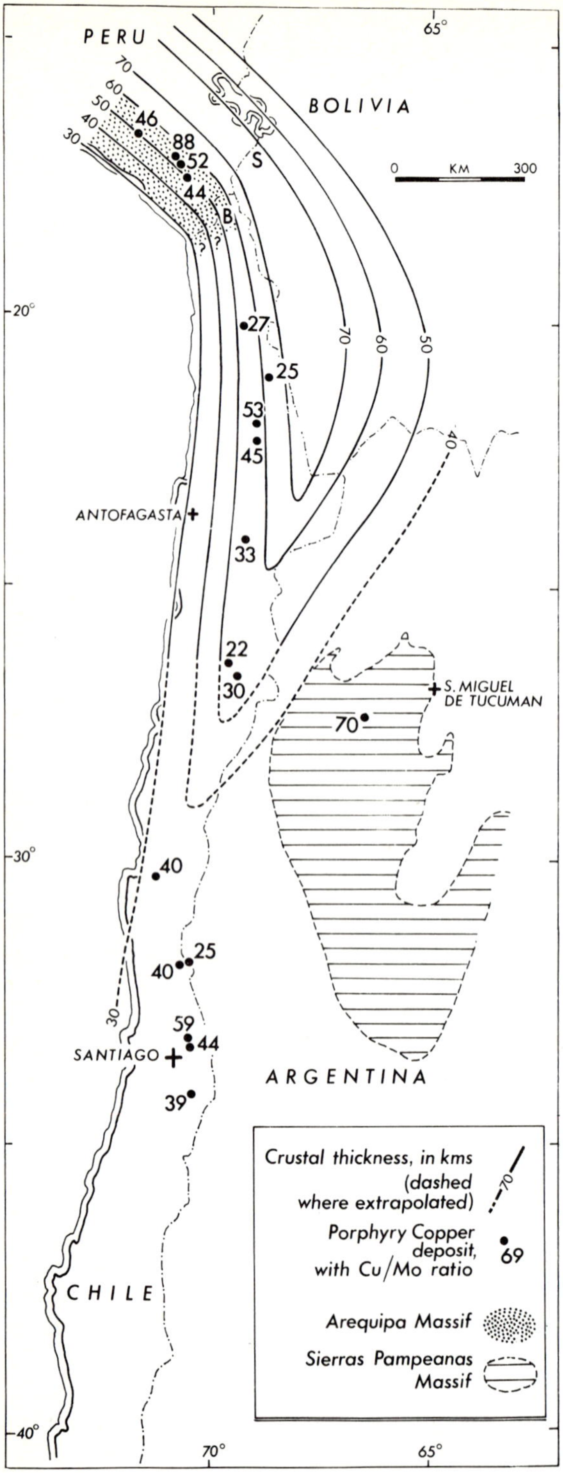

Fig. 4. Outcropping Precambrian massifs (from Shackleton et al. 1979, Ramos and Pesce 1982), crustal thicknesses (mainly from James 1971a but influenced south of lat. 27°S by Cummings and Shiller 1971), and Cu/Mo ratios of principal porphyry copper deposits (from Table 2) in the central Andes. *B* = Belén Schist Formation; *S* = San Andrés granite

littoral (James 1971a, Cummings and Shiller 1971). Much of the increase of crustal thickness to the extreme of 50 to 70 km in northern Chile-southern Peru-western Bolivia is believed to have been a consequence of voluminous Meso-Cenozoic magmatism (James 1971b).

In the central Andes of northern Chile-southern Peru-western Bolivia, Miocene to Recent volcanic rocks have been shown to possess intial $^{87}Sr/^{86}Sr$ ratios >0.705. These relatively high initial ratios, combined with other isotopic and trace-element criteria, are interpreted to signify that mantle-derived magmas underwent small degrees of crustal contamination (Thorpe et al. 1981, James 1982). In southern Peru, at least part of the contaminant has been shown to be 2000-m.y.-old granulites in the upper crust (Tilton and Barreiro 1980, James 1982). The significantly lower (<0.705) initial ratios for andesitic volcanic rocks from central and southern Chile, where the continental crust is <40 km thick, are taken to indicate less crustal involvement in magma genesis, and to imply that contamination farther north is due to passage of magmas through an unusually thick (>50 km) crust (Harmon et al. 1984).

Cu/Mo Ratios

It is difficult to obtain precise hypogene Cu/Mo ratios for much of the porphyry copper mineralization in the central Andes (Fig. 2) because drilling has been insufficient to obtain representative Cu and Mo grades. Satisfactory Cu/Mo ratios are available for only 18 major deposits, including all the operating mines (Table 2). The grade information is based on assay of 3-m lengths of split core obtained from extensive programmes (generally >30000 m) of diamond drilling conducted by the mining companies concerned to determine ore reserves.

Although the data in Table 2 are taken to be reasonably reliable, say within ±25% of the true Cu/Mo ratios, two main problems are inherent in the provision of average hypogene Cu and Mo values for entire porphyry copper deposits.

First, porphyry copper deposits are commonly characterized by an inhomogeneous, sometimes zoned, distribution of metals. This is particularly the case for molybdenum, which may occur as higher-grade concentrations: in a halo to a copper-enriched core, as at Bajo de La Alumbrera (Sillitoe 1979) and Potrerillos (H. Flores, personal communication, 1983); in hydrothermal breccia bodies, as at Quebrada Blanca (Hunt et al. 1983), El Pachón (A. N. Lencinas, personal communication, 1983) and Disputada (Warnaars 1983); in halos to breccias at El Teniente (A. Enrione, personal communication, 1983) and Los Pelambres (Atkinson et al. 1984); and associated with a phase of overprinted sericitic alteration adjoining the West Fissure at Chuquicamata (Ambrus 1978). However, although inhomogeneities in Mo distribution may be large, they are effectively averaged out when metal values become available for entire deposits. It may be seen from Table 2 that average Mo values range from 0.01 to 0.025 percent, except for Quellaveco and Bajo de La Alumbrera with averages of 0.008 and 0.007%, respectively.

Table 2. Hypogene copper and molybdenum contents of selected central Andean porphyry copper deposits[a]

Deposit	% Cu	% Mo	Cu/Mo
Santa Rosa	0.65	0.014	46
Quellaveco	0.7	0.008	88
Cuajone	0.7	0.0135	52
Toquepala	0.6	0.0135	44
Cerro Colorado	0.4	0.015	27
Quebrada Blanca	0.5	0.02	25
El Abra	0.8	0.015	53
Chuquicamata	0.9	0.02	45
La Escondida	0.5	0.015	33
El Salvador	0.55	0.025	22
Potrerillos	0.75	0.025	30
Andacollo	0.6	0.015	40
Los Pelambres	0.6	0.015	40
El Pachón	0.4	0.016	25
Río Blanco	1.0	0.017	59
Disputada	0.7	0.016	44
El Teniente	0.86	0.022	39
Bajo de La Alumbrera	0.49	0.007	70

[a] All data from unpublished company sources, except for those from Andacollo (Llaumett et al. 1975) and Quebrada Blanca (Hunt et al. 1983).

Second, and more problematic, is the effect of supergene copper enrichment, which modified most central Andean porphyry copper deposits during the Cenozoic. In deposits such as Cuajone, Toquepala, Cerro Colorado, Chuquicamata, La Escondida, Quebrada Blanca, El Salvador and Potrerillos, characterized by high-grade supergene enrichment, average hypogene grades, commonly less well constrained by drilling, were estimated for depths below the enrichment blankets. Where supergene enrichment is more localized, as at Los Pelambres, El Pachón, Disputada and El Teniente, an attempt was made to subtract the supergene contribution from the overall average copper grades. Copper values are therefore of variable accuracy, with deposits such as El Abra, Río Blanco and Bajo de La Alumbrera, which underwent essentially no supergene enrichment, providing the most reliable figures. Hypogene copper contents vary from a low of 0.4% to a high of 1.0% (Table 2).

From Table 2 and Fig. 4 it may be appreciated that all Cu/Mo ratios fall in the range from 22 to 88. Based on Table 2, the average Cu/Mo ratio for the 18 central Andean porphyry copper deposits is 43.4. Bajo de La Alumbrera, the deposit with the second highest Cu/Mo ratio (70), is the only central Andean porphyry copper deposit with a high gold content. Its average gold grade of 0.62 ppm (Sillitoe 1979) is more than 2.5 times that of Potrerillos and Andacollo, and at least 5 times that of all other deposits listed in Table 2.

Metallogenic Implications

Cu/Mo ratios are not available for any of the Paleozoic or Late Cretaceous porphyry-type occurrences. Nevertheless, it may be reasonably concluded from Figs. 3 and 4 that the Cu/Mo ratios of central Andean porphyry copper deposits are unrelated to the age of emplacement (Fig. 5), their position relative to the Pacific littoral (and the Peru-Chile trench), and their latitudinal position within individual sub-belts. Since systematic variations in crustal constitution are to be expected both along and across Andean-type margins, and during their evolution, Fig. 4 implies that the Cu/Mo ratios are probably independent of the central Andean continental crust. The possibility that Fig. 4 reflects a random spatial variation of one or more critical crustal parameters is reduced by differences in Cu/Mo ratios of as great as 44 between closely spaced deposits in southern Peru.

Furthermore, reference to Figs. 4 and 6 reveals that Cu/Mo ratios appear to bear no relation to crustal thickness or, as far as can be determined, to the distribution of outcropping Precambrian crust. Deposits in southern Peru and northern Chile with Cu/Mo ratios from 22 to 88 (average 42.2 for 11 deposits) are underlain by crust from 50 to 60 km thick, whereas those in central Chile and

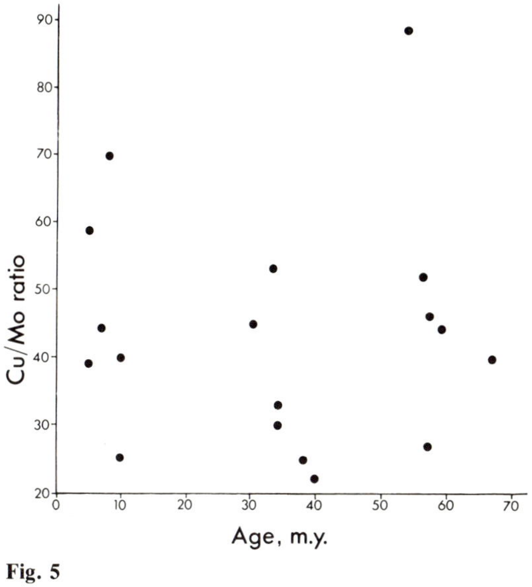

Fig. 5

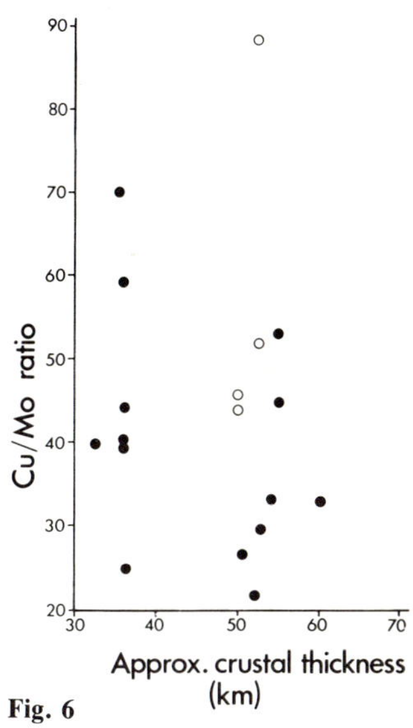

Fig. 6

Fig. 5. Plot of Cu/Mo ratio (from Table 2) vs K – Ar age (from Fig. 3) for selected porphyry copper deposits in the central Andes. Lack of any correlation is apparent. Age of Cerro Colorado assumed same as nearby Mocha deposit

Fig. 6. Plot of Cu/Mo ratio (from Table 2) vs approximate crustal thickness (estimated from Fig. 4) for selected porphyry copper deposits in the central Andes. No correlation is apparent. No Cu/Mo ratios available for deposits underlain by crust from 40 to 50 km thick. Deposits emplaced in Arequipa massif denoted by *open circles*

westermost Argentina with Cu/Mo ratios from 25 to 59 (avergae 41 for 6 deposits) are underlain by crust <40 km thick. It is particularly noteworthy that deposits confined to the 2000-m.y.-old Arequipa massif, where the crustal thickness is now (but perhaps not during Paleocene porphyry copper emplacement) some 50 km thick, are not significantly more molybdenum-rich than those generated farther south, where known 2000-m.y.-old rocks are lacking. Furthermore, Bajo de La Alumbrera, underlain by the Precambrian Sierras Pampeanas massif (Fig. 4), is one of the most molybdenum-poor deposits in the central Andes. The emplacement of porphyry copper deposits (with similar Cu/Mo ratios) into granulite-facies upper crust in southern Peru and into essentially unmetamorphosed volcanic and sedimentary sequences elsewhere in the central Andes fails to support the general applicability of Plant et al.'s (1983) hypothesis that porphyry copper generation is inhibited by granulite-facies (anhydrous) metamorphic host rocks.

It should also be stressed that Cu/Mo ratios for Cenozoic deposits emplaced in northern Chile and southern Peru through thickened crust, as denoted by Miocene to Recent volcanic rocks with initial $^{87}Sr/^{86}Sr > 0.705$, are not systematically higher than those for deposits in central Chile, where volcanic rocks possess initial Sr-isotopic ratios <0.705.

It is concluded that there is no discernable correlation between the Cu/Mo ratios of central Andean porphyry copper deposits and the thickness, age, lithology, or other observable or measurable characteristics of the continental crust into which they were emplaced. This conclusion militates against a crustal source for Cu and Mo, and possibly other elements, in central Andean porphyry copper deposits. The data are consistent with derivation of Cu and Mo from a subcrustal source region. However, since crustal parameters, apart from thickness, pertain only to upper crustal levels, the lower continental crust cannot be discounted as a metal source. Observed variations in Cu/Mo ratios (Table 2) are perhaps attributable to detailed differences in the conditions of magma generation and evolution, given broadly similar starting products.

Discussion

Karig and Kay (1981) recognized a correlation between $^{87}Sr/^{86}Sr$ and $^{206}Pb/^{204}Pb$ ratios for pelagic sediments and island-arc volcanic rocks. They concluded, contrary to many previous workers, that pelagic sediments overlying the oceanic plate are selectively subducted, and contribute to arc magmatism. Although the bulk of calc-alkaline magma is considered to be generated by partial melting of the mantle wedge above the subduction zone (Fig. 1), a substantial sediment flux is required to account for the contents of K, Rb, Sr, Pb and REE in arc magmas. Combined isotopic and trace-element data for arc magmas from many parts of the world also strongly support a significant contribution from the downgoing slab (e.g., Hawkesworth 1982).

If it is accepted that the metal contents of porphyry copper deposits in calc-alkaline magmatic arcs, such as the central Andes, are largely independent of crustal parameters, then an adequate subcrustal source could be provided by subducted pelagic sediments (cf., Sillitoe 1981). Representative layer-1 sediments on the Nazca plate, west of the central Andes, contain up to 1000 ppm Cu and 300 ppm Mo (Field et al. 1983). Although subducted oceanic basalts and the peridotitic mantle wedge, both characterized by much lower Cu and Mo contents than pelagic sediments, cannot be ruled out as sources for these two metals, they do not appear to supply much of the trace Pb found in porphyry copper deposits (Sillitoe and Hart 1984). The lead-isotopic signatures of porphyry copper mineralization from the northern Andes of Colombia (Sillitoe and Hart 1984) fall in broadly the same isotopic fields as those determined for Disputada (Tilton 1979), El Salvador and Toquepala (Tilton et al. 1981) in the central Andes, and are not incompatible with a pelagic-sediment source. However, since pelagic-sediment and continental-crustal leads are difficult to distinguish, Pb-isotopic ratios do not provide a discriminant between crustal and subcrustal Pb sources (cf., McNutt et al. 1979), except under special circumstances (Sillitoe and Hart 1984).

It was noted above that small amounts of ^{87}Sr, ^{207}Pb, ^{18}O and other isotopes may have been supplied as contaminants to subduction-related magmas that ascended through continental crust >50 km thick in northern Chile, southern Peru and adjacent areas (Thorpe et al. 1981, James 1982, Harmon et al. 1984). However, this analysis has revealed no compelling reason to suspect a significant contribution of Cu or Mo from this exceptional thickness of crustal rocks, although it should be pointed out that the continental crust of the Altiplano region may well have been substantially thinner during Paleocene porphyry copper emplacement in southern Peru. However, as emphasized by Cobbing and Pitcher (1983), crustal thickening was probably essentially complete prior to Eocene magmatism, so that the Late Eocene-Early Oligocene porphyry copper deposits in northern Chile were emplaced through the thickened crust.

Available ^{87}Sr/^{86}Sr initial ratios for most central Andean porphyry copper deposits (Table 3), including Toquepala, El Salvador and Disputada, all approximate 0.704, thereby confirming that Sr, although known to be a possible crustal contaminant, is entirely of subcrustal origin. It is therefore extremely unlikely that Cu and Mo in these deposits possess anything other than a subcrustal source.

Andesitic volcanic rocks cogenetic with porphyry copper emplacement at Bajo de La Alumbrera, northwestern Argentina (Figs. 2 and 3), have yielded

Table 3. Available ^{87}Sr/^{86}Sr initial ratios for central Andean porphyry copper deposits

Deposit	^{87}Sr/^{86}Sr (mean value)	Reference
Toquepala	0.7044	Beckinsale et al. (1985)
El Salvador	0.7041	Gustafson and Hunt (1975)
	0.7033	Halpern (1978)
Disputada	0.7041	Halpern (1979)
Bajo de La Alumbrera	0.7068[a]	McNutt et al. (1975)

[a] Determined on comagmatic volcanic rocks only.

$^{87}Sr/^{86}Sr$ initial ratios of 0.706 and 0.7075 (McNutt et al. 1975, Table 3), which may be taken (as they were by Tilton et al. 1981) to suggest an addition of radiogenic Sr from subjacent sialic crust, although this is <40 km thick (Fig. 4). Contrary to expectation, however, the Bajo de La Alumbrera deposit possesses the highest Au content of any porphyry deposit in the central Andes (see above), and the second highest Cu/Mo ratio in Table 2, a combination of features that renders it the least likely candidate to have gained a significant metal contribution from sialic crust. Bajo de La Alumbrera therefore demonstrates that a porphyry copper metal budget can lack the generally accepted signs (high Mo/low Au) of an upper crustal source even where sialic crust appears to have supplied an ^{87}Sr contaminant. A similar conclusion was reached by Damon et al. (1983) for porphyry Cu – Mo deposits throughout Mexico, where those emplaced through Precambrian sialic basement gave a mean $^{87}Sr/^{86}Sr$ initial ratio of 0.7075, significantly higher than a mean value of 0.7048 for those farther south in accreted terranes lacking sialic basement.

Acknowledgments. The assistance of Sergio Barassi, José Cabello, Mario Chabert, Héctor Flores, Andrés Lencinas, Joy Merz, Dorian Nicol and Enrique Tidy in the collection of Cu and Mo grades for central Andean porphyry copper deposits in gratefully acknowledged. Comments on the manuscript were provided by Nigel Grant and Richard Thorpe.

References

Alfaro HG (1980) Prospección de pórfidos cupríferos en la zona central-sur de Chile. In: Minería de cobres porfídicos, vol II. Anales del Congreso Cincuentenario. Ins Ing Min Chile, Santiago, pp 295 – 313

Ambrus J (1978) Chuquicamata deposit. In: Sutulov A (ed) International molybdenum encyclopaedia 1778 – 1978, vol I. Intermet, Santiago, pp 87 – 93

Araneda GR (1982) El Indio, yacimiento de oro, plata y cobre. Coquimbo, Chile. Minerales (Santiago) 37(160):5 – 13

Atkinson WW Jr, Souviron A, Vehrs TI (1984) Geological results of recent exploration of the porphyry copper deposit at Los Pelambres, Chile. Geol Soc Am (Abstracts with Programs) 16:434

Beckinsale RD, Sanchez-Fernandez AW, Brook M, Cobbing EJ, Taylor WP, Moore ND (1985) Rb – Sr whole-rock isochron and K – Ar age determinations for the Coastal Batholith of Peru. In: Pitcher WS, Atherton MP, Cobbing EJ, Beckinsale RD (eds) Magmatism at a plate edge: The Coastal Batholith of Peru. Blackie, London, pp 177 – 202

Burnham CW (1981) Convergence and mineralization – Is there a relation? Geol Soc Am Mem 154:761 – 768

Caminos R, Cingolani CA, Hervé F, Linares E (1982) Geochronology of the pre-Andean metamorphism and magmatism in the Andean Cordillera between latitudes 30° and 36°S. Earth Sci Rev 18:333 – 352

Cobbing EJ, Pitcher WS (1983) Andean plutonism in Peru and its relationship to volcanism and metallogenesis at a segmented plate edge. Geol Soc Am Mem 159:277 – 291

Cobbing EJ, Ozzard JM, Snelling NJ (1977) Reconnaissance geochronology of the crystalline basement rocks of the Coastal Cordillera of southern Peru. Geol Soc Am Bull 88:241 – 246

Coney PJ, Jones DL, Monger JWH (1980) Cordilleran suspect terranes. Nature (London) 288:329 – 333

Cummings D, Shiller GI (1971) Isopach map of the earth's crust. Earth Sci Rev 7:97 – 125

Dalmayrac D, Lancelot JR, Leyreloup A (1977) Two-billion-year granulites in the late Precambrian metamorphic basement along the southern Peruvian coast. Science 198:49 – 51

Damon PE, Shafiqullah M, Clark KF (1983) Geochronology of the porphyry copper deposits and related mineralization of Mexico. Can J Earth Sci 20:1052 – 1071

Field CW, Rye RO, Dymond JR, Whelan JF, Senechal G (1983) Metalliferous sediments of the East Pacific. In: Shanks WC III (ed) Cameron volume on unconventional mineral deposits. Am Inst Min Metall Petrol Eng, New York, pp 133 – 156

Griffiths JR, Godwin CI (1983) Metallogeny and tectonics of porphyry copper-molybdenum deposits in British Columbia. Can J Earth Sci 20:1000 – 1018

Gustafson LB, Hunt JP (1975) The porphyry copper deposit at El Salvador, Chile. Econ Geol 70:857 – 912

Halpern M (1978) Geological significance of Rb – Sr isotopic data of northern Chile crystalline rocks of the Andean orogen between latitudes 23° and 27°South. Geol Soc Am Bull 89:522 – 532

Halpern M (1979) Strontium isotope composition of rocks from the Disputada copper mine, Chile. Econ Geol 74:129 – 130

Harmon RS, Barreiro BA, Moorbath S, Hoefs J, Francis PW, Thorpe RS, Déruelle B, McHugh J, Viglino JA (1984) Regional O-, Sr-, and Pb-isotope relationships in late Cenozoic calc-alkaline lavas of the Andean Cordillera. J Geol Soc (London) 141:803 – 822

Hawkesworth CJ (1982) Isotope characteristics of magmas erupted along destructive plate margins. In: Thorpe RS (ed) Andesites: orogenic andesites and related rocks. John Wiley & Sons, New York, pp 549 – 571

Hervé F, Davidson J, Godoy E, Mpodozis C, Covaccevich V (1981) The late Paleozoic in Chile: Stratigraphy, structure and possible tectonic framework. An Acad Bras Ciênc 53:361 – 373

Hollister VF (1975) An appraisal of the nature and source of porphyry copper deposits. Min Sci Eng 7:225 – 233

Hunt JP, Bratt JA, Marquardt LJC (1983) Quebrada Blanca, Chile: An enriched porphyry copper deposit. Min Eng 35:636 – 644

James DE (1971a) Andean crustal and upper mantle structure. J Geophys Res 76:3246 – 3271

James DE (1971b) Plate tectonic model for the evolution of the central Andes. Geol Soc Am Bull 82:3325 – 3346

James DE (1982) A combined O, Sr, Nd and Pb isotopic and trace element study of crustal contamination in central Andean lavas, I. Local geochemical variations. Earth Planet Sci Lett 57:47 – 62

Jensen ML (1971) Provenance of Cordilleran intrusives and associated metals. Econ Geol 66:34 – 42

Karig DE, Kay RW (1981) Fate of sediments on the descending plate at convergent margins. Philos Trans R Soc London Ser A 301:233 – 251

Laznicka P (1976) Porphyry copper and molybdenum deposits of the USSR and their plate tectonic settings. Inst Min Metall Trans B 85:B14 – 32

Lehmann B (1978) A Precambrian core sample from the Altiplano/Bolivia. Geol Rundsch 67:270 – 278

Llaumett C et al. (1975) El yacimiento cobre porfídico "Andacollo", Provincia de Coquimbo, Chile. Rev Geol Chile 2:56 – 66

Lowell JD (1974) Regional characteristics of porphyry copper deposits of the Southwest. Econ Geol 69:601 – 617

Lowell JD (1980) Metallogenesis and porphyry deposits of North America and the Pacific region. In: Mineral resources. Bur Rech Geol Min Mem 106:40 – 47

McNutt RH, Caelles JC, Farrar E, Haynes SJ, Zentilli M (1975) Initial $^{87}Sr/^{86}Sr$ ratios of plutonic and volcanic rocks of the central Andes between latitudes 26° and 29°S. Earth Planet Sci Lett 27:305 – 313

McNutt RH, Clark AH, Zentilli M (1979) Lead isotopic compositions of Andean igneous rocks, latitudes 26° to 29°S: Petrologic and metallogenic implications. Econ Geol 74:827 – 837

Oyarzún MJ, Frutos J (1980) Metallogenesis and porphyry deposits of the Andes (southeastern Pacific region). In: Mineral resources. Bur Rech Geol Min Mem 106:50 – 62

Pacci D, Munizaga F, Hervé F, Kawashita K, Cordani UG (1980) Acerca de la edad Rb – Sr Precámbrica de rocas de la Formación Esquistos de Belén, Depto. de Parinacota, Chile. Rev Geol Chile 11:43 – 50

Plant JA, Simpson PR, Green PM, Watson JV, Fowler MB (1983) Metalliferous and mineralized Caledonian granites in relation to regional metamorphism and fracture systems in northern Scotland. Inst Min Metall Trans B 92:B33 – 42

Ramos A, Pesce AH, and collaborators (1982) Mapa geológico de la República Argentina, 1:2500000. Subsecretaría de Minería, Buenos Aires

Shackleton RM, Ries AC, Coward MP, Cobbold PR (1979) Structure, metamorphism and geochronology of the Arequipa massif of coastal Peru. J Geol Soc (London) 136:195–214

Sillitoe RH (1972) A plate tectonic model for the origin of porphyry copper deposits. Econ Geol 67:184–197

Sillitoe RH (1979) Some thoughts on gold-rich porphyry copper deposits. Miner Deposita 14:161–174

Sillitoe RH (1981) Regional aspects of the Andean porphyry copper belt in Chile and Argentina. Inst Min Metall Trans B 90:B15–36

Sillitoe RH, Hart SR (1984) Lead-isotopic signatures of porphyry copper mineralization in oceanic and continental settings, Colombian Andes. Geochim Cosmochim Acta 48:2135–2142

Thorpe RS, Francis PW, Harmon RS (1981) Andean andesites and crustal growth. Philos Trans R Soc London Ser A 301:305–320

Tilton GR (1979) Isotopic studies of Cenozoic Andean calc-alkaline rocks. Carnegie Inst Washington Yearb 78:298–304

Tilton GR, Barreiro BA (1980) Origin of lead in Andean calc-alkaline lavas, southern Peru. Science 210:1245–1247

Tilton GR, Pollak RJ, Clark AH, Robertson RCR (1981) Isotopic composition of Pb in central Andean ore deposits. Geol Soc Am Mem 154:791–816

Titley SR, Beane RE (1981) Porphyry copper deposits. Part I. Geologic settings, petrology, and tectogenesis. Econ Geol 75th Anniversary Vol, pp 214–235

Warnaars FW (1983) Copper tourmaline breccias at Los Bronces, Chile. Soc Min Eng AIME Trans 272:1902–1911

Westra G, Keith SB (1981) Classification and genesis of stockwork molybdenum deposits. Econ Geol 76:844–873

Zimmermann JL, Kihien Collado A (1983) Détermination par la méthode K/Ar de l'âge des intrusions et des minéralisations associées dans le porphyre cuprifère de Quellaveco (sud ouest du Pérou). Miner Deposita 18:207–213

Metallogenic Zoning of Volcano-Plutonic Belts and Porphyry-Copper Mineralization

V. A. EVSTRAKHIN, A. I. KRIVTSOV, and I. F. MIGACHYOV[1]

Abstract

Volcano-plutonic belts (VPB) that host porphyry-copper deposits are grouped into basaltic and andesitic types. Basaltic (eugeosynclinal) VPB give rise to island-arc barrier zones (fore-arc). Andesitic VPB originate in an orogenic-activation regime and, depending on the character of the basement, are subdivided into epicratonic, epimiogeosynclinal, and epieugeosynclinal types. Ore genesis products within the VPB are consistent with vertical and lateral-temporal metallogenic zoning (MZ). In basaltic VPB, copper and gold porphyry-copper deposits are associated with copper massive sulfide ores and are replaced in time by veined polymetallic, and in places, skarn mineralization. MZ of andesitic VPB is controlled by metallogeny of their basement combined with molybdenum-, copper-molybdenum porphyry, and molybdenum-copper deposits (depending on the substratum) as well as with later porphyry Sn, Sn – W, W – Mo, W, Au – Ag ores. The metallogenic style of the VPB is determined by the formation of porphyry deposits. The metallogenic zoning of the VPB is exemplified in some provinces of the Soviet Union.

The formational approach to a specific metallogeny that is much favored now implies that mineral deposits are natural components of certain geologic formations. Appraising and correlating the existing metallogenic concepts, Smirnov (1981) clearly gives preference to the formational approach, pointing out not only its prospects, but its appreciable practical significance as well. Recognition of ore-bearing geologic formations and their geotectonic and paleotectonic settings, serves as a basis for metallogenic models, that, in the long run, enable us to define the position of the mineralization in the spatial and temporal series of the products of the tectono-magmatic cycle and to outline boundaries of metallogenic areas.

Applied to porphyry-copper deposits, the approach was extensively employed by such authors as Vlasov (1978), Krivtsov (1977, 1983), Pavlova (1978), Popov (1977), and others. It was clearly demonstrated that porphyry-copper deposits are related to specific geostructures: volcano-plutonic belts known in geosynclinal folded systems of different ages (from Phanerozoic to Recent). It is obvious that regimes that are responsible for the origin of the volcano-plutonic

[1] Ministry of Geology, Moscow, USSR

Geology and Metallogeny of Copper Deposits
(ed. by G. H. Friedrich et al.)

belts do not belong exclusively to the Kimmerian or Alpine epochs, but manifest themselves in much earlier periods of the Earth's crust development.

At the same time, American and English authors, using the Laramide and Pacific provinces as examples, look upon the geotectonic position of porphyry-copper deposits from the plate-tectonic viewpoint. Spatial and temporal siting of such deposits are analyzed, based on the position and tilting of the associated subduction zones, the interposition of plates, and the speed and directions of their displacements. Most models are based on the reconstruction of a subduction paleozone setting, while the formational analysis of the environments of ore origin, which are available for direct observations, are neglected.

Independent of the time of formation, volcano-plutonic belts (VPB) hosting porphyry-copper mineralization are looked upon as specific geostructures that occupy certain lateral-temporal positions in conjugate geotectonic elements (Krivtsov 1977, 1983). We subdivide VPB into the basaltic (island-arc) type that terminates preorogenic evolution of eugeosynclines, and the andesitic type forming in orogenic-activation regimes at the flanks of contemporaneous geosynclinal systems (Krivtsov 1977, 1983, Krivtsov and Ageyeva 1979).

According to Vlasov (1978) belts of both types are conjugate with other geotectonic elements that differ in their regimes. The East Asian type of orthogeosynclines, in particular, consist of the following lateral elements: ocean, trench, Benioff zone, barrier zone of the island-arc (fore-arc, on the ocean side) basaltic VPB, inner zone of the island-arc with massive sulfide formations, median massif, back-arc basin with a miogeosynclinal regime, epicontinental andesitic VPB, and continent with local troughs with terrigenous filling (see Fig. 1). It should be noted that rear troughs, in case they acquire specific features of the secondary (regenerated) geosynclines, hamper interpretations of such systems.

Being different in regime, the aforesaid geostructures are conjugate in time and share a common genesis that is usually interpreted through a single subduction zone associated with it. On the other hand, continent-ocean migration of such series have been established for the Cimmerides and Alpinedes of the Eastern Soviet Union, the Hercynides of the Urals, and the Hercynides and Alpinedes of the Caucasus. These relationships make it possible to superimpose andesitic VPB upon any of the previously evolved elements of such geostructures. Accordingly, depending on the composition of the substratum, we group the andesitic VPB into epicratonic, epimiogeosynclinal, and epieugeosynclinal ones (Krivtsov 1977, 1983).

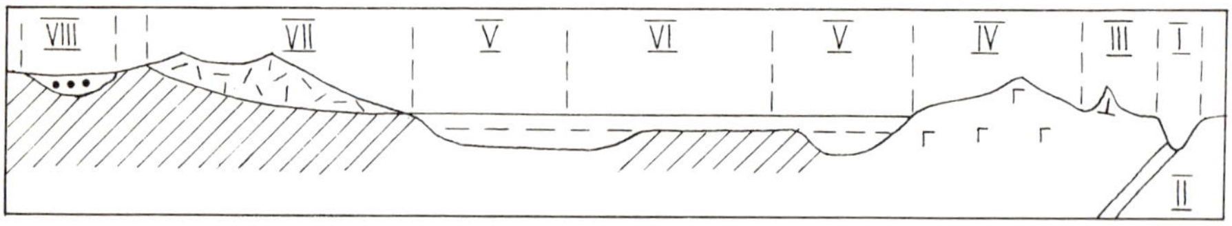

Fig. 1. Geotectonic elements of the East Asian type of orthogeosynclinal systems. Trench (*I*), Benioff zone (*II*), Barrier zone (fore-arc, basaltic VPB) of island-arc (*III*), inner zone of island-arc (*IV*); back-arc basin (miogeosynclinal) (*V*); median mass (*VI*); epicontinental andesitic VPB (*VII*), continental local trough (*VIII*). Continent is shown by *diagonal lines*

Intracontinental orthogeosynclinal systems, well-studied through the paleozoids of Kazakhstan, also involve some conjugate geostructures with a different evolution regime. As shown in Fig. 2, within the paleozoids, axial-rift troughs exposing oceanic crust, are fringed on one or two sides by the secondary geosynclines produced as a result of the expansion of the geosynclinal regime onto the sialic frame. Being most remote from the rifting axis, andesitic VPB are superimposed both on the secondary-geosynclinal complexes, and on their old basement. In the general scheme of the system (orthogeosynclinal, secondary geosyncline, andesitic VPB, foredeep, continent), there is lack of the basaltic VPB (though they are not prohibited). In the above scheme series, each succeeding element is somewhat younger than the previous one and is partially overlapped in space. The latter fact is not characteristic of marginal-continental systems. Nevertheless, intracontinental andesitic VPB also is subdivided into these three above mentioned types.

The differences in the positions of the basaltic and the andesitic VPB, as well as the nature of their basements, affect significantly not only the geochemistry of the associated porphyry-copper deposits, but also the general metallogeny of the VPB.

As shown in Fig. 3, the basaltic VPB that evolved on the oceanic crust and was composed of undifferentiated mafic lavas are relatively poor in ore-genesis

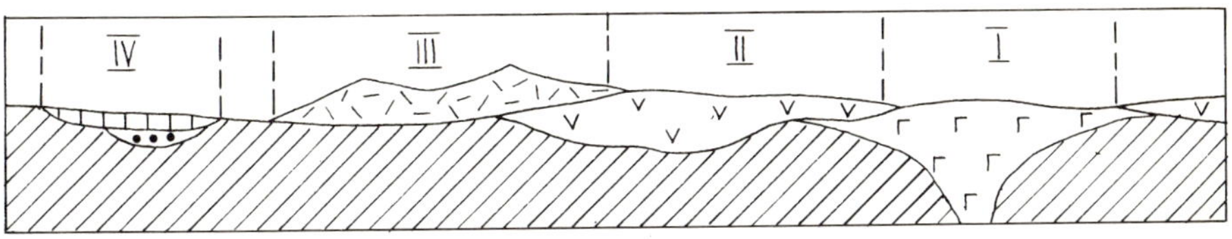

Fig. 2. Geotectonic elements of intracontinental orthogeosynclinal systems. Primary geosyncline (rift trough) (*I*); secondary geosyncline (*II*); andesitic VPB (*III*); foredeep (*IV*). Continent is shown by *diagonal lines*

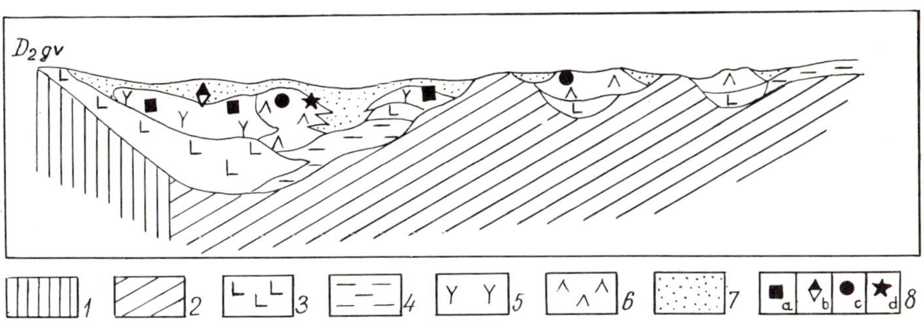

Fig. 3. Basic paleo-reconstructed section of the southern-central Urals. *1* Precambrian rocks; *2* Pre-Middle Ordovician sedimentary and volcanogenic-sedimentary sequences; *3* basaltic rocks of the undifferentiated formation associated with coaly-shaly and silicious rocks; *4* coaly-shaly and siliceous rocks; *5* volcanics of contrasting and successive massive sulfide-bearing formations; *6* volcanics of the basalt-andesite-basalt formation of basaltic belts; *7* sediments including flyschoids and turbidites; *8* deposits; *a* zinc-copper massive sulfide; *b* iron-manganese; *c* porphyry copper; *d* veined polymetallic

products. Porphyry-copper and gold-porphyry-copper deposits proper (Titley 1975), related to gabbro-diorite formations, are predominant in such belts. Previously evolved massive-sulfide ores are localized in differentiated formations of the inner zones of the island-arc outside the basaltic VPB-hosting porphyry-copper ores. Younger ore-genesis products within the VPB consist of veined polymetallic mineralization that locally can be related to the periphery of the porphyry-copper systems. The monotony of the metallogeny and simplicity of the basaltic VPB can apparently be attributed to the primitive composition of their substratum that undoubtedly supplies some amount of the ore-forming constituents. The fact that basaltic VPB's are incorporated into a lateral-temporal series of geostructures makes it possible to explain the confining at such deposits to structures similar in the nature of zoning.

Andesitic VPB that evolve on a basement varied in character and composition are distinguished both in the geochemistry of porphyry-copper deposits and the diversity of other ore-genesis products. Porphyry-molybdenum deposits are common to the initial stages of epicratonic VPB; porphyry-copper-molybdenum deposits are typical of epimiogeosynclinal VPB; and porphyry molybdenum copper ones, of epieugeosynclinal. These deposits are related to granitoid, monzonitoid, and granodiorite plutonogenic formations, respectively.

Porphyry-molybdenum-copper and porphyry-copper deposits of andesitic belts are incorporated into a temporal series of metallogenic zoning of two ranks. The first rank zoning, which has predominant lateral expression, is attributed to the andesitic VPB that is conjugate with eugeosynclines and pericratonal basins. In the eugeosynclines, preorogenic metallogeny with predominant massive-sulfide ores is classical. In the second case, stratiform polymetallic deposits can evolve. The second-rank zoning that manifests itself in VPB structures and has mainly vertical expression is associated with deposits attributed to different ages, viz., from ore occurrences localized at the VPB basement to deposits formed at the late stages of VPB evolution (see Fig. 4).

Metallogeny of the VPB basement is determined by the style of its evolution. The sialic substratum of epicratonic VPB is typically poor in minerals. Characteristic granitophilic elements are inherited by metallogenic processes at the period of VPB emplacement. Predominantly terrigenous, carbonate-terrigenous, and carbonate complexes of the basement of epimiogeosynclinal VPB are not infrequently hosts to stratiform polymetallic deposits. Being superimposed by porphyry-copper systems, the latter undergo skarnization and regeneration as a result of which veined deposits, conjugate with stratiform ones, come into being. Volcanogenic formations of the basement of epieugeosynclinal VPB are characterized by preorogenic metallogeny with typical massive-sulfide, iron, and manganese volcanogenic ores. So far only some examples can be given to illustrate the involvement of this metallogenic material into the processes of formation of porphyry-copper systems, though skarn magnetite ores are known from the periphery of porphyry-copper systems in VPB of various ages.

Later periods of VPB emplacement (after porphyry-copper systems have formed) produced ore-genesis material ranging greatly in its composition (see Fig. 4). All the deposits occur within the VPB, yet it has been so far difficult to admit the possibility of the generation of vertical series in some parts of the VPB.

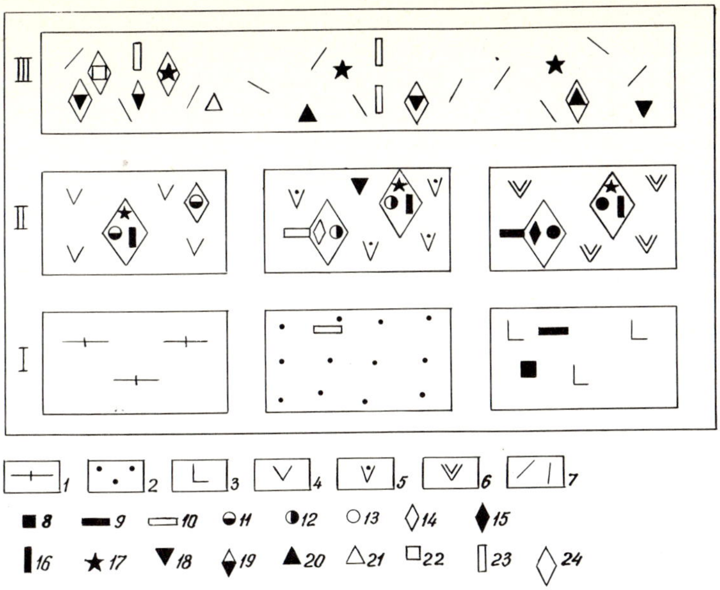

Fig. 4. The structure and metallogenic zoning of andesitic volcano-plutonic belts (VPB). **I** Substratum of the belts: *1* rigid sialic (cratonic); *2* miogeosynclinal; *3* eugeosynclinal; *4–7* volcano-plutonic associations (VPA) of VPB. **II** Favorable for porphyry-copper mineralization and preceding VPA: *4* epicratonic zones; *5* epimiogeosynclinal zones; *6* epieugeosynclinal zones; *7* late VPA **III**. *8–23* Deposits; *8–10* stratiform deposits of belts' substratum; *8* zinc-copper massive-sulfide; *9* iron ore; *10* lead-zinc. VPB Deposits; *11* porphyry molybdenum (with copper); *12* porphyry-copper-molybdenum; *13* porphyry-molybdenum-copper; *14* skarn lead-zinc; *15* iron-copper skarn; *16* veined polymetallic; *17* gold-silver and gold-bearing; *18* tin; *19* tin-tungsten; *20* molybdenum-tungsten; *21* tungsten; *22* molybdenum-uranium; *23* niobium-tantalum; *24* deposits (of the system) with porphyry appearance

Thus, the following geologic-economic series of deposits are recognized in basaltic volcano-plutonic belts: copper-zinc, massive-sulfide, iron-manganese-porphyry copper + veined polymetalllic. In andesitic belts, these are as follow: massive-sulfide, base-metal-lead zinc stratiform, iron ore, porphyry-copper + skarn copper iron ore, gold, silver, tin + tin-tungsten + molybdenum-tungsten + tungsten, niobium-tantalum.

There foregoing lateral-temporal, vertical-temporal, and lateral-vertical-temporal series of metallogenic zoning are available for observation, as a rule, in lateral expression. This seems to be borne out of the fact that the vertical extension of zoning as well as the real depth of exposure of a deposit are not comparable in values. The general characteristics of the VPB metallogenic zoning is exemplified by some provinces of the Soviet Union.

The products of early and late geosynclinal ore genesis in the region of the Urals are incorporated into various elements of eugeosynclinal space, which rules out their spatial conjunction (see Fig. 3). Basaltic belts in the Magnitogorsk intracontinental eugeosyncline give rise to extended volcanogenic island-arcs of barrier zones and fringe inner areas of contrasting and successive massive-sulfide ore-bearing formations. Basaltic rocks of the undifferentiated formation commonly make up the basement of the belts. Porphyry-copper mineralization here is preceded by the formation of zinc-copper massive-sulfide deposits associated with earlier eugeosynclinal formations. Porphyry-copper mineralization within the studied areas of the Urals is represented by the porphyry-copper (without molybdenum) type proper. The productive volcano-plutonic association involves essentially sodium volcanogenic basalt-andesite-basalt and plutonogenic gabbro-diorite-plagiogranite formations. Intrusive bodies are controlled by ruptures bounding barrier zones, paleovolcanic vents, and these apparently belong to "subvolcanic intrusions". Porphyry-copper deposits are related to quartz-diorite and diorite-porphyry massifs. Ore mineralization occurs in apical, endo- and exocontact parts of the such intrusions, viz., in albitization, silicification, chloritization, and sericitization zones of rocks regionally metamorphosed in prehnite-pumpellyite facies. Iron-manganese deposits of the Near-Magnitogorsk group are commonly associated with the terminations of massive-sulfide formations. However, they can occur in the basalt-andesite-basalt formation of basaltic belts. The formation is known with host-veined polymetallic mineralization traced in the Irendyk and Imennovsk zones of the Urals. These areas are consistent only in the gabbro-diorite quartz-diorite formation that hosts porphyry-copper mineralization.

Metallogenic zoning of intracontinental andesitic volcano-plutonic belts conjugate with geosynclines can be exemplified by the Late Paleozoic belt of the northern Near-Balkhash region of Kazakhstan. As shown in Fig. 5, from southwest and northeast, it fringes the northern Balkhash geosyncline that evolved into a eugeosyncline in Cambrian – Late Ordovician time and into miogeosyncline, in Late Ordovician – Early Famennian time. In the Famennian-Tournasian along the periphery of the geosyncline, secondary geosynclines evolved. Temporally (after the Visean) and laterally (southwest and northeast) the secondary geosynclines give way to a volcano-plutonic belt. Depressions with terrigenous sedimentation evolved along the periphery of the belt.

The substratum of the belt is known to host Early Paleozoic massive-copper sulfide ore occurrences. Middle-Late Devonian massive-sulfide – base metal and Later (Famennian-Tournasian) lead-zinc stratiform mineralization is related to Famennian-Tournasian secondary geosynclines. Gold-silver mineralization confined to the latest dacite-liparite components of effusive rock sequences is related to the volcanic rocks favorable for porphyry-copper mineralization. Copper-magnetite ore-bearing skarn bodies are traced at the contact of ore-bearing intrusions hosting porphyry-copper mineralization. Molybdenum-tungsten and niobium-tantalum mineralization is associated with plutonic rocks of later, Permian dacite-liparite, and trachidacite-trachiliparite volcano-plutonic association. Porphyry-copper mineralization of the region is represented by three groups of deposits related to epicratonic (essentially molybdenum deposits of the Karatasak area), epimiogeosynclinal (the Borlinsk area), and epieugeosynclinal (molybdenum-copper deposits of the Kounrad area) environments, respectively.

It is important to emphasize some conservatism in the succession of metallogenic processes.

In eu- and miogeosynclines, typical of preorogenic ore formation processes was volcanogenic-sedimentary ore accumulation, irrespective of the ores accu-

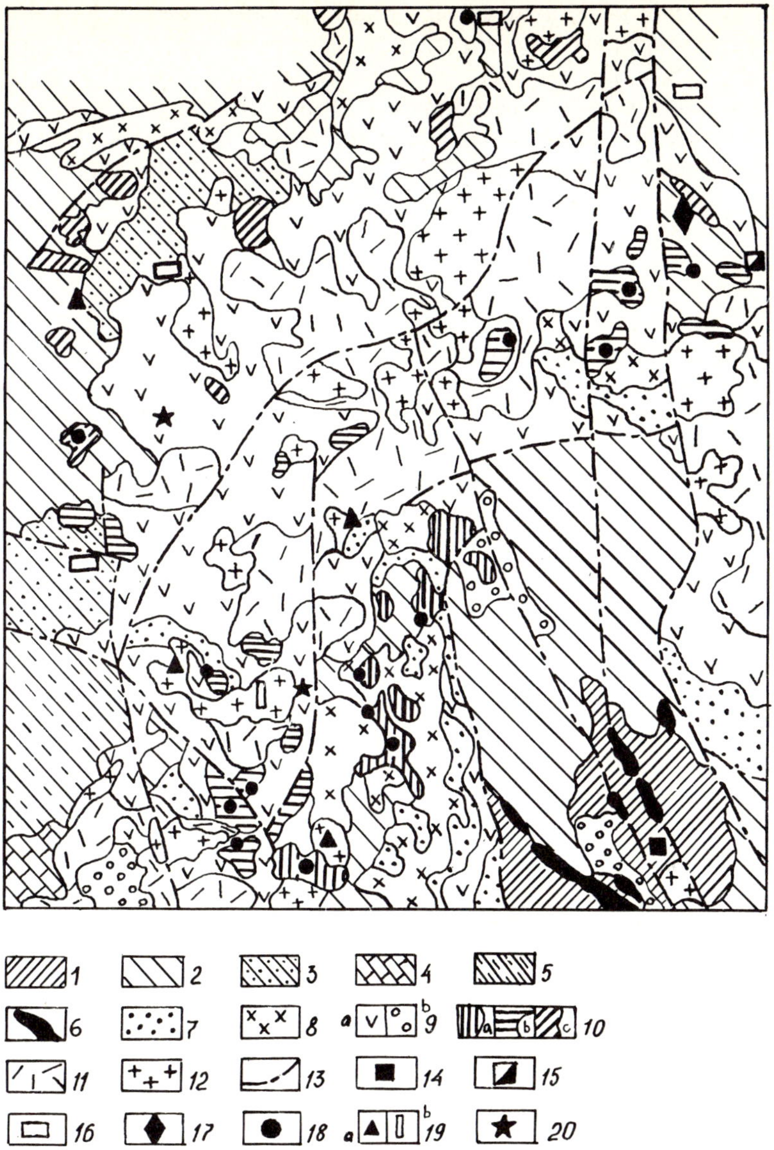

Fig. 5. The position of porphyry-copper deposits in metallogenic zoning of the andesitic belt in the northern Near-Balkhash region (a fragment) of Kazakhstan. *1–6* Caledonian and Early Hercynian basement of the volcano-plutonic belt: *1* eugeosynclinal complexes; *2* miogeosynclinal complexes; *3* molasse; *4* gneiss; *5* granitoids; *6* ultramafic rocks; *7* volcanogenic and sedimentary sequences for the Famennian-Tournaisian secondary geosynclines; *8* intrusions of the gabbro-plagiogranite secondary geosynclinal formation; *9–12* volcano-plutonic belt rocks: *9–10* favorable volcano-plutonic association: volcanics $C_{1-2}-C_2$ (*9a*) and synchronous sediments (*9b*). *10* Ore-bearing granodiorite massifs: Sodium (*a*); potassium-sodium (*b*); potassium (*c*). *11–12* Postore associations: *11* volcanics $C_{2-3}-P_1$; *12* granitoids $C_{2-3}-P$ and younger; *13* mayor faults. *14–20* mineralization; *14* copper massive-sulfide; *15* massive-sulfide copper-zinc metal; *16* lead-zinc, stratiform; *17* skarn iron ore; *18* porphyry-molybdenum copper; *19* gold-silver; *20* molybdenum tungsten; *21* niobium-tantalum

mulated (sulfide, iron-oxide, manganese). Meanwhile, in the case of andesitic VPB, preferable ore-genesis processes were those producing porphyry-copper type systems.

As is known, ore matter is accumulated in porphyry-copper deposits as veinlets and impregnations in intrusive bodies and at their exocontacts, breccia pipes as well as veins, in the intrusive frame. The similarity of the style of ore accumulation at the early and late epochs of the VPB emplacement manifests itself primarily in the intensity of mineralization of the first two types.

For foreign VPB, Sutherland-Brown et al. (1971), Sillitoe et al. (1975), and some other authors showed that some tin and tungsten deposits belong to the porphyry type. All elements of such deposits are the ones common to porphyry systems, though of a different qualitative expression.

Evstrakhin and Itsikson (1980), who purposefully studied the position and structure of deposits related to andesitic VPB in the Soviet Union demonstrated the constancy of the "porphyry style" of accumulation for some metals: tin, tin and tungsten, uranium, and gold and silver.

These researchers identify porphyry deposits of the volcanogenic level with metals being accumulated in breccia pipes, necks, and volcanic intrusions. The group includes tin, uranium-molybdenum, and gold-silver porphyry deposits (see Fig. 6).

Tin-tungsten, molybdenum-tungsten, and tungsten-porphyry deposits are related to deeper plutonic levels, and occur in hypabyssal intrusions as well as at their exocontacts (see Fig. 7).

Depending on the formational conditions of mineralization, on the way an ore-magmatic system was opened or closed, the products of early and main stages of ore genesis can occupy different positions with respect to ore-bearing porphyry intrusions. They can be far away from their apical parts, occurring as veinlets or veined zones (tin deposits of a supra-intrusive zone, tungsten-tin deposits) or, on the contrary, be concentrated in a stock or nearby one, giving rise to veinlet-impregnated mineralization (porphyry-copper and porphyry-molybdenum deposits, and tin-bearing greisens of an intrusive zone). Later mineralization occurs in the peripheral and the upper parts of ore-magmatic systems and controls the mineralogical-geochemical zoning of deposits.

Metallogenic specialization of the substratum of the belts plays a crucial role in formation of some types of porphyry deposits as much as in the case of

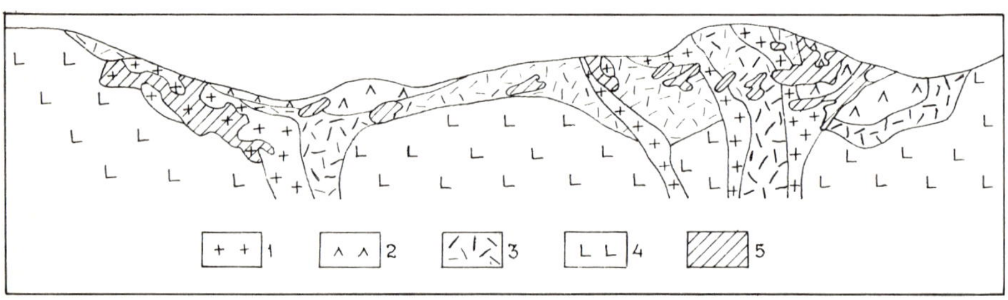

Fig. 6. Ore zones and generalized geologic sections of a molybdenum-uranium deposit of the USSR. *1* Granite porphyry; *2* quartz porphyry; *3* clastic quartz porphyry; *4* andesite dacite lavas; *5* ore mineralization

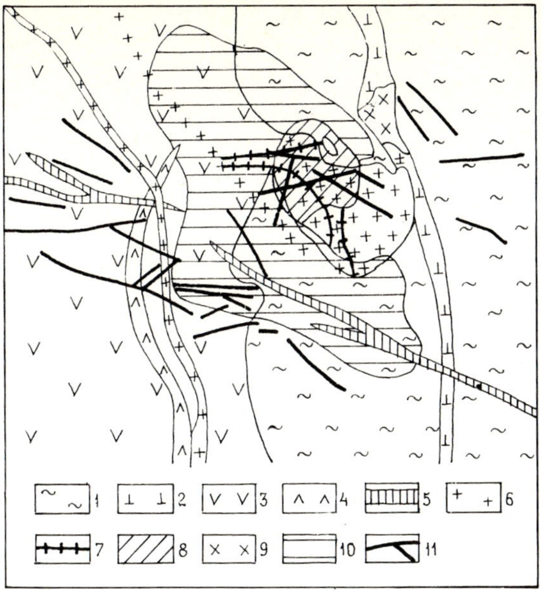

Fig. 7. Ore zones and generalized geologic plan of the molybdenum-tungsten deposit of the USSR. *1* Porphyrites, schists, limestones Early Cambrian; *2* ultrabasite Cambrian; *3* diorites, granodiorites, Early Paleozoic; *4* plagiogranites Early Paleozoic; *5* bostonites Triassic; *6* granite porphyry Triassic-Early Jurassic; *7* quatz-molybdenite veins; *8* molybdenite stockwork; *9* granite porphyry; *10* tungsten stockwork

porphyry-copper deposits. Taking the Sikhote-Alin folded system as an example, Kulish and Usenko (1982) showed that widely spread sandy-shaly formation with higher tin contents played a significant role in tin potential of granitoid and in tin concentrations. Granitoids intruding these terrains are commonly tin-bearing.

The continuity of the metallogeny of substratum has been so far established only for some VPB. Thus, in particular, epicratonic zones of VPB host copper-molybdenum porphyry, molybdenum porphyry, tin and tin-tungsten porphyry, uranium porphyry, and gold-silver-porphyry deposits. Epimiogeosynclinal zones, along with porphyry copper, host tin porphyry deposits.

The foregoing series certainly do not encompass the whole metallogeny of volcano-plutonic belts, but their identification is important for metallogenic researches, viz, for outlining zones favorable for porphyry-copper mineralization.

References

Evstrakhin VA, Itsikson MI (1980) Porphyry type ore deposits. Razved Okhr Nedr 1:9 – 16 (in Russian)

Krivtsov AI (1977) Types of areas of porphyry-copper mineralization. Geol Rudn Mestorozhd 4:46 – 57 (in Russian)

Krivtsov AI (1983) Geological fundamentals of prognostication and prospecting for porphyry-copper deposits. Nedra, Moscow, 254 pp (in Russian)

Krivtsov AI, Ageyeva ST (1979) Petrological features of volcanoplutonic association of porphyry copper mineralization areas. Sov Geol 7:36 – 49 (in Russian)

Kulish EA, Usenko SF (1982) Granites and tin ore formation in the Sikhote-Alin folded system. Collected abstracts, Tbilisi, IAGOD VIth Symp, pp 79 – 80

Pavlova IG (1978) Porphyry-copper deposits. Nedra, Leningrad, 275 pp (in Russian)
Popov VS (1977) Geology and genesis copper- and molybdenum porphyry deposits. Nauka, Moscow, 201 pp (in Russian)
Sillitoe RH, Halls C, Grant JN (1975) Porphyry tin deposits in Bolivia. Econ Geol 70:913 – 927
Smirnov VI (1981) The trends in metallogenics. In: Distribution of mineral resources. Nauka, Moscow, XIII:8 – 20 (in Russian)
Sutherland-Brown A, Cathro RJ, Panteleev A, Ney CS (1971) Metallogeny of the Canadian Cordillera. Can Inst Min Metall Bull 64-709:37 – 61
Titley SR (1975) Geological characteristics and environment of some porphyry copper occurrences in the Southwestern Pacific. Econ Geol 70:768 – 784
Vlasov GM (1978) Young geosynclines of the Pacific belt, their volcanogenic and ore formations. Nauka, Moscow, 177 pp (in Russian)

Geological and Structural Conditions of Localization of the High-Grade Ores of Porphyry Copper Deposits

I. M. Golovanov, E. I. Nikolayeva, and M. A. Khazikhin[1]

Abstract

A generalized geological-structural modeling of porphyry copper deposits reveals the common features of formation of the high-grade ores of the inner parts of stockworks. The common internal structure of the ore stockworks has been established. The ore stockworks form a series of bodies concentric in plan, whose copper concentration decreases outward from the centre. Studies of the Almalyk area show that the morphology of ore stockworks and high-grade ore parts of chalcopyrite-bornite ores is determined by the form of the stock roof, particularly by the presence of ore-localizing negative structures – depressions, cavities and trenches. The so-called structure of "an egg in a wineglass" is the prototype of high-grade core sections of porphyry copper ores enveloped by the tongues of porphyritic stocks at Almalyk. Low-grade ore envelopes the high-grade ores, smoothing their outline and filling up the irregularities of the stock roof.

Porphyry copper deposits are considered to be the most important type of copper deposit. In developed capitalist countries and in developing countries, these deposits made up 35.5% of the total mineral reserves in 1963, and their share increased to 65.1% in 1980. For a long time porphyry copper deposits have played the leading role in the balance of the world's reserves and output of copper. The deposits determine the strategy of development of the world's extractive copper industry for the near future (Krivtsov 1983, Pervago 1978, and others).

The classical high-grade ores (copper content 3 – 5%) of porphyry copper deposits are characteristic of the zone of secondary sulphide enrichment. The average composition of the primary ores is notable for the low amount of copper (0.3 – 0.6%) and accompanying elements – molybdenum, gold and silver.

Generalization of the data on the structure of deposits in the USSR, USA, Canada and Iran and elaboration of the geological-structural models of porphyry copper deposits show the common features of the high-grade hypogene ore sections of porphyry copper stockworks. The most important initial elements in a model appear to be a porphyritic ore-generating granitoid stock and an ore-

[1] Central Asian Scientific Research Institute of Geology and Mineral Resources, Ministry of Geology, Tashkent, USSR

Geology and Metallogeny of Copper Deposits
(ed. by G. H. Friedrich et al.)

bearing stockwork. In Almalyk the core structure of the latter is determined first of all by the morphology of the roof of the porphyritic stock and by the composition and structure of the enclosing rocks. A common inner structure of ore stockworks has been established for many ore deposits. The stockworks have been found to form a series of bodies concentric in plan, with copper concentration decreasing from the centre.

The localization of stockwork bodies of copper ore (especially those of high-grade ores) is, clearly, in negative structures ("rims", depressions, cavities) of the intrusives in the Kounrad and Aktogai deposits in Kazakhstan, Erdenetuin-Obo in Mongolia (Krivtzov 1983), and the Almalyk group of deposits in Uzbekistan (Golovanov 1982).

Porphyry copper deposits of the Almalyk area may be considered typical of porphyry copper ore formation. The history of their complex investigation and industrial exploitation numbers 50 yrs. During this period a lot of work has been done at the deposits – from prospecting to exploitation. The deposits were explored to a depth exceeding the pinching out of mineralization. Within the limits of the ore field there are ore bodies considerably stripped by erosion, as well as those unstripped.

The detailed information on the vertical section of the Almalyk deposits and the similarity of their composition to that of the majority of the world-known porphyry copper manifestations allowed us to take these deposits as a standard for determining the localizing conditions of porphyry copper mineralization and for making a complex model of the ore-metasomatic and magmatic system.

The geological-structural specificity of the area is determined by the development of two systems of faults. The first system is represented by northwest-trending faults that have no conspicuous manifestation in the modern erosion section. These faults control the distribution of porphyritic stocks. The second system conspicuously is represented by sublatitudinal long-lived faults. They are manifested in the modern erosion section by cataclastic zones.

Effusive quartz porphyries (Lower Devonian), diorite and monzonites (Middle Carboniferous) and the rocks of porphyritic stocks (Upper Carboniferous) are considered to be the main ore-bearing rocks of the Almalyk deposits.

Judged from their petrographic and petrochemical parameters, the porphyritic stock rocks are quartz monzonite porphyries. They appear to be the last link in the incomplete magmatic association gabbro-diorite-diorite-monzonite (syenodiorite)-quartz monzonite porphyry. Hyperalkalinity of the porphyritic stocks is ensured by an alkali sum ($Na_2O + K_2O$) of 8.4%, with K_2O dominating over Na_2O ($K_2O/Na_2O = 1.56 - 2.13$). Biotite content is higher and amphibole content is lower in depth and also in the inner parts of stocks. Clinopyroxene appears also in depth.

Morpho-structural modeling was shown to the rather informative for studying the relationship between magmatism and mineralization. To make such models the authors used the data from exploratory drill holes.

Three-dimensional block diagrams (models) of fields with ore concentrations and ore-bearing stocks are dissected to show sets of parallel vertical sections passing across the strike of the ore bodies (Fig. 1). The ore bodies are correlated on

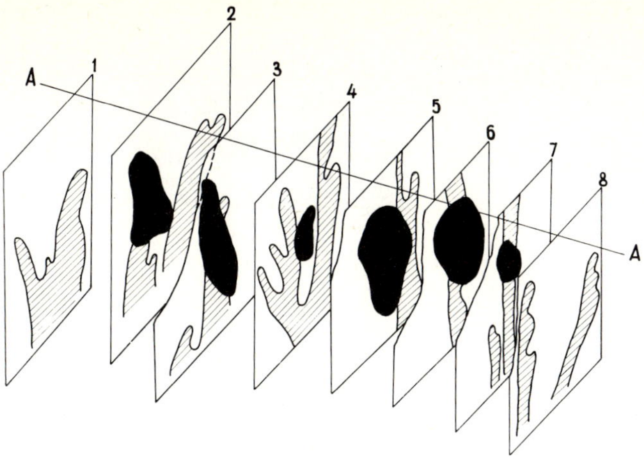

Fig. 1. Geological sections of a porphyry copper deposit. *Hachures:* porphyry stock, *black:* stockwork of high-grade ore

the maps of various hypsometric levels and on longitudinal vertical sections. A similar method has been used in the Mount Morgan deposit (Cornelius 1969). The group of geologists from the Central Research Institute of Geological Exploration, under the leadership of Krivtzov (1983), also used this method.

To make the models of the bodies, various outlines of copper concentration have been used. The external outline corresponds to low-grade ores (0.2% Cu) and the internal contours represent high-grade bornite-chalcopyrite ores (up to 1% Cu) (Fig. 4).

The analysis of morpho-structural models that included five units of the Almalyk ore field showed that all the deposits investigated, though different in size, had the same structure of ore bodies and porphyritic intrusives and the same dimensional relationship between these two.

The roof morphology of the porphyritic stock is considered to be the decisive factor in the localization of copper ores at Almalyk. The ore stockwork of any deposit has a certain correspondence to the host body of quartz monzonite porphyry. It should be noted that the bigger deposits are known to have bigger stocks. The smaller deposits are characterized by the simplest morphology of the apical part of the stock and its copper ore bodies, respectively. The form of stocks and ore bodies becomes more complicated in big units (Dalnee, Kalmakyr).

The apical parts of all ore-generating intrusives in the Almalyk ore area are characterized by negative structures, such as depressions, pits and trenches, that are bordered by tongues (apophyses) of the intrusives. These negative structures are represented by the subsidence domes, the latter are formed when the stock edges have already consolidated, but the middle part reduces in size while cooling, and then shrinks. Very frequently the tongues bordering the depressions form solid, ribbed, multidissected walls while emerging, but sometimes they look like the tentacles of an octopus. The simplest possible stock structure reminds one of a crater with sloping walls.

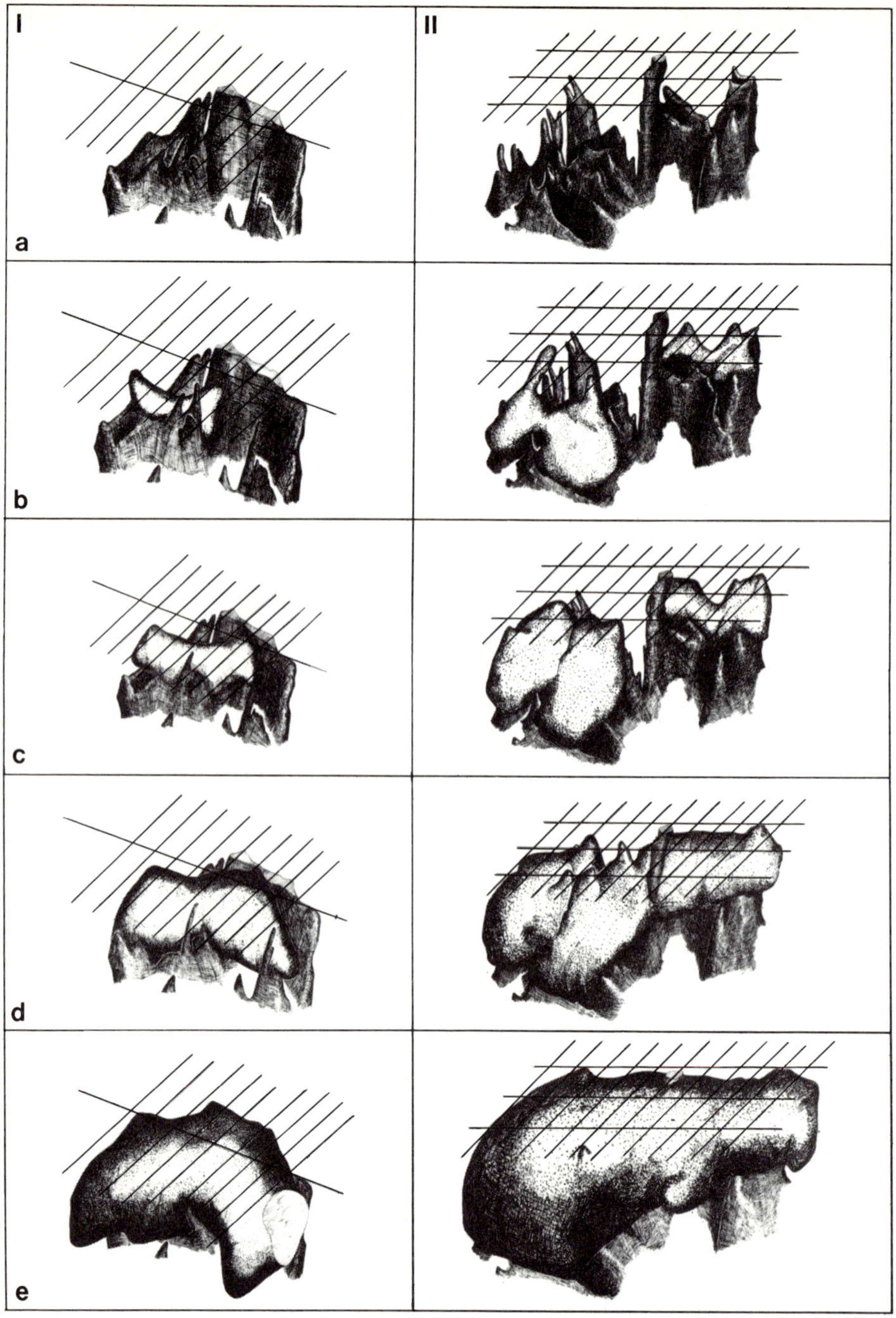

Fig. 2a – e. Models of porphyry stocks and ore stockworks outlined by various copper content. *Dark to light shading* stock; *dark-bordered stippling* ore-bearing stockwork. *I* Morphologically simple deposit; *II* morphologically complicated deposit. **a** Porphyry stock; **b** stockworks of rich ores in stock depressions; **c, d** stockworks of intermediate copper content; **e** stockworks of poor copper ores

High-grade stockwork ore is situated directly in the stock roof depression, and the shape of the ore body imitates the relief of the stock roof surface (the so-called structure of an egg in a wineglass). Ore bodies with lower copper concentrations gradually expand and cover the tongues, walls and upper parts of the stock. The form of intermediate stockworks is similar to that of the central part, but has a smoother profile (Fig. 2, I). The approximate geometry of the ore bodies defines them as ellipsoids of revolution or elliptical paraboloids of revolution with similar or equal relations between short axes. The porphyritic stock and its ore bodies are sure to have a most regular and isometric form if the intrusive was emplaced in an isotropic medium. In this case, in plan, the contour of the stock and its ore bodies approximates a circle. The presence of enclosing rocks of diverse composition as well as tectonic dislocation, distorts the form of the intrusive and its ore bodies. The Almalyk deposits have several types of ore-containing rocks. Stockwork mineralization is usually associated with monzonites (syenodiorites), sometimes with diorites, and very rarely with effusive quartz porphyries (Fig. 3). The roof morphology pf large intrusives is characterized by the presence of sags. Sometimes these sags are localized on different hypsometric levels. The bottom of large depressions is complicated by swells and crests. The morphology of ore bodies, particularly that of their high-grade core areas, is determined by the complicated relief of the stock surface. If there are several sags in the intrusive roof, each negative structure has a corresponding section of high-grade ore. Swells, crests and tongues in the sag bottom prevent the even development of high-grade ore in the sags. Low-grade ores form a body of complicated morphology that includes also high-grade cores. While expanding, this body envelopes the upper part of the stock and its tongues (Fig. 2, II).

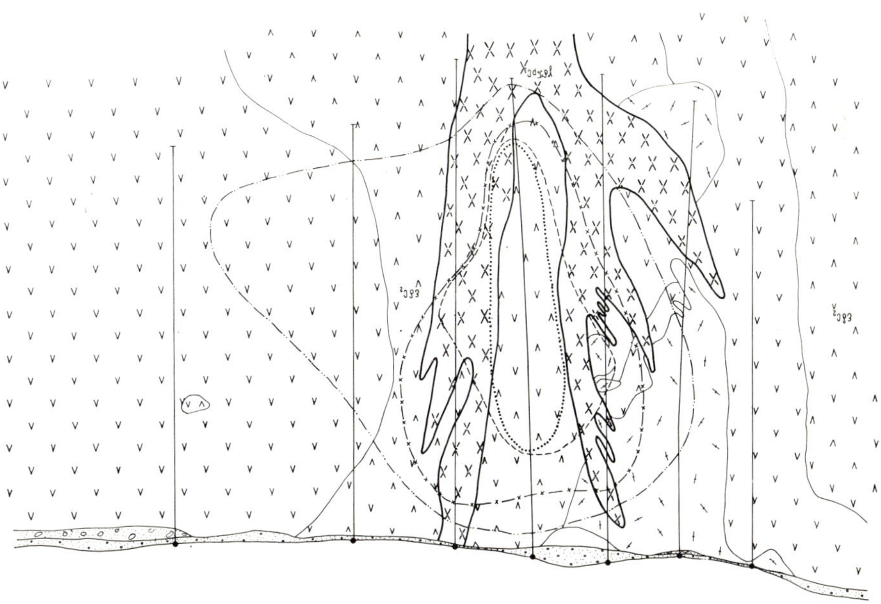

Fig. 3. Typical geological section of a porphyry copper deposit (for legend see Fig. 4)

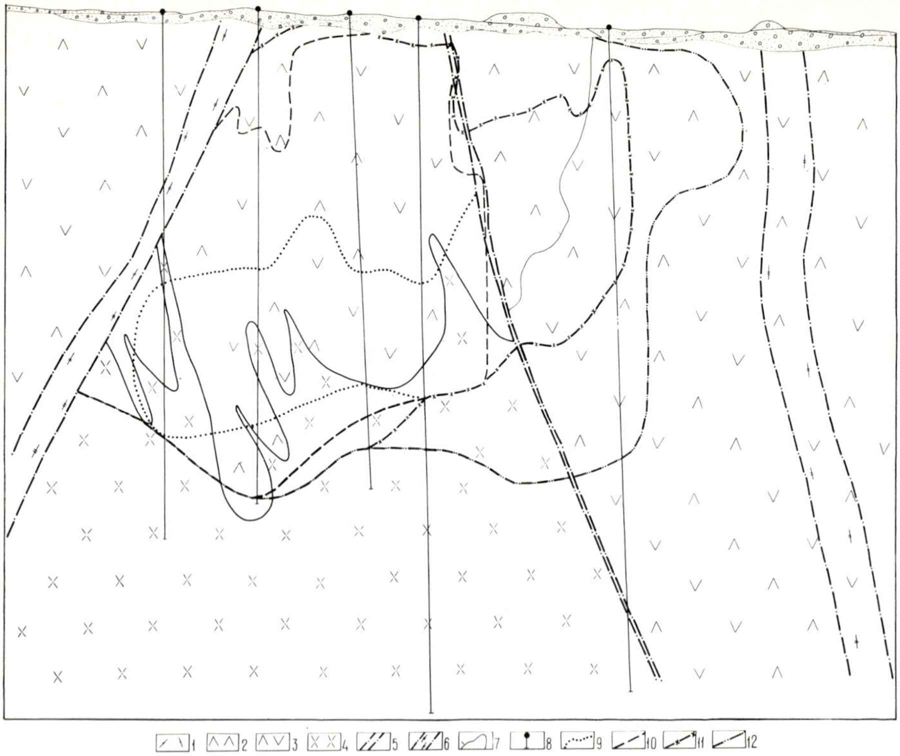

Fig. 4. Typical geological section of a porphyry copper deposit in a wedge-shaped tectonic block. *1* Quartz porphyries; *2* diorites, gabbro-diorites; *3* monzonites (syenodiorites); *4* quartz monzonite porphyries; *5* faults; *6* faults with fracture zone; *7* rock boundary; *8* drill hole; *9, 10, 11, 12* contours of ore stockworks according to copper content; *9* rich copper ores; *10, 11* intermediate content; *12* poor copper ores

The level of erosion determines to a great extent the morphology of the deposit. The Almalyk deposits, except the Kalmakyr, appear to be practically uneroded. The Kalmakyr deposit has been stripped only to one-third. Thus, earlier the Kalmakyr ore zone structure was described as a ring put on a stock; actually the stock is the bottom swell of a big downwarp in a large intrusive.

It is important to note that the negative structures of the stock roof coincide with faults, paired to give a wedge-shaped block (Fig. 4). These negative structures were shown to be the main factor determing the location of porphyry copper mineralization. The localization of core parts of high-grade copper ores directly in the roof downwards of the fault-broken intrusives determines the location of ore and the metasomatic zonation of the porphyry copper system.

The composition and zonation of metasomatites in porphyry copper deposits (Fig. 5) vary, depending on:

1. Localization of the ore-metasomatic system relative to large faults.
2. Localization of the ore-metasomatic system in negative structures of the upper part of the stock; i.e., morphology of subsidence dome, inhomogeneity of the enclosing rocks, and stock configuration.

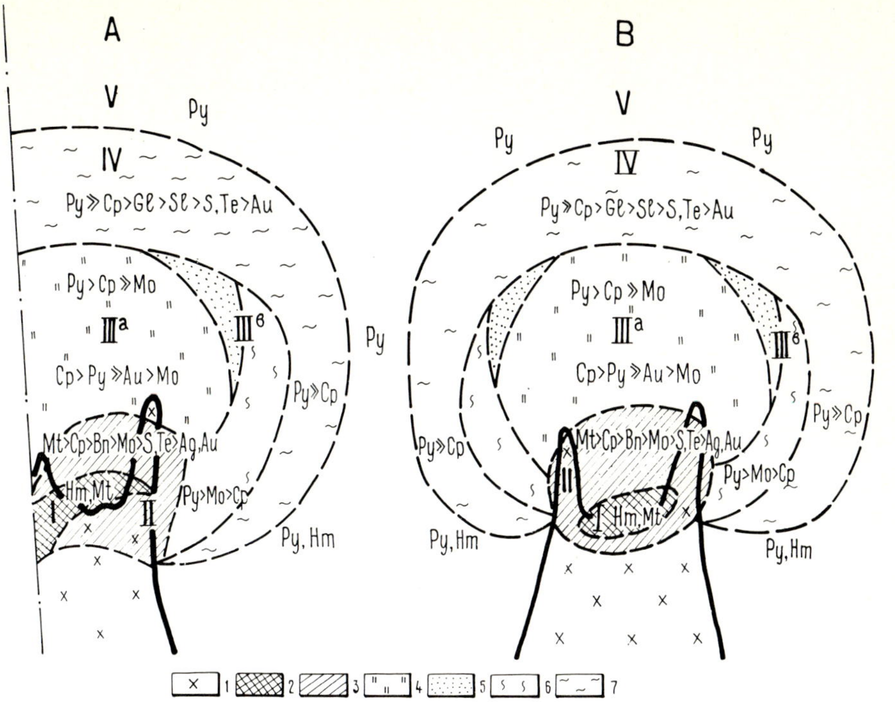

Fig. 5 A, B. Ore-metasomatic zonation model of a porphyry copper deposit. **A** Section perpendicular to the ore-bringing fault plane; **B** section parallel to the fault plane. Zones: *I* central; *II* inner; *IIIa, IIIb* intermediate; *IV* fringe; *V* peripheral; *1* porphyry stock; *2* silicic alteration; *3* K-feldspar alteration; *4* phyllic alteration; *5* argillic alteration; *6* propylitic alteration I; *7* propylitic alteration II. Abbreviations: *Py* pyrite; *Cp* chalcopyrite; *Mt* magnetite; *Hm* hematite; *Bn* bornite; *Mo* molybdenite; *Gl* galena; *Sl* sphalerite; *S, Te* sulphides, tellurides; *Au, Ag* native gold and silver

3. Physico-mechanical characteristics (composition, density, jointing, porosity, etc.) of the enclosing rocks.

The metasomatic composition naturally changes relative to an ore-generating fault and to the quartz monzonite porphyry stock. At their intersection, the fault and the stock determine a focal plane for metasomatic zonation. The zones of silicic, K-feldspar, and phyllic metasomatites enclosed in a propylitic aureole replace one another while moving away from the given plane, progressively advancing into the enclosing rocks. The metasomatic system is pyritizing, and pyritization is especially intense in the external phyllic zone, which is anhydrite-bearing at the lowermost levels. The external lateral zones are more persistent at depth than the internal ones.

Mineralization of subeconomic character occupies a certain place in the general zonation of the system. Usually it is localized much higher than pay ores. Primary ore zonation is expressed in the change of molybdenum mineralization to copper-molybdenum, copper, polymetallic, or gold-silver mineralization. These changes go from the centre to peripheral parts as well as vertically from the bottom to the top. Zonation is expressed relative to the contact of the porphyritic intrusion and ore-localizing faults.

Changes in the mineral composition of the ores are determined by the same zonation as that in the metasomatites: ore-zonation is three-dimensional and coaxial. Five zones of various mineral compositions can be identified.

1. Central zone – silicified barren core asymmetrically placed downwards from the system centre to the stock body.
2. Internal zone of maximum grade ores – high-grade copper core with gold and silver.
3. Intermediate zone – lower grade, but thicker copper-molybdenum ores.
4. External zone – low-grade copper-molybdenum ores with lead, zinc and silver.
5. Peripheral zone with lack of copper mineralization and presence of veined gold-silver metallization.

Dimensionally, central zone ore mineralization corresponds to certain facies of metasomatic zoning. Impregnations of hematite and magnetite (the "bottom" of the ore-metasomatic system plunging within a stock body) are associated with the silicified core. Magnetite-chalcopyrite-bornite ore with gold, silver and a number of additional minerals – sulphides, sulphosalts, tellurides and native elements (high-grade copper ore core of internal zone) (Fig. 5) – is associated with orthoclase-rich rocks and lowermost levels of phyllic alteration.

Judged from the amount and concentration of long-lived veinlets, the high-grade copper ore zone may be called the zone of maximum ore genesis. Mineral associations in this part of the stockwork are very diverse. They are characterized by a wide range of ore-depositing conditions.

The veinlets, from early to late, vary in composition. They may include magnetite, pyrite, molybdenite, chalcopyrite, chalcopyrite-bornite-tetrahedrite-gold with chalcopyrite, digenite, sulphides and sulphosalts of tin, bismuth, nickel, silver, with native copper, bismuth, tin, silver and others; sphalerite-chalcopyrite-gold and many other veinlets (Nikolayeva 1980).

The deposition of the inner zone mineral complex started under conditions of relatively high temperature, as evidenced by decomposition of bornite-chalcopyrite solid solution; then the veinlets underwent repeated opening. Thus, mineralization was renewed and early minerals were recrystallized and desulphurized. At the last stage of ore formation the ores underwent the influence of low-temperature solutions with high-oxidation states: the formation of hypogene goethite, chalcocite, digenite, covelline and also low-temperature minerals – stromeyerite, and others – took place. Recycled minerals are also found there. The stage is characterized by the following processes: diffusion regrouping – metamorphogenic decomposition of copper sulphide solid solutions, and dissociation of electrum, resulting in the appearance of native silver and gold of high fineness. These processes and some others indicate the extreme permeability and active ore genesis of the so-called bowl, plunging within the deep negative structure of the stock roof. The inner zone of the stock roof partially involves phyllic and early propylitic zones passing outward from the stock centre. This inner part has its own zonation: the lower section is represented by a hypogene oxidation zone (locally absent) with bornite-bearing ores; the middle section by associations in which chalcopyrite dominates, the external edges by enrichment in molybdenite.

The central high-grade copper core is known to have high ratios of copper : molybdenum, gold : molybdenum, and silver : molybdenum. It is also distingusihed by having a maximum-positive-paired correlations coefficient for copper with gold and silver (up to 0.7 – 0.8) and a negative one for molybdenum.

Langton and Williams (1982) described similar geological-structural conditions for the localization of high-grade ores and metasomatites of the Dos Pobres deposit in Arizona. The core sections of high-grade copper ores are represented by bornite-chalcopyrite ores. Copper, gold, silver and K_2O have common or contiguous zonation centres. Molybdenum-enriched sites are situated within the limits of the 0.4% copper contour. Copper, gold, silver and K_2O zonation centres are near the Foothill fault.

The generalized model of ore and metasomatic zonation of similar distribution has been also described for the Christmas deposit by Koski and Cook (1982), for the Bingham deposit by Einaudi (1982) and for some others. According to Einaudi's scheme the deep core section of copper is represented by chalcopyrite-bornite and is situated under the zone of intensive potassium feldspathization. Approaching the peripheral parts, the ores change to chalcopyrite-pyrite and then to pyrite-chalcopyrite.

According to John (1978), the bornite-chalcopyrite zone of the Bingham deposit appears to be the central part of the high-grade copper ore body containing equal amounts of bornite and pyrite. Chalcopyrite occurs together with bornite and substitutes for it. Jambor and Beaulne (1977) reported the predominance of bornite over other sulphides in the central part of the high-grade core zone, e.g. for the Valley Copper deposit in Canada.

Potassium metasomatism (potassium feldspar, biotite and anhydrite) and development of the bornite-chalcopyrite-digenite association are connected with the early-, late-magmatic stage in the Chuquicamata deposit in Chile (Soto 1981). Rock transformation and appearance of the pyrite-bornite and pyrite-bornite-chalcopyrite associations took place during the second main phase of the hydrothermal stage. The central part of the El Abra deposit in Chile is characterized according to Ambrus (1977), by a high content of copper and a low content of sulphur in biotite-potassium feldspar metasomatites produced by diorites. The high copper content is provided here by bornite-chalcocite-chalcopyrite and bornite-chalcopyrite associations.

The review presented here shows that the established morphostructural peculiarities of ore-bearing intrusives and stockwork bodies, and the zonation distribution of ore and metasomatic complexes, appear to be common features for a number of porphyry copper deposits. The detailed geological investigation of other deposits (their morphostructural constructions in particular) will make it possible to determine similar structural characteristics for high-grade core zones of porphyry copper deposits. The foregoing data on the structure of ore-generating porphyry intrusives, copper-ore stockwork bodies and their dimensional position, ore zonation, and metasomatism may be recommended for application to certain stages of geological prospecting and exploration and probably also to exploitation.

References

Ambrus G (1977) Geology of the El Abra porphyry copper deposit. Chile. Econ Geol 72:1062 – 1085

Cornelius KD (1969) The Mount Morgan mine – a massive gold-copper pyritic replacement deposit. Econ Geol 64:885 – 902

Einaudi MT (1982) Description of skarns associated with porphyry copper plutons. In: Advances in geology of the porphyry copper deposits, Southwestern North America. Univ Ariz Press, Tucson, Arizona, USA, pp 139 – 183

Golovanov IM (1981) Geological-genetic model of concentric zonation of a porphyry copper deposit. Genetic models of endogenic ore formations. Abstracts of reports of All-Union conference. Novosibirsk, p 68 (in Russian)

Golovanov IM (1982) Conditions of ore localization in concentric zonal porphyry copper systems. Genesis of ore deposits. Rep IAGOD VIth Symp (Abstr), Tbilisi, pp 202 – 203 (in Russian)

Jambor JL, Beaulne JM (1977) Exploration possibilities for porphyry deposits in the central part of Highland Valley, British Columbia. Pap Geol Surv Can 1:101 – 106

John EC (1978) Mineral zones of the Utah copper orebody. Econ Geol 73:1250 – 1259

Koski RA, Cook DS (1982) Geology of the Christmas porphyry copper deposit, Gilta County, Arizona. In: Advances in geology of the porphyry copper deposits, Southwestern North America. Univ Ariz Press, Tucson, Arizona, USA, pp 353 – 374

Krivtzov AI (1983) Geological basis for porphyry copper deposit forecasting and exploration. Nedra, Moscow, pp 256 (in Russian)

Krivtzov AI, Yudin IM (1976) Hypogenic zonation of porphyry copper formation deposits. Geological methods of search and exploration of ore deposits. Mem All-Union Inst Miner Res Econ, p 44 (in Russian)

Langton JM, Williams SA (1982) Structural, petrological and mineralogical controls for the Dos Pobres orebody: Lone Star Mining District, Graham County, Arizona. In: Advances in geology of porphyry copper deposits, Southwestern North America. Univ Ariz Press, Tucson, Arizona, USA, pp 335 – 352

Melnikova LV, Poletayev AI (1981) Mineralogical zonation of porphyry copper deposits and the possibility of using it for search and exploration (after the example of Aktogai). Prospects of copper ore provinces in Kazakhstan. Alma-Ata, pp 9 – 17 (in Russian)

Nikolayeva EI (1980) Paragenesis of gold and silver in porphyry copper deposits of the Almalyk ore field. Notes of Uzbek Branch. Mem All-Union Miner Soc 33:16 – 23 (in Russian)

Pavlova IG (1978) Porphyry copper deposits. Nedra, Leningrad, p 273 (in Russian)

Pervago VA (1978) Geology and economics of porphyry copper deposits. Nedra, Moscow, p 170 (in Russian)

Popov VS (1977) Geology and genesis of porphyry copper and porphyry molybdenum deposits. Nauka, Moscow, p 204 (in Russian)

Shayakubov TSh, Golovanov IM, Rakhubenkov AT (1983) The Dalnee porphyry copper deposit. Nedra, Moscow, p 110 (in Russian)

Soto PH (1981) Alteración y mineralización primaria en Chuquicamata (Chile). Tecniterrae 8, 43:27 – 44

Elements Determining in the Geological-Structural Model of the Erdenetuin-Obo Copper-Molybdenum Ore Field, Mongolia

M. ZHAMSRAN[1], V. I. SOTNIKOV[2], A. P. BERZINA[2], D. GARAMZHAV[1], and YU. A. SARYANOV[1]

Abstract

Copper-molybdenum mineralization in Mongolia is localized within the limits of regional structures referred to as the superimposed metallogenic belts. The latter is spatially controlled by the late Paleozoic (in Southern Mongolia) and late Paleozoic-early Mesozoic (in Northern Mongolia) volcanic belts, elongate in a sublatitudinal trend. The through-going transverse fault system of the northwest trend has played an important role in forming ore-concentrating structures. Intersection of transverse faults and ones that trend along the volcanic belt account for the development of a block fabric and highly permeable zones that are favourable for the existence of long-lived magmatic centers that host the ore fields.

The Erdenetuin-Obo ore field is located in the Erdenet-Tsagansuburgin ore-concentrating structure that stretches throughout Mongolia and on into the USSR (in the north) and into China (in the south). Distribution and localization of mineralization within the ore field are determined by three fault systems: northwest, north-northwest (submeridional) and west-southwest (sublatitudinal). Of these fault systems, the first one was of the most significance. It is considered to have provided channels for fluid flow concentration at a level necessary to produce large volumes of ores.

Nowadays, Mongolia is among the leading copper-molybdenum provinces of the world due to the discovery of the Erdenetuin-Obo (Khasin et al. 1977) and Tsagan-Suburga (Sotnikov et al. 1980) deposits as well as of numerous copper-molybdenum prospects in different regions of the country. At the Erdenetuin-Obo deposit, the most efficient mining and concentrating plant in Asia is functioning.

Copper-molybdenum mineralization in Mongolia is localized within the limits of regional structures of the superimposed metallogenic belt type: North Mongolian, Central Mongolian and South Mongolian. Their formation is related to the development of extrageosynclinal tectonic-magmatic processes, and they are spatially controlled by the late Paleozoic and late Paleozoic-early Mesozoic volcanic belts (of the same names, respectively), elongate in sublatitudinal trend

[1] Institute of Geology and Mining Industry of the MN, Ulan Bator, MNR

[2] Institute of Geology and Geophysics, Siberian Branch of the USSR Academy of Science, Novosibirsk, 630090, USSR

Geology and Metallogeny of Copper Deposits
(ed. by G. H. Friedrich et al.)

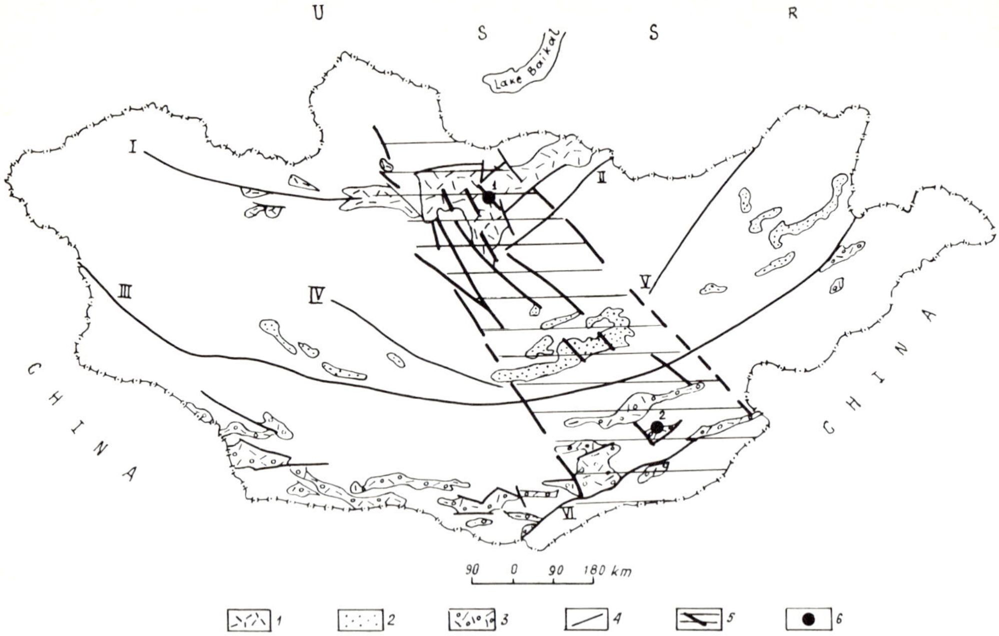

Fig. 1. Late Paleozoic-early Mesozoic volcanic belts in Mongolia and the Erdenet-Tsagansuburgin ore-concentrating structure. The volcanic belts (*1*–*3*): *1* North Mongolian ($PZ_3 - MZ_1$); *2* Central Mongolian ($PZ_3 - MZ_1$); *3* South Mongolian (PZ_3); *4* the fault systems (*I*–*VI*): *I* Khangai; *II* Mongolo-Okhotsk; *III* Main Mongolian; *IV* Bayankhongor; *V* Kerulen; *VI* Gobi-Tien-Shan; *5* the Erdenet-Tsagansuburgin ore-concentrating structure; *1, 2* the copper-molybdenum deposits: Erdenetuin-Obo (*1*) and Tsagan-Suburga (*2*)

(Fig. 1). On the whole, volcanic belts inherit the orientation of the Hercynides on which they are superimposed. Similar orientation is observed in structures with a Caledonian or a Proterozoic basement. All these considerations suggest the connection of volcanic belts with deep long-lived faults, the latter periodically becoming active.

The North Mongolian volcanic belt is traced by a system of extensive faults of east-north-east trend on the east and of west-north-west ones on the west (Fig. 1). The fault system on the west of the North Mongolian belt is designated as the Khangai lineament, and the one on the east represents the western flank of the Mongolo-Okhotsk lineament extending from the USSR. Late Paleozoic volcanic rocks of the South Mongolian belt are concentrated in two linear zones, one of which is controlled by the Main Mongolian lineament, and the other by the Gobi-Tien-Shan system of faults (Yarmolyuk and Kovalenko 1981). The Central Mongolian volcanic belt is limited to the north by the Bayan-Khongor and Kerulen faults, and to the south by the Main Mongolian lineament. According to their importance, the magma-controlling faults are deep and regional ones that governed the development of orogenic complexes (Geology of the Mongolian People's Republic 1973). These faults, where they reached the zones of deep-seated magma generation, where the principal magma-delivering structures. Volcanic belts can be regarded as common regional ore-controlling structures.

Through-going transverse structures of nortwest (in Eastern Mongolia) and northeast (in Western Mongolia) trends played an important part in forming ore-concentrating structures. The greatest one is the Erdenet-Tsagansuburgin ore-concentrating structure (Fig. 1) that is traced across all Mongolia from north to south (from the USSR frontiers up to the Chinese ones) and that crossses geosynclinal structures of different ages and the Proterozoic basement. This ore-concentrating structure includes the majority of ore prospects of the North Mongolian, Central Mongolian and South Mongolian belts; the Erdenetuin-Obo and Tsagan-Suburga deposits occur along them. The structure is bounded by deep long-lived zones of basement faults. These are combination of faults, traced over tens of kilometres, zones of crumpling, crushing and high jointing as well as individual, usually echelonlike fractures tens to hundreds of metres long. In many cases these tectonic dislocations are cryptic, revealed only during large-scale mapping and structural investigations of deposits and prospects.

The combination of both transverse faults, involved in the system of through-going ore-concentrating structures, and of others trending along the volcanic belts, gave rise to the development of block structures as well as of highly permeable zones. These regions are characterized by intense occurrence of late Paleozoic (late Paleozoic-early Mesozoic) volcanoplutonic association. The most intense, long-developing, and diverse magmatism occurred within the limits of the ore fields. The latter spatially fix centres of continuous endogene activity. Evidently, an ore field can be regarded as a taxonomic unit, most completely summing up all features of evolution of an ore-magmatic process in a concrete geological situation. As a typical example of such an ore field, picting the conditions of occurrence of copper-molybdenum mineralization in geological structures of Mongolia, one can study the Erdenetuin-Obo ore field.

Copper-bearing mineralization of the Erdenetuin-Obo ore field is clearly controlled by the zone of drawn-together northwest faults, arranged in an echelon pattern from the Tola river up to the lower and middle reaches of the Eigiin-gol river. Localized in the Erdenet-Tsagansuburgin ore-concentrating structure, this zone is known as the Erdenet zone (Fig. 2) and not only determines the arrangement of ore-metasomatic rocks, but also influences both localization and morphology of intrusive massifs and dikes, different in age and contained within the framework of Lower-Upper Permian volcanic rocks.

The Erdenet zone is traced along the contact of the regional blocks (Fig. 2), differing in composition and interval of movement (Mossakovsky and Tomurtogoo 1976). It has been suggested (Gretskaya and Mossakovsky 1969, Mossakovsky and Tomurtogoo 1976) that the block structure was best developed in the Middle Jurassic period, but its elements began to form earlier. Being formed initially in the Caledonides; the faults continued to develop during the accumulation of the Permian volcanogenic series and were regenerated at the end of the Permian – at the beginning of the Triassic period. The block structure is mainly delineated by a combination of faults of three directions: northwest, northeast and north-northeast. Of these, the fault and highly fractured NW-trending system is most clearly defined.

To achieve efficiency in exploration the most promising areas on the whole are those with a combination of the above fault systems showing maximum activ-

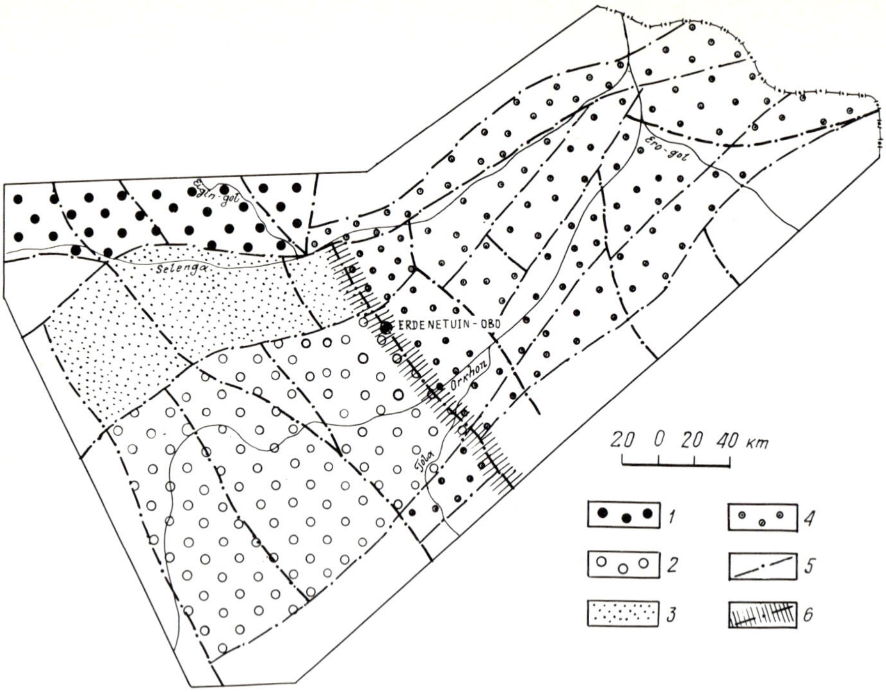

Fig. 2. Block pattern of the northern part of the Erdenet-Tsagansuburgin structure. The blocks (*1–4*): *1* Northern; *2* Central; *3* Southern; *4* Eastern; *5* faults; *6* the Erdenet ore-bearing zone

ity and highest permeability. However, those structures most responsible for the total character of distribution of arrangement of copper-molybdenum mineralization within the framework of the Erdenetuin-Obo ore field remain the northwest dislocations (Fig. 3). The largest of them are represented by faults and uplifts with an amplitude up to several hundreds of metres. This most clearly manifested fault system appeared to be the main ore-concentrating structure providing maximum drainage for ore-bearing fluids and highest metal concentrations in relatively local areas.

The Erdenetuin-Obo ore field is subdivided into three areas: the Erdenetuin-Obo deposit, the Central area and the Southern area (Fig. 3). In the deposit (Fig. 4), magmatic rocks of the Selenginsk complex mainly occur. They are represented by granodiorite, granite and gabbro-diorite. Granosyenitelike rocks in which generation metasomatic processes played an important role are widespread. In pendants of the granitoid massif, relicts of Riphean-Cambrian gneisses, amphibolites and crystalline schists, often considerably K-feldspathized survive. It should be emphasized that the occurrence of potassium metasomatites is the most characteristic feature of the rocks exposed in the ore-bearing district, especially close to the ore manifestations and in the highly permeable zones.

Within the deposit area, various shapes of porphyry bodies of the ore-bread-ing magmatic complex are widespread. These consist of granodiorite and plagio-granite porphyries, quartz-diorite porphyries, and more rarely, of granite and

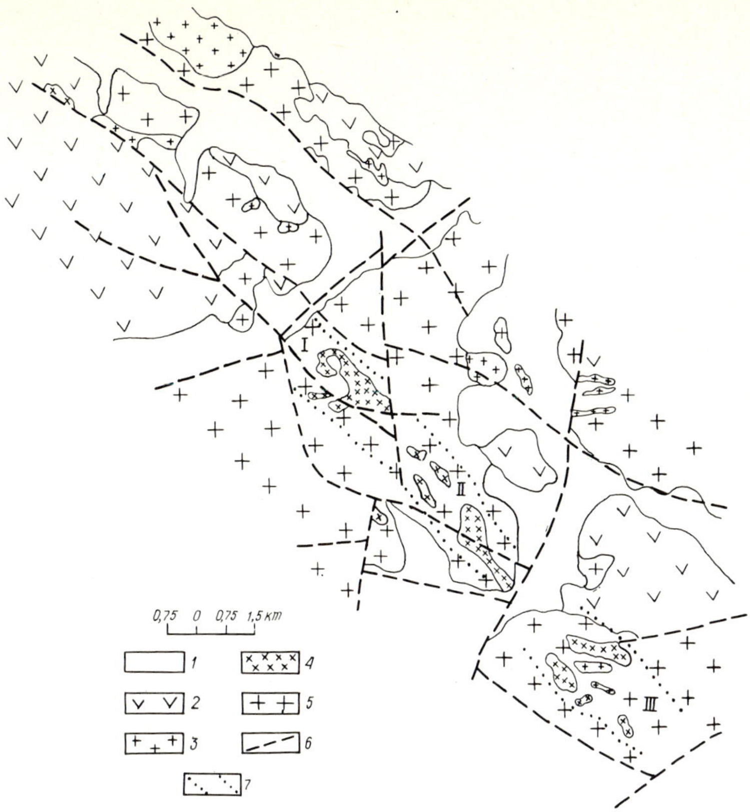

Fig. 3. Geological scheme of the Erdenetuin ore field. *1* Quaternary sediments; *2* volcanic rocks, P_{1-2}; *3* leucocratic granite, T_1; *4* porphyry (the ore-bearing magmatic complex), P_2-T_1; *5* granitoid (the Selenginsk magmatic complex), P; *6* faults; *7* the mineralization areas (*I–III*) *I* Erdenetuin-Obo deposit; *II* Central area; *III* Southern area

granosyenite porphyries with transitions to dacite and rhyolite-dacite in intrusive endocontacts. Pre-ore, intra-ore and post-ore porphyries are recognized. On the whole, the porphyries of different intrusive phases are similar in composition. Pre-ore porphyries are mainly represented by northwest-trending stocks. Among intra-and post-ore porphyry bodies, dikes of the north-northwest strike predominate. Stocklike porphyry bodies are continuously accompanied by near-contact explosion breccias composed of granitoid fragments. The latter are cemented by finely-crushed biotitized and K-feldspathized material derived from the same granitoids.

In the central portion of the deposit, a porphyry body rather hydrothermally altered was mapped. Being generally stretched out in a northwest direction which is consistent with the total orientation of the Erdenet fault zone, the stocklike body about 2100 m long and 900 to 1000 m wide is of complex shape with numerous embayments into the host rocks as apophyses and satellites about 250 to 500 m long and 10 to 100 m wide. The stock dimensions decrease with depth,

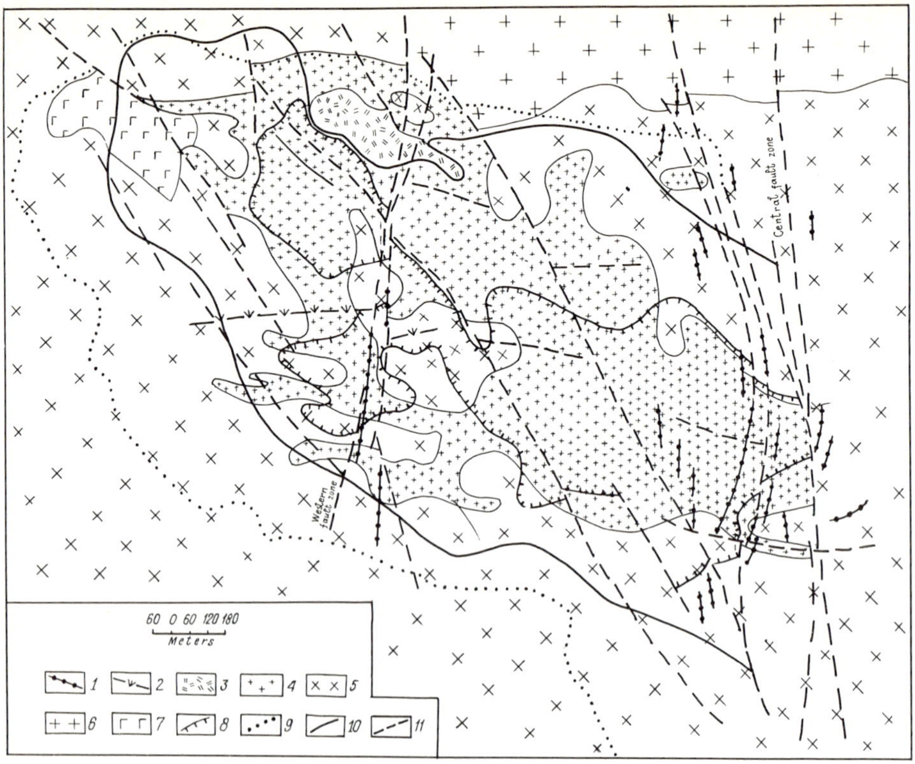

Fig. 4. Geological scheme of the Erdenetuin-Obo deposit. *1, 2* Dikes, T_2-J_1: *1* andesitic and trachyandesitic porphyry; *2* syenodiorite; *3* stock of felsite porphyry, T_2-J_1; *4* porphyry. The Selenginsk magmatic complex (*5–7*): *5* granodiorite; *6* granite; *7* gabbro-diorite; *8, 9* Outline of hydrothermal alteration: *8* strong silicification; *9* sericitization and silification; *10* outline of ore body; *11* faults

its morphology becomes somewhat simpler changing to a piplelike body with a northeast pitch. In the east the stock is cut off by the Central fault.

Dikes and small isometric bodies of fine-grained leucocratic granites, microgranites and granite-aplites are less developed in the deposit. They are widespread in the southwestern portion of the deposit and continue into the western part of the Central area (Fig. 3). Leukocratic rocks belonging to the post-ore phase are hydrothermally altered (mainly poor K-feldspathic and bleached) only at certain places, mainly in the zones of high permeability and early mineralization.

In the deposit area, rocks of the Late Triassic-Early Jurassic age are represented by a feldsite porphyry stock with a northwest orientation as well as by dikes of andesitic and trachyandesitic porphyry that fill northwest and especially north-northwest to north-northeast tectonic zones. Syenodiorite porphyry dikes predominantly have a west-southwest strike.

The structural pattern of the deposit is determined by a combination of the three main systems of faults: generally northwest, west-southwest and north-northwest. In general, the most important is the northwestern system that is a component of the large, long-lived Erdenet fault zone – the main structural element of the ore field. This fault system accounts for the localization of the stock- and dikelike bodies of porphyries as well as for the generation of the zone of high jointing and permeability to which ore-metasomatic rocks of the deposit are confined.

The north-northwest fault system is represented by the two large zones, i.e. the Central and Western ones. The former is developed most intensely, being represented by a series of large sutures and highly jointed zones marked by the andesite porphyry dikes. This seems to have prevented ore-metasomatic processes from spreading to the east. The latter, together with the registered north-northwest orientation of quartz-sulphide veinlets, testify to the development of faults of this direction during the pre-ore period also. Most intensive renewal of the Central zone evidently took place in the Late Triassic when, in its eastern part, numerous dikes of andesite porphyry were involved in a large north-northwest dike belt stretching far beyond the deposit. Along most joints of the Central zone, post-ore movement with occurrence of mylonites and clay gouge took place. The Western fault zone, which is similar to the Central one, but differing from it by the lesser intensity of the tectonic processes, together with northwest faults, controlled the localization of the Late Triassic felsite neck in its northern part.

A west-southwest fault system is traced by apophyses of the porphyry stock, a number of quartz-sulphide veins as well as post-ore dikes of syenodiorite porphyry. This system is of minor significance and has influenced the formation of the deposit structure much less than the two fault system described above.

Vein-disseminated ore mineralization is confined to the zone of hydrothermally altered rocks elongated northwest for over 2.8 km, its width being 0.3 to 1.3 km. Spatially, it is related to the porphyry stock that penetrates into the granitoid stocks of the Selenginsk complex for over 300 – 500 m. Ore mineralization is observed at the depth of 1000 m without signs of pinching out. Towards the periphery of the ore body, the succession of vein-disseminated mineralization is followed for more massive veins. Quartz-sulphide veinlets have strikes that follow the three directions of jointing (Khasin et al. 1977): northwest (305 – 330°), north-northwest to north-northeast (355 – 20°) and west-southwest (270°). Veins of the first two directions dominate. The veinlets range in width from parts of a mm to 2 – 3 cm and usually are traced along the same trend over tens of centimetres and their direction often changes. The veins most persistent in trend are characteristic of the periphery of the deposit. In the central parts of the ore body, especially in the near-contact zone of the porphyry stock, high-grade breccia ores, irregular in morphology, occur. Spatial zones of pre-ore explosion breccias (including those connected with the emplacement of the porphyry stock) as well as areas of high jointing with maximum occurrence of joints of the three above-mentioned directions are filled by breccia ores. As a whole, the mineralized zone can be considered as a zone of intense pre-ore brecciation and jointing favourable for development of ore-metasomatic processes and concentrated ore deposition.

The intensity of hydrothermal alteration of the host rocks decreases from the centre to the periphery of the deposit. An inner zone of strong silicification, an intermediate one of sericitization and silicification and an outer one of chloritization have been distinguished. In the zone of chloritized rocks, early K-feldspar alteration is observed, whereas in other metasomatic zones the K-feldspars are destroyed by later, superimposed metasomatic processes. Relicts of K-feldspar granitoids create a discontinuous fringe around the zone of quartz-sericite metasomatites. Hydrothermally altered rocks are characterized by different ore-bearing features. The copper and molybdenum contents in different types of metasomatites are respectively: strongly silicified rocks (not more than 0.11% Cu and 0.009% Mo), moderately silicified and sericitic ones (0.4 – 0.8% Cu and 0.022 – 0.026% Mo), slightly silicified and sericitic ones (0.2 – 0.3% Cu and about 0.01% Mo). In fresh and poorly chloritized rocks, copper and molybdenum are present in negligible amounts. The ore body structure is complicated by post-ore dikes of syenodiorite, andesite and trachyandesite porphyry.

According to Krivtsov et al. (1981), the Erdenetuin-Obo deposit by relationship between the ore body and the porphyry intrusion, are assigned to the group of disconformable deposits the formation of which is supposed to have taken place in a subsurface environment. The ore-magmatic system is considered to have been open and to have been developed at low pressure and temperature as well as in a changeable deformation pattern. Here, ore deposition corresponds to the period when the porphyry stock was practically crystallized. The authors believe that in the present case the ore-generating portion of the ore-magmatic system was deeper than the level of the intrusion and crystallization of a porphyry stock. The latter give rise to the possibility of mineralization occurring at several levels. Under these conditions the ore-controlling role of tectonic dislocations is clearly shown to be a part of the regional structures.

Structural factors in the distribution and localization of ore mineralization within the framework of the Erdenetuin-Obo deposit are also determinants for most prospects in the northern part of the Erdenet-Tzagansuburgin structure. These prospects show in miniature the structural conditions of ore formation that are characteristic for certain parts of the deposit discussed. The degree of development of ore mineralization very much depends on the intensity of occurrence of pre-ore-concentrating and ore-controlling structures. In general, the combination of transverse northwest (north-northwest at a later stage) dislocations and west-southwest faults consistent with the general trend of the volcanic belt are the decisive factors for the formation of ore concentrations for all prospects in the Erdenetuin-Obo ore field as well as in other areas that make up the system of the Erdenet-Tzagansuburgin ore-concentrating structure. It should be noted, in particular, the channeling role of transverse ore-concentrating structures in providing loci for the upward flowing ore fluid allowing a level favourable for the formation of large volumes of ores.

Acknowledgments. The manuscript was reviewed by Drs. Richard H. Sillitoe and K. F. Clark whose comments have greatly improved the work and whose contributions are gratefully acknowledged.

References

Geology of the Mongolian People's Republic, vol II (1973) In: Khasin RA, Borsakovskii YuA, Zonenshain LP (eds) Magmatism, metamorphism, tectonics. Nedra, Moscow, 750 pp (in Russian)

Gretskaya TA, Mossakovsky AA (1969) Some questions of the stratigraphy and structural relation between the Carboniferous and the Triassic in northern Mongolia. Akad Nauk SSSR Izv Ser Geol 2:56 – 71 (in Russian)

Khasin RA, Marinov NA, Khurts Ch, Yakimov LI (1977) The Erdenetuin-Obo copper-molybdenum deposit in northern Mongolia. Geol Rudn Mestorozhd 6:3 – 15 (in Russian)

Krivtsov AI, Migachev IF, Shishakov VB (1981) Morphogenetic type designs of ore bodies at copper porphyry deposits. Sov Geol 10:28 – 41 (in Russian)

Mossakovsky AA, Tomurtagoo O (1976) The Upper Paleozoic of Mongolia. Sovmestnaya Sovetsko-Mongol'skaya Geologicheskaya Ekspeditziya, Trudy, vol 15. Nauka, Moscow, 126 pp (in Russian)

Sotnikov VI, Berzina AP, Zhamsran M, Myagmar L (1980) The Tsagan-Suburga molybdenum-copper deposit (MNP). Geol Rudn Mestorozhd 3:34 – 46 (in Russian)

Yarmolyuk VV, Kovalenko VI (1981) The role of largest faults in the location of continental volcanic rocks of MNR. Akad Nauk SSSR Doklady 256, 2:436 – 440 (in Russian)

Genetic Aspects of the Recsk Mineralized Complex, Hungary

Cs. Baksa[1]

Abstract

The following conclusions can be drawn concerning the genesis of the Recsk deposit:

1. The mineralization belongs to the Paleogene volcanic arc along the Balaton-Darno line.
2. The main phases of mineralization, culminating with porphyry copper formation, are products of hydrothermal events related to a diorite porphyry intrusion (a_3 stage).
3. The two younger (post-intrusion) volcanic cycles have resulted in late-stage near-surface mineralization, which is considered to be a product of remobilization.

In comparison with other porphyry copper mineralization it may be suggested that the Recsk deposit has an affinity with the island-arc-type magmatism. It shows similarities with the diorite model in the zoning of the intrusive rocks and the ore distribution. Morphogenetically it may be termed a "conformable" deposit. Its classification was simplified by the completeness of the ore depositional sequence, although the sequence of superimposed ore-forming stages created difficulties for interpretation.

Introduction

More than 130 yrs have elapsed since exploration and mining commenced in the Recsk ore district. Numerous studies and papers dealing with several genetic aspects of the Recsk and Parádfürdö ore occurrences have been published in this period (Pantó 1951) A new chapter in the history of Recsk exploration opened in 1958 with a deep drilling program, the results of which were summarized in 1975 (Baksa 1975a, b, Cseh Németh 1975, Csillag 1975, Csongrádi 1975) (Fig. 1).

The common factor of genetic interpretation in these papers was the acceptance of a relationship between the igneous activity and the repeatedly

[1] National Ore and Mineral Mining Co., 1062 Budapest, Hungary

Geology and Metallogeny of Copper Deposits
(ed. by G. H. Friedrich et al.)

activated Darnó megastructural zone (Zelenka 1973, 1974). This ideas was reflected by the paleogeographic reconstructions, suggesting a genetic link between the different types of mineralization and the Eocene Priabonian tectonic evolution along this zone. The Darnó zone merges with the Balaton line southeastward. Several other volcanic areas are known along this line, among which is the NE part of the Velence Mts., where the igneous complex shows age relations and chemistry similar to Recsk. All these occurrences are situated on the northern boundary of the Igal-Bükk eugeosyncline (Wein 1969, 1978), in the Central Mountains belt and include different types of mineralization.

An updated synthesis of this volcanic belt was given by Csillag et al. (1980), who postulated the uniform island-arc nature of these Paleogene volcanic occurrences. This volcanic arc and other earlier volcanic zones in the Carpathian system (Banat-Timok area in Rumania and Yugoslavia, Vardar zone in Yugoslavia) can be interpreted uniformly, based on the changes in the convergence style of the Africa and European plates and the several microplates between them (Hadži et al. 1977). This approach also explains the similiar mineralization and petrologic features of these deposits, and the relative age differences.

Geological Setting of the Recsk Deposit

Recently several new aspects concerning the relationship between the tectonic evolution of the Darnó zone and the Recsk ore deposit were revealed (Balla et al. 1980; Zelenka et al. 1983). Field observations in the SW part of the Bükk Mts., in the Darnó area and evaluation of drill core from Recsk have called into question the earlier theories concerning the links between the Central Mountain belt and the Recsk basement. The Mesozoic rocks at Recsk were in a similar manner overturned and folded, and exhibit similar lithological properties to the Mesozoic on the SE side of the Darnó zone. Thus Zelenka and co-workers (1983) have concluded that the Darnó zone was not a paleogeographic boundary in the Mesozoic period, as was previously stated.

The age of the igneous activity can be quite accurately identified from both stratigraphic and radiometric data. The limestones and marls beneath and on top of the volcanic sequence belong to the Nummulites fabianii horizon, of the Priabonian (Upper Eocene) stage (determination by Báldi). The K/Ar age of relatively fresh andesite from the Recsk Andesite Formation falls in the 34.9 – 35.7 (±2.5 – 4.9) m.y. range (determinations by Balogh).

The first volcanic products were the submarine effusive (+ to a lesser extent explosive) products assigned to the a_2-type andesites (this is a biotite-hornblende andesite of the first stratovolcanic stage in local, informal mine terminology), which accumulated on the Triassic basement, or on earlier Priabonian sediments (Fig. 2). The a_2 andesites are barren, no post-volcanic hydrothermal activity is linked to this stage. The ore indications in these volcanics are later products. The effusive centre of the a_2 andesites is not known, only the extent of the volcanics can be contoured. This coincides with the position of the diorite porphyry

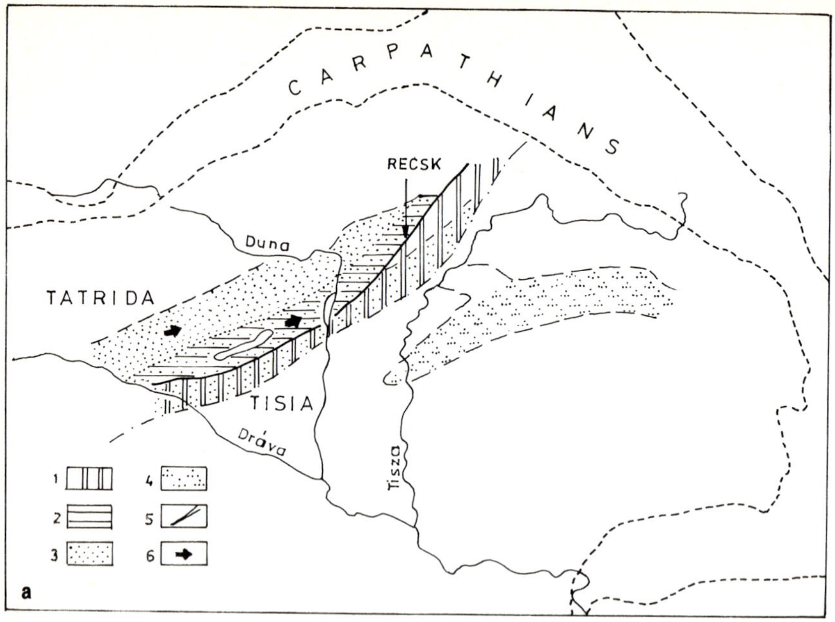
C A R P A T H I A N S
RECSK
Duna
TATRIDA
TISIA
Dráva
Tisza
1
2
3
4
5
6
a

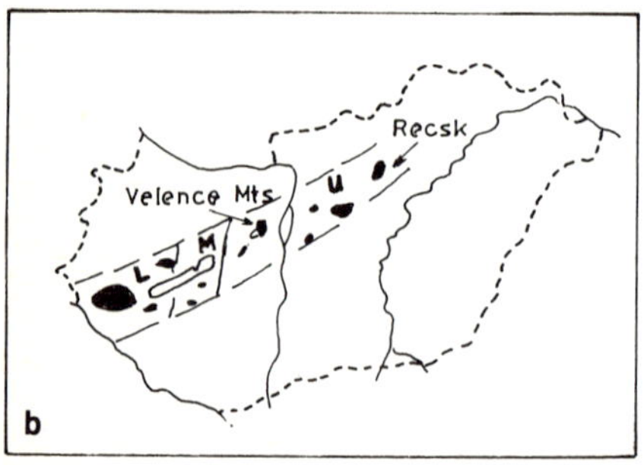
Recsk
Velence Mts
L
M
U
b

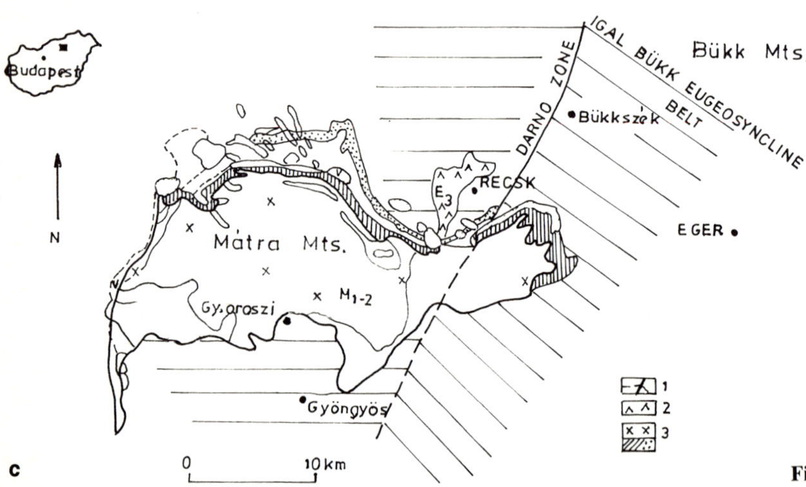
Budapest
N
IGAL BÜKK EUGEOSYNCLINE BELT
Bükk Mts.
DARNO ZONE
Bükkszék
RECSK
E_3
EGER
Mátra Mts.
M_{1-2}
Gy.oroszi
Gyöngyös
1
2
3
0
10 km
c

Fig. 1a – c

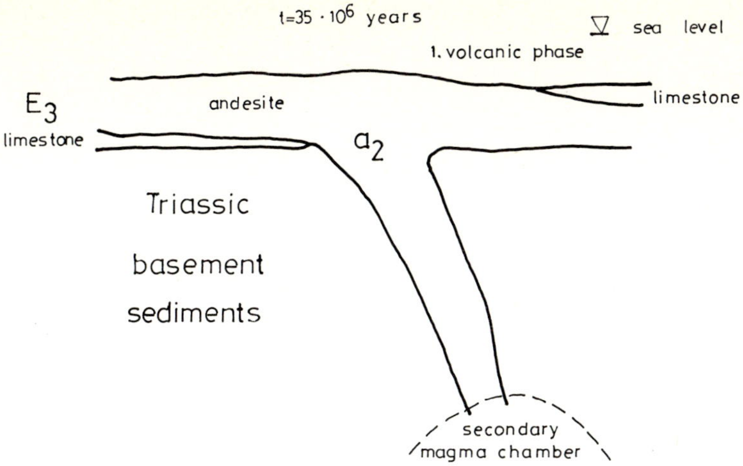

Fig. 2. The first phase of the Priabonian (Upper Eocene) biotite-amphibole andesite volcanism, intruding and accumulating on the Triassic basement

intrusion, which comprises the second magmatic phase (Fig. 3). It can thus be that the a_3-type diorite (this is local, informal mine terminology also) porphyry stock was emplaced by the partial or total destruction of a secondary magma chamber and accompanying vent for the a_2 effusives, and by the assimilation of an equal volume of Triassic host rocks. The a_3 intrusive body is 3 km long and 600 – 800 m wide, with an elongate shape in a N – S direction. It follows the morphology of the uplifted basement horst. The a_3 intrusive body is lithologically complex.

The chemical and lithological properties of the a_3 intrusives are similar to the diorite porphyries of eugeosynclinal belts (trench). This is in agreement with the setting described in the island-arc reconstruction of Csillag et al. (1980). Thus these characteristics (chemical and lithological properties) and mineralogical parameters (alteration and zonation) link the Recsk occurrence to the diorite model of Hollister (1978).

The Triassic wall rocks were both mechanically and chemically affected by a_3 intrusive emplacement. Due to the destruction by later processes, only scarce data are available for the alteration event synchronous with a_3 emplacement. Typical hornfels are not abundant. The movements within and adjacent to the intrusion, which created the fracturing necessary for the later flow of hydrothermal fluids, are related to the cooling history of the system below 800 °C.

Fig. 1 a – c. a The regional setting of the Paleogene volcanic arc in the Carpathian Basin. *1* Oceanic lithospheric remnants of the Igal Bükk eugeosyncline belt; *2* the Paleogene volcanic arc; *3* Eocene shelf and shoreline-facies marine sedimentation; *4* Cretaceous-Paleogene flysh zone; *5* Balaton-Darno line; *6* the direction of Eocene marine transgression and volcanic activity (after Csillag et al. 1980). **b** Space and time relationships of the Paleogene volcanism indicating the ages of the first volcanic products: L = Lower Eocene; M = Middle Eocene; U = Upper Eocene (after Csillag et al. 1980). **c** Tertiary volcanics of the Mátra Mts. *1* Basement formations; *2* Upper Eocene biotite-hornblende andesites; *3* Miocene volcanics

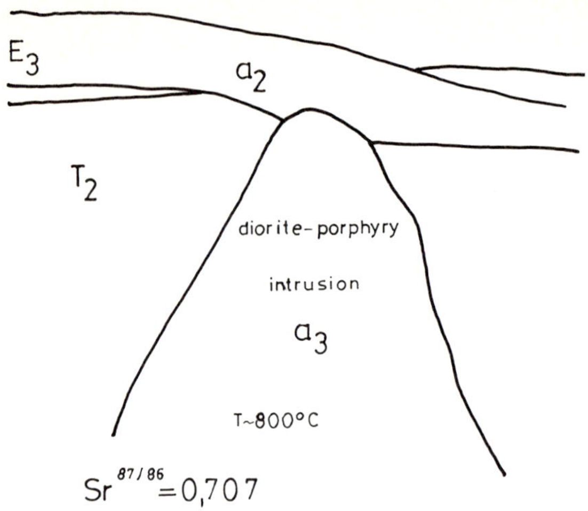

Fig. 3. Idealized section of the diorite porphyry intrusion that ascended from the secondary magma chamber during the second magmatic cycle

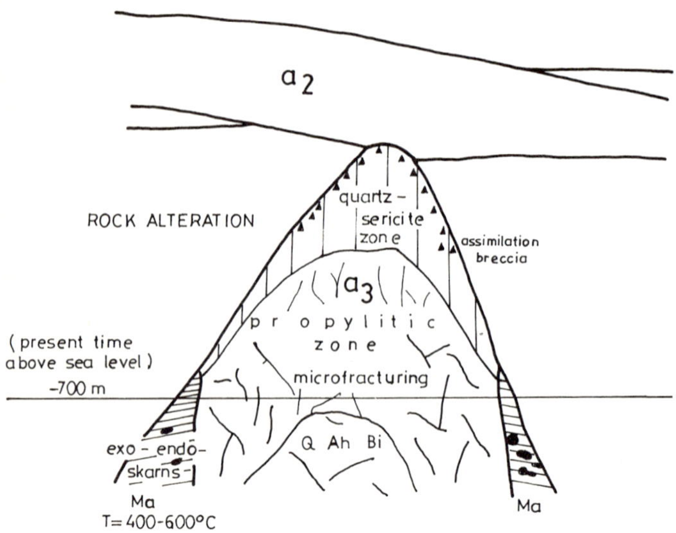

Fig. 4. Alteration zoning in the diorite porphyry intrusion. Q = quartz; Ah = anhydrite; Bi = biotite; Ma = magnetite

Localized stresses developed by the conversion of heat to mechanical energy, and produced zones of tension, microfracturing and brecciation. These zones provided favourable loci for alteration and mineralization.

The most typical alteration, which is zoned, is plotted on Fig. 4. The alteration was studied by Csillag (1975). A 150–200 m wide skarn zone has developed on the contacts of the intrusion below −700 m (above sea level), and consists of calc-silicate assemblages, which were formed during contact metasomatism. Both endo- and exoskarns can be distinguished. The zonal pattern, with certain generalizations, is similar to the diorite model of island-arcs (Hollister 1978).

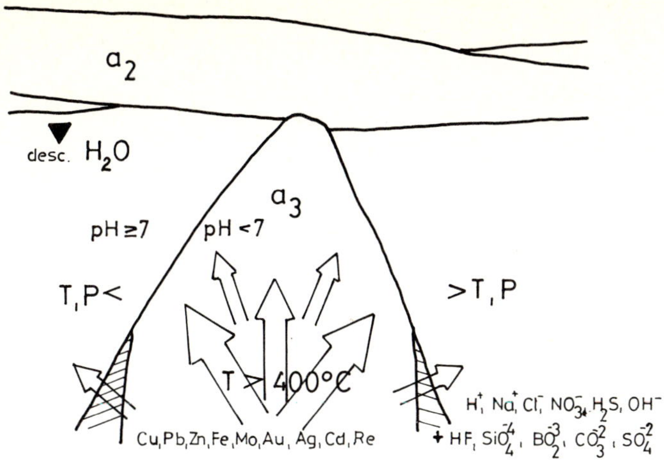

Fig. 5. Hypothetical compositions and flow paths of the fluids related to the cooling of the intrusion and the physicochemical parameters of the system

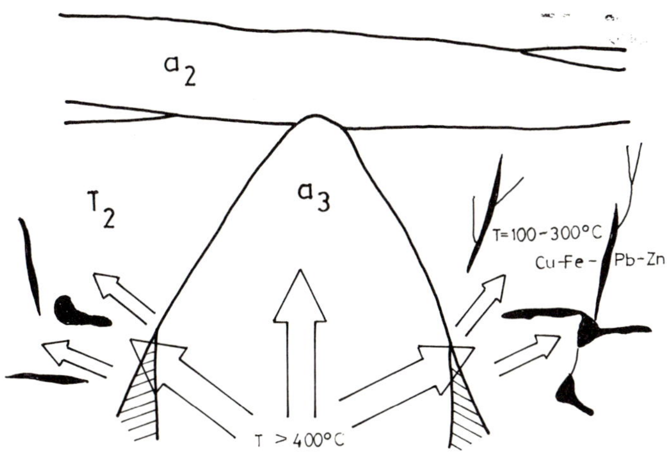

Fig. 6. Vein-type and irregular hydrothermal replacement polymetallic ore deposits in the Triassic wall rocks of the intrusion. The first stage of hydrothermal ore deposition

Ore Formation

Ore formation began with the emplacement of the a_3 diorite porphyry intrusion. The alteration processes, which were related to the cooling intrusion, first affected its peripheral zones and the adjacent wall rocks, and progressively invaded the inner zones of the intrusion. Later alteration phases were superimposed on the earlier ones, and a complex system was developed.

The space-time model of the Recsk deposit can be summarized as follows:

A. Pre-hydrothermal stage:

1. Contact-metasomatic magnetite deposits of the skarn zone (Fig. 4).

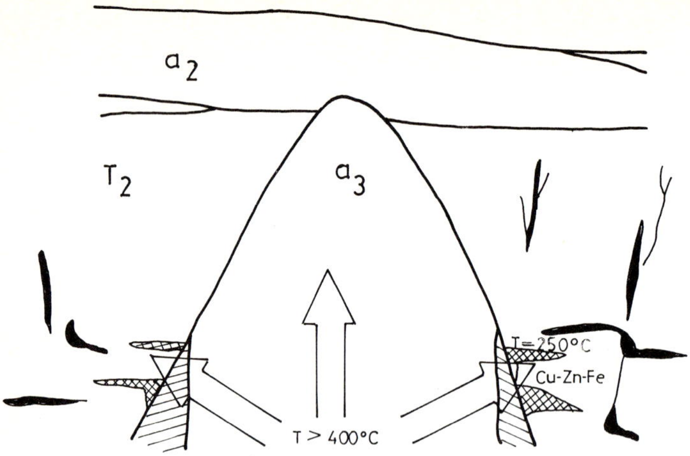

Fig. 7. Hydrothermal replacement polymetallic and pyrite deposits on the outer boundary of the skarn zone, along lithological contacts in Triassic wall rocks. The second stage of hydrothermal ore deposition

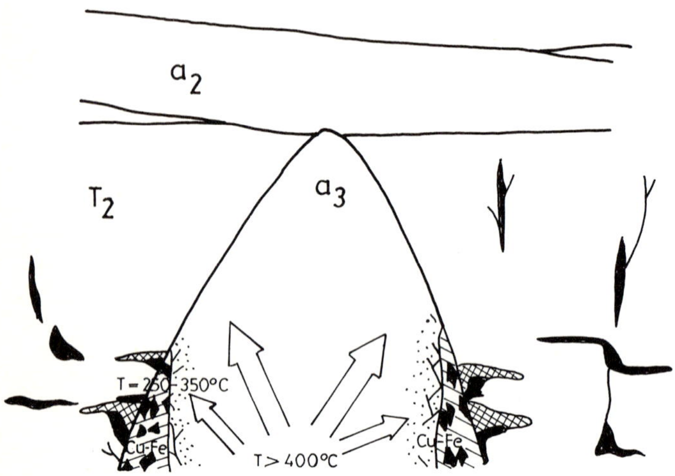

Fig. 8. Hydrothermal replacement chalcopyrite-pyrite deposits in veins, lenses and pods in the skarn zone, replacing calc-silicates. The third stage of hydrothermal ore deposition

B. Deposits related to hydrothermal activity of the diorite porphyry intrusive:

1. Hydrothermal vein-type and replacement Pb – Zn – Cu – Fe deposits (Fig. 6).
2. Hydrothermal stratabound replacement Cu – Zn – Fe and pyrite deposits (Fig. 7).
3. Hydrothermal replacement Cu – Fe deposits in the exo- and endoskarns of the skarn zone (Fig. 8).
4. Porphyry-type Cu – Fe – Mo mineralization of the intrusion (Fig. 9).

C. Deposits emplaced in the late stages. Due to remobilization effects of later volcanic phases:

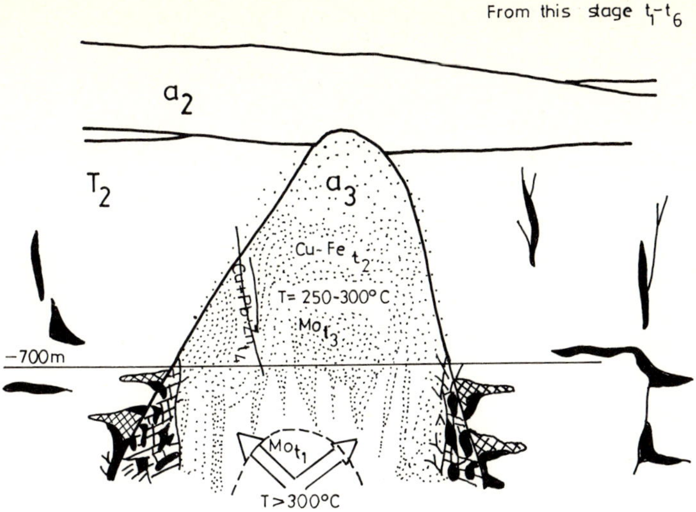

Fig. 9. Hydrothermal disseminated and vein-type Cu – Mo deposits (porphyry copper), formed in stockworks in the intrusion. The fourth stage of hydrothermal ore deposition. Also shown are the relative timing (t_1 t_2 t_3 t_4) of the formation of the main paragenetic types in the whole complex in this stage

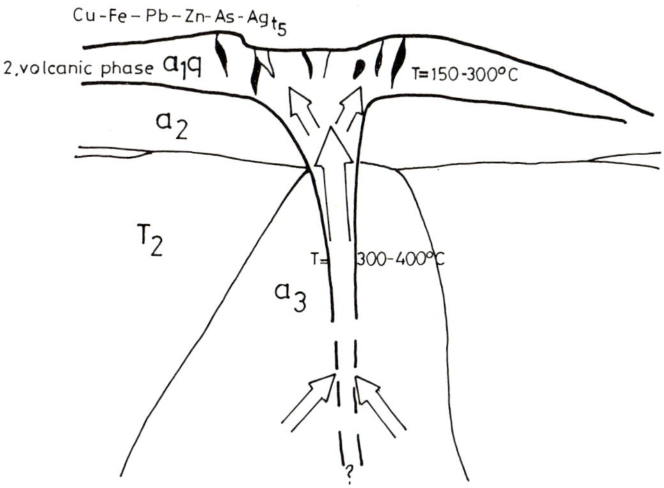

Fig. 10. Hydrothermal polymetallic ore deposits in silicified zones produced during the eruption phase of the a_1q type biotite-hornblende andesite as the second phase of stratovolcano activity. The fifth stage of hydrothermal ore formation

1. Hydrothermal metasomatic Cu – Fe – Pb – Zn – As – Ag ores related to silicified zones in the a_1q phase (local, informal mine terminology of the second stratovolcanic stage, Fig. 10).
2. Hydrothermal stockwork, massive sulphide and volcanic-sedimentary Cu – Fe – Au – Ag – Sb – As ores related to the a_1 phase (this is a hornblende-

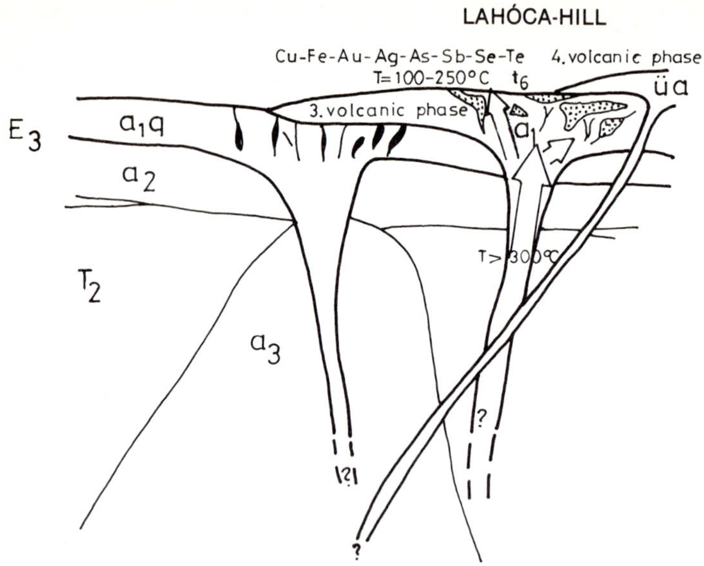

Fig. 11. The Cu – Fe – Au – Ag – As – Sb – Se – Te volcano-sediment deposits generated by the a_1-type biotite-hornblende andesite eruptions during the third phase of stratovolcano activity. The sixth stage of hydrothermal ore formation. The post-ore *üa*-type pyroxene-biotite-hornblende andesite is the fourth phase of stratovolcano activity

biotite andesite of the third stratovolcanic stage in the local, informal mine terminology (Fig. 11). The *üa* phase on Fig. 11 is the fourth stratovolcanic stage without mineralization).

These phases ($a_1 q$, a_1 and *üa*) are post-intrusion stratovolcanic stages.

This sequence of events indicates the temporal evolution and genetic development at Recsk: the different stages are depicted in Figs. 4 to 11. As stated above, the mineralization comprises a number of different, superimposed parageneses, which developed as functions of continuously changing heat fluxes of the igneous phases, which, in turn, caused circulation of fluids. In the hydrothermal stage related to a_3 intrusion, mineralization adavanced from the outside of the intrusion inward. In the subsequent volcanic phase, the formation of the mineralization was governed by local, though similar, *PT* and chemical factors.

The temperatures of the various mineralization stages (Figs. 4 to 11) were taken from the results of decrepitation studies (determinations by Csillag and Kovách).

The last and most important mineralization event related to the diorite porphyry intrusion was the porphyry copper stage. During this stage, the minerals filled stockworks in the diorite porphyry. The ore-grade contours and the alteration zone boundaries closely follow the retreating path of the thermal fronts, which generated the fluid circulation. The structural parameters of the rocks (i.e. fracturing in the intrusion, lithological boundaries, and pre-ore fissures in the sedimentary wall rocks) have played on important role in the shaping of these contours. The grade contours of the porphyry mineralization are not related to lithological parameters. Two types of the porphyry mineralization can be distinguished according to their shape. "Conformable" deposits are those in which grade contours are subparallel to the shape of the intrusion, whereas "disconformable" deposits are those in which the symmetry axis of the ore body does not fit with that of the intrusion. Cu and Mo distributions coincide in the disconformable type, but differ in the conformable type. The Recsk porphyry copper mineralization shows the following characteristics:

1. The grade contours are subparallel to the geometry of the intrusion;
2. Cu and Mo show a negative correlation;
3. The centre of the Mo reserves coincides only in part with the centre of the Cu reserves in the explored part of the intrusive body.

Comparing these data with the classification of Popov (1977), the Recsk deposit to belongs to the category of conformable deposits in a morphogenetic aspect. The formation of this geometry is promoted by the covered and quasi-closed nature of the system.

The detailed description of ore-genetic data and the parageneses of each ore-forming stage is beyond the scope of this study. Csongrádi (1975) and Baksa (1975a, b) have discussed the paragenetic sequences in the deep-level and near-surface mineralization, respectively.

Acknowledgments. This study would have been impossible without the constructive assistance and cooperation of geologists working on the Recsk exploration for more than 10 yrs. Special thanks are due to J. Földessy who helped with separation of the different magmatic cycles, and to J. Csillag for providing laboratory results and valuable discussions concerning the study.

References

Baksa Cs (1975a) New data on the enargite-luzonite-pyrite massive sulphide deposits, North from Lahóca-Hill, Recsk, Hungary. Földtani Közlöny, Bull of the Hungarian Geol Soc 105:58–74 (in Hungarian)

Baksa Cs (1975b) The subvolcanic andesite body of Recsk and its dikes. Földtani Közlöny, Bull of the Hungarian Geol Soc 105:612–624 (in Hungarian)

Baksa Cs, Cseh Németh J, Földessy J, Zelenka T (1980) The Recsk porphyry and skarn copper deposit, Hungary. International Symposium Belgrade. In: Jankovic S, Sillitoe RH (eds) European Copper Deposits; proceedings of an international symposium. Special Publications – Society for Geology Applied to Mineral Deposits (SGA) 1:73–76

Balla Z, Baksa Cs, Földessy J, Havas L, Szabó I (1980) The tectonic setting of the ophiolites in the Bükk Mts., Hungary. Geologica Carpathica 31 4:465–493

Cseh Németh J (1975) Deep-seated base metal ore occurrence of Recsk: geological pattern of ore accumulation. Földtani Közlöny, Bull of the Hungarian Geol Soc 105:692–708 (in Hungarian)

Csillag J (1975) Rocks transformed upon magmatic effect in the Recsk area, Hungary. Földtani Közlöny Bull of the Hungarian Geol Soc 105:646–671 (in Hungarian)

Csillag J, Földessy J, Zelenka T, Balázs E (1980) The plate tectonic setting of the Eocene Volcanic Belt in the Carpathian Basin. In: Bisztricsányi E, Szeidovitz S (eds) Proceeding of the 17th Assembly of the European Seismological Commission. Budapest, publishing House Hung Acad Sci 1:589–599

Csongrádi J (1975) Characterization of the deep-seated base metal ore mineralization of Recsk on the basis of ore-microscopic analyses. Földtani Közlöny, Bull of the Hungarian Geol Soc 105:672 – 691 (in Hungarian)

Földessy J (1975) The stratovolcanic andesite formation of Recsk. Földtani Közlöny, Bull of the Hungarian Geol Soc 105:625 – 645 (in Hungarian)

Hadži E, Aleksic V, Pantic N, Kalenic M (1977) The plate movements in South-Eastern Europe during the Alpine cycles. In: Jankovic S et al. (eds) Metallogeny and Plate Tectonics in the Northeastern Mediterranean; proceedings of an international IGCP symposium 1:231 – 248. International Symposium, Belgrade

Hollister VF (1978) Geology of the porphyry copper deposits of the Western Hemisphere: New York, Soc Mining Engineers AIME 219 p

Krivtcov AI (1977) Types of the regions of porphyry copper ore mineralization. Geologija Rudnüh Mesztorozsgenij XIX-4:44 – 57 (in Russian)

Lowell JD, Guilbert JM (1970) Lateral and vertical alteration-mineralization zoning in porphyry ore deposits. Econ Geol 65:373 – 408

Pantó G (1951) Ore deposit of Lahóca at Recsk, Hungary. Földtani Közlöny, Bull of the Hungarian Geol Soc 81:146 – 152 (in Hungarian)

Popov VSz (1977) Geology and genetic model of porphyry copper deposits. Nauka, Moscow (in Russian)

Sillitoe RH (1972) A plate tectonic model for the origin of porphyry copper deposits. Econ Geol 67:148 – 197

Sillitoe RH (1973) The tops and bottoms of porphyry copper deposits. Econ Geol 68:799 – 815

Wein Gy (1969) Tectonic review of the Neogene covered areas of Hungary. Acta Geol Acad Sci Hung 13:399 – 436

Wein Gy (1978) Development of Carpathian Basin. General Geological Review, Issued occasionally by the Section for General Geology of the Hung Geol Soc 11:5 – 28 (in Hungarian)

Zelenka T (1973) New data on the Darnó megatectonic zone. Acta Geol Acad Sci Hung 17:155 – 162

Zelenka T (1974) Megatectonic and magmatogenic history of the Northeastern Mátra Mountains. Acta Geol Acad Sci Hung 18:377 – 385 (in Russian)

Zelenka T, Baksa Cs, Balla Z, Földessy J, Földessyné JK (1983) The role of the Darnó line in the basement structure of NE Hungary. Bratislava, Geologica Carpathica 31.4:53 – 69

Ore-Magmatic Systems of Copper-Molybdenum Deposits

V. I. SOTNIKOV and A. P. BERZINA[1]

Abstract

Ore deposition concludes a complex cycle of transport and evolution of materials throughout the crust (20 – 30 km thick) where magmatic and hydrothermal associations, derivatives of a single ore-magmatic system, are being formed. Ore formation takes place in or near porphyry stocks considered as cupolas of intermediate magmatic chambers, the crystallization of which markedly influenced the formation of the flow of ascending fluid as well as a total temperature rise. The zone of magmatic cupolas (stocks) is characterized by more intensive ascending fluid flow. Here, favorable conditions for magmatic-hydrothermal systems exist. Conditions of mineralization in the hypabyssal environment become more complex because of the anatectic chamber, the crystallization of which causes an additional fluid flow into the upper levels of the crust. The development of the ore-magmatic system results from the interaction of subcrustal mafic magma and fluids separating from it with a magmatic melt formed in the crust. Fluids of two types existed in forming copper-molybdenum deposits. The magmatic system includes concrete intrusive bodies of an ore-bearing magmatic complex and the rock volumes adjacent to them. The meteoric-hydrothermal fluid system does not evolve in close juxtaposition with these intrusives.

Porphyry crystallization took place within the 1230 – 800 °C temperature range, explosive breccia formation from 750 – 600 °C to 300 – 200 °C, hydrothermal alteration and mineralization from 730 °C to 200 °C.

Nowadays complex copper-molybdenum deposits with large reserves of relatively low-grade ores offer one of the main sources of copper and molybdenum as well as of a number of accompanying rare and precious metals. That is why both theoretical and practical interest in these deposits exists all over the world.

As to the systematics of copper-molybdenum deposits, the authors suggest the use of ore-association analysis. The concept of ore association is widely applied in Soviet geology and is considered by investigators to be any group of deposits related in the common mineral composition and conditions of their formation. An ore association is characterized by specific geologic-genetic

[1] Institute of Geology and Geophysics, Siberian Branch of the USSR Acad. Sci., Novosibirsk, USSR

Geology and Metallogeny of Copper Deposits
(ed. by G. H. Friedrich et al.)

features common to given groups of deposits (Kuznetsov 1972, Konstantinov 1973, Shcheglov 1980). The main task of the analysis is to elucidate the principal features of ore mineralization that enable one to determine the particular ore association and its boundaries and to demonstrate that it has been correctly established. In the USSR, for the last 20 – 30 yrs, the ore-association analysis has developed for both the scientific and practical study of deposits and has influenced the development of theoretical principles of metallogeny and prognosis.

In contrast to the ore-association analysis, other systematics are based on separate features of mineralization: geotectonical position (Sillitoe 1972, Pavlova 1983, Pokalov 1972, Krivtsov 1977), chemical features of ore-bearing magmatism (Westra and Keith 1981, Mutschler et al. 1981), geochemical profile of mineralization, and accompanying alteration (Rekharskii and Distler 1965). One can see from these works that the deposits, different in some features, are similar in others. That is why copper-molybdenum deposits are considered by some researchers as one group, but are distinguished by others into porphyry-copper and porphyry-molybdenum groups.

Revealing in some cases essential differences among each other, on the whole, all copper-molybdenum deposits show rather stable common geologic-genetic and physicochemical parameters. Their differences are of regional appearance and are attributed to the specific features of development of the ore-magmatic systems that depend on the type of geologic setting, the location of the magmatic chamber, the relationship among the different sources of fluids and metals, the patterns of the fluid-transporting zone, hydrothermal alteration, ore deposition, and other parameters. That is why all deposits in question should be grouped into one class – copper-molybdenum association that is subdivided into molybdenum, copper-molybdenum, and copper subassociations (Sotnikov 1966, Sotnikov et al. 1977, 1985).

In this paper, the development of ore-forming, particularly the conditions of formation of large-volume ore concentrations, are considered on the data obtained in the study of copper-molybdenum deposits in the principal ore districts of the USSR and Mongolia and also on the basis of the generalizations advanced in many publications on these problems. The authors have taken into account a variety of the geologic situations in which ore mineralization occurs, different ages, and the geochemical distinctive features of the deposits. The principal assumptions concern the copper-molybdenum association as an integral unit.

Copper-molybdenum deposits are located mainly along regional structures referred to as superimposed ore belts, which are related to the evolution of extra-geosynclinal tectonic-magmatic processes and on the whole controlled by volcano-plutonic belts. Location of ore districts within these volcano-plutonic belts is controlled by the intersection of regional tectonic structures, coinciding with the direction of volcanic belts, and large through-going zones of transverse dislocation. These extend far downward (actually into the mantle), such zones create great possibilities for the connection of the ore-forming process with deep-seated layers of the Earth. Further, they offer very efficient permeable zones to permit the ascent of mantle material and the accumulation of fluid flow up to a level sufficient for intense magmatic and ore-bearing processes to develop. These belts contain centers of repeated endogenous activity where magmatic rocks of various ages, morphology, and genesis have been emplaced.

Copper-molybdenum deposits refer to the class of magma-related deposits associated with the volcano-plutonic series. The ore mineralization shows close spatial and temporal relation to porphyry magmatism (stocks and dikes). Because of their small sizes, difference in age, and a certain independence with respect to large host plutons. porphyry bodies generally cannot both energetically and by volume of separating fluids account for the formation of large-scale copper-molybdenum deposits. The latter seem to be possible only if porphyries are an integral part of larger magmatic systems, developed over a long period of time. All these points are taken into account in interpreting the relationship of mineralization with porphyry magmatism. Ore deposition concludes a complex cycle of transport and evolution of materials throughout the crust 20 – 30 km thick where magmatic and ore-metasomatic rocks, derivatives of a single ore-magmatic system, are being formed (Fig. 1). Porphyry stocks are supposed (Kalinin et al. 1980) to be cupolas of intermediate magmatic chambers, the crystallization of which markedly influenced the formation of ascending fluid flows as well as providing a total temperature rise in the zone of ore deposition.

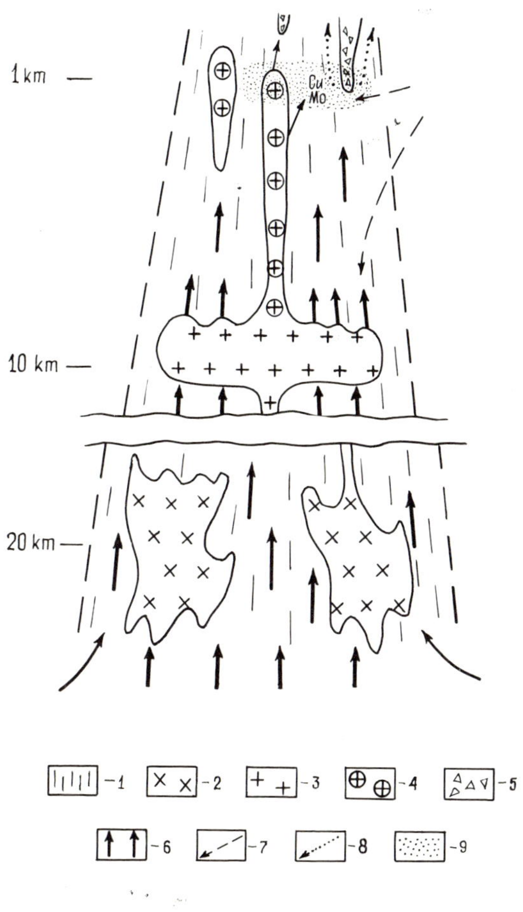

Fig. 1. The ore-magmatic system of copper-molybdenum deposits. *1* Ore-concentrating structure of high permeability; *2* primary anatectic magma chamber; *3* intermediate magma chamber; *4* porphyry; *5* explosive breccia; *6* fluid flow (from mantle and magma chambers); *7* meteoric water; *8* exhausted hydrothermal solutions; *9* ore-bearing zone

Magmatism developed in high permeability structures (under the conditions of drastic decrease in external pressure) are considered to be extremely favorable for the separation of late- and post-magmatic ore-bearing fluids. Conditions of mineralization in the hypabyssal environment become more complex because of the presence of a primary anatectic magmatic chamber, the crystallization of which during the regressive period of the development of the magmatic system causes the production of additional fluid in the upper levels of the crust.

According to the mathematical model of a magmatic-hydrothermal system with magmatic chambers at different levels (Fig. 2a), the zone of the magmatic cupolas (stocks) are characterized by more intense ascending flows especially in the early stages of crystallization. Here, favorable conditions for magmatic-hydrothermal systems are being formed. Near-contact zones of the cupolas are noted to show the maximum temperature gradients (Fig. 2b) that are suitable for the development of the ore-bearing process. After the melt crystallized, the stocks themselves exhibited high-contractional porosity (especially in the peripheral zones), and they can be treated as structures draining any ascending flows. Metasomatic processes play an important role in forming high-potassium porphyries under the conditions of high-fluid saturation of an ore-magmatic system (Artyshev and Vasil'ev 1982, Sotnikov et al. 1983, Spatz 1979). Usually such metasomatic processes are characteristic of ore-magmatic systems with elevated activities of potassium and fluorine.

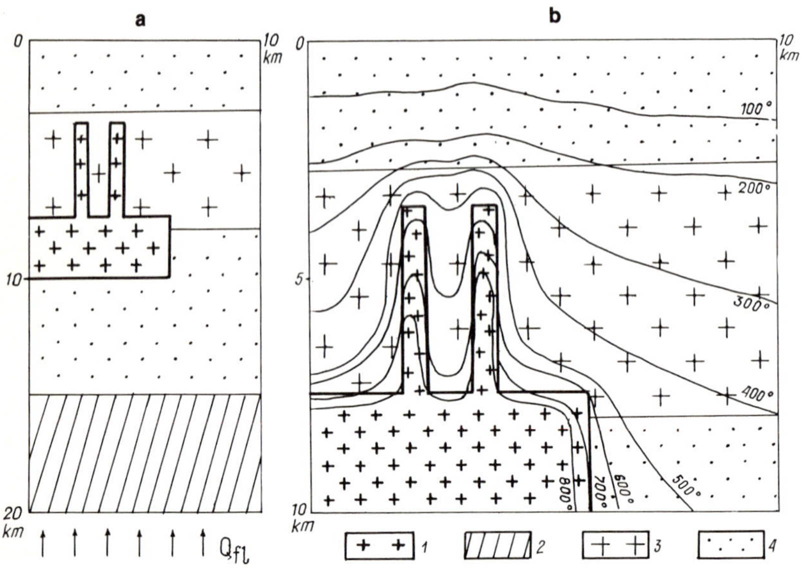

Fig. 2a, b. Model of magma system (**a**) and temperature distribution in supra-intrusive zone of an intermediate magmatic chamber (**b**). Q_{fl} (the magnitude of mantle fluid flow) = 10^{-8} g/(cm^2 s^{-1}). The intermediate chamber is 2.5 km high and 6 km wide. Melt crystallization temperature, 800°C. Before formation of the intermediate magmatic chamber, the heat flow was $1.25 \cdot 10^{-6}$ cal/(cm^2 s^{-1}). The temperature distribution is shown at 10000 yrs after formation of the intermediate magmatic chamber. *1* Intermediate chamber and its cupolas (stocks of porphyries); *2* deep zone of partial melting; *3* granitoid pluton hosting porphyry stocks and mineralized rocks; *4* volcanic-sedimentary and metamorphic rocks

At present, rather abundant isotopic data are available (Taylor 1982, Tugarinov et al. 1974, Ford and Green 1977, Johan and Le Bel 1980, Sotnikov and Berzina 1979), which indicate that crust and mantle matter take part in different quantitative combinations in the ore-magmatic processes of copper-molybdenum deposits. These data, along with the often observed occurrence of more mafic rocks in the early stages of development of the ore-bearing magmatic complexes, suggest that the ore-magmatic system resulted from interaction of subcrustal mafic magma and fluids separating from it with a magmatic melt being formed in the crust. A clear trend of potassium enrichment in the late rocks of the ore-magmatic complexes and the common development of potassium metasomatism are caused either by a continuous flow of deep fluids or by fluid interaction and the redistribution of mobile components (potassium in particular) in the boundary region between the alkaline-mafic and secondary silicic melts (Dobretsov and Dobretsov 1983). These processes are most fully realized in the zones of elevated and lasting tectonic-magmatic activities. Due to the combination of differently directed deep-fault structures, favorable conditions are created for the mantle fluids to penetrate the zone of magma generation (Korzhinskii 1952, Perchuk 1973, Beus and Golubev 1979, Banks and Page 1980) with intensive heating of the crust basement and fluid concentration; the deep basic magma ascent to the crust being possible in the same zones.

Such characteristics as the geotectonic, lithologic, and geochemical, among others, of certain units of the crust involved in the tectonic-magmatic processes can cause certain differences in the nature of the volcano-plutonic series that affect the development of the ore-bearing magmatic system, as well as the geologic and geochemical features of the ore districts (Kalinin and Sotnikov 1982). To the west of Southern Mongolia, the copper-molybdenum-bearing volcano-plutonic belt is emplaced in an oceanic domain and is represented by andesite associations with subordinate monzonite-diorite-granodiorite ones. Ore mineralization is essentially copper-rich (often auriferous) with slight quartz-sericite alteration. High chlorine activity was registered especially at early stages of the magmatic process. A "diorite" model (Hollister 1978) of the formation of copper-molybdenum deposits appear to have been realized in these regions. In activated old structures of essentially the mafic type located in Northern Mongolia (Sotnikov et al. 1985) and Middle Asia, USSR (Pavlova 1978), the portion of granitoids among magmatic rocks increases. Higher Mo/Cu ratios are characteristic of the ores in these areas. Strong quartz-sericitic alteration is widespread in the better mineralized centers. Development of an ore-bearing process is considered to conform to a "quartz-monzonite" model (Lowell and Guilbert 1970). In the zones of activated old median massifs and platform margins, including a prominent granite-metamorphic series (Kuznetskii Alatau, Transbaikalia), dacite-rhyolite-granodiorite, dacite-rhyolite-granodiorite and rhyolite-granite associations, usually with significant predominance of plutonic rocks, enjoy preferential development within the framework of volcano-plutonic belts (Sotnikov et al. 1977). Mineralization of copper-molybdenum and essentially molybdenum profile with the appearance of tungsten in a number of cases is typical. K-feldspar alteration is dominant. The magmatic-hydrothermal process is characterized by higher fluorine activity. In a number of cases these deposits begin to acquire the features of greisen in edones. A "granite" model, essentially applicable to molybdenum deposits (Clark 1972), and partly a "quartz-monzonite" model are realized in these areas. In the zones of activation of old consolidated structures (the Aldan Shield) copper-molybdenum mineralization is likely to manifest itself in connection with the development of alkaline magmatism (Kochetkov 1982).

Fluids of two types are suggested to have existed in forming copper-molybdenum deposits (Sotnikov and Berzina 1979, Sotnikov et al. 1983). The magmatic system includes concrete intrusive bodies of an ore-bearing magmatic complex and their adjacent areas (within the scope of a given fluid system, the processes of explosive brecciation and those of high-temperature potassium alteration usually take place). Meteoric-hydrothermal fluid systems evolve outside a close spatial connection with these intrusives (to a great extent it is the region of the development of sericitization, silicification, and chloritization and, in many cases, of the concentrated ore deposition). In the former fluid system, magmatic sources are sharply dominant (often with preferential juvenile component participation). The latter is provided by the components of deep-seated (including juvenile) and meteoric sources. In this system a meteoric part for a number of components can dominate appreciably. In particular, in cases of wide and intensive hydrothermal alteration, the host rocks can offer one of the principal sources of ore-forming substances. Hydrothermal alteration occurring within the limits of the influence of the meteoric-hydrothermal fluid system and usually characterized with an exothermal effect (Sotnikov et al. 1977, Norton and Cathles 1982) could appreciably contribute to the energy budget of the ore-magmatic system, facilitating in a number of cases spontaneous development of these processes. It should be noted that the zones of hydrothermally altered rocks, which, in contrast to the initial (unaltered) ones, are characterized by elevated values of thermo- and temperature conductivity coefficients and also by higher permeability, can offer good heat conductors and a suitable medium for moving fluids.

Ore formation in copper-molybdenum deposits (occurring in connection with the development of pulsed multiphase porphyry magmatism) is considered to have been a complex process of multistage development characterized by repeated occurrence of ore-hydrothermal rocks. The ore-forming process, in general, is characterized by a regressive development with a step-to-step temperature decrease and by the replacement of the oxide mineralization that is most typical for the period of the early explosive brecciation and high-temperature mineral deposition, by a sulfide one. At an early stage a large-scale "metasomatic steaming" of the host rocks occurs, with huge volumes of disseminated mineralization generated. During this preliminary stage the conditions potentially favorable for the subsequent formation of the large-volume ore concentrations are being created. Later on, during the main productive stage, mobilization of disseminated metals takes place. Ore components also enter in the solutions generated by deep-seated zones of the ore-magmatism, the efficiency of which depends greatly on the activity of Cl complexes. The latter considerably influence both the extraction and the transportation of ore components by magmatic fluids. Insignificant redeposition of ore substance also took place during the later

Table 1. Some data resulting from an investigation of melt and fluid inclusions in feldspar, quartz, apatite, and zircon[a]

Process	T_{hom} °C[b]	P_{bar}[c]	State of aggregation
Crystallization of ore-bearing melt	1230 – 800		Melt
Explosive brecciation	750 – 250	2500 – 1100	Gas
K-feldspathization	730 – 400	1800 – 1600	Mainly gas
Ore-formation	470 – 220	1500 – 600	Gas, liquid
Sericitization	410 – 330		Liquid
Argillization	360 – 320		Liquid

[a] Generalized after Berzina and Sotnikov (1979).
[b] Determination of temperature of melt inclusion as described in Roedder (1979).
[c] Syngenetic inclusions of essentially water and essentially carbon dioxide, formed by boiling of mixed $H_2O - CO_2$ solutions and three-phase inclusions with halite were used for estimating the pressure. See: Roedder and Bodnar (1980).

stages of the formation of the deposits in connection with final stages of ore-bearing magmatism. Economically polyascendent deposits where multiple magma intrusions (accompanied by explosive brecciation) took place, are the most favorable for prospecting.

Due to the multistage development of ore-magmatic systems, the ore-forming process in copper-molybdenum deposits is characterized by a complex evolution of physicochemical parameters (Piznyur 1983, Sotnikov et al. 1979). Some of these are shown in Table 1. The formation of porphyry intrusives of the ore-bearing magmatic complexes took place within the 1200 – 800 °C temperature range. Feldspar and quartz phenocrysts crystallized within the 1230 – 1150 °C and 1110 – 865 °C temperature range, respectively. The lower limit of homogenization of the primary melt inclusions in quartz phenocrysts was found to be within a narrow temperature range, i.e., 960 – 950 °C in most of the dike- and stocklike intrusive bodies studied. Only in a few cases was the temperature lowered to 865 °C. The homogenization temperatures of both secondary and pseudo-secondary melt inclusions is usually lower than the minimum homogenization temperature or primary ones; their lowest values nearing those of the highest temperatures of homogenization of inclusions from explosive breccias and early high temperature metasomatites.

Fluids of an early stage of intrusive activity that were fixed in primary melt inclusions of quartz phenocrysts are represented by nitrogen. In the composition of fluids coexisting with the melt during the formation of secondary melt inclusions along with nitrogen, a high content of carbon dioxide and halides is noteworthy (Fig. 3). The maximum effect of these components in a fluid manifests itself later during the accessory mineral crystallization with carbon dioxide and halides dominating during zircon and apatite deposition, respectively.

Fluid-explosive breccias form mainly pipelike bodies composed of host rock fragments. A cement is dominated by finely crushed and grated material of these rocks. Newly formed minerals, with biotite and magnetite being prevalent, are

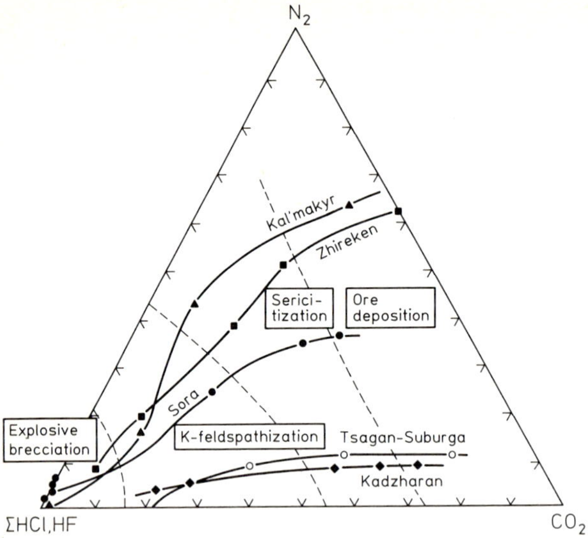

Fig. 3. Composition of the gas phase of fluid inclusions in the quartz of mineralized and altered rocks. The *curves* show a change of the gas composition in a trend from high- to middle-temperature rocks (from *left to right*)

observed as well. As a result of later superimposed hydrothermal processes, the cement of the breccias (to a lesser extent, the fragments) is altered. The temperature range of explosive-breccia formation varied from 750 – 600 °C (in the early stages) to 300 – 250 °C (in the late stages). The highest pressures (registered during the thermobarometric investigations) reached 2000 – 2500 atm. On the whole, the early explosive brecciation occurred somewhat earlier than the formation of porphyry bodies. Violent degassing of the magma was accompanied with the separation of the fluids enriched in halogens (mainly HCl and ferrochlorides). Breccias were formed as a result of multiple explosions; during relatively quiescent stages (when the feeding channel was blocked), the multiple extrusion of breccias ceased, and ejections consisted of finely crushed and ground rocks. The pressure decrease following the ejection produced the heterogeneity of the solutions. The conditions favorable for the inflow of meteoric waters from the surroundings were being developed.

The important role of chlorine during the process of early explosive brecciation is the reflection of its geochemical history in a magmatic process connected with its escaping from the magmatic chamber during the crystallization of the melts. Magmatic melts could have been degassed both during crystallization and emplacement of intrusives, when, with the existence of high permeable zones, melts boiling up resulted in major fluid separation. The latter cases were more favorable for the evolution of explosive brecciation as well as for the formation of wide zones of disseminated ore mineralization. In a number of deposits, carbon dioxide (along with halogens) were important components in the composition of the fluids and accounted in large part, for the formation of the explosive breccias (Fig. 3).

Ore-alteration rocks were generated in the broad temperature range: 730 – 600 °C in the early period (K-feldspathization, albitization, early silicification) to 450 – 200 °C (the main ore-forming stage as well as sericitization, silicification, argillization) and still lower in the final stage. The pressure ranged from 1500 to 100 – 80 atm. The solutions were as a rule, of alkali-chloride character and, at relatively low temperatures, alkali-chloride-carbonate composition. High-temperature fluids were characterized by a gaseous state, although due to the high pressure obtained they had the density of liquids. The medium- and low-temperature mineralization was deposited as if from liquids. The fluids responsible for the formation of high-temperature K-feldspathic rocks were characterized by a large content of halogens. At lower temperatures, the contribution of carbon dioxide and nitrogen in the fluid composition increased, the highest increase occurring after the solutions became truly liquids. The latter as compared to gaseous fluids usually contained higher concentrations of essentially all constituents. Concentrated ore deposition took place from the solutions with a relatively high carbon dioxide contribution.

Under conditions of relatively shallow depths and highly permeable structures, mineral-forming solutions boiled and as a result, the salt concentration in the liquid phase increased. In a number of cases, typical brine-melts appeared. The regions where highly concentrated solutions existed, a more concentrated ore deposition was observed. Less concentrated solutions that migrated to the peripheral zones of the deposits (ore field), are of an essentially carbonate composition. Under favorable circumstances they could dump their ore loads to form mineralization of a different geochemical profile (in particular, polymetallic and gold-silver ores).

The development of ore-magmatic systems in ore fields of copper-molybdenum deposits is accompanied by a considerable concentration of ore substance in a relatively limited volume; this predetermines the following most important conditions of ore formation: the concentration of a fluid (hydrothermal) flow in certain structures, responsible not only for the morphology, but also for the very possibility of the formation of ore bodies; the presence of a mechanism of ore deposition within a relatively narrow depth range.

It has already been emphasized that draining structures of different types and orders play a significant role in determining the concentration of fluid flows. Analyzing the conditions of the formation of large-volume veinlet-disseminated ores, which mostly characterize copper-molybdenum deposits, it is necessary to take into account that ore-forming solutions must be delivered to the region of the widestspread jointing (to microjoints often not connected directly to each other). Along with the explanation of the fluid mechanism developed by Burnham (1982), one can use capillary transport, the significance of which markedly decreases at temperatures above 300°C. Special conditions arising in the process of the formation of stockwork copper-molybdenum deposits that fix the elevated tectonic activity sites are also likely to stimulate the solutions transfer into the region of relatively high temperatures and pressures. Here, the relaxation of tension results in discompaction of the rocks and in the increase of their volume through the appearance of tiny cracks and pores. The process of mass relaxation of tectonic tensions results in the creation of an ultrasonic field in which the capillary transfer of fluids rises drastically since it is no longer limited by high temperature and pressures.

Thermo-osmotic transfer of the substance could widely develop in the region of a meteoric-hydrothermal system under conditions of nonhomogeneous heating, considerable humidifying of huge volumes of rocks, their variable porosity, and due to the presence of the two thermodynamic phases of interstitial water – the free and bound ones (Dudarev and Sotnikov 1982). Thermo-osmosis presents the process of formation of heat slip flow of bound (film) water in a temperature gradient field (Lykov 1954). This film liquid flow is usually turned to a heater. Here, as result of evaporation, the thickness of the water film is decreased and the liquid can transfer from a thicker film into a thin one. Thus, under nonuniform heating conditions in a porous environment, migration of a substance from a bound water zone into a thermo-osmotic zone and further to a heater takes place. Experimental work suggests that along with film liquid (water), alkalies and metals, including copper and molybdenum, are transferred effectively due to this thermo-osmotic phenomenon. Transfer effect rises with increase of porous medium dispersion. In this case, higher-temperature bound water, when compared with lower temperature, is characterized by enhanced transporting capability. A continuous thermo-osmotic flow (and what is of paramount importance, the flow to the source of heating) can reach 500 – 600 °C even at a distance far enough from the region of heating granted a certain amount of free fluid is available. This is comparable with the maximum temperatures of the process of ore formation in copper-molybdenum deposits.

Ore deposition on the geochemical temperature barrier in the region of maximum temperature gradients is considered as a possible mechanism of ore-substance deposition, thus providing for the formation of large copper-molybdenum deposits. Computation of a similar model (Kalinin et al. 1981, Sotnikov et al. 1983) revealed that the pattern, on the whole, is consistent with peculiarities of ore formation in copper-molybdenum deposits. During the progressive stage of development of a temperature field in the intrusive, the geochemical temperature of a temperature field in the intrusive, the geochemical temperature barrier situated in the supraintrusive region at a distance from the contact (the ore-substance concentration in the contact zone is fixed in case of highly concentrated solutions only) is gradually shifting upward and simultaneously growing narrower. This is the time when the area (ore bodies) rich in copper and molybdenum are being formed and, on the whole, instead of being scattered in a large-volume mineralization (by "metasomatic steaming") occur in a locally concentrated one. During the regressive stage, the geochemical temperature barrier is lowered, and here late mineral associations are depositing. The latter seems to account for superimposition of the late polymetallic and gold-silver mineralization onto the early copper-molybdenum one of higher temperature. It has been verified that economic concentrations in the geochemical temperature barrier zone can be formed in the presence of the fluid flow only. The presence of a lateral flow of meteoric waters introduces certain complications into the pattern of ore deposition. The lateral flow acts as an additional temperature barrier facilitating the formation of a new zone of ore deposition in the lower portion of the horizontally permeable structure in its point of intersection with the vertical one. On the whole, the zone of lateral flow of meteoric waters showing minimum solubility of ore components can offer a region of maximum ore deposition.

To conclude, it is necessary to point out that some elements of the ore-magmatic system discussed have been investigated and evidence has been produced to a lesser degree; for others, the evidence is much better, which is why they are not equally reliable. The greatest difficulties naturally arise in the genetic model for deep portions of the ore-magmatic system. And here we anticipate the analysis of the dynamic models of endogenous processes on the basis of recent geologic and geophysical evidence on deep-seated structures of the crust, experimental and theoretical developments of both modern petrology and the theory of ore formation and use of the latest computational equipment.

Acknowledgments. The authors are grateful to Drs. Richard H. Sillitoe and Richard L. Nielsen for their helpful reviews of the manuscript.

References

Artyshev SA, Vasil'ev BD (1982) Quartz-albite metasomatites of magmatic stage of median depth in Kuznetsk Alatau. In: Metasomatism and ore formation. Tezisy dokladov 5th Vsesojuznoi Konf, 23 – 25 November, Leningrad, pp 33 – 34 (in Russian)

Banks NG, Page NJ (1980) Some observations that bear on the genesis of porphyry-copper deposits. In: Ridge JD (ed) Proc 5th Q IAGOD Symp, vol I. Stuttgart, pp 49 – 73

Berzina AP, Sotnikov VI (1977) Physicochemical conditions of endogene processes in copper-molybdenum deposits in Central Asia. Econ Geol 72:25 – 36

Beus AA, Golubev VS (1979) Geochemical model of evolution of continental lithosphere. Akad Nauk SSSR Izv Seriya Geol 3:108 – 121 (in Russian)

Burnham C Wayne (1982) Magmas and hydrothermal fluids. In: Barnes HL (ed) Geochemistry of hydrothermal ore deposits. Mir, Moscow, pp 71 – 121 (in Russian) (original edition in English)

Clark KF (1972) Stockwork molybdenum deposits in the Western Cordillera of North America. Econ Geol 67:731 – 758

Dobretsov GL, Dobretsov NL (1983) On problem of genesis of alkaline-salic rocks. Geol Geofiz 2:3 – 11 (in Russian)

Dudarev AN, Sotnikov VI (1982) Thermo-osmotic filtration of solutions in rocks. Nauka, Novosibirsk, 108 pp (in Russian)

Ford JH, Green DC (1977) An oxygen and hydrogen isotope study of the Panguna porphyry copper deposit, Bougainville, Papua New Guinea. Geol Soc Aust J 24:63 – 80

Hollister VF (1978) Geology of the porphyry copper deposits of the Western Hemsiphere. Amer Inst Mini Metall Petrol Eng, New York, 219 pp

Johan Z, Le Bel (1980) Géochimie isotopique du soufre et sa contribution à la compréhension de la genèse des minéralisations type porphyry cuprifère. Mem Bur Rech Geol Min 99:151 – 161

Kalinin AS, Sotnikov VI (1982) New aspects of associational analysis of endogenic ore deposits. Sov Geol Geophys (Geol Geofiz) 23, 5:29 – 37 (in English)

Kalinin AS, Sotnikov VI, Berzina AP (1980) Hydrodynamic and temperature model of magmatogene hydrothermal system of molybdenum deposits (modelling by means of electronic computer). Akad Nauk SSSR Doklady 252, 5:1230 – 1234 (in Russian)

Kalinin AS, Sotnikov VI, Kolonin GR, Berzina AP (1981) A model of geochemical temperature barrier of copper-molybdenum deposit (modelling by means of electronic computer). Akad Nauk SSSR Doklady 259, 4:962 – 965 (in Russian)

Kochetkov AYu (1982) On new type of porphyry copper mineralization (Aldan shield). Akad Nauk SSSR Doklady 267, 2:443 – 445 (in Russian)

Konstantinov RM (1973) Principles of associational analysis of hydrothermal ore deposits. Nauka, Moscow, 215 pp (in Russian)

Korzhinskii DS (1952) Granitization as magmatic replacement. Akad Nauk SSSR Izv Ser Geol 2:56 – 69 (in Russian)

Krivtsov AI (1977) Types of porphyry copper areas. Geol Rudn Mestorozhd 4:3 – 20 (in Russian)
Kuznetsov VA (1972) Ore associations. Geol Geofiz 6:3 – 14 (in Russian)
Lowell JD, Guilbert JM (1970) Lateral and vertical alteration-mineralization zoning in porphyry ore deposits. Econ Geol 65:373 – 408
Lykov AV (1954) Transfer in capillary-porous bodies. Gostekhizdat, Moscow, 269 pp (in Russian)
Mutschler FE, Wright EG, Ludington S, Abbott J (1981) Granite molybdenite system. Econ Geol 76:874 – 897
Norton D, Cathles L (1982) Thermal aspects of ore deposition. In: Barnes HL (ed) Geochemistry of hydrothermal ore deposits. Mir, Moscow, pp 481 – 496 (in Russian)
Pavlova IG (1978) Porphyry copper deposits. Nedra Leningrad, 275 pp (in Russian)
Pavlova IG (1983) Geologo-genetic models of porphyry molybdenum-copper deposits. In: Kuznetsov VA (ed) Genetic models of endogene ore associations, vol I. Nauka, Novosibirsk, pp 127 – 134 (in Russian)
Perchuk LL (1983) Thermodynamical conditions of deep petrogenesis. Nauka, Moscow, 318 pp (in Russian)
Piznyur AV (1983) An evolution model of ore-bearing fluids as the basis of local evaluation and prediction of ore mineralization. In: Kuznetsov VA (ed) Genetic models of endogene ore associations, vol I. Nauka, Novosibirsk, pp 148 – 153 (in Russian)
Pokalov VT (1972) Genetic types and prospecting criteria of endogene molybdenum deposits. Nedra, Moscow, 271 pp (in Russian)
Rekharskii VI, Distler VV (1965) On feldspar-quartz association in molybdenum deposits. Geol Rudn Mestorozhd 4:19 – 26 (in Russian)
Roedder E (1979) Origin and significane of magmatic inclusions. Bull Miner 102:487 – 510
Roedder E, Bodnar RJ (1980) Geologic pressure determination from fluid inclusion studies. Annu Rev Earth Planet Sci 8:263 – 301
Shcheglov AD (1980) Principles of metallogenic analysis. Nedra, Moscow, 430 pp (in Russian)
Sillitoe HR (1972) A plate tectonic model for the origin of porphyry copper deposits. Econ Geol 67:184 – 197
Sotnikov VI (1966) An experience of systematics of tungsten and molybdenum mineralization in the Altai-Sayan geosyncline region. In: Kuznetsov VA (ed) Endogene ore associations in Siberia and the Far East. Nauka, Moscow, pp 115 – 123 (in Russian)
Sotnikov VI, Berzina AP (1979) Fluid and metal sources of copper-molybdenum deposits. In: Kuznetsov VA (ed) Nature of fluids and metal sources of endodene deposits. Nauka, Novosibirsk pp 15 – 32 (in Russian)
Sotnikov VI, Berzina AP, Nikitina EI, Proskuryakov AA, Skuridin VA (1977) The copper-molybdenum ore-association. Nauka, Novosibirsk, 422 pp (in Russian)
Sotnikov VI, Berzina AP, Shugurova NA, Motorina IV (1979) Physicochemical parameters of the processes of the generation of the copper-molybdenum association. In: Kuznetsov VA (ed) Main parameters of nature processes of endogene ore formation, vol I. Nauka, Novosibirsk, pp 209 – 220 (in Russian)
Sotnikov VI, Kalinin AS, Berzina AP (1983) Genetic model of the copper-molybdenum association. In: Kuznetsov VI (ed) Genetic models of endogene ore formation, vol I. Nauka, Novosibirsk, pp 112 – 127 (in Russian)
Sotnikov VI, Berzina AP, Zhamsran M, Garamzhav D, Bold D (1985) Copper associations in Mongolia. Nauka, Novosibirsk
Spatz DM (1979) Potassium-argon age of the Cerro Colorado porphyry copper deposit, Panama. Econ Geol 74:693 – 695
Taylor HP Jr (1982) Oxygen and hydrogen isotopes of hydrothermal mineral deposits. In: Barnes HL (ed) Geochemistry of hydrothermal ore deposits. Mir, Moscow, pp 200 – 237 (in Russian)
Tugarinov AI, Voinkov DM, Grinenko LN, Pavlenko AS (1974) Isotope composition and sulphur sources of copper-molybdenum prospects in Mongolia. Geokhimiya 2:171 – 178 (in Russian)
Westra G, Keith SB (1981) Classification and genesis of stockwork molybdenum deposits. Econ Geol 76:844 – 873

Part III
Deposits of Volcanic-Hydrothermal Association

Base Metal Deposits in the Iberian Pyrite Belt*

G. K. STRAUSS and K. G. GRAY[1]

Abstract

The Spanish-Portuguese Pyrite Belt covers a large area in the SW part of the Iberian Peninsula. Known reserves of massive pyrite and cupreous stockwork ore exceed 1000 million tons.

The stratiform sulphide deposits and accompanying manganese mineralization are of synsedimentatry-exhalative origin and occur in a Lower Carboniferous, geosynclinal, volcanic-sedimentary rock sequence, strongly folded during the Hercynian Orogeny. Epigenetic stockwork ores are in close relationship with volcanism and often represent the feeder channels of the massive sulphide sediments.

Most important ore types mined are massive "crude" pyrites, lead-zinc-rich "complex" pyrites, high-copper "banded" ores, copper-rich stockwork ores, disseminated "porphyry"-type copper ores and gold-silver-rich gossans. The main active mines are described and beneficiation of these ores is discussed.

Introduction

The Iberian Pyrite Belt represents one of Europe's most prominent metallogenic provinces, as well as one of its oldest mining districts. It occupies the SW part of the Iberian Penisula, extending approximately E–W over 250 km, from the Province of Seville, Spain to the west coast of Portugal, covering some 8000 km^2.

Proved mineral reserves are in excess of 1000 million tons and since the middle of the last century some 250 million tons of massive pyritic ores have already been mined. Other ore types, geologically closely associated in time and space, include low-grade copper and copper-zinc mineralizations and gold-silver-bearing gossans, both of great economic importance. Numerous small-sized manganese deposits were worked in the past.

Mining and smelting were carried out by the Tartessians, Phoenicians, and Romans. Following this period most of the mines lay dormant until their sudden

[1] Compañia Española de Minas de Tharsis, S.A., Minas de Tharsis (Huelva), Spain

* Manuscript submitted 12-2-1984.

Geology and Metallogeny of Copper Deposits
(ed. by G. H. Friedrich et al.)

resurgence in the middle of the 19th century when extraction of copper from the secondary enriched cementation zones of the rediscovered pyrite ore bodies became the principal objective. Copper was produced either by direct smelting of high-grade ores or by open-air roasting of cupreous pyrites and subsequent leaching and recovery of copper via cementation of scrap iron. In the second half of last century and during some 15 yrs. Spain was thus the leading copper district in the world – until the much bigger producers in the USA and Chile started to gather momentum.

At the same time, and with the depletion of the Sicilian sulphur deposits, the enormous Spanish pyrite reserves became an alternative source as raw material for the sulphuric acid production demanded by the ever-growing European chemical industry. The base metals contained in the "cinders" (residues from pyrite roasting), mainly copper and zinc, were recovered by newley developed hydrometallurgical processes in several Northern and Central European plants. The remaining "purple ore" (iron ore, freed of its base metal contents) was used as feedstock for blast furnaces.

The Spanish-Portuguese pyrite industry, the world's third largest after the USSR and P.R. China, nowadays produces some 2500000 tons per year (Tpy) of crude pyrites. Some 2600000 Tpy of complex pyrites are mined to produce copper, lead, and zinc concentrates with flotation pyrites as a by-product. Another 6000000 Tpy of low-grade copper ores are mined and gold and silver are produced from widespread gossans.

General Geologic Setting

The Iberian Pyrite Belt forms the central part of a Devonian-Carboniferous Geosyncline which represents the southernmost tectonic segment of the Hercynian "Iberian Meseta". It consists of Upper Paleozoic eugeosynclinal rocks, pyritic ores and manganese mineralizations and stretches as an arcuate belt across southwestern Spain and southern Portugal (Fig. 1).

A general description of the Pyrite Belt geology has been given among others, by Strauss (1965, 1970); Schermerhorn (1971) and Strauss et al. (1977). According to Strauss (1965, 1970), the rocks of this belt are grouped into three units (from bottom to top):

a) The basal *Slate-Quartzite Group* ("Phyllite-Quartzite Group", Schermerhorn 1971) is very uniform and consists principally of phyllitic slates and quartzites. At its top occur rare quartzite conglomerates and some limestone lenses which have been dated by conodonts to be of Upper Devonian (Famennian) age (Höllinger 1958, Van den Boogaard 1967, Van den Boogaard and Schermerhorn 1975).
b) The *Volcanic-Sedimentary Complex* ("Volcanic-Siliceous Complex", Schermerhorn 1971) overlies conformably the Slate-Quartzite Group. Its thickness ranges from less than 50 m up to some 800 m and its lithology is strongly variable. Out of the three, this is the only group with a wide variety of felsic to

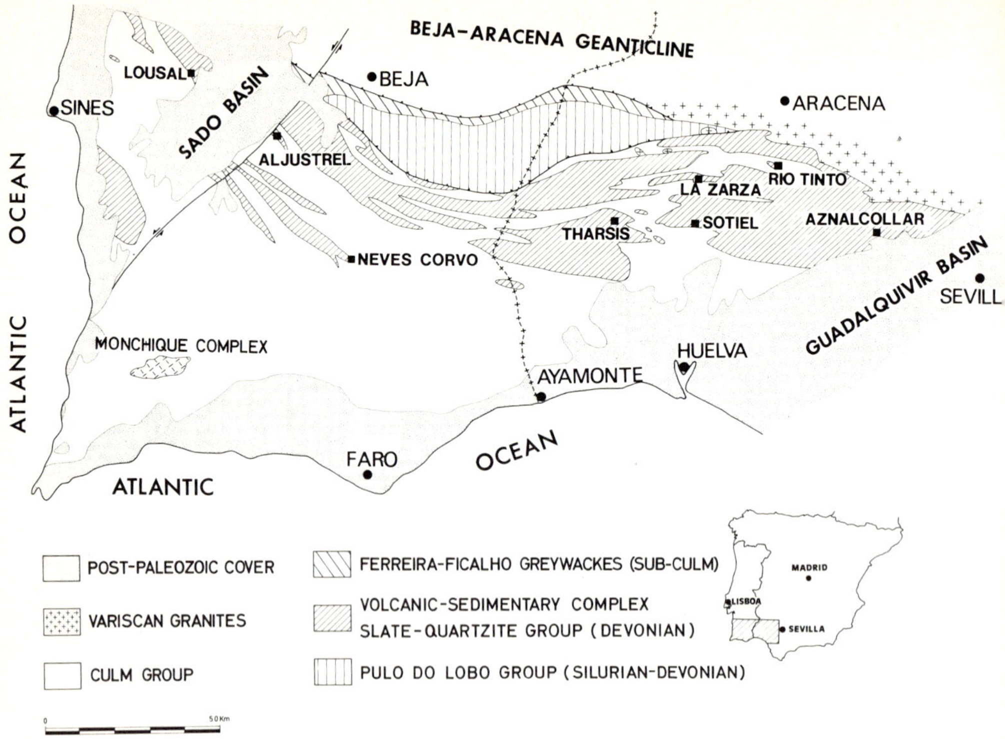

Fig. 1. Geology of the Pyrite Belt

mafic volcanics and it exclusively contains all the stratabound, volcanogenic sulphide and manganese mineralizations. Its age ranges from Tournaisian to Lower Visean (Lower Carboniferous).

c) The Volcanic-Sedimentary Complex is conformably overlain by *Culm*, a thick, well-stratified, monotonous succession of alternating slates and greywackes. It represents the flysch phase of the Pyrite Belt Geosyncline and reaches from Upper Visean to Lower Westphalian. Its thickness probably changes from only some 500 – 1000 m in Spain, to several thousands of metres in the western, Portuguese part.

During the *Hercynian Orogeny* (Middle Westphalian, Schermerhorn 1971) the whole Paleozoic rock sequence was intensively folded, with a prevailing general E – W direction in Spain, changing into a SE – NW strike in Portugal. Mostly narrow, isoclinal, and S – SW overturned fold structures were developed with accompanying thrust faults and, at the same time, a coaxial slaty cleavage was produced. Following the main folding phase, transverse faults were developed, striking preferentially NE – SW and with occasionally important displacements. A low-grade Hercynian regional metamorphism took place during the final stages, somewhat later than the main folding phase.

Post-Paleozoic cover rocks are predominantly *Tertiary sediments*, mainly sands and conglomerates.

Volcanic-Sedimentary Complex

The Volcanic-Sedimentary Complex (VS) is a heterogeneous rock sequence with a rapidly varying thickness and quickly changing volcano-sedimentary facies in both lateral and vertical directions. Felsic and mafic volcanics occur in lenses and "beds" up to several hundred metres thick and are closely related to the volcanogenic sulphide and manganese mineralizations, and a sedimentary framework consisting predominantly of slates (Fig. 2).

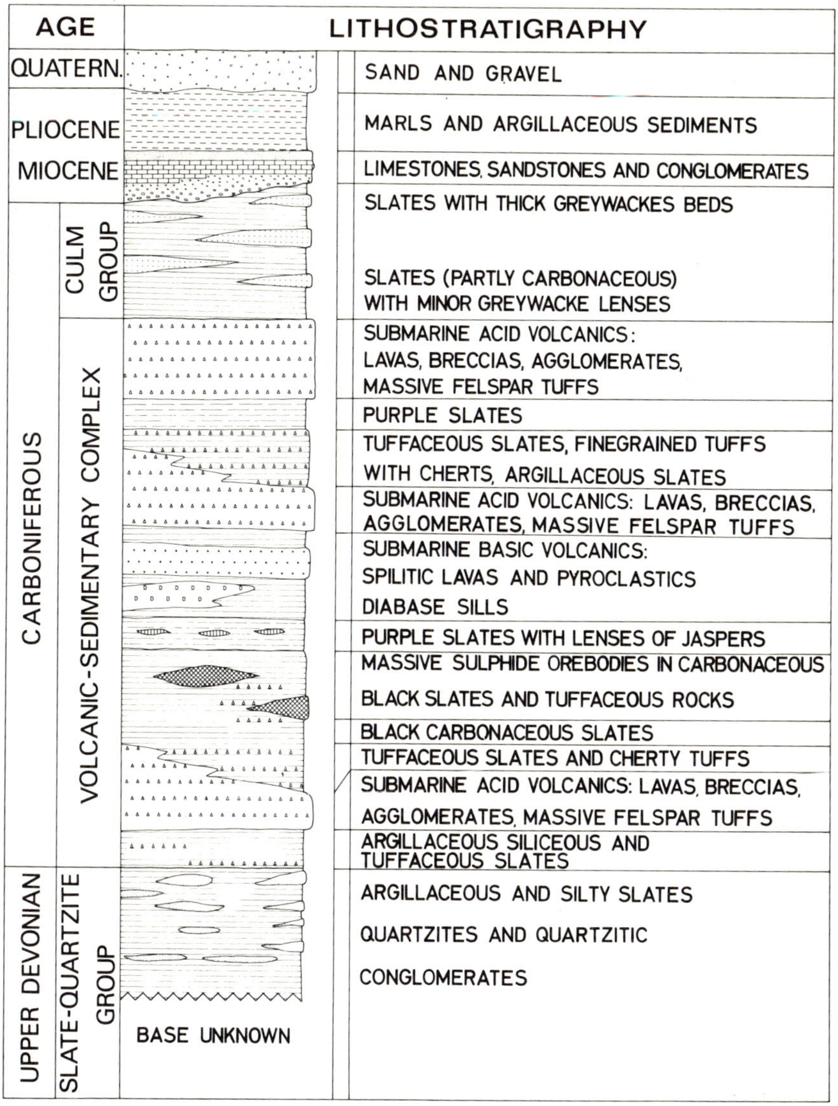

Fig. 2. Lithostratigraphy of the VS Complex

Lithology

At the beginning of the VS, the sedimentation of the Slate-Quartzite Group was abruptly terminated by the widespread eruption of felsic tuffs produced by *acid explosive volcanism* at many different centres forming thick accumulations of pyroclastics of all grain sizes and frequently interbedded with shales, thus indicating a submarine deposition. Massive acid pyroclastics are grouped around eruptive centres, lined up along fissural volcanic lineaments parallel to the general strike. These eruptive centres are outlined by volcanic breccias, agglomerates and isolated lava flows and volcanic chimneys have occasionally been recognized. Laterally these volcanics interfinger rapidly with sediments such as black shales, radiolarian cherts, etc. and/or decrease in grain size to form well-bedded medium- to fine-grained tuffs or tuffaceous shales (Strauss and Madel 1974, Strauss et al. 1977). The felsic volcanism is predominantly of soda- or potash-rich quartz-keratophyre or sodic rhyolite composition with albite being the principal felspar phase. Potash-felspar megacryst-bearing pyroclastics are rare.

The *basic volcanics* consist mainly of intrusive albite diabases, together with less common spilitic pyroclasts, pillow lavas and tuffs, representing submarine effusions. Their silica contents vary between 40 and 66% SiO_2, indicating intermediate compositions (Schermerhorn 1970). Ultrabasic cumulates are known in a few localities of the Spanish part of the Pyrite Belt.

Extrusive spilites occur always in the immediate vicinity of acid volcanic lineaments. Their lithostratigraphic position is, almost in every case, on top of a preceding acid volcanic eruption cycle (Bernard and Soler 1974, Strauss and Madel 1974), although the stratigraphic position of the extrusive mafics differs between the eastern and western part of the Pyrite Belt.

The *intrusive albite diabases*, either as dykes or, more frequently, as sills, are spaced at greater distances from acid pyroclastics and have intruded all formations of the VS Complex.

The *sedimentary facies* of the VS is largely dominated by slates, typically dark to black shales often with a high content of carbonaceous matter and, in part, with finely disseminated iron oxides and sulphides that weather characteristically in pink to reddish tints.

Cherts and slaty cherts represent phases of stronger siliceous precipitation. They often contain radiolaria and when abundant, even pass into radiolarites. Dark grey or red jaspers are another typical sedimentary rock of the VS Complex, usually occurring above felsic volcanite series and often associated with manganeses ores.

Coarse detrital sediments occur locally in the upper part of the VS, while limestones form a very minor part of the VS assemblage and occur locally as thin layers or small lenticular bodies, often in a sideritic-dolomitic composition. These do, however, often occur close to the sulphide horizons and can form useful geologic marker horizons.

The *strata-bound massive sulphide deposits* are always connected, in time, to the different submarine explosive acid volcanisms and specifically to the final, waning stage of each eruptive cycle. In space they are confined to the vicinity of the volcanic centres, either deposited directly on massive acid pyroclastics (autochtonous ores) or sometimes deposited at some distance and interbedded with black shales and tuffs (allochthonous ores) (Schermerhorn 1970, Strauss and Madel 1974, Strauss et al. 1977).

The *stockwork ores* are of epigenetic origin and were generally the feeder channels of the overlaying massive sulphide sediments.

The *manganeses ores* are always associated with grey and red ferruginous jaspers and purple-violet tuffaceous shales that make the "manganese formation" (Strauss 1965, 1970). This ore facies (silicates, carbonates and oxides) is either contemporaneous with the sulphide deposits or, as in most cases, somewhat later. It was formed during greater time spans and occupies more extensive areas than the sulphide facies.

Ore Deposits

The geology and metallogeny of the Iberian pyrite deposits have been described and discussed in a great number of papers (Strauss 1965, 1970, Strauss and Madel 1974, Strauss et al. 1977 and references cited therein) and will not be repeated at length here. Several types of mineralizations are known, of which the best known and economically most important ones are the *syngenetic-sedimentary, massive, polymetallic pyritic ores.* They always occur in massive, lenticular-shaped ore bodies of varying dimensions and can vary in size from less than 100000 to more than 100000000 tons. Their principal mineral is always pyrite, together with varying amounts of chalcopyrite, sphalerite and galena, and lesser amounts of arsenopyrite, pyrrhotite, tetrahedrite, lead sulfosalts, cassiterite and many other minor minerals. Principal gangue consists of finely disseminated or banded intercalations of the host rocks, together with silica, carbonates and to a lesser extent barytes (Strauss 1965, 1970).

The most important facts with respect to the geologic environment and formation of these ore deposits can be summarized as follows:

1. The massive sulphide ores, together with disseminated sulphides and stockwork-type mineralizations occur exclusively in the Volcanic Sedimentary Complex (Fig. 3).
2. Massive sulphide ores and manganese mineralizations are both sedimentary and are closely linked with submarine, acid-alkaline volcanism.
3. In the case of multiple acid eruptive cycles along one volcanic lineament, only one seems to be "productive" with regard to economical sulphide mineralizations.
4. The sedimentary sulphide concentrations, both massive and disseminated, are always located near the borders of the volcanic extrusive centres and at a geologically well-defined distance, where black shales are interbedded with acid pyroclastic rocks. (The manganese mineralization, both laterally and vertically, is distributed more widely than the much more restricted sulphide facies).
5. The concentration of sulphides into massive ore lenses is explained by the inflow of sulphide mud and/or detrical sulphide material into topographical depressions around submarine volcanos.

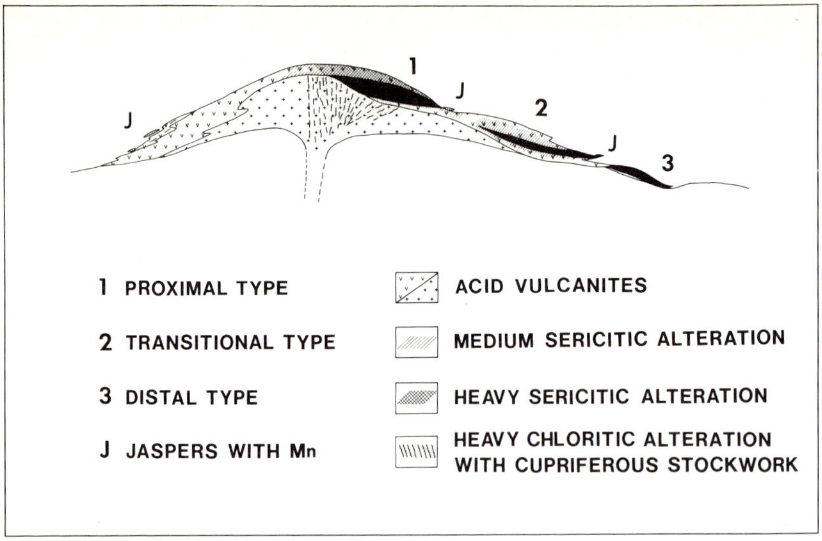

Fig. 3. Types of mineralization

6. Sulphide ores with massive, fine-grained textures, were generally deposited on coarse porphyritic pyroclastics in the immediate vicinity of the corresponding extrusive centre. Sulphide ore bodies with a layered, internal structure and detrital sedimentary textures were deposited in a more shaly environment at greater distances from the eruptive centres.
7. Clastic sedimentary ore textures indicate resedimentation of sulphide material. This implies that previously formed, already solidified pyritic deposits were submitted to submarine erosion and/or flows, their material being transported into newly formed depressions.
8. The close relationship between massive sulphide ores, manganese ores and submarine acid-alkaline volcanism clearly implies a genetical link between this type of volcanic activity and the supply of sulphur and metals: The volcanism was associated with hydrothermal activity venting on the seafloor, where iron and base metal sulphides and manganese silicates and carbonates were precipitated as chemical sediments.

Synsedimentary disseminated sulphide mineralization is widespread and occurs usually in the hanging wall of massive sulphide ore bodies, sometimes as isolated ore bodies. They consist mostly of pyrite, chalcopyrite and sphalerite, finely disseminated and sometimes layered, within a sedimentary environment, consisting mostly of black shales, tuffs and/or tuffites. Their origin can be interpreted as the waning phase of volcanic supply of metals and sulphur into basins where sedimentation of shaly and tuffaceous material was more intensive than that of the sulphides.

Closely related to the synsedimentary massive sulphide ores is the *epigenetic stockwork-type sulphide mineralization* occurring underneath autochthonous massive pyritic deposits, predominantly in their highly chloritized, sericitized and silicified footwall rocks. Their sulphide grade varies usually from less than 10% to over 50% pyrite equivalent (<5 – 25% S), but a slow transformation into massive pyritic ores is known at several mines. They consist mainly of hydrothermally altered volcanics and their sulphides occur within the brecciated host rock (breccia ore), in scattered veins and veinlets (stockwork ore) and finely disseminated (disseminated ores). By far the most widespread sulphides are pyrite, chalcopyrite and sphalerite. This type of mineralization is interpreted as of epigenetic origin with respect to the host rocks, and as the hydrothermal feeder channels of the sedimentary sulphides (Williams 1967, Rambaud and Perez 1969, Bernard and Soler 1974, Strauss and Madel 1974, Garcia Palermo 1980, Williams et al. 1975).

The *syngenetic, volcano-sedimentary manganese mineralization* forms an integral and most typical part of the VS Complex and represents together with accompanying jaspers, cherts and purple slates one of the most outstanding marker formations within the Pyrite Belt. Over 500 small manganese ore bodies are known, none of which, however, ever reached any economic importance. The primary ores consist mainly of rather low-grade silicates and carbonates and only the near surface oxide cappings were worked in the past.

Pyritic Ores

Ore Types

The great variety of pyritic ores can be subdivided due to their chemical and mineralogical composition and industrial use, on the one hand, and to their formation, on the other.

Massive Pyrite is a typical polimetallic sulphide ore with pyrite as the predominant mineral component, always accompanied by a great variety of other base metal sulphides, mainly chalcopyrite, galena and sphalerite, together with minor amounts of arsenopyrite, tennantite, complex Cu-, Pb-, As-, Sb-, Hg-, Ag-sulphides, cassiterite, gold and many other ore minerals. Main gangue components, which as a rule do not exceed 5% of the total, are finely distributed or layered host rock, quartz, carbonates and sometimes baryte. A common feature of all massive pyritic ores is that the mineral components are extremely fine-grained and with intense intergrowth (Strauss and Gray 1982).

The massive pyrite ore is known on the international market as "crude pyrite" or "standard pyrite" and contains typically: S 45 – 48%; Fe 40 – 43%; Cu 0.6 – 1.0%; Pb 0.5 – 1.0%; Zn 1.0 – 2.5%; As 0.3 – 0.4%; Au 0.5 – 1.2 g/t; Ag 20 – 30 g/t. This ore type has been mined for more than a century at Rio Tinto, Tharsis, La Zarza and many other smaller operations.

Since the early 1970s *complex pyrites* are of growing importance. They are found in most pyrite mines and occur usually as interbedded layers and lenses within crudes pyrites, or in their hanging and footwalls. The typical features of these massive, "complex" pyrites, compared to the "standard" ones, are their somewhat lower sulphur content, higher lead and zinc values and sometimes rather high contents of arsenic, antimony, mercury, silver and tin. The extremely fine intergrowth of their mineral components as well as the occurrence of complex sulphide minerals make their separation and beneficiation by flotation very difficult. Typical values of these complex pyrites are: S 40 – 45%; Fe 34 – 41%; Cu 0.4 – 1.5%; Pb 1 – 2%; Zn 2.5 – 5%; As 0.4 – 0.8%; Sn 0.01 – 0.06%; Au 0.5 – 1.0 g/t; Ag 40 – 60 g/t.

Banded copper ores were discovered at Neves-Corvo, Portugal and are locally known as "banded ores". Compared to normal pyrites they are very rich in copper (8% Cu) and silver (44 g/t), low in zinc (0.40% Zn) and contain only traces of lead.

Typical and widespread are massive *low grade pyrites*, known in Spain as "azufrones". They consist of massive pyritic ores which either contain a large amount of finely disseminated host rock as gangue or occur as thin layers interbedded with the country rock. These ores are economically of little importance, except when they contain higher amounts of base metals, usually copper.

A special type of massive, low-grade pyrite is the *carbonate ore*, principally found in Tharsis where it occurs interstratified with normal crude pyrites. A typical feature is the abundance of brecciated pyrite-clasts in a siderite matrix as gangue.

Closely related to these snygenetic-sedimentary, massive pyritic ores occur epigenetic, disseminated and/or stockwork mineralizations, often underneath autochthonous massive pyrite ore bodies and sometimes as isolated units, but always in hydrothermally altered, chloritized, sericitified and silicified, acid pyroclasts.

A typical example of this group are the *disseminated copper ores* of Cerro Colorado/Rio Tinto which have in the last decade become one of the most important copper sources of Western Europe. Pyrite-chalcopyrite- and quartz-chlorite veinlets and stringers in strongly, hydrothermally-altered felsic tuffs contain about 0.64% Cu.

Massive stockwork copper ores are known and have been extracted for many years at Alfredo/Rio Tinto. Thick chalcopyrite-pyrite veins and veinlets cross the strongly chloritized and silicified tuffaceous host rock which sometimes is so shattered that it forms a breccia within a sulphide matrix. Copper and zinc values both average about 1.5% and sulphur amounts to 20 – 30%. Very similar is the "siliceous copper ore" of La Zarza. The "stockwork" ores and "brecciated" ores of Neves-Corvo, Portugal seem to differ from the other ones only by their much higher copper values (1.5 – 4.0% Cu).

Mines

The Spanish-Portuguese Pyrite Belt hosts a total of eleven operating mines with another two being developed (Fig. 4).

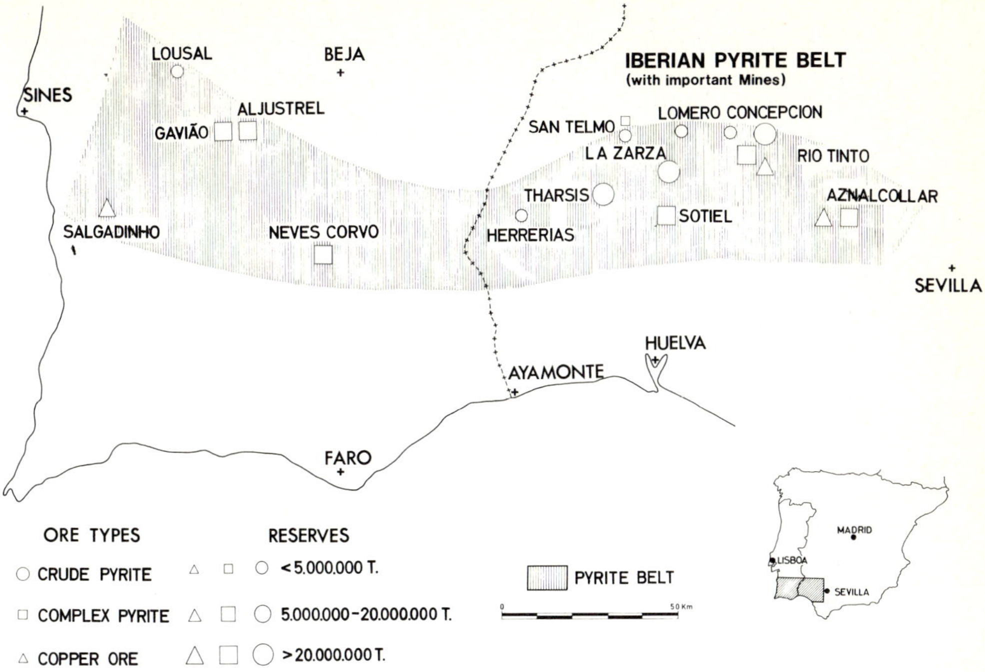

Fig. 4. Mines of the Pyrite Belt

The biggest and in all ways the most important ore field is *Rio Tinto*, originally formed by a single stratiform sheet of massive pyritic ores, 4500 m long, 1500 m wide, and up to 80 m thick, which was later folded into an anticline with most of the pyrite in the flat crest of the fold being gossanized (Cerro Colorado) (Fig. 5). The massive pyrite ore bodies all lie in a thick sequence of quartz-keratophyric and rhyolitic lavas and tuffs and show widespread pyritic and cupriferous-pyritic stockwork mineralizations in their foot walls. These consist of felsic pyroclasts, which are hydrothermally altered (chloritized, sericitized and silicified).

Over 140 million tons of crude and cupreous ores have been mined with 50 million tons remaining in the San Dionisio ore body, where nowadays 500000 Tpy are extracted by open pit. At "San Antonio" 12 millions tons of complex ore have been proved by underground workings, with 1.6% Cu, 1% Pb and 2% Zn. In the "Alfredo" stockwork ore-deposit reserves of 15000000 tons grading 1.7% copper are mined underground at a rate of 600000 Tpy. Reserves of disseminated copper ores at the "Cerro Colorado" open pit amount to over 150 million tons, grading 0.60% Cu and are mined at a rate of over 4 million Tpy. Total production of copper in concentrates is over 30000 Tpy from a 20000 Tpd (tons per day) capacity mill. Up to 4 tons of gold and 45 tons of silver are produced every year from a cyanide mill treating 2 million Tpy of gossan which contains 2.0 g/t Au and 50 g/t Ag.

At *Tharsis* several big ore bodies are known, the most important ones being North Lode, S. Guillermo and Sierra Bullones, with a combined strike extension of 1500 m, average thickness 80 m and reaching to over 400 m depth (Figs. 6 and 7). Proved reserves are over 100 million tons, consisting mainly of massive crude pyrites with 46.5% S, 0.7% Cu, 0.8% Pb, 1.8% Zn and rather high gold values (1.2 g/t Au). The stratiform pyrite lenses are enclosed by black slates with some tuffaceous interbeds and thin layers of dolomitic-ankeritic limestones and spilites in their hanging wall, overlain by a thick sequence of felsic tuffs. Approximately 800000 Tpy of crude pyrites are extracted from the North Lode open pit.

La Zarza consists of a large single ore body, 3000 m long, locally over 100 m thick and over 400 m deep (Fig. 8). Some 50 million tons of massive pyrites were extracted since modern mining started with more than the same amount remaining in this 400000 Tpy underground mine. The massive pyrites average 47% S, 0.7% Cu, 0.6% Pb and 1.5% Zn and are conformably interbedded within felsic tuffs, tuffites, black slates and cherts. They are locally underlain by "siliceous copper ores", a stockwork mineralization of yet unknown dimensions, averaging over 1.5% Cu, 1.4% Zn and 28% S. Disseminated pyrites and low-grade stockwork ores with only 0.1 – 0.6% Cu occur in the same formation, as well as chloritized pipelike structures with no mineralizations at all.

Aznalcollar mines a large ore body containing 45 million tons of massive pyritic complex ores, grading 0.44% Cu, 1.77% Pb, 3.33% Zn, 1 g/t Au and 67 g/t Ag. It is overlain by "pyroclast ore", finely disseminated and thin stockwork veinlets of pyrite, chalcopyrite and minor sphalerite in chloritized, felsic tuffs. Reserves of 47 million tons average 0.58% Cu, 0.40% Zn and 10 g/t Ag. Both ore bodies are interbedded in a thick series of volcano-sedimentary rocks, consisting mainly of acid tuffs, tuffites and black shales (Fig. 9). The open pit operation started in 1980 and present production is at a rate 2000000 Tpy complex ore and 1600000 Tpy of "pyroclastics".

At *Sotiel Coronada* a massive complex pyrite ore body is enclosed in black shales with some limestone beds and tuff intercalations. Approximately 42 million tons of complex ore grading 0.44% Cu, 1.36% Pb, 4.25% Zn and 39 g/t Ag are now prepared for underground operation. Eventually 600000 Tpy of ore are expected to yield some 38000 Tpy of zinc concentrates (50% Zn), 11000 Tpy of copper concentrates (23% Cu) and 500000 Tpy of fine-grained pyrite concentrates as a by-product, half of which will be used locally for sulphuric acid production.

At *Lousal, Portugal*, known workable ore reserves are limited and only 40000 Tpy of low sulphur pyrites are extracted. Relatively high cobalt values are typical for this ore body (0.1% Co) (Fig. 10).

At *Aljustrel*, six ore bodies contain over 200 million tons of crude and complex pyrites (Fig. 11). They occur at the top of a sequence of coarse-grained and fine-grained tuffs, siliceous shales and jaspers. In the northern part, a single ore body, over 60 m thick and containing more than 100 million tons of pyritic ores, was cut and displaced by a Hercynian wrench fault, forming the separate masses Feitais and Estacao. The southern ore zone includes the four remaining deposits, extending over a strike length of over 4000 m. These ores are more cupreous than the ones of the northern zone, averaging 1.1% Cu in the Moinho deposit and

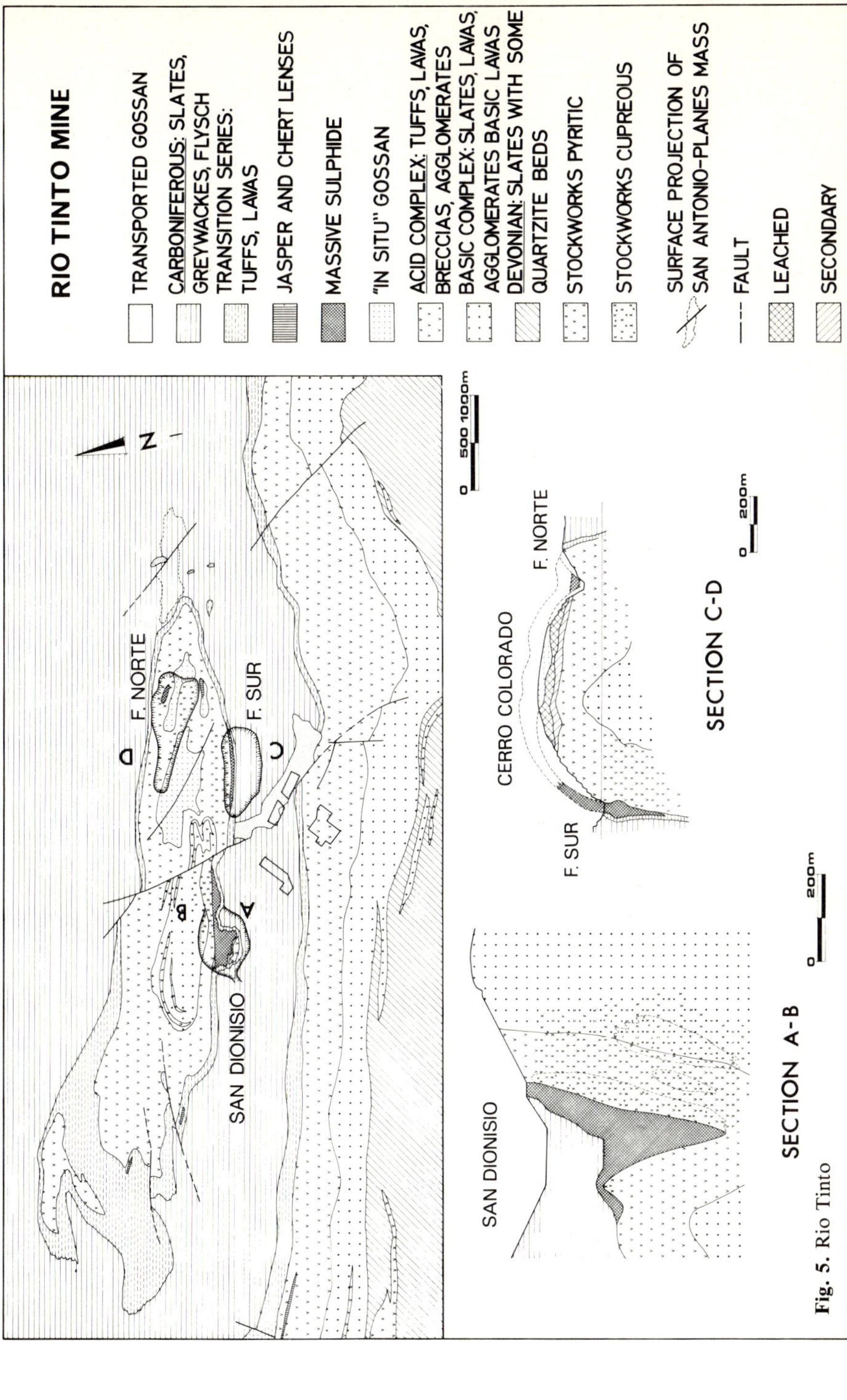

Fig. 5. Rio Tinto

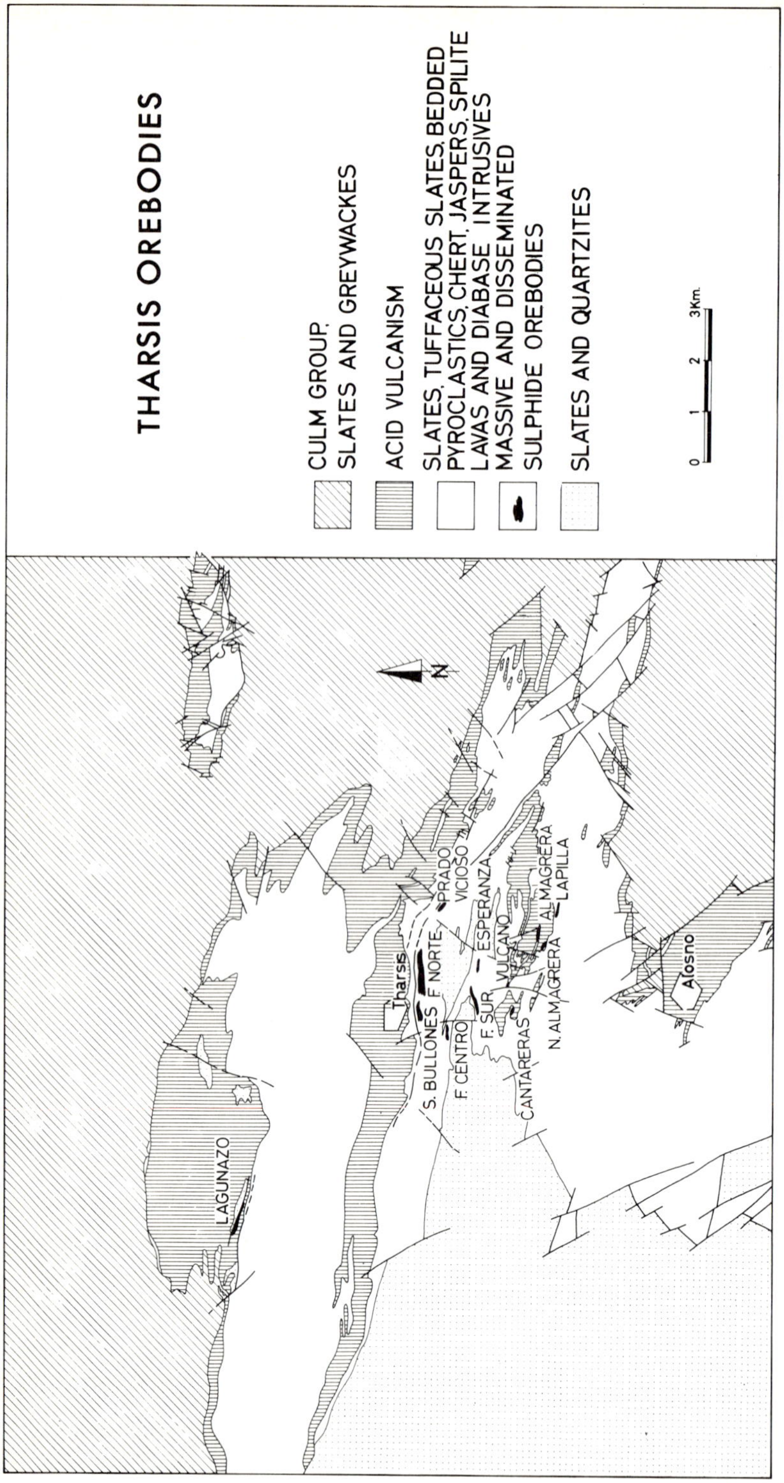

Fig. 6. Tharsis ore bodies

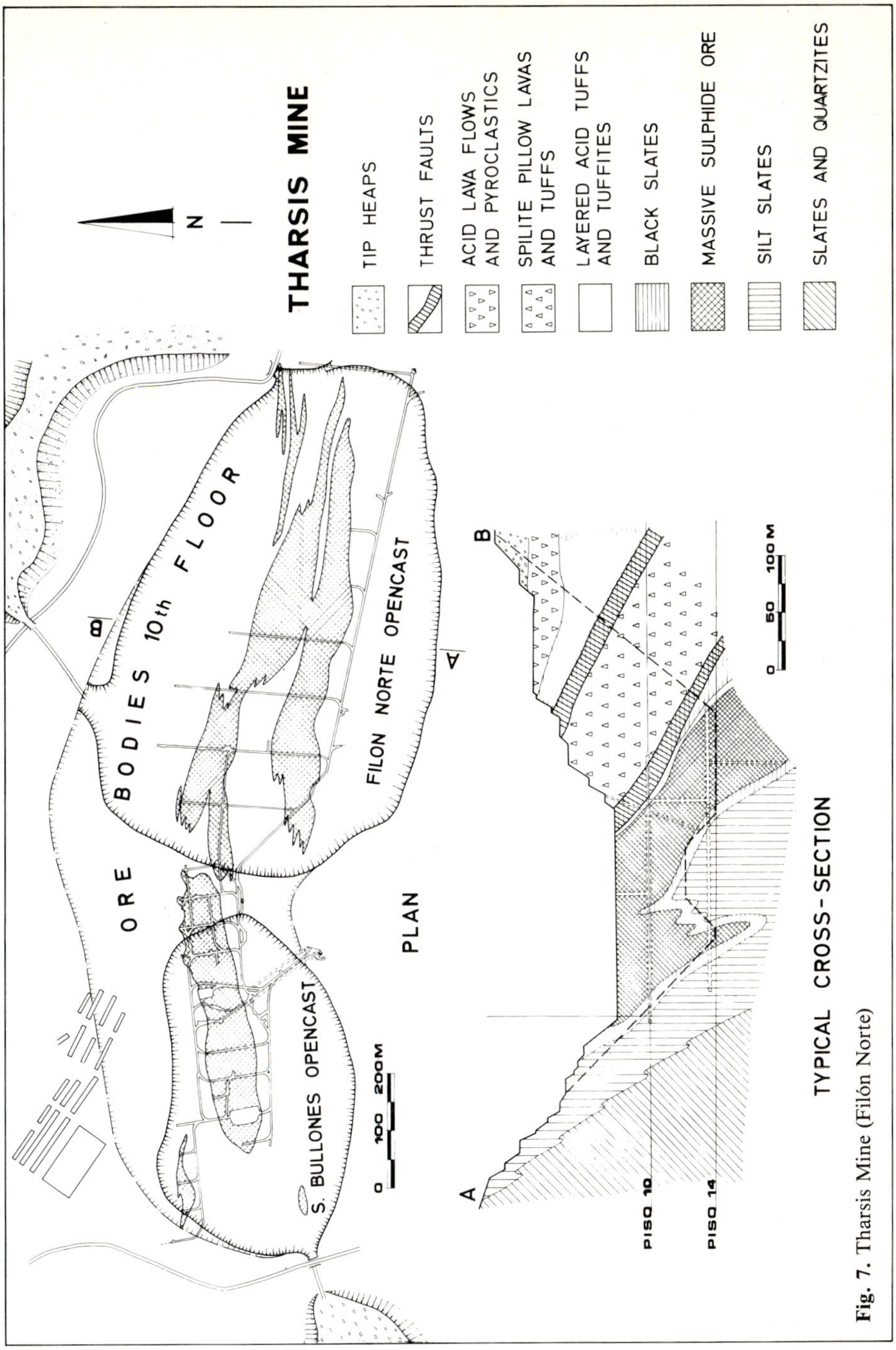

Fig. 7. Tharsis Mine (Filón Norte)

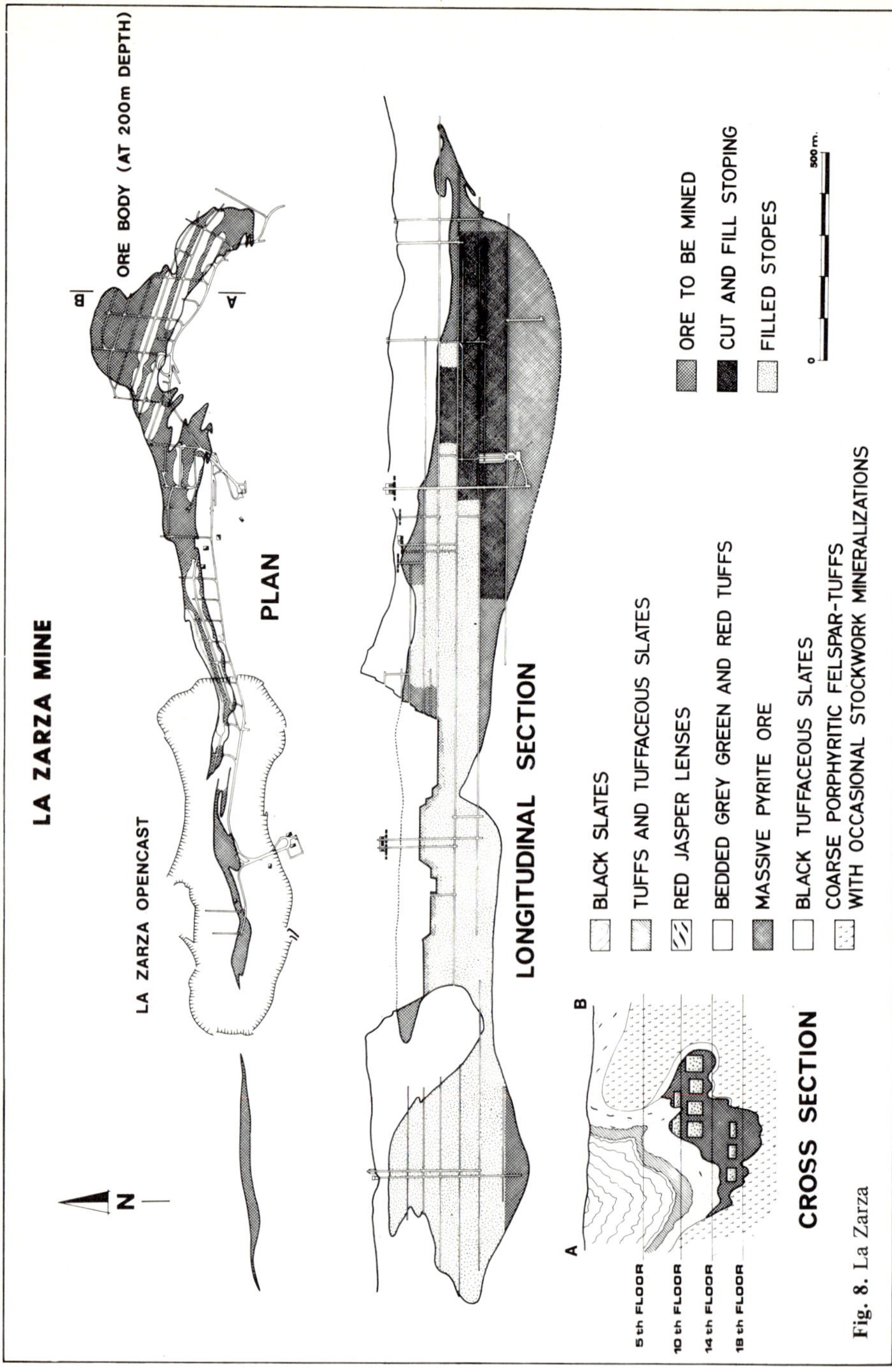

Fig. 8. La Zarza

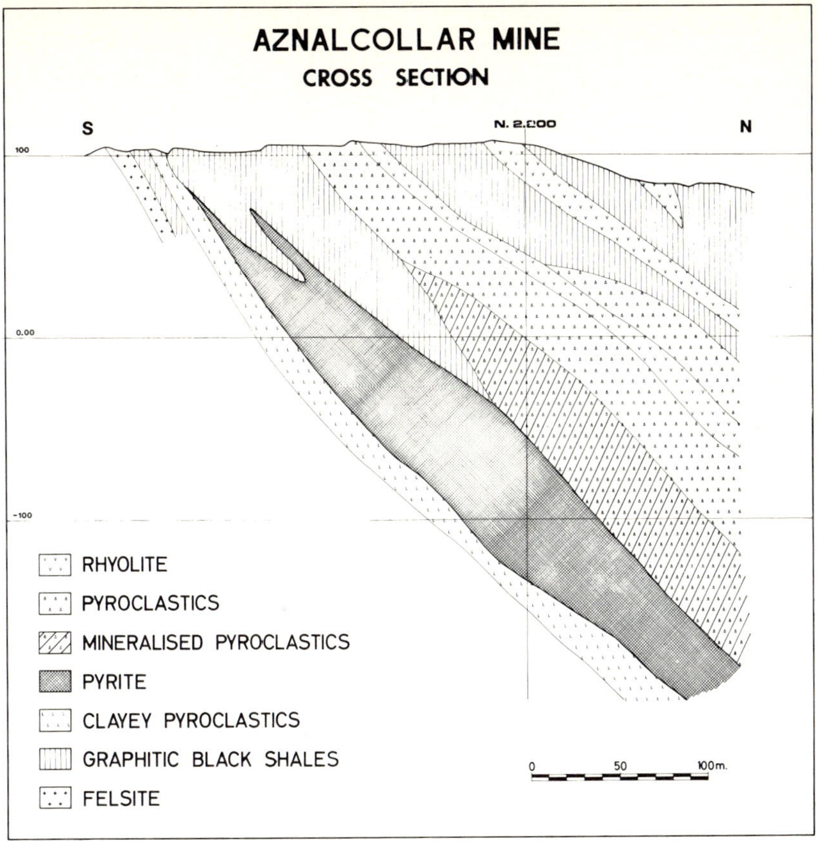

Fig. 9. Aznalcollar

1.5% Cu at Gaviao. Moinho and Feitais are worked underground and some 250000 Tpy of complex pyrite ores are produced, with an average grade of 45% S, 0,86% Cu, 1.42% Pb, 3.18% Zn, 0.7% As and 40 g/t Ag and are used for sulphuric acid production.

The most spectacular *Neves-Corvo* ore deposits were discovered in 1977 40 km SE of Aljustrel and under 300 to 500 m deep bedrock. Since then four flat lying ore bodies have been outlined, Neves, Corvo, Graca and Zambujal, with estimated total reserves of 100 million tons of massive pyritic and stockwork ores (Fig. 12). The ore deposits are interbedded in a thick sequence of acid volcanics, tuffites, black shales and beds of greywackes, siliceous shales, jaspers and limestones and overlain by Culm greywackes.

The most striking discoveries were 34 million tons of high-grade copper ores. Banded ores, massive, synsedimentary copper ores, form a 1.5 to 4 m and locally up to 10-cm-thick layer on top of the massive pyrites and acid pyroclastic rocks, grading usually about 8% Cu and 44 g/t Ag, but only 0.4% Zn and traces of Pb. The felsic pyroclastic rocks in the foot wall of the massive pyrite ores are strongly chloritized, sericitized and silicified and show an epigenetic stockwork mineral-

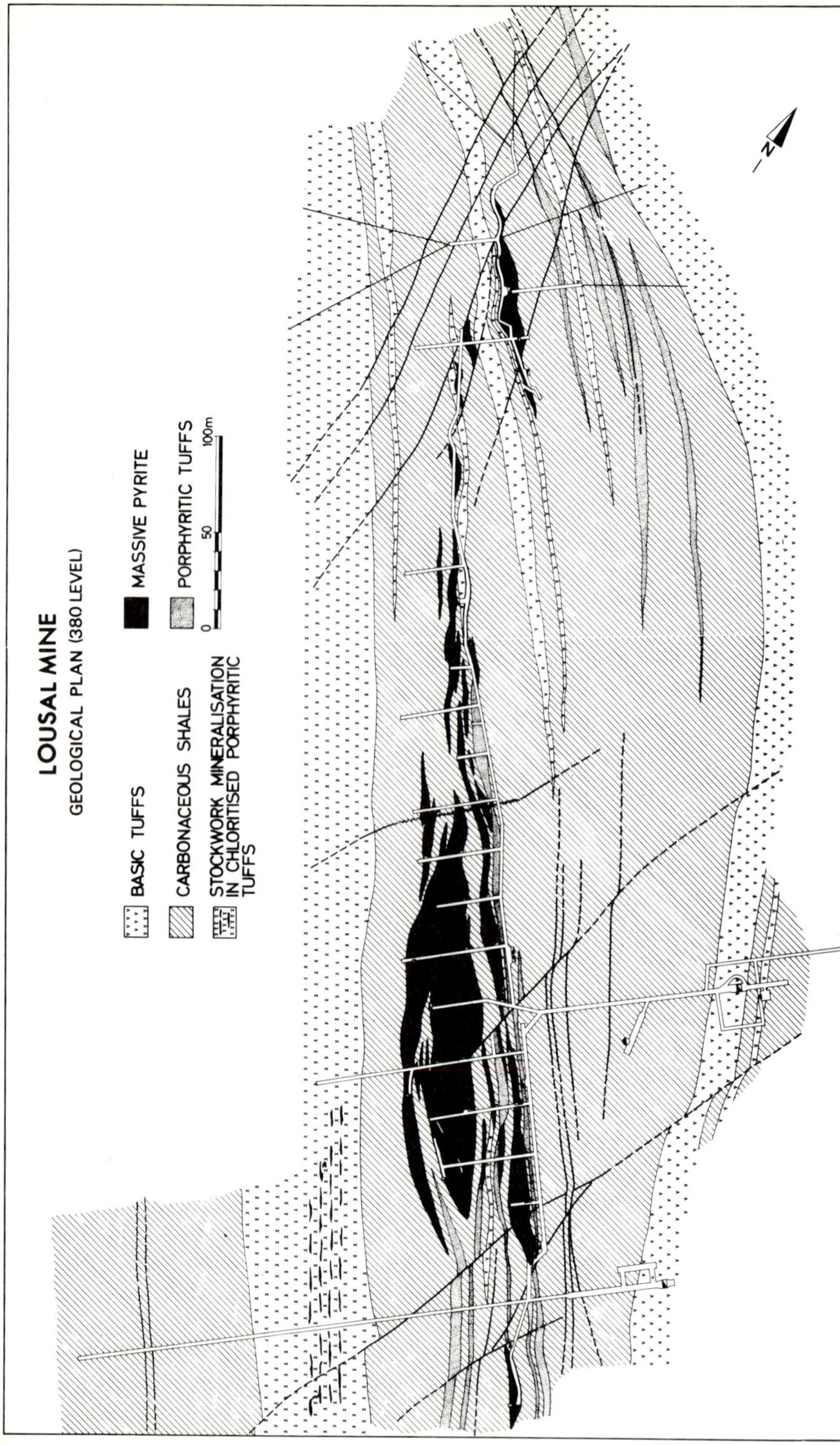

Fig. 10. Lousal

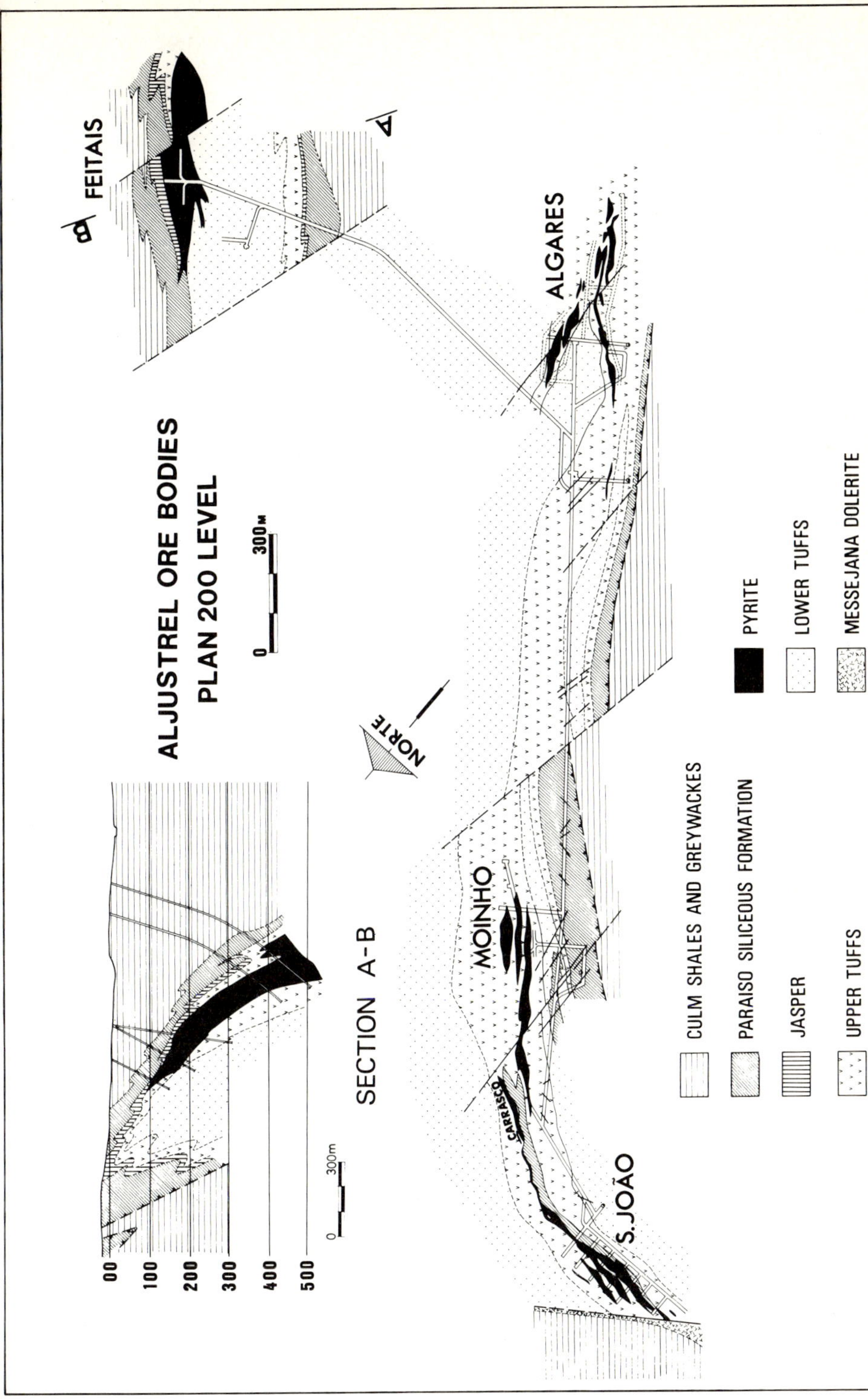

Fig. 11. Aljustrel

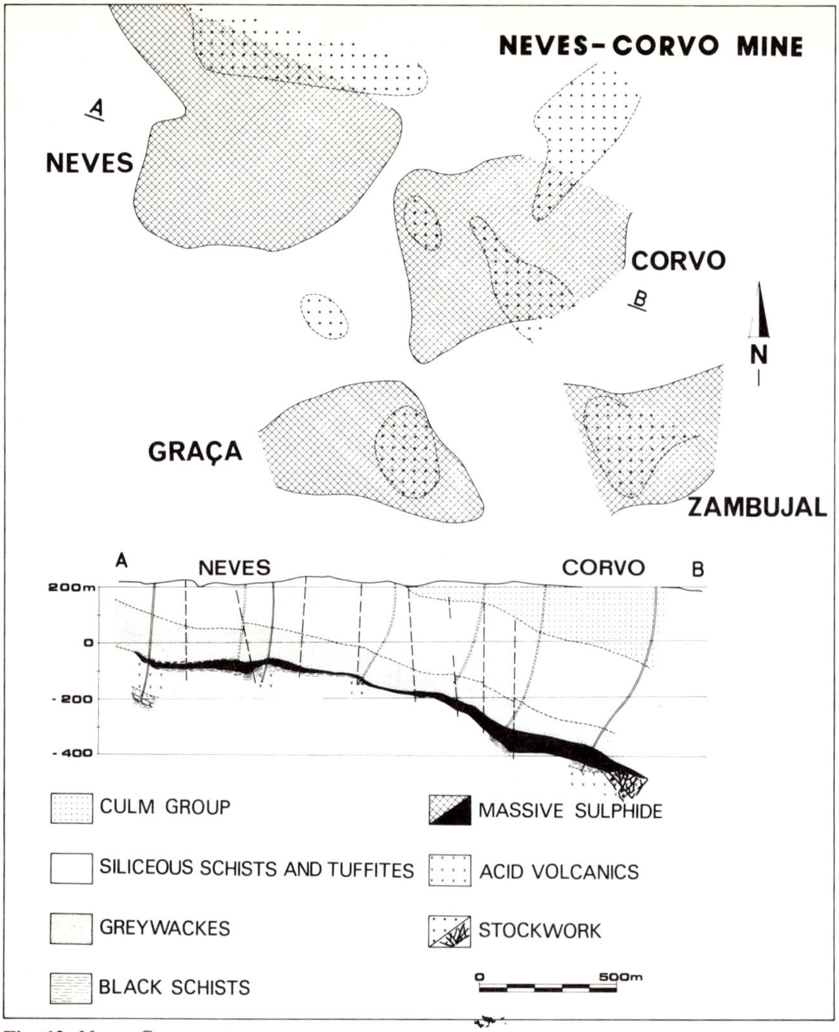

Fig. 12. Neves-Corvo

ization consisting mainly of chalcopyrite and pyrite. These massive stockwork ores average 1.50 – 4.0% Cu and 10 g/t Ag and similar breccia ores grade 2.5 – 3.5% Cu and up to 2% Zn (Albouy et al. 1981). An underground mining operation is being prepared and should be on stream in 1988/89 on a basis of 1 000 000 Tpy of ore producing some 60 000 Tpy of copper in concentrates.

Ore Beneficiation

The Iberian pyrite mining industry is based on many different types of ores and, similarly, several ways to use their contents of sulphur, base metals, gold and silver have been used or are being investigated.

Crude pyrites was until a few years ago the most important ore type produced in the pyrite belt and, for many years, constituted the back-bone of the European sulphuric acid industry. At present some 2.5 million tons per year of massive, crude pyrites are produced, crushed and screened, and sold mostly to the local sulphuric acid industry together with some exports. After roasting there remain 0.7 tons of residues ("cinders") per ton of pyrite, some of which are then treated for the recovery of base and precious metals and the production of iron ore (purple ore), suitable as a blast furnace feed.

Although there are several approaches to the treatment of *complex ores*, the only one being used industrially at this time is differential flotation with the production of copper, lead and zinc concentrates together with large quantities of very fine residual pyrite concentrates (Aznalcollar and Sotiel mines/Spain). The complexity of the mineral association and the extremely fine intergrowth require an ultrafine milling (80% $< 5/20\ \mu$) and even so make the production of prime quality base-metal concentrates difficult.

Alternative routes for the beneficiation of these complex ores, which are still under investigation, can be subdivided into hydrometallurgical and pyrometallurgical approaches and the following are representative (Strauss and Gray 1982):

1. Flotation of bulk concentrates followed by recovery of the base and precious metal values by hydrometallurgical leaching processes under varying physical and chemical conditions. Examples of these are the Sherrit Gordon and Comprex pressure leaching processes, on the one hand, and the Elkem Spigerverket and Minemet ferric chloride leach processes, on the other.
2. Roasting of the ore under sulphatizing conditions allowing the recovery of mainly, copper, zinc and silver by a simple water or dilute sulphuric acid leach. The roasting can be carried out in two steps with the production of elemental sulphur in the first, if so desired. Examples of this route are Outokumpu's Kokkola operation (Finland), and the Auxini (Spain) and Sapec (Portugal) processes under investigation.
3. Direct pyrometallurgical (smelting) routes such as the Kivcet process are also under consideration.

Another important source of base metals, mainly copper, are the *massive stockwork copper ores* and *disseminated copper ores*. Being both of epigenetic-hydrothermal origin, although also fine-grained, they are less so than the massive ores, with less intimate mineral intergrowth characteristics and minor occurrence of complex Cu – Pb – As – Sb – Hg minerals. As a result, milling to 80% less 325 Mesh is usually sufficient and relatively good liberation and clean concentrates of copper, zinc and pyrite are achieved.

The *gold-silver gossan ores* are submitted to cyanide leach processes and precious metals are recovered by zinc-dust precipitation.

References

Albouy L, Conde LN, Follerini F, Leca T, Morikis A (1981) Les gisements de sulfures massif polymetalliques de Neves-Corvo (Baixo Alentejo, Sud Portugal) Chron Rech Mines (Paris) 460:5 – 31

Bernard AJ, Soler A (1974) Aperçu sur la province pyriteuse Sud-Ibérique. In: Bartolomé P (ed) Gisements stratiformés et provinces cuprifères. Liege Soc Geol Belg, pp 287 – 315

Boogaard M van den (1967) Geology of the Pomaro region (South Portugal). Thesis, Univ Amsterdam

Boogaard M van den, Schermerhorn LJG (1975) Conodont faunas from Portugal and southwestern Spain. Script Geol Leiden 28:1 – 43

Carvalho D (1979) Geologia, metalogenia e metodologica da investigacao de sulfuretos polimetalicos do Sul de Portugal. Comun Serv Geol Portg (Lisboa) 65:169 – 191

Garcia Palermo F (1980) Caracteres geologicos y relaciones morfologicas y genéticas de los yacimientos del Anticlinal de Rio Tinto. Thesis, Univ Salamanca

Höllinger R (1958) Beitrag zur Kenntnis der Geologie im SW der Provinz Huelva. Thesis, Univ Münster

Kinkel AR Jr (1962) Observations on the pyrite deposits of the Huelva district, Spain, and their relation to volcanism. Econ Geol 57(7):1071 – 1080

Rambaud Perez F (1969) El sinclinal carbonifero de Rio Tinto (Huelva) y sus mineralizaciones asociadas. Mem Inst Geol Min Esp (Madrid) 71:1 – 229

Routhier P, Aye F, Boyer C, Lécolle M, Molière P, Picot P, Roger G (1979) La ceinture sub-ibérique à amas sulfures dans sa partie espagnole médiane. Mem BRGM 94:265 pp

Schermerhorn LJG (1970) The deposition of volcanics and pyritite in the Iberian pyrite belt. Min Deposits 5(3):273 – 279

Schermerhorn LJG (1971) An outline stratigraphy of the Iberian pyrite belt. Bol Inst Geol Min Esp (Madrid) 82(3/4):239 – 268

Schermerhorn LJG (1980) Copper deposits of the Iberian peninsula. In: Jankovic S, Sillitoe RH (eds) European copper deposits. Fac Min Geol, Belgrade, SGA Spec Publ 1:295 – 303

Solomon M, Walshe JL, Garcia Palermos F (1980) Formation of massive sulphide deposits at Rio Tinto, Spain. Trans Inst Min Metall London Sect B 89:B16 – B24

Strauss GK (1965) Zur Geologie der SW-Iberischen Kiesprovinz und ihrer Lagerstätten, mit besonderer Berücksichtigung der Pyritgrube Lousal/Portugal. Thesis, Ludwig-Maximilians-Univ, München

Strauss GK (1970) Sobre la geologia de la provincia pyritifera del Suroeste de la peninsula ibérica y de sus yacimientos, en especial sobre la mina de Lousal/Portugal. Mem Inst Geol Min Esp (Madrid) 77:266 pp

Strauss GK, Gray KG (1981) Complex pyritic ores of the Iberian peninsula and their beneficiation, with special reference to Tharsis Company Mines. Inst Mine Metall London 79 – 87

Strauss GK, Gray KG (1982) The international outlook for Spanish pyrites. Ind Min London (Abstr) 175:29

Strauss GK, Madel J (1974) Geology of massive sulphide deposits in the Spanish-Portuguese pyrite belt. Geol Rundsch 63(1):191 – 211

Strauss GK, Madel J, Fernandez-Alonso F (1977) Exploration practice for strata-bound volcanogenic sulphide deposits in the Portuguese-Spanish pyrite belt. In: Klemm DD, Schneider HJ (eds) Time- and strata-bound ore deposits. Springer, Berlin Heidelberg New York, pp 55 – 93

Vazquez Guzman F (1978) Depositos minerales de Espana. Bol Inst Geol Esp (Madrid) 1958 pp

Williams D (1962) Further reflections on the origin of the porphyries and ores of Rio Tinto, Spain. Trans Inst Min Metall (London) 71(5):265 – 266

Williams D, Stanton RL, Rambaud F (1975) The Planes-San Antonio pyritic deposit of Rio Tinto, Spain: its nature, environment and genesis. Trans Inst Min Metall London Sect B 84:B73 – B82

Copper-Pyrite and Pyrite Base Metal Deposits of the Caucasian Region

N. K. KURBANOV[1]

Abstract

This paper analyzes metallogenic specialization of two pairs of pyrite-bearing structural-formational zones[2] belonging to the Caucasian segment of the Mediterranean fold belt. These zones evolved during the Alpine tectono-magmatic cycle along the northern and southern framing of the Transcaucasian microcontinent. In the northern pair (the Great Caucasus meganticlinorium) the pyrite-bearing zones are in the central trough and the south continental rise, and in the southern pair (the Lesser Caucasus meganticlinorium), they are in the rift ophiolitic zone and the island arc. A clustering concentration pattern of base-metal mineralization with centrifugal and centripetal types of localization of pyrite, porphyry and vein deposits has been revealed.

Introduction

The present-day classification of the pyrite-bearing provinces represented by various types of geosynclinal systems is based on the character of volcanic or volcano-sedimentary rocks filling in the geosynclines and paleotectonic setting, which determines the types and evolution of ore-bearing formations, their succession and the ore grade (Borodayevskaya et al. 1979, Kurbanov 1982, Hutchinson 1973).

The Caucasian segment of the Mediterranean fold belt is noted for a variety of pyrite deposits associated with geosynclinal systems of different age and tectonic style. In the long period of polycyclic evolution (from Riphean to Neogene) the major recent structures of the Caucasian region – Great and Lesser Caucasus meganticlinoria separated by the Transcaucasian median mass (microcontinent) – underwent progressive rejuvenation of the geosynclinal regime, folding, and metallogeny in a southward direction (Fig. 1).

[1] TsNIGRI, Varshavskoje Shosse, 58, Moscow, M-430, USSR

[2] A structural-formation zone lies within the fold area and differs from the neighboring zones in features of sedimentation, structure, magmatism, determined by tectonic regime and a number of physicogeographical factors (such as climatic) characteristic of the zone during its evolution. The zone is usually bounded by deep faults. The term has been used extensively of late.

Geology and Metallogeny of Copper Deposits
(ed. by G. H. Friedrich et al.)

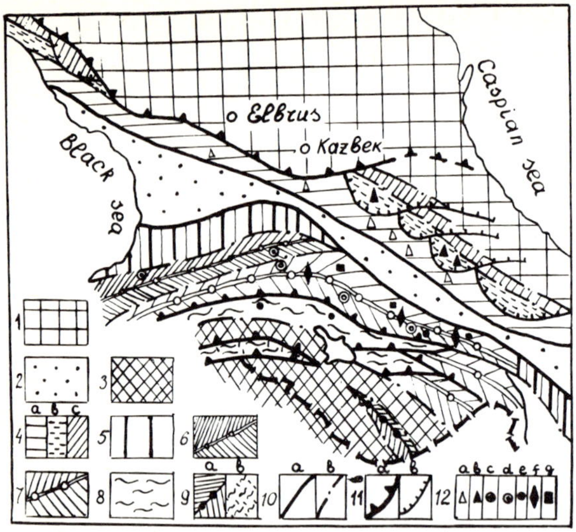

Fig. 1. Location of copper-lead-zinc deposits in the Alpine zones of the Caucasus. *1* Scythian craton; *2* Transcaucasian median mass; *3* Iranian median mass; *4* structural-formational zones of the Great Caucasus terrigenous eugeosyncline: *a* Pshekhi-Dagestan zone of early stabilization; *b* Bzyb-Tfan trough zone (central); *c* Chkhalta-Sarybash zone of the continental rise. *5* Adjar-Trialeti and Talysh superimposed troughs; *6* cretaceous volcanogenic island arc of Pontid; *7* Jurassic Somkheti-Kafan island arc; *8* Sevan-Akerin ophiolitic zone; *9* activated areas of the Iranian median mass: *a* Ordubad volcanogenic trough; *b* Vedi ophiolitic zone of the Cretaceous. *10* Boundaries of structural-formational megazones (*a*) and zones (*b*). *11* Alpine-type regional (*a*) and local (*b*) trusts. *12* Commercial-genetic types of deposits: *a* stratiform pyrite-base metal; *b* pyrite-chalcopyrite-pyrrhotite; *c* copper-pyrite; *d* pyrite-barite base metal; *e* sulfur-pyrite; *f* vein base metal; *g* copper-porphyry

An analysis of the geodynamics and formational composition of the Mesozoic rocks of the Great and Lesser Caucasus shows that in the Alpine tectono-magmatic cycle, which was most prolific in pyrite ores, that these were independent systems, i.e. terrigenous (Great Caucasus) and volcanogenic (Lesser Caucasus) eugeosynclines that evolved on a consolidated subplatform base as secondary linear tension zones (Kurbanov 1982, Tvalchrelidze 1977, Khain 1975). The essential difference in the tectono-magmatic evolution of the above megazones predetermined in many ways the variety of commercial-genetic types of base metal mineralization within them.

Genetic Models and Formative Conditions of Pyrite Deposits of the Great Caucasus Terrigenous Eugeosyncline

The Alpine terrigenous eugeosyncline of the Great Caucasus meganticlinorium, which hosts the base metal-bearing massive sulfide province, emerged in the north of the region, along the margin of the craton (Russian platform) and microcontinent (Transcaucasian median mass) at the onset of the Lower Jurassic within a narrow strip of the oceanic crust. It consists of three structural-formational zones: (1) the northern, early stabilized Pshekhi-Dagestan zone adjacent to the craton; (2) the central Bzyb-Tfan trough zone of the rift type; (3) the southern Chkhalta-Sarybash zone of the continental rise, which developed along the active margin of the microcontinent and was transformed subsequently into a volcano-terrigenous island arc. These structural-formational zones are separated by longitudinal long-lived synsedimentary fault zones (the Great Caucasus and Kekhnamedan zones) related probably to the Benioff paleozones.

All the known pyrite deposits of copper, lead, and zinc are confined exclusively to the central and southern structural-formational zones that are complicated by allochthonous overthrust sheets and display complex structure.

Within the *central Bzyb-Tfan trough zone* (see Table 1) the ore-bearing rocks are the slate, slate-graywacke, and terrigenous-flysch sedimentary formations, as well as two unimodal volcanic formations: the early (J_1, pl) siliceous-picrite-spilite-basalt, and the late ($J_2 t-a$) spilite-diabase ones.

The ore-bearing terrigenous and volcanic formations within the trough zone compose two polycyclic regressive rhythmostratigraphic complexes[3]: *the lower* (J_1, pl – J_2) 4600-m-thick complex, and *the upper* (J_2a – bt), 3900-m-thick complex. Both complexes are noted for clay (with sulfide disseminations and concretions) and volcanic units that accumulated in the linear synsedimentary paleobasins.

Volcanic formations are accompanied by belts (the Kazbek and Kakhetian) of diabase stocks and dikes, bounding the trough zone across the strike.

In terms of petrography and petrochemistry, basaltic rocks of siliceous-picrite-spilite-basalt formation are comparable with oceanic tholeiites (similar to tholeiites of the present-day rift of the Red Sea), whereas those of spilite-diabase formation are correlatable with their subvolcanic equivalents similar to basaltic rocks of unimodal formations of volcanogenic eugeosynclines (Fig. 2).

The above-described formations of the trough zone are crumpled into linear and isoclinal folds showing an S-vergence and complicated by co-folded reverse-thrust and overthrust faults. A thick crush zone with abundant dikes and minor intrusions of volcano-plutonic associations, namely, basalt-dacite-rhyolite-plagiogranite association ($T_2 bt$), is traceable in the present-day structures along the southern margin of the trough zone.

Three rhythmostratigraphic complexes are recognized in the *southern Chkhalta-Sarybash zone* (see Table 1): (1) the Sinemurian-Pliensbachian (1 300 m) complex that incorporates a layer of basal conglomerate – extrusions and tuff of andesite-rhyodacite calcareous-alkaline potassium-sodium formation (Sinemurian) and the ore-bearing slate-greywacke formation of the continental rise (Toarcian); (2) the Toarcian-Aalenian (1 500 m) a complex composed of terrigenous-flysch (Toarcian) and an ore-bearing slate-greywacke (Aalenian) unit containing lava flows and extrusive-tuffaceous facies related to the contrasting basalt-rhyolite suboceanic formation of sodium series; and (3) the Bajocian-Bathonian (2600 m) complex, comprising volcanics and hypabyssal intrusions of potassium-sodium basalt-andesite-dacite-rhyolite-plagiogranite volcano-plutonic

[3] Rhythmocomplexes in the Soviet geologic literature mean alternating sets of volcanic and sedimentary formations that evolved at the same stage in a similar tectonic setting.

Table 1. Ore-bearing structural-formational zones of the Great Caucasus

Rhythmo-strati-graphic complex	Southern zone of active continental rise (Chkhalta-Sarybash)					Central trough (rift) zone (Bzyb-Tfan)				
	Magmatic formations		Sedimentary formations		Localization levels of stratiform ore bodies	Magmatic formations		Sedimentary formations		Localization levels of stratiform ore bodies
	Composition	Thick-ness (m)	Composition	Thick-ness (m)		Composition	Thick-ness (m)	Composition	Thick-ness (m)	
Bajocian-Bathonian	Basalt-andesite-dacite-rhyolite gabbro-plagio-granite	500 – 1400	Terrigenous-flyschoid	700 – 1200	–	Gabbro-plagio-granite (dikes and minor intrusions)	–	Terrigenous-carbonate-flyschoid	1300	–
Toarcian-Aalenian	Basalt-rhyolite	200 – 700	Slate-gray-wacke (Aalenian) and ter-rigenous-flyschoid (Toarcian)	1500	Upper – in shales of the upper parts of Aalenian; lower – in shale of the upper part of Toarcian	Splite-dia-base (Lower Aalenian) Splite-dia-base (Toarchian)	300 – 500 550 – 600	Terrigenous-flyschoid: rhythmic alternation of sandy-shale flysch-oid members (50 – 10 m) and shale beds (100 – 150 m)	3900	Over spilite flows, in shales of the lower parts of the Upper Aale-nian. Over spilite flows, in shales of the lower parts of Toarcian
Sinemurian-Pliensh-achian	Andesite-rhyodacite	150 – 200	Slate-gray-wacke: alter-nation of graywacke sandstone members (50 – 70 m) and shale beds	1300	In the up-per parts of shales	(Siliceous)-picrite-basalt	350	Slate: shale, phyllite with graywacke members (30 – 50 m) and rhythm-ically thin-bedded silty sandstone	4600	In the up-per parts of shales, over basalt flows

association and a sandy-siltstone terrigenous-flysch formation developed mainly in the southeastern segment of the southern slope of the Great Caucasus.

Magmatism in both zones was terminated by the emplacement of dikes and minor intrusions of gabbro-diorite formation (J_3-K_1). In terms of petrography and petrochemistry, the products of magmatism are related to calcareous-alkaline series of ensialic island arcs (Fig. 2).

Typical of the ore-bearing formations of rhythmocomplexes is the accumulation of clay facies (enriched by C_{org} up to 5 – 6% against 0.1 – 0.9% background value, as well as by numerous impregnations and concretions of pyrite and pyrite-siderite) in local synsedimentary basins frequently controlled by magma- and ore-feeding faults and bounded laterally by vast sand uplifts ("sand barriers").

The zone are noted for large anticlines (with recumbent southern, relatively gently inclined northern limbs and heavily block-faulted cores), and narrow synclines. The cores and northern limbs of the anticlines are frequently complicated by local injection structures, high-order folds with steep hinges and co-folded reverse-thrust faults. Many of them emerged as synsedimentary inversion uplifts in place of the above-mentioned local paleobasins.

The majority of investigators (Borodayevskaya et al. 1979, Kurbanov 1982, Tvalchrelidze 1977) relate copper-lead-zinc mineralization of the Great Caucasus

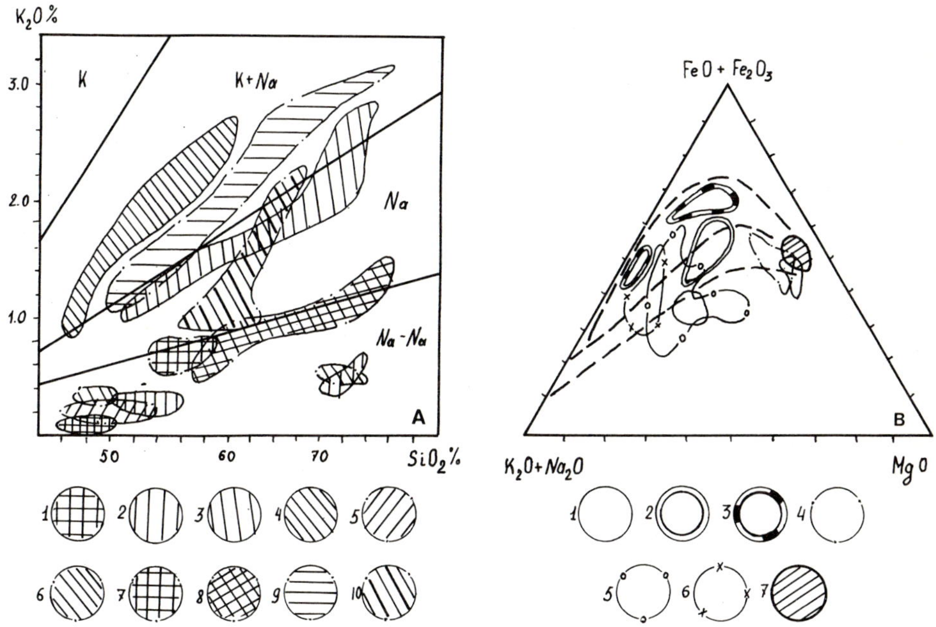

Fig. 2A, B. Volcano-plutonic associations of the Great and Lesser Caucasus. **A** Variation diagram showing potassium to siliceous acid ratio. Volcanic formations of the Great Caucasus: *1* unimodal; *2* bimodal; *3* trimodal; *4* gabbro-diorite-plagiogranite Volcanic formations of the Lesser Caucasus; *5* uni modal; *6* bimodal basalt-rhyolite; *7* bimodal andesite-basalt; *8* trimodal; *9* gabbro-plagiogranite; *10* cretaceous volcanic trimodal. **B** The ratio of the sum of alkali, ferrum, and magnesium, Great Caucasus: *1* unimodal; *2* bimodal; *3* trimodal, Lesser Caucasus: *4* unimodal; *5* bimodal; *6* trimodal; *7* field of undifferentiated basalt magma

terrigenous eugeosyncline to two major ore types – stratiform pyrite and hydrothermal vein. Two subtypes, namely, pyrite-chalcopyrite-galena-sphalerite and chalcopyrite-pyrrhotite are distinguished within the pyrite ore type. The former subtype developed in two stages: an early, stratified hydrothermal-sedimentary (essentially pyrite or pyrrhotite, with sulfides of copper, lead, and zinc) and a late, hydrothermal-metasomatic (the most abundant in base metals) stage that emerged after early geosynclinal deformations of stratified ores and the emplacement of minor intrusions of plagiogranite bodies. The later ores of the hydrothermal-vein type, which form after the intrusion of late geosynclinal gabbrodiorite formation, are represented by two paragenetic associations of different ages: the early, quartz-chlorite-chalcopyrite-pyrrhotite association and the late, quartz-carbonate-pyrite base-metal one. The combination of ores of the pyritic subtype with those of the hydrothermal-vein type gave rise to polygenetic combined mineral deposits of different ages within the province.

Localization of polygenetic and monogenetic deposits known from the region follows a zonal pattern. Thus, the majority of deposits in the Bzyb-Tfan zone are made up of pyrite ores of chalcopyrite-pyrite-pyrrhotite composition, whereas those of the Chkhalta-Sarybash zone are represented by pyrite base-metal ores. Superimposed veinlet-disseminated ores of quartz-chlorite-chalcopyrite-pyrrhotite composition are widespread in the trough zone, whereas quartz-carbonate-pyrite base-metal ores are common in the southern zone of the active continental rise. Such a localization pattern of the mineralization makes it possible to identify two metallogenic zones: the Bzyb-Tfan and Chkhalta-Sarybash zones whoses geological boundaries coincide with the structural-formational zones of the same names.

The second localization of stratiform pyrite ores is determined by their confinement to three stratigraphic levels (see Table 1): Pliensbachian, Toarcian, and Upper Aalenian, and within these levels – by their restriction to stratified ore-bearing layers (with relatively high concentrations of iron sulfides ranging from 1% to 15% and more). These layers occupy certain regular positions in shale units of the transgressive and regressive series of rhythmostratigraphic complexes.

The analysis of the localization of the deposits shows that the commercial concentration of the stratiform pyrite mineralization within the ore-bearing layers is associated with clay facies that accumulated in the above-mentioned local paleobasins under the conditions of a reducing regime. The central parts and active sides of the basins are complicated by magma- and ore-feeding faults and the products of submarine hydrothermal exhalations, such as sulfide ores of solid-massive, bedded, mottled, and other structural types. Pyrite is the main mineral in these ores; the amount of sphalerite, chalcopyrite, and galena is varied.

As has been established during prospecting for stratiform ores, the massive deposits change laterally and vertically to sulfide rhythmites represented by alternating sulfide and clay interbeds ("ore flysch"). The latter interbeds are, in their turn, replaced laterally by facies composed of lenticular-streaky pyrite accumulations, and then by stratified layers of pyrite concentrations.

Owing to the emplacement of subintrusions associated with a persistent early island-arc ore-generating formation, the stratiform pyrite ores were subjected to the most essential alterations of the first stage, which consisted of the introduction (and, possibly, regeneration and redeposition as well) of sulfides of the productive associations that rcplaced metasomatically the sulfides of the earlier ores. In time, this process coincided with the early geosynclinal stage of folding and faulting (development of linear folds, reverse-, and overthrust faults, and flow cleavage) that were responsible for boudinage and recrystallization of ores accompanied by regeneration and redeposition of the ore matter along conjugate faults and resulted in vein and veinlet-disseminated systems.

The most intense metamorphism of the stratiform deposits is connected with hypabyssal intrusions of the late island-arc gabbro-diorite formation that were accompanied by considerable regeneration of the ore matter and transformation of chalcopyrite-sphalerite-pyrite ores from many mineral deposits of the Bzyb-Tfan zone into chalcopyrite-pyrrhotite ores with halos of veinlet-disseminated quartz-chlorite-chalcopyrite-pyrrhotite and quartz-carbonate-pyrite-sphalerite-galena ores.

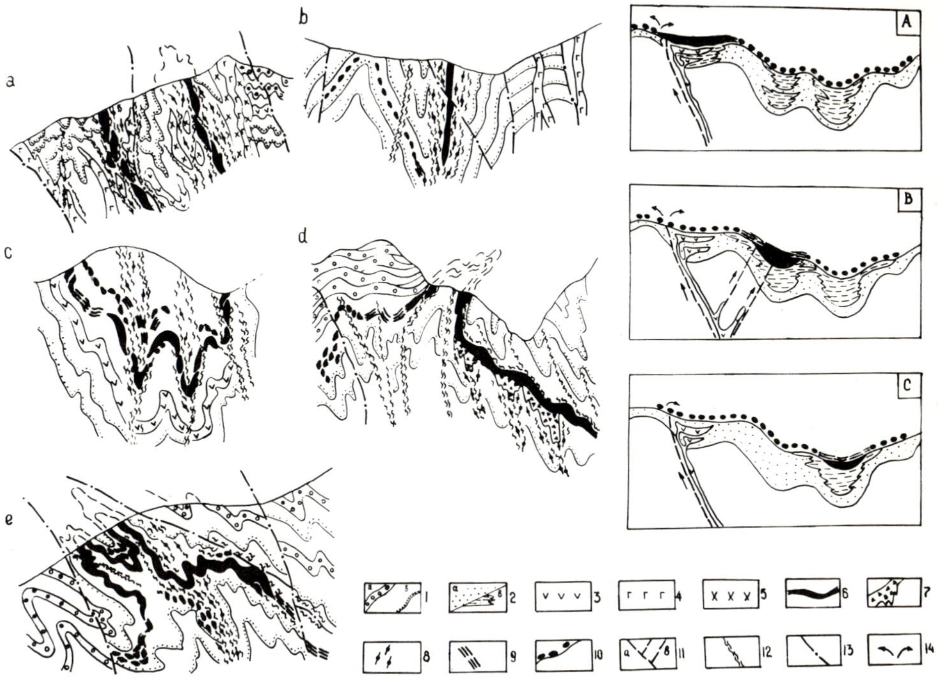

Fig. 3. Type models showing various settings in which the Great Caucasus copper-pyrite deposits were formed. Accumulation of hydrothermal-sedimentary ores in the local basins: **A** above ore-generating source; **B** near the source; **C** at a distance from the source. Location in the present-day structures: **a** in the fold zones; **b** in axial zones of linear folds; **c** in synclinal folds; **d** in inversion uplifts; **e** in inversion uplifts complicated by injection structures. *1* Sandstone beds (**a**), pockets of intercalating silty sandstone and shale (**b**); *2* sandy (**a**) and clayey (**b**) facies; *3* spilite and basalt lava flows and their feeding channels; *4* minor intrusions and dikes of gabbro-diorite formation; *5* subvolcanic intrusions of andesite, rhyodacite, and rhyolite; *6* beds of stratiform pyrite ores of mottled-breccioid structure; *8* veinlet-disseminated ores; *9* rhythmites (ore flysch); *10* sulfide concretions; *11* magma- and ore-feeding channels: *a* main, *b* feathering. *12* Schistosity and crush zones; *13* reverse-thrust and thrust faults; *14* direction of outpouring of mineral solutions

The analysis of the localization of the pyrite and hydrothermal-vein base-metal mineralization has shown that against the background of linear metallogenic zones, the stratiform mineral deposits occur as clusters. These ore districts within the longitudinal metallogenic zones are confined to relatively uplifted inversion cross-cut blocks characterized by greater thickness of the ore-bearing sedimentary formations. Major uplifts, in their turn, consist of several conjugate local ore-enclosing inversion uplifts of various types; these are the most promising ore fields. The latter frequently form compact ore clusters controlled by regional longitudinal long-lived magma- and ore-feeding faults.

Taking all this into account, the type model of the setting in which stratiform monogenetic deposits were formed, for the region under review, would be the accumulation of these deposits in local paleobasins: (1) directly above the ore-generating source; (2) near the source, where they are controlled by long-lived magma- and ore-feeding faults; and (3) at a distance from the ore-generating source. It is noteworthy that with the distance from the ore-generating source the role of biogenic-sedimentary sulfides (Kurbanov 1982) increases significantly (Fig. 3A, B). However, the present-day models of the targets for prognosis and exploration are determined not only by initial conditions of their formation, but also by settings that depend on a combination of these and on the intensity of their alteration during subsequent tectonic-magmatic processes that produced polygenetic deposits of different ages.

Genetic Models and Formative Conditions of Pyrite Barite Base-Metal Deposits Hosted by the Lesser Caucasus Volcanogenic Eugeosyncline

The volcanogenic eugeosyncline of the Lesser Caucasus occupies a geotectonic position similar to that of the Great Caucasus, since it is situated between the Afro-Arabian craton in the south and Transcaucasian microcontinent in the north. Initiated as a secondary geosyncline only in the Middle Jurassic (Kurbanov et al. 1978, Tvalchrelidze 1977, Khain 1975), this metallogenic province differs considerably from the above-described one in its tectonic style, role of magmatism and, therefore, metallogenic specialization and the set of commercial-genetic types of base-metal deposits.

Three structural-formational zones, namely: (1) the northern Somkheti-Kafan zone; (2) the central Sevan-Akerin zone; and (3) the southern Miskhan-Zangezur zone are recognized in the Alpine volcanogenic eugeosyncline of the Lesser Caucasus (Kurbanov et al. 1978, Tvalchrelidze 1977, Khain 1975).

The Miskhan-Zangezur zone consists of rigid blocks (Daralagez, Miskhan-Zangezur) activated in the Jurassic, Cretaceous, and Paleogene and confined to the northern framing of the Iranian median mass. These blocks include the Cretaceous allochthon of the Vedi ophiolitic zone (Khain 1975). The Alpine base-metal metallogeny of the zone is characterized by syngenetic, frequently stratiform, pyrite ores associated with extrusions and volcanic domes of the Eocene basalt-andesite-dacite-rhyolite calcareous-alkaline volcanic formation.

The Sevan-Akerin zone which is an eastern continuation of the Tavr-Anatolia ophiolitic zone, represents a linear rift tension zone that emerged (most probably in the Bajocian) on the oceanic crust and was transformed subsequently into a melange suture structure with numerous allochthonous slabs of the Cretaceous limestone, ophiolitic rocks, and olistostrome units of Oxfordian-Tithonian age. As to the endogenic mineralization, most typical of the early geosynclinal stage is the development of chromite ore occurrences and hydrothermal-sedimentary copper-pyrite mineralization of the Ergani-Maden type hosted by volcanics of the unimodal tholeiite-siliceous-spilite-basalt formation.

The set of the igneous rocks, their geotectonic evolution, and their metallogenic specialization make it possible to identify the Jurassic Somkheti-Kafan zone and the Cretaceous Pontid zone, replacing it an echelon in the west, as a single belt of the ensialic volcanogenic island arc that formed on a thick and active southern margin of the Transcaucasian microcontinent and that is a surface expression of an old Benioff zone (Fig. 1).

Three volcano-plutonic associations are recognized within the island arc. Each of these is characterized by its own metallogenic specialization. *The earliest, Middle Jurassic association* is represented (Fig. 2) by the Early Bajocian unimodal spilite-basalt formation developed upon the Toarcian-Aalenian sandy-shale unit and reaching its maximum thickness in the *Southern subzone*; here, it grades vertically into volcanic products of the suboceanic contrasting basalt-rhyodacite formation of the sodium series ($\Sigma_{ex} = 15-20\%$)[4]. In the *Northern subzone* these products are replaced by facies change to thick (up to 2900 m) central-type volcanics ($\Sigma_{ex} = 75-80\%$) of calcareous-alkaline basalt-andesite-dacite-rhyolite formation of the sodium series, filling there separate volcano-tectonic uplifts and depressions. This Middle Jurassic volcanism is terminated by the Bathonian basalt-andesite formation of the potassium-sodium calcareous-alkaline series and by the emplacement of hypabyssal polyphase intrusions of gabbro-plagiogranite formation.

The successive, *Upper Jurassic volcano-plutonic association* is represented by two formations of a transitional sodium and sodium-potassium calcareous-alkaline series: the earlier, andesite-dacite, and the later, gabbro-diorite-granodiorite-adamellite-banatite formations (Fig. 2) developed exclusively in the superimposed troughs.

The *Upper Cretaceous early island-arc volcano-plutonic association* (Figs. 1 and 2) identified in the Bolnisi volcano-tectonic depression of the extreme south-east termination of the Pontid branch of the island arc (Figs. 1 and 4) is compositionally referred to calcareous-alkaline magmatites of the sodium and potassium-sodium series (Fig. 2); products of trimodal volcanic formation are concentrated in the depression proper, while those of plutonic gabbro-plagiogranite-granodiorite formation – along its periphery, in uplifts of the old basement (Khram and Lok massifs) that terminate the depression (Fig. 4).

It should be stressed that the geotectonic evolution regime of the above volcano-plutonic associations was centripetal in the volcano-tectonic uplifts, being centrifugal in the depressions (Fig. 4). This being so, the volcanic activity in

---

[4] Σ_{ex} = Coefficient of explosiveness.

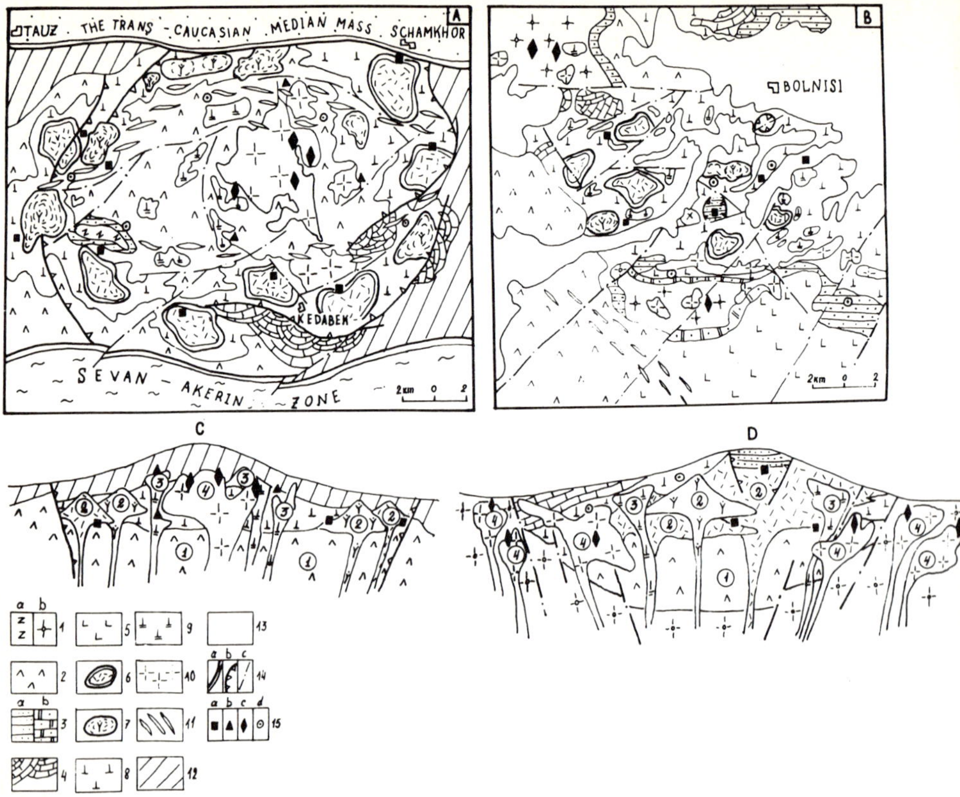

Fig. 4A – D. Type models for ore districts of the Lesser Caucasus island arcs. **A** Ore districts with centripetal tectonic regime of ore-bearing-volcano-plutonic associations and the related base metal deposits: **A** in plan, **C** in the section. **B** The same, with centrifugal tectonic regime: **B** in plan, **D** in the section. *1* Basement rocks: *a* schists; *b* gneiss. *2* Andesite-basalt and andesite volcanics; *3* terrigenous (*a*) and volcano-terrigenous (*b*) sediments; *4* reef limestone; *5* andesite-dacite volcanics; *6* rhyolite and rhyodacite extrusions; *7* rhyolite and rhyodacite volcanic domes; *8* rhyolite lava and tuff; *9* rhyolite and rhyodacite subvolcanic intrusions; *10* hypabyssal intrusions of gabbro-plagiogranite formation; *11* trachyandesite, trachydacite, trachybasalt dike belts; *12* Upper Jurassic-Lower Cretaceous undifferentiated volcano-sedimentary formations; *13* quaternary plateau basalt; *14* boundaries of island arc (*a*), subsidence-calderas (*b*), and synvolcanic faults (*c*); *15* commercial-genetic types of deposits: *a* pyrite, *b* pyrite-chalcopyrite-enargite, *c* copper-porphyry, *d* vein-quartz base metal

the volcano-tectonic uplifts started with the formation (Fig. 4) of a vast caldera filled with andesite-basalt and andesite volcanics that subsequently were replaced along the periphery by semiring structures of volcanic domes and multivent extrusions of rhyolite-dacite and rhyolite. The successive rhyolite subintrusive facies intruded the central parts of the uplifting caldera, the formation of which was terminated by the intrusion of gabbro-plagiogranite into the apical part of the uplift.

In contrast, despite the similar character of the volcanism in the volcano-tectonic depressions, their tectonic style is reversed: the evolution of silicic and moderately silicic volcanism was centrifugal and was terminated by the emplacement of major polyphase intrusions along the periphery of a vast volcano-tectonic depression showing a mosaic structural pattern.

The known base-metal ores of the Lesser Caucasus province are attributed to three main commercial-genetic types, viz., pyrite, porphyry, and vein types that are noted for higher concentrations of tungsten, tin, molybdenum, gold and silver. All the listed base-metal ore types are related spatially and genetically to the Bajocian-Bathonian and Upper Cretaceous volcano-plutonic associations of the Somkheti-Kafan and Pontid island arcs (Fig. 1). The pyrite deposits include (Kurbanov 1982, Tvalchrelidze 1977): (1) iron-pyrite and pyrite-chalcopyrite hydrothermal-sedimentary deposits (the Cyprus type), associated with lavas and jaspers of unimodal mafic rock formation; (2) iron-pyrite synvolcanic deposits (the Urals type) associated with contrasting formation; and (3) synvolcanic copper-pyrite deposits associated with the Bajocian basalt-andesite-dacite-rhyolite sequences and localized in subsidence calderas, vents and slopes of extrusions, and volcanic domes (the Lesser Caucasus type); (4) chalcopyrite-enargite (with luzonite) pyrite postintrusive deposits linked with copper-porphyry ores; (5) polygenetic copper-pyrite and barite base-metal deposits of various located in the Bolnisi area and attributed to the Kuroko type (Tvalchrelidze 1977, Kurbanov et al. 1978, Borodayevskaya et al. 1979) that are localized in subsidence-calderas, volcanic domes, and volcano-tectonic depressions filled with volcanics of Upper Cretaceous andesite-dacite-rhyolite formation (Fig. 5).

Permanently, following the emplacement of polymodal intrusions of gabbro-granodiorite-plagiogranite sequences and their porphyry series, within the fields of high-temperature (with cordierite) quartzites, stockwork zones of island-arc (Kurbanov et al. 1978) copper-porphyry ores were developed.

According to the above-described evolution regime of volcano-tectonic uplifts and depressions, which correspond to ore districts (ranging in size from 350 km^2 to 400 km^2), the localization pattern of the various types of deposits is centripetal and/or centrifugal. Pyrite deposits are concentrated in the volcano-tectonic uplifts along the periphery of the ore districts (Fig. 4); closer to the center of the uplifts, they are replaced by pyrite ores superimposed on subintrusive rhyolite facies. These, in turn, are replaced in the endocontact halo of the gabbro-plagiogranite intrusions by pyrite-chalcopyrite-enargite ore deposits. Finally, copper-porphyry deposits are concentrated in the center of the uplifts. These deposits are localized directly among plagiogranites (Fig. 4). In the volcano-tectonic depressions, the localization patterns of the above deposits within the ore districts are reversed (Fig. 4).

Taking into account the above-stated facts the following conclusions can be drawn:

1. Pyrite deposits of the Caucasian region are associated genetically and spatially with conjugate pairs of structural-formational zones, viz., frontal troughs and island arcs that emerged along the southern and northern bordes of the Transcaucasian microcontinent as ensimatic linear tension zones and activated ensialic heterogenous blocks of the Earth's crust.
2. Hydrothermal-sedimentary copper-pyrite deposits are associated with unimodal mafic rock formations in rift trough zones, while in the island arcs they

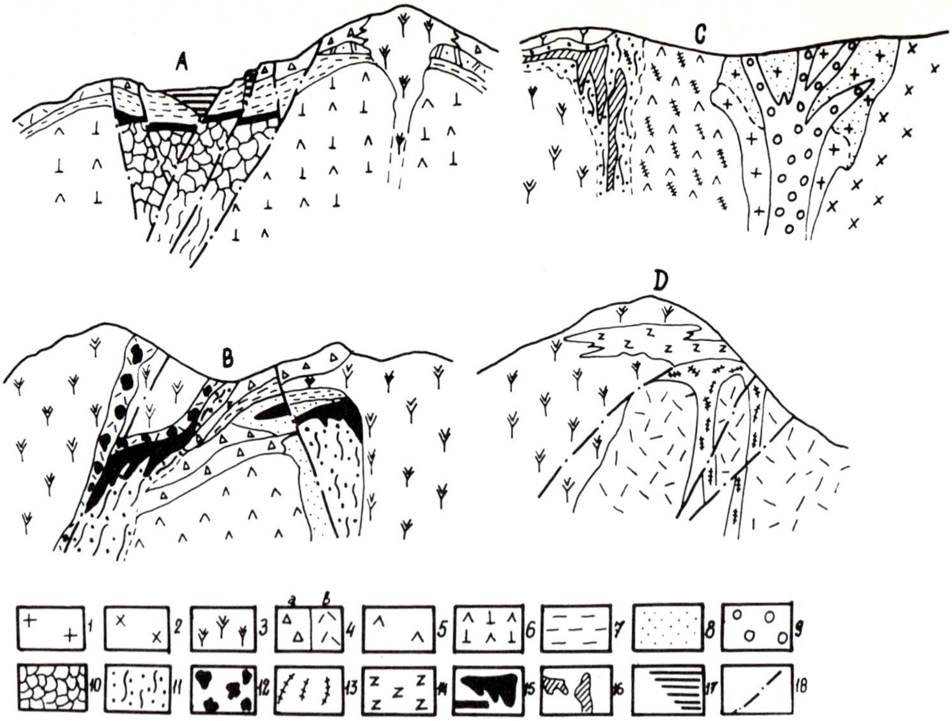

Fig. 5. Formative conditions of the Lesser Caucasus copper-lead-zinc deposits of various commercial-genetic types. *1* Plagiogranite; *2* granodiorite-porphyry; *3* rhyolite and rhyodacite; *4* rhyolite tuff (*a*) and ignimbrite (*b*); *5* andesite-basalt; *6* andesite-dacite and dacite; *7* tuffstone, tuffite and tuffmudstone; *8* copper-porphyry ores of the outer zone; *9* copper-porphyry ores of the inner zone; *10* copper-pyrite ores of the breccioid structure; *11* veinlet-disseminated ores of quartz-pyrite-chalcopyrite and quartz-pyrite-enargite composition; *12* ore clasts (fragments of pyrite ores) in tuff; *13* veinlets of late-quartz base metal ores; *14* barite ores; *15* stratiform and metasomatic copper-pyrite ores; *16* massive pyrite-chalcopyrite-enargite ores; *17* barite-base metal ores; *18* faults

are connected with bi- and trimodal types, manifesting a well-pronounced polygenetic nature, different ages, and a copper-pyrite section.

3. The clustering and zonal localization pattern of the copper-pyrite deposits is associated with the tectonic style of individual transverse blocks of the ensialic island arcs. Two types, viz., centripetal (in the volcano-tectonic uplifts) and centrifugal (in the depressions) are recognized in the volcanogenic arcs; in active continental rises only one type is developed – vast inversion uplifts consisting of local uplifts that appeared in the place of compensated paleobasins.

References

Borodayevskaya MB, Gorzhevsky DI, Krivtsov AI, Tvalchrelidze GA, Yakovlev GF (1979) Geotectonic and paleotectonic types of pyrite-bearing provinces. In: World massive sulfide deposits. Nedra, Moscow, pp 15 – 26 (in Russian)

Hutchinson RW (1973) Volcanogenic sulfide deposits and their metallogenic significance. Econ Geol 68:1223 – 1246

Khain VE (1975) The main stages of the Caucassus tectono-magmatic evolution: geodynamic interpretation practice. Geotektonika 1:13 – 27 (in Russian)

Kurbanov NK (1982) The main stages of formation of combined copper-pyrite deposits and their correlation with evolution stages of the Alpine Great Caucasus terrigenous geosyncline. Tr TsNIGRI Moscow Iss 168:3 – 18 (in Russian)

Kurbanov NK, Kulakov VV, Zaryanov YuP, Antonov VA (1978) Jurassic-Cretaceous igneous formations of the north-eastern Lesser Caucasus. Sov Geol 5:99 – 113 (in Russian)

Tvalchrelidze GA (1977) Metallogeny of the main volcanic belt types. Nedra, Moscow (in Russian)

The Hidden Mineralogical and Geochemical Zoning and the Ore-Forming Conditions of Copper and Copper-Lead-Zinc Massive Sulfide Deposits

N. I. EREMIN[1]

Abstract

The hidden mineralogical and geochemical zoning (HMGZ) of the repeated and nonrepeated structural types is characterized by examples of different massive sulfide deposits of Karelia, South Urals, Rudni Altai, and Japan. This zoning is detected in different scale orders by means of microanalysis through the composition variations of the ore-forming sulfides: pyrite, sphalerite, and tennantite-tetrahedrite. Is is shown that in the case of primary origin of such zoning its nonrepeated type corresponds to volcanic-sedimentary ore deposition and repeated type – to a hydrothermal replacement one. It is established that the variations of Co/Ni values or Co-content in pyrite, Fe-content in sphalerite, As-, Sb-, Ag-content in tennantite-tetrahedrite in cross-sections of the individual ore bodies (ore intervals, rhythms) are mainly due to alteration of temperature and sulfur activity during ore formation.

Introduction

Hidden mineralogical and geochemical zoning (HMGZ), as a regular space variation in composition of minerals within a geological unit, is extremely characteristic of volcanic-associated massive sulfide deposits. This group of deposits is very important economically and genetically; it includs pure pyrite, Cu-pyrite, Cu – Zn-pyrite, and polymetallic-pyrite (i.e., Cu – Pb – Zn-pyrite) types of mineralization. In contrast to sediment-hosted massive sulfide deposits, which are mainly Pb – Zn in composition, almost all the types given above include economic contents of copper (in addition to a general, but fluctuating content of zinc and sometimes of lead). An exception is the highly deformed and metamorphosed group of deposits which are characterized by morphological transformation and newly formed, mainly Zn and Pb – Zn ore segregations (i.e., metamorphism differentiates in the transformed ore bodies as a result of ore material redeposition by hydrothermal-metamorphic solutions).

[1] Department of Mineral Deposits, Geological Faculty, Moscow State University, Moscow, 119899, USSR

Geology and Metallogeny of Copper Deposits
(ed. by G. H. Friedrich et al.)

A substantial part of the ores of the group of deposits considered here usually forms massive stratiform bodies and consists predominantly of pyrite (or pyrrhotite) with variable content of base metal sulfides and various gangue minerals. The scale of HMGZ in such bodies may be of four orders: in ore bodies as a whole (first order); in individual intervals composed of a single mineralogical type of ore (second order); in hand specimens of specific texture (third order); and in individual mineral grains (fourth order). Nonrepeated and repeated structural types of HMGZ are the usual ones in the first and second orders. The nonrepeated type is characterized by a polarity of mineral composition from the bottom of the ore body (ore interval) to its top, with the formation of a layerlike structure. The repeated type reflects a symmetrical or asymmetrical concentric change of mineral composition from the central part of the ore body (ore interval) to its peripheral, marginal parts. According to geological data, nonrepeated and repeated types of HMGZ in weakly metamorphosed deposits are characteristic of volcanic-sedimentary and hydrothermal-replacement ore bodies, respectively. The identical structural types of HMGZ in strongly metamorphosed and dislocated deposits are of metamorphic origin.

In most cases the third order HMGZ is indicated in layered ores by a rhythmic variation of chemical composition of some specific mineral corresponding usually to the size of its grains. Sometimes rhythmic structure is determined for ore intervals (HMGZ of the second order) in weakly metamorphosed ore bodies. The fourth order HMGZ corresponds to the concentric phase heterogeneity identified sometimes in weakly metamorphosed volcanic-sedimentary ore bodies. It indicates diffusion, interrupted growth of crystals, exsolution, and other phenomena accompanying crystallization of ores with the participation of hydrothermal and metamorphic solutions.

The indicator minerals for HMGZ are pyrite, sphalerite, and tennantite-tetrahedrite typical of the ores; the variations of chemical elements in these minerals are established by means of microanalysis.

For the reader's convenience all deposits considered here are listed in Table 1, as well as the types of HMGZ exhibited in each of them.

HMGZ in the Pure Pyrite and Cu-Pyrite Ores

In the massive, pure pyrite and Cu-pyrite ore bodies that are typical of the Karelian, Cyprus, and some Urals types of deposits, the HMGZ is normally established through the Co/Ni value or Co-content of the pyrite.

The steep lens of the Parandovskoe deposit (East Karelia) is composed of recrystallized, metamorphosed (amphibolite grade), primary volcanic-sedimentary pyrite ores of Proterozoic age containing hydrothermal-metamorphic veinlets of pyrite-pyrrhotite composition with scattered grains of chalcopyrite and sphalerite. Abundant crystalline-granular pyrite, resulting from metamorphic recrystallization of an early colloidal variety of the mineral, shows the regular fluctuation of its composition in the cross-section of the ore body. The mean Co/Ni values of 39 individual analyses were calculated separately for presumably

Table 1. List of deposits considered, their location, type, and HMGZ exibited

Types	Deposits		HMGZ[a]				
	Name	Location	A	B	C	D	E
Cu – (Zn) (Cyprus, Karelian)	Parandovskoe	East Karelia, USSR	pp	nrp	1	py	Co/Ni
	Shimokawa	Hokkaido, Japan	Cu	nrp	1	py	Co/Ni
			Cu	cnc	4	py	Co
Cu – Zn (Urals)	Osennee	South Urals, USSR	Cu	nrp	1	py	Co/Ni
	Uchali	South Urals, USSR	Cu	nrp (rhm)	3	py	Co/Ni
	Gai	South Urals, USSR	Cu	rp	1	py	Co
	Oktjabrskoe	South Urals, USSR	Cu – Zn	rp	1	sph	Fe
Cu – Pb – Zn (Altai, Kuroko)	Tishinskoe	Rudni Altai, USSR	pp	nrp (rhm)	3	py	Co/Ni
			pm	rp	1	sph	Fe
			pm	rp	1	tt	As, Sb, Ag
	Gusljakovskoe	Rudni Altai, USSR	pm	rp	1	sph	Fe
			pm	rp	1	tt	As, Sb, Ag
	Nikolaevskoe	Rudni Altai, USSR	pm	nrp	2	tt	As, Sb
			pm	cnc	4	tt	As, Sb, Ag
	Shakanai	North Honshu, Japan	Cu	nrp (rhm)	2	py	Co/Ni
	Hanawa	North Honshu, Japan	Cu	rp	1	py	Co/Ni

[a] **A** Ore: *pp* pure pyrite; *Cu* copper pyrite, yellow; *Cu – Zn* copper-zinc pyrite; *pm* polymetallic pyrite, Kuroko. **B** Structural type: *rp* repeated; *nrp* nonrepeated; *rhm* rhythmic; *cnc* concentric. **C** Order; **D** indicator mineral; **E** indicator variable (value) in the mineral

upper (top), middle, and lower (bottom) parts of the ore body; they are as follows: 2.5; 1.9; 1.2 respectively, and the gradient (change of this value for 1 m of thickness) of this ratio is no more than 0.07. Such a noncontrasting and nonrepeated HMGZ of the first order is a primary one, reflecting volcanic-sedimentary origin of the ore body; its weakening occurred during metamorphic recrystallization accompanied by decrease of Co- and Ni-content in the newly formed crystalline-granular pyrite compared with the initial colloidal one (Sergeeva 1973).

The ores of the Devonian Cu-pyrite Osennee deposit (South Urals) form massive subhorizontal beds and consist of pyrite with chalcopyrite, rare sphalerite, and magnetite. The deposit is of volcanic-sedimentary origin (Potapenko et al.

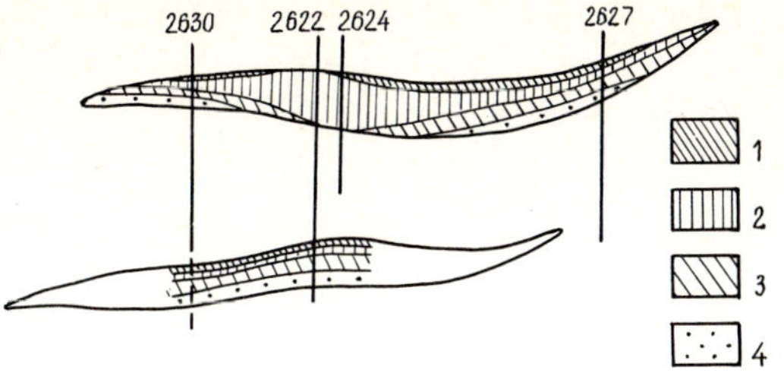

Fig. 1. A section of the ore bodies, Osennee deposit (nonrepeated HMGZ of the first order through the variation of the Co/Ni value in pyrite, see Table 2). Co/Ni values: *1* more than 2.0; *2* 2.0–0.5; *3* 0.5–0.2; *4* up to 0.2

Table 2. Variation of the Co/Ni value in pyrite, Osennee deposit, Cu-pyrite ore (data of laser microanalysis, Co- and Ni-content in $n \times 10^{-3}\%$)

Bore hole (m)	Depth (m)	Co	Ni	Co/Ni	Gradient of Co/Ni
2630	89.2	78.0	5[a]	15.6	
	89.5	20.0	5	4.0	
	91.0	11.0	5	2.2	1.2
	91.9	2.5	5	0.5	
	102.1	2.0	5	0.4	
2622	89.0	18.0	5	3.6	
	91.0	33.0	10	3.3	0.1
	97.5	19.0	6.8	2.8	
	148.0	32.0	8.5	3.8	
	153.0	18.5	5	3.7	0.6
	154.0	9.0	21.0	0.4	
2624	89.0	27.0	5	5.4	
	90.5	18.0	5	3.6	
	92.7	23.0	5	4.6	0.3
	101.0	9.5	5	1.9	
	102.0	9.0	5	1.8	
2627	88.0	79.5	5	15.9	
	96.0	14.0	5	2.8	1.2
	101.0	2.0	5	0.4	

[a] In all these cases (when the Ni values are set at 5) the density of a selected analytical line (Ni 1 3414.77) measured by high resolution microphotometer (G-11 Karl Zeiss, Jena, GDR) on the sensitive photoplate (ZU-2 ORWO) was distinctly and stably somewhat higher as compared with neighboring background

1973), and its metamorphic level corresponds to greenshist-facies. In vertical drill holes of the two ore bodies (Fig. 1) there is a marked decrease of the Co/Ni value in pyrite from the top of each body to the bottom. The gradient of this ratio increases sharply toward the flank margins. In general, the cross-section of each ore body shows nonrepeated, marked HMGZ (Table 2) which reflects the volcanic-sedimentary genesis of this deposit.

The Mesozoic Cu-pyrite Shimokawa deposit (Japan) of the Cyprus type (Eremin 1983) is an inclined, flat bed of volcanic-sedimentary sulfides (Bamba

1977) located among the spilitized basalts. In some places it is complicated by subsequent hydrothermal replacement mineralization and wall rock alteration (both in the foot wall and hanging wall); the metamorphic transformation is a very weak one, manifested mostly in dislocations. The principal ore minerals are pyrite, chalcopyrite, pyrrhotite, and sometimes sphalerite. Cobalt is of economic value and most of it is concentrated in the pyrite. The distribution of this metal in the grains themselves is irregular within the range of 0.0X to 0.X wt%; in some cases it forms concentric HMGZ of the fourth order. The high Co- and Ni-content and the irregular distribution of these metals in the pyrite grains have caused a significant, often irregular variations of the Co/Ni value in the cross-sections of the bedded ore body. Nevertheless in several traverses from the top to the bottom of the ore body, the smoothed values of this ratio show a distinct monotonous increase in the range of 7.2 – 25.0, i.e., here the HMGZ of the first order is revealed (Eremin 1983). It should be pointed out that the degree of contrast of this zoning is very high and its gradient in different traverses is identical (2.8 – 2.9). Together with the volcanic-sedimentary nature of pyrite, the latter confirms the primary origin of this HMGZ which was preserved in some parts of the ore body.

For the three deposits mentioned above, the nonrepeated HMGZ of the first order is a primary one and corresponds to volcanic-sedimentary deposition of the ore bodies; its gradients decrease with increasing metamorphic grade and increasing geological age.

Intervals of pure pyrite and Cu-pyrite ores occur also in the complex Cu – Zn-pyrite and polymetallic-pyrite ore bodies of the Urals, Altai, and Kuroko types. In such cases it is possible to find HMGZ through Co/Ni values and Co-content in pyrite, but now they are of the second and third orders.

At the southern flank of the Cu-pyrite ore body of the weakly metamorphosed Uchali deposit (South Urals) there are rhythmically-layered, volcanic-sedimentary textures in the pure pyrite ores. This layering is due to the difference in the grain size of pyrite and the total quantity of the mineral (graded bedding). This picture is complicated by discordant and rare concordant diagenetic veinlets of pyrite 1 – 2 mm thick. In some places the layers are distorted with atectonic dislocations of block movement and slump folding. Each rhythm combines two neighboring layers with a distinct border. The bottom layer, 0.5 – 0.6 cm thick, contains up to 60 – 65% of pyrite and the average size of its grains is 0.1 – 0.2 mm. The top layer up to 0.5 cm in thickness contains an apparently lesser quantity of pyrite (40 – 50%) and the size of its grains ranges between 0.00X and 0.1 mm. The principal gangue mineral is quartz. Though the fluctuations in Co- and Ni-content for individual grains of pyrite are irregular, the decrease of Co/Ni values from bottom to top of each rhythm (with the greatest range in one of them 36.0 – 1.0) is clearly seen, i.e., the nonrepeated rhythmic HMGZ of the third order is recorded.

Similar pure pyrite layered ores are recorded at the highly metamorphosed and dislocated Tishinskoe deposit (Rudni Altai). In addition to intensive folding of varied scale, shearing and cleavage of the host rocks, and the elongate form of the ore body, the coarse-grained, mainly sphalerite-galena layers located in the margins of the massive ores are here very clearly seen (as a result of remobiliza-

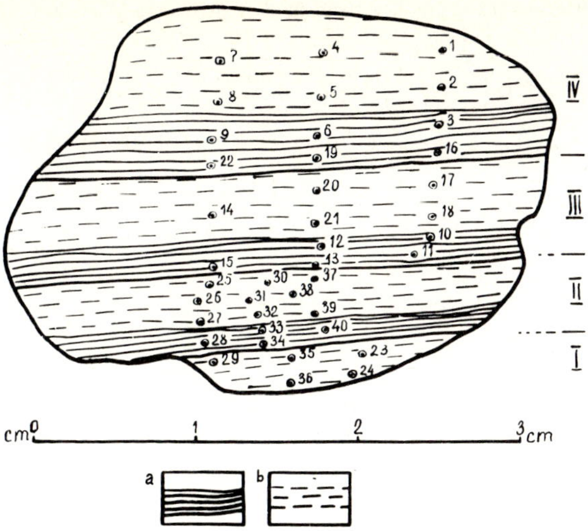

Fig. 2. A specimen of pure pyrite rhythm-layered ore, Tishinskoe deposit: *I – IV* rhythms; *a* bottom parts of rhythms; *b* top parts of rhythms (rhythmic HMGZ of the third order through the variation of the Co/Ni value in pyrite; see Table 3, the sample numbers in this Table correspond to those in figure)

tion of ore material). The impregnated (semimassive) pure pyrite ores under consideration form thin, elongated, and folded concordant lenses outside the main ore body among the quartzite and quartz-sericite-chlorite rocks. The layering in these ores is due to the alternation of sericite-chlorite and quartz-sericite-chlorite layers of differing pyrite content and regular change in the pyrite grains. Each single layer is 0.5 – 3.0 cm thick with a regular decrease both in grain size in the range of tenths to hundredths of a millimeter (graded bedding) and in total quantity of pyrite from the bottom to the top. This is the cause of the rhythms (Fig. 2).

In this case, as with the Uchali deposit, the distinct decrease in Co/Ni values for pyrite grains from bottom to top of each rhythm is also established (Table 3); this corresponds to HMGZ of the third order. The disseminated mode of occurrence of pyrite among the hydrothermal-sedimentary rocks is the reason for preservation of both the pyrite grain size and the content of metals in them, notwithstanding their recrystallization during metamorphism. This is the reason for the almost equal degree of contrast of the zoning in question in these two deposits.

Very impressive nonrepeated HMGZ of the second order is shown in the Miocene Shakanai deposit (Japan) of the Kuroko type. The fourth ore body of this deposit is a subhorizontal layer of black (polymetallic-pyrite) ores in its upper part and yellow (Cu-pyrite) ores in the bottom part. Bore hole No. 756 cuts this body and reveals four intervals of yellow ore. The regular variation of Co/Ni values in individual grains of pyrite is apparently shown; for each interval this ratio sharply increases from top to bottom (Table 4). The thickness of each interval of neighboring, massive pyrite-chalcopyrite ores is up to 1.5 m. These intervals are separated from each other by layers of highly altered acid tuffs. This nonrepeated HMGZ corresponds to (i.e., agrees with) the volcanic-sedimentary origin of these ores; this idea is supported by the majority of investigators of the Kuroko-type ore deposits (e.g., Matsukuma and Horikoshi 1970, Tatsumi and Watanabe 1971, Lambert and Sato 1974, Smirnov 1976, Large 1977).

Table 3. Variation of the Co/Ni value in pyrite, Tishinskoe deposit, pure pyrite ore (data of laser microanalysis, Co- and Ni-content in arbitrary numbers)

Sample Nos.	Co	Ni	Co/Ni	Sample Nos.	Co	Ni	Co/Ni	Sample Nos.	Co	Ni	Co/Ni
7	4	3	1.3	4	3	3	1.0	1	2	1	2.0
8	5	2	2.5	5	3	3	1.0	2	2	1	2.0
9	5	3	1.7	6	5	2	2.5	3	5	1	5.0
22	4	2	2.0	19	5	1	5.0	16	8	1	8.0
				20	3	3	1.0	17	3	5	0.6
14	3	2	1.5	21	2	2	1.0	18	4	3	1.3
				12	4	2	2.0	10	3	2	1.5
15	5	1	5.0	13	5	2	2.5	11	5	2	2.5
25	7	6	1.1	37	2	2	1.0	30	4	9	0.5
26	7	6	1.1	38	2	2	1.0	31	4	9	0.5
27	5	2	2.5	39	3	1	3.0	32	3	2	1.5
				40	4	1	4.0	33	3	1	3.0
28	8	2	4.0					34	4	1	4.0
29	4	2	2.0	23	3	2	1.5	35	4	3	1.3
				24	3	1	3.0	36	6	4	1.5

Table 4. Variation of the Co/Ni value in pyrite, Shakanai deposit, intervals of yellow ore in bore hole No. 756 (data of laser microanalysis, Co- and Ni-content in $n \times 10^{-3}\%$)

Intervals (from the stratigraphic top to the bottom)	Sample Nos.	Distance from the bore hole collar (m)	Co	Ni	Co/Ni	Gradient of Co/Ni in the interval
I	244	17.7	4.7	5.9	0.8	
	245	17.2	5.8	4.1	1.4	2.3
	246	16.8	5.2	1.8	2.9	
II	247	16.4	4.0	8.8	0.4	
	248	15.9	4.0	2.7	1.5	1.3
	249	15.5	5.0	3.2	1.6	
III	253	6.9	4.0	7.8	0.5	2.8
	254	6.4	3.4	1.8	1.9	
IV	256	5.2	4.3	9.2	0.5	
	257	4.8	4.3	3.2	1.3	2.6
	258	4.4	3.4	1.3	2.6	

The repeated structural type of HMGZ established through Co/Ni values or Co-content in pyrite is uncommon for the massive pure pyrite and Cu-pyrite ores referred to.

The steep dipping Cu-pyrite bodies of the deep horizons at the Gai deposit (South Urals) are of the hydrothermal replacement origin as compared with the uppermost volcanic-sedimentary Stergnevaja lens. They mark the ore-controlling structure (Eremin 1983). The mode of the Co-content distribution (Ni is not detected) in pyrite (up to 90% in the ore) corresponds to HMGZ of the first order, which is seen in the cross-section of three ore lenses. In one lens this zoning is a relatively simple, concentric one with the maximum Co-content (more than 30 units) in the form of an elongated, slightly curved stripe along the central axis of this lens. Each of the two other lenses has two parallel maxima (complicated concentric zoning). The obtained temperatures of crystallization in the range 300 – 120 °C for these ores (using the distribution of Co-content between coexisting pyrite and chalcopyrite, Bezmen et al. 1978) mark the concentric-zoned structure of a paleothermal field. The elongated maxima of this structure are traced concordantly in the general direction of the ore lenses coinciding with the maxima in distribution of the Co-content discussed above.

A similar picture of the space distribution of Co/Ni values in pyrite has been established in the Motojama ore body of the Kuroko-type Hanawa deposit (Japan). This is a subvertical pipelike body of massive yellow ore consisting of two neighboring shoots named Motojama Shinko Sho and Higashi Daijiti. Detailed sampling of these shoots along the walls and the roofs of workings at four horizons has been carried out. At the planes of each shoot the complicated concentric structure of HMGZ (Eremin 1983) is indicated by one to two isometric maxima, with several isolines of the Co/Ni value (5; 10; 15; 20; 25). Bearing in mind the primary origin of the analyzed crystalline-granular pyrite, we conclude that the genesis of this HMGZ of the first order is primary. It is suggested that the yellow ore of the Motojama ore body was formed by hydrothermal replacement of an ore-controlling vent structure just below the seafloor.

Thus, the repeated structural type of HMGZ established through the distribution of Co-content or Co/Ni values in pyrite of the Gai and Hanawa deposits corresponds to hydrothermal replacement deposition of these ores.

HMGZ in the Cu – Zn-Pyrite and Polymetallic-Pyrite Ores

In the Cu – Zn-pyrite and polymetallic-pyrite ore bodies HMGZ of both structural types is established on the basis of the Fe-content in sphalerite and the fluctuation in composition of tennantite-tetrahedrite. It is manifested most impressively in some deposits of the Kuroko, Altai, and, rarely the Urals types.

The regular space fluctuation of Fe-content in sphalerite during the formation of the repeated type of HMGZ is characteristic of the first ore body of the Oktjabrskoe deposit (South Urals). This body consists of massive sulfide ore and is located in volcanogenic rocks of Paleozoic age. Its top contact is concordant with the layering of host rocks, but its bottom one is complicated, with steep dipping apophyses. In vertical cross-section, the body is of mushroom shape, and in longitudinal ones it is inclined 15 – 20°. The principal ore minerals are pyrite, chalcopyrite, and sphalerite. The maximum Fe-content in sphalerite (more than 1.5%) is found in the central, thickest part of the body, whereas the minimum Fe-content (up to 0.25%) is found on its flanks and marginal parts. The picture is clearly seen both in the cross- and longitudinal-vertical sections (Eremin 1983). According to geological data sphalerite mineralization is the latest one at the deposit, being formed as a result of the reaction between Zn-containing hydrothermal solution and Cu-pyrite ores (Malachov et al. 1974).

The increase in Sb- and Ag-content in tennantite-tetrahedrite toward the top of the ore bodies provides evidence of a distinct nonrepeated HMGZ in the black ore of a majority of Kuroko deposits (Matsukuma and Horikoshi 1970). A decrease in Fe-content in sphalerite (HMGZ of the first and second orders) is noted in the same direction for these ores at the Shakanai deposit (Kajiwara 1970). It was found also (Urabe 1974) that the Fe-content in sphalerite decreased toward the top of complex bodies composed of several types of sulfide ores (HMGZ of the first order).

The principal ore constituents at the Gusljakovskoe deposit (Rudni Altai) are syngenetic pyrite followed by polymetallic mineralization with sphalerite, galena, chalcopyrite, and tennantite-tetrahedrite. Tennantite-tetrahedrite compositions across the subvertical ore zone of mainly impregnated sulfides are concentrically zoned, i.e., the more Ag-rich tetrahedrite varieties of the mineral are located in the central part, while tennantite varieties occur in the marginal parts of the ore zone (the Sb-content fluctuates in the range of 26.37 – 4.23%; the Ag-content, in the range of 5.98 – 0.15%; and the As-content, in the range of 4.67 – 18.36%). The distribution of the Fe-content in sphalerite is also characteristically zoned: the most Fe-rich varieties of the mineral are located in the axial part of the ore zone. Concentric HMGZ of the first order, defined by the composition of tennantite-tetrahedrite and sphalerite, definitely corresponds to the structure of the paleothermal field identified by the decrepitation of sphalerite. The axis of temperature maximum (300 – 360 °C) coincides also with the central part of the ore zone (Zolotarev et al. 1975). According to geological and geochemical data, the polymetallic mineralization followed hydrothermal alteration which in turn succeeded pyrite impregnation (pyrite is recrystallized in the zone of alteration).

The volcanic-sedimentary Kreshenskoe ore body (Nikolaevskoe deposit, Rudni Altai), which is weakly metamorphosed and, hence, likely to be preserved in its original position, is comparable to the Kuroko deposits in the mode of occurrence of different ore types. The dominant massive ores of this body are divided (mainly according to the data of chemical sampling) into the following types (from the bottom to the top): pure pyrite and Cu-pyrite, Cu – Zn-pyrite, and polymetallic-pyrite. There are no distinct variations in the composition of tennantite-tetrahedrite, which is all of tennantite variety at this deposit (in contrast to the Kuroko deposits). Nevertheless, in the polymetallic-pyrite interval (as in the Kuroko ores) nonrepeated HMGZ of the second order is marked. Here we have a decrease in Sb-content (from 0.11 to 0.04%) and a corresponding increase in As-content (from 14.02 to 20.32%). In these ores there are also single grains of tennantite-tetrahedrite with concentric-zoned structure inside each grain due to the internal chemical heterogeneity (HMGZ of the fourth order).

The main ore body of the Tishinskoe deposit mentioned above consists of dominant veinlet-impregnated ores and massive polymetallic-pyrite ores. There is a regular fluctuation in composition of sphalerite and tennantite-tetrahedrite in vertical cross-section (Eremin 1983). The maximum Fe-content in sphalerite (up to 2.40%) is in the central part of the body at the level of the thickest massive ore. This content decreases to 0.10% toward the flanks and both (upper and lower) ends of the body. Thus, in this cross-section of the vertical body the elongated concentrically-zoned distribution of Fe-content in sphalerite is concordantly outlined with maxima in the upper horizons, which are richer in the massive ores.

In this cross-section the same structure is outlined by the variation in composition of the tennantite-tetrahedrite: the maxima of Sb- and Ag-contents are found in those grains of the mineral that are located in the thickest portion of the massive ore as mentioned above. The isolines of mineral composition form a concentric asymmetric pattern extending along the axes of the ores body. Toward the flanks the composition of the mineral becomes more arsenical. The range in mineral composition is very broad: As 1.00 – 18.33%; Ag 7.05 – 0.15%; Sb 31.01 – 0.94%. This is responsible for the strong zoning. In many respects this HMGZ resembles that described for the Gusljakovskoe deposit. This possibly reflects the similar condition of ore formation at both deposits. But it is quite possible that at the Tishinskoe deposit this zoning may result from metamorphic transformation of the ore body.

The Ore-Forming Conditions

In the cases when the HMGZ is primary, it reflects the conditions of deposition of the ore bodies. Relicts of such zoning could be found even in highly metamorphosed deposits.

Keepling in mind the more common occurrence of the nonrepeated, sometimes rhythmic HMGZ of the different orders as compared with the repeated one, it could be concluded that most of the pure pyrite and Cu-pyrite ore bodies have been formed in a volcanic-sedimentary manner on the seafloor. In this case (Shimokawa, Uchali, Tishinskoe, and Shakanai deposits) we see more often a decrease in the Co/Ni values in pyrite toward the top of such bodies (intervals, rhythms) and more seldom – an increase in this ratio (Parandovskoe, Osennee deposits). A decrease of temperature toward the top of the ore bodies (intervals, rhythms) is assumed to occur during volcanic-sedimentary ore deposition. This has been proved particularly for the Kuroko deposits (Large 1977).

Following Bezmen and Tichomirova (1975) it was shown by the author (Eremin 1983) that a decrease in temperature of chlorine-bearing solutions of constant composition causes a decrease in the Co/Ni values in coexisting pyrite below the vaesite-millerite equilibrium line in the diagram $\log a_{S_2} - T$, or an increase in this ratio above this line. In other words, the volcanic-sedimentary ore deposition at the Parandovskoe and Osennee deposits took place under conditions of relatively high activity of sulfur.

One can always a decrease in the Co/Ni values (or Co-content) in pyrite from the center to the margins in the hydrothermal replacement Cu-pyrite ore bodies (the deep horizons of the Gai deposits, the Motojama body of the Hanawa deposit); this is also in accordance with similar lowering of the temperature of mineralizing solutions during their diffusion away from the axis of the ore-controlling fissure structure. For the Gai deposit this is confirmed by the paleothermal zoning structure. Consequently, in the case of both these deposits the formation of pyrite, as a basic part of the ores, occurred at the comparatively low sulfur activity below the equilibrium line vaesite-millerite in the diagram $\log a_{S_2} - T$.

The successive volcanogenic-sedimentary deposition of the black ore, which was deposited after the yellow ore at the Kuroko deposits, is confirmed by geological data and correlated with the nonrepeated HMGZ established through the decrease of Fe-content in sphalerite and increase of Sb- and Ag-content in tennantite-tetrahedrite toward the top of the ore bodies. Since the deposition of these ores occurred during a decrease in temperature, the observed decrease in the Fe-content in sphalerite, as it follows from the diagram $\log a_{S_2} - T$ (Barton and Skinner 1967), points out the increase in the sulfur potential (although the sulfur activity was lowered, the vector of the ore formation was shifted toward equilibria with participation of more sulfide sulfur). Deposition of black ore in this manner conforms with the HMGZ identified by the compositional fluctuations of the tennantite-tetrahedrite. In the diagram $\log a_{S_2} - T$, in the area of relatively low sulfur activity, tennantite and Sb-sulfosalts are the stable phases, whereas under increasing sulfur activity, the antimony combines more readily in the form of tetrahedrite and arsenic in the form of enargite. The direct correlation between antimony and silver is due to the crystallochemical peculiarities of the mineral.

In the hydrothermal replacement Cu – Zn-pyrite and polymetallic-pyrite ores (Oktjabrskoe, Gusljakovskoe and, possibly, Tishinskoe deposits), decrease in the Fe-content in sphalerite related to the lowering of the temperature of the solution diffusing away from the ore-controlling fissure structure is also seen. But the change of the tennantite-tetrahedrite composition here is opposite the change observed in the volcanic-sedimentary black ore. The more tetrahedrite-rich varieties of the mineral with high Sb- and Ag-content are located in the high-temperature (central parts of the bodies; the tennantite varieties, in the low-temperature (marginal) parts. It is possible that in this case the determining factor of the HMGZ formation was likely to be more the different mobility of As and Sb ions during metasomatism than the change in temperature and sulfur activity in the diffusing solution. It is interesting to note that for these ores the heterogeneity of the tennantite-tetrahedrite grains themselves (HMGZ of the fourth order) is not typical. But this phenomenon is more or less ordinary for the volcanic-sedimentary black ores in Kuroko deposits, and is found in individual cases in the polymetallic-pyrite ores of the Nikolaevskoe deposit.

Acknowledgments. The author is grateful to Academician V. I. Smirnov and to all the staff members of the Department of Mineral Deposits (Moscow State University) for support in this work. Special thanks to Mrs. T. S. Maximova for assistance in translation of the manuscript into English. The author also wishes to acknowledge Dr. A. D. Genkin and all foreign colleagues for helpful suggestions and criticism.

References

Bamba T (1977) Metallogenic province of Hokkaido. J Fac Sci Hokkaido Univ Ser IV, 17:695 – 708

Barton PB Jr, Skinner BJ (1967) Sulfide mineral stabilities. In: Barnes HL (ed) Geochemistry of hydrothermal ore deposits. Holt, Rinehart and Winston, Inc, New York, pp 236 – 333

Bezmen NI, Tichomirova VI (1975) The effect of temperature on the distribution of Co and Ni between the Fe-sulfides and the solutions of different composition. Geochemia 11:1691 – 1697 (in Russian)

Bezmen NI, Eremin NI, Narazauly IG, Pozdnyakova NV, Sergeeva NE (1978) Pyrite-chalcopyrite geothermometer: cobalt distribution. Geochemia 3:384 – 389 (in Russian)

Eremin NI (1983) The differentiation of the volcanogeneous sulfide ore-formation. MGU, Moscow, 256 pp (in Russian)

Kajiwara Y (1970) Some limitations on the physico-chemical environment of deposition of the Kuroko ore. In: Tatsumi T (ed) Volcanism and ore genesis. Univ. Tokyo Press, Tokyo, pp 367 – 380

Lambert IB, Sato T (1974) The Kuroko and associated ore deposits of Japan: a review of their features and metallogenesis. Econ Geol 69:1215 – 1236

Large RR (1977) Chemical evolution and zonation of massive sulfide deposits in volcanic terrains. Econ Geol 72:549 – 572

Malachov AA, Vorobjev VV, Malachov DA et al. (1974) On the complex processes of formation of the Oktjabskoe Cu-pyrite deposit (South Urals). Geol Rudn Mestorogd 3:40 – 50 (in Russian)

Matsukuma T, Horikoshi E (1970) Kuroko deposits in Japan. A review. In: Tatsumi T (ed) Volcanism and ore genesis. Univ Tokyo Press, Tokyo, pp 153 – 179

Potapenko BP, Trebuchin VS, Ageeva ST (1973) Dombarovskii ore region. Trudi ZNIGRI 105:178 – 188 (in Russian)

Sergeeva NE (1973) Pyrite from the massive sulfide deposits of Karelia. Vestn MGU Geol 5:100 – 105 (in Russian)

Smirnov VI (1976) The geology of mineral deposits, 3rd edn. Nedra, Moscow, 688 pp (in Russian)

Tatsumi T, Watanabe T (1971) Geological environment of formation of the Kuroko-type deposits. In: Proc IMA-IAGOD Meet, Tokyo-Kyoto, 1970, IAGOD Vol Spec Iss 3. Inst Assoc Genesis Ore Deposits, Tokyo, pp 216 – 220

Urabe T (1974) Iron content of sphalerite coexisting with pyrite from some Kuroko deposits. In: Ishihara S (ed) Geology of Kuroko deposits. Min Geol Spec Iss 6. Soc Min Geol Jpn, Tokyo, pp 377 – 384

Zolotarev VG, Eremin NI, Dergatchev AL, Polenkov AI (1975) The correlation of metasomatic, geochemical and thermal zoning at the polymetallic-pyrite Gusljakovskoe deposit (Rudni Altai). In: Magakjan IG (ed) Metasomatism and massive sulfide ore-formation. Erevan, pp 111 – 113 (in Russian)

Zoning of Massive Sulphide Deposits and Their Origin

N. S. SKRIPCHENKO, V. I. SHCHEGLOV, G. V. RYABOV, and V. L. ANDREYEV[1]

Abstract

Vertical and lateral zoning of massive sulphide deposits is well known. Vertical zoning is characteristic of volcanic-hosted deposits (e.g. Kuroko), whereas lateral zoning occurs in sediment-volcanic-hosted ore (e.g. Zhairem). A dynamic model of such zoning would belong to the "barrier reaction" type. An oxide-acidic model is herein proposed for the volcanic-hosted and a reduced-carbon model for the sediment-volcanic-hosted deposits. The first model relates to the boundary of the metalliferous solutions/oxidized water system. Oxidized sulphur components in this system range from SO_2 to S^{2-}. The acidity and ratio of S^{2-}/SO_4^{2-} are increased and decreased, respectively, from the base to the upper part of the ore section. The result of these variations is the following sequence (from bottom to top): sulphide-silicate metasomatites, pyrite-, copper- and lead-zinc-pyrite massive ores. The form and size of these ore types is caused by the dynamics of the hydrothermal flow and the size of the recycling zone. According to the reduced-carbon model, the ore is deposited from the hydrothermal brine into depressions on the seafloor. Deposition is controlled by the brine/seawater contact, where sulphate-reduction takes place. The laterally concentric emplacement of lead, zinc and pyrite ores is caused by depletion of the sulphidic ion and concentration of the metallic ions in the brine pool.

Massive sulphide copper, zinc and lead deposits are hosted by basalt-carbonaceous-argillaceous megaformations. Ferrous disulphides and some baryte are the dominant minerals of their ores. Ferric and manganese oxides are minor constituents. As a rule, the classification of massive sulphide deposits reflects the composition of their ores and host rocks (Borodaevskaya et al. 1979). If not taken into consideration the morphology-combined multimodula ore bodies, the inner structure of the massive sulphide bodies may be classified by only two types of zoning: the vertical and lateral (horizontal). This zoning is the basis for subdivision of the massive sulphide deposits into two genetic classes: vertically and laterally zoned deposits. Examples of the vertical zoning type are the Kuroko deposits (Fig. 1), and of the horizontal type, the ore bodies of the Zhairem deposits (Fig. 2). The genetic models of both types of zoning are examined below.

[1] The Novocherkassk Polytechnical Institute, Mining-Geological Department, Novocherkassk, USSR

Geology and Metallogeny of Copper Deposits
(ed. by G. H. Friedrich et al.)

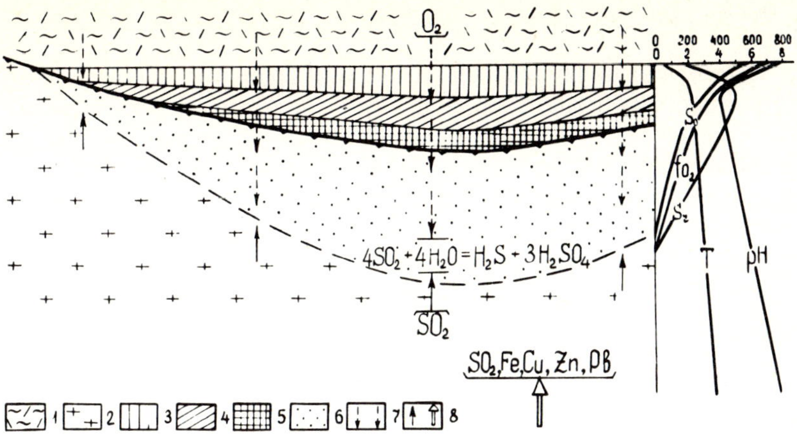

Fig. 1. Structure and genetic model of vertically-zoned massive sulphide deposits of the Kuroko type. *1* Unaltered rocks of the post-ore time; *2* volcanic rocks; *3, 4, 5, 6* massive lead-zinc, copper, pyrite and disseminated silicate-pyrite ores, respectively; *7* descending exogenic flow saturated with oxygen; *8* ascending ore-bearing flow. The diagram shows the limits of pH changes and the limits in mineral formation temperatures (T °C), and also suggests trend curves for concentrations of sulphide (S_r) and sulphate (S_0) and for oxygen fugacity (fO_2)

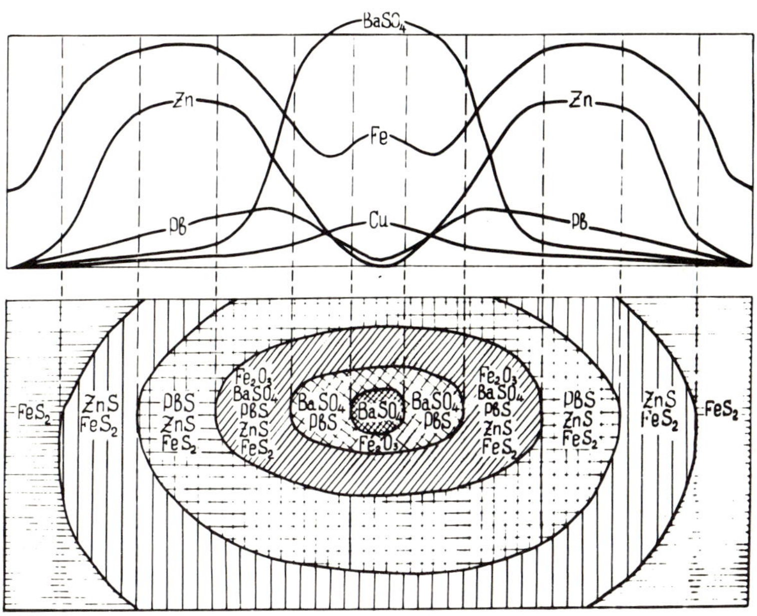

Fig. 2. Characteristic lateral zoning in Zhairem-type ore deposits. The ore-mineral composition of zones is shown on the horizontal projection of the deposit (*below*). The curves of the *upper diagram* indicate variations in metal content in a cross-section of the ore body

Vertical Zoning

The complete zoning of this type is represented by a column where pyritization metasomatite, essentially pyrite, copper- and lead-zinc-pyrite ores, occur in succession upwards (Fig. 1). The data characterizing the structural and genetic features of the vertical zonal ore deposits are given below.

The thickness of sulphide mineralization in sections through the deposits is very small (from a few to 100 m) in comparison to the distance from the source of ore-bearing solutions. The lithological varieties in ore-bearing sections do not control the ore zone location or the degree of differentiation. The ore-hosted rocks are represented by lava, subvolcanic, pyroclastic-sedimentary and sedimentary complexes. The local ore-controlling forms of paleorelief are defined as deeps in basins with characteristic iron-oxide and, more rarely, manganese syngenetic mineralization accompanying the massive sulphide ores. The data confirm the suggestion that massive sulphide ores were deposited in basinal depressions saturated with oxygen.

Stratiformlike mineral-component differentiation of the ore bodies and asymmetric location of various genetic types of ores (metasomatic in the lower and hydrothermal-sedimentary in the upper parts of the deposit) define the ore deposits as a genetically combined type (Skripchenko 1966, Smirnov 1968). The ratio of thickness of hydrothermal-sedimentary to hydrothermal-metasomatic types of ores varies widely. Hydrothermal-sedimentary ore is characterized by a high sulphide: lithogenic mineral ratio (usually from 1 to 10). The degree of lithogenic components leached and the accompanying sulphide deposition, especially pyrite, in the metasomatic part of section increases from bottom to top.

Vertical zoning belongs to the facial zoning type according to Smirnov (1982). The origin of this zoning is explained by ore sedimentation at the time of discharge of hydrothermal solutions into exogenic basins. The projections of the pyrite metasomatite of the ore footwall defines the discharge area. A number of workers (Skripchenko 1972, Smirnov 1960, 1968) prove the convergent character of massive sulphide ores. Isostructural sulphide aggregates generated under different $P-T$ conditions are associated in cross-sections of ore deposits. Mineral thermometric measurements in massive sulphide ores established the limit from 40° to 350° (Bogysh and Trufanov 1983, Smirnov et al. 1968). These data demonstrate that temperature fluctuations are not the only factor in the process of massive sulphide ore formation and differentiation, although it may have an indirect influence (Tvalchrelidze and Yaroshevich 1981, Large 1977, Large and Both 1980). The main cause of vertical differentiation and sulphide deposition, in general, is the existence of intensive chemical reactions at the contact of concentrated hydrothermal solutions and exogenic water. Reactions become active as a result of the high oxidizing potential of exogenic water.

Sulphide profiles of ore formation, characterized by deposition of large masses of iron disulphide, raise a question about the nature of the sulphide ion. Data from Holland and Malinin (1982), Stolajarov (1980) and Barsukov et al. (1981) show that the main, and possibly the only, form of sulphur present in endogenic hydrothermal systems during their generation is SO_2. Consequently,

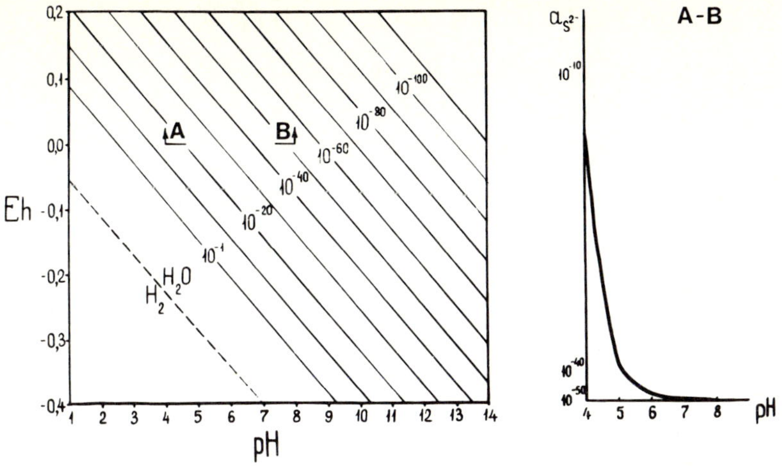

Fig. 3. Change in S^{2-} concentrations depending on pH and Eh of aqueous solutions (T = 25 °C, P = 1 atm, $S^{2-} + SO_4^{2-} = 0.1$ m). The diagram on the *right* is the section *AB* of Eh-pH diagram (Skripchenko 1966)

the massive sulphide deposits genetic model may be based on the assumption that the primary ore-bearing solution contains SO_2 as the main sulphur component. In accordance with Holland and Malinin's data (1982), cooling of hydrothermal solutions is accompanied by the reaction: $4\,SO_2 + 4\,H_2O = H_2S + 3\,H_2SO_4$. A result of this reaction is an increase in the fugacity of the sulphide ion $S^{2-}(S_r)$ as well as the sulphate ion $S^{6-}(S_0)$. This reaction may be supported by the thermodynamic calculation of the S^{2-}/SO_4^{2-} ratio on the Eh-pH diagram (Skripchenko 1966). As indicated in Fig. 3, S^{2-} concentrations increase with decreasing pH of the aqueous solution. At a constant isopotential (Eh) level, the S^{2-}/SO_4^{2-} ratio increases from pH_{max} to pH_{min} (Fig. 3).

Decrease in pH in the discharge zone of ore-forming hydrothermal solutions results from interaction of these solutions and the surface oxygen-saturated waters. The possible geochemical parameters in a section of the reaction zone are given in Fig. 1. This section of mixed zone, also known as the ore-forming zone or "worked" zone according to Pospelov (1973), is subdivided into three parts corresponding to three types of mineralization (from bottom to top): (1) Metasomatic silica-pyrite ores with low copper content; their formation is related to growth of the ratio S_r/S_0 and intensified leaching of wall rocks caused, in general, by a progressive increase in acidity. If the process proceeds according to the "recycling" model, the shape and thickness of the metasomatite zone will vary (Fig. 4). (2) Weakly or non-differentiated massive copper and essentially pyrite ores. These ores are formed in a zone characterized by maximum acidity and concentration of S_r. (3) Differentiated ores with high copper, zinc and lead content. They are formed in the upper section of the worked zone where, it is suggested, there exists a deficit of S_r. The high content Cu, Pb and Zn in these ores is explained by variations in the bond strength of metal and sulphide ions. The bond strength of the non-ferrous metal ions is more than that of the ferrous

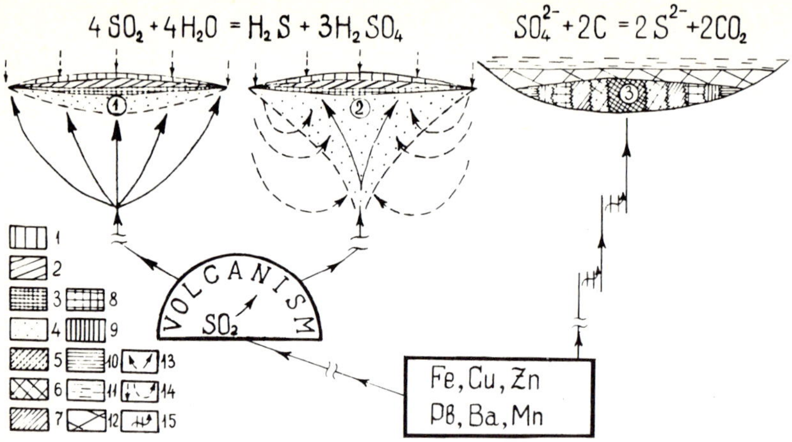

Fig. 4. Genetic relations and zoning of massive sulphide deposits of the Kuroko type (*1* ascending flow; *2* recycling flow), and zoning of pyrite deposits of the Zhairem type (*3*). Explanations: *1* polymetallic; *2* copper; *3* pyrite; *4* silicate-pyrite ores; *5* baryte; *6* baryte-galena; *7* baryre-galena-sphalerite; *8* galena-sphalerite; *9* sphalerite; *10* pyrite ores; *11* seawater; *12* hydrothermal brine; *13* flow of endogene ore-bearing solutions; *14* direction of flow of exogenic oxygen-saturated solutions; *15* zones of inhibited flow of endogenic solutions

ion. Solutions containing only minor amounts of the sulphide ion must precipitate Pb – Zn – Cu sulphides only.

The morphological and thickness parameters of ore bodies are defined by a number of factors; among these are local seafloor relief and the speed and form of the hydrothermal solution flow. The major control on ore body morphology is the structure of metal-bearing solution channels. Two structural varieties of simple ore bodies (1 and 2) are given in Fig. 4. Interpretation of the second variety assumes the possibility of lowering of mixing front to considerable depths. This is caused by the variation in the morphology of the hydrothermal ore-solution flow. Mixings, of the recycling nature, promote an increase in the thickness of the pyritization metasomatite zone. The latter changes its form from a gently sloping, conformable lenticle (in the first variety) to vertical and funnel-shaped (in the second variety). One can observe morphological variations of the pyritization metasomatite zone in various ore bodies within the limits of a single ore field and even within the limits of one multimodular ore body.

Lateral Zoning

One can observe lateral zoning best in Zhairem type deposits occurring in black siliceous-clay-lime flyschoid strata *series*. In this case vertical sections of the ore bodies are homogeneous. Ore zones replace each other in concentric fashion outward from the centre as follows: siliceous-baryte, baryte-galena, galena-sphalerite, sphalerite, pyrite (Fig. 2). Ore bodies and ores are characterized as follows: (1) (non-metallic minerals) : (ore minerals) ratio gradually increases from the centre to the periphery. The ore-hosted section of flyschoid is thinly bedded. As a rule, ore layers are characterized by high metal concentrations. (2) Absence or rare occurrence of footwall alteration zone. (3) Competing sulphide-oxide association among iron mineral assemblages. One can observe an increase of sulphides in ores from the centre to the periphery of ore bodies. The central zone is the site of baryte concentrations, and in some cases anhydrite and goethite. Copper, lead and zinc are represented only by sulphides. (4) Lateral zoning is the facies type of zoning. This conclusion is emphasized by the absence of the mutual crossing of outer ore zones by the inner one and also the same disposition of ore zones from centre to periphery in the different ore bodies (Skripchenko et al. 1971, Skripchenko 1980).

Though the general nature of Zhairem type deposits is of hydrothermal-sedimentary type, the ore minerals (textural) relations are controlled by diagenetic transformations. These transformations include crystallization and local regeneration of the primary ore silt deposits. Framboidal pyrite applies to the earliest form. Other ore minerals (sphalerite, galena, baryte) are the results of late crystallization in the order: (1) baryte; (2) sphalerite; (3) galena. Because the sedimentation of iron took place in oxide form and non-ferrous metal as sulphides, the diagenetic pyrite and Cu – Pb – Zn-sulphide aggregates are observed as different layers.

Primary ore sediments formed under conditions of sulphide-ion deficit, hence, only the sulphides of non-ferrous metals were deposited. The prevailing form of iron deposition was as an oxide. Pyritization of sediments occurred in thin layers during the diagenetic stage. Goethite was converted to Fe-disulphide, but thick iron-oxide beds were not pyritized.

Normal interbedding of carbonic-clay-silica-carbonate sediments, sulphide and oxide ores limits the number of genetic models for Zhairem deposits to that of precipitation from a hydrothermal brine in basin deeps on the seafloor. The main pyrite form in the ores is framboidal pyrite. As investigations showed (Berberian 1981, Skripchenko 1980), this early, diagenetic form is related to biogenic sulphate reduction.

The textural-mineralogical and genetic features outlined above have been interpreted in a genetic model suggested by Skripchenko (1983) for ore sediments of the Atlantis II Deap in the Red Sea. In accordance with this model (Fig. 5), the generation of non-ferrous metal sulphides occurs in the contact zone between the hot hydrothermal brine and normal seawater saturated by oxygen. The contact zone is known as the zone of biogenic sulphate reduction (Bischoff 1974) and functions as a filter preventing the dispersal of non-ferrous metals because of their great affinity for sulphide ion. The iron precipitate is oxygen; that of baryte, sulphate. Other elements, especially manganese, disperse into the sea. The lateral distribution of non-ferrous metals is determined by differences in their affinity for S^{2-} and, according to this criterium, should be deposited in the order: copper, lead and zinc outward from the centre of the periphery. The zone of zinc accumulation is shifted towards the periphery relative to the lead and baryte zones of the central part of the Zhairem type ore deposits (Fig. 2).

Two factors, (1) rhythmically-interrupted accumulation of ores and *carbonic* flyschoid and (2) relatively high concentration of organic mass, are favourable

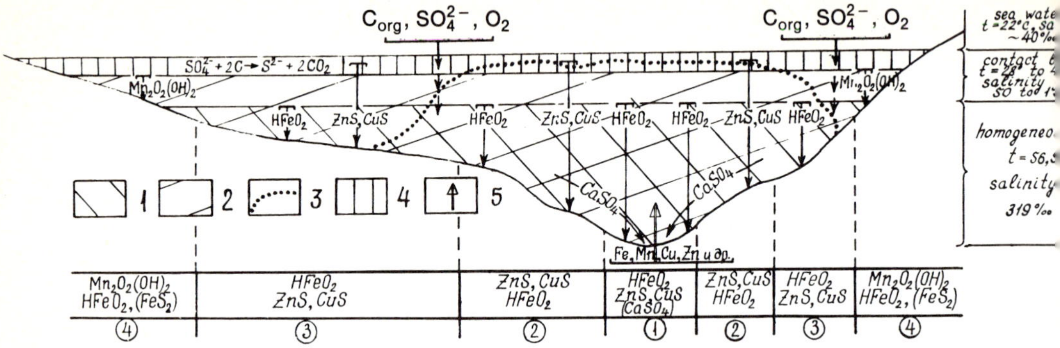

Fig. 5. The sedimentation model for mineralization in the Atlantis II deep (Skripchenko 1983). *1, 2* Brine zones with maximum metal content: *1* iron (over 87 ppm); *2* manganese (over 77 ppm). *3* Upper limit for zinc over 7 ppm. *4* Zone of biogenic sulphate reduction. *5* Brine in flux in axial part of depression. *1*–*4* (*in circles*) distribution of ore types on deep bottom: *1* iron-copper-zinc; *2* rich copper-zinc; *3* poor copper-zinc; *4* iron-manganese. In *parantheses:* the minor mineral occurrences (pyrite and anhydrite)

conditions for biogenic iron disulphide formation. Deposition of pyrite in outer ore zones is explained by the possibility of biogenic sulphate reduction in sediments only in the outer belt of the ore sedimentation deep. The high-temperature and chemically corrosive brine in hydrothermal deeps does not promote the possibility of intensive sulphate reduction in the sediment in the inner ore zones. The lithogenetic component of ores of the inner zone in Zhairem deposits is generally silica.

Conclusion

Two genetic classes of massive sulphide deposits, vertical and horizontal zoning types accordingly, are characterized by identical mineral and elemental compositions. The hydrothermal-sedimentary phase of their formation is described by two models: acidic oxygen and carbonic reduction. The vertically zoned group is related to middle and final stages of basaltic volcanic activity; they are also developed in thick carbonic argillaceous formations of shale according to Tvalchrelidze (1982). High (ore material):(non-metallic mineral) ratios in the ore bodies and their association with rocks formed under rapid sedimentation rates suggests that this deposit type is the result of deposition from highly concentrated ore-bearing fluids (Skripchenko 1966, Henley and Thornley 1979).

The laterally-zoned group of deposits is formed under conditions of low sedimentation and ore deposition rates. As a result of these conditions rhythmic layers of ore and lithogenic deposits accumulate. In addition, the non-ferrous metal content of these deposits is large.

The sulphur sources for the two types of massive sulphide deposits are different. In the first group of deposits, sulphur is introduced by the ore solutions themselves; in the second group, it is taken from seawater.

Several possible sources of metals necessary for ore formation and different models of their delivery to the site of ore deposition are indicated in Fig. 4. In some cases, activation of ore source occurs during the process of active volcanism; in other cases, it is connected with the development of tectonic fractures in the Earth's crust. Consequently, one may assume a magmatic sulphur source in massive sulphide deposits of the first group. Movement of concentrated ore-bearing fluids is also intensified by active volcanism. Where the source rock is cut by fractures, ore-bearing fluids pass through the rock stratum less intensively and, as a rcsult, ore accumulation will be defined by the dynamics of sedimentary processes.

References

Barsukov VL, Grigoryan SV, Ovchinikov LN (1981) Geochemical methods of ore deposits search. Nedra, Moscow, 318 pp (in Russian)

Berberian TK (1981) Icosahedron – a geometrical model of ordered pyrite framboids. Pap Acad Sci USSR 258, 2:456–458 (in Russian)

Bischoff JL (1974) The sediments of Red sea thermal brine (mineralogy, chemical composition and genesis) – In: Modern hydrothermal ore deposition. Mir, Moscow, pp 157–193 (in Russian)

Bogysh IA, Trufanov VN (1983) Temperature regime and zoning of ore formation of polygenetic massive sulphide lodes (Northern Caucasus). Geol Ore Dep 4:71–79 (in Russian)

Borodaevskaya MB, Gorzhevsky DI, Krivtsov AI, Ruchkin GV, Skripchenko NS, Tvalchrelidze GA, Yakovlev GF (1979) Massive sulphide deposits of World. Nedra, Moscow, p 284 (in Russian)

Henley RW, Thornley P (1979) Some geothermal aspects of polymetallic massive sulphide formation. Econ Geol 74:1600–1612

Holland HD, Malinin SD (1982) Solubility and occurrences of non-metallic minerals. In: Geochemistry of hydrothermal ore deposits. Mir, Moscow, pp 370–404 (in Russian)

Large RR (1977) Chemical evolution and zoning of massive sulphide deposit in volcanic terrains. Econ Geol 72:549–572

Large RR, Both RA (1980) The volcanogenic sulphide ores at Mount Chalmers, eastern Queensland. Econ Geol 75:992–1009

Pospelov GL (1973) Paradoxes geologic physical essence and metasomatoz mechanism. Nauka, Novosibirsk, 355 pp (in Russian)

Skripchenko NS (1966) Volcanogenic-sedimentary ore deposition. Nedra, Moscow, 292 pp (in Russian)

Skripchenko NS (1972) Hydrothermal sedimentary sulphide ores of basaltoid formations. Nedra, Moscow, 241 pp (in Russian)

Skripchenko NS (1980) Hydrothermal-sedimentary polymetallic ores of limeshale formations. Nedra, Moscow, 215 pp (in Russian)

Skripchenko NS (1983) Deposition and differentiation of ore silts at the Atlantis II deep of Red sea. Geol Ore Dep 1:74–86 (in Russian)

Skripchenko NS, Rozhnov AA, Lytkin VA (1971) Zoning of ore deposits of Zhairem group polymetallic deposits (Central Kazakhstan). Geol Ore Dep 5:3–11 (in Russian)

Smirnov VI (1960) Convergence of the massive sulphide deposits. Vestnic Moscow State Univ Series 4 Geol 2:19–26 (in Russian)

Smirnov VI (1968) Massive sulphide deposits. In: Genesis of endogenic ore deposits. Nedra, Moscow, pp 586–647 (in Russian)

Smirnov VI (1982) Geology of mineral deposits. Nedra, Moscow, 688 pp (in Russian)

Smirnov VI, Borodayev VS, Starostin VI (1968) Massive sulphide ores and deposits of Japan. Geol Ore Dep 1:17–31 (in Russian)

Stolyarov YuM (1980) Hypogenic sulphide-sulphate zoning of porphyry-copper deposits. Geol Ore Dep 3:47–57 (in Russian)

Tvalchrelidze GA (1982) Metallogeny of two main types of the Earth's crust development. Geol Ore Dep 2:3–13 (in Russian)

Tvalchrelidze AG, Yaroshevich VZ (1981) Genesis of stratiform massive sulphide deposits. Nedra, Moscow, 131 pp (in Russian)

Physico-Chemical Conditions of Base Metal Sulphide Ore Formation

A. G. TVALCHRELIDZE[1]

Abstract

Primary base metal massive sulphide deposits are formed as a rule in seafloor depressions. Ore formation is preceded by stratification of seawaters into oxiding and reducing zones. Hydrothermal sedimentary pyritic bodies are formed in two stages. During sedimentogenesis, monosulphide iron and non-ferrous thin sediments are accumulated under thermodynamic disequilibrium conditions. During diagenesis, sea-floor sulphide sediments are crystallized as a result of their interaction with pore solutions. The latter, according the thermodynamic data, are characterized by the buffer ratio of sulphur in oxidized and reduced forms. The process of diagenesis results in standard base metal zonation. If pyrite-bearing fluids are characterized by increased total activity of sulphur, then such a zonation is complicated by an additional zonal distribution of sulphides and sulphates. The disappearance of anhydrite at the barite-polymetallic level is due to the spontaneous evolution of the external physical and chemical parameters against a background of temperature decrease and displacement of the invariant barite-calcite-anhydrite equilibrium towards to barite crystallization field.

Introduction

During the past 25 yrs. a large volume of scientific literature has confirmed the hydrothermal-sedimentary origin of base metal massive sulphide deposits as proposed by Oftedahl (1958), Smirnov (1960) and Anderson (1969). The geological conditions and environment necessary for the hydrothermal-sedimentary formation of pyritic massive sulphide deposits were discussed by Smirnov (1968), and various aspects of this genetic theory have also been discussed in numerous publications by Watanabe (1965), Borodajevskaya et al. (1976), Dzotsenidze (1969), Gilmour (1971), Hutchinson (1973, 1980) and others.

The basic geochemical differences between hydrothermal-sedimentary and metasomatic sulphide ore formation were first discussed by Skripchenko (1969),

[1] Caucasian Institute of Mineral Resources, 85 Paliashvili St., 380062 Tbilisi, USSR

Geology and Metallogeny of Copper Deposits
(ed. by G.H. Friedrich et al.)

who has shown that synsedimentary sulphide ore deposition has a long and complex history, involving sedimentation and diagenetic processes. These geochemical differences are not adequately taken into account by some workers. For instance, Plimer and Finlow-Bates (1978) suggested that primary iron sulphide minerals (pyrite and pyrrhotite) are directly deposited from seawater pyrite in an oxidizing environment, pyrrhotite in a reducing environment. Current research into genetic models for stratiform massive sulphide deposits tends to adopt one of two approaches:

1. Investigations of the physical aspects of hydrothermal discharge, including the hydrodynamics of buoyant plumes (Turner and Gustafson 1978, Solomon and Walshe 1979a, b). Generally, these investigations consider the mineralogical zoning typical of these deposits to be the result of the dynamics of progressive mixing of hydrothermal solutions with seawater.
2. Investigations of the chemical consequences of such a mixing. Generally, these investigations consider the mineralogical zoning to be due to chemical gradients established by the mixing of hydrothermal solutions with seawater (Ridge 1974, Large 1977) or the interaction between ore solutions of possible magmatogenic affiliation with convecting, entrained shallow pore fluids of marine origin (Spooner 1980).

In this paper we try to formulate a third geochemical approach.

Hydrodynamics of Interaction Between Bottom Brine and Hydrothermal Solution

Numerous paleotectonic reconstructions of pyrite deposits in volcanic and sedimentary rocks (Yakovlev 1984) have shown that primary synsedimentary pyrite formation proceeds in bottom depressions. Even in paleovolcanic areas where base metal sulphides are related to rhyolite domes (for instance, in the Urals-type provinces, Tvalchrelidze 1984), massive ore bodies lie directly in volcanic craters and other small negative structures in the central parts of the paleotopographic highs. These structures have sharp contacts with country rocks (Sato 1977, Tvalchrelidze et al. 1982) and are characterized by specific conditions of chemical-sedimentary lithogenesis (Skripchenko 1972). Recent investigations by Sato (1972), Rickard (1973) and Tvalchrelidze (1981) have shown that syngenetic sulphide ore formation occurs under reducing conditions.

Commonly, the level of the oxidizing-reducing margin of seawaters is fixed by the appearance of banded iron and manganese formations. Such a margin may be situated in the upper part of negative structures. If so iron and/or manganese oxide minerals are deposited directly at the roof of sulphide ore bodies. In other cases the oxide sediments are distal to, but always younger than, sulphide ores.

As shown by Sato (1972), Turner and Gustafson (1978) and Solomon and Walshe (1979a, b), the hydrothermal solutions supplied into the described negative structures do form anaerobic brine pools due to the different densities

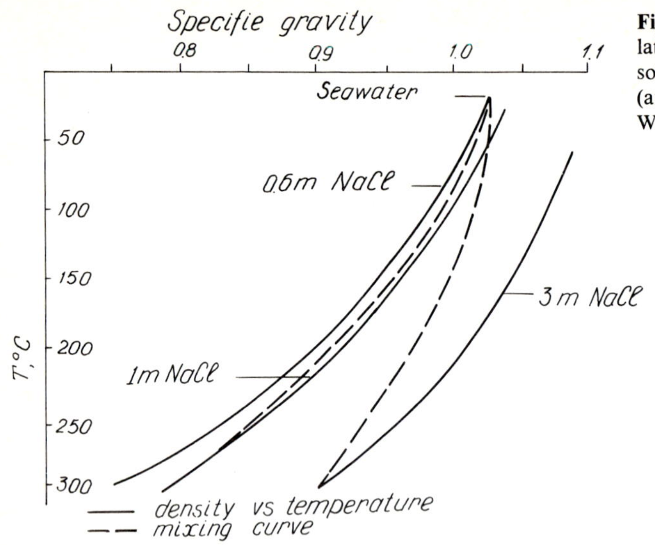

Fig. 1. Density vs temperature relationship of seawater and model solutions of different salinity (after Sato 1972, Solomon and Walshe 1979a)

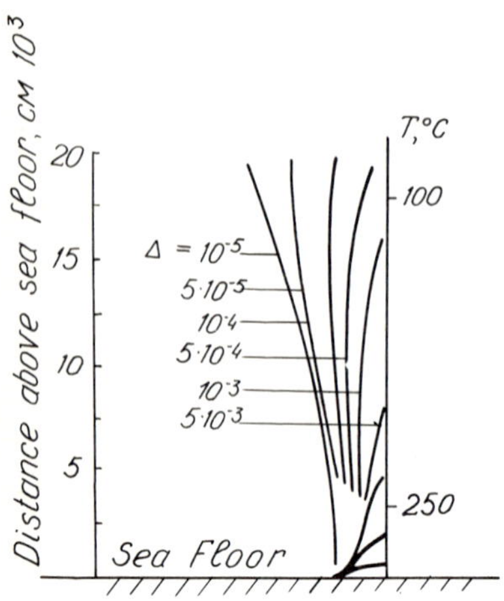

Fig. 2. Reconstruction of the hydrothermal flow as a function of the buoyancy main factor (after Solomon and Walshe 1979a, simplified)

of seawater and hydrothermal fluids. We have recently suggested (Tvalchrelidze et al. 1982) that bottom brines have the same chemical composition as a hydrothermal fluid except that their lower temperatures (about 55° – 60°C) result in oversaturation with respect to ore minerals. Furthermore, any subsequent hydrothermal discharge into a brine pool that has achieved a state of stable stratification would result in the development of buoyant plumes due to the difference in densities between the high-temperature hydrothermal solutions and the cooled brines of the pool (see Fig. 1).

Investigators of hydrodynamic phenomena (Solomon and Walshe 1979a, b) postulate a progressive decrease in temperature vertically upwards along the axis of a buoyant plume formed in this manner. Such a model is illustrated in Fig. 2.

The marginal stream lines are expressed by a buoyancy main factor ("buoyancy parameter") Δ:

$$\Delta = g(\varrho_0 - \varrho_1)/\varrho_1 , \tag{1}$$

where: ϱ_0 = density of plume fluid, ϱ_1 = density of seawater, g = gravitational constant.

The buoyant plume is predicted to rise a height of 500 m. Such hydrodynamic hypotheses deserve careful consideration because they are suggested to be a fundamental cause of the characteristic zoning of pyritic deposits (Solomon et al. 1980). However, any error in the vertical temperature gradient may cause significant errors in the genetic model. We have tried to solve this problem from heat balance considerations.

The heat amount (Q) which is supplied in an ore-forming medium from a hydrothermal flow in hydrodynamically stable conditions can be determined by the classical equation:

$$Q = C_p m \Delta T , \tag{2}$$

where: C_p = thermal heat capacity; ΔT = temperature difference between a stream and its medium, which averages 100 °C, according to our (Tvalchrelidze 1981) investigations; and $m = vt$, where v is mass transfer during the time of stream stabilization (t).

On the other hand, as the stream has the form of an inverted cone:

$$m = \frac{\varrho_0 + \varrho_1}{6} Sh , \tag{3}$$

where: h = the height of the cone axis, S = the area of the cone base. Then:

$$h = \sqrt[3]{\frac{6Q}{\pi C_p \sin^2_\alpha(\varrho_0 + \varrho_1) \Delta T}} , \tag{4}$$

where α is the inclination angle of the cone edge.

Using in Eqs. (2), (3) and (4) maximal values $v = 5 \cdot 10^5\ \mathrm{g\ s^{-1}}$, $t = 100$ s, $\alpha = 10°$ (Turner and Gustafson 1978, Solomon and Walshe 1979a), $\varrho_0 = 1.1099\ \mathrm{g\ cm^{-3}}$, $\varrho_1 = 1.0495\ \mathrm{g\ cm^{-3}}$ (Tvalchrelidze et al. 1982) and tabular data of $C_p = 0.950\ \mathrm{kcal \cdot l^{-1}}$ it could be shown that the maximum height to which a buoyant hydrothermal stream would rise would not exceed 50 m. Because the real values of the given parameters should be much lower than those used in calculation, it could be concluded that commonly the real height would average less than 10 m. So the real buoyant hydrothermal streams are thought to be unable to provide mineralogical zoning of pyrite deposits.

Ore Deposition

Based on experimental (Tvalchrelidze and Yaroshevich 1981) and empirical data (Shanks and Bischoff 1977, Mossman and Heffernan 1978), it was suggested (Tvalchrelidze 1984) that polymetallic sediments before their crystallization consist of monosulphide species of iron and non-ferrous metals. So, ore bodies of the pyritic type must be formed as a result of diagenesis. It is well known that decrease in temperature is the main cause of ore precipitation. Because the temperature of the brine is much lower than that of the hydrothermal fluids, this brine would be entirely oversaturated with respect to ore minerals (see Fig. 3), and consequently suspensions of sulphide species would be formed over a range of heights above the floor of the brine pool. The kinetics of sedimentation by such a mechanism are governed by Stokes' law, which in the present case can be expressed as

$$V = \frac{2r^2(\varrho_2 - \varrho_0)}{9\eta} g \,, \tag{5}$$

where: V = sedimentation rate, r = radius of sulphide species in the suspension, ϱ_2 = density of dispersive phase, η = brine viscosity.

Using the recently published values (Tvalchrelidze et al. 1982) of these parameters, it can be shown that a 4-m-thick ore body could have formed within 1000 yrs. In other words the rates of hydrothermal-sedimentary sulphide accumulation are many times greater than the rates of normal marine clastic sedimentation, and result in sulphide deposits that contain little detrital matter. As a result of continued sulphide accumulation in a brine pool, the sediment-brine interface would tend to come closer to the oxidizing-reducing boundary of the brine pool surface. So new portions of hydrothermal fluids would discharge into the oxidizing zone, giving rise to the formation of oxide manganese and iron sediments.

Another feature of sulphide ore formation is as follows. Accumulating sulphide sediments contain pore solutions, which of course must be over-

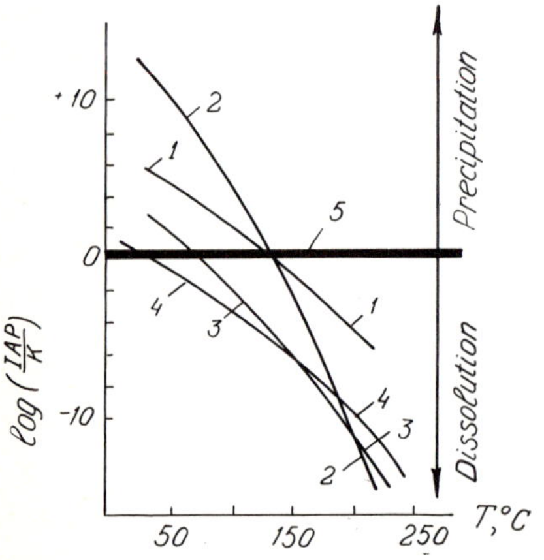

Fig. 3. Temperature dependence of saturation of a hydrothermal fluid (after Tvalchrelidze and Yaroshevich 1981). *IAP* Total ion activity of a given solution ($\log a_S = -3.9$; for other thermodynamic data see Table 1); *K* equilibrium ion activity. *1* Pyrite; *2* chalcopyrite; *3* sphalerite; *4* galena; *5* equilibrium saturation line

saturated with respect to ore components. However, deposition of ore species takes place only over the sediment-brine interface. The fundamental cause of this event, as has been suggested recently (Tvalchrelidze and Yaroshevich 1981, Tvalchrelidze et al. 1982), is the protionization of water molecules in the micropore environment and their extremely high compression. Such a water with distorted properties (water dissipative state, according to Blokh 1984) forms super-high concentrated solutions which can exist only in a micropore environment because their saturation does not correspond to the external thermodynamic conditions.

These phenomena have three important consequences:

1. Since the super-high concentrated solutions, as mentioned, can exist only in pores among sulphide species, sulphide sediments and their pore waters are relatively independent and may be theoretically considered as a half-isolated system.
2. The metastable state of pore solutions preserves sulphide sediments from recrystallization. So, for some time hydrothermal activity would cause convective flow of pore solutions. This migration of pore fluids redistributes ore components, which would tend to homogenize the chemical composition of the sulphide deposit. This is why all the conceptions that consider the geochemical zonation of stratiform sulphide deposits to be due to the chemical evolution of the ore-forming solution and to be a primary depositional pattern (Large 1977) are open to criticism.
3. Crystallization of ores and consequent hypogene zonation are diagenetic processes, which are controlled by interaction between sulphide sediments and pore solutions. It must be noted that the diagenetic nature of hypogene zonation of base metal sulphide ores was suggested by Govett and Whitehead (1974), but geochemical aspects of this process have not received proper attention. It is this aspect of the problem that is discussed below.

Crystallization of Sulphide Sediments and Formation of Monoascending Zonation

We have recently shown (Buadze and Tvalchrelidze 1980) that during diagenesis the crystallization of sulphide minerals in sediments takes place under specific conditions. Briefly, since the sulphur of monosulphides such as FeS or H_2S have a 2^- oxidation state (Rickard 1975), the diagenetic crystallization of disulphide minerals, in which sulphur has a 1^- oxidation state, requires partial oxidation of the bulk sulphur content. It has been suggested that, in the diagenesis of sediments, the oxidizing agent is sulphur (Berner 1972) or polysulphide (Rickard 1975) which has been formed as the result of bacterial reduction of sulphate. It is clear that such conditions are not applicable to hydrothermal-sedimentary ores, because, as indicated by detailed isotopic studies (Tvalchrelidze 1984), there is only minimal involvement of biogenic sulphur. Consequently, most of the pyrite and other disulphide minerals must have been formed without any significant influence of bacterial metabolism.

Wikjord et al. (1976) suggested oxidizing-reducing reactions resulting in the liberation of hydrogen and consequent pyrite crystallization. Fabriol and Maurel (1975) believed that under certain conditions pyrite could form as a result of the reaction of FeS with sulphate ion:

$$7\,FeS + SO_4^{2-} + 8\,H^+ \rightarrow 4\,FeS_2 + 3\,Fe^{2+} + 4\,H_2O\,. \tag{6}$$

Simple thermodynamic conclusions, however, indicate that the concentration of ferrous ion produced by the reaction would not allow the reaction to proceed to the right.

Berner (1970) considered the reaction with sulphate ion and hydrogen sulphide:

$$FeS + 3\,H_2S + 2\,H^+ + SO_4^{2-} \rightarrow FeS_2 + 3\,S^0 + 4\,H_2O\,. \tag{7}$$

Under the conditions of elevated temperature of diagenesis (Tvalchrelidze 1981), however, the elementary sulphur is unstable:

$$4\,S^0 + 4\,H_2O \rightarrow 3\,H_2S + 2\,H^+ + SO_4^{2-}\,. \tag{8}$$

Summing up reactions (7) and (8):

$$4\,FeS + 3\,H_2S + 2\,H^+ + SO_4^{2-} \rightarrow 4\,FeS_2 + 4\,H_2O\,. \tag{9}$$

Thus, pyrite formation from iron monosulphide needs a buffer ratio of sulphur in oxidized and reduced forms. Figure 4 indicates that the pyrite crystallization field lies on the univariant equilibrium line between hydrogen sulphide and sulphide ion. If such a diagram were supplemented with mineral equilibria of non-ferrous metals with the solution (see Fig. 5), then the bulk thermodynamic picture of base metal ore formation would be manifested. Activities of the main components of the model solution are given in Table 1. Table 2 contains some additional thermodynamic data. The method of calculation was published recently in detail (Tvalchrelidze et al. 1982, Tvalchrelidze and Yaroshevich 1981).

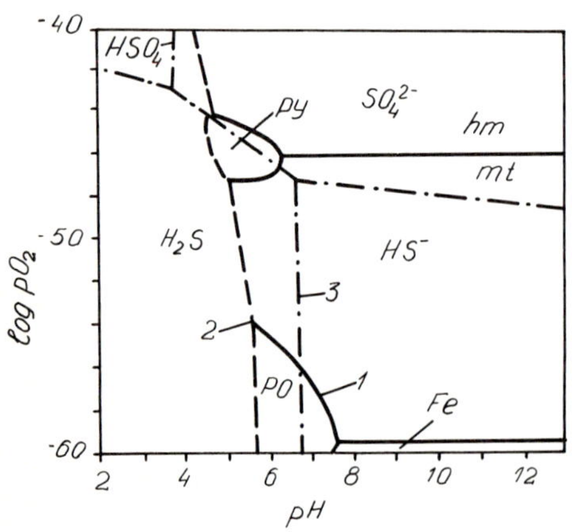

Fig. 4. Pyrite crystallization field and relation of sulphur dissolved species under $T = 150\,°C$; $\log a_S = -5.5$. Equilibria of minerals with model solution are calculated using Table 1 (after Tvalchrelidze 1981). *hm* = Hematite; *mt* = magnetite; *po* = pyrrhotite; *py* = pyrite. Univariant equilibrium lines: *1* between minerals; *2* between minerals and a solution; *3* between sulphur dissolved species

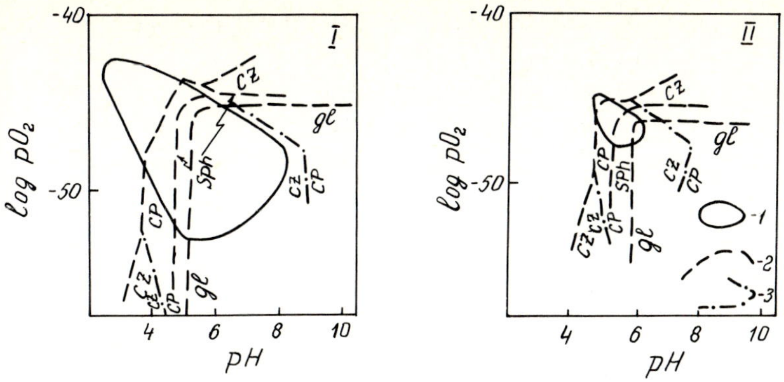

Fig. 5. Fragment of total thermodynamic diagram illustrating the consequence of crystallization of sulphide minerals from a model solution under $T = 150\,°C$; $\log a_{Fe^{2+}} = -3.7$; $\log a_{Cu^{2+}} = -17.0$; $\log a_{Zn^{2+}} = -9.4$; $\log a_{Pb^{2+}} = -11.0$; **I** $\log a_S = -3.9$; **II** $\log a_S = -5.5$ (after Tvalchrelidze et al. 1982). *cz* = Chalcocite; *cp* = chalcopyrite; *gl* = galena; *sph* = sphalerite. *1* Pyrite crystallization field; *2* univariant equilibria between minerals and the model solution; *3* univariant equilibria between minerals. Thermodynamic properties of the model solution are given in Table 1. Examples of some additional simple reactions are given in Table 2. For the full thermodynamic picture, see Tvalchrelidze et al. (1982)

Table 1. Activities of some components of the model solution (after Tvalchrelidze and Yaroshevich 1981)

Component	Temperature, °C					
	25	50	100	150	200	250
	$-\log a_i$					
H_2CO_3	2.4			2.6		2.7
HCO_3^-	3.3			5.0		
$CaHCO_3^+$	3.5			3.9		5.7
$MgHCO_3^+$	4.4			3.1		3.1
$NaSO_4^-$	2.1			2.2		2.6
SO_4^{2-}				5.1		3.9
$FeOH^+$	3.8			4.5		5.3
HSO_4^-	6.3			5.6		5.3
$ZnCl_4^{2-}$	4.8			5.3		5.8
$ZnCl_3^-$	5.2			5.7		6.2
$CuCl_2$	6.2			6.7		7.2
$CuCl$	7.0			7.3		7.8
$PbCl_4^{2-}$	6.5			6.7		7.2
$PbCl_3^-$	6.7			7.2		7.7
Pb^{2+}		11.0	11.4	11.0	12.5	13.2
Zn^{2+}		6.3	7.8	9.4	11.6	13.9
Fe^{2+}		3.3	3.5	3.7	3.9	4.1
Cu^{2+}		16.4	17.5	17.0	18.5	20.7
H^+		5.3	4.8	4.3	3.9	4.0
H_2S [a]		4.3/5.9	3.9/5.5	3.9/5.5	3.8/5.4	4.4/4.6
HS^- [a]		5.8/7.4	6.0/7.6	7.1/8.7	8.3/9.9	10.0/11.0
S^{2-} [a]		13.6/15.2	13.3/14.9	14.2/15.8	15.8/16.9	16.6/18.2

[a] *Note:* in nominator – the component activity below $-\log a_{H_2S}^{150} = 3.9$; in denominator – the component activity below $-\log a_{H_2S}^{150} = 5.5$.

Table 2. Equilibrium constants of some reactions (after Tvalchrelidze and Yaroshevich 1981)

Reactions	Temperature, °C						
	50	100	150	200	250	300	350
	$-\log K$						
$PbS \rightarrow Pb^{2+} + S^{2-}$	26.77	24.05	22.00	20.43	19.21	13.35	17.58
$ZnS \rightarrow Zn^{2+} + S^{2-}$	24.60	22.68	21.17	19.97	19.02	18.30	17.79
$FeS \rightarrow Fe^{2+} + S^{2-}$		17.19			15.21	14.88	
$CuFeS_2 \rightarrow Cu^{2+} + Fe^{2+} + 2S^{2-}$	62.74	57.30	53.01	49.71	47.00	44.95	43.34
$FeS_2 + H_2O \rightarrow Fe^{2+} + 2S^{2-} + 2H^+ + \frac{1}{2}O_2$	80.81	71.97	64.64	61.19	58.07	56.12	
$H_2S \rightarrow 2H^+ + S^{2-}$	19.87	19.05	18.90	19.34	20.24	21.68	23.41
$H_2S \rightarrow H^+ + HS^-$	6.81	6.95	7.51	8.36	9.59		

The diagrams need some clarification. We have mentioned that the pore water dissipative state results in the crystallization of pyrite due to the interaction between pore fluids and monosulphides. This process does not change the buffer ratio of the sulphur dissolved forms, but would decrease the total activity (and concentration) of sulphur. As simple calculations have shown, after the crystallization of pyrite and other disulphide minerals $\log a_s$ would decrease from -3.9 to -5.5. As long as the new hydrothermal flow preserved the initial concentration (activity) of sulphur in the lower part of the sulphide stratum, a gradual decrease of sulphur activity would be observed in the ascending section of the crystallizing sediments. So the first diagram (see Fig. 5, I) represents the thermodynamics of the bottom part of a sulphide body, while the second one (see Fig. 5, II) refers to its top.

It is well known that non-ferrous sulphides commonly occur only in the upper part of base metal sulphide ore bodies, where they are crystallized simultaneously with pyrite. As the pyrite crystallization field here is narrow (see Fig. 5, II) and $\log pO_2$ could not range widely, the zonal deposition of chalcopyrite, sphalerite and galena would be due to a neutralization of pore solutions.

The reason for such a neutralization could be explained (Tvalchrelidze 1981) by the results of experiment (see Table 3). Analysis of Table 3 indicates:

1. Pyrite does not crystallize from FeS without oxidation of H_2S.
2. The neutralization of solutions proceeds only *after* pyrite crystallization (compare pH values in Fig. 4 and Table 3!) and only in solutions containing $(NH_4)_2S$. So, the neutralization of solutions must be due to partial hydrolysis resulting in NH_4OH formation.

Theoretically, as Na, the main component of any hydrothermal solution (Skinner and Barton 1973), is a stronger base than ammonium, temperature decrease must immediately cause neutralization of fluids. Practically, neutralization is prevented both by the buffer ratio of the dissolved forms of sulphur and by a stable distribution of metals in chloride complexes (see Table 1). The latest experimental and thermodynamic investigations of hydrolysis under hydrothermal conditions (Govorov 1977, Massalovich 1975, etc.) have shown that the hy-

Table 3. Experimental study of hydrothermal crystallization of pyrite under $T = 150\,°C$, $P = 500$ bar, experiment duration = 35 days

Experiment number	Autoclave load	pH		Stable phase	Metastable phases
		Primary	Final		
T – 41	14.68 ml $(NH_4)_2S$ + 4 ml 0.9 m $FeSO_4$	7	10	Monoclinic pyrrhotite	Mackinawite
T – 42	200 mg FeS + 15 mg S + 16.68 ml H_2S	5.5	1.5	Pyrite + sulphur	Monoclinic pyrrhotite + marcasite
T – 43	250 mg FeC_2H_4 + 18.68 ml H_2S	5.5	4	Monoclinic pyrrhotite	
T – 44	200 mg FeS + 19.62 ml H_2S	5.5	2	Monoclinic pyrrhotite	X-ray amorphus FeS
T – 45	200 mg FeS + 15 mg S + 18.68 ml H_2O	7	1	Pyrite	Sulphur + marcasite
T – 46	14.68 ml $(NH_4)_2S$ + 4 ml 0.9 m $FeSO_4$ + 15 mg S	7	12	Pyrite	Marcasite + monoclinic pyrrhotite

Note: Solutions, the concentrations of which are not shown, are saturated under $T = 25\,°C$ and $P = 1$ bar.

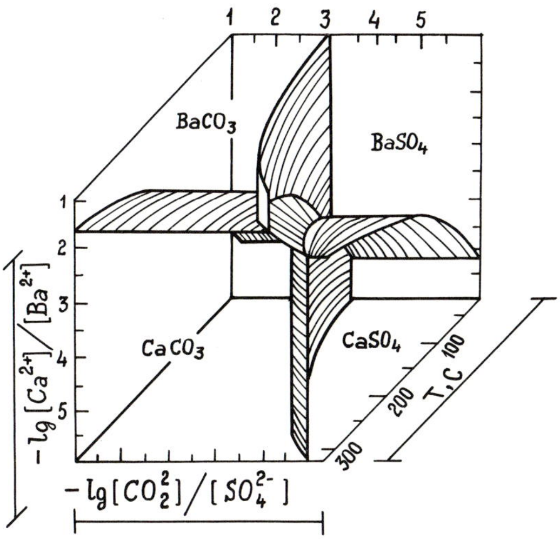

Fig. 6. Phase relations in the system Ba – Ca – S – O (after Gomelauri 1978, slightly modified)

drolysis would be active only after a decrease of sulphur activity and a displacement of equilibria between the chloride complexes of iron and non-ferrous metals.

The above considerations indicate that zonal deposition of polymetallic sulphides, being governed by neutralization of pore solutions, would be active only in the upper parts of a crystallizing ore body.

Thus, although the primary sedimentation of sulphide deposits takes place under conditions of general chemical disequilibrium, the final form of the deposit, particularly with regard to bulk mineralogy and zonal patterns, is determined by diagenetic processes and not by the evolution of ore-forming solutions. This is the main difference between our genetic model and that of Large (1977). These diagenetic processes take place under chemical conditions that are buffered within narrow ranges of various chemical parameters. Buffering within narrow ranges explains the chemical and mineralogical uniformity of massive base metal sulphide deposits throughout the world. A single exception to the general rule is presented by Kuroko-type deposits, where the usual mineralogical zonation is complicated by the presence of sulphate zones. In general terms, such a zonation is expressed by the disappearance of anhydrite at the barite-polymetallic sulphide zone boundary. Briefly, this phenomenon may be explained in two ways: (1) by higher total sulphur activity compared to the model described above; and (2) by the tendency of the anhydrite phase to dissolve from a barite-calcite-anhydrite equilibrium assemblage as temperature decreases below 250 °C (see Fig. 6).

Acknowledgment. I would like to express my deepest gratitute to my anonymous opponent who has read and edited the manuscript of this paper and supplied a number of important and valuable remarks.

References

Anderson CA (1969) Massive sulfide deposits and volcanism. Econ Geol 64:129 – 146

Berner RA (1970) Sedimentary pyrite formation. Am J Sci 268:1 – 23

Berner RA (1972) Sulfate reduction, pyrite formation and the oceanic sulphur budjet. In: Dyssen D, Jagder J, (eds) The changing chemistry of the oceans. Almqvist & Wiksell, Stockholm, pp 347 – 361

Blokh AM (1984) Mobilization of matter in superhigh-concentrated solution as a result of the dissipative state of water in microspaces of the Earth's interior. In: Janelidze TV, Tvalchrelidze AG (eds) Proc 6th IAGOD Symp, vol I. Schweizerbart, Stuttgart, pp 59 – 67

Borodajevskaya MB, Krivtsov AI, Shiray EP (1976) Peculiarities of early geosyncline volcanism and ore formation in pyrite-bearing different provinces. Geol Ore Dep 1:20 – 28 (in Russian)

Buadze VI, Tvalchrelidze AG (1980) Geological, geochemical and physico-chemical conditions of formation of stratiform non-ferrous deposits in sedimentary rocks of the Caucasus. In: Ridge JD (ed) Proc Fifth Quadrennial IAGOD Symposium, vol 1. Schweizerbart, Stuttgart, pp 245 – 260

Dzotsenidze GS (1969) Role of volcanism in the formation of sedimentary rocks and ores. Nedra, Moscow (in Russian)

Fabriol R, Maurel C (1975) Réactions chimiques entre pyrrhotite et solutions d'acide sulfurique: étude expérimentale et thermodynamique. Bull Soc Miner Crystallogr 98:227 – 234

Gilmour P (1971) Strata-bound massive pyritic sulfide deposits – a review. Econ Geol 66:1239 – 1244

Gomelauri AI (1978) Temperature conditions of barite formation in the Georgian deposits and their examination in autoclaves. In: Ermakov NP (ed) Thermobarogeochemistry of the Earth's crust and ore formation. Nauka, Moscow, pp 127 – 130 (in Russian)

Govett GJS, Whitehead RES (1974) Origin of mineral zoning in stratiform sulfides; a hypothesis. Econ Geol 69:551 – 556

Govorov IN (1977) Thermodynamics of ion-mineral equilibria and the mineralogy of hydrothermal deposits. Nauka, Moscow (in Russian)

Hutchinson RW (1973) Volcanogenic sulfide deposits and their metallogenic significance. Econ Geol 68:1223 – 1247

Hutchinson RW (1980) Massive base metal sulphide deposits as guides of tectonic evolution. In: Strangway DW (ed) The continental crust and its deposits. Geol Assoc Can Spec Pap 20:659 – 684

Large RR (1977) Chemical evolution and zonation of massive sulfide deposits in volcanic terrains. Econ Geol 72:549 – 572

Massalovich AM (1975) Ore transport during pyrite deposition in hydrothermal systems. Geol Ore Dep 2:59 – 63 (in Russian)

Mossman DJ, Heffernan KJ (1978) Simulated low-grade metamorphism in metalliferous mud from the Read Sea Atlantis II deep. Econ Geol 73:1150 – 1154

Oftedahl C (1958) A theory of exhalative-sedimentary ores. Geol Fören Stockholm Förh 80:1 – 19

Plimer IR, Finlow-Bates T (1978) Relationship between primary iron sulphide species, sulphur source, depth of formation and age of submarine exhalative-sulphide deposits. Miner Deposita 13:399 – 410

Rickard DT (1973) Limiting conditions for synsedimentary sulfide ore formation. Econ Geol 68:605 – 617

Rickard DT (1975) Kinetics and mechanism of pyrite formation at low temperatures. Am J Sci 275:636 – 652

Ridge JD (1974) Notes on boiling of ascending ore fluids and the position of volcanic-exhalative deposits in the modified Lindgren classification. Geology 6:287 – 288

Sato T (1972) Behaviours of ore-forming solution in seawater. Min Geol 22:31 – 42

Sato T (1977) Kuroko deposits: their geology, geochemistry and origin. In: Volcanic processes in ore genesis. Inst Min Metall, London, pp 153 – 161

Shanks WC, Bischoff JL (1977) Ore transport and deposition in the Red Sea geothermal system: a geochemical model. Geochim Cosmochim Acta 41:1507 – 1519

Skinner BJ, Barton PB (1973) Genesis of mineral deposits. Annu Rev Earth Planet Sci 1:183 – 211

Skripchenko NS (1969) On the stability of sulphides and oxides during low temperature mineragenesis. Geochemistry 3:301 – 308 (in Russian)

Skripchenko NS (1972) Hydrothermal-sedimentary sulphide ores in basaltic rocks. Nedra, Moscow

Smirnov VI (1960) Convergence of pyrite deposits. Moscow State Univ Bull Geol 2:19 – 26 (in Russian)

Smirnov VI (1968) Pyrite deposits. In: Smirnov VI (ed) Genesis of endogenous ore deposits. Nedra, Moscow, pp 586 – 647

Solomon M, Walshe JL (1979a) The behavior of massive-sulphide ore solution entering seawater and the development of zoned deposits. Bull Miner 102:463 – 470

Solomon M, Walshe JL (1979b) The formation of massive sulfide deposits on the sea floor. Econ Geol 74:797 – 813

Solomon M, Walshe JL, Garcia-Palomero P (1980) Formation of massive sulphide deposits at Rio Tinto, Spain. Trans Inst Min Metall B89:16 – 24

Spooner ETC (1980) Cu-pyrite mineralization and seawater convection in oceanic crust – the ophiolitic ore deposits of Cyprus. In: Stangway DW (ed) The continental crust and its mineral deposits. Geol Assoc Can Spec Pap 20:685 – 704

Turner JS, Gustafson LB (1978) The flow of hot saline solution from vents in the sea floor – some implications for exhalative massive sulfide and other deposits. Econ Geol 73:1082 – 1100

Tvalchrelidze AG (1981) Geochemistry of hydrothermal-sedimentary sulphide ore formation. Geol Ore Dep 2:1 – 16 (in Russian)

Tvalchrelidze AG (1984) Ore-Bearing hydrothermal systems of stratiform pyrite deposits. In: Janelidze TV, Tvalchrelidze AG (eds) Proc 6th IAGOD Symp, vol I. Schweizerbart, Stuttgart, pp 246 – 257

Tvalchrelidze AG, Yaroshevich VZ (1981) Genesis of stratiform base metal massive sulphide deposits. Nedra, Moscow (in Russian)

Tvalchrelidze AG, Yaroshevich VZ, Narozauli IG (1982) A genetic model for the Urup copper massive sulfide deposit. Int Geol Rev 24:11 – 24

Watanabe T (1965) Submarine volcanism and ore deposition. J Geol Soc Jpn 71:332 – 336

Wikjord AG, Rummery TE, Doern FE (1976) Crystallization of pyrite from deoxygenated aqueous sulphide solution at elevated temperatures and pressures. Can Miner 4:571 – 573

Yakovlev GF (1984) Volcanogene structures of mineral deposits. Nedra, Moscow (in Russian)

On the Genesis of Barite-Associated with Volcanogenic Massive Sulfides, Fukazawa Mine, Hokuroku District, Japan

S. I. KALOGEROPOULOS[1,2] and S. D. SCOTT[1]

Abstract

The barite bed at Fukazawa mine overlies a massive black ore zone and underlies a ferruginous chert bed (tetsusekiei). It is characterized by simple mineralogy which consists predominantly of barite with small but variable amounts of pyrite, sphalerite, chalcopyrite, galena, hematite, quartz and traces of intimately intergrown chlorite-sericite. Chemical and textural evidence indicates that barite formed by the mixing of a circulating barium-rich hydrothermal solution with cold seawater within an unconsolidated pyritic tuff. Growth of barite displaced the very fine silicate matrix through a process of fluidization of the tuffaceous bed during hydrothermal discharge. Variations in the intensity of the hydrothermal activity resulted in vertical fluctuations of the redox boundary providing for replenishment of the barite environment with cold seawater which passed on to the barite both its sulfur and strontium isotopic characteristics. The barite bed was affected by later hydrothermal solutions, the activity of which continued even after the deposition of hanging wall rocks. As a consequence, equilibrium thermodynamics cannot be applied to describe the chemical behaviour of the bulk dynamic system.

Introduction

Kuroko-type ore deposits occur in Miocene volcanic and sedimentary rocks of the Green Tuff Region in Japan. The Hokuroku district located in northern Honshu (Fig. 1) is the most productive Kuroko-type ore field in the Green Tuff Region (Sato 1974, Ohmoto and Skinner 1983). Tuffs and tuff breccias in the Green Tuff Region are also the hosts of relatively small lenticular massive barite deposits of Kuroko type (Igarashi et al. 1974). Thin (generally less than 1 m), essentially monomineralic, massive and sandy barite ores also form the uppermost zone of massive sulfide Kuroko mineralization. The barite bed in the Tsunokakezawa No. 1 ore body of the Fukazawa mine, the subject of this study,

[1] Department of Geology, University of Toronto, Toronto, Ontario, M5S 1A1, Canada
[2] Present address: Institute of Geology and Mineral Exploration, 70 Messoghion Street, Athens 608, Greece

Geology and Metallogeny of Copper Deposits
(ed. by G.H. Friedrich et al.)

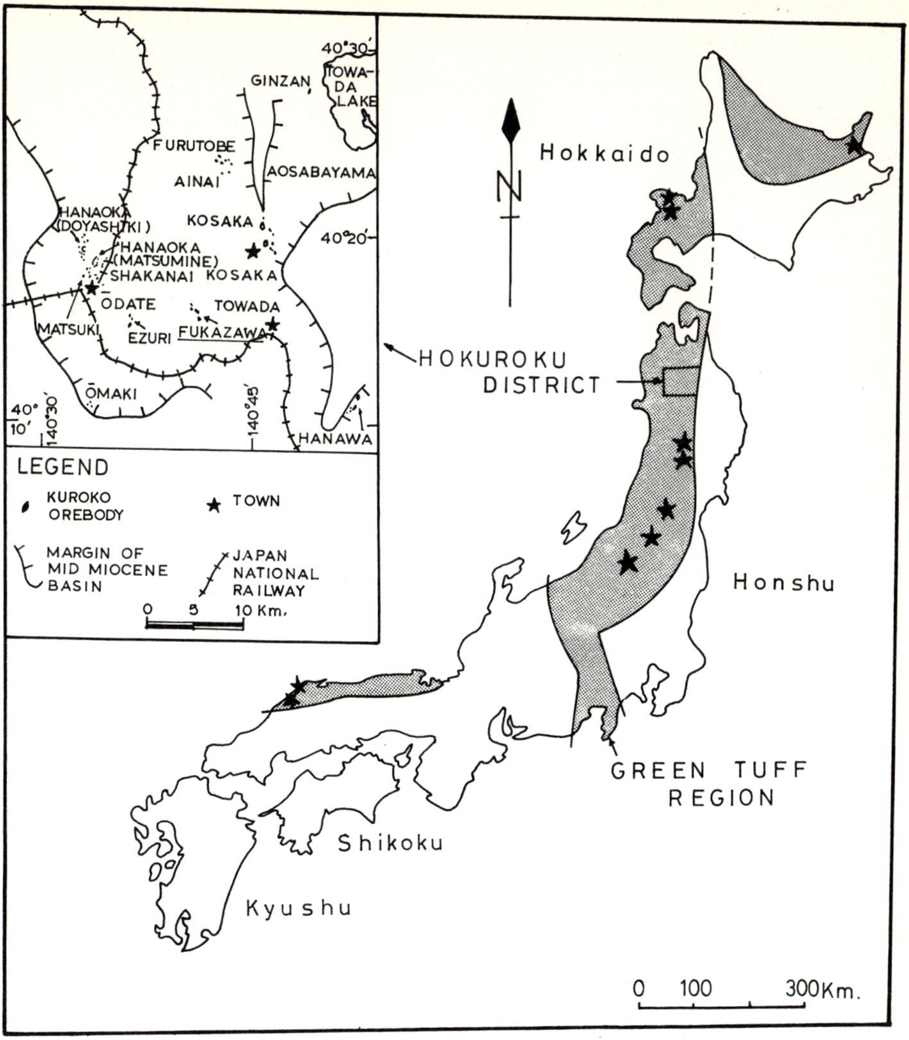

Fig. 1. Location of Kuroko deposits in relation to the Green Tuff Region (from Hashimoto 1977). The *inset* shows the distribution of the Kuroko deposits in the Hokuroku district. The studied barites are from the Fukazawa mine (*underlined*)

conformably overlies a massive black ore zone and underlies a bedded cherty tuff that is variably enriched in iron (tetsusekiei; Fig. 2). Chemical (Shikazono personal communication 1979) and isotopic (Farrell 1979, Farrell and Holland 1983, Kusakabe and Chiba 1983) studies on barite from the Tsunokakezawa No. 1 ore body have contributed to an understanding of the genesis of the barite zone. This paper describes a study utilizing mineralogy, fluid inclusions, stable isotopes, and the relation to the overlying tetsusekiei zone to elucidate that genesis further.

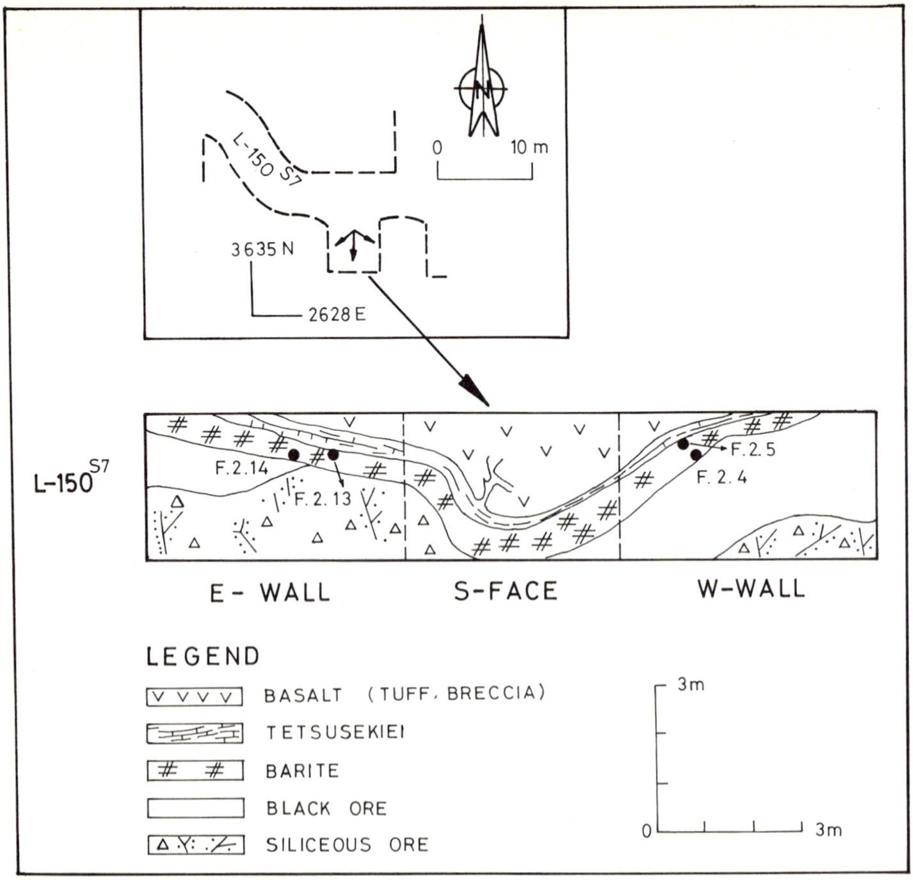

Fig. 2. A part of Tsunokakezawa No. 1 ore body on level − 150 at stop 7 (-150^{s7}). The massive barite zone lies immediately above the massive black ore and below tetsusekiei (ferruginous cherty tuff). *Numbers* represent samples referred to in the text. (Courtesy of Dowa Mining Co. Ltd., Japan)

Mineralogy and Geochemistry

The barite zone at Fukazawa is characterized by a simple mineralogy consisting predominantly of barite with small but variable amounts of pyrite, chalcopyrite, sphalerite, galena, hematite, quartz, and traces of intimately intergrown chlorite – sericite. The sphalerite contains minute inclusions of "chalcopyrite disease" which is thought by Barton (1978) to be produced by CuS-rich solutions reacting with FeS in sphalerite.

Pyrite

Two textural types of pyrite have been observed in the barite ore zone: (1) euhedral disseminated clusters; and (2) euhedral to anhedral pseudomorphs after barite.

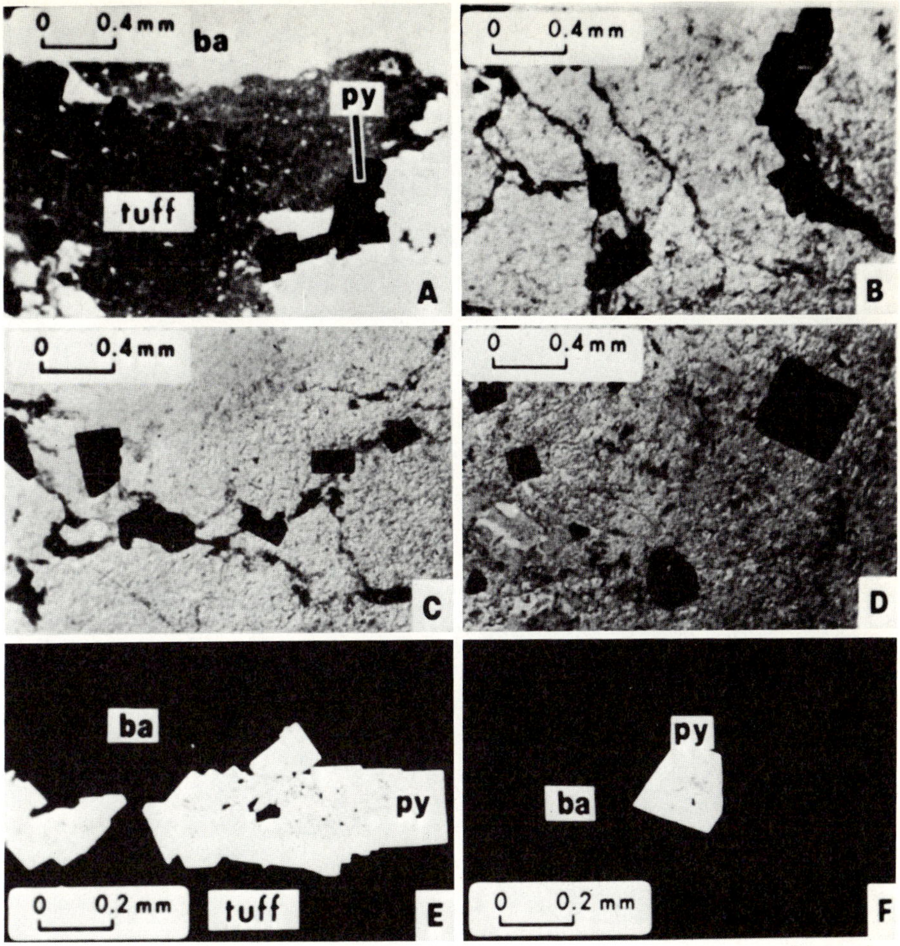

Fig. 3A – E. Textures exhibited by pyrite in the barite zone of Tsunokakezawa No. 1 ore body. All photomicrographs are in plane polarized light. **A** Euhedral pyrites (*black*) are exclusively associated with the tuff component, and both are entrapped by barite. Transmitted light. **B** and **C** Intermediate stages of tuff displacement and entrapment of pyrite by barite. The tuff is *dark grey* and occupies the irregular interstices between large aggregates of barite. Transmitted light. **D** The clay matrix of the tuffaceous component is completely displaced, and pyrite is concentrated and recrystallized within barite. Note that regions originally occupied by tuff can still be traced by the contrast in relief of barite and by the distribution of pyrite. Transmitted light. **E** Euhedral pyrites within tuff both of which are enclosed by barite. Reflected light. **F** Euhedral pyrite inclusion within barite. Note the similarity in internal texture between this type of pyrite and that of **E**. Reflected light

Euhedral pyrite grains are usually disseminated in clusters whithin the barite zone (Fig. 3D). However, whenever tuffaceous (silicate) material is enclosed by the massive barite, the pyrites are exclusively associated with the tuff (Fig. 3A). Figures 3A – D depict pyrites whose host is progressively changing from a tuffaceous material to barite via displacement of the former. All these pyrites have identical textures (Fig. 3E, F). These features suggest that the clusters of euhedral

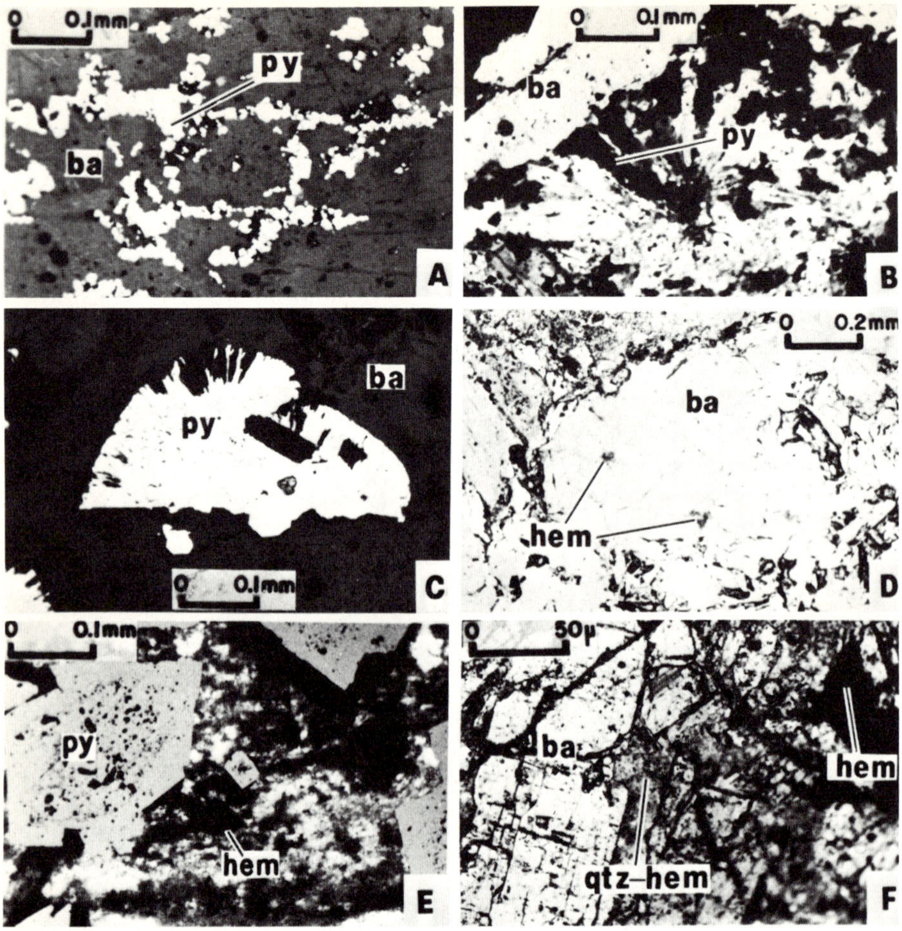

Fig. 4A – F. Textures exhibited by pyrite in the barite zone of Fukazawa (Tsunokakezawa No. 1 ore body) and Furutobe mines. **A** Pyrite replacing recrystallized barite along its cleavages. Reflected light. **B** Fanlike barite blades, partially replaced by pyrite. Plane transmitted light. **C** Pyrite pseudomorphs after fanlike barite. Plane reflected light. **D** Hematite inclusions within recrystallized barite. Hematite has cubic and triangular outlines. Plane transmitted light. **E** Hematite pseudomorphs after pyrite in tetsusekiei overlying barite zone. Partially plane transmitted and reflected light. **F** Hematite pseudomorph after pyrite and quartz-hematite both interstitial to barite. Plane transmitted light

pyrites were enclosed in the barite as the unconsolidated tuffaceous matrix was progressively displaced.

Pyrite replaced barite along its cleavages (Fig. 4A) and pseudomorphed fanlike barite blades (Fig. 4B, C). A probable reaction is

$$2BaSO_4 + 2H^+ + Fe^{2+} \rightleftharpoons 2Ba^{2+} + FeS_2 + H_2O + \tfrac{7}{2}O_2 .$$

The released barium may have been adsorbed by clays and/or precipitated as barite in the overlying zone.

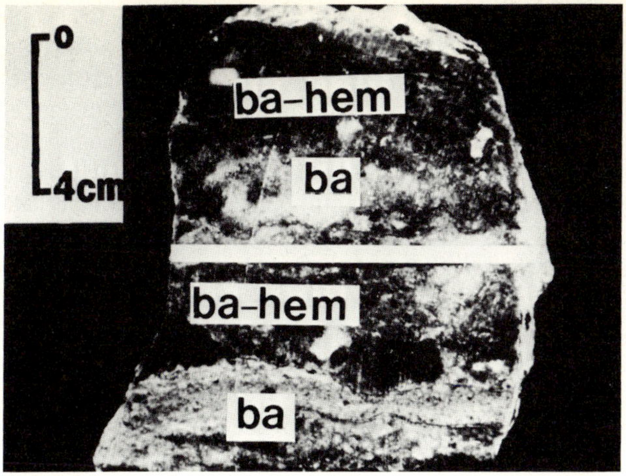

Fig. 5. Alternating hematite-rich bands in barite

Hematite

Scattered inclusions of hematite, some apparently pseudomorphous after pyrite, occur within recrystallized barites (Fig. 4D). At Fukazawa, there is evidence that hematite has pseudomorphically replaced pyrite in the overlying pyritic tetsusekiei as well. Therefore hematite inclusions are probably incorporated into barite by the same process as was described for the "displacement" type of pyrite.

Hematite also occurs interstitially to barite crystals (Fig. 4F) in certain hematite-rich bands (Fig. 4E). There are two possible explanations for this phenomenon:

1. Pyrite grains within the barite zone may have been oxidized without pseudomorphic textures being preserved.
2. The Fe^{2+}/Fe^{3+} ratio in the ascending hydrothermal fluid may have been so far removed from the stability field of any sulfide that at some level within the barite zone the solutions yield interstitial hematite. In case (2) any pyrite initially present within the appropriate parts of the barite zone would also be expected to be oxidized. The presence of hematite pseudomorphs after pyrite together with quartz-hematite intergrowths, both interstitial to barite (Fig. 4F) indicates that the second case is the more likely. Moreover, the observation of alternating hematite-rich and sulfide-bearing barite layers (Fig. 5) suggests that the redox boundary was fluctuating apparently due to fluctuations in the intensity of the hydrothermal activity.

Barite

Fanlike blades and sheafs of barite, euhedral to subhedral tubular grains, and occasionally, single barite grains combining these textural features, are observed in the barite zone. Veins of barite crosscut the barite beds.

Experimental work has shown that $(Ba, Sr)SO_4$ is unreactive in terms of Ba – Sr exchange between aqueous and solid phases. This means that during the course of precipitation the solid does not readily re-equilibrate in response to changes in the composition of the aqueous phase and that chemical equilibrium is established only between the liquid and the last increment of barite precipitated (Doerner and Hoskins 1925, Gordon et al. 1954, Hanor 1968). This type of equilibrium has been referred to as "surficial equilibrium" by Barton et al. (1963).

Recognition of texturally and chemically distinct stages of precipitation with respect to Sr content can provide information concerning the mechanism(s) of barite precipitation and the variation of the $m_{Sr^{2+}}/m_{Ba^{2+}}$ ratio in solution. This is true only if the initial precipitation follows a prescribed path. For instance, if a (Ba, Sr)-rich solution suddenly encounters an excess SO_4^{2-} presumably barite will be "dumped" and the $m_{Sr^{2+}}/m_{Ba^{2+}}$ ratio will match the initial ratio in the solution. Actually, sulfur isotopes of Kuroko barites suggest that the major precipitating mechanism of these barites is the mixing of a barium-bearing hydrothermal fluid and cold seawater (Bryndzia et al. 1983, Kusakabe and Chiba 1983).

The internal texture of single barite grains, regardless of whether they are fine (<0.5 mm) or coarse-grained (0.5 – 2.5 mm) is typically characterized by two zones with different textural features (Fig. 6).

1. A write clean variety, referred to as zone 1; and
2. A greyish-white, dusty-looking variety of very fine-grained (<0.2 mm) barite aggregates, referred to as zone 2. Zone 2 has morphological characteristics that could be associated with higher degrees of supersaturation than those of zone 1. The most likely way to achieve this supersaturation would be to mix (Ba, Sr)-rich hydrothermal fluid with cold SO_4^{2-}-bearing seawater. This inference can further be supported by the sulfur isotopes of the Kuroko barites (Kusakabe and Chiba 1983). Single barite grains exclusively representing zone 1 or zone 2, as well as composites of 1 and 2 in zone patterns such as zone 1 (centre)/zone 2 (rim), zone 2 (centre)/zone 1 (rim), and zone 1 (centre)/zone 2 (intermediate)/zone 1 (rim) have been identified in the barite ore. Evidently, the last zone pattern represents a complete evolution of barite growth.

The observation of (1) transitional textural stages from fan- and sheaflike barites to tabular barites with identical internal textural features (Fig. 6B – D), and (2) the incorporation of a number of textural units which characterize single barite grains (i.e. zones 1 and 2) within individual but larger braik grains (Fig. 6D, E), suggest that the composite barite grains were most likely produced by recrystallization of the fan- or sheaflike aggregates. Green et al. (1983) have shown that the hydrothermal activity in the Fukazawa ore continued even after the deposition of the hanging wall rocks. Thus, the recrystallization textures observed in the barite zone may have been caused or aided by this later hydrothermal solution.

Both textural types of barite (clean zone 1 and dusty-looking zone 2) were analyzed for their strontium content by energy dispersive electron microprobe. In addition, qualitative spectral scans were performed on single grains in order to

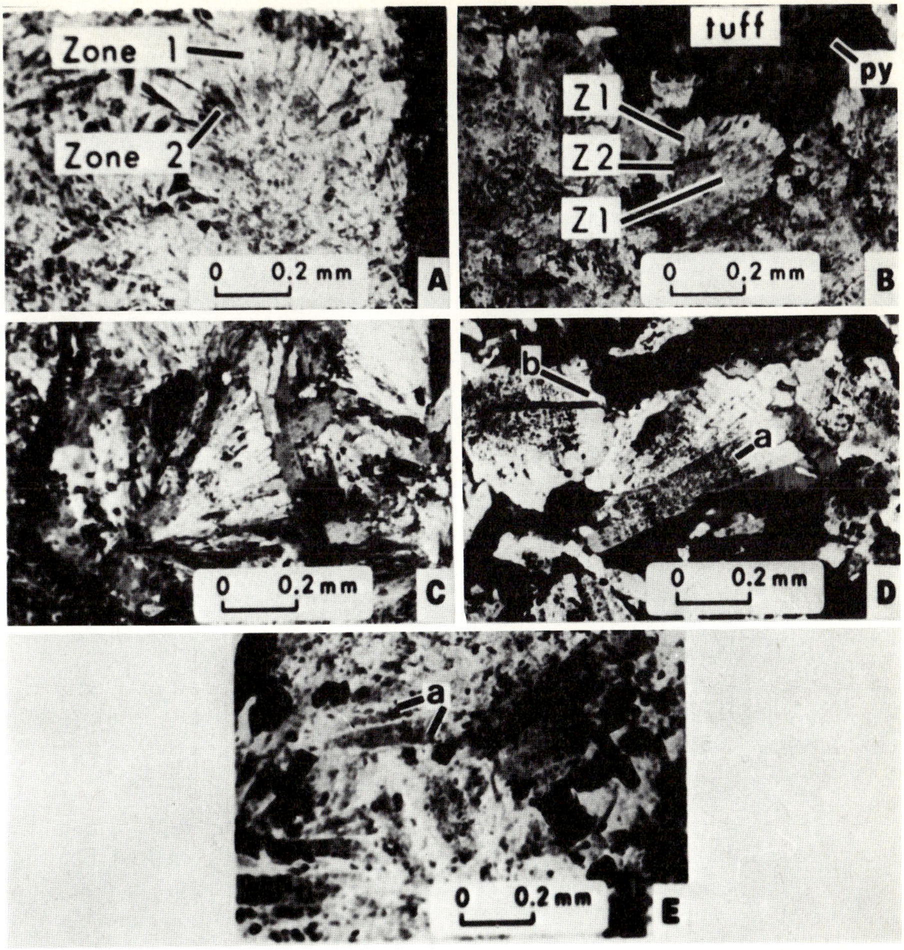

Fig. 6A – E. Textural types of barite from Tsunokakezawa No. 1 ore body, Fukazawa mine. **A** Fanlike barite blades showing a typical dusty-looking fine-grained zone 2 (Z_2), and a clean coarser-grained zone 1 (Z_1). Plane transmitted light. **B** A polycrystalline barite overgrowth (*zone 1*) is succeeded inwards by *zone 2* and a central *zone 1*. Plane transmitted light. **C** Fanlike barite blades of textural zone 1 with an almost perfect tabular outline. Transmitted light, crossed nicols. **D** Zoned barites (Z_1 and Z_2). Grain (*a*) is composite like those in **A** and **B**. The rim of (*b*) is monocrystalline. Transmitted light, crossed nicols. **E** Randomly oriented barite grains (as indicated by their zonation) some of which (*a*) form a single grain, apparently resulting from recrystallization. Transmitted light, crossed nicols

characterize the variations in internal composition. Details of the analytical procedures are in Kalogeropoulos (1982, Appendix I).

Table 1 summarizes the strontium variations along barite blades which form constituent units of fanlike barite aggregates. These barite blades in turn may consist of two, three or four growth zones which can texturally be classified as zone 1 or 2. These textural zones possess different strontium contents, high values coinciding with the texturally dusty-looking (fine-grained) barite stage of zone 2.

Table 1. Mean SrO (wt% ± 1 s[a]) concentrations of texturally-zoned fanlike barite blades from the massive barite ore zone of the Tsunokakezawa No. 1 ore body (Fukazawa mine)[b]

Sample	Early zone 1	Intermediate zone 2	Late zone 1
F.20.6	0.97 ± 0.13(7)	2.14 ± 0.30(9)	0.95 ± 0.07(2)

[a] 1 s = One standard deviation.
[b] Numbers of analyses are in parentheses.

Table 2. Mean SrO (wt% ± 1 s[a]) concentrations of texturally-zoned tabular and vein barites from the massive barite ore zone of the Tsunokakezawa No. 1 ore body (Fukazawa mine)[b]

Sample and textural zonation	Core	Intermediate	Edge
F.20.6A Zone 1	0.20 ± 0.17(3)	0.70(2)	0.98 ± 0.13(3)
F.2.13A Zones 1-2-1	1.14 ± 0.27(14)	2.63 ± 0.34(15)	1.02 ± 0.29(24)
F.20.6B Zones 2-1	2.52 ± 0.33(6)	1.26 ± 0.15(5)	1.08 ± 0.23(5)
F.20.6C (veinlet) Zone 1	1.12 ± 0.19(3) (early)		0.77 ± 0.06(3) (late)

[a] 1 s = One standard deviation.
[b] Numbers of analyses are in parentheses.

Table 2 summarizes the variations in concentration of strontium (from the core, through the intermediate zone to the rim) in texturally-zoned tabular and vein barite. The average strontium content of these barites varies from 0.2 to 2.6 wt% SrO (total range 0.1 – 3.2 wt% SrO). They exhibit three characteristic zonation trends:

1. An increase in SrO from 0.20 at the core to about 1.0 wt% at the edge (Table 2, texturally zone 1);
2. An increase of SrO from about 1.1 wt% at the core to 2.6 wt% at an intermediate zone which texturally coincides with zone 2, and a subsequent decrease at the edge where SrO concentrations approach those at the core (1.0 wt% SrO; Table 2; textural zonation 1-2-1); and
3. A decrease from about 2.5 wt% SrO at the core to a value of about 1.1 wt% at the edge (Table 2; textural zonation 2-1).

A late stage barite veinlet crosscutting the bedding of the massive barite ore zone shows a decrease in SrO from 1.1 wt% in an early formed zone to 0.77 wt% in a later zone (Table 2; texturally zone 1). Taken together, these data for tabular and vein barites show a consistent systematic change in composition which is illustrated in Fig. 7. Early formed barite contains about 0.2 wt% SrO, followed by intermediate barite at 2.6 wt% SrO and late stage barite at about 0.8 wt% SrO.

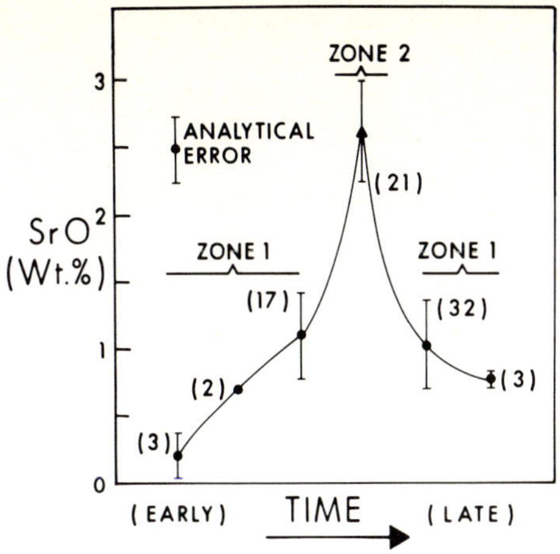

Fig. 7

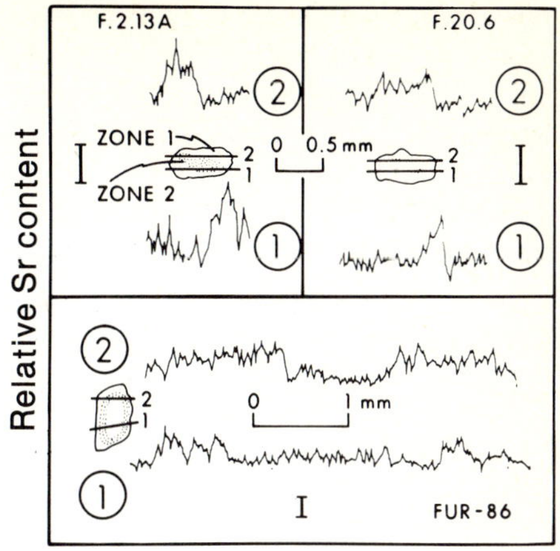

Fig. 8

Fig. 7. Evolution of SrO (wt%) content in tabularlike barite grains. Note the similarity in trend and concentration level with those in Table 1. The *bars* incorporate one standard deviation of the mean and the analytical error

Fig. 8. Spectral scans for Sr(Kα) of two barite grains from Fukazawa (F.2.13A and F.20.6) and one grain from Furutobe (FUR-86). Note the sharp increase in strontium coinciding with textural zone 2. *Vertical bars* represent a relative SrO content of 1 wt%

Comparing Table 1 and Fig. 7 one can clearly see not only similar zonation trends, but also identical SrO concentrations for both the barite blades and single tabular barite grains. These data together with textural evidence suggest that the coarse-grained tabular barites most likely evolved from the fanlike or tabular sheafs by a process of isochemical recrystallization.

Spectral scans for Sr across single grains of barite (Fig. 8) reveals that the transition from zones of low to high Sr content is generally abrupt and that the strontium-rich zones represent a small fraction of the total barite volume. Consequently, the chemical and the concomitant textural changes during barite precipitation, whatever their cause, are characterized by an ability to change abruptly and by a short residence time in the system.

Sulfur Isotopes

Three pyrite samples from the barite ore zone have $\delta^{34}S$ values of +3.8 and +3.9‰. These lie within the range of $\delta^{34}S$ (−0.6 to +6.05‰; Kalogeropoulos and Scott 1983) for pyrites from the overlying tetsusekiei bed at Tsunokakezawa No. 1, suggesting a similarity in their environments of formation.

If pyrites are assumed to have attained sulfur isotopic equilibrium with the coexisting barites ($\delta^{34}S$ = +21.5‰; Kusakabe and Chiba 1983), a wide range of temperatures, all over 330 °C, are obtained from the sulfur isotopic geothermometer (Sakai 1968, Nakai 1970, Ohmoto 1972). These temperatures are in disagreement with the tight cluster (150° – 210 °C) obtained from oxygen isotopic geothermometry for the pair barite-water (Kusakabe and Chiba 1983). Therefore, sulfur isotopic geothermometry based on sulfate-sulfide equilibria does not appear to give a reliable estimate of temperature, probably because pyrite and barite from the barite ore zone were not precipitated together, as we have shown above.

Fluid Inclusions

Barites from the bedded barite of Tsunokakezawa No. 1 ore body have two-phase (vapour + liquid) fluid inclusions of variable size (10 – 30 μ) and shape (tabular, ovoid, irregular). Vapour to vapour + liquid ratios range from less than 0.05 to almost 1. Because of the possibility of leakage along cleavage planes during heating, fluid inclusions were subjected to freezing experiments before they were homogenized. All measurements were made on the Linkam TH600 stage (Macdonald and Spooner 1981, details of the analyses are in Kalogeropoulos 1982, Appendix II). The salinity of the 14 inclusions examined is quite constant [4.6 ± 0.4(1 s)[3] wt% eq. NaCl] and falls within the 3 to 6 wt% equivalent NaCl obtained for barites from the black ore of Kosaka, Furutobe, Ainai and Shakanai Kuroko deposits (Lu 1970, Watanabe 1970, Tokunaga and Honma 1974).

The coexistence of vapour-rich with liquid-rich inclusions could be a result of boiling, leakage or necking down. All of the fluid inclusions studied homogenized into the liquid phase and at highly variable temperatures (Table 3). Boiling can thus be eliminated as an explanation.

There is no unequivocal evidence that the inclusions have not leaked. However, leakage of a fluid from any inclusion that has not exsolved a daughter crystal will not affect the salinity estimates derived from that inclusion (although it will affect homogenization temperatures).

The morphological and microthermometric properties of fluid inclusions in barite from beds in the Fukazawa mine and from the active geothermal field of Milos Island (Greece) which shows similar characteristics to Kuroko (Kalogero-

Table 3. Range of homogenization temperatures, mean (± 1 s[a]) and range of salinity in wt% eq. NaCl for barite from Milos (Greece) and the bedded barite of Fukazawa mine (Japan)

Deposit	Sample no.	T_H(°C) Range	T_F(°C)	Salinity wt% eq. NaCl
Milos	MA	169.6 – 304	n.d.[b]	–
Fukazawa	F.2.4	130.8 – 302	– 2.79 ± 0.23 (– 2.4 to – 3.2)	4.61 ± 0.4 (4.02 to 5.25)

[a] 1 s = One standard deviation.
[b] n.d. = not determined.

[3] 1s = One standard deviation.

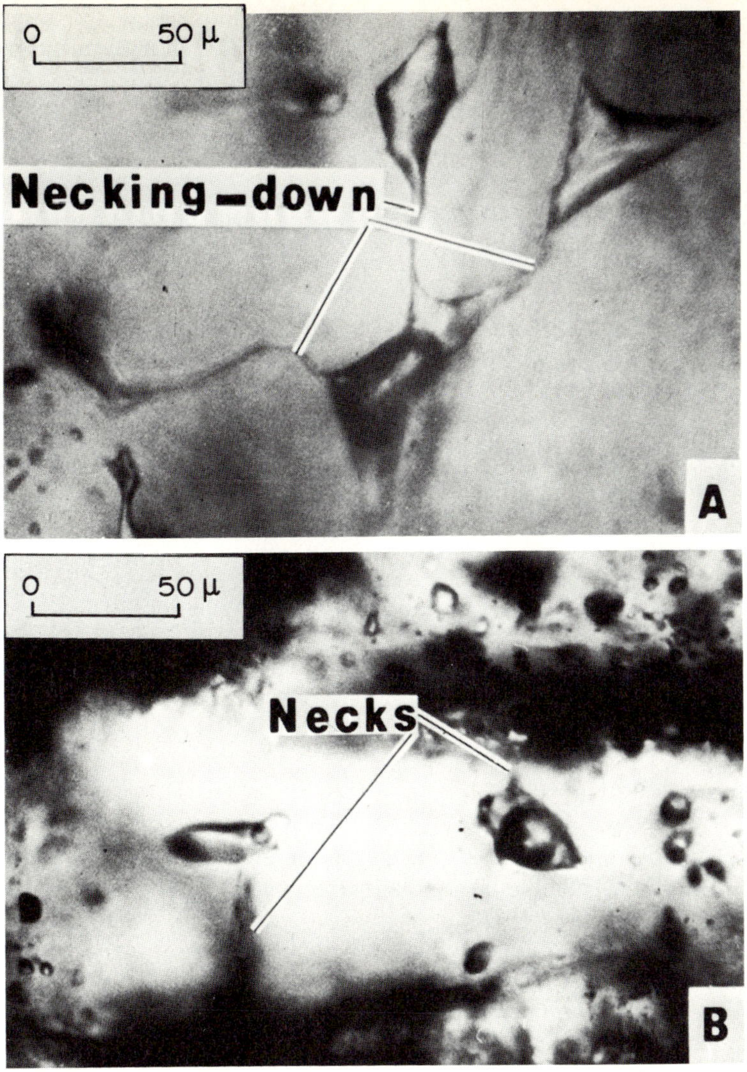

Fig. 9A, B. Fluid inclusions in barite from hydrothermal deposits. **A** Milos, Greece. Note the process of necking-down. **B** Bedded barites, Fukazawa mine. Note the slightly camouflaged necks

poulos and Mitropoulos 1983) are compared in Fig. 9A, B and Table 3. Both sets of inclusions share characteristics of variable size, vapour/liquid ratio and homogenization temperature. The Milos inclusions, however, show indisputable evidence of necking down (Fig. 9A). The evidence is not so obvious in the Fukazawa inclusions, but these also can be seen, in places, to have necked down (Fig. 9B).

In conclusion, fluid inclusions in barite from the Fukazawa mine have certainly undergone necking down and possibly some leakage. Neither process will affect the validity of the salinity estimates, but either would render homogenization temperatures useless for the purpose of geochemical modelling. The latter point is well illustrated by the consistently large ranges of homogenization temperatures obtained from inclusions in barite from several Kuroko deposits (Tokunaga and Honma 1974, this chapter, Table 3).

Parameters Affecting the Distribution of Strontium in Barite and Estimation of Strontium Content in Solution

Textural and chemical zonation on the scale of single barite grains precludes the use of equilibrium thermodynamics to describe a chemically dynamic system as a whole. However, certain individual subsystems may be treated using equilibrium theory. An example of such a subsystem is an individual, texturally and chemically homogeneous zone of a single barite grain. Changes in the strontium content of each zone can be used to trace variations in the strontium composition of the coexisting fluid from which it was precipitated.

The partitioning of strontium between $(Ba, Sr)SO_4$ and a coexisting fluid is described by:

$$(BaSO_4)^{bar} + (Sr^{2+})^{soln} \rightleftharpoons (SrSO_4)^{bar} + (Ba^{2+})^{soln}$$

$$K_{(P,T)} = \frac{a^{bar}_{SrSO_4}\, a^{soln}_{Ba^{2+}}}{a^{bar}_{BaSO_4}\, a^{soln}_{Sr^{2+}}} = \frac{X^{bar}_{SrSO_4}\, m^{soln}_{Ba^{2+}}\, \gamma^{bar}_{SrSO_4}\, \gamma^{soln}_{Ba^{2+}}}{X^{bar}_{BaSO_4}\, m^{soln}_{Sr^{2+}}\, \gamma^{bar}_{BaSO_4}\, \gamma^{soln}_{Sr^{2+}}} \tag{1}$$

$$K^{bar}_{Sr} = \left(\frac{X_{SrSO_4}}{X_{BaSO_4}}\right)^{bar} \div \left(\frac{m_{Sr^{2+}}}{m_{Ba^{2+}}}\right)^{soln} \tag{1a}$$

where: $K_{(P,T)}$ = equilibrium constant, K^{bar}_{Sr} = distribution coefficient of Sr in barite, a = activity, X = mole fraction, m = molality (mol kg^{-1} H_2O), and γ = activity coefficient.

The partitioning of strontium in barite is a function of temperature, pressure, $(m_{Sr^{2+}}/m_{Ba^{2+}})$ in solution and the mixing properties of the $BaSO_4-SrSO_4$ solid solution. Moreover, experimental work (Doerner and Hoskins 1925, Gordon et al. 1954) has shown that the rate of crystal growth may affect the partitioning behaviour of strontium between solid and liquid phases. All of these vary with the degree of mixing between hydrothermal solution and cold seawater at the time of barite precipitation.

If the fluid was relatively stationary in a closed system and barite precipitated slowly, the $m_{Sr^{2+}}/m_{Ba^{2+}}$ ratio in the fluid phase is expected to increase gradually with barite precipitation (Gordon et al. 1954). This would result in a progressive strontium enrichment from early to late barites as they were precipitated out of this solution. However, such a mechanism can be ruled out for the Fukazawa barites by the observations that: (1) the relative increase of Sr in the barites studied in sharp rather than gradual (Fig. 8); and (2) the trend of zonation in barites (Figs. 7 and 8) is that of an initial increase and subsequent decrease. Therefore, the mineralizing solution must have been circulating. During the course of such circulation the hydrothermal solution has been mixing with cold seawater. This process is further strengthened by evidence from the sulfur isotopic data of the Kuroko barites and their morphological characteristics, both of which are mentioned earlier in this text.

Barites of the textural zone 1 from the barite bed contain less strontium than those of the textural zone 2 (Table 1). This observation may represent either (1) precipitation of the textural zone 1 barites from a hydrothermal solution containing a low concentration of strontium or (2) a composition closer to "equilibrium" when thermodynamics began to rule over kinetics. Since no study of kinetics of barite precipitation has to our knowledge, been made to date we tend to incline towards accepting case (1). This speculation is further supported by the observation of Shikazono (personal communication 1979) that barites from stratigraphically lower zones of the Fukazawa deposit contain less strontium than those from the barite bed.

Equation (1a) shows that the strontium content of barite is a function of the distribution coefficient K_{Sr}^{bar} and of the $m_{Sr^{2+}}/m_{Ba^{2+}}$ ratio in solution. K_{Sr}^{bar} decreases with decreasing temperature (0.4 at 300 °C and 0.16 at 200 °C; Starke 1962). Thus, if a low-strontium ascending hot mineralizing solution were to mix with cold seawater the effect of temperature and that of changing strontium content would be opposite. Any barite precipitated will have a strontium content determined by the balance of these two parameters.

The evolution in concentration of strontium in solution from which the Fukazawa barites were precipitated can be calculated from the following reorganization of Eq. (1a):

$$(m_{Sr^{2+}})^{soln} = \frac{1}{K_{Sr}^{bar}} (m_{Ba^{2+}})^{soln} \cdot (X_{SrSO_4}/X_{BaSO_4})^{bar} . \tag{2}$$

Even though the X_{SrSO_4}/X_{BaSO_4} ratio in the individual stages of barite precipitation is known accurately from their strontium content, both the $(m_{Ba^{2+}})$ in solution and $1/K_{Sr}^{bar}$, which have opposite effects on the $(m_{Sr^{2+}})$ in solution upon mixing, are indeterminate because temperature and salinity data for each particular zone of single barite grains are lacking. Consequently, certain assumptions were made in order to estimate the strontium content in solution:

1. Barites from the Tsunokakezawa No. 1 ore body of Fukazawa mine formed from a hydrothermal solution with an average temperature of 250 °C (Tokunaga and Honma 1974, Pisutha-Arnond and Ohmoto 1983) and salinity of 0.8 *m* NaCl (see proper section); and
2. $m_{Ba^{2+}} = m_{SO_4^{2-}}$ during the texturally different stages of barite precipitation.

Figure 10A summarizes the means and ranges of estimated concentrations of strontium in solution from which bedded barites and barite veins were formed (Kalogeropoulos 1982). Figure 10B shows strontium isotopic data for the above types of barites and for Miocene volcanic rocks (Farrell 1979). Both the isotopic and compositional data indicate a greater involvement of seawater during precipitation of the bedded barites than for the veins. The ranges in Fig. 10 indicate that the barite veins precipitated from a solution with rather uniform strontium concentration and isotopic composition. On the other hand, the bedded barites formed from a solution with variable strontium isotopic values and strontium

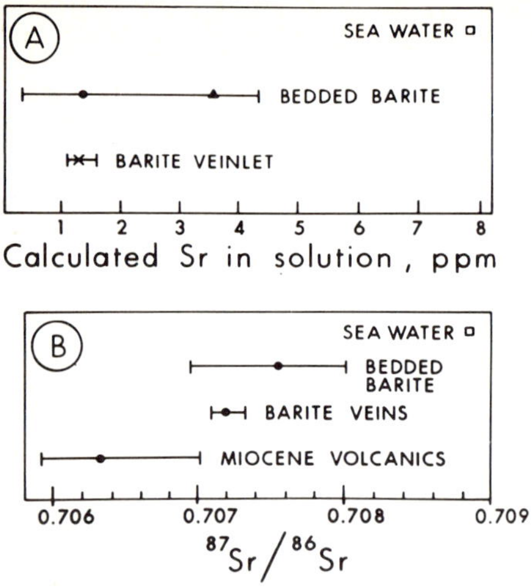

Fig. 10. A Means and ranges of calculated strontium in solution (ppm) for a barite veinlet (*x*) and for bedded barite zones 1 (●) and 2 (▲). The value of seawater (□) is from Krauskopf (1967). **B** Means and ranges of $^{87}Sr/^{86}Sr$, ratios for Miocene volcanics, Fukazawa barites (veins and bedded) (Farrell 1979) and contemporaneous seawater (Dasch and Biscaye 1971)

content, apparently resulting from variable degrees of mixing between ascending hydrothermal solution and seawater. The addition of strontium from seawater during precipitation of the fine-grained zones (zone 2) must have resulted in a concomitant enrichment of these zones in ^{87}Sr. Therefore, the strontium isotopic values in each barite sample will be dependent upon the relative abundance of such growth zones.

Conclusions: A Model for Formation of the Barite Zones

Since the solubility of barite in hydrothermal solutions is quite low (Blount 1977), there are two plausible ways of transporting and depositing barite in the Kuroko deposits: (1) oxidation of a hydrothermal fluid containing barium and reduced sulfur; and (2) mixing of a barium-rich hydrothermal fluid with seawater. Sulfur isotopes of Kuroko barites regardless from which part of a deposit they come (Bryndzia et al. 1983, Kusakabe and Chiba 1983), are consistent with a derivation of the barite-sulfur directly from seawater sulfate. Consequently, mixing of an ascending barium-rich hydrothermal fluid with seawater (case 2) is the most probable mechanism for the precipitation of the Kuroko barites. A subordinate amount of barite may have precipitated by process (1). The active participation of seawater during the course of barite formation is further evidenced by the oxidation of pyrite to hematite, and by the fine-grained (dusty-looking) barites characterized by higher strontium content and $^{87}Sr/^{86}Sr$ ratios (Farrell 1979).

The strontium content of barites precipitating out of a stationary solution in a closed system is expected to increase progressively with time. However, the observed sharp rather than gradual increase in strontium content within single barite grains (Fig. 8), and the initial increase followed by a decrease of strontium in zones barites (Fig. 7) require that the mineralizing solution was circulating.

Textural evidence indicates that bedded barites at Fukazawa have formed within a pyritic tuff (Fig. 3) by displacing the silicate matrix. This displacement seems to have been caused by the circulating mineralizing fluid and probably through a process of fluidization of the unconsolidated tuffaceous bed. An unconsolidated bed fluidizes when the rate of fluid flow passing through it reaches a point at which the bed expands slightly, and the particles move apart and are supported in the upward flowing hydrothermal solution. As the tuffaceous bed expands during fluidization the redox boundary is displaced upwards (Fig. 11, stage A). Seawater sulfate persisting probably metastably within the reducing zone (Malinin and Khitarov 1981) is consumed by barite formation as the front

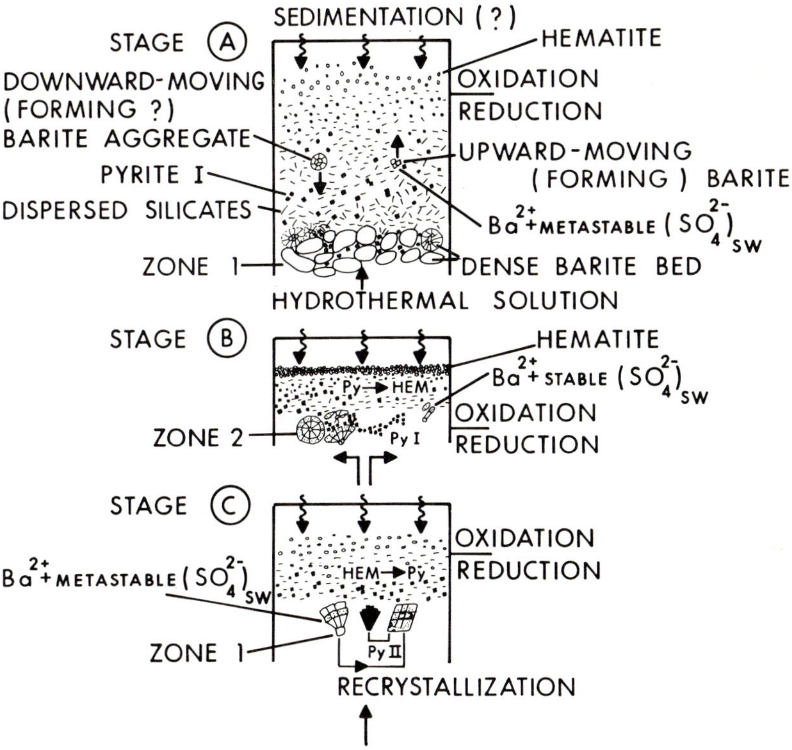

Fig. 11A – C. Stepwise development of the barite ore zone of Kuroko-type ore deposits (not to scale). High sedimentation rates or relatively massive barite at the base of the pyritic tuff result in channelling of the hydrothermal fluid to the seafloor and consequently much compositional variation in the overlying tuffaceous unit. During stage **A**, ascending barium-rich hydrothermal solutions mix with seawater sulfate. They form barite of the textural type represented by zone 1, within a fluidized pyritic tuff (see text for details). The fluid flow rates during this stage are expected to be lower than those of massive sulfide formation, but high enough to initiate and sustain fluidization of the tuffaceous bed. Stage **B**, represents the collapse of the fluidized bed as a result of decline of fluid flow rates. During this stage pyrite oxidizes to hematite and fine-grained barite (i.e. textural zone 2) forms. These barites have higher strontium contents than those measured in textural zone 2, and strontium isotopic values approaching that of seawater. Stage **C** is characterized by channelling of the ascending hydrothermal solutions due to the already deposited barite and tetsusekiei beds. These solutions result in the sulfidation of hematite, replacement of barite mainly by pyrite and overall recrystallization

of the barium-rich hydrothermal solution advances upwards. A certain amount of barium will also be taken up by the clay fraction of the system. These initially formed barites may be suspended or transported upwards depending on their mass. As they progressively grow to a size that cannot be further supported by the drag force of the upwelling hydrothermal solution, these barites begin to settle, forming a dense barite bed (textural zone 1) at the base of the fluidized pyritic tuff (Fig. 11, stage A).

There are at least three lines of evidence which support a sudden, short, but nevertheless major influence of seawater on the barite ore environment: (1) the sharp increase of strontium concentration in single barite grains (Fig. 8); (2) the smaller volume of the strontium-rich (fine-grained barite) relative to the strontium-poor (coarser-grained barite) zones (Fig. 6); and (3) the sulfur isotopic ratios of all analyzed barites (Kusakabe and Chiba 1983) which approach that of contemporaneous seawater (+22‰). This significant participation of seawater may have occurred during a stage of waning intensity of the hydrothermal activity when incursion of seawater into the environment of barite growth is possible (Fig. 11, stage B). During this stage the tuffaceous bed is no longer fluidized and becomes static; the redox reaction involving pyrite and hematite (Fig. 4D – F) merely provides a chemical indication of the incursion of oxygenated seawater and this seawater also provides sulfate to precipitate fine-grained barite (i.e. textural zone 2) from sulfate-deficient ascending hydrothermal solutions. These vertical fluctuations of the redox boundary apparently due to fluctuations in the intensity of the hydrothermal activity are also demonstrated at Fukazawa by the intervening hematite-rich bands in sulfide-bearing barite (Fig. 5). This mechanism provides for replenishment of the barite environment with seawater sulfate that is initially stable (downward flow of seawater) and subsequently metastable (upward flow of hydrothermal fluid) as indicated by the replacement of barite by sulfides. The seawater passes on to the barite not only its sulfur (Kusakabe and Chiba 1983), but also its strontium isotopic characteristics (Farrell 1979), thus explaining the inconsistency in stratigraphic relationships between strontium isotopes and strontium contents of barites discussed previously. When the hydrothermal system re-establishes itself, barites continue to form (equivalent to zone 1) and re-crystallize (Fig. 6). This is followed by a stage of pyrite (py II) deposition as pseudomorphs after barite (Fig. 4A, C and Fig. 11, stage C).

The mechanism of barite precipitation within an unconsolidated pyritic tuff which is a precursor of the tetsusekiei bed has contributed significantly to the nature of the overlying tetsusekiei, which acquires its final chemical and textural characteristics during the conclusion of the ore-forming process (Kalogeropoulos and Scott 1983). This may coincide with a stage of copper-mineralization (or Cu-remobilization) evidenced by the "chalcopyrite disease" exhibited by sphalerites from the barite ore zone. Because the hydrothermal fluid-seawater system has large ranges of temperature and chemistry existing simultaneously in different parts of the system and because the system as a whole evolved chemically and thermally in a complex fashion punctuated by repetitive alternating incursions of sea- and hydrothermal waters, equilibrium thermodynamics cannot treat this system as a whole.

Our future research to elucidate the genesis of these barites further should be focussed on the following two major aspects. The first, involves an oxygen and sulfur isotopic study of the two different textural types of barites, as they were identified in this work (i.e. zone 1 and zone 2); the second, and most important, involves a mathematical modelling of the depositional processes.

Acknowledgements. Our studies on barite were conducted as part of the U.S.-Japan-Canada Co-operative Research Project on the Genesis of Volcanogenic Massive Sulfide Deposits (Ohmoto and Skinner 1983). We acknowledge, in addition to our colleagues on the project, the assistance of and valuable discussions with the following people on specific aspects of this investigation: L. T. Bryndzia, C. Cermigniani, M. P. Gorton, E. T. C. Spooner, University of Toronto; and N. Shikazono of Tokyo University. Special thanks are extended to the Dowa Mining Co. and particularly to K. Hashimoto and J. Date. Moreover, we wish the express our sincere thanks to Drs. P. B. Barton Jr. and D. J. Vaughan for their constructive criticism which improved the text of this manuscript.

Funding was provided to S. D. S. by the Natural Sciences and Engineering Research Council of Canada operating grant A7069; the Ontario Geoscience Research Fund grant number 3-706-111-30; the Natural Science and Engineering Research Council of Canada; and a Japanese Society for the Promotion of Science Exchange Fellowship; and to S. I. K. by University of Toronto Fellowships and Ontario Graduate Scholarships.

References

Barton PB Jr (1978) Some ore textures involving sphalerite from the Furutobe mine, Akita Prefecture, Japan. Min Geol 28:293 – 300

Barton PB Jr, Bethke M, Toulmin P III (1963) Equilibrium in ore deposits. Min Soc Am Spec Pap 1:171 – 185

Blount CW (1977) Barite solubilities and thermodynamic quantities up to 300 °C and 1400 bars. Am Miner 62:942 – 957

Bryndzia LT, Scott SD, Farr JE (1983) Mineralogy, geochemistry and mineral chemistry of siliceous ore and altered footwall rocks in the Uwamuki No. 2 and No. 4 deposits, Kosaka mine, Hokuroku District, Japan. Econ Geol Monogr 5:507 – 522

Dasch EJ, Biscaye PE (1971) Isotopic composition of Sr in Cretaceous-to-Recent pelagic forminifera. Earth Planet Sci Lett 11:201 – 204

Doerner HA, Hoskins WM (1925) Co-precipitation of radium and barium sulfates. J Am Chem Soc 47:662 – 675

Farrell CW (1979) Strontium isotopes of Kuroko deposits. PhD Thesis, Harvard Univ, Cambridge, Mass

Farrell CW, Holland HD (1983) Strontium isotope geochemistry of the Kuroko deposits. Econ Geol Monogr 5:320 – 328

Gordon L, Reimer CC, Burit PB (1954) Distribution of strontium within barium sulfate precipitation from homogeneous solution. An Chem 26:842 – 846

Green GR, Ohmoto H, Date J, Takahashi T (1983) Whole-rock oxygen isotope distribution in the Fukazawa-Kosaka area, Hokuroku District, Japan. Econ Geol Monogr 5:395 – 411

Hanor JS (1968) Frequence distribution of compositions in the barite-celestite series. Am Miner 53:1215 – 1222

Hashimoto K (1977) The Kuroko deposits of Japan – geology and exploration strategies. Assos Ing Min Metal Geol Mex Mem Tech 13:25 – 88

Igarashi T, Okabe K, Yajima J (1974) Massive barite deposits in West Hokkaido. In: Ishihara S (ed) Geology of the Kuroko Deposits. Min Geol Spec Iss 6:39 – 44

Kalogeropoulos SI (1982) Chemical sediments in the hanging wall of volcanogenic massive sulfide deposits. PhD Thesis, Univ Toronto, Can

Kalogeropoulos SI, Mitropoulos P (1983) Geochemistry of barites from Milon island (Aegean Sea), Greece. N Yearb Miner Mh 1:13 – 21

Kalogeropoulos SI, Scott SD (1983) Mineralogy and geochemistry of tuffaceous exhalites (tetsusekiei) of the Fukazawa mine, Hokuroku District, Japan. Econ Geol Monogr 5:412 – 432

Krauskopf BK (1967) Introduction to geochemistry. McGraw-Hill, New York, p 721

Kusakabe M, Chiba H (1983) Oxygen and sulfur isotope composition of barite and anhydrite from the Fukazawa deposit, Japan. Econ Geol Monogr 5:292 – 301

Lu LE (1970) In: Aoki K, Sato T, Tatsumi T (eds) Kuroko deposits and Towada volcanoes. IMA-IAGOD Meet 1970, Tokyo-Kyoto, Guideb 3, 53 pp

Macdonald AJ, Spooner ETC (1981) Calibration of a Linkam TH 600 programmable heating-cooling stage for microthermometric examination of fluid inclusions. Econ Geol 76:1248 – 1258

Malinin SD, Khitarov NI (1981) Fluid inclusion study of stockwork siliceous orebodies of Kuroko deposits at the Kosaka mine, Akita, Japan. Min Geol 28:349 – 360

Nakai N (1970) Isotopic and chemical equilibrium of $CH_4 - CO_2$ and pyrite-anhydrite in geothermal areas in Japan. Int Symp Hydroch Biochem, Tokyo (Abstr), p 22

Ohmoto H (1972) Systematics of sulfur and carbon isotopes in hydrothermal ore deposits. Econ Geol 67:551 – 578

Ohmoto H, Skinner BJ (eds) (1983) The Kuroko and related volcanogenic massive sulfide deposits. Econ Geol Monogr 5:604

Pitsutha-Arnond V, Ohmoto H (1983) Thermal history, and chemical and isotopic compositions of the ore-forming fluids responsible for the Kuroko massive sulfide deposits in the Hokuroku district of Japan. Econ Geol Monogr 5:523 – 558

Sakai H (1968) Isotopic properties of sulfur compounds in hydrothermal processes. Geochem J 2:29 – 49

Sato T (1974) Distribution and geological setting of the Kuroko deposits. In: Ishihara S (ed) Geology of the Kuroko Deposits. Min Geol Spec Iss 6:1 – 9

Starke R (1962) Die Strontiumgehalte der Baryte. Freiberg Forschungsh C:150

Tokunaga M, Honma H (1974) Fluid inclusions in the minerals from some Kuroko deposits. In: Ishihara S (ed) Geology of the Kuroko Deposits. Min Geol Spec Iss 6:385 – 389

Watanabe M (1970) Kuroko deposits and Towada and Hakkoda Volcanoes. IMA-IAGOD Meet, Tokyo-Kyoto, Guideb 3, 53 pp

Part IV
Sediment-Hosted Deposits

Diagenetic Features at White Pine (Michigan), Redstone (N.W. Territories, Canada) and Kamoto (Zaire)*. Sequence of Mineralization in Sediment-Hosted Copper Deposits (Part 1)

A. C. BROWN and F. M. CHARTRAND[1]

Abstract

Disseminated copper and copper-iron sulfide mineralization at the White Pine (Michigan), Redstone (N.W. Territories, Canada) and Kamoto (Shaban Province, Zaire) deposits are interpreted to have been emplaced during early diagenesis by an influx of dissolved copper into initially pyritic basal units of grey-bed host rocks. This concept is supported by the position and configuration of the mineralized zones, by numerous textural features, by the zoning of Cu- and Fe-bearing sulfides as well as affiliated metals (e.g. Pb, Zn, Co and Mo), and by comparisons with modern analogs of strata hosting stratiform copper deposits. Most evidence supports an early diagenetic introduction of copper from underlying coarse-grained red beds. Plausible sources of copper include the red beds themselves, or pene-exhalative solutions which "exhale" into the red beds. Preliminary quantitative data now available suggest that the red beds may have provided sufficient copper to mineralize basal units of the overlying grey beds; the tectonic setting and timing of mineralization also encourage further research into the pene-exhalative model.

Introduction

The years 1960–1962 were pivotal years in our evolving understanding of the genesis of stratiform copper deposits. With few exceptions, it was quite normal until that time to interpret the fine-grained sedimentary-like distribution of cupriferous sulfide minerals along bedding as evidence of a sedimentary origin, or else in another school of thought, to attribute these sedimentary-like features to magmatic hydrothermal replacements, i.e. to selective telethermal mineralization à la Lindgren and Gratton.

In 1960, W. S. White of the U.S. Geological Survey instigated several important studies of the well-preserved, unmetamorphosed White Pine (Michigan) deposit when he reported details on the distribution of base-metal sulfide minerals

* Contribution of the Mineral Exploration Research Institute, Montreal, Canada.

[1] Ecole Polytechnique, Dept. of Mineral Engineering, P.O. Box 6079, Sta. "A", Montreal, Quebec, H3C 3A7, Canada

Geology and Metallogeny of Copper Deposits
(ed. by G.H. Friedrich et al.)

in the excellent exposures at the White Pine mine (White 1960). He described a vertical zoning of Cu,Fe-bearing sulfides, with the ore-zone chalcocite (+native copper and silver) grading abruptly into the overlying pyritic zone through an orderly array of chalcocite→bornite→chalcopyrite→pyrite; this narrow transition zone reportedly formed a single undulating blanket-like surface cross-cutting bedding at gentle angles. Furthermore, White noted the occurrence of minor amounts of CdS immediately above this bedding-transgressive Cu–Fe transition. The idea of cupriferous fluids moving upward through the host sediments during early diagenesis captured the attention of Wm. C. Kelly at the University of Michigan, and studies designed to test this concept were set in motion.

In 1962, Bartholomé commented on the Cu–Fe zoning found in the Kamoto deposits (Shaban Province, Zaire), and described a mineralogy and zoning of Cu,Fe-bearing sulfides which parallels the array studied independently at White Pine. He illustrated in careful detail the fine-scale replacement of pyrite by progressively more cupriferous sulfides, and noted the gently cross-cutting nature of the mineral zones grading from a basal chalcocite-bornite zone to an overlying pyritic zone in the Kamoto district. As a result, an upward influx of cupriferous solutions was proposed for Kamoto as well.

From those years onward, research into the genesis of stratiform copper deposits has been predominantly oriented toward testing and refining this model which calls for an early diagenetic influx of copper into formerly iron sulfide-rich host sediments. The present paper is not intended to be a compilation of observations and interpretations supporting this hypothesis – summaries of that sort can be found in Rentzsch et al. (1976), Brown (1978, 1980a,b, 1981a), Gustafson and Williams (1981) and others. It is interesting to note, however, that the early diagenetic model has survived tests of its accuracy during more than two decades of detailed investigations of many different stratiform copper deposits by many researchers.

Secondly, it is pertinent to point out that much progress has been made since the early 1960s. Until that time, the study of an ore deposit was typically aimed at descriptions and interpretations of the ore minerals, commonly without great attention to the nature and origins of enclosing host rocks. The latter was due in good part to the all too pervasive pre-1960 perception that the origins of the host rocks belonged to pre-ore history and that the prime prerequisite of the host rock was that firstly, it be structurally deformed to allow entry of magmatic ore solutions and that secondly, it be chemically reactive at the local and microscopic scale so as to precipitate metals from invading hydrothermal fluids. Today stratiform copper mineralization is seen to be intimately related in time and space to the origins of the enclosing host rocks. In fact, mineralization may be at least indirectly interactive with environments prevailing in the contemporaneous hydrosphere, biosphere and atmosphere during deposition of the host sediment (Brown and Chartrand 1983).

Current evidence for the diagenetic[2] emplacement of copper may be "convincingly for" or at least "permissive", and it is significant that the observations made thus far belong to these categories and do not conflict with the diagenetic

[2] Footnote 2 see page 392.

hypothesis. For example, the replacement of pyrite (or a more primitive iron sulfide), the zoning of sulfides and the gently cross-cutting nature of the sulfide zones form strong arguments, alone and together, for a post-sedimentary influx of copper from a source outside of the host sediments.

Other studies, such as those on the derivation of copper from adjacent red beds (e.g. Lustwerk and Rose 1983, Rose et al. 1984) or pene-exhalative sources (Brown 1981b, 1984), have described "permissive" evidences for the diagenetic influx model – permissive in the sense that the evidence does not conflict with the diagenetic concept.

In some cases, one may qualify such evidence as being "complementary" or even "essential". For example, the diagenetic influx model requires adequate transport mechanisms to introduce copper to the fine-grained portions of the host rocks, and to bring the copper to the host rocks from plausible distant external sources. Calculated solubilities of metals in interstitial brines provide needed assurance that the metal transport can be adequate. At the same time, it is reassuring to find that trace amounts of copper within large volumes of red beds could release copper to these brines in significant quantities. These observations on sources and transport of copper are complementary to the diagenetic mineralization model, and in fact the existence of such evidence (or some adequate alternative such as the pene-exhalative model) is essential to form a complete source-transport-deposition model for the diagenetic emplacement of copper in the case of stratiform copper deposits.

Types of Evidence for the Early Diagenetic Origin of Stratiform Copper Deposits

The precise evidence for the diagenetic influx of copper can only be properly appreciated by direct reference to the extensive literature on stratiform copper deposits. The following brief summary is limited to the identification of the *types* of evidence currently available on the White Pine, Redstone and Kamoto deposits:

1. *Position and Configuration of the Mineralized Zone.* The cupriferous zone at White Pine, Redstone and Kamoto is located at the base of a sequence of organic-rich sediments. The beds overlying the cupriferous zone are pyritic, whereas the underlying beds are red, hematitic sediments. The location of the cupriferous

---

[2] The usage of the term "diagenetic" with regard to the post-sedimentary emplacement of copper in stratiform copper deposits may be accepted, on the one hand, and disputed, on the other. Gustafson and Williams (1981) properly limit usage of the term to cases where the copper is redistributed within the sediments during diagenesis. That concept clearly conflicts with the current model for stratiform copper deposits in which copper is interpreted to have been introduced from sediments other than those currently mineralized. However, if one expands one perception of the sediments involved to include adjacent strata (e.g. underlying red beds), then it may be stated that the copper was redistributed diagenetically within this single sedimentary sequence.

The term diagenetic is also acceptable if it clearly applies to a period of time characterized by diagenetic processes; in such cases, time may be denoted by modifiers such as "early" or "late".

zone in the basal reduced beds of an originally iron sulfide-rich stratigraphic unit lying directly above oxidized sediments would be consistent with a concept whereby copper was chemically screened from upward moving ore solutions.

The upper limit of the cupriferous zone also transgresses bedding as if it resulted from an upward advancing mineralization front. In the case of White Pine, the mineralization front is generally abrupt, whereas in the case of Redstone and Kamoto, the transgressive nature of the mineralization front is shown by a general trend from copper-rich to copper-poor sulfides extending over broad zones, each of which nevertheless transgresses bedding in a gentle manner. Moreover, at White Pine, there is an inverse correlation between the amount of copper in the lowermost beds of the mine section and the height of the top of the cupriferous zone in the mine section. This correlation strongly suggests that the top of the cupriferous zone is stratigraphically low where most copper was screened out of an upward migrating ore solution which added approximately equal quantities of copper across the base of the host sediments.

2. *Zoning of Metal Sulfides.* The stratiform copper deposits of White Pine, Redstone, Kamoto and other areas are typically zoned, both with respect to sulfide minerals of the Cu–Fe system, and with respect to other associated minor-to-trace metals. Among Cu–Fe sulfides, the least soluble sulfides such as chalcocite and bornite occur close to the site of copper influx into the reduced, sulfide-bearing host rocks, whereas the more soluble chalcopyrite and the still more soluble iron sulfides are located further downstream in the path of the upward circulating ore solution. The replacement of the more soluble iron-rich sulfides by the less soluble copper-rich sulfides has been repeatedly confirmed in reports on various textural features at White Pine, Redstone, Kamoto and elsewhere.

Other affiliated metals show similar distributions according to their solubilities relative to the Cu–Fe sulfides. For example, cadmium, lead and zinc are clearly located downstream from copper at White Pine, as would be predicted from their solubilities relative to copper and iron in sulfide-rich host rocks. Recent examinations of lead and zinc at Redstone reveal that these minor metals are also distributed downstream from copper-bearing sulfides, whereas trace amounts of molybdenum sulfide at Redstone are located at the base of the mineralized zone (i.e. at the site of ore solution *influx*), again predictable from the lower solubility of molybdenite relative to copper sulfides.

The essential point from these observations on metal sulfide zoning is that the sequence of metal zones is consistent with an influx model, generally pointing to an upward circulation of ore fluids from adjacent red beds during post-sedimentary time. However, one may ask when, within reasonable post-sedimentary time, did the fluids enter and mineralize the reduced, sulfide-rich host rocks?

3. *Timing of Copper Mineralization Relative to Early Diagenesis of the Host Rocks.* Sedimentological studies have provided more precision on the timing of mineralization relative to the normal diagenesis of the enclosing host rocks. The data are derived from detailed petrographic analyses and from comparisons with diagenetic features observed in modern sediments resembling these host rocks.

The Redstone deposit has been studied in this manner, and a close analogy is drawn here with the sedimentation and early diagenesis of intratidal and supratidal (sabkha) sediments deposited within a series of marginal-marine basins which may be of intracontinental-rift origin.

Copper mineralization at Redstone can be identified as a post-sedimentary event within the normal early diagenesis of microbial-laminated carbonate beds of peritidal origin. The sequence of events through sedimentation and early diagenesis of the carbonate beds may be summarized as follows (from Chartrand and Brown 1984):

Stage I. Syndiagenetic Stage: Dolomitization of initial sabkha carbonates; formation of blade-like gypsum; transformation of gypsum to anhydrite; formation of disseminated pyrite under anoxic conditions in decaying microbial mats.

Stage II. Very Early Diagenesis: Disseminated pyrite euhedra of stage I are rimmed by epitaxial calcite; gypsum-anhydrite blades are replaced by quartz, calcite and feldspars. Quartz overgrowths consist of more than one generation of silica.

Stage III. Ore-Stage Diagenesis: Copper-bearing sulfides (chalcopyrite, bornite, digenite, chalcocite) replace quartz, pyrite and remnants of anhydrite; cupriferous sulfides also form pseudomorphs after blade-like gypsum-anhydrite.

Stage IV. Post-Ore Diagenesis: Neomorphic calcite closes the pores of the mineralized carbonate, possibly arresting the process of mineralization by impeding further influx of ore solution.

The importance of this paragenetic sequence among the early diagenetic minerals at Redstone lies in the observation that mineralization was intimately related to early diagenesis, but is nevertheless a post-sedimentary event superimposed on the host-rock carbonates. These microbial-laminated carbonates formed under normal sabkha conditions and proceeded through normal sabkha diagenesis including the transformation of sulfate phases and the formation of iron sulfide in the presence of abundant organic matter. However, the normal diagenetic evolution of the sabkha beds was modified by the introduction of copper to the carbonates during stage III diagenesis. As mentioned above, the zoning of Cu, Pb, Zn and Mo sulfides indicates that the direction of copper influx was from the underlying red beds.

We are currently examining the Zairian mineralization in the same manner, and we find many parallels with the Redstone mineralization because the initial sedimentary environment was also of the sabkha type, characterized by microbial-laminated carbonate beds. The proportions of lithotypes, their dimensions and the concentrations of metals at Kamoto are quite different from those at Redstone, but the essential lithologies and environmental conditions are similar.

Some important lithologic differences relating to sedimentation and early diagenesis deserve attention, however, because these features probably formed close to the period of early diagenetic mineralization and should limit conditions within the host rocks during mineralization. For example, the Zairian carbonates are highly magnesian while those of Redstone are predominantly calcitic. In addition, an enormous volume of authigenic silica has been added to the Zairian sedi-

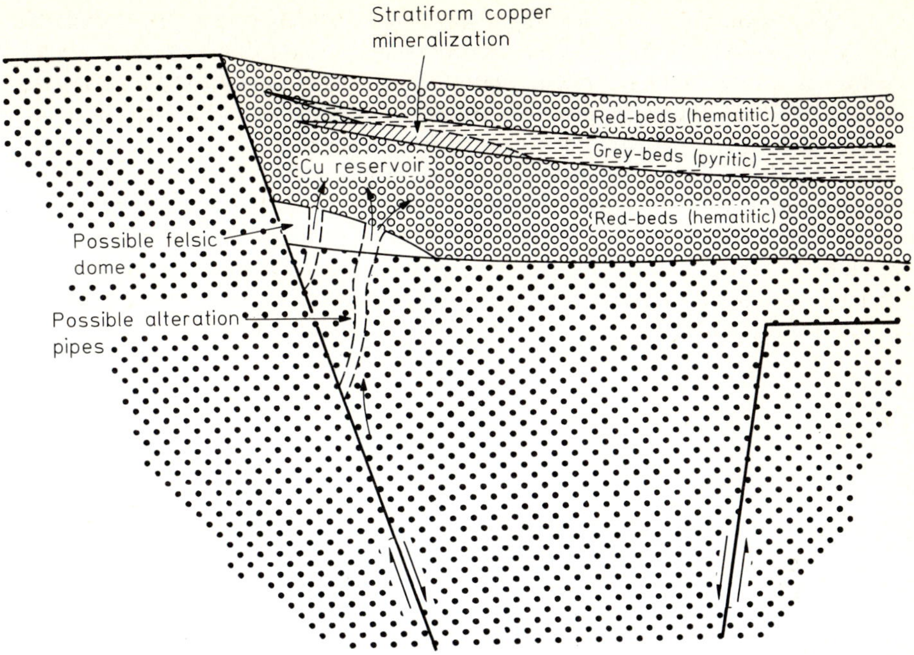

Fig. 1. The fundamental features of a pene-exhalative model which could generate a reservoir of cupriferous solutions in permeable red-bed strata underlying initially iron sulfide-bearing and/or sulfate-bearing grey beds (from Brown 1984)

ments, suggesting anormal diagenesis in the early history of these copperbelt host rocks (Bartholomé et al. 1972). Also the Shaban copperbelt ores are cobaltiferous as well as cupriferous, whereas the Redstone mineralization is not cobaltiferous.

Speculations on Ore-Forming Models

Although stratiform copper deposits are commonly considered monometallic (with, in some cases, accessory economic silver or cobalt), a complete description of the total added post-sedimentary mineralization commonly includes mention of significant concentrations of other metals such as Pb, Zn, Mo, etc. The potential sources of these associated metals, as well as the ore metals, may be summarized in a very limited number of hypotheses consistent with an emplacement of metals during early diagenesis of the host rocks and with an influx from underlying red beds.

The derivation of metals from the red beds and associated lithologies, is a very attractive option. Some very interesting data on this hypothesis are presented in Lustwerk and Rose (1983) and Rose et al. (1984), and no further discussion is required in the present paper.

A more speculative model which has certain merits is the pene-exhalative model (Brown 1981b, 1984). This concept is supported in part by general similarities in the tectonic setting of many stratiform copper deposits (including White Pine and Redstone) and the tectonic model presently adopted by many geologists in the case of sediment-hosted Pb-Zn deposits of possible submarine exhalative origin [see, for example, Large (1980, 1983), Finlow-Bates (1980), Russell et al. (1981) and Russell (1983)]. The tectonic setting is that of intra-cratonic rifting and rapid filling of the downfaulted blocks by red beds and variable amounts of mafic flows. In brief, it may be possible to generate low-temperature cupriferous brines which "exhale" into the red beds, and then mineralize overlying basal portions of favourably reducing, sulfide-rich grey-beds (see Fig. 1).

Closing Comment

As suggested initially in this paper, we are not yet sufficiently advanced in our understanding of stratiform copper deposits to make definitive decisions on their complete genesis. However, we do have certain guidelines such as those described here on the timing of mineralization relative to diagenesis, and on the directions of ore-solution flow. As for long-range transport and ultimate sources of metals, we are beginning to obtain significant quantitative data, some of which may eliminate the need for a complex pene-exhalative model. However, in principle, the discovery of other adequate models does not exclude the possibility of another potentially adequate model, and at this time the so-called pene-exhalative concept should receive additional thought and research in parallel with other attractive models.

References

Bartholomé P (1962) Les minerais cupro-cobaltifères de Kamoto (Katanga Ouest). I Pétrographie. Stud Univ Louvanium, Fac Sci Kinshasa 14, p 40

Bartholomé P, Evrard P, Katekesha F, Lopez-Ruiz JM, Ngongo M (1972) Diagenetic ore-forming processes at Kamoto, Katanga, Republic of the Congo. In: Amstutz GC, Bernard AJ (eds) Ores in sediments. Springer, Berlin Heidelberg New York, pp 21–41

Brown AC (1978) Stratiform copper deposits – evidence for their post-sedimentary origin. Miner Sci Eng 10:145–163

Brown AC (1980a) The diagenetic origin of stratiform copper deposits. In: Ridge JD (ed) Proc 5th Q IAGOD Symp, pp 81–89

Brown AC (1980b) Stratiform copper deposits and pene-exhalative environments. Geol Soc Am, Abstracts with Programs 13:418

Brown AC (1981a) The timing of mineralization in stratiform copper deposits. In: Wolf KH (ed) Handbook of stratabound and stratiform ore deposits, vol IX. Elsevier/North Holland Biomedical Press, Amsterdam New York, pp 1–23

Brown AC (1981b) Stratiform copper deposits and pene-exhalative environments. Geol Soc Am, Abstracts with Programs 13 no. 7:418

Brown AC (1984) Alternative sources of metals for stratiform copper deposits. Precambr Res 25:61–74

Brown AC, Chartrand FM (1983) Stratiform copper deposits and interactions with co-existing atmospheres, hydrospheres, biospheres and lithospheres. Precambr Res 20:533 – 542

Chartrand FM, Brown AC (1984) The diagenetic origin of stratiform copper mineralization, Coates Lake, Redstone copperbelt, NWT, Canada. Econ Geol 80:325 – 343

Finlow-Bates T (1980) The chemical and physical controls on the genesis of submarine exhalative orebodies and their implications for formulating exploration concepts. A review. Geol Jahrb D40:131 – 168

Gustafson LB, Williams N (1981) Sediment-hosted stratiform deposits of copper, lead and zinc. In: Skinner BJ (ed) Economic Geology. 75th Anniversary Vol, pp 139 – 178

Large DE (1980) Geological parameters associated with sediment-hosted, submarine exhalative Pb-Zn deposits: an empirical model for mineral exploration. Geol Jahrb D40:59 – 129

Large DE (1983) Sediment-hosted massive sulphide lead-zinc deposits: an empirical model. In: Sangster DF (ed) Sediment-hosted stratiform lead-zinc deposits. Miner Assoc Can, Short Course Handb 8:1 – 29

Lustwerk RL, Rose AW (1983) Source and segregation of transition metals during diagenetic formation of the Redstone stratiform copper deposit, Mackenzie Mts., NWT, Canada. Geol Soc Am, Abstracts with Programs 15, 6:632

Rentzsch J, Schirmer B, Rölling G, Tischendorf G (1976) On the metal source of non-ferrous mineralizations in the Zechstein basement (Kupferschiefer type). In: The current metallogenic problems of Central Europe. Geol Inst, Warsaw, pp 171 – 188

Rose AW, Ohmoto H, Lustwerk RL, Smith AT, Hoy L (1984) Sequence of mineralization in sediment-hosted copper deposits: Part 3 – Geochemical aspects of the Catskills Formation (Pennsylvania) and Redstone (Canada). Int Geol Congr, Moscow (Abstracts)

Russell MJ (1983) Major sediment-hosted exhalative zinc + lead deposits: formation from hydrothermal convection cells that deepen during crustal extension. In: Sangster DF (ed) Sediment-hosted stratiform lead-zinc deposits. Miner Assoc Can, Short Course Handb 8:251 – 282

Russell MJ, Solomon M, Walshe JL (1981) The genesis of sediment-hosted zinc + lead deposits. Miner Dep 16:113 – 127

White WS (1960) The White Pine copper deposit (discussion). Econ Geol 55:402 – 410

Diagenetic Sulphide Mineralization Within the Stratiform Copper-Cobalt Deposit of West Kambove (Shaba-Zaire). Sequence of Mineralization in Sediment-Hosted Copper Deposits (Part 2)

J. CAILTEUX[1]

Abstract

The stratiform copper-cobalt mineralization in the Shaban Mines Group took place in intertidal sediments enclosing the stromatolitic masses of the R.2.1.3 unit and around similar masses corresponding to the R.2.3.1 unit.

The cristallization of primary sulphides displays a vertical zoning with pyrite at the base and at the top of the ore bodies. They form micro- and macrocycles suggesting a sequence of copper and cobalt mineralization. This mineralization results from a diagenetic process involving a mixing of stagnant interstitial water with brines which were enriched in metals and which migrated through still porous sediments. Secondary reactions with the brines enriched previously formed sulfides with additional copper and/or cobalt.

Introduction

The metallic stratiform mineralization of the Roan Supergroup in Shaba is composed mainly of copper and cobalt sulphides distributed principally along two mineralized units of the Mines Group (R.2)[2] (Table 1). These units have been studied in detail at Kamoto (Kolwezi) by Bartholomé and his colleagues (1962, 1963, 1969, 1972, 1974).

Prospecting by drilling on the West-Kambove deposit in the heart of the Shaban copperbelt (Fig. 1) has identified copper-cobalt mineralization similar to that of the classic copperbelt ore bodies (Cailteux 1983) in the lower R.2.3 unit (the highest formation of the Mines Group). The presence of a number of oxidized deposits associated with this stratigraphic horizon has been known for some time, as in the case at Etoile, Kakanda, Kamatanda, Kamfundwa, Luisha, Shamitumba, etc. The mineralization is concentrated in an alteration pockets between residual dolomite mounds, but its origin has never been carefully studied. Oosterbosch (1962) and Francois (1973, 1974) noted only rare and irregular occurrences of pyrite as well as copper-cobalt anomalies in the lower R.2.3, and it has been supposed that the metallic concentrations were produced by a

[1] GECAMINES, P.R.D., Bureau d'Etudes Géologiques, Likasi, Shaba, Zaire

[2] For terminology see Table 1.

Geology and Metallogeny of Copper Deposits
(ed. by G.H. Friedrich et al.)

Table 1.

Kundelungu Supergroup	K.s	Upper Kundelungu Group	
	K.i	Lower Kundelungu Group	
	R.4	Mwashya Group	
	R.3	Dipeta Group	
Roan			R.2.3
Supergroup	R.2.	Mines Group	R.2.2
			R.2.1
	R.1	R.A.T Group	
Unconformity			
Kibarian Basement			

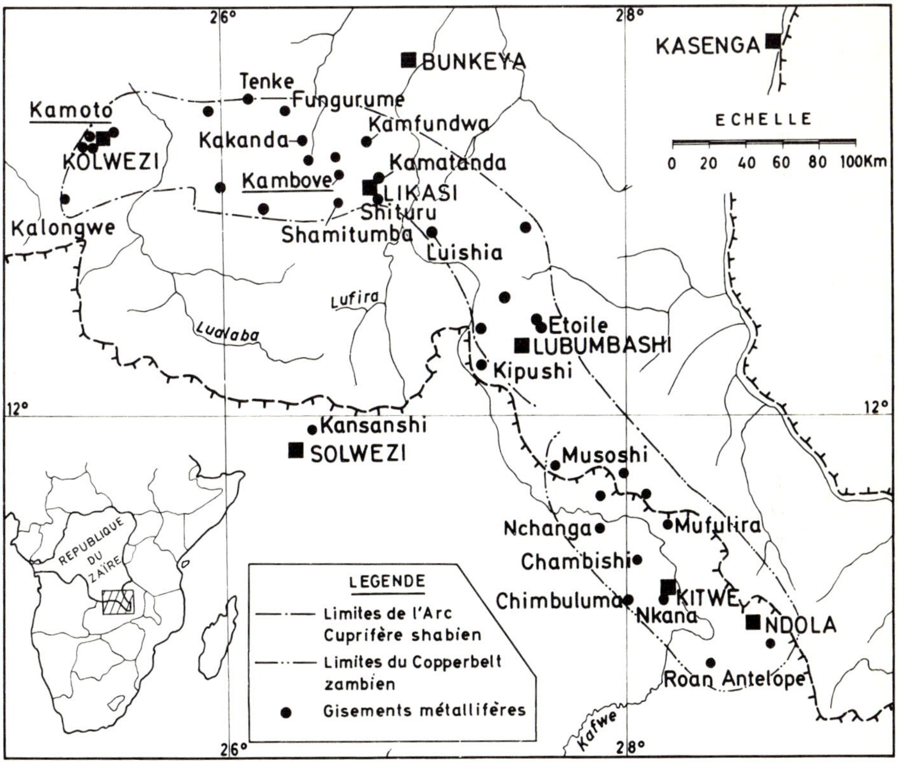

Fig. 1. General distribution of metalliferous deposits in the Central African Copperbelt

supergene enrichment of minor amounts of background sulphides, as in the case of the R.4 unit in Shituru (Lefebvre 1974).

Our objective is to examine the sulphide mineralization of West-Kambove in the regions beyond recent supergene alterations in order to compare the mineralization of the R.2.3 unit with that of usual mineralized bodies of the copperbelt and to draw genetic conclusions from these observations.

Sediments of the Kambove Mines Group and the Distribution of Copper-Cobalt Mineralization

At Kambove, the Roan Supergroup lies on the Upper Kundelungu Group (K.S.2.) which crops out further north (Figs. 2 and 3). As is usual in Shaba, the Roan occurs as "écailles" ("scales" or mega-breccia fragments) of Mines Group (R.2) strata floating between the crushed R.A.T. Group unit (R.1) and Dipeta Group (R.3) [3].

It is possible to recognize three very distinct entities within the sediments of the Mines Group:

1. A dolomitic unit (R.2.1) including stratified dolomites and *laminites,* with a massive stromatolitic dolomite at the top and a chloritic microsandstone with pyroclastic characteristics at the base;
2. A shaly dolomitic unit (R.2.2);
3. A talc-dolomite unit (R.2.3) including stromatolitic massive dolomites laminites, talc dolomites and chloritic dolomitic shale bands in its upper part.

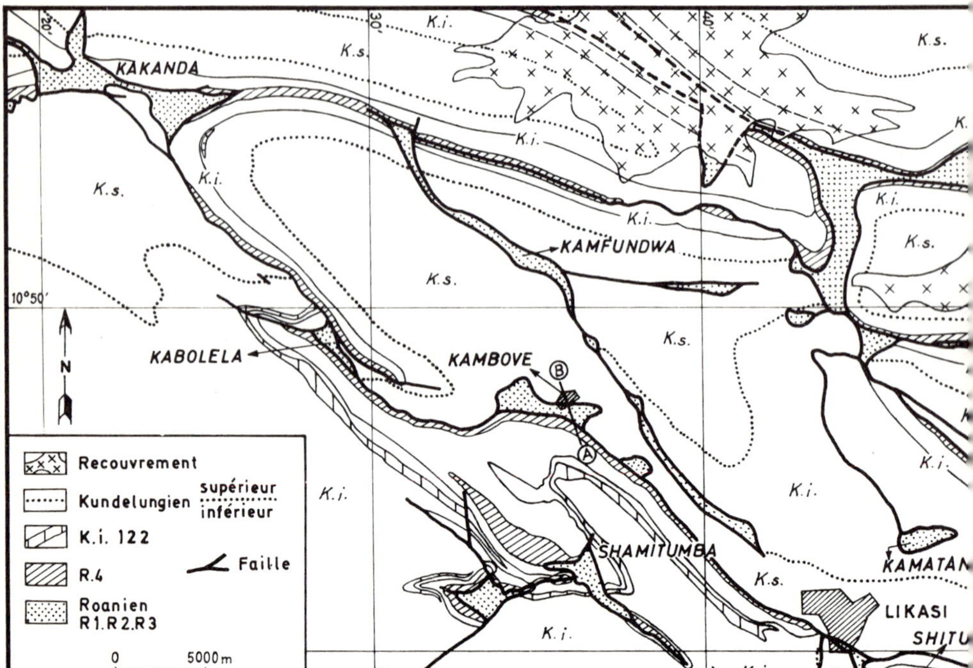

Fig. 2. General geology in the central portion of the Shaban Copperbelt (after François)

[3] For more complete geographic and stratigraphic information, the reader should consult Cahen (1954, 1974), Demesmaeker et al. (1962), Bartholomé et al. (1972) and François (1973, 1974).

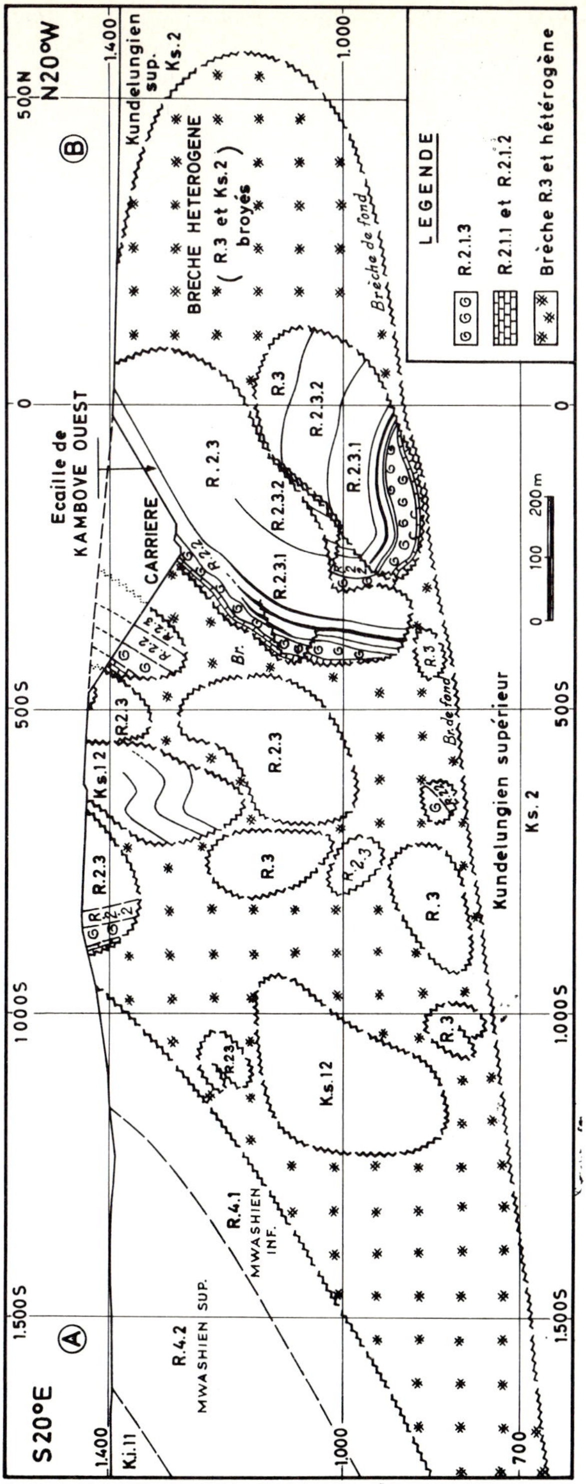

Fig. 3. Vertical profile across the Roan of Kambove along the line *A – B*

The R.2.1 and R.2.3 beds are similar in composition and constitute units characteristic of an intertidal environment at the edge of salt-saturated lagoons (Cailteux 1977a, 1977b, 1983). The intervening R.2.2. sediments were formed in a more marine or open lagoonal environment than the R.2.1 or R.2.3 beds.

The West-Kambove deposit is located asymmetrically on the northern flank of a syncline. The mineralization is situated mainly below and above a stromatolitic unit (R.2.1.3) in what is called the lower ore body (in R.2.1.1, in R.2.1.2 and at the base of R.2.1.3) and the upper ore body (at the top of R.2.1.3, in R.2.2.1 and at the base of R.2.2.2).

However, the upper portions of R.2.2 and the R.2.3 also contain sulphides, commonly in considerable quantities. In R.2.2, bituminous horizons (2.2 and 3.2) may contain chalcopyrite and more rarely carrollite, except where they are pyritic. At the base of R.2.3, the sulphides are principally cupriferous; they occupy the dolomitic shales, and the laminites enclose the stromatolitic masses (R.2.3.1.1 and R.2.3.1.2). Within the laminites of the R.2.3.1.2 unit, as well as in the talc-dolomite unit of R.2.3.1.3, both copper and cobalt sulphides form lenticular ore bodies rich in carrollite, chalcopyrite and bornite. Finally, sulphides can also be found in R.2.3.2 in some dolomitic shales.

Mineralography of the Sulphide Assemblages

Sulphide Zoning

The primary sulphides[4] of the classic copperbelt ore bodies generally display a vertical zoning which can be readily observed in horizontal planes on the northern flank of the West-Kambove syncline, as well as in adjacent poorly mineralized deposits[5] (Fig. 4). From the stratigraphic bottom to the top, within the lower mineralized body, a pyrite-dominant zone lies below the chalcopyrite-bornite zone which is itself overlain by the carrollite-rich zone. The three zones overlap to a considerable extent. Locally, chalcopyrite and bornite are not developed and the carrollite zone overlaies directly the pyritic zone.

Within the upper mineralized body, the zoning is similar but inverted. Carrollite is situated near the base and is overlain by bornite-chalcopyrite and then by pyrite. In poorly mineralized deposits, however, pyrite appears at the base of the ore body as well, and is overlain by co-existing chalcopyrite, bornite and carrolite.

In R.2.3, the laminite unit R.2.3.1.2 is especially characterized by a zonal distribution of sulphide minerals grading from richer to poorer cupriferous sediments:

pyrite → chalcopyrite → bornite → chalcopyrite → pyrite

These zones form cycles with ticknesses which vary from a few centimetres to one or more metres. Figure 5 describes a complete cycle with a tickness of about 17 cm, including two partial cycles.

---

[4] "Primary" sulphides are those which precipitated before or during diagenesis.

[5] "Poorly mineralized" deposits are those containing less than 2% copper.

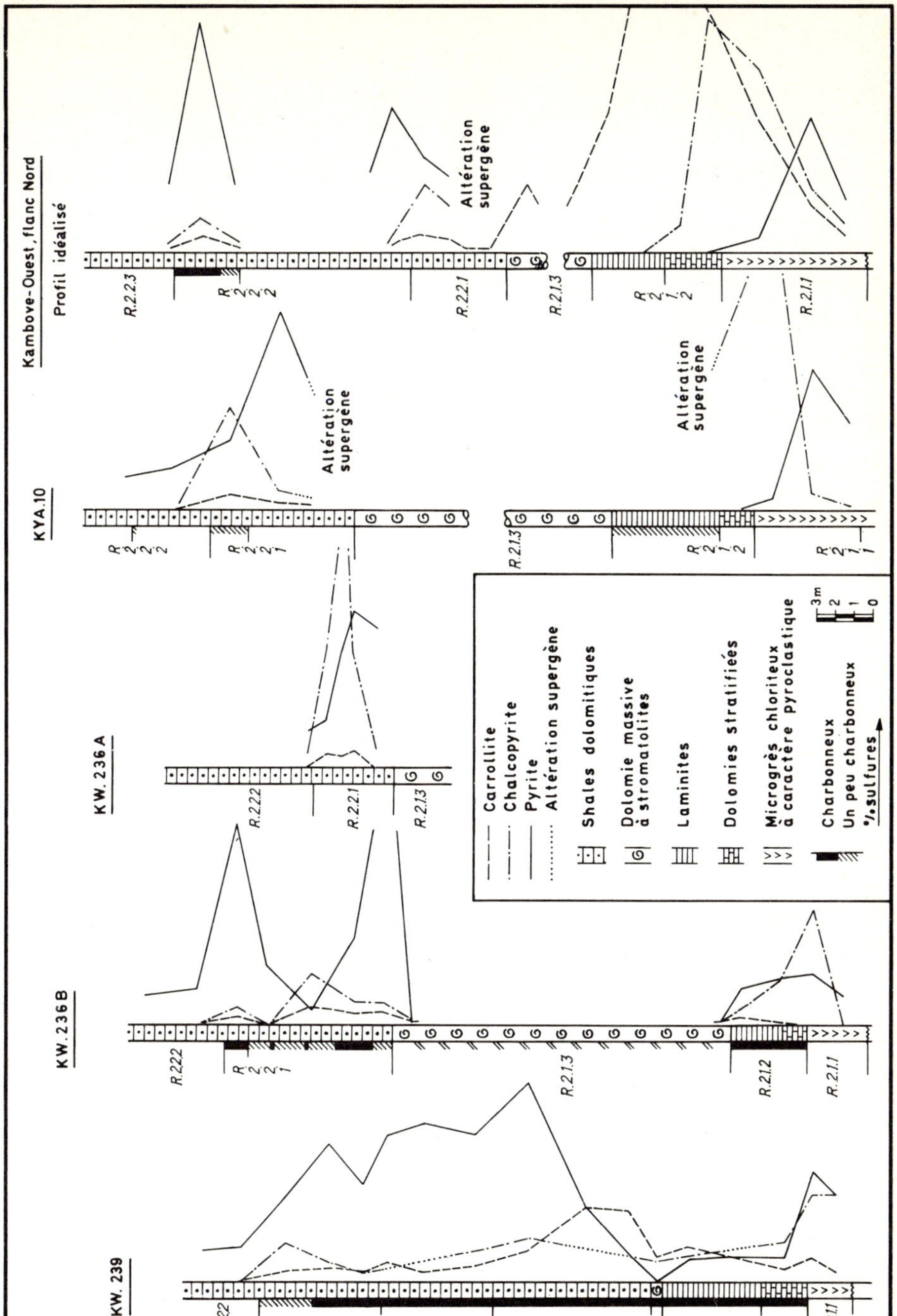

Fig. 4. Profiles of primary sulphide minerals in lower Roan strata at the West Kambove deposit (North limb) and in nearby poorly mineralized deposits

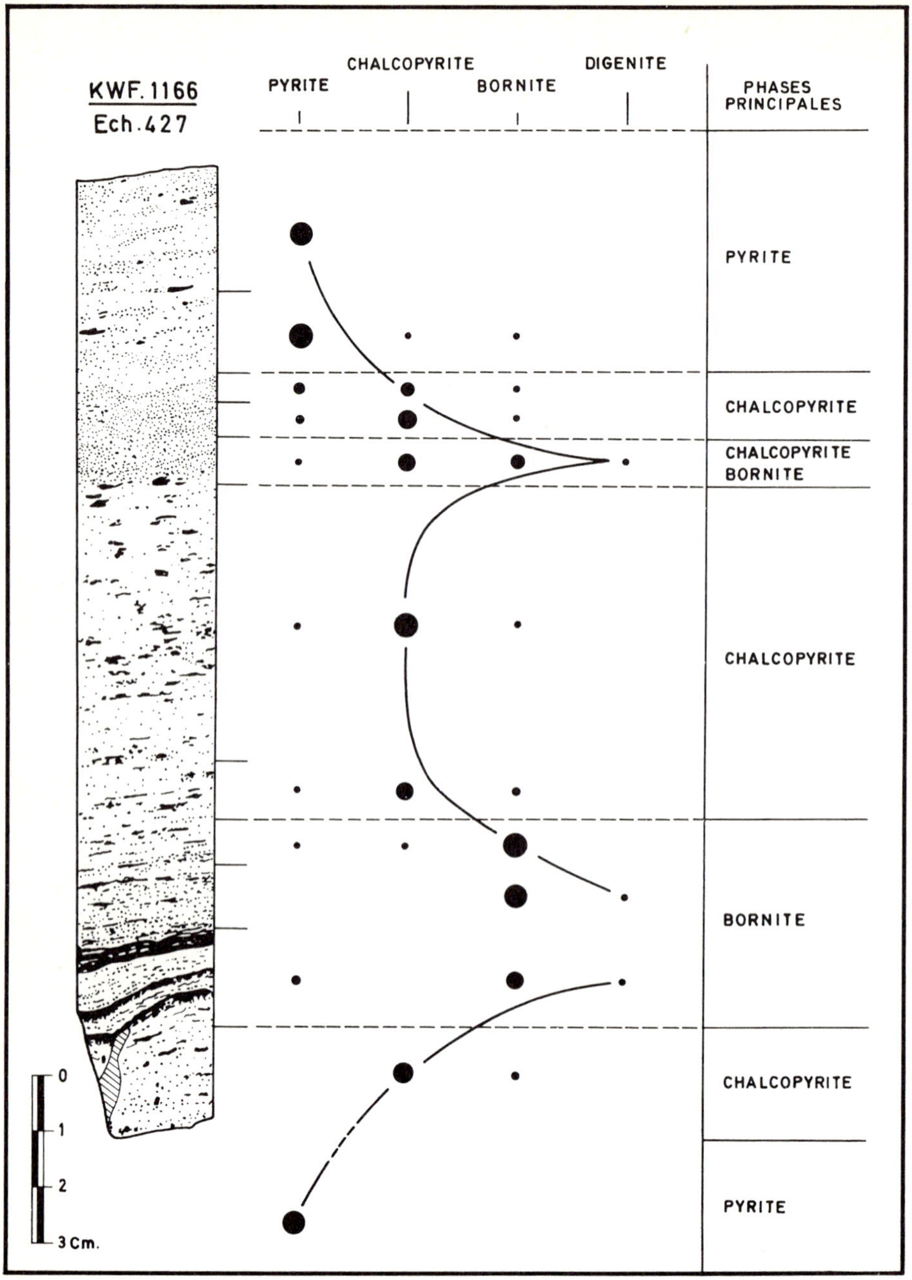

Fig. 5. A complete micro-cycle of zoned primary sulphides in the laminite bed of unit R.2.3.1.2

Primary Paragenesis

Pyrite forms small grains, or large patches of poecilitic pyrite encircling gangue minerals. In association with copper- and cobalt-bearing minerals, pyrite is enclosed by these economic sulphides (Fig. 6b, c and e), but in some cases it contains them (Fig. 6a). The growth of pyrite took place especially around bornite, porphyroblastic carrollite, etc. (Fig. 6b).

Chalcopyrite and bornite comprise the primary cupriferous mineralization. Bornite is not consistently present throughout the mineralized layers, but it is consistently associated with chalcopyrite, in some cases as exsolutions in chalcopyrite.

Digenite and chalcocite (or djurleite) are also primary where they are in exsolution with bornite (Fig. 6a). They are found in the bornite-rich beds and account for the strongest copper mineralization at West-Kambove.

Cobalt associated with copper forms *carrollite.* This mineral is most commonly present in isolated poecilitic grains encircling chalcopyrite and bornite, or in association with chalcopyrite alone. In the R.2.3 unit, it is found mainly in the the lower talc-rich layers.

A Secondary Diagenetic Sulphide Assemblage

During diagenesis, previously formed sulphides (or those formed during still earlier diagenetic processes) reacted with the surrounding solutions and were at least partially transformed to new mineral assemblages.

Bornite has been transformed locally into digenite on the periphery of grains (an enrichment in copper with an accompanying loss in iron), and digenite has been altered in turn to chalcocite. These associations are *not supergene* in origin because they are found as inclusions within later primary sulphides.

In more cobaltiferous strata, the borders of chalcopyrite and bornite grains are replaced by carrollite. In this reaction, digenite is produced as a result of the liberation of copper from chalcopyrite or bornite in the presence of cobalt; the digenite produced rims on and veins within the chalcopyrite and bornite. As for the iron from chalcopyrite and bornite, it has been removed in solution, or more rarely, it has been reprecipitated as pyrite within carrollite (Fig. 6b, d, e).

The reactions produced can be outline as follows:

1. *For chalcopyrite:*

$$\underset{\text{chalcopyrite}}{2\,CuFeS_2} + 2\,Co \rightarrow \underset{\text{carrollite}}{CuS \cdot Co_2S_3} + Cu^{\vee} + 2\,Fe^{\vee}$$

$$\underset{\text{chalcopyrite}}{5\,CuFeS_2} + 13\,Cu \rightarrow \underset{\text{digenite}}{2\,CuS \cdot 4\,Cu_2S} + 5\,Fe^{\vee}$$

2. *For bornite:*

$$\underset{\text{bornite}}{3\,Cu_5FeS_4} + 6\,Co \rightarrow \underset{\text{carrollite}}{3\,CuS \cdot Co_2S_3} + 12\,Cu + 3\,Fe^{\vee}$$

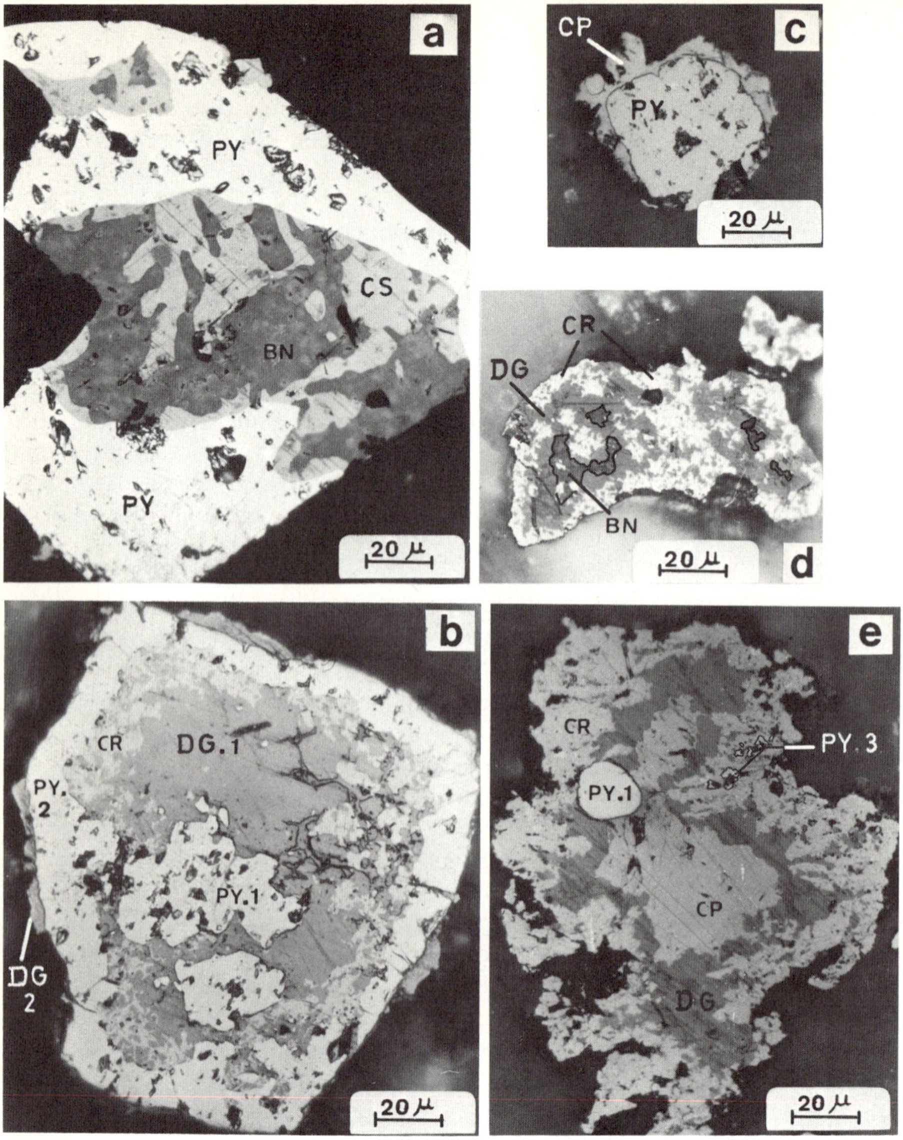

Fig. 6. a Pyrite (*PY*) enclosing the primary assemblage bornite (*BN*) – chalcocite (*CS*). R.2.3.1.2, Drill hole KWf. 1035, Sample 1161. **b** The association digenite-chalcocite (*DG. 1*) enclosing grains of pyrite (*PY. 1*); carrollite (*CR*) formed from a reaction between copper sulfides and introduced cobalt. A later pyrite (*PY. 2*) surrounds this early assemblage and is in turn partially enclosed by late copper sulphides (digenite-chalcocite, *DG.2*). R.2.2.1.1, Drill hole KWf. 1138, Sample 1511. **c** Deposition of chalcopyrite (*CP*) as a rim around a grain of pyrite (*PY*). R.2.1.1, Drill hole KWf. 1150, Sample 1311. **d** Remnant bornite (*BN*) from a grain initially transformed to carrollite (*CR*) and digenite (*DG*). R.2.2.1.1, Drill hole KWf. 1166, Sample 840. **e** Chalcopyrite (*CP*) partially transformed to carrollite (*CR*) and digenite (*DG*). Remnants of digenite remain in the carrollite and the liberated iron is precipitated as pyrite (*PY. 3*). R.2.1.1, Drill hole KWf. 1150, Sample 1291

$$\underset{\text{bornite}}{5\,Cu_5FeS_4} + 11\,Cu \rightarrow \underset{\text{digenite}}{4\,CuS \cdot 4\,Cu_2S} + 5\,Fe^{\vee}$$

In some cases, digenite is transformed into chalcocite, possibly with the aid of cobalt, and as a result the assemblage appears to be enriched in copper:

$$\underset{\text{digenite}}{CuS \cdot 4\,Cu_2S} + 2\,Co \rightarrow \underset{\text{carrollite}}{CuS \cdot Co_2S_3} + S^{\vee} + 8\,Cu^{\vee}$$

$$\underset{\text{digenite}}{CuS \cdot 4\,Cu_2S} + Cu \rightarrow \underset{\text{chalcocite}}{5\,Cu_2S}$$

The carrollitization of copper sulphides is especially intense in the lower mineralized body which is the richest in cobalt. In the talc-rich layers of R.2.3, this process is clearly linked to the crystallization of carrollite porphyroblasts because certain strata without porphyroblasts show the chalcopyrite to be intact.

In reference to digenite, Dimanche (1974) makes a distinction between natural digenite, the stability of which is assured only by an iron content of less than 1%, and a digenite solid-solution metastable at ordinary low temperatures. Dimanche showed by microprobe analyses that the digenite-type minerals which he identified at Kipushi and Kamoto were too poor in iron (less than 0.1 wt%) to belong to the group of natural digenites, and that these grains are probably composed of anilite. In contrast, the same mineral, associated with carrollite at Kamoto, contains a minor amount of cobalt (0.3 wt%).

Carrollite porphyroblasts enclose gangue minerals (chlorite, dolomite and quartz), primary sulphides (pyrite, chalcopyrite and bornite) and secondary paragenetic sulphides. It is therefore possible to observe bornite inclusions which have reacted further toward digenite-chalcocite closer to the borders of porphyroblastic grains than in the core regions. In the same way, the outer edge of chalcopyrite inclusions has been transformed into digenite and carrollite (commonly confused with porphyroblastic carrollite).

Locally (especially in the upper mineralized body), chalcopyrite and bornite porphyroblasts have developed in a similar fashion. For example, chalcopyrite porphyroblasts may have bornite inclusions intact in the core and bornite inclusions partially or totally transformed into digenite-chalcocite at the periphery. These chalcopyrite porphyroblasts may also have undergone subsequent carrollitization. In the same way, bornite porphyroblasts contain grains of digenite-chalcocite, in some cases ringed with carrollite and still containing remnants of the initial bornite.

So, there is no doubt that the crystallization of the porphyroblasts started in the primary stage of mineralization and then continued during the secondary diagenetic reactions involving primary and early secondary copper sulphides.

A late diagenetic copper mineralization (especially bornite, but in some cases chalcopyrite) formed overgrowths on carrollitic porphyroblasts and filled the microfractures in the carrollite.

Discussion

Numerous observations on the Kamoto and Kambove deposits sufficiently demonstrate that from the top of the R.1 up to and including the R.2.3 unit, the sediment-hosted deposit is of a tidal type (Bartholomé et al. 1972; Cailteux 1977a, b, 1983). The evidence of evaporite minerals and textures (gypsum, anhydrite, evaporite swellings, collapsed salt structures, etc.) in the Mines Group, of high-salinity fluid inclusions (Pirmolin 1970), and of neoformed minerals (dolomite, magnesite, etc.) strongly support arguments in favour of the presence of brines and, therefore, tend to confirm a diagenetic model for metal enrichment. The above-described observations permit a definition of the mechanism.

Firstly, there is no doubt that the primary mineralization of R.2.3 has an origin similar to that of the classic mineralized bodies of the Shaban copperbelt. The sulphide paragenesis is identical and the sedimentary environment is similar.

The early emplacement of iron sulphides is well established from observations on the bituminous layers of R.2.2, at the base of R.2.3, as well as in poorly mineralized classic lodes where pyrite generally forms grains of the framboidal type accompanied to a minor extent by inclusions of chalcopyrite.

In the second stage of mineralization, a major chemical reaction occurred between stagnant interstitial waters present within the sediment and a metal-rich brine (copper, cobalt) of higher pH and higher Eh circulating within more porous layers. The reaction transformed early pyrite into copper sulphides and copper-cobalt sulphides.

At West-Kambove, there is no systematic replacement or destruction of pyrite; commonly only a simple coating of pyrite by cupriferous sulphides is observed.

A non-framboidal pyrite identified in the diagenetic succession is closely associated with the copper-cobalt mineralization. In some cases, it includes primary sulphides (including secondary diagenetic products) or constitutes growth rings around the primary sulphides.

The primary paragenesis indicates that the emplacement of the sulphides required a certain length of time during which it underwent a secondary reaction with the brines. Several mineral phases formed in succession and sulphide zones were developed. The sequence from one phase to the following phase caused the progressive replacement of previously formed sulphides (or of those in the process of being formed). This process led to the "carrollitization" of cupriferous sulfides as well as the transformation of bornite into digenite where high-grade copper occurs.

Electron microprobe analyses show that grains of pyrite are zoned with slightly cobaltiferous outer edges (Bartholomé et al. 1971, Cailteux 1974). Probably the origin of cobalt in the pyrite rims is related to the enrichment of cobalt in copper sulfides.

The transformations are well known for copper. For cobalt, the emplacement hypothesis is supported by laboratory experiments on copper-cobalt sulphide ores. Among many types of metallurgical leaching studies (see, for example, Dimanche et al. 1979), it is especially interesting to note the results obtained from experiments using K and Na chlorides in concentrated solutions (F. Dimanche, J. Frenay, personal communication 1984). The authors note that chalcopyrite and bornite transform into digenite, which is then transformed rapidly into covellite. Carrollite is changed slowly into a cobaltiferous digenite, which then gives rise to a cobaltiferous covellite.

The vertical zoning of sulphides within classical mineralized bodies, (i.e.

$$\text{Pyrite} \rightarrow \text{Sulphides of (Fe}-\text{Cu), (Cu}-\text{Co), (Fe}-\text{Cu)} \rightarrow \text{Pyrite,}$$

exhibiting a symmetrical distribution of zones with regard to the R.2.1.3 unit), disagrees with the concept of an influx of a brine from an underlying conglomerate (e.g. a channel conglomerate) situated at the base of the Mines Group. Such conglomerates do not exist in the Kambove region. Furthermore, one should find a zoning similar to that at Kamoto, with chalcocite at the base, followed by bornite, chalcopyrite and then pyrite at successively higher stratigraphic levels. However at Kamoto, can one be sure that a supergene process has not altered the primary profile of the sulphides? For the R.2.3 unit, the lenticular configuration of ore bodies, the distribution of the ore lenses and the apparent delicacy of the sulphides within homogeneous sediments dozens of metres thick (laminites) suggest that such a placement mechanism would be improbable. As well, it is not reasonable to contemplate the downward circulation of a brine through the R.2.1.3 unit and into the underlying beds. The occurrence of pyrite in the poor of the mineralized bodies contradicts that concept. On the other hand, a comparable relationship does not exist between the sulphide mineralization and stromatolitic beds within the R.2.3 unit.

In conclusion, rather than propose a late inflow of ore solutions along fractures in the sediment, it is more probable, as suggested by Brown (1978), that the brine migrated through the sediment itself and formed ore bodies during early diagenesis when the sediment was still sufficiently porous and permeable. Figure 7 illustrates this concept for the metal emplacement in Shaba.

In addition, it is highly probable that changes in pH and Eh conditions at the interface between the metalliferous brine and the interstitial pore solution, combined with chemical (e.g. salinity) – biological variations and changes in the metal content of the brine, caused a variable precipitation of sulphides within the sediment as well as secondary reactions. Annels et al. (1983) confirmed this hypothesis following studies of the Zambian Copperbelt: they concluded that the increases in cobalt in the sediment and the Cu/Co ratio in carrollite are related to increases in the quantity of carbonate in the host rock, and also result from modifications in the activity of sulphur due to variations in the production of biologically-reduced sulphur.

As to the origin of the Mines Group metals in Shaba, a possible hypothesis would be that of leaching under oxidizing conditions of a pyroclastic level such as that found at the top of the R.A.T. Group which underlies the mineralized bodies. The abnormally high background level of metal in this type of sediment could have been taken into solution during diagenesis and accumulated in the basin waters (Cailteux 1983). It is then possible that this water, already enriched in ionic metals, could have been trapped in the sabkha-zone and formed metalliferous brines.

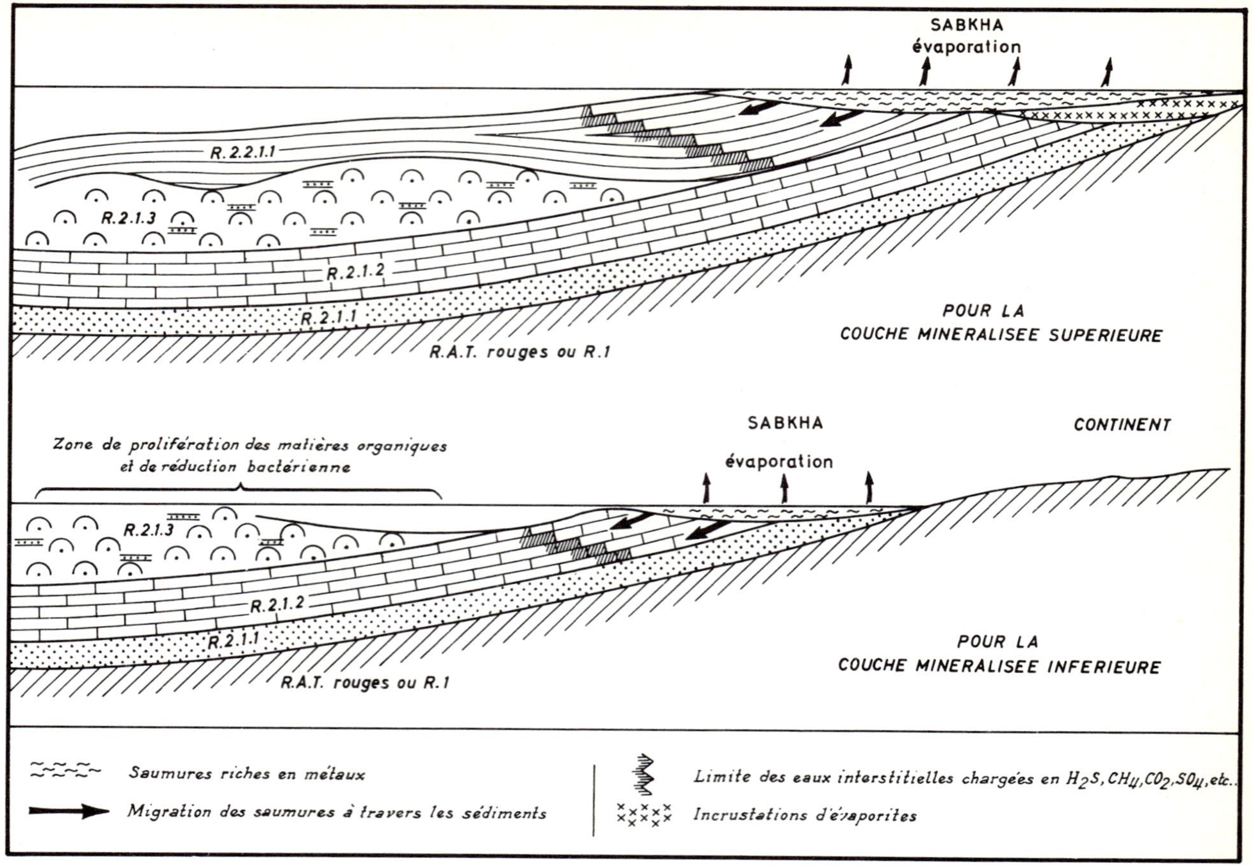

Fig. 7. Genetic model for the emplacement of stratiform metal deposits in Shaba

Acknowledgements. The management of Gécamines has authorized the publication of this work, and I wish to express my appreciation for this permission. I would also like to acknowledge Prof. F. Dimanche of the Université de Liège for the many useful discussions that we have had on this topic.

References

Annels AE, Vaughan DJ, Craig JR (1983) Conditions of ore mineral formation in certain Zambian Copperbelt deposits with special reference to the role of cobalt. Min Dep 18:71 – 88

Bartholomé P (1962) Les minerais cupro-cobaltifères de Kamoto, Katanga-Ouest (Shaba). I. Petrographie – II. Paragenese. St Univ Lovanium, Fac Sci 14, 16. Léopoldville (Kinshasa)

Bartholomé P (1963) Sur la zonalité dans les gisements du Copperbelt de l'Afrique Centrale. Symposium. Problems of postmagmatic ore déposition. V. 1. Prague, 317 – 321

Bartholomé P (1969) Kamoto (Katanga) et White Pine (Michigan), deux gisements stratiformes de cuivre. Acad Roy Sc d'Outre-Mer Bull Séances (Bruxelles) 397 – 410

Bartholomé P, Katekesha F, Lopez-Ruiz J (1971) Cobalt zoning in microscopic pyrite from Kamoto, Republic of the Congo. Min Dep 6:167 – 176

Bartholomé P, Evrard P, Katekesha F, Lopez-Ruiz J, Ngongo M (1972) Diagenetic ore forming processes at Kamoto, Katanga, Rep. of the Congo. In: Amstutz GC, Bernard AJ (eds) Ores in sediments. Springer, Berlin Heidelberg New York, pp 21 – 41

Bartholomé P (1974) On the diagenetic formation of ores in sedimentary beds, with special reference to Kamoto, Shaba, Zaire. In: Bartholomé P (ed) Gisements stratiformes et provinces cuprifères. Liège, pp 203 – 213

Brown AC (1978) Stratiform copper deposits – evidence for their postsedimentary origin. Min Sci Eng Vol 10; 3:172 – 181

Cahen L (1954) Géologie du Congo Belge. Vaillant-Carmanne, Liège

Cahen L (1974) Geological background strata to the copper-bearing strata of Southern Shaba (Zaire). In: Bartholomé P (ed) Gisements stratiformes et provinces cuprifères. Liège, pp 57 – 77

Cailteux J (1974) Les sulfures du gisement cuprifère stratiforme de Musoshi, Shaba, Zaire. In: Bartholomé P (ed) Gisements stratiformes et provinces cuprifères. Liège, pp 267 – 276

Cailteux J (1977) Particularités stratigraphiques et pétrographiques du faisceau inférieur du Groupe des Mines au centre de l'Arc cuprifère Shabien. Ann Soc Geol Belg T 100:55 – 71

Cailteux J (1977) La succession stratigraphique du C.M.N. (ou R.2.3.) au centre de la sous-province cuprifère shabienne. Ann Soc Geol Belg T 100:73 – 85

Cailteux J (1983) Le "Roan" shabien dans la région de Kambove (Zaire). Etude sédimentologique et métallogénique. Thesis Univ Liège. Belgium

Demesmaeker G, Francois A, Oosterbosch R (1962) La tectonique des gisements cuprifères stratiformes du Katanga. In: Lombard J, Nicolini P (eds) Gisements stratiformes de cuivre en Afrique. 2ème partie. Lusaka. Assoc Serv Geol Afr Paris 47 – 115

Dimanche F (1974) Paragenèse des sulfures de cuivre dans les gisements du Shaba (Zaire). I. Kipushi, II. Kamoto. In: Bartholomé P (ed) Gisements stratiformes et provinces cuprifères. Liège, pp 185 – 201

Dimanche F, Frenay J, Herman JC (1979) Lixiviation ammoniacale d'un minerai cupro-cobaltifère sulfuré. Ann Soc Geol Belg T 102:199 – 211

Francois A (1973) L'extrémité occidentale de l'Arc Cuprifère shabien. Etude géologique. Gécamines. Likasi, Shaba, Zaire

Francois A (1974) Stratigraphie, tectonique et minéralisations dans l'Arc cuprifère du Shaba (Rép. du Zaire). In: Bartholomé P (ed) Gisements stratiformes et provinces cuprifères. Liège, pp 79 – 101

Lefebvre JJ (1974) Minéralisations cupro-cobaltifères associées aux horizons pyroclastiques situés dans le faisceau supérieur de la Série de Roan à Shituru Shaba, Zaire. In: Bartholomé P (ed) Gisements stratiformes et provinces cuprifères. Liège, pp 103 – 122

Oosterbosch R (1962) Les minéralisations dans le système de Roan au Katanga. In: Lombard J, Nicolini P (eds) Gisements strafiformes de cuivre en Afrique. 1ère partie. Assoc Serv Geol Afr Paris. 71 – 136

Pirmolin J (1970) Inclusions fluides dans la dolomite du gisement stratiforme de Kamoto (Katanga Occidental). Ann Soc Geol Belg T 93:193 – 202

Geochemical Aspects of Stratiform and Red-Bed Copper Deposits in the Catskill Formation (Pennsylvania, USA) and Redstone Area (Canada). Sequence of Mineralization in Sediment-Hosted Copper Deposits (Part 3)

A. W. ROSE, A. T. SMITH, R. L. LUSTWERK, H. OHMOTO, and L. D. HOY[1]

Abstract

Noneconomic red-bed copper occurrences in the Devonian Catskill Formation are localized by concentrations of reducing plant fragments, commonly in the basal parts of fining-upward fluvial (or tidal?) cycles. The occurrences are within and near thick zones of red mudstone to fine sandstone interpreted to have been deposited in relatively inactive parts of an extensive alluvial plain in an arid environment. Regionally, the occurrences are most abundant adjacent to areas of major sediment input, and in more detail are controlled in part by sandstone beds which were more mature and probably more permeable than the rest of the formation. The copper minerals were emplaced after earliest diagenesis, but before induration and formation of quartz overgrowths. Thin zones of transgressive marine sandstone are present near most occurrences. Copper has been redistributed within the Catskill, being greatly depleted from nearly all red (oxidized) sediments, but showing highly variable values in green and gray (reduced) sediments. Mobilization of copper from the red beds into chloride-bearing pore fluids of marine or evaporative origin is proposed, followed by flow of these fluids through the more permeable sandstones until organic reductants were encountered which precipitated the copper by bacterial sulfate reduction. The flow was driven by compaction, sea-level changes, and gravity flow down the alluvial plain. Red beds near the Coates Lake stratiform copper deposit in the Redstone area show similar depletion of copper.

Introduction

The source of copper and other metals in red-bed copper and stratiform copper deposits, and the geological-geochemical controls and processes forming these ores, are topics of continuing discussion (Gustafson and Williams 1981). The intent of this paper is to summarize evidence on these problems from geochemical studies of red-bed copper-uranium occurrences in the Devonian of Pennsylvania and the Precambrian Redstone stratiform copper deposit in the Northwest

[1] Department of Geosciences, The Pennsylvania State University, University Park, PA 16802, USA

Geology and Metallogeny of Copper Deposits
(ed. by G. H. Friedrich et al.)

Territories, Canada. Though neither of these areas currently produces copper, the Redstone deposit would be economic in a more accessible area, and both areas serve to illustrate the processes and relations in producing districts.

Red-Bed Copper-Uranium Occurrences in the Upper Devonian Catskill Formation, Pennsylvania

The Catskill Formation is a complex clastic wedge of dominantly fluvial red and gray sediments that prograded to the west during erosion of the Acadian Highlands along the east coast of the United States (Fig. 1). The formation thins rapidly westward, from about 2400 m near Lehighton at the eastern limit of the exposures to less than 800 m in central and north-central Pennsylvania. The Catskill overlies and to the west thins and interfingers into non-red marine shales and turbidites of Middle and Upper Devonian age. The Paleozoic sediments were extensively folded in late Paleozoic and now crop out along narrow belts in much of the region.

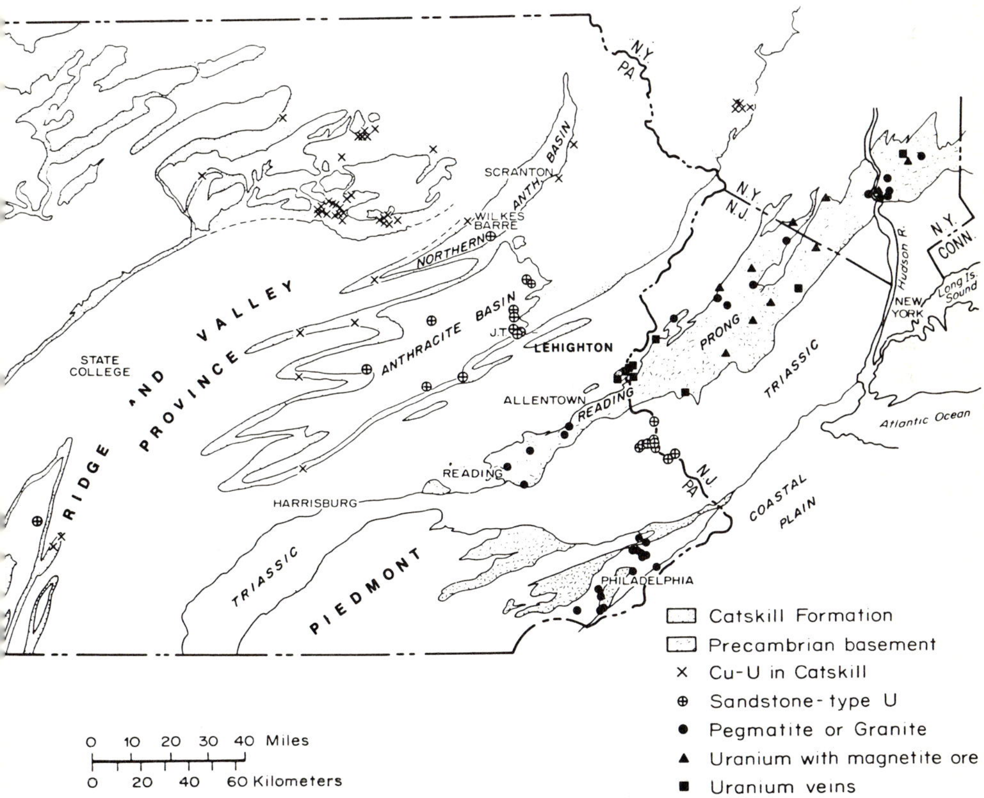

Fig. 1. Distribution of Catskill Formation, red-bed copper-uranium occurrences, sandstone-type uranium occurrences, other uranium occurrences, and geologic provinces in Pennsylvania and adjacent states

The Catskill Formation has been divided into four magnafacies utilizing repeating patterns of individual facies (Smith and Rose 1985). These magnafacies show an overall progradational pattern (Fig. 2). Over most of the area, the lower part of the Catskill (Magnafacies A) is composed of red shale and fine-grained sandstone containing plant fragments and root casts, interlayered with gray to green fine-grained sandstones and shales occasionally containing marine fossils. This magnafacies (A) is interpreted as interbedded nonmarine and marine sediments deposited in an intertidal to supratidal environment.

The middle part of the Catskill (Magnafacies *B*) is typically composed of fine, red sandstone to mudstone in fining-upward cycles, interpreted as the deposits of relatively small meandering streams on an alluvial plain. The basal zones of some cycles are green and locally contain plant fragments. Caliche in mudstone horizons near the top of some cycles suggests an arid climate. Occasional thin tongues of marine sandstone within Magnafacies B are recognized by green color, bi-directional ripples and cross-beds, flaser bedding, planar bedding and occasional marine fossils. These tongues apparently reflect transgressions of marine (or at least tidal) conditions across the alluvial plain and appear to be of considerable importance in forming the deposits, as discussed later.

The upper Catskill contains two magnafacies. Magnafacies C is composed of extensively cross-bedded, gray-green sandstones and conglomeratic sandstones deposited by northwestward-flowing braided streams, and is limited to three areas which are interpreted as loci of sediment input from a source area that was uplifted contemporaneously (Fig. 3). Magnafacies D overlies Magnafacies B or C, and is composed of thick, fining-upward cycles with a much higher proportion of sand than Magnafacies B. The siltstone and mudstone in the upper part of these cycles are red; the lower sandstones are commonly gray or green. This magnafacies is interpreted as the deposit of relatively large meandering streams.

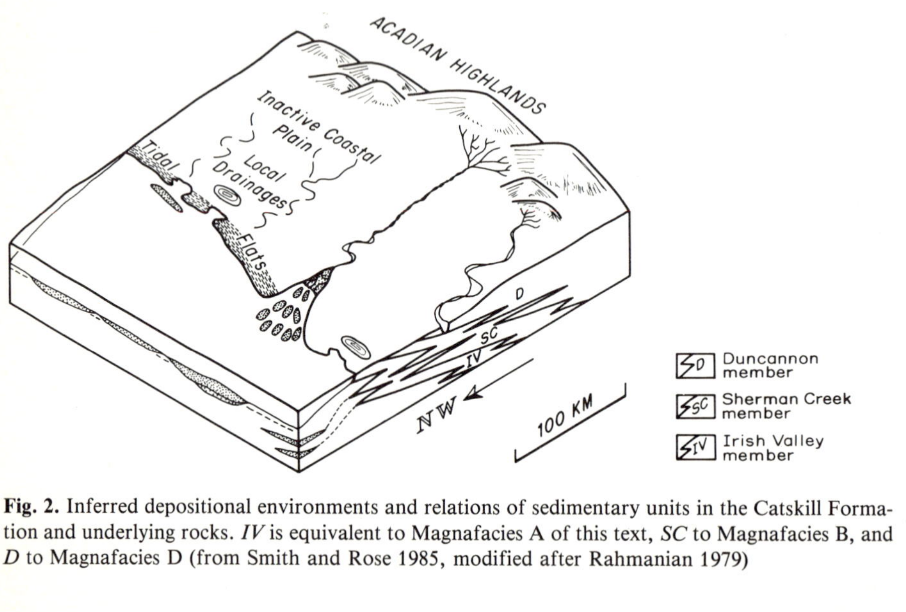

Fig. 2. Inferred depositional environments and relations of sedimentary units in the Catskill Formation and underlying rocks. *IV* is equivalent to Magnafacies A of this text, *SC* to Magnafacies B, and *D* to Magnafacies D (from Smith and Rose 1985, modified after Rahmanian 1979)

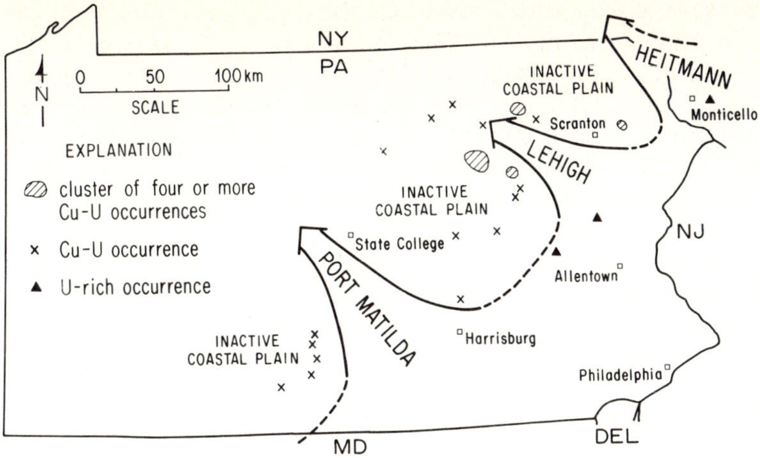

Fig. 3. Location of red-bed copper-uranium occurrences and sandstone uranium occurrences relative to sediment input areas (Magnafacies C) in the Catskill Formation (after Smith and Rose 1985)

The red-bed copper-uranium occurrences are lens-shaped to tabular and typically occupy zones of plant fragments either in the basal sandy to conglomeratic portions of fining-upward cycles, or in thin gray siltstone units. The mineralized zones range in size from 0.1 to 1 m thick and extend 3 to 10 m along bedding. Chalcocite and digenite furnish grades of 0.05 to 1% copper and are accompanied by small amounts of uranium (20 – 1000 ppm), only rarely as recognizable uraninite, along with enrichments in arsenic, lead, and vanadium. The copper and uranium are localized by concentrations of plant fragments, which apparently acted as reducing agents.

Several lines of evidence indicate a low temperature of copper deposition. Sulfur isotopes of sulfides range from –34 to –18‰ (CDT), averaging –27‰, suggestive of biogenic sulfate reduction at temperatures less than 100 °C. The grain size of copper minerals and the gangue minerals is very fine, typical of sediments rather than hydrothermal processes. Bornite from the occurrences exsolves chalcopyrite upon heating above 75 °C, indicating formation as a meta-stable phase at low temperatures (McCauley 1961). The age relations of the sulfides indicate a diagenetic origin, as discussed later.

The red-bed copper-uranium occurrences are mainly in Magnafacies B, with some in Magnfacies A and D. They are lacking in Magnafacies C, but tend to be concentrated adjacent to the input areas marked by the distribution of Magnafacies C.

An additional local control on the distribution of Cu minerals is provided by relatively coarse-grained bodies of sandstone which probably had high permeability prior to late Paleozoic deformation. Most sandstones of the Catskill are "dirty", low-rank graywackes with a high proportion of phyllite grains and fine-grained matrix, but the sandstones near ore bodies are significantly different. "Mineralized" sandstones collected within or immediately adjacent to mineralized zones were compared with sandstones of similar grain size collected nearby, but outside the mineralized zones. The mineralized sandstones characteristically have 70 to 75% framework grains (mainly quartz) and only 20 to 30% ductile grains (rock fragments, micas, matrix) compared to 55 to 65% framework grains and 30 to 45% ductile grains in the "nonmineralized" sandstones (Smith 1983). Some of these mineralized sandstones are marine as discussed below. The higher permeability of the mineralized sandstones prior to deep burial and deformation would have allowed preferential flow through these zones.

Of 30 localities examined in detail, 23 have exposures of sandstones interpreted as "marine bars" or "barrier sandstones" (Smith 1980, Smith and Rose 1985) within 10 m above the Cu occurrence. At many of the remaining seven occurrences, exposures are poor and such sediments could be concealed. Also, most of these seven occurrences are near other occurrences that do exhibit marine sandstone facies. This correlation with thin tongues of marine sediment appears highly significant in view of the effect of Cl-bearing waters on copper solubility.

The solubility of copper as Cu^{2+} or complexes of common natural anions with Cu^{2+} is very low at pH values greater than 6, whereas very few stream or groundwaters have pH values less than 6 (Rose 1976). In addition, Fe oxides are very effective adsorbents of copper at pH values above about 6 (Benjamin and Leckie 1981). Thus, in a red-bed depositional environment, one would not expect much copper mobility.

However, in the presence of chloride under intermediate oxidation conditions, strong complexes of Cu^+ with Cl^- are formed, and the solubility of copper is greatly increased (Fig. 4), so that high solubilities extend to pH 8 or higher.

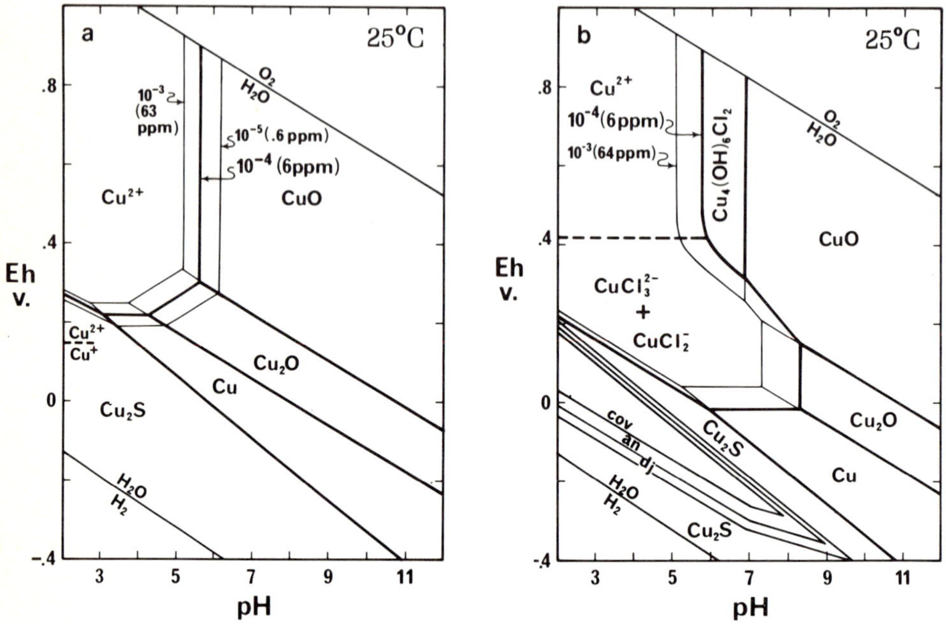

Fig. 4a, b. Solubility of copper at 25 °C as a function of pH, Eh, and Cl^-, showing increase at pH 6–8 in the presence of Cl. **a** No chloride, $\sum S = 10^{-4}$ m; **b** 0.5 m Cl, $\sum S = 10^{-4}$ m, *cov* = covellite; *an* = anilite; *dj* = djurleite (after Rose 1976)

The ingress of marine waters during the occasional transgressions provides a convenient source of Cl-rich solutions. If small amounts of organic matter or Fe^{2+}-bearing clay or chlorite were present in the initial sediment, O_2 would be consumed and the pore fluid would reach an appropriate oxidation state. Waters of compaction released from underlying marine sediments, or evaporation of fresh waters in an arid environment are other possible sources of Cl. Uranium is also relatively soluble in solutions of intermediate oxidation state, especially if some carbonate is present in solution.

The abundance of copper in the Catskill and related rocks provides more direct evidence for the source of metals. So far as is known, the source area for Catskill detritus was similar to the source for the underlying Middle Devonian marine sediments, the main difference being the environment of deposition and the resulting color (Sevon et al. 1978, Willard 1939). Essentially all sediments in this sequence are low-rank graywackes and equivalent shales and mudstones.

Figure 5 shows copper contents of a set of Catskill and pre-Catskill rocks. The Catskill samples were collected in a previous study of uranium (Pirc 1979, Pirc and Rose 1981, Rose and Wright 1980), and were intended to be representative of the various colors and grain sizes of Catskill sediments. The pre-Catskill rocks were collected for this study. In the reducing marine pre-Catskill rocks, copper contents are mostly in the range 15 – 30 ppm and show a possible correlation with Fe content. Changes with grain size appear to exist, but are relatively minor. These samples are believed to represent the copper content of detritus entering the Catskill sedimentary basin. In contrast, in the Catskill, red (oxidized) sediments are generally much lower in copper, mostly 0 – 5 ppm, and gray to green (reduced) sediments vary widely, with a few values over 400 ppm.

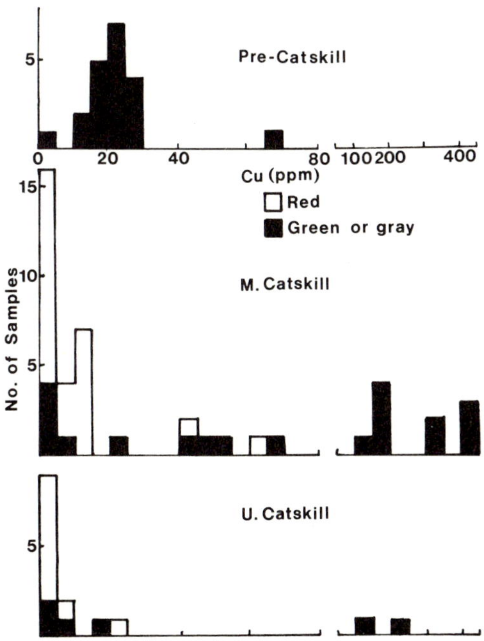

Fig. 5. Abundance of total copper (HF-$HClO_4$ digestion) in red and green-gray sandstones and shales of the Catskill and pre-Catskill Formations near Lehighton, Pottsville, and Harrisburg, PA

There is no obivous correlation with grain size or with major elements in the samples; high values are found in gray-green rocks ranging from mudstone to medium-grained sandstone, and low values in red rocks ranging from mudstone to fine sandstone.

These data indicate major redistribution of copper in the Catskill during deposition and early to middle diagenesis. Much copper appears to have been lost from the red beds. Uranium behaves somewhat similarly, being enriched in some reducing horizons, though not to the extent shown by copper. The rocks are now highly indurated and impermeable so the redistribution must have occurred before appreciable burial and compaction. In the exposures near Lehighton from which these samples were collected, no red-bed copper occurrences were previously recognized, but copper at levels of 100 to 400 ppm is present in numerous locations, even though not visually evident.

Loss of 20 ppm copper from a 30-m-thick section of sediment (25 ppm initial to 5 ppm final) could furnish copper for a layer 0.3 m thick with 0.2% copper, which is a common concentration in the red-bed copper occurrences (Fig. 6). This calculation assumes that all transport is normal to bedding. Lateral migration from a large volume of sediment into a similar small volume could create higher grades or thicker mineralized zones. Thus, ample copper is available to form the red-bed copper occurrences by depletion of copper from nearby sediment.

The time of copper mobilization is also indicated by the age relations of minerals in the red-bed copper occurrences. In the earliest diagenetic stage, quartz and other detrital grains are coated by hematite and locally by fine clay that apparently infiltrated into the stream bed (Walker et al. 1978, Loucks et al. 1977). In reducing environments manganiferous calcite cement has formed, and can be seen to surround the clay coats in a few instances. A middle diagenetic stage is represented by chlorite as an alteration of rock fragments and biotite and as radiating and fibrous crystals, apparently formed in relatively large pore spaces before compaction, but later than calcite because it is not found in calcite-cemented zones. Copper sulfides accompany some chlorite and also appear to have filled relatively large pores prior to compaction, or have replaced organic matter. A late diagenetic stage is characterized by quartz overgrowths which cover clay coats and are not found with copper minerals. In sediments of the Frio

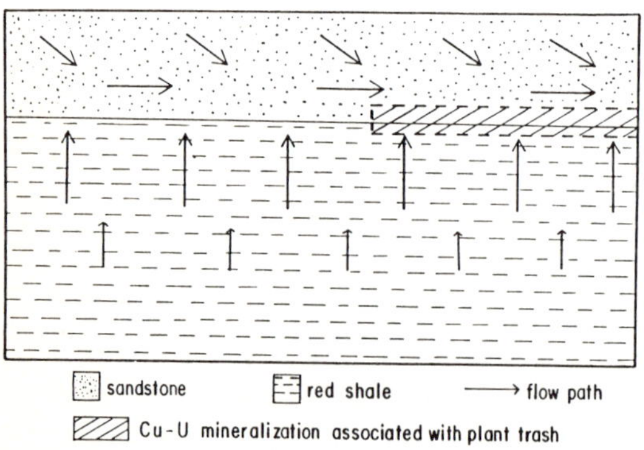

Fig. 6. Model for formation of red-bed copper-uranium occurrence in the Catskill Formation. *Arrows* indicate flow of copper-bearing pore fluid out of red shale and along bedding in more permeable sandstone, and localization of copper in zone of plant trash with dimensions approx. 0.5 m thick ×5 m×5 m

Formation on the Gulf Coast of the United States, quartz overgrowths are first observed at depths of 1 to 2.5 km (Loucks et al. 1977) and are assumed to be formed at a similar stage in the Catskill. A final stage is hematite along fractures, perhaps related to recent weathering.

Based on the above age relations, copper was emplaced after earliest diagenesis (after hematite, clay coats, and calcite cement), but probably before quartz overgrowths and before compaction and induration of the sediments.

Flow of pore fluids during diagenesis could be driven by compaction, gravity-driven flow of groundwater down the alluvial fan, or small changes in sea level as demonstrated by the thin marine tongues. The flow caused by these processes would be concentrated in the network of relatively coarse-grained sands deposited in channels of the larger streams. These permeable sands may be more abundant near the sediment-input areas.

In summary, the following factors appear to control the formation of the red-bed copper occurrences in the Catskill Formation:

1. Plant fragments act as reductants to localize sulfide precipitation.
2. Chloride-bearing solutions tend to mobilize copper; these solutions might be seawater emplaced during small marine transgressions, pore fluids from underlying compacting marine sediments or evaporated fresh water in the arid coastal plain.
3. The process of forming red beds appears to have liberated copper, perhaps by oxidation of the iron-bearing silicates initially hosting copper, and possibly affected by the behavior of the resulting Fe-oxides.
4. Permeability of sediments; the coarser, better-sorted sands deposited in channel bottoms of the larger streams on the alluvial plain, and the thin tongues of sand in the marine transgressions, would be locations of maximum flow.
5. Copper content of sediments; the copper content of the relatively "dirty" Catskill sediment probably was slightly higher than many other sedimentary sections. Also, data in the literature suggests that copper is concentrated in the finer sediments with higher Fe (Wedepohl 1964).
6. Hydraulic gradients causing flow of pore fluid, as dicussed above.

Redstone Area

The stratiform to lenticular copper occurrences in late Precambrian sediments of the Redstone area in the MacKenzie Mountains, Northwest Territories, Canada are described in a companion paper by Brown and Chartrand, this Vol., and by Ruelle (1982) and Lustwerk and Rose (1983). The hosts of ore are one or more beds of reducing marine algal carbonate of sabkha and shallow marine origin. These units are underlain by 1000 m or more of Redstone River Formation, composed of red siltstone deposited on an evaporative, alluvial plain and/or continental sabkha, as indicated by common gypsum-anhydrite as cement and locally as thin beds of gypsum-anhydrite in the siltstone. The pore fluids in these sediments clearly were saline with elevated chloride content during part of their diagenetic history.

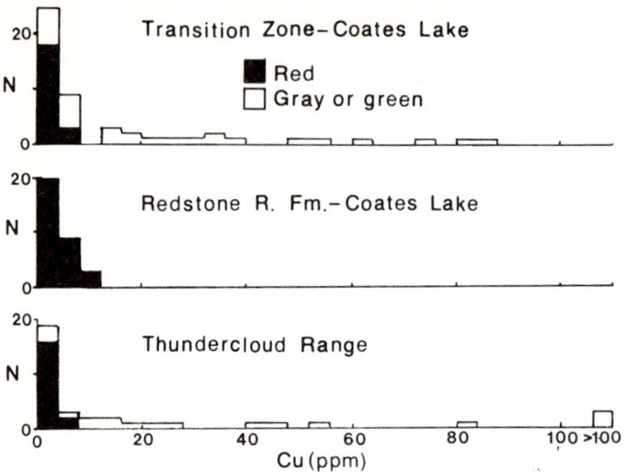

Fig. 7. Abundance of total copper in siltstone and sandstone from the Redstone River and Transition Zone near the Coates Lake stratiform copper deposit, Redstone Belt, MacKenzie Mts., NWT. The ore occurs within the lower Transition Zone and is underlain by 1000+m of Redstone River Formation. The Thundercloud area, 40 km to the south, contains a thinner section of the same rocks in exposures along the margin of the basin

Figure 7 shows analyses for copper in numerous samples of siltstone and sandstone from the Redstone River Formation and the Transition Zone containing the ore. Over half of all samples and about 75% of the red samples contain <2 ppm copper. Some gray to green samples contain a wide range of higher values. Most other transition elements (zinc, nickel) show fair to good correlation with iron, and it appears that copper in detritus entering the basin probably shared this correlation. Although some sandstones with low iron content may have originally contained only a few ppm copper, the average siltstone with 3% to 8% iron probably contained 15 to 50 ppm, based on worldwide averages and estimates of the type of detritus entering the basin. The pattern is thus similar to that found for the Catskill; namely, that copper is extensively redistributed during diagenesis of sediments near "sedimentary" copper deposits; most of these rocks, especially red beds, are greatly depleted in copper, and appear to have been the source of copper in the ore. This type of copper depletion may be a guide to favorable areas for sedimentary copper deposits.

Acknowledgments. The research described above has been supported by a grant from the National Science Foundation and by the Mineral Conservation Section of the Pennsylvania State University. We are indebted to Prof. A. C. Brown, Mr. Frank Chartrand, and Dr. John Ruelle for discussion and assistance with the Redstone studies and to Prof. E. G. Williams and Dr. R. L. Slingerland for ideas and discussion on the Catskill studies.

References

Benjamin MM, Leckie JO (1981) Multiple-site adsorption of Cd, Cu, Zn and Pb on amorphous iron oxyhydroxide. J Colloid Interf Sci 79:209–221

Gustafson LB, Williams N (1981) Sediment-hosted stratiform deposits of copper, lead, and zinc. In: Skinner BJ (ed) Economic geology 75th anniversary, vol Econ Geol Publ Co, pp 139–178

Loucks RG, Bebout DG, Galloway WE (1977) Relationships of porosity formation and preservation to sandstone consolidation history – Gulf Coast Lower Tertiary Frio Formation. Gulf Coast Assoc Geol Soc 27:109–120

Lustwerk RL, Rose AW (1983) Source and segregation of transition metals during diagenetic formation of the Redstone stratiform copper deposit, MacKenzie Mts., NWT, Canada. Geol Soc Am (Abstracts with Programs) 15:632

McCauley JF (1961) Uranium in Pennsylvania. Penn Geol Sur (4th) Bull M43, 71 pp

Pirc S (1979) Uranium and other elements in the Catskill Formation of east-central Pennsylvania. Ph D Thesis, Penn State Univ 300 pp (unpublished)

Pirc S, Rose A (1981) Uranium anomalies in paleoaquifers near sandstone-type uranium deposits in the Devonian Catskill Formation of Pennsylvania. J Geochem Explor 15:219–231

Rahmanian VDK (1979) Stratigraphy and sedimentology of the Upper Devonian Catskill and uppermost Trimmers Rock Formations in central Pennsylvania. Ph D thesis, Penn State Univ, 329 pp (unpublished)

Rose AW (1976) The effect of cuprous chloride complexes in the origin of red-bed copper and related deposits. Econ Geol 71:1036–1048

Rose AW, Wright RJ (1980) Geochemical exploration models for sedimentary uranium deposits. J Geochem Explor 13:153–179

Ruelle JC (1982) Depositional environments and genesis of stratiform copper deposits of the Redstone Copper Belt, Mackenzie Mountains, NWT. Geol Assoc Can Spec Pap 25:701–737

Sevon WD, Rose AW, Smith RC, Hoff DT (1978) Uranium in Carbon, Lycoming, Sullivan and Columbia Counties, Pennsylvania. Guidebook, 43rd Field Conf Penn Geol, Harrisburg, PP, pp 7–22

Smith AT (1980) Stratigraphic and sedimentologic controls for copper and uranium in red-beds of the Upper Devonian Catskill Formation in Pennsylvania. MS Thesis, Penn State Univ, 216 pp (unpublished)

Smith AT (1983) Geology of the red-bed Cu-U occurrences in the Upper Devonian Catskill Formation, Pennsylvania. Ph D Thesis, Penn State Univ, 240 pp (unpublished)

Smith AT, Rose AW (1985) Relation of red-bed copper-uranium occurrences to the regional sedimentology of the Catskill Formation in Pennsylvania. In: Woodrow DL, Sevon WD (eds) Geol Soc Am Spec Pap 201:183–197

Walker TR, Waugh B, Crone AJ (1978) Diagenesis in first-cycle desert alluvium of Cenozoic age, southwestern U.S. and northwestern Mexico. Geol Soc Am Bull 89:19–32

Wedepohl KD (1964) Untersuchungen am Kupferschiefer in Nordwest-Deutschland. Geochim Cosmochim Acta 28:305–364

Willard B (1939) The Devonian of Pennsylvania. Penns Geol Sur (4th) Rep G19, 481 pp

Stratabound Copper Deposits in East South-Central Alaska: Their Characteristics and Origin

M. K. SOOD, R. J. WAGNER, and H. D. MARKAZI[1]

Abstract

Six stratabound Kennecott-type copper deposits are being studied to evaluate the characteristics of the ore-host rock system. Three (Mountain Grill, Radovan, and Clear-Porcupine) are in the Middle to Upper Triassic Nikolai Greenstone – a thick sequence of subaerial amygdaloidal tholeiite flows – and the other three (Binocular, Nelson, and Peavine) are in the Upper Triassic Chitistone Limestone, a platformal limestone and dolomite unit.

X-ray diffraction, petrographic and chemical studies have led to the identification of ten ore minerals in the deposits, the most abundant being digenite, chalcocite, djurleite, covellite, bornite, and chalcopyrite. Minor arsenic and antimony sulfide phases are also present. Mineral deposits in greenstones have both Cu and Cu – Fe sulfides in contrast to the essentially Cu-sulfides in limestones.

The mineralization is proposed to have formed in two stages. Hydration-dehydration reactions associated with alteration-metamorphism of the Nikolai Greenstone liberated copper from the metal oxides and the mafic minerals. Copper was carried by the circulating fluids along the existing fracture-fluid pathway system, reacting and precipitating ore minerals in them. The weak and widely distributed mineralization in the belt is probably of this type. Greenstone alteration was probably coincidental with Cretaceous accretionary orogeny as shown by K-Ar systematics (Silberman et al. 1981).

Cenozoic deformation induced a steep fracture system that enhanced the porosity and permeability of the rocks, thus focusing the meteoric water circulatory system for effective and efficient solution and transport of metals. Activity in such a system at this stage was probably driven by a thermal charge from the Tertiary intrusives. Lack of extensive alteration of the Chitistone Limestone and of recrystallization at the ore vein-limestone contacts suggests that the ore solutions were at relatively low temperatures. The mineral associations and textures suggest that the upper temperature that prevailed in the Triassic ore-host rock system was about 200 °C.

[1] Department of Earth Sciences, Northeastern Illinois University, Chicago, IL 60625, USA

Geology and Metallogeny of Copper Deposits
(ed. by G.H. Friedrich et al.)

Introduction

The Triassic age Nikolai Greenstone and Chitistone Limestone belt lying in the central part of the McCarthy Quadrangle, Alaska, (141° – 144° long and 61° – 62° lat), is a host to stratabound copper mineralization. Of particular interest have been the limestone-hosted copper deposits at Kennecott considered to be of low-temperature hydrothermal origin (Bateman and McLaughlin 1920, Jensen and Bateman 1981). These are large, high-grade deposits characterized by simple mineralogy, limited host-rock alteration, and sparsely distributed gangue minerals.

There are a number of other copper occurrences in this Triassic belt that have affinities with those at Kennecott, but have not been systematically studied. This paper describes the geologic and mineralogical characteristics of six Kennecott-type copper deposits both from greenstone (Mountain Grill, Radovan, and Clear-Porcupine) and limestone (Binocular, Nelson and Peavine), with special emphasis on the alteration-metamorphism of the Nikolai Greenstone, the character of the ores, thermal regimes in the ore-host rock system, and the structural focus of mineralization. The objective is to develop a model for the genetic evolution of the widely scattered copper mineralization in the McCarthy Quadrangle and to study its relation to the accretionary tectonics in east-south-central Alaska.

Location

The McCarthy Quadrangle is located about 260 miles east of Anchorage and extends to the Canadian border in east south-central Alaska (Fig. 1a). The generalized geology of the northwest trending Nikolai Greenstone and Chitistone Limestone belt and the locations of the copper deposits are shown in Fig. 1b.

Geologic Setting

Lithounits in the McCarthy Quadrangle constitute Upper Paleozoic to Cenozoic age stratigraphy (Fig. 2).

The Upper Paleozoic andesitic island-arc sequence and argillites and limestones of the Skolai Group are unconformably overlain by Nikolai Greenstone, Chitistone Limestone, basinal Nizina Limestone, and the Triassic-Jurassic McCarthy Formation. Jones et al. (1977) termed this cohesive lithostratigraphic setting the allochthonous Wrangellia terrane that comprises the major part of the McCarthy Quadrangle. The superjacent Cenozoic units comprise largely silicic and felsic shallow-level intrusives, and andesitic to rhyolitic rocks of the Wrangell Lava (Fig. 2).

Southern Alaska – an example of collisional tectonics – is composed of many disparate terranes generally becoming progressively younger to the south

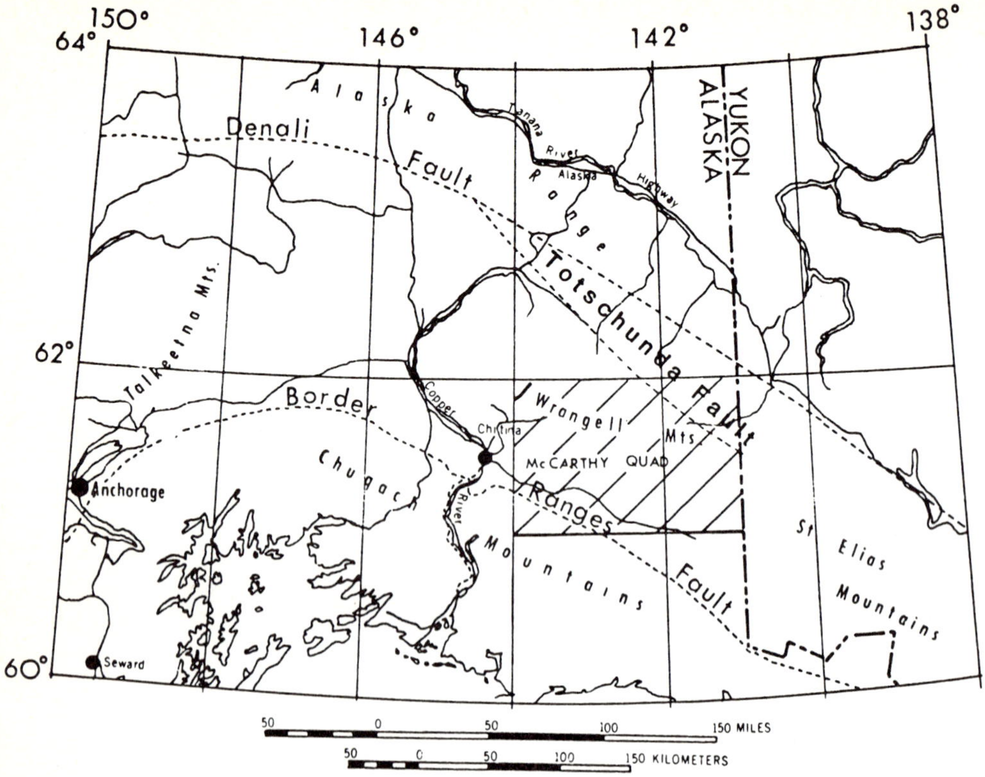

Fig. 1a. Location of the McCarthy Quadrangle in relation to the major structural features in southern Alaska (modified from MacKevett et al. 1977)

Fig. 1b. Scheme of the 15-min-quadrangles, the location of the mineral prospects of the present study, and the distribution of the Nikolai Greenstone and Chitistone Limestone belt in the McCarthy Quadrangle (based on MacKevett 1978, and MacKevett et al. 1977)

Tertiary intrusives

Jurassic Chitina Valley batholith

Chitistone Limestone

Triassic

Nikolai Greenstone

······ Fault

—— Geologic boundary

△ Mineral prospects

1. Radovan
2. Nelson
3. Mountain Grill
4. Binocular
5. Peavine
6. Clear-Porcupine

K. Kennecott

B4. 15′ quadrangle

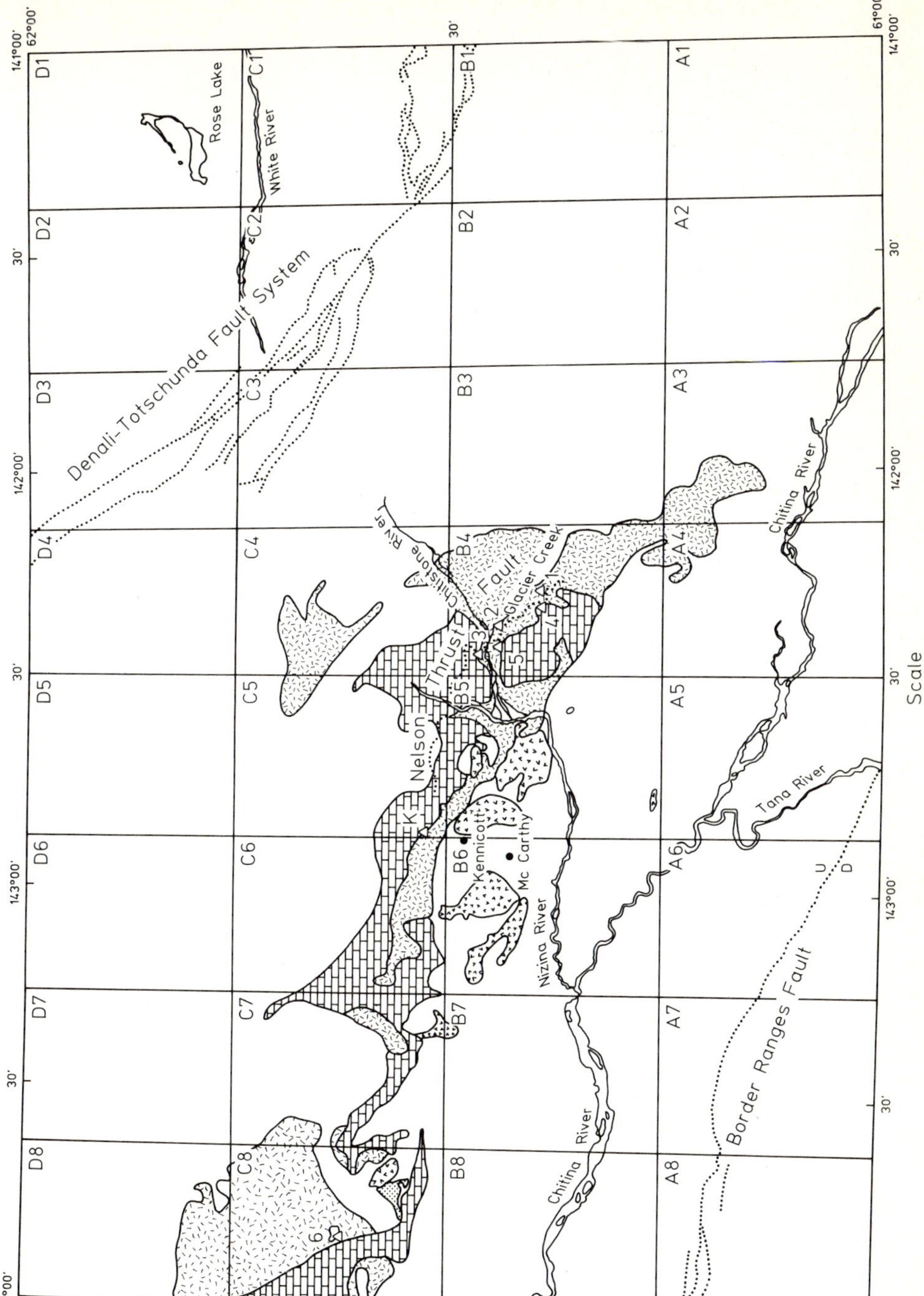
Rose Lake
White River
Denali-Totschunda Fault System
Chitistone River
Nelson Thrust Fault
Glacier Creek
Kennicott
Mc Carthy
Nizina River
Chitina River
Tana River
Border Ranges Fault
Scale
D1
D2
D3
D4
D5
D6
D7
D8
C1
C2
C3
C4
C5
C6
C7
C8
B1
B2
B3
B4
B5
B6
B7
B8
A1
A2
A3
A4
A5
A6
A7
A8
141°00'
142°00'
143°00'
144°00'
62°00'
61°00'
30'

Fig. 1b

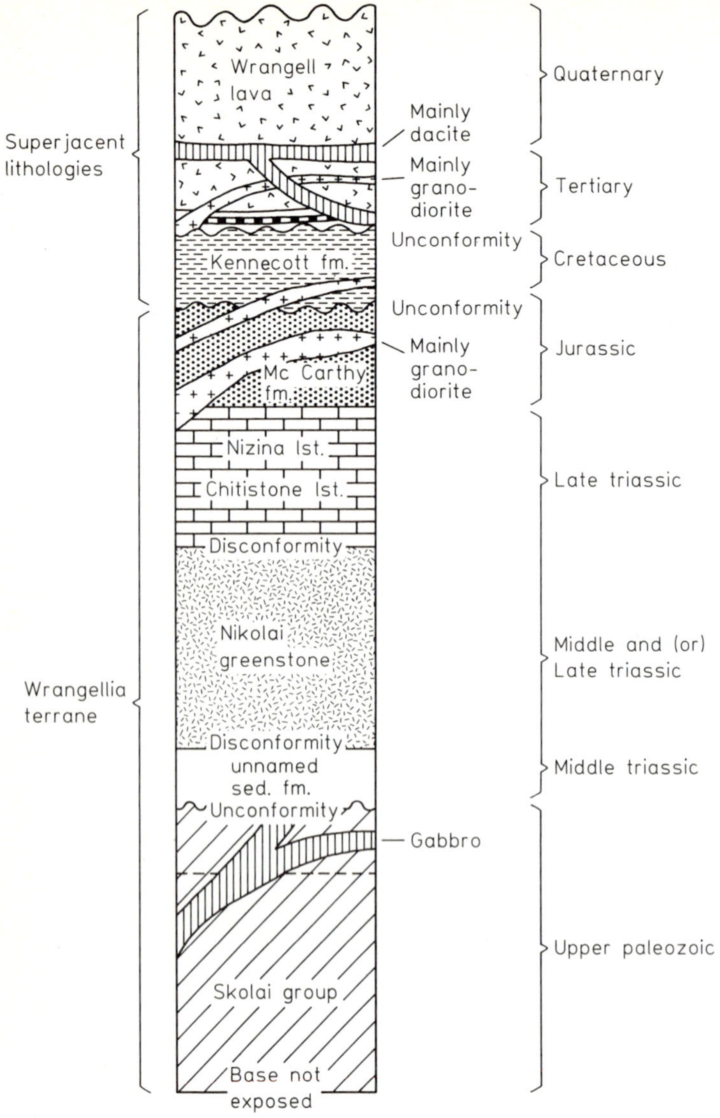

Fig. 2. Generalized lithostratigraphic relations in the McCarthy Quadrangle (based on MacKevett 1978, Silberman et al. 1981)

(Fig. 3). Csejtey et al. (1980) consider Wrangellia terrane a part of the accretionary continental margin of western North America. It joined the geologically and stratigraphically similar Peninsular terrane (Csejtey et al. 1978, Jones and Siberling 1979), occurring to the west of the Quadrangle, in a tectonic coupling in the Middle Jurassic. This new coherent terrane, named the Talkeetna superterrane by Csejtey et al. (1980), was emplaced in Alaska during the Middle Cretaceous accretionary orogeny. The Wrangellia part was juxtaposed in the

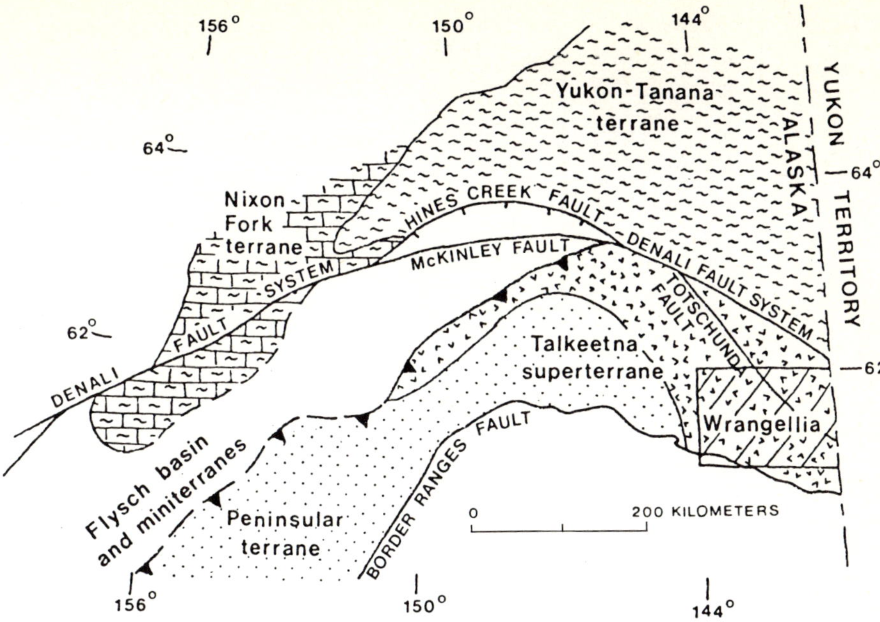

Fig. 3. Tectonostratigraphic terrane map of Southern Alaska. Location of the McCarthy Quadrangle is shown by the *box* (modified from Csejtey et al. 1982)

northeast against the Precambrian Yukon-Tanana terrane along a right lateral strike-slip (or transform) Denali fault system – a tectonostratigraphic suture (Csejtey 1976, Csejtey et al. 1982).

In the northeastern part of the McCarthy Quadrangle, the Totschunda fault is a conjugate part of the eastern portion of the Denali fault system (Fig. 3) that developed in the Cenozoic deformational phase along the reactivated Cretaceous fault system (Richter and Matson 1971) as a postaccretionary feature (Csejtey et al. 1982).

The high-angle reverse or a thrust fault in the central part of the McCarthy Quadrangle, herein referred to as the Nelson thrust fault (Fig. 1b), is a prominent feature that affects the Triassic stratigraphy. It is probably related to the Cretaceous accretionary process. Along its trend, the Nelson thrust is disrupted by steep-dipping, northeasterly and easterly faults. Kennecott-type mineralization is largely localized within these younger, northeasterly fracture systems.

In the southwestern part of the Quadrangle, the Cretaceous or Tertiary age Border Ranges fault (Fig. 3) is a high-angle reverse fault that grades into a thrust fault to the west of the Quadrangle (Mackevett 1978).

Lithologic Characteristics of the Triassic Units

The Triassic Nikolai Greenstone (dark) and Chitistone Limestone (light gray) form an impressive and cohesive geologic unit whose contact is marked by a

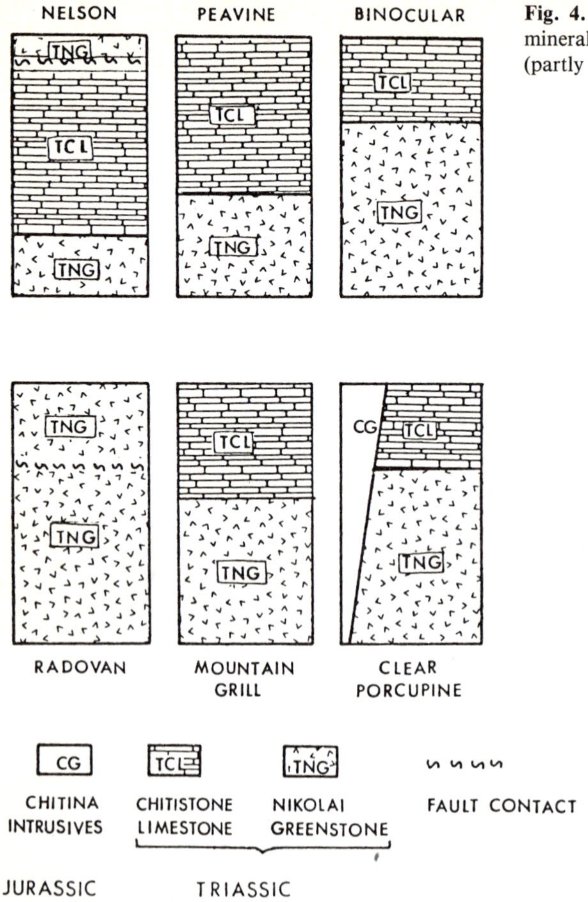

Fig. 4. Lithostratigraphic relations at the mineral prospects of the present study (partly based on MacKevett 1978)

lithologic color contrast along its entire length in the Quadrangle. The stratigraphic relations at each prospect of this study are shown in Fig. 4.

Nikolai Greenstone

The Nikolai Greenstone is comprised of a series of 15-cm to 4-m-thick subaerial basaltic flows which, due to their overlapping relations, appear to be continuous. Mackevett (1978) estimated the cumulative thickness for the unit to be about 4 km and reported that pillowed structures are present in the basal portions in the northeastern part of the Quadrangle. However, in the study area the unit is massively bedded and abundantly amygdaloidal.

At Radovan, the Nikolai unit is intensely fractured and altered along the fractures; at Peavine and Mountain Grill prospects, it is affected by faulting and overturned folding (Plate I) and at Clear-Procupine, it has been contact metamorphosed by the Jurassic Chitina Valley granodiorite.

Plate I. Faulting and overturned folding in the Triassic lithounits at the Mountain Grill Prospect can be seen in the central part of the picture which is taken looking north. Nikolai Greenstone (*darker*) is seen in the *left central part*. In other places, the greenstone/limestone contact is cohesive and structurally conformable

Petrographically, the Nikolai rocks display the effects of pervasive, low-temperature alteration-metamorphism, but without penetrative deformation. Primary minerals (augite, labradoritic plagioclase, relict olivine, hypersthene, magnetite, and ilmenite) and textures (fine-grained volcanic, porphyritic, intergranular, and subophitic) are preserved to varying degrees (Plate IIa). Plagioclase is a common phenocryst. A variety of minerals fill the amygdules (Plate IIb).

The representative mineralogy (Plate IIa, b, c) of the unit based on the examination of about 100 samples is given below:

Primary Minerals (Plate IIa, b, c)

Augite	40 to 50%
Plagioclase (An 35 to An 65)	35 to 45%
Olivine Hypersthene Pigeonite	Not always present. Up to 5%
Magnetite Ilmenite Apatite	Up to 10%

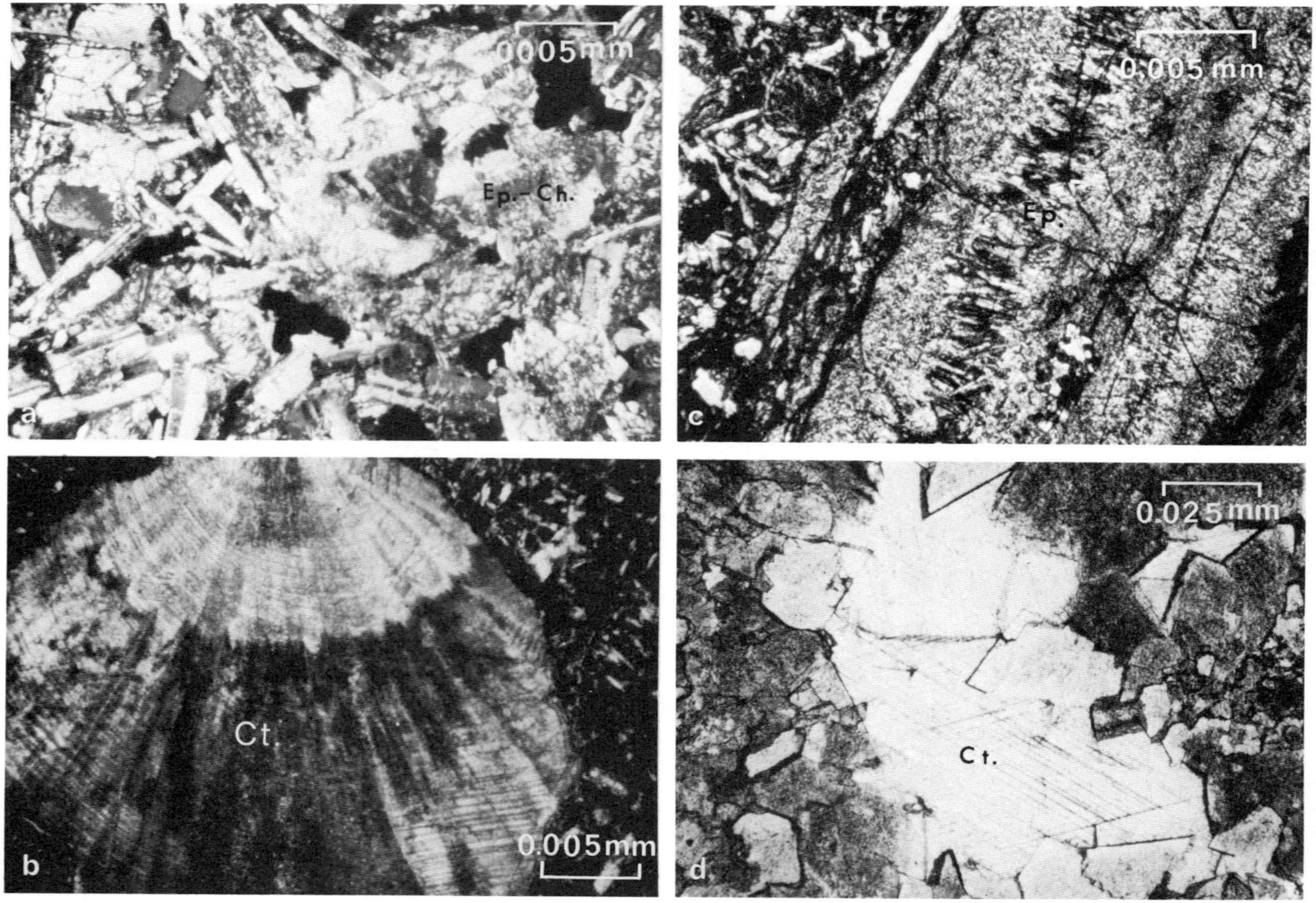

Plate II. **a** Nikolai Greenstone in partially altered state, but the volcanic texture is preserved. Chlorite (*Ch*) and epidote (*Ep*) are the principal products of alteration-metamorphism in this sample. **b** Calcite (*Ct*) filling an amygdule in Nikolai Greenstone. **c** Epidote (*Ep*)-filled fracture in

Secondary and Alteration Minerals (Plate IIa, c)

Kaolinite	Sericite	May total up to 30%
Chlorite	Prehnite	
Pumpellyite	Calcite	
Quartz	Epidote	
Hematite	Mica	
Laumontite		

Amygdule Minerals (Plate IIb)

Chlorite	Natrolite
Calcite	Thomsonite
Quartz	Native copper

The average chemical composition of the Nikolai rocks is given in Table 1. The rocks are largely quartz/hypersthene normative, but do contain normative olivine/nepheline. Major and minor element chemistry of less altered samples shows a relatively high degree of compositional uniformity. Their chemical relation in terms of $Al_2O_3 - MgO - FeO$ show ocean island-continental transitional tectonic affinities.

Chitistone Limestone

The Chitistone Limestone generally forms thick beds and overlies the Nikolai Greenstone as a cohesive structurally conformable unit. At Nelson (Fig. 4), the

Table 1. Average chemical compositions of Nikolai Greenstone, McCarthy Quadrangle, Alaska

Oxides	Wt%					
	Radovan[a] Prospect	Peavine[a] Prospect	Mountain[b] Grill Prospect	Nelson[a] Prospect	MacKevett and Richter (1974)	Muller et al. (1974)
SiO_2	51.12	49.05	54.78	50.28	50.30	49.30
Al_2O_3	13.22	16.10	17.49	14.27	15.23	14.46
Fe_2O_3	4.54	4.95	2.3	6.49	5.46	4.26
FeO	8.94	7.19	6.73	6.46	6.72	6.76
MgO	6.51	8.34	3.89	7.37	7.25	8.01
CaO	9.27	9.24	12.75	10.09	9.87	10.92
Na_2O	2.97	2.59	0.04	2.53	3.26	3.64
K_2O	0.17	0.49	0.22	0.40	0.50	tr
TiO_2	2.62	1.65	1.64	1.72	1.47	1.30
P_2O_5	0.27	0.16	0.13	0.18	0.17	0.10
MnO	0.22	0.17	0.09	0.15	0.19	0.18
Total	99.85	99.94	100.50	99.94	99.98	99.98
Volatiles	4.99	4.26	17.26	3.60	3.60	3.90

Analysis by K. Ramlal.

tr = trace

[a] Average of 3 analyses.

[b] Single analysis of epidotized sample.

Chitistone is overlain by Nikolai Greenstone along the Nelson thrust contact; at Peavine and Mountain Grill prospects the overturned folding and reverse faulting has resulted in the limestone wrapping around the Nikolai Greenstone (Plate I). At the Clear-Porcupine prospect, Chitistone is metamorphosed to skarn and marble near its contact with the Jurassic intrusives (MacKevett 1978). At Binocular, Peavine, and Nelson prospects, the limestone is abundantly fractured with most trending almost north-south and northeasterly. Calcite and dolomite veins, hairlike to several inches wide, commonly occur criss-crossed, parallel, and less frequently braided. Breccia and collapse features have been observed at Nelson.

The Chitistone unit is predominantly a limestone. Dolomite, dolomitic limestone, and chert are better developed in the stratigraphically higher parts. For the most part, it shows mineralogical and chemical purity within the lithofacies.

The main minerals are calcite [micrite sparry (Platte IId) fibrous, crystalline, and intracrystalline varieties] and dolomite [limpid or clear (Plate IId), fine-grained matrix, granular and coarse-grained]. Clays, quartz, pyrite, and chert constitute the minor minerals.

Comprehensive studies of the stratigraphic section at Nelson Prospect show that the unit is comprised of many microlithofacies conforming to the classification of Folk (1962), Dunham (1962), and as amplified by Wilson (1975). Lithofacies characteristics for the stratigraphically lower 125 m compare well with lithofacies characteristics of the Chitistone unit at the Donoho Peak section at Kennecott (Armstrong and MacKevett 1982) and the section at Green Butte (Armstrong et al. 1969). Of particular importance is the fact that gray-lime mudstone facies in the lower 60 m seems to be the most continuous horizon of the Chitistone and, therefore, a good marker bed in the Quadrangle.

Table 2. Chemical composition of Chitistone Limestone, McCarthy Quadrangle, Alaska

Oxides	Wt%				
	Nelson[a]			Binocular[b]	Peavine[a]
	Dolomite	Dolomitic limestone	Limestone		
SiO_2	0.62	2.07	1.16	3.0	nd
Al_2O_3	0.12	0.50	0.37	<0.10	nd
FeO	0.09	0.13	0.15	1.62	0.16
MgO	18.57	2.04	0.74	17.20	0.50
CaO	33.07	50.77	53.46	24.36	53.69
MnO	0.05	0.47	0.03	nd	0.01
H_2O total	tr	0.78	0.44	1.15	0.19
CO_2	46.38	42.85	42.86	40.42	42.49
S (ppm)	78	nd	tr	nd	nd

Analysis by K. Ramlal.
tr = trace
nd = not determined
[a] Average of 3 analyses.
[b] Single analysis.

Argillaceous limestones are prominent in the stratigraphically lower parts and these contain burrows, crinkles, pelletoids, peloids, and other sedimentary structures. The more dolomitic upper parts of this platform sequence resemble the "zebra bed" described for the stratigraphic section at Kennecott (Armstrong and MacKevett 1982). The Chitistone Limestone was deposited in subtidal to supratidal environments with possible sabkha affiliation.

The $CaCO_3/MgCO_3$ ratios (1.39 to 71.78) for Chitistone correspond to dolomites, dolomitic limestones, and limestones.

Chemical composition for the Chitistone unit is given in Table 2. Some dolomitic samples from the Nelson prospect show sulfur contents of 78 ppm.

Table 3. Geological characteristics of copper mineralization in Nikolai Greenstone, McCarthy Quadrangle, Alaska

Characteristics	Radovan Prospect	Mountain Grill Prospect	Clear-Porcupine Prospect
Lithologic	Intensely fractured; calcite, chlorite, and laumontite; common fracture fillings	Faulted, overturned folding; veins of epidote	Fractured; affected by Jurassic grandiorite intrusion
Alteration (Plate II)	Quartz, chlorite, pumpellyite, K-feldspar, mica, and hematite	Calcite, chlorite abundance appears to have "bleached" the host rock	Quartz, calcite sulfides near the Jurassic intrusive; epidote
Form and control of ore	Veins up to 45 cm wide; disseminations; limited reaction; ores confined to northeasterly fractures	Veins 5 to 10 cm wide; native Cu in epidote veins; ores localized to shear zones	Veins up to 10 cm wide; disseminations; native Cu in amygdules; ore along northeasterly fracture system
Ore mineralogy	(See Plate IIIa, b)		
Major	Digenite, chalcocite, bornite, chalcopyrite	Chalcocite, digenite, bornite	Bornite, chalcopyrite, digenite
Minor	Covellite, enargite, stibnite	Chalcopyrite, covellite	Covellite
Secondary	Malachite	Malachite	Malachite
Texture	Massive; replacement features of bornite by other phases	Massive/granular; some sulfide and host rock replacement	Massive covellite replaces bornite along fracture
Mineral associations	Bornite-digenite-chalcopyrite	Chalcopyrite-digenite-chalcocite	Bornite-chalcopyrite-digenite
Metal content of ores			
Cu (Wt%)	45 to 71	4[a]	55 to 58
Fe (Wt%)	2 to 8	nd	10 to 15
Ag (ppm)	50 to 200	<3	131 to 300

[a] Grade of disseminated ore in Nikolai Greenstone.
nd = not determined

Mineralization

The distribution of known copper occurrences in the McCarthy Quadrangle defines a northwesterly trend in conformity with the Triassic belt (Fig. 1b). Geologic characteristics of mineralization in the Nikolai Greenstone at Radovan, Mountain Grill, and Clear-Porcupine prospects are summarized in Table 3 and for those occurring in Chitistone Limestone are summarized in Table 4.

Table 4. Geologic characteristics of copper deposits associated with Chitistone Limestone, McCarthy Quadrangle, Alaska

Characteristics	Nelson Prospect	Binocular Prospect	Peavine Prospect
Lithologic	Platformal carbonates, abundant criss-cross and irregular calcite, dolomite fillings	Fractured dolomite, lime mudstone, veined	Lime mudstone, dolomite facies brecciated and solution features
Alteration	Dolomitization, minor silicification, recrystallization on fracture walls	Rusty stains from pyrite weathering	Hematite stains, cherty zones; some recrystallization
Form and control of ore	Swell and pinch reach 60 cm in width, stockwork system, only very limited bedding control	Fracture control of ore veins, disseminated, minor bedding control	Up to 30 cm wide, steep fracture fillings
Ore mineralogy	(See Plate IIIc and d)		
Major	Digenite, chalcocite, djurleite, covellite, bornite	Digenite, chalcopyrite	Chalcopyrite, digenite, pyrite
Minor	Enargite	Covellite	Covellite
Secondary	Malachite, azurite, cuprite	Malachite, azurite	Malachite, azurite
Texture	Massive, fine to coarse granular; bornite-chalcopyrite, digenite-chalcocite-djurleite exsolution; bornite replaced by digenite	Massive, fine-grained; mutual replacement features in ores	Massive, granular; replacement between principal phases; covellite alteration
Mineral associations	Bornite-chalcopyrite digenite-djurleite-chalcocite	Chalcopyrite-digenite	Chalcopyrite-digenite-covellite
Metal content of ores			
Cu (Wt%)	49 to 76	10[a]	15[a]
Fe (Wt%)	0.2 to 0.3	nd	nd
Ag (ppm)	190 to 400	14	11

[a] Grade of disseminated ore in Chitistone Limestone.
nd = not determined

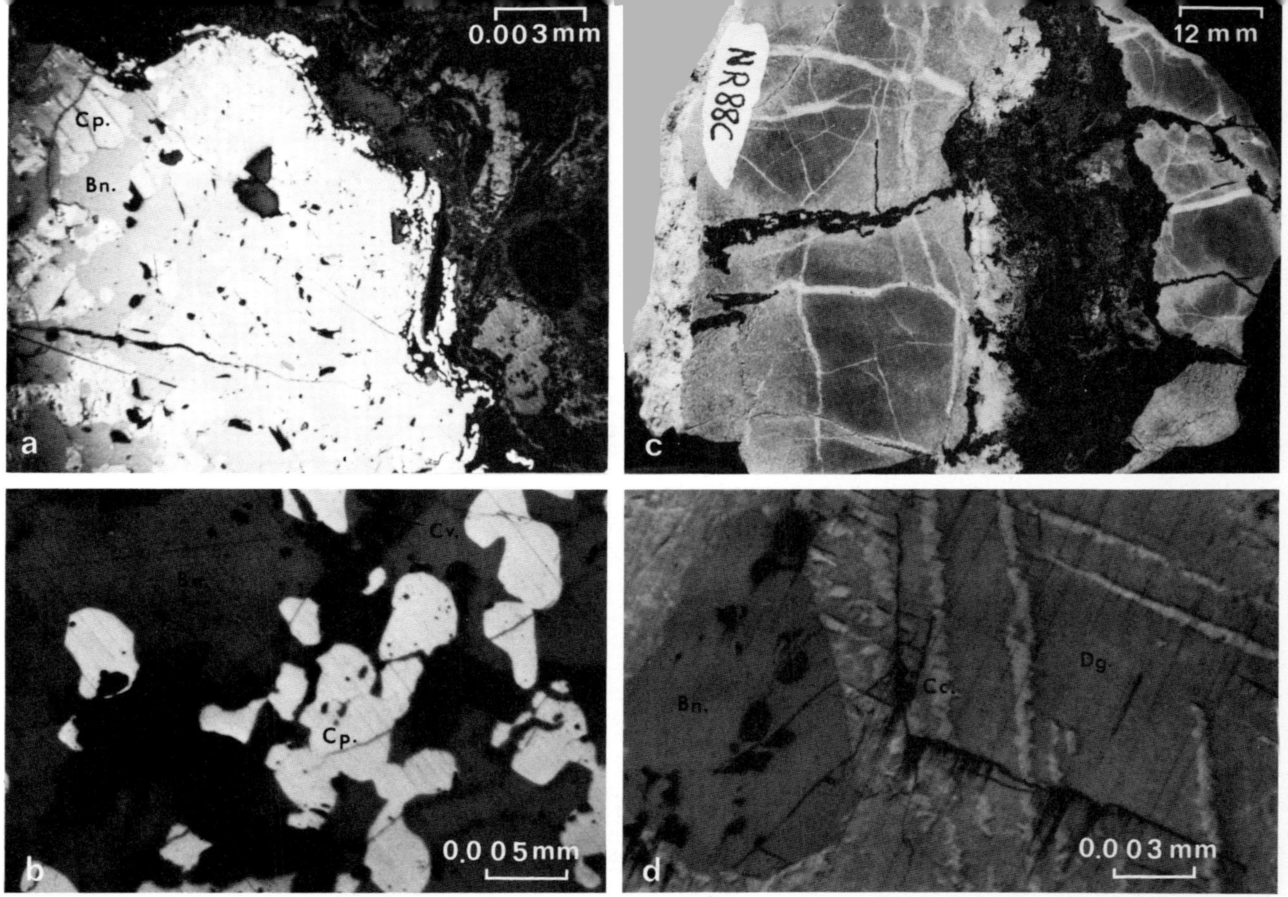

Plate III. a Bornite (*Bn*) and chalcopyrite (*Cp*) showing replacement relations in ores in Nikolai Greenstone at Radovan. Note the embayed margins with silicate vein material. **b** Bornite (*Bn*) and chalcopyrite (*Cp*) replacement relations in the sulfide ores at the Clear-Porcupine prospect. **c** Mineralization in Chitistone Limestone. Note the distinct ore vein-host rock relations. **d** Digenite (*Dg*) chalcocite (*Cc*) and bornite (*Bn*) in the ores at the Nelson prospect. Note the exsolution relations

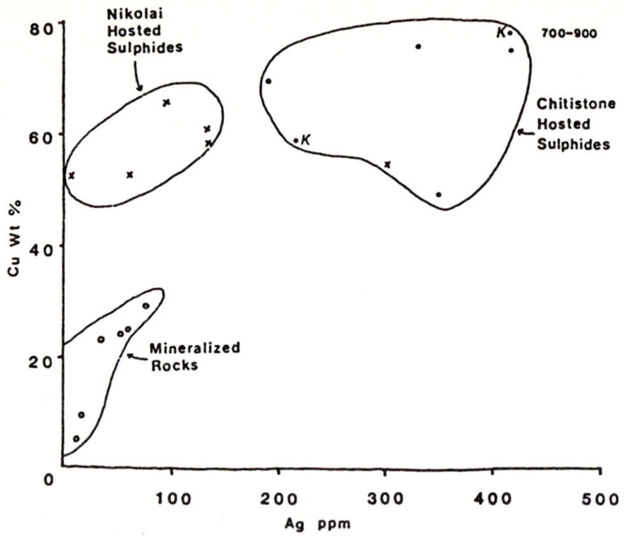

Fig. 5. Relation of copper and silver in the Kennecott-type ore deposits (Kennecott data from Bateman and McLaughlin 1920)

Their comparison brings out the following points:

1. In both the units, the structural control of mineralization predominates.
2. The fracture system is relatively continuous and prominent in the Chitistone.
3. Digenite is more abundant than chalcocite in the Kennecott-type ores.
4. Copper sulfides, digenite, and chalcocite are present in both the units, but bornite and chalcopyrite are relatively abundant in the ores in greenstone, particularly at Clear-Porcupine. (Plate IIIb)
5. There is no pervasive deformation of the ores.
6. Ore vein-host rock contacts are generally distinct, sharp, or with low degree of alteration reaction.
7. Secondary alteration minerals, malachite, and/or azurite, occur at every prospect.
8. Ore textures and structures indicate precipitation from fluids.
9. Silver content parallels copper content (Fig. 5).

Discussion

The ore-host rock system in Nikolai Greenstone is a representative of low-temperature, hydrothermal alteration-metamorphism. Rock-water interaction through hydrothermal circulation at the ocean ridges is well recognized and established for the alteration of ocean crust (Rise Project Group 1980, 1981, Mevel 1981). Nikolai Greenstone has such alteration features in common; it lacks penetrative deformation as shown by the preservation of primary textures and mineralogy in the rocks and by crystallization of secondary hydrous minerals in veins, amygdules, and as pseudomorphs of magmatic minerals.

Recently, Seyfried (1982) and Reed (1983) have given comprehensive experimental data and calculations, respectively, for the seawater-basalt reaction to

300 °C. Many minerals found in metabasalts are also produced in seawater-basalt reactions. Reed (1983) concludes that the seawater-basalt ratio greather than 30 favors the formation of chlorite (hydration in natural systems), whereas a ratio of less than 30 promotes the crystallization of epidote (dehydration in natural systems at depth). Therefore, the dehydration reactions in the lower parts of a lava pile could make significant contributions to the hydration in the upper parts. Jolly (1974) has correlated the mobilization and enrichment of copper in the Keweenawan basalts to the hydration and the precipitation of native copper-epidote to the dehydration of the basalts. Depth-dependent, dehydration-hydration reactions in volcanic piles, therefore, become important as they affect chemical and elemental redistributions. The variation in the fluid composition as it penetrates the rock is probably responsible for the diversity observed in the mineralogy of altered mafic volcanics.

Chlorite, pumpellyite, quartz, calcite, laumontite, and epidote mineralogy found in the veins is also found in altered Nikolai Greenstone. Despite their somewhat irregular distribution, such hydrous mineral assemblages form representative paragenesis of low-temperature alteration-metamorphism of mafic volcanics (Mevel 1981). Using the fluid pressure vs temperature data by Nitsch (1971) and Liou (1979) on the stability relations for the prehnite-pumpellyite (chlorite + albite) reaction, we conclude that the Nikolai assemblages formed around 250 °C and pressure of 1 to 2 kb – the pressure-temperature conditions of prehnite-pumpellyite facies metamorphism (Coombs 1960, Turner 1981).

The presence of primary bornite-chalcopyrite association and the identification of a high chalcocite phase in the ores of this study are also in conformity with temperature estimates of 200 to 300 °C.

Table 5. Selected trace metal contents in epidote-quartz veins in Nikolai Greenstones, McCarthy Quadrangle, Alaska

	(ppm)				
	Radovan Prospect[a]		Kennecott[b]		
	RMS-8	RMS-9	9	18	E3
Width of vein (cm)	10	25	10	<3	30
Ag	3	<3	2	0.5	30
Co	18	47	15	30	15
Cr	74	150	70	150	300
Cu	10830	264	5000	300	>20000
Ni	20	83	30	50	15
Pb	nd	nd	10	ND	10
Zn	36	76	10	15	10
As	<50	nd	20	ND	ND

[a] Analysis by K. Ramlal.
[b] From Silberman et al. (1981).
ND = not detected
nd = not determined

Silberman et al. (1981), on the basis of oxygen isotopic fractionation studies on the Nikolai (and the epidote-quartz mineral veins in it) estimate the temperature for hydrothermal alteration-metamorphism of Nikolai to be about 200 °C.

At Radovan, laumontite-ore intergrowth is an important indicator that alteration and ore formation, at least in Nikolai Greenstone, was a related process. Ore vein-host rock contacts, for the most part, are sharp.

In Table 5, the trace element contents of epidote veins from the Nikolai Greenstone are compared. There are obvious similarities in the trace metal contents of epidotes both at Kennecott and those of the present study. This would indicate to us that the Nikolai unit in the belt, or at least in these two places, experienced similar thermal conditions and chemical changes.

The ore-host rock system in Chitistone Limestone displays several important characteristics:

1. Ore veins are in intimate and sharp contact with the unaltered limestone and have matching walls suggesting fracture fillings by relatively cold solutions.
2. Ore vein-host rock contacts are embayed and separated by thin, calcite/dolomite walls indicating reaction and accompanying precipitation/recrystallization (Plate IIIc).
3. Stockwork arrangement of closely-spaced fracture fillings.
4. Limited or minor disseminations and bedding replacement.

Dolomitization is the principal alteration and is expressed on all scales, but commonly discordant to bedding. It is probably premineralization and of hydrothermal origin.

Model for Mineralization

Undoubtedly, the Nikolai Greenstone due to its intrinsically high copper content (MacKevett et al. 1981), is the most likely source of the metal for the Kennecott-type mineralization in the McCarthy Quadrangle. Copper is known to concentrate in magnetite (200 to 880 pm), ilmenite (90 to 300 ppm), and mafic silicate minerals (50 to 320 ppm) (Cornwall 1956, Duncan and Taylor 1968).

The following processes are considered to have been important in the concentration of copper from such minerals into Kennecott-type deposits. Hydration associated with the alteration-metamorphism of the Nikolai Greenstone rocks released copper. With progressive hydration, copper ions were mobilized and enriched in fluids in the fractures as water was fixed in the structure of the mafics.

Our studies of the mineralogy and chemistry of the veins and the Nikolai suggest that the alteration characteristics of greenstones were analogous to those described by Sinclair (1977) at White River in Yukon and by Silberman et al. (1981) at Kennecott. Sinclair (1977) envisaged that the fluids which caused the alteration and mineralization in Triassic greenstones were metamorphic waters mixed with connate waters derived from the Upper Paleozoic sedimentary rocks. Silberman et al. (1981) favors the modified connate water concept on the basis of their oxygen isotopic values of +1 per mil for vein epidote in the Nikolai Greenstone. We consider that this thermal-alteration and ore-forming event was of a regional nature and that the chemistry of the fluids that affected the greenstone of the present study were similar to those proposed by Sinclair (1977).

Dehydration-hydration reactions in mafic volcanic piles can result in pervasive mobilization, redistribution, and concentration of elements during prehnite-pumpellyite facies conditions. Development of chlorite, pumpellyite, and epidote occurs first on the fracture walls before fluids diffuse into the homogeneous interiors of the rock. The intensity and nature of fractures and fluid pressure control the extent of the alteration.

We propose that dehydration-hydration reactions in this early stage may have been responsible for the widespread, but low-grade disseminated, thin-fracture filling, and native copper mineralization in the Nikolai. Particularly important in this respect are the ore-laumontite intergrowths and native copper-epidote associations observed in the Radovan area. The relatively iron-rich (see Tables 3 and 4) character of the ore is also an important feature of this stage.

Lack of extensive alteration particularly near the veins in Chitistone Limestone suggest that the ore solution at least when they arrived there were not very hot or acidic (Sood and Wagner 1981). MacKevett et al. (1981) favor a low-temperature (100 °C) meteoric water hydrothermal circulatory system for the ore-forming process at Kennecott.

Our studies of the Clear-Porcupine prospect indicate that the Kennecott-type mineralization is younger than Jurassic as it is not affected by the Jurassic Chitina Valley batholith. On the basis of K-Ar systematics, Silberman et al. (1981) reported an age of 112 ± 11 m. y. for metamorphism of the Nikolai Greenstone, coincident with the Cretaceous accretionary event.

Structural relations are also important for timing of mineralization. The Nelson thrust fault occurs close or near to the centers of intense mineralization (Fig. 1b). It is probably related to the late phase of the Createaceous accretionary event or the later Cenozoic deformation that activated the Denali system in its eastern portion forming the younger Totschunda fork (Fig. 3). The study area is characterized by three principal fault sets, i.e., northerly, northeasterly, and easterly. The Nelson thrust at Radovan, is displaced by steep-dipping northeasterly faults, and at the Mountain Grill area by an easterly fault set. Copper mineralization is localized by such (probably Cenozoic) fracture systems. They are better developed in the Chitistone Limestone than in the Nikolai Greenstone. The expanded fracture system at this stage, further enhanced the porosity and permeability of the units, thus focusing the meteoric water circulatory system for effective and efficient solution and transport of metals. We conclude that formation of the Kennecott-type mineralization initiated with the main phase of the Cretaceous accretion, but occurred mainly in its late phase, perhaps continuing into the Tertiary.

The proximal relations of the mineralization to the Tertiary intrusives, both at Kennecott and the prospects of this study in the McCarthy B-4 Quadrangle suggests that the Tertiary igneous activity may have thermally charged and driven the system. Although data are incomplete in this respect, we have, however, noticed high potassic alteration zones (see also MacKevett and Radtke, 1966) and relatively high As and Sb contents in the ores in the limestones.

The size of the deposits may be more a function of the availability of the fracture system and cavities which are preferentially developed in limestone.

Conclusion

It is concluded that Kennecott-type mineralization in southeastern Alaska, though of a different character in the Nikolai Greenstone and the Chitistone Limestone units is the result of a continuous mineralizing epoch. It is, most likely, the last major chemical event affecting the Triassic rocks may, therefore, be Late Cretaceous to Tertiary age.

Dehydration-hydration reactions of prehnite-pumpellyite facies metamorphism in the volcanic pile contributed to the ore-solution chemistry. The upper temperature prevalent in the ore-host rock system, as indicated by mineral associations, was about 250 °C. Prominent mineralization in the Triassic belt is a function of the disposition of the fracture system and the efficiency of solution and transport by water circulatory systems thermally charged by the Tertiary igneous activity.

The control of Kennecott-type mineralization by a fundamental process of ore formation operating on a regional dimension strongly favors the presence of large mineralization centers. The alteration characteristics of the Nikolai Greenstone, particularly the MgO and K_2O contents, and the fracture analyses in the Triassic units should be utilized in exploration for additional deposits.

Acknowledgments. This study has been supported by grants (to MKS) by Geneva Pacific Corp. (now Testing Sciences, Inc.), Glenview, IL, USA. The help of C. Witt and F. Megiris with the drafting of the diagrams and of M. McGuigan for typing the manuscript is gratefully acknowledged.

References

Armstrong AK, MacKevett EM Jr (1982) Stratigraphy and diagenetic history of the lower part of the Triassic chitistone limestone, Alaska. US Geol Surv Prof Pap 1212A:A1 – A26

Armstrong AK, MacKevett EM Jr, Silberling NJ (1969) The Chitistone and Nizina limestones of part of the southern Wrangell Mountains, Alaska – a preliminary report stressing carbonate petrography and depositional environments. US Geol Surv Prof Pap 650-D:-49-D62

Bateman AM (1932) Notes on Kennecott-type copper deposits, Glacier Creek, Alaska. Econ Geol 27:297 – 306

Bateman AM, McLaughlin DH (1920) Geology of the ore deposits of Kennecott, Alaska. Econ Geol 15:1 – 80

Coombs DS (1960) Lower grade mineral facies in New Zealand. Rep Int Geol Congr, 21st Session Norden 13:339 – 351

Cornwall HR (1956) A summary of ideas on the origin of native copper deposits. Econ Geol 54:351 – 373

Csejtey B Jr (1976) Tectonic implications of Late Paleozoic volcanic arc in the Talkeetna Mountains, Southcentral Alaska. Geology 4:49 – 52

Csejtey B Jr, Nelson WH, Jones DL, Silberling NJ, Dean RM, Morris MS, Lanphere MA, Smith JG, Silberman ML (1978) Reconnaissance geologic map and geochronology, Talkeetna Mountains Quadrangle, northern part of Anchorage Quadrangle and southwest part of Healy Quadrangle, Alaska, Scale 1:250,000. US Geol Surv Open File Rep 78-558-A, 62 pp

Csejtey B Jr, Foster HL, Nokleberg WJ (1980) Cretaceous accretion of the Talkeetna Superterrane and subsequent development of the Denali fault in Southcentral Alaska. Geol Soc Am (Abstracts with Programs) 12:409

Csejtey B Jr, Cox DP, Evarts RC (1982) The Cenozoic Denali fault system and the Cretaceous accretionary development of Southern Alaska. J Geophys Res 87:3741 – 3754

Duncan AR, Taylor SR (1968) Trace element analyses of magnetites from andesitic and dacitic lavas from the Bay of Plenty, New Zealand. Contrib Mineral Petrol 20:30 – 33

Dunham RJ (1962) Classification of Carbonate Rocks according to depositional textures. In: Ham WE (ed) Classification of carbonate rocks. AAPG Mem, pp 108 – 122

Folk RL (1962) Spectral division of limestone types Symp. Classification of carbonate rocks. In: Ham WE (ed) Classification of carbonate rocks. AAPG Mem, pp 62 – 84

Jensen ML, Bateman Am (1979) Economic mineral deposits. John Wiley & Sons, New York, 593 pp

Jolly WT (1974) Behavior of Cu, Zn, and Ni during prehnite-pumpellyite rank metamorphism of Keweenawan basalts, Northern Michigan. Econ Geol 69:1118 – 1127

Jones DL, Silberling NJ (1979) Mesozoic stratigraphy, the key to tectonic analysis of Southern and Central Alaska. US Geol Surv Open File Rep 79-1200, 41 pp

Jones DL, Silberling NJ, Hillhouse JW (1977) Wrangellia – A displaced terrane in northwestern North America. Can J Earth Sci 14:2565 – 2577

Liou JG (1979) Zeolite facies metamorphism of basaltic rocks from the Taiwan ophiolite. Am Mineral 64:1 – 14

MacKevett EM Jr (1976) Mineral deposits and occurrence in McCarthy Quadrangle, Alaska. US Geol Surv Misc Field Stud Map M F 773-B Scale 1:250000, 2 sheets

MacKevett EM Jr (1978) Geologic map of McCarthy Quadrangle Alaska. US Geol Surv Misc Invest Ser Map I-1032 Scale 1:250000

MacKevett EM Jr, Radtke AS (1966) Hydrothermal alteration near the Kennecott copper mines, Wrangell Mountains area, Alaska – A preliminary report. US Geol Surv Prof Pap 550-B:B165 – B168

MacKevett EM Jr, Richter DH (1974) The Nikolai Greenstone in the Wrangell Mountains, Alaska, and nearby areas. Geol Assoc Can, Cordilleran Section, Program and Abstracts 13:14

MacKevett EM Jr, Albert NRD, Barnes DF, Case JE, Robinson K, Singer DA (1977) The Alaskan mineral resource assessment program: background information. US Geol Surv Cir 739, 23 pp

MacKevett EM Jr, Armstrong AK, Potter RW, Silberman ML (1981) Kennecott-type copper deposits, Wrangell mountains, Alaska, an update and summary. In: Silberman ML et al. (eds). Symp on Mineral Deposits of the Pacific Northwest. US Geol Surv Open File Rep 81-355:51

Mevel C (1981) Occurrence of pumpellyite in hydrothermally altered basalts from the Vema fracture zone (Mid-Atlantic Ridge). Contrib Mineral Petrol 76:386 – 393

Muller JE, Northcote KE, Carlisle D (1974) Geology and mineral deposits of Alert-Cape Scott map area, Vancouver Island, British Columbia. Geol Surv Can Pap 74-8:77 pp

Nitsch H (1971) Stabili-tats-beziehungen von prehnit-pumpellyite-haltingen paregenesen. Contrib Mineral Petrol 30:240 – 260

Reed MH (1983) Seawater-basalt reaction and the origin of Greenstones, and related ore deposits. Econ Geol 78:466 – 485

Richter DH, Matson NA (1971) Quaternary faulting in the eastern Alaskan Range. Geol Soc Am Bull 82:1529 – 1540

Rise Project Group (1980) East Pacific Rise: Hotsprings and geophysical experiments. Science 207:1421 – 1433

Rise Project Group (1981) Crustal processes of the mid-oceanic ridge. Science 213:31 – 40

Seyfried WE (1982) Hydrothermal alteration of basalt by seawater under seawater-dominated conditions. Geochim Cosmochim Acta 46:985 – 1002

Silberman ML, MacKevett EM Jr, Connor CL, Mathews A (1981) Metallogenic and tectonic significance of whole rock potassium – Argon ages of the Nikolai Greenstone, McCarthy Quadrangle, Alaska. In: Silberman ML et al. (eds). Symp on Mineral deposits of the Pacific northwest. US Geol Surv Open File Rep 81-355:53 – 73

Sinclair AJ (1977) The White River Copper deposit, Southwestern Yukon. Geol Assoc Can (Program with Abstract) 2:49

Sood MK, Wagner RJ (1981) Copper mineralization at Glacier Creek, McCarthy Quadrangle, Alaska. Geol Soc Am Abstr (Northcentral Section) 13:317

Turner FJ (1981) Metamorphic petrology: mineralogical, field and tectonic aspects, 2nd edn. McGraw-Hill, New York, 451 pp

Wilson JL (1975) Carbonate facies in geologic history. Springer, Berlin Heidelberg New York, 471 pp

Major Element Geochemistry of the Host Rocks in Some Sediment-Hosted Copper Deposits

B. MOINE[1,2], L. GUILLOUX[3], and D. AUDEOUD[2,4]

Abstract

The chemical compositions of fine-grained sedimentary rocks provide important genetic indications when their mineralogical significance is brought out. The chemical approach is essential when the primary features have been obliterated by metamorphic crystallization. Recent progress in the geochemistry of shales and marls from evaporite-bearing series is of special interest with regard to sediment-hosted copper deposits.

Two major chemical characteristics are obvious in the environment of the Ore-Shale deposits in the Zambian Copperbelt based on comparison of 378 analyses on systematically chosen samples from drill holes and cross-sections in Konkola, Chingola, Chambishi, Mindola, Rokana, and Luanshya (metamorphism from greenschists to lower amphibolite facies) with shales and marls from common platform series. Firstly, the relatively high Mg (and Li) contents are characteristic of magnesian clay minerals, in the early members of evaporitic sequences. This is an ubiquitous feature whereas anhydrite is only of local occurrence. Secondly, primary (premetamorphic) fine-grained rocks with very high K-feldspar and comparatively low quartz contents (around 63% feldspar, 10% quartz and 27% chlorite-rich clays) are inferred from extremely high K_2O concentrations (often $\geqslant$10%) and K/Al ratios in the Ore-Shale Formation. These compositions cannot be obtained by sedimentary processes only and a large part of the feldspar is probably of diagenetic/hydrothermal origin.

In Shaba (Kamoto, Kambove) the Cu and Co concentrations are carbonate-hosted. However, very uncommon chemical compositions are exhibited by the Mg-chlorite-quartz-dolomite rocks of the underlying RAT formation (32 analyses): high Mg, Li and low alkali contents. They could partly derive from felsic-volcanic glass altered by reaction with Ca- and Mg-rich and alkali-poor

[1] Laboratoire de Minéralogie, UA67 du CNRS, Université Paul Sabatier, 39 Allées Jules Guesde, 31400 Toulouse, France

[2] Laboratoire de Pétrographie, UA805 du CNRS, Université Claude Bernard, 27–43 Bd du 11 Novembre, 69622 Villeurbanne Cédex, France

[3] Département Gîtes Minéraux, Bureau de Recherches Géologiques et Minières B.P. 6009, 45060 Orléans Cédex, France

[4] Centre de Recherches sur la Géologie de l'Uranium B.P. 23, 54501 Vandoeuvre-Nancy Cédex, France

Geology and Metallogeny of Copper Deposits
(ed. by G.H. Friedrich et al.)

brines. Such brines have been observed in fluid inclusions related to U-mineralization which occurs at the top of the formation.

In contrast, the German Kupferschiefer (Mansfeld district, 22 analyses) does not show major element compositions different from those of common black shales.

Introduction

The chemical study (major and selected trace elements) of the rocks hosting stratiform Pb, Zn and Cu mineralization, in particular of the free-grained sediments and metasediments such as in the African Copperbelt and Kupferschiefer types of deposits, provides interesting information regarding metallogeny. Such quantitative descriptions of argillites, marls and other fine-grained sediments usefully complement the mineralogical observations with sedimentological interpretation in view. As illustrated by numerous studies in metamorphic domains (Shaw 1956, de La Roche 1965, 1968, 1974, Moine 1971), these chemical data are necessary for the precise identification of the sedimentary origin of metamorphic rocks in which the original mineralogy has been changed and the primary structures have been modified.

Regarding metallogeny, one might expect information on the following:

1. The sedimentary maturity of the host rocks.
2. Characteristics indicating a specific depositional environment, especially a restricted (evaporitic) environment as many authors have pointed out the frequent association of Pb, Zn and Cu stratiform deposits with evaporitic series (Renfro 1974). Several recent works (Trauth 1974, Kulke 1976, 1978, Jarousse et al. 1978 and Moine et al. 1981) have defined the chemical characteristics of clays and marls of the evaporitic series thus contributing to the identification of these series in the metamorphic domains.
3. The possible contamination by volcanogenic materials.
4. The possible diagenetic/hydrothermal alteration which can be assumed in the case of compositions unexplained by sedimentary or volcanic processes.

This approach is illustrated here by the results of the chemical study on the Copperbelt of Central Africa, i.e. the Ore-Shale of Zambia and the bottom of the Mine series of Shaba. These results are included in the work of Guilloux (1982) and Audeoud (1982) who carried out geochemical studies in cooperation with the first author (B.M.).

The Ore-Shale of Zambia

Samples and Chemical Data – Metamorphism

Many papers have been devoted to the geology of the Zambian Copperbelt in particular the composite studies by Mendelsohn (1961), Garlick and Fleischer

(1972) and Fleischer et al. (1976). The stratiform Cu mineralization is located in the Lower Roan Group (around 1 000 m.y. old) directly overlying the Kibarian (around 1 300 m.y.) gneissic and granitic basement. In the area to the southwest of the Kafue anticline (Fig. 1), Annels et al. (1983) summarized the lithological succession of the Lower Roan as follows:

1. Hangingwall: argillites and quartzites with local dolomite interbeds;
2. Ore-Shale Formation: siltstones, argillites, carbonaceous argillites and impure dolomites;
3. Footwall: footwall conglomerates, quartzite and argillite, lower conglomerate, quartzite (from top to bottom).

The mineralization is located in the Ore-Shale (Ore-Shale Alignment), then, towards the southwest, in the basal arenites. To the northeast of the Kafue dome, the Ore-Shale is no longer observed, but its probable arenite equivalent (Binda and Mulgrew 1974) contains extensive mineralizations (Mufulira).

Figure 1 shows the location of the eight series of samples taken in the drill holes and cross-sections in the Ore-Shale Alignment. Chingola and Rokana are located in the carbonaceous facies of the Ore-Shale, whereas the other deposits are in the littoral facies according to the distinctions made by Annels (1974, 1983; Fig. 1). Table 1 presents concise information on the distribution of the 378 samples between the various formations and on the respective thickness sampled. Detailed descriptions of the profiles and rocks are given by Guilloux (1982). Chemical analyses were performed on each sample for major and trace elements (including Cu, Pb, Zn and U) and an X-ray diffraction study, for the semi-quantitative determination of the mineralogical composition. The main data of the chemical study of major elements are presented in this paper. The analyses were carried out by standard techniques in the Société Nationale Elf-Aquitaine (SNEA) laboratories in Pau and some additional data were obtained at the Centre de Recherches Pétrographiques et Géochimiques in Nancy-Vandoeuvre (see footnote, Table 1).

Although terms applied to sedimentary rocks are used here (siltstones, argillites, etc.) the Zambian Copperbelt formations are distinctly metamorphic. In

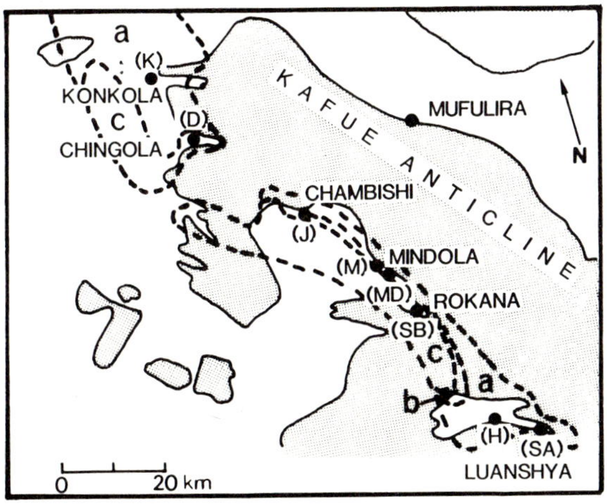

Fig. 1. The Ore-Shale Alignment, Zambian Copperbelt. The basemen is shown as *dashed. a, b* and *c*: respectively, littoral, transition and carbonaceous facies of the Ore-Shale (after Annels et al. 1983). The sampled districts are shown with the corresponding index letters in *parentheses*

Table 1. Distribution of the samples from the Ore-Shale Alignment, Zambia (see also Fig. 1)

	HW[a]	ORE	FW
Konkola (K)		22 (21.3)	15 (33)
Chingola (D)	24 (257.4)[c]	22 (18)[b]	17 (12.4)
Chambishi (J)	15 (15.4)	14 (16)	19 (54.5)
Mindola (M)	7 (36.2)	31 (29.6)	7 (5.5)
Mindola (MD)	3 (16.3)	28 (22.4)	7 (43.7)
Rokana (SB)	8 (36.3)	13 (11.9)	15 (55.6)
Luanshya (H)	10 (14.6)	22 (30.9)	7 (20.6)
Luanshya (SA)	11 (20.2)	25 (53.4)	19 (39.2)
Total	78	177	106

[a] HW, ORE and FW: Hangingwall, Ore and Footwall Formation, respectively.
[b] 22(18) means 22 samples along 18 m, etc.
[c] Plus 17(235) Upper Roan.

particular, the rocks of the Ore-Shale mainly occur as fine-grained micaschists which may contain carbonates, talc or tremolite. Drysdall et al. (1972) have shown that the grade of metamorphism increases from the biotite zone in the northwest to the epidote-actinolite zone in the southeast. According to the various mineral associations of mineral phases and some compositions of phengites in the ubiquitous quartz-phengite-biotite-K. feldspar paragenesis (experimental work by Massonne 1981 and Massonne and Schreyer 1983a), the approximate metamorphic conditions can be defined as follows: $P = 2$ kb, $T = 420°C$, $X_{CO_2} = 0.5$ at Chingola and Konkola; $P = 3$kb, $T = 460°C$, $X_{CO_2} = 0.2$ at Chambishi and Rokana; $P = 6.5$ kb, $T = 550°C$, $X_{CO_2} \leqslant 0.2$ at Luanshya. The medium to high pressure conditions observed at Luanshya are consistent with the presence of Roan metasediments with talc, kyanite and phlogopite in the Solwezi area, to the northwest of Zambia (Vrana and Barr 1972, Audeoud, unpublished observations, experimental work by Massonne and Schreyer 1983b).

Magnesian Characteristics

Table 2 presents typical chemical compositions mostly of the rocks of the Ore-Shale Formation (*n°* 3 to 11).

The Ore Shale Formation is divided, in places, into impure dolomite, ore-shale and hangingwall argillite, but these units do not seem to correspond to major geochemical and lithological variations. These divisions cannot be taken into account in a very composite study. Table 2 shows that the rocks of the Ore-

Table 2. Some typical chemical compositions from the Ore-Shale Alignment[a]

Wt%	1 K28	2 MS11[b]	3 K9	4 D55	5 D61	6 J19	7 MO8	8 M38	9 SA35	10 MS30[b]	11 MD5	12 7413	13 7339	14 FB1	15 SH
SiO_2	72.0	80.10	59.6	65.3	62.6	57.0	56.43	63.7	55.5	55.46	62.15	53.4	45.02	61.48	58.10
Al_2O_3	11.9	8.05	16.6	15.3	14.5	18.1	14.41	8.5	17.9	17.36	16.94	17.35	18.56	17.60	15.40
Fe_2O_3t	1.4	2.81	3.7	4.3	2.8	4.1	3.7	2.4	6.1	5.0[c]	2.48	4.51	6.10	3.65	6.72[d]
MgO	2.2	0.36	6.3	1.7	7.5	5.6	2.31	10.2	7.0	4.54	3.23	9.20	8.92	1.40	2.44
CaO	tr	1.13	0.5	tr	tr	0.5	4.2	3.1	0.2	1.53	0.01	1.43	1.64	0.47	3.11
Na_2O	0.2	0.50	0.2	0.1	0.2	1.0	0.72	1.3	0.4	0.34	0.41	0.12	0.66	0.22	1.30
K_2O	8.7	5.57	10.2	5.2	6.3	11.5	9.58	4.2	9.9	9.88	10.18	7.13	4.92	12.0	3.24
TiO_2	0.7	0.21	0.9	0.9	0.8	0.9	0.8	0.4	0.9	0.63	1.12	1.05	1.01		0.65
I.L	12	(1.03)	1.4	5.0	4.0	1.3	6.75	4.7	1.8	(2.89)	1.84	4.96	12.27		(8.68)
Total	98.3	99.76	99.4	97.8	98.7	100.0	98.90	98.5	99.7	97.63	98.36	99.15	99.10		99.64
CO_2		0.48								1.45				0.07	2.63
C	0.02		0.16	1.43	0.91	0.03		1.3	0.01						0.80
H_2O		0.48								0.39					5.00
S		0.07	0.1	0.04	0.04	0.01	(2.7)	0.2	0.03	1.05					0.25

[a] *1, 2*: Footwall Formation. *3* and *6* to *11*: Ore-Shale Formation. *4, 5*: Lower Banded Shale. See Table 1 for provenance. *12, 13*:: Argillites from evaporitic series (Triassic, Aquitaine Basin, France; in Jarousse 1978). *14*: Potassic shale from the Fucoïd Beds, Scotland (Bowie et al. 1966). *15*: Mean composition of the shales, after Clarke (1924). Analyses *1, 3, 4, 5, 6, 8* and *9* by SNEA (Pau). The quantity of Si, Al, Fe, Ca, K and Ti was determined by X-ray fluorescence and of Na and Mg by atomic absorption after calcination at 1050 °C and measurement of ignition loss. Total S was measured by combustion in the Leco furnace and total C(CO_2 + organic C) by combustion and coulometry. Analyses *7, 11, 12* and *13* by CRPG (Nancy) by automatic optical emission spectrometry with microwave plasma excitation.

[b] MS11, MS30: Analyses from Luanshya quoted by Mendelsohn (1961).

[c] Fe_2O_3 = 1.4%; FeO = 3.3%

[d] Fe_2O_3 = 4.02%; FeQ = 2.45%

Shale are different from the common shales (analysis 15) by their frequently high magnesium and very high potassium contents.

In sedimentary rocks, high Mg concentrations are related to carbonates (dolomite when Mg is associated with Ca and more rarely magnesite) or to magnesian clays (Mg-smectites, chlorite, corrensite, sepiolite). Various geochemical and mineralogical studies (Trauth 1974, Kulke 1976, 1978, Jarousse et al. 1978, Moine et al. 1981) showed that many evaporite series, in particular the carbonate platform series such as the Triassic of Western Europe and North Africa contain argillites and marls rich in magnesian clays and consequently showing very specific chemical compositions. For the study of metamorphic rocks, these chemical characteristics are presented in the [Al – (Na + K)] – Mg – Ca diagram of Fig. 2, a variation of the Al – Mg – Ca diagram of Moine et al. (1981) for the study of feldspar rocks. Figure 2a shows the distinct difference between the compositions of common sediments where the predominant clays are illites, montmorillonites and kaolinite associated with calcite and dolomite (+ quartz + feldspars) and those of argillites and marls of the evaporite series dominated by the association of magnesian clays, illites and dolomite with a calcic tendency towards anhydrite.

Figure 2b shows the distribution of all the samples of the Ore-Shale Alignment. They are remarkably similar to the evaporitic series. Only the Chingola samples, mainly those of the Lower-Banded Shale, equivalent here of the Ore-Shale, correspond to predominant illite sediments (analysis 4, Table 2). The location of a sample along the (Al – Na – K) – Mg line depends only upon the relative quantity of the aluminous and magnesian clays and not upon the total quantity of clays in the initial rock. Consequently, arkoses with low clay proportion, buth with high chlorite/illite ratio are grouped with the chlorite-rich argillites (for instance, rocks 1 and 3, Table 2 are projected on rather close points). The importance of magnesian clay is very clear for the Ore-Shale (Table 2, analyses 12 and 13 of evaporitic argillites in comparison with analyses 3, 5, 6, 8 and 9), and less so for most rocks of arenaceous origin. This evaporitic characteristic is supported by the high Li concentrations (up to 600 ppm) noted in some samples. At Mufulira, some of the analyses given by Malan (1964) of the argillite associated with the stromatolite layer of inter *B/C* deposits reveal similar Mg-rich compositions.

It has been accepted for many years that the sediments of the Lower Roan were deposited in an evaporitic environment. Anhydrite was observed in Mindola, Chambishi, Mufulira, etc. and Annels (1974) reported the formation of sulphides from anhydrite. Other minerals such as magnesian tourmaline (X_{Mg} = 0.63 and 0.74 in two analyses made by L. Guilloux) and scapolite also provide evidence of an evaporitic heritage in metamorphic rocks; the high quantity of albite in certain districts (Mindola, Mulfulira) may have the same significance. The identification in the overall Ore-Shale Alignment of the magnesian clay heritage, reveals the very general character of these evaporitic trends. These aluminous magnesian clays probably were not formed by direct precipitation, but rather by diagenetic transformation of illites in the basin.

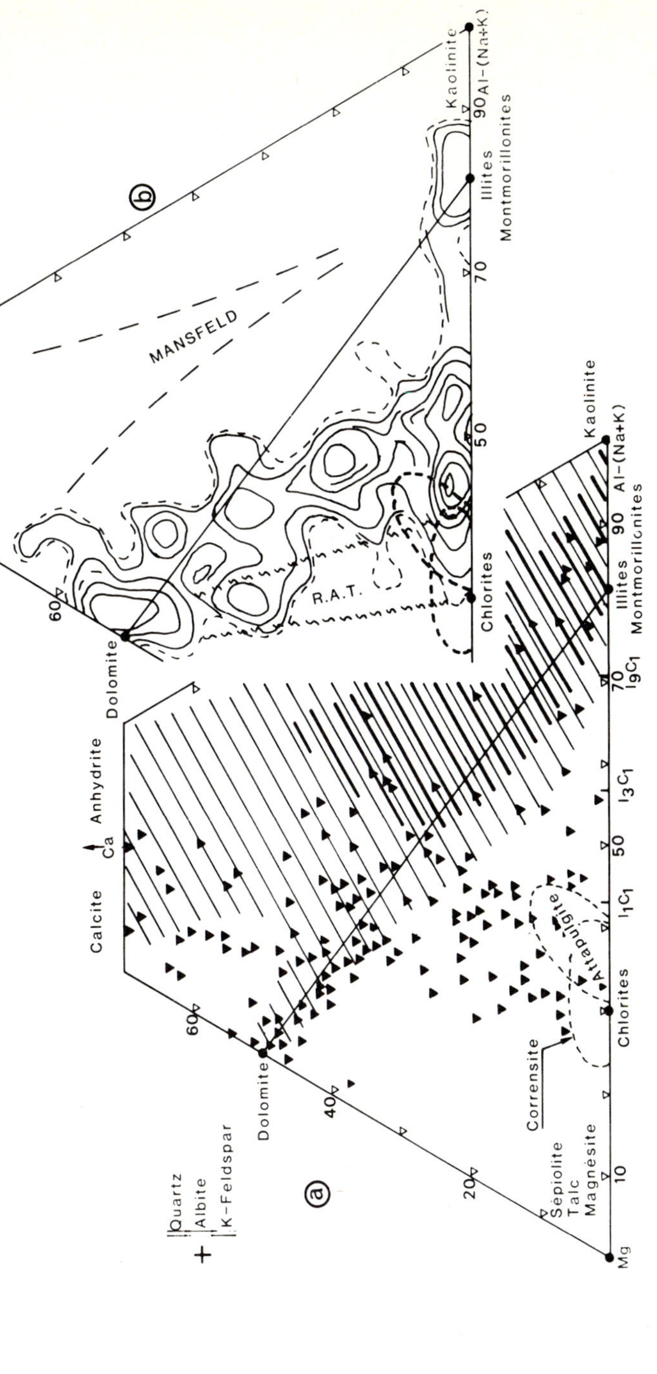

Fig. 2a, b. (Al − (Na + K)) − Mg − Ca (milliatoms) for the identification of the (magnesian) evaporitic trend. **a** Evaporite argillites, marls and carbonates (*black triangles*; data from Jarousse, 1978) compared with common sedimentary rocks (*hatched*; compilation by CRPG). Clay minerals from Weaver and Pollard 1973. I_1C_1, I_3C_1 and I_9C_1: illite/chlorite (mol) = 1/1, 3/1 and 9/1, respectively. **b** Ore-Shale Alignment of Zambia (density curves), RAT Formation from Shaba and Mansfeld district

Potassic Characteristics

Table 2 shows that the K_2O concentration of the Ore-Shale rocks frequently exceed 10% which is quite unusual for rocks termed "argillites" and "siltstones" (Pettijohn 1957). However, the Na_2O concentrations are low except in some albite-enriched rocks.

The Si/3 Al – (Na + K)/Al diagram (Fig. 3) exhibits a distinct grouping of the samples apart from the common sedimentary compositions. The initial sedimentary rocks were quartz poor and very rich in K-feldspar. Only the Lower-Banded Shale of Chingola is projected on the shale domain. The arenites (mostly the Footwall Formation) contain feldspar-rich arkose compositions.

In the K/(Al – Na) – FM/(Al – Na) diagram (Fig. 4), quartz, carbonates (except magnesite) and albite exert no influence and the rock distribution is examined in comparison with the relative proportions of the various clays and K-feldspars. The rocks studied here also show unusual compositions with a dense grouping near the chlorite-K-feldspar line and only the Chingola rocks exhibit predominantly illite normal compositions. Some samples are shifted towards high FM/(Al – Na) values, probably because of magnesite. The arkosic rocks of the Footwall Formation are grouped near the K-feldspar pole. However, to our knowledge, the K-feldspar and magnesian chlorite association defined by the original Ore-Shale rocks has no equivalent either in the common sediments or in the evaporitic series. The argillites and marls of the evaporitic series often contain, with magnesian clays, illites very rich in K and small quantities of authigenic K-feldspar (Moine et al. 1981), but the K/(Al – Na) values seldom exceed 0.5. The origin of the very particular composition of the Ore-Shale poses a difficult problem requiring a more thorough discussion.

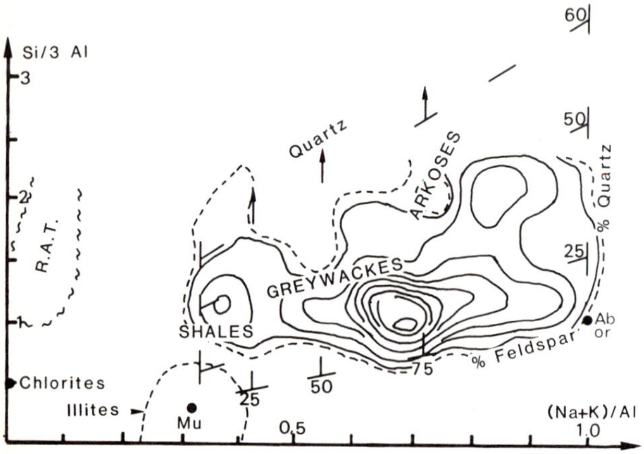

Fig. 3. Si/3 Al vs (Na + K)/Al (milliatoms) exhibiting the low quartz content and the high K-feldspar content of the precursors of the Ore-Shale compared with common sedimentary rocks (cf. legend Fig. 2b)

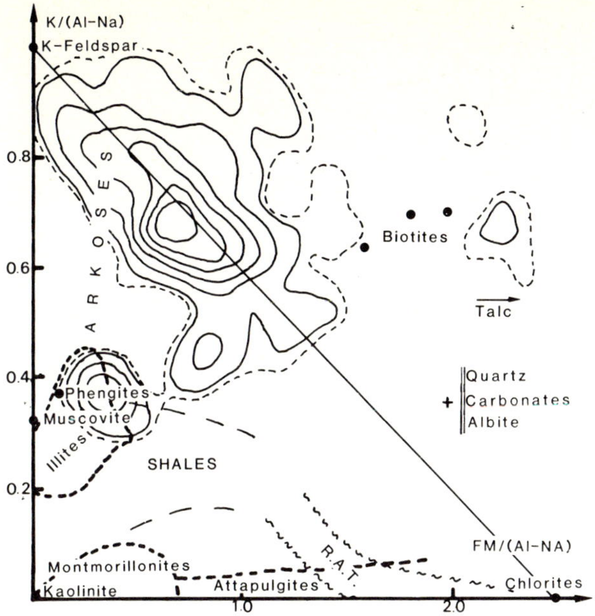

Fig. 4. K/(Al − Na) vs FM/(Al − Na) (milliatoms) showing the unusual K-feldspar-chlorite association in the precursors of the Ore-Shale. *FM*: Fe − 0.5 S + Mg + Ca − C so that FM = 0 for pyrite, calcite and dolomite (cf. legend Fig. 2b)

Origin of the Shale

The pre-metamorphic mineralogy of the Ore-Shale may be determined by distributing the elements between the following mineral phases: quartz, albite, K-feldspar, illite, chlorite and dolomite. The average composition given by Weaver and Pollard (1973) was taken for the illite and the $(Mg, Fe)_5Al_2Si_3O_{10}(OH)_8$ formula for chlorite. Table 3 shows the calculated results for some rocks. Very high K-feldspar contents, low quartz and clay contents with high chlorite/illite ratios are noted. Only the Chingola rock (D55) has a mineralogical composition similar to that of the average shales (SH)

The relative proportions of quartz, feldspars and clays in the Konkola, Chambishi and Chingola Lower-Banded Shales samples are presented in Fig. 5. The position of samples (numbers) of Table 2, the arkose domain and the position of the greywackes and the shales are also plotted in this diagram. The rocks of Chambishi and Konkola Ore-Shale fall in a specific domain away from the common sediments. A high concentration of the samples may be noted around an average point corresponding to 10% quartz, 63% feldspar and 27% clay. Although these rocks are not argillites, the term is used because of the presence of fine-grained metamorphic rocks richer in phyllites (micas) and quartz than the initial rocks.

No intense modifications of the chemical compositions took place during regional metamorphism. With certain exceptions, the metamorphic recrystalliza-

Table 3. Calculated mineralogical composition of the precursors of some Ore-Shale micaschists[a]

	D55	H29	MD5	K9	SH
Illite	47.6	31.4	17.3	6.7	45.0
Chlorite	4.5	20.4	10.3	21.8	11.7
Albite	0.9	0.9	3.8	1.9	13.8
K-feldspar	12.8	30.5	55.6	59.2	2.0
Quartz	34.2	16.8	13.0	10.4	27.5
Clay min.	52.1	51.8	27.6	28.5	56.7
Feldspar	13.7	31.4	59.4	61.1	15.8

[a] These samples are quoted in Table 2 except H29, a sample from Luanshya. Illite to quartz = 100; quartz to feldspar = 100.

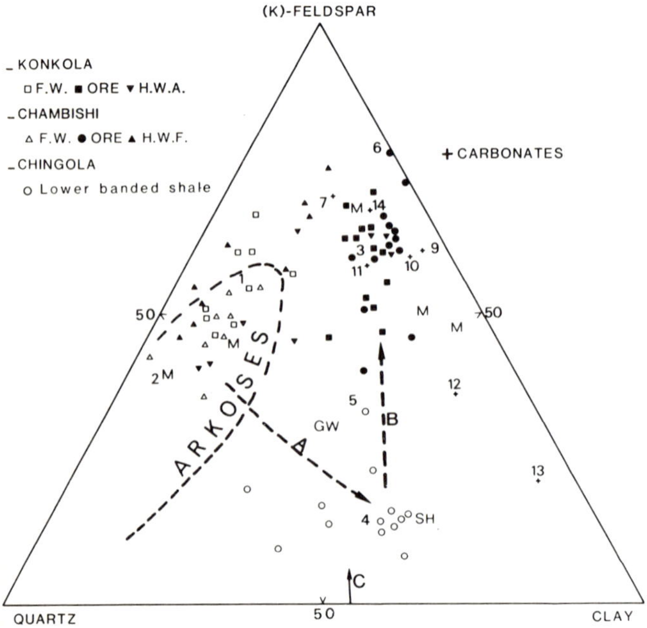

Fig. 5. Calculated proportions of quartz, clay minerals (illite + chlorite) and feldspar in the precursors of the environment of the mineralization in some Ore-Shale districts, Zambia. Abbreviations: *FW*: Footwall Formation; *ORE*: Ore Formation; *HWA*: Hangingwall Argillite; *HWF*: Hangingwall Formation. *GW, SH*: Mean composition of graywackes and shales, respectively. *M*: chemical analyses from Mendelsohn 1961. *1* to *14*: analyses from Table 2. *A, B, C*: see text

tions are mainly isochemical owing to the low porosity limiting the fluid quantity and preventing their circulation. The characteristics described here are by no means related to the metamorphic grade.

The Footwall Formation arenites are arkoses and conglomerates unusually rich in K-feldspar (Fig. 5), and one may suggest that the K-feldspar of the Ore-Shale is mainly detrital. But a sedimentary process has to be envisaged producing a considerable K-feldspar enrichment in the fine-grained fraction with a feldspar/quartz ratio increasing from 1 to 6. This would be a trend contrary to that generally occurring and difficult to explain. An argument invalidating this hypothesis is given in the $TiO_2-Al_2O_3$ diagram (Fig. 6). In the argillaceous and marly series a correlation is observed between TiO_2 which is mainly contained in the clay minerals and Al_2O_3 which measures the quantity of these minerals mixed with varied quantities of quartz and carbonates. The trend determined by many samples of the French Pyrenees Paleozoïc is given as an example (study in progress by B. Moine and F. Saupé, CRPG, Nancy). The rocks of the Ore-Shale are located on the argillaceous trend and, for these elements, no considerable difference is observed between the Chingola rocks of known argillaceous origin and the Ore-Shale of Konkola and Chambishi (and other districts) whose origin is yet to be determined. The rocks of the Footwall Formation show a lower TiO_2/Al_2O_3 ratio, and the presence of 15% clay and 20% feldspar instead of 35% quartz could not shift them in the Ore-Shale domain.

The above observations also do not corroborate the hypothesis of a contamination by acidic volcanic material (cinerites) whose composition may be compared to that of the Tuff Marker Beds, very rich in K-feldspar, describe by Croxford (1964) in the Mount Isa deposit (Fig. 6). Moreover, no evidence of volcanic activity was noted in the Zambian Lower Roan.

If the Ore-Shale is of argillaceous origin, its high K-feldspar content has therefore been acquired after the sedimentation by reaction with solutions in equilibrium with this mineral. During this transformation, Ti and Al may have remained inert which is commonly the rule in metasomatic processes (Moine et al. 1982). Illite, the dominant mineral of the initial shales of Chingola may have reacted according to the following equilibrium:

$$\underset{\text{(muscovite for illite)}}{KAl_2Si_3AlO_{10}(OH)_2} + \underset{\text{(quartz)}}{6\,SiO_2} + 2\,K^+ = \underset{\text{(K-feldspar)}}{3\,KAlSi_3O_8} + 2\,H^+$$

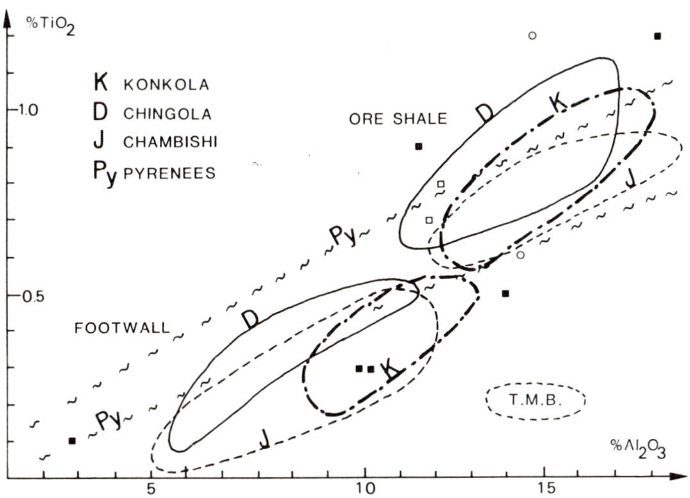

Fig. 6. TiO_2 vs Al_2O_3 (wt%) for the Ore-Shale districts shown in Fig. 5. *TMB*: Tuff Marker Beds from Mount Isa (Croxford 1964) (cf. legend of Fig. 5)

This reaction consumes quartz and makes the mineralogical composition of the argillaceous rocks evolve according to the *B* trend, parallel to *C* in Fig. 5. Located in space between sedimentation and metamorphism, this transformation would be diagenetic in the broad sense of the word. It may occur during early or during burial diagenesis. The temperature is not determined, but hydrothermal conditions are more consistent with the intensity of the transformations undergone than the low temperature conditions. The formation of the magnesian phyllites discussed above is also diagenetic. As this phenomenon is widely developed in the evaporitic series it is likely of early diagenesis and could not be contemporaneous with feldspathization.

Intense feldspathization of argillaceous rocks has seldom been described. Moine (1974) has used the example of the Fucoïd Beds of Scotland to explain the origin of the feldspar-rich micaschists observed in the carbonate platform series of the Upper Proterozoic of Central Madagascar. In the Fucoïd Beds (Cambrian) adularia-rich shales and potassic marls (analysis 14, Table 2) are intercalated between quartzites and dolostones. Discussions on potassic characteristics have permitted to compare the hypotheses of a volcanic or diagenetic origin. A sound argument in favour of the diagenetic hypothesis was furnished by the determination of the crystallization age of the adularia (K/Ar datation) at around 100 m.y. after that of the sedimentation (Bowie et al. 1966, 1967). It is noteworthy that the two above examples are situated like the Zambian Ore-Shale, in restricted environments.

Finally, the chemical study of the Ore-Shale shows that these rocks cannot be considered like ordinary, fine-grained, detrital sediments affected by regional metamorphism. We agree on this point with Darnley's conclusions (1960) to which specialists have not taken much notice. Without entering upon the metallogenic problems, our conclusion is in agreement with Annels' work (1974) emphasizing the role of diagenesis in the formation of some sulphides.

In contrast, Guilloux (1982) has shown the Mansfeld Kupferschiefer to be shales and calcareous-dolomitic marls with a mainly illitic argillaceous fraction. Its relationship with the evaporitic conditions (some samples contain anhydrite) does not particularly affect the geochemistry of the silicates (cf. domain of Fig. 2b) by comparison with the common platform series. Also the Kupferschiefer is much more rich in organic carbon and, as underlined by Wedepohl (1971), the Kupferschiefer of Germany and Poland is characterized not only by high Cu grades, but also by high Zn and Pb grades exceeding 0.5%. This is not the case in the Zambian Ore-Shale where Pb and Zn concentrations around 50 to 100 ppm are observed.

The Shaban "RAT"

In Shaba, the RAT (argillaceous, talc-bearing rocks) forms the bottom of the Roan (*R*) series. Francois (1974) distinguishes from bottom to top:

1. RAT Group (R_1) formed of "micro-sandstones or hematite-bearing chlorite pelites, lilac to purplish red" or red RAT. Its bottom is unknown. In Kolwezi, R_1 is identified over 235 m;

2. Mines Group (R_2) containing the copper and cobalt mineralization. It is composed of two dolomitic formations enclosing a schistose-sandstone formation. At the bottom of the first dolomitic formation, layer $R_{2.1.1.}$ is composed of rocks similar to R_1, but grey-coloured (grey RAT) where iron occurs in the form of pyrite. The grey RAT are the host rocks of uranium mineralization (Shinkolobwe, Kamoto, Kambove).

The separation between grey and red RAT is based on the presence of a sedimentary conglomerate between the two formations. However, this separation is not constant. Cailteux (1977, 1983) shows that a continuous transition is often observed (like in Kambove) with a variation of the deposition environment, oxidizing (red RAT), then reducing (grey RAT).

The present study is concerned with the upper part of the RAT: 32 samples from 42 m of red RAT and 3 – 5.6 m of grey RAT at Kamoko and Kambove-west (see footnote, Table 4). These rocks mainly consist of chlorite, quartz and dolomite. Grains of scaled or splintered quartz (Cailteux 1977) or with corrosion pits (Audeoud 1982) are of probable volcanic origin. Chlorite forms a commonly predominant matrix that is mainly sheridanite with X_{Mg} around 1 (microprobe analyses by D. Audeoud).

The chemical compositions are also very homogeneous and their variations correspond to those of the quartz + chlorite and dolomite proportions (Figs. 2 and 3, Table 4). The three analyses presented in Table 4 clearly show the dominant characteristics of the RAT. These are highly magnesian (chlorite and dolomite), iron-poor and extremely alkali-poor rocks. The Li contents in the most chloritic material are high (several hundred ppm). A cobalt anomaly (a few

Table 4. Some typical chemical composition of the "red RAT" from Kamoto (1) and Kambove (2, 3)

Wt%	1[a] RT16	2 KB52	3 KB50	4[b] 3535	5[c] AT1
SiO_2	62.1	45.9	28.10	45.44	54.86
Al_2O_3	11.0	15.7	6.02	15.32	7.56
Fe_2O_3t	5.9	1.6	0.52	5.45	3.69
MgO	13.5	20.8	20.99	14.19	10.22
CaO	0.2	0.9	14.81	0.34	0.62
Na_2O	tr	tr	0.04	1.04	0.03
K_2O	tr	tr	0.04	4.43	0.53
TiO_2	0.6	1.0	0.46	0.60	0.43
P_2O_5	0.09	0.05	0.17	–	0.44
I.L	6.5	13.0	27.67	11.70	20.1
Total	99.89	98.95	98.82	98.55	98.48
Li	390	666	157	320	
Co	132	659	317	24	63

[a] Analyses 1 and 2 by SNEA, the others by CRPG (see Table 2). The sample RT16 is from the RT drill hole (N° 1328/Gécamines), at 38 m under the grey/red RAT limit. The samples KB50 and KB52 were taken from a cross-section in a gallery at Kambove-West, at respectively 0.1 and 5.7 m under the same limit.

[b] Argillite from evaporite series Triassic, Aquitaine Basin, France, in Jarousse (1978).

[c] Attagpulgite argillite (Eocene phosphate-bearing series, Senegal; sample M. Seyler).

hundred ppm) is constant. Cailteux (1983) has published 15 new chemical analyses of the RAT from Kambove. The two sets of data are very similar.

The highly magnesian character and prominence of Li are quite consistent with our observations in the evaporitic series (Moine et al. 1981). Figure 2a and b clearly shows that the RAT are located in the argillite and dolomitic marl domain of the evaporitic series. However, a major problem is the nearly total lack of alkalies contrasting with the abundance of K and the relatively high K/Al values in the untransformed evaporite series (cf. analyse 4, Table 4). Various interpretations may be proposed.

In an isochemical hypothesis, clay and attapulgite marls as those identified in the Cenozoïc cover of the West African Shield (analysis 5, Table 4) may be considered as the initial sediment. Several observations contradict this interpretation. The attapulgite (with sepiolite) is mainly associated with calcite in marls thus showing a trend different from that observed in Fig. 2b. These rocks contain low, but appreciable K_2O contents (analyses 5, Table 4), whereas most RAT show no K_2O at all.

Above all, attapulgites are definitely richer in SiO_2 than chlorites. Their metamorphic recrystallization should lead to the formation of chlorite + quartz rocks. But the Si – Al – (Mg + Fe) diagram of Fig. 7 shows that the RAT, in spite of the contribution of detrital quartz, are not as siliceous as attapulgites and palygorskites. The argillaceous fractions which originated the RAT should be less siliceous than these rocks and should necessarily be plotted near the chlorites (Fig. 7).

The presence of few mineral phases (here, mainly three) and very low concentrations of mobile elements such as the alkalies are often the signature of exchanges between fluids and minerals (metasomatic alteration in the broad sense of the word). This confirms the two hypotheses put forward to explain the formation of the RAT: chloritization of detrital illite (Katekesha 1975) and chloritization of fine-grained pyroclastic material such as cinerites (Cailteux 1977, 1983).

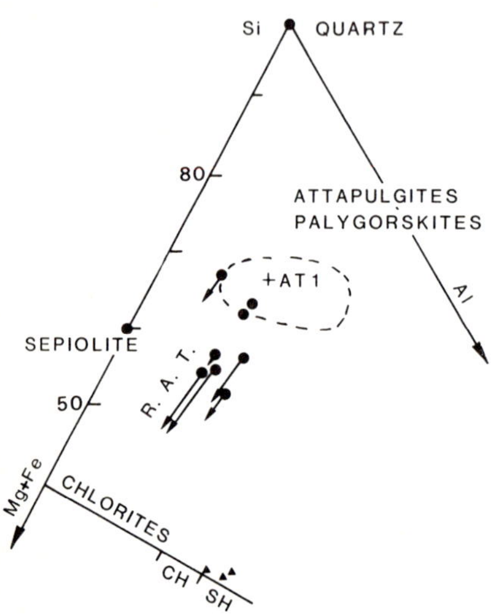

Fig. 7. Si – Al – (Mg + Fe) (milliatoms) showing the discrimination between the RAT from Shaba and attapulgites-palygorskite compositions (*circled* and *ATl*, Table 4). Each rock from the RAT is plotted as a *large dot* (assuming primary dolomite) and also as an *arrowhead* (assuming primary calcite). *Triangles*: some chlorite compositions

The chloritization of white micas may form chloritites in hydrothermal conditions. However, this transformation should be evidenced by structures, such as zonings intersecting the bedding. This chloritization is accompanied by a considerable increase of the volume of the mineral which may be compensated by the dissolution of some quartz (Moine et al. 1982). On the contrary, the petrographic observations show well-preserved detrital quartz. The formation of aluminous, magnesian phyllites (chlorite, corrensite) from detrital illites may occur during early diagenesis as is the case in the evaporitic environments. However, as we have already seen, this transformation seldom produces only chloritic rocks with the total leaching of potassium. As a rule, high concentrations of K are related to the presence of illites or authigenic feldspar.

Cailteux (1977, 1983) suggests that the top of the RAT is of pyroclastic origin. He bases his hypothesis in particular on the observation of pyroclastic material in certain parts of Shaba (Lefebvre and Cailteux 1975). The precise nature of this volcanism is yet to be determined, but the presence of quartz and low Fe, TiO_2 (usually from 0.3 to 0.6%) Cr, etc. contents suggests acidic volcanism. The chloritization of the feldspathic ashes may have taken place in a sedimentary environment (analogy with the "Tonstein") or under hydrothermal conditions during burial diagenesis. In this case, the moderate volume increase (13% for K-feldspar) may have been compensated by the partial dissolution of the quartz. The RAT have low SiO_2 contents, usually lower than 60%. The SiO_2 contents may have been fairly higher (up to 70%) in the initial quartz-feldspathic material.

Audeoud (1982) presents the results of the quantitative analyses of REE by neutron activation in six RAT samples. The spectra are varied, contrasting with the rock mineralogical and chemical homogeneity. It seems that mobilization may have taken place between different layers and that REE may have been trapped in the monazite. The nearly total absence of an europium anomaly does not confirm the feldspathic heritage (plagioclase). The spectra which appear to be the less disturbed are rather identical to those of the shales. However, no clear conclusion can be definitely reached on the origin of the RAT.

Whatever the precise origin of the RAT – argillaceous or volcano-sedimentary – their mineralogical and chemical compositions have probably been achieved in a restricted environment. This is in agreement with petrographic observations such as dolomitic "patches" which are explained by the transformation of anhydrite nodules (Cailteux 1977, 1983).

Audeoud (1982) carried out a study by microthermometry and by Raman spectroscopy with the MOLE microprobe of the fluid inclusions contained in the dolomitic veinlets of the RAT and the identical inclusions associated with the anchimetamorphic recrystallization. They are mainly aqueous with very high salinities i.e. more than 60% of dissolved salts (Pirmolin 1970, has also shown high salinities in the Kamoto dolomite). These salinities are mainly due to $MgCl_2$ and $CaCl_2$, whereas the NaCl and KCl concentrations are comparatively low. The origin of these brines probably lies in the series itself. They are quite consistent with the RAT composition characterized by chlorite, dolomite and the absence of alkali minerals. The metallogenic interest of these observations need not to be emphasized.

Conclusions

The approach in this study has been to define the primary mineralogical composition of metasediments from their chemical composition, thus revealing two main features of the African Copperbelt:

1. The rocks which host the mineralization were deposited in an early evaporitic environment of great extent whose signature is former magnesian clays.
2. Characteristics related to the alkali: high amounts of K in the Zambian Ore-Shale, and in contrast, very low K and Na concentration in the Shaban RAT, define original mineralogical compositions which do not correspond to sedimentary or volcano-sedimentary material. They are very likely the signature of pronounced diagenetic or hydrothermal transformations. The role played by the reaction between solutions and minerals is supported in the RAT by the observation of fluid inclusions whose composition is consistent with the mineralogical characteristics.

The role of the sedimentation environment and the post-sedimentary process in the formation of the silicates is thus demonstrated and may contribute towards a better understanding of the formation conditions of the associated sulphide minerals. With this in mind, further studies should be carried out to determine the precise relations between the sulphide content and the geochemistry of metasediments at various levels of observation. Variations of the chemical characteristics in unmineralized zones should be more thoroughly investigated than they have been so far. The major variations observed in the compositions of the Shaban RAT and formations of the same age in Zambia have to be explained. They may be due partly to paleogeographic factors and partly to the nature of the materials which underwent diagenetic transformation (for example, the role of volcanic material in Shaba). These lithologies are likewise quite distinct from the Kupferschiefer of northern Europe.

Finally, this study strengthens the view that the African Copperbelt deposits can no longer be regarded as strictly syngenetic sedimentary deposits. In spite of their apparent control by sedimentary structures, they are most likely related to diagenetic or hydrothermal processes involving evaporitic brines and an unknown source of metals.

Acknowledgements. Full facilities for sampling were given to L. Guilloux by the mining companies: Nchanga consolidated Copper Mines Limited and Roan Consolidated Mines in Zambian; Gécamines in Zaïre; Kombinat of Sangerhausen in G.D.R. The study by L. Guilloux has received financial support by the Délégation Générale à la Recherche Scientifique et Technique and by the Bureau de Recherches Géologiques et Minières and technical support by the Société Nationale Elf-Aquitaine and by the Centre de Recherches Pétrographiques et Géochimiques, Nancy. The study by D. Audeoud was supported by the Centre de Recherches sur la Géologie de l'Uranium, Nancy. The manuscript was prepared with the assistance of the Laboratoire de Minéralogie, Toulouse. Mrs. J. Guilloux is thanked for the translation of the text into English.

References

Annels AE (1974) Some aspects of the stratiform ore deposits of the Zambian copperbelt and their genetic significance. In: Bartholomé P (ed) Gisements stratiformes et provinces cuprifères. Soc Geol Belg, Liège, pp 235–254

Annels AE, Vaughan DJ, Craig JR (1983) Conditions of ore mineral formation in certain Zambian Copperbelt deposits with special reference to the role of cobalt. Miner Depos 18:71–88

Audeoud D (1982) Les minéralisations uranifères et leur environnement à Kamoto, Kambove et Shinkolobwe (Shaba, Zaïre). Petrographie, géochimie et inclusions fluides. These 3eme Cycle, Lyon, 211 pp

Bartholomé P (1974) On the diagenetic formation of ores in sedimentary beds with special reference to Kamoto, Shaba, Zaïre. In: Bartholomé P (ed) Gisements stratiformes et provinces cuprifères. Liège, Soc Géol Belg, Liège, pp 203–213

Binda PL, Mulgrew JR (1974) Stratigraphy of copper occurrences in the Zambian Copperbelt. In: Bartholomé P (ed) Gisements stratiformes et provinces cuprifères. Soc Géol Belg, Liège, pp 215–233

Bowie SHU, Dawson J, Gallagher MJ, Ostle D (1966) Potassium-rich sediments in the Cambrian of Northwest Scotland. Trans Inst Ming Metal GB 75, 714:125–145

Bowie SHU, Dawson J, Gallagher MJ, Ostle D (1967) Potassium-rich sediments in the Cambrian of Northwest Scotland. Report of discussion at Nov. 66, general meeting, contributed remarks and author's reply. Trans Inst Ming Metal GB 76, 723:60–69

Cailteux J (1977) Particularités stratigraphiques et pétrographiques du faisceau inférieur du Groupe des Mines au centre de l'arc cuprifère shabien. Ann Soc Geol Belg 100:55–71

Cailteux J (1983) Le "Roan" shabien dans la région de Kambove (Shaba, Zaïre). Etude sédimentologique et métallogénique. These, Univ Liège

Croxford NJW (1964) Origin and significance of volcanic potash-rich rocks from Mount Isa. Trans Inst Ming Metal GB 74:33–43

Darnley AG (1960) Petrology of some Rhodesian Copperbelt orebodies and associated rocks. Trans Inst Ming Metal GB 69:137–173, 371–398, 540–569

Drysdall AR, Johnson RL, Moore TA, Thieme JG (1972) Outline of the geology of Zambia. Geol Mijnbouw 51:265–276

Fleischer VD, Garlick WC, Haldane R (1976) Geology of the Zambian Copperbelt. Handbook of stratabound and stratiform ore deposits. In: Wolf KH (ed) Elsevier, New York, 6:223–352

Francois A (1974) Stratigraphie, tectonique et minéralisations dans l'Arc cuprifère du Shaba (République du Zaïre). In: Bartholomé P (ed) Gisements stratiformes et provinces cuprifères. Soc Geol Belg, Liège, pp 79–101

Garlick WG, Fleischer VD (1972) Sedimentary environment of Zambian copper deposition. Zambia issue. Geol Mijnbouw, 51:277–298

Guilloux L (1982) Etude chimique des séries porteuses de quelques grands gisements de type Kupferschiefer. Conséquences métallogéniques. Thèse d'Etat Lyon. Sciences de la Terre, Nancy, Mém 43

Guilloux L, Pellisonnier H (1974) Les gisements de schistes, marnes et grès cuprifères. In: Gisements stratiformes et provinces cuprifères. Soc Geol Belg, Liège, pp 35–53

Hewitt DA (1973) Stability of the assemblage muscovite-calcite-quartz. Am Miner 58:785–791

Hoschek G (1973) Die Reaktion Phlogopit + Calcit + Quartz = Tremolit + Kalifeldspat + H_2O + CO_2. Contrib Miner Petrol 39:231–237

Hoschek G (1980) Phase relations of a simplified marly rock system with application to the Western Hohe Tauern (Austria). Contrib Miner Petrol 73:53–68

Jacobs GK, Kerrick DM (1981) Devolatilization equilibria in H_2O-CO_2 and H_2O-CO_2-NaCl fluids: an experimental and thermodynamic evaluation at elevated pressures and temperatures. Am Miner 66:1135–1153

Jarousse J (1978) Contribution au problème de l'identification d'anciennes séries évaporitiques dans les ensembles métamorphiques. Approche principalement géochimique. These 3eme Cycle, Univ Claude Bernard, Lyon-I, 232 pp

Jarousse J, Moine B, Sauvan P (1978) Etude chimique de séries évaporitiques en vue de leur identification dans les ensembles métamorphiques. CR Acad Sci Ser D 286:1057–1060

Katekesha WM (1975) Conditions de formation du gisement cuprocobaltifère de Kamoto Principal (Shaba, Zaïre). Mem Doct, Fac Sci Univ Liège, 237 pp

Kerrick DM (1974) Review of metamorphic mixed-volatile (H_2O-CO_2) equilibria. Am Miner 59:729 – 762

Kulke H (1976) Die salinare Trias des Atlassystems (Nordwest Afrika). Faziesverteilung, Tektonik, Morphologie, Petrographie, Mineralogie und Geochemie, 213 S. Habil Schr Fak Geowiss, Univ Bochum (unpublished)

Kulke H (1978) Tektonik und Petrographie einer Salinarformation am Beispiel der Trias des Atlassystems (Nordwest Afrika). Geotekt Forsch, 55 I-II:1 – 58

La Roche H de (1965) Sur l'existence de plusieurs faciès géochimiques dans les schistes paléozoïques des Pyrénées Luchonnaises. Geol Rundsch Dtsch 55:274 – 301

La Roche H de (1968) Comportement géochimique différentiel de Na, K, et Al dans les formations métamorphiques et plutoniques. CR Acad Sci Ser D 267:39 – 42

La Roche H de (1974) Geochemical characters of the metamorphic domains: survival and testimony of their premetamorphic history. Sci Terre, T XIX, 2:101 – 117

Lefebvre JJ, Cailteux J (1975) Volcanisme et minéralisations diagénétiques dans le gisement de l'Etoile, Shaba, Zaïre. Ann Soc Geol Belg, Liège, 98(1):177 – 195

Malan SP (1964) Stromatolites and other algal structures at Mufulira, Northern Rhodesia. Econ Geol 59:397 – 415

Massonne HJ (1981) Phengite: eine experimentelle Untersuchung ihres Druck-Temperatur-Verhaltens im System $K_2O-MgO-Al_2O_3-SiO_2-H_2O$. Dissertation, Ruhr-Univ Bochum

Massonne HJ, Schreyer W (1983a) A new experimental phengite barometer and its application to a variscan subduction zone at the southern margin of the rhenohercynicum. 2nd E.U.G. meeting, Strasbourg, 1983. In Terra Cognita, 3:187

Massonne HJ, Schreyer W (1983b) Stability of the talc-kyanite assemblage revisited. 2nd E.U.G. meeting, Strasbourg, 1983. In Terra Cognita, 3:187

Mendelsohn F (1961) The geology of the Northern Rhodesian Copperbelt. Mac Donald, London, 523 p.

Moine B (1971) Caractères de sédimentation et de métamorphisme des séries précamrbiennes épizonales à catazonales du centre de Madagascar (région d'Ambatofinandrahana). Sci Terre Fr Mem 31:293

Moine B, Sauvan P, Jarousse J (1981) Geochemistry of evaporite-bearing series: a tentative guide for the identification of metaevaporites. Contr Mineral Petrol 76:401 – 412

Moine B, Gavoille B, Thiebault J (1982) Géochimie des transformations métasomatiques à l'origine du gisement de talc et chlorite de Trimouns (Luzenac Ariège, France). I. Mobilité des éléments et zonalité. Bull Minéral 105:62 – 75

Pirmolin J (1970) Inclusions fluides dans la dolomite du gisement stratiforme de Kamoto (Katanga Occidental). Ann Soc Geol Belg 93(2):397 – 406

Petitijohn FJ (1957) Sedimentary rocks. Harper and Row, New York, 718 p

Puhan D, Johannes W (1974) Experimentelle Untersuchung der Reaktion Dolomit + Kalifeldspat + H_2O = Phlogopit + Calcite + CO_2. Contr Miner Petrol 48:23 – 31

Renfro AR (1975) Genesis of evaporite-associated stratiform metalliferous deposits – a sabkha process. Econ Geol 69:33 – 45

Rentzsch J (1974) The Kupferschiefer in comparison with the deposits of Zambian Copperbelt. In: Gisements stratiformes et provinces cuprifères. Soc Geol Belg, Liège, pp 395 – 418

Shaw DM (1956) Geochemistry of pelitic rocks. Part III: Major elements and general geochemistry. Geol Soc Am Bull 67:919 – 934

Skippen G (1974) An experimental model for low pressure metamorphism of siliceous dolomitic marble. Am J Sci 274:487 – 509

Slaughter J, Kerrick DM, Wall VJ (1965) Experimental and thermodynamic study of equilibria in the system $CaO-MgO-SiO_2-CO_2-H_2O$. Am J Sci 275:143 – 162

Trauth N (1974) Argiles évaporitiques dans la sédimentation carbonatée continentale et épicontinentale tertiaire. Sci Geol Fr Mem 49:195 p

Vrana S, Barr MWC (1972) Talc-kyanite-quartz schist and other high-pressure assemblages from Zambia. Mineral Mag 38:837

Weaver CE, Pollard LD (1973) Developments in sedimentology. The chemistry of clay minerals. Elsevier, Amsterdam, 213 p

Wedepohl KH (1971) "Kupferschiefer" as a prototype of syngenetic sedimentary ore deposits. Soc Mining Geol Japan, Spec Issue 3:268 – 273

Zechstein Copper-Bearing Shales in Poland. Lagoonal Environments and the Sapropel Model of Genesis

C. HARAŃCZYK[1]

Abstract

The palaeogeographic situation of the Fore-Sudetic Monocline during the first Zechstein evaporative subcycle, evolution of the metalliferous sedimentation and subsequent subaerial geological processes are illustrated on a redox map, geochemical profiles and diagram. The sequence of the sedimentation of the seaward and lagoonward slopes of the lagoonal barrier explains the differences between copper- and lead-bearing shale sedimentation environments. Subaerial abrasion, redeposition, oxidation, cementation and lagoonal brine infiltration contributed to the redistribution of metals in the just-formed sediments before Werra main dolomite sedimentation closed the first evaporative subcycle.

Introduction

Kupferschiefer as a stratigraphic term has for years been used in the Mansfeld Ore District in Germany. According to Chronica Mansfeldensi written by Spangenberg in 1572, mining operation (Bergbau) started in 1199 in this ore district. However, the first geognostic modern descriptions we owe to Freiesleben (1815), a pupil of the Werner school. These were followed by papers of several generations of geologists; let us mention only Schneiderhöhn (1921), Marowsky (1969), Rentzsch and Knitzschke (1968) and Wedepohl (1971), namely, authors who introduced new methods of laboratory investigations, and created the concept of Kupferschiefer as a prototype of the sedimentary ore bed.

The discovery of new copper-rich areas on the southern margin of the Zechstein basin in Poland has greatly stimulated research on these sedimentary ore deposits. The Fore-Sudetic Monocline in Lower Silesia is now not only the largest European copper deposit, producing more than 350000 tons of copper and 700 tons of silver yearly, but it is the area of most complex and holotypical development of sediments. Therefore, it must also be regarded as Locus Classicus, the prototype of euxinic sedimentary ore beds.

Results of the recent investigations carried out here substantiate a revision of some traditional concepts concerning this prototype. This review aims to high-

[1] Institute of Geological Sciences, Jagellonian University, 2a Oleandry St., Kraków, Poland

Geology and Metallogeny of Copper Deposits
(ed. by G.H. Friedrich et al.)

light some revised aspects of the genesis of these sediments, indeed, also to explain the genesis of the richness of this unique Zechstein ore district. These aspects are: Kupferschiefer as a sediment formed under regressive conditions of the first Zechstein subcycle, the development of euxinic sediments in the lagoonal area, the enrichment processes due to subaerial exposure, and finally, the sapropelic model of genesis acting in lagoonal environment.

Early Zechstein Sea

The Zechstein Sea was formed when the northernmost rim of a chain of interlinked sub-sea level inland drainage basins was breached or overflown by a relative rise of sea level (Smith 1980), thus creating an inland sea, in places initialy more than 250 m deep. It is contended that the rim remained as a barrier near sea level and exerted a strong influence on the subsequent depositional history of the whole basin (Smith 1980). A complex Zechstein inland drainage basin the bottom of which lies well below contemporary sea level was thus immediately established, drowning existing Lower Permian red sandstone desert deposits, including aeolian sand dunes and rocky hills as well as the extensive plains of the basin floor and widespread marginal peneplains (Smith 1980). The Zechstein inland basin overlapped and in much of the area it might also be genetically related to the foreland rift basins formed after Hercynian orogenic collision and filled with Rotliegend sediments. Lorenz and Nicholls (1976) and in this year also Jowett (1984) published an excellent plate-tectonic explanation of the taphrogenic Permian evolution of Central Europe. The explanation is important also for the discussion of the provenance of metals for the metalliferous Zechstein sediments.

Palaeogeography of the Fore-Sudetic Monocline

The ingression of the Zechstein Sea came from the northwest along the line Głogów-Polkowice (Fig. 1). Shortly after ingression and stabilization of the sea it extended about 70 km farther toward Wrocław, as is evidenced by a thin bed of basal limestone (Eisentraut 1939, Neuhaus 1942, Rydzewski 1969). Formation of the Zechstein metalliferous sediments is associated with the first regression of the sea. A lagoonal topography was created when the sea retreated to the north, leaving the Rudna lagoon, a set of barrier reefs around it (see Fig. 1) and a narrow seaway from Głogów to just west of Polkowice extending toward Szklary. Three main Palaeotopographic zones may be distinguished here (Fig. 1). These are the western highs of oxidized sediments (Rote Fäule) extending westward from Sieroszowice (Rydzewski 1969, Harańczyk 1970, Oszczepalski 1982), the narrow Głogów-Polkowice-Szklary depression which ran southward in the central part of the area, and the eastern coastal elevation extending eastward from Lubin (Kłapciński 1964). Within this area a lagoon barrier separated the sea from the adjacent northern marginal basin which because of increasing regres-

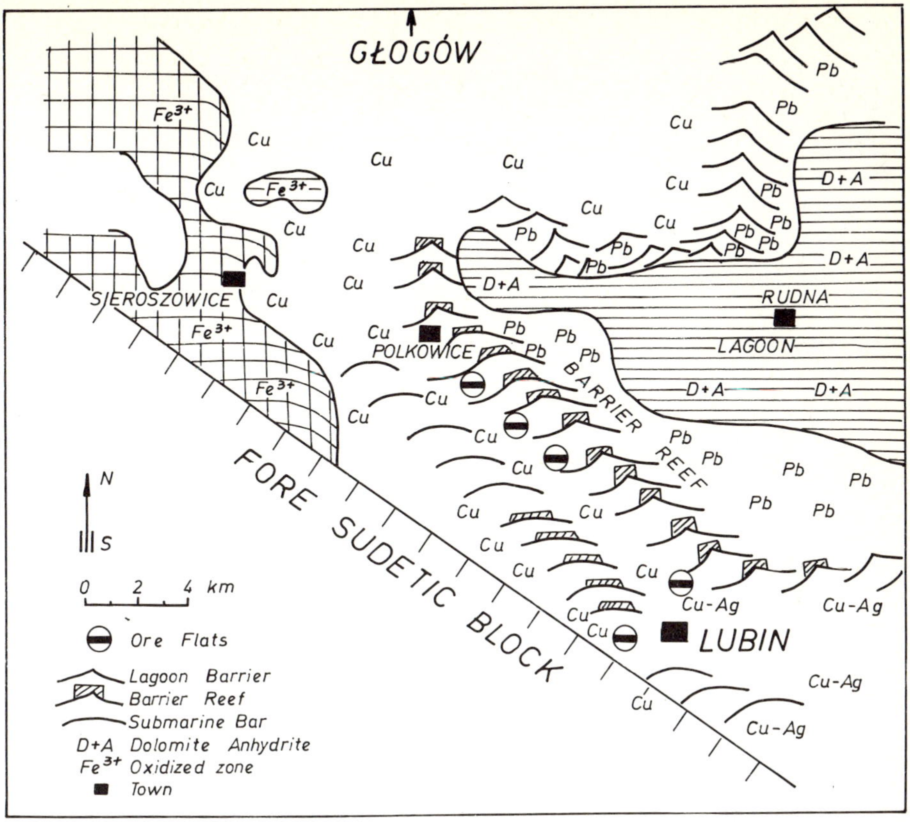

Fig. 1. Palaeogeographic situation of the Fore-Sudetic Monocline during the first Zechstein regression subcycle

sion and isolation evolved into a lagoon, named Rudna Lagoon after the main locality (Figs. 1 and 2; see also Harańczyk 1970). More than 150 km^2 of this lagoon have been explored by drilling, however, its eastern extension is still unknown (Fig. 3). The southwestern marginal part of this lagoon is accessible in the stopes of the copper mines. Apart from the main lagoonal barrier representing a zone of sand bars and elevations topped by reef material, some sandstone bars which are also partly covered with a flat reef and are free from any copper-bearing shales, due to erosion and reworking of the previously existing shales, are recognized in the southwestern foreground of the lagoon (Fig. 4). These bars did not separate any isolated basin and are surrounded by flats covered by copper-bearing sediments. Only the lagoonal barrier separates areas of the copper- and lead-bearing shales. In the lagoon, the lead-bearing shales are underlain by a 2- or 3-cm-thick layer of the copper-bearing shales formed at an early stage of regression before isolation of the Rudna Lagoon and evaporative regime, and also before stratification of the lagoonal brines was established. Rudna Lagoon was a deeper lagoon than the shallow lagoonal pans at Mansfeld depicted by Rentzsch (1974).

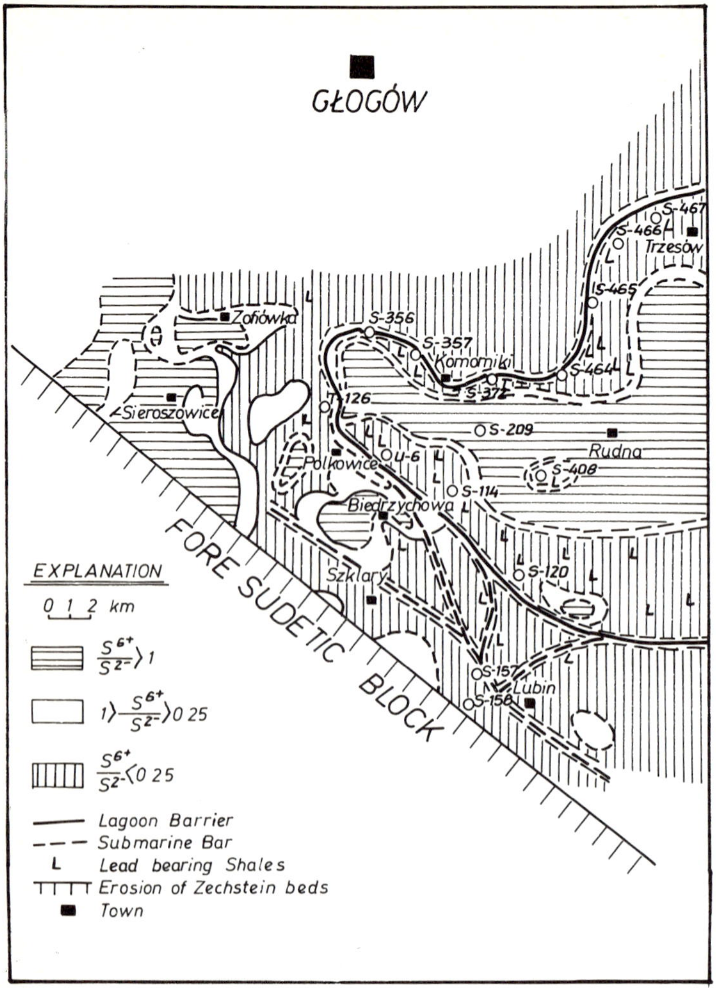

Fig. 2. Geochemical redox map of the Fore-Sudetic Monocline

Rudna Metallogenic Subcycle

The first Zechstein evaporative subcycle began shortly after ingression of the sea and it is included in the Werra cyclothem. The most complete evaporative sediment sequence is known from the locality of Rudna. Ingression of the Zechstein sea is marked in places by basal conglomerates, and by basal limestone or dolomite formed under oxidizing conditions. These sediments delimit the largest extension of the Zechstein Sea (Eisentraut 1939, Kłapciński 1964, Rydzewski 1969). They are followed by sediments formed during stagnation associated with growing regression of the sea. Let us consider data substantiating the existence of the first evaporative subcycle:

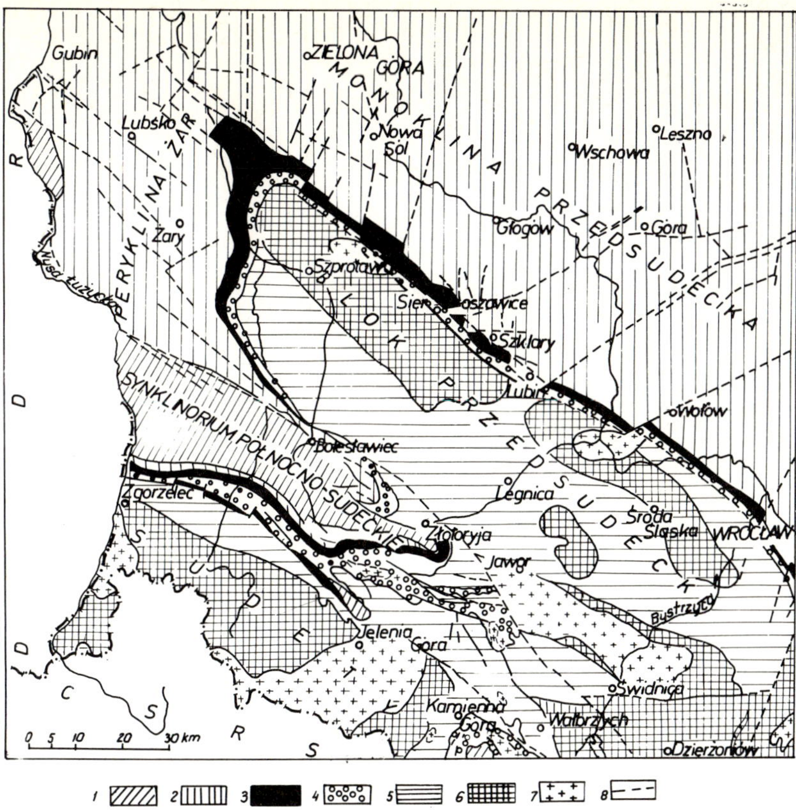

Fig. 3. Geological sketch map of Fore-Sudetic area according to J. Krasoń and J. Sokołowski, showing strata older than Cenozoic. *1* Cretaceous; *2* Triassic; *3* Zechstein; *4* Lower Permian red beds (Rotliegend); *5* Palaeozoic; *6* Precambrian; *7* Hercynian igneous rocks; *8* dislocations

1. The shoreline marked by transgressive basal limestone (locally dolomite) was shifted almost 70 km westward in the flat peneplain area between Wrocław and Lubin.
2. According to Smith (1980) in England, marl slate with high metal content is restricted to near-shore deposition in steep relief, however, in a flat marginal peneplain, it is 40 – 50 km distant from the extensive shoreline.
3. A typical subcycle sediment sequence ranges in thickness from 2 m on the internal slopes of the lagoonal barrier (Figs. 5 and 8) to almost 20 m in the central part of Rudna Lagoon (Fig. 9). The Rudna subcycle sediment sequence within the lagoonal area includes shales, marly dolomites and dolomites with anhydrite and limestones with anhydrite from bottom to top; and it is a typical regression sequence with increasing evaporation. The subcycle sediment sequence is overlain by stylolithic dolomite. The sediment sequence in the open sea was described by Bereś et al. (1973).
4. Preserved lagoonal relief of more than 30 m is observed in the mine stopes of the Lubin, Polkowice and Rudna mines in the Fore-Sudetic Monocline. The

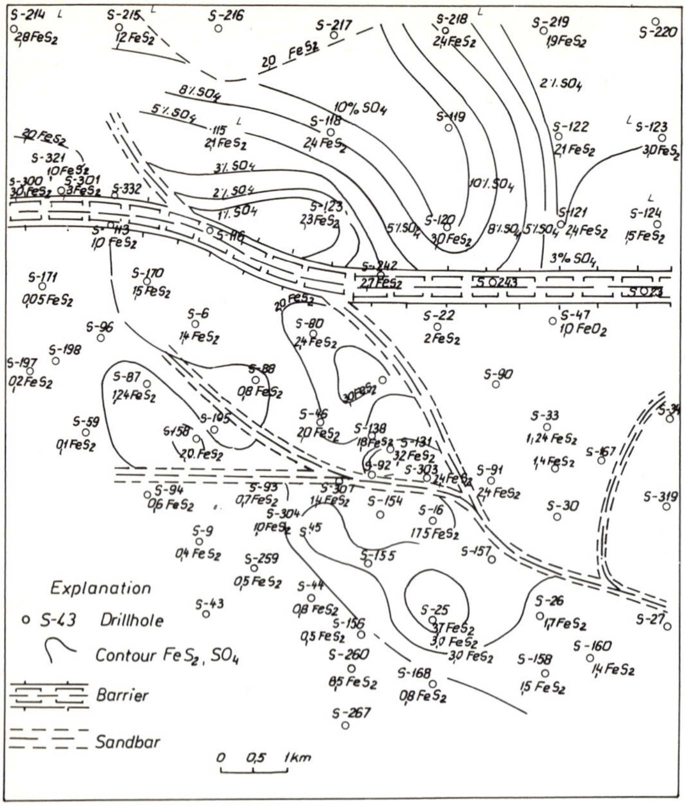

Fig. 4. Fragment of the map shown in Fig. 2 displaying area of subaerially exposed shales consequently enriched in pyrite

difference is measured between the flat regressive reef capping the lagoonal barrier and the tidal flats in the fore-shore which in their shallower parts are free from shales, due to abrasion, and in the lower part are covered by the copper-bearing shales burrowed by scavengers, and finally the lowest shoreline marked by lenses of redeposited copper and organic-rich shales. Also the thickness of the mineralized zones in the sandstones indicates a ca. 50-m-high barrier in the southeastern part of the lagoon (Harańczyk 1970).

Taking into consideration the above-mentioned data, the fall of the sea level during the Rudna subcycle is estimated to have been at least 70 m. It is highly probable that it prevailed in the whole Zechstein basin. The Rudna subcycle lasted a few hundred thousand years as inferred from the maximal thickness of the lagoonal sediment sequence consisting mainly of seasonal rhythmic sediments.

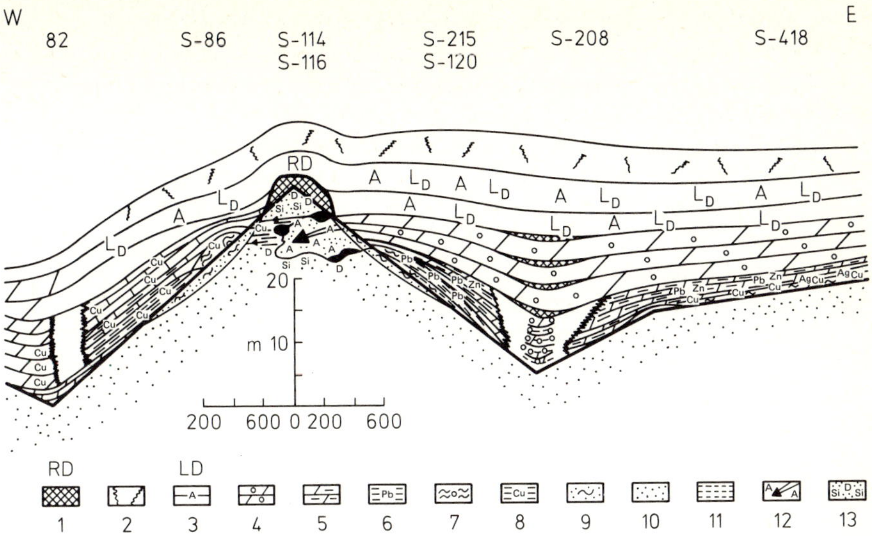

Fig. 5. Distribution of sediments around the Rudna lagoonal barrier. *1* Reef dolomite; *2* Stylolitic dolomite; *3* dolomitic limestone with anhydrite inlayers; *4* dolomite with anhydrite nodules; *5* pyrite-bearing argillaceous dolomite; *6* lead-zinc-bearing argillaceous dolomite; *7* redeposited copper-rich (smearing) shale; *8* copper-bearing shale (Kupferschiefer); *9* bioturbidite sandstones hosting detrital biota of the reef talus; *10* Weissliegend sandstone; *11* the same with sulphide diffusion laminae; *12* zone of sandstones with matrix replaced by anhydrite-hosting nests of massive ore; *13* regenerated and/or dolomitized sandstone

Evolution of the Sedimentation During the First Subcycle

The lagoonal relief of the surface of the Weissliegend sandstones in the Fore-Sudetic Monocline was only partly inherited from desert topography. Subaqueous formation of sandstone bars is suggested for most of the Weissliegend in the area (Krasoń and Grodzicki 1965, Konstantynowicz 1965, Metzler 1975, Jerzykiewicz et al. 1976). An influence of synsedimentary tectonics is also probable as the southern barrier runs parallel to the Sudetian tectonic direction. During a short period of sea-level stability shortly after transgression, the top of the lagoonal barrier and some other bars were in the euphotic zone and active reef growth, locally with a stenohaline fauna, flourished. The reef is fairly flat – rarely more than 2 m thick because of the regression of the sea caused the top of the reef to approach sea level, limiting upward growth and encouraging lateral expansion. Bivalves and crynoids are dominant, stromatolite structures are also seen.

The next stage of the subcycle is marked by a sharp fall of sea level leaving all reefs subaerially exposed. Separation of the Rudna Lagoon was established and early dolomitization of the reef by refluxing lagoonal brines began.

In the subsequent stage a continuing fall of sea level initiated the growing isolation of the lagoon, and the ensuing abrasion and washing out of the copper-bearing shales lying on the steep, seaward slopes. Redeposition of the disintegrated material produced copper- and organic-rich shales (smearing shales), forming lenses along the lowest shoreline. Weathering of the exposed shales was associated with cementation of the barrier sandstones by descending waters (compare the barrier cross-section, Harańczyk 1970). Tidal flats in the fore-shore of the barrier are covered with shales burrowed by scavengers. Cementation of these shales has produced ore flats, a unique ore mineral form (Harańczyk 1984a). The burrowed shales from the tidal flats are also extremely rich in fish fossils. Parenthetically, it should be noted that Freiesleben (1815) was right in saying that the fish died of poisonous copper concentrations in the seawater, however, it was true only for the zone adjacent to the supratidal flats where the oxidation of the subaerially-exposed shales proceeded. Deposition of lead- and zinc-rich shales in the Rudna Lagoon on the internal barrier slopes and of nodular carbonates with intercalations of anhydrite in the central part developed, indicating that lagoonal brines were already strongly concentrated and stratified.

During the lowest sea level, total isolation of the lagoonal basin produced brines which invaded barrier sandstones and reef carbonates, replacing the matrix and in places whole sandstone beds by anhydrite or gypsum and dolomite. The replacement front is marked by regeneration of quartz grains and is followed by massive concentrations of chalcocite, forming the richest ore in the district.

Full recovery of the sea level and uniform sedimentation of a beige stylolitic dolomite or dolomitic limestone brought to a close the first regression and the Rudna subcycle of the Zechstein Sea in the Fore-Sudetic Monocline (Figs. 5 and 6). Because the relief was not completely compensated by sedimentation, the barrier formed a cupola and steplike faults developed along it.

The above presented evolution precludes a sabkha model suggested by Renfro (1974) and others discussed by Lagny (1980).

Subaerial Ore Processes

Emergence of the metalliferous shales and prolonged contact with atmospheric oxygen induced ore processes important for the redistribution of metals and their concentration or dispersion. In the Fore-Sudetic Monocline, this process developed on a large scale depending on the exposure time and contrast of the relief. The following ore processes are recognized as important for metal concentration:

1. Wave abrasion and washing out of the shale and sandstone beds lying on the steep seaward slopes of the lagoonal barrier and on the slopes of submarine bars causing redeposition (Błaszczyk 1982) and formation of lenses of shale along the contemporaneous shoreline (Figs. 5 and 7). The redeposited shale consists mostly of disintegrated shale matter which lost its rhythmic sedimentation structure and sometimes its organic carbon and copper contents, ranging up to 20%. The disintegrated shale matter fills also scavenger trails.
2. Short, subaerial exposure of copper-bearing shales induces their partial oxidation and the migration of metals mainly to the basal layers producing enrichment in pyrite (Fig. 4) and formation of a contact layer of pure chalco-

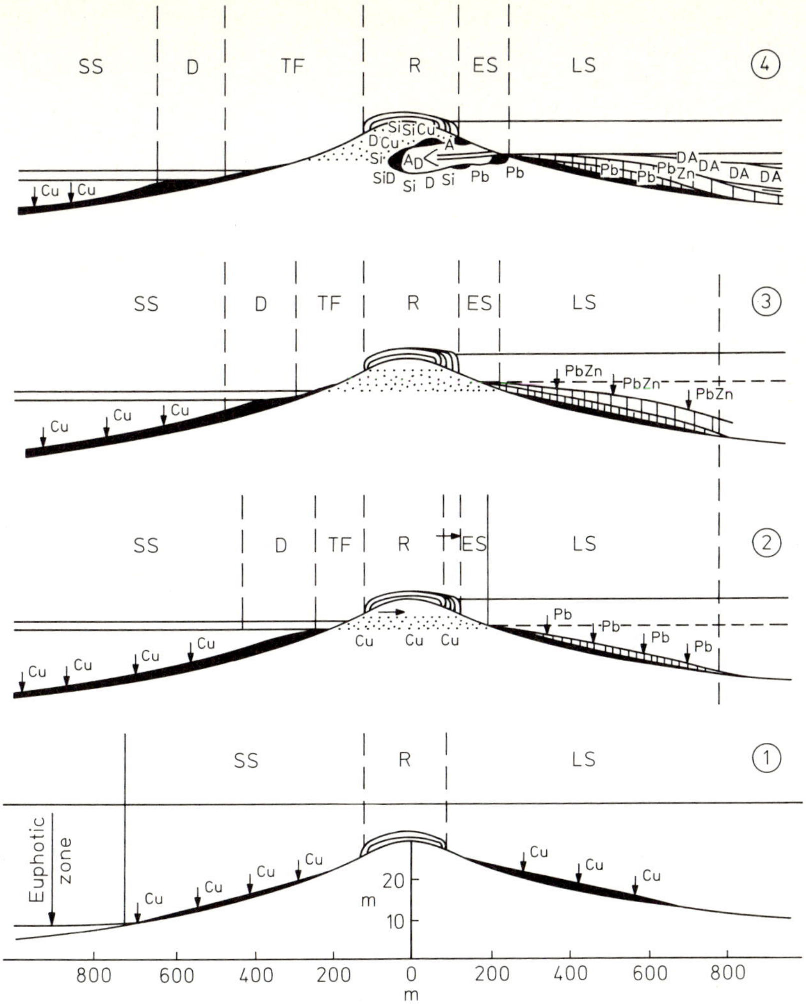

Fig. 6. Evolution of sedimentation in the zone of lagoonal relief during the first regression of the Zechstein Sea. Symbols of sedimentation zones: *SS* sea sedimentation; *D* redeposition of sediments; *TF* tidal flats; *R* barrier reef; *ES* lagoonward slope of the barrier; *LS* lagoonal sedimentation; *1 – 4* stages of evolution. D_A dolomite with anhydrite nodules

cite (Fig. 10). Covellite, digenite, native silver and rare cobalt minerals frequently contribute to this layer (see also Harańczyk 1972).

3. Advanced oxidation of shales and descendent migration of solutions is the most frequent cause for the genesis of the sandstone ore extending only under subaerially-exposed shale areas of bars, barrier, highs and elevations. The form of the mineralized sandstone zone was described in an earlier paper (Harańczyk 1970). The sulphides disseminated in the sandstone, form Liesegang-like parallel-asymmetrical diffusion bands built of sulphide grains replacing the sandstone matrix.

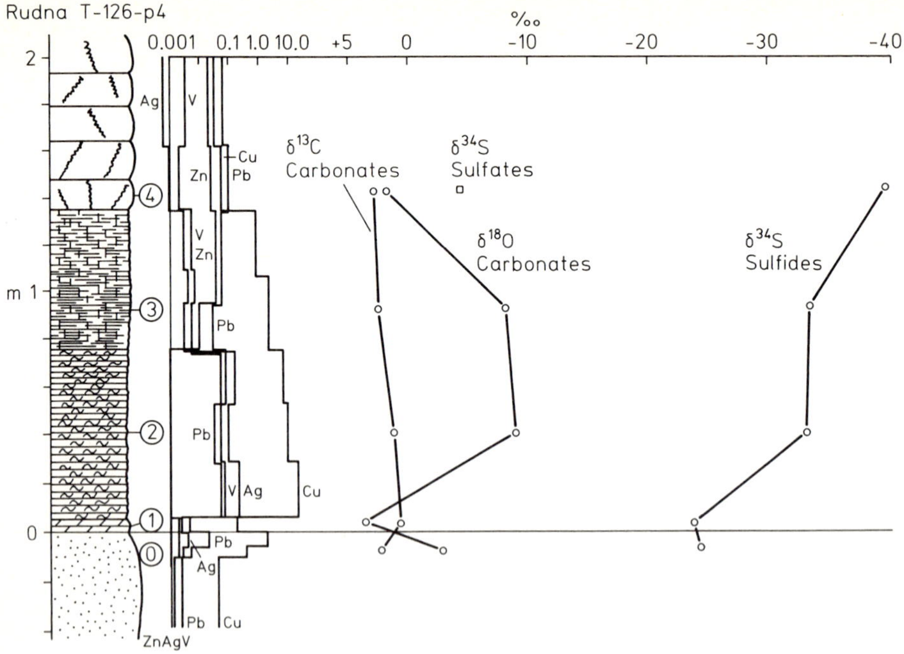

Fig. 7. Sediment sequence of the sea shoreline, the location where lenses of redeposited shales are formed. Localization is shown in Fig. 5. Additional explanations see Fig. 8

4. Surficial oxidation of the scavenger-burrowed shale beds lying on the tidal flats in the fore-shore of the lagoonal barrier has induced the formation of the ore flats within plywoodlike fissile shales. Some scavenger burrows are filled with slightly diagenised sediments rich in foraminifera and other minute fossils and subsequently replaced mimetically by sulphides, the same as those that make ore flats (Fig. 11). Further details are given in another paper (Harańczyk 1984a).
5. Prolonged emergence of the black shales occurring on eminences and highs caused complete oxidation of the shales producing the so-called Rote Fäule, which means oxidized hematite-stained red shales. The metal content of the shales was dispersed, or it was concentrated laterally in adjoining unweathered shales. Since Freiesleben's time (1815) the contacts of Kupferschiefer and Rote Fäule have been regarded as the most important sites for rich ore.
6. Infiltration of the lagoonal barrier sandstones by concentrated lagoonal brines induced the migration of metals, earlier distributed in the sandstones. As a result irregular ore bodies of massive chalcocite (rarely galena) developed in parts of the barrier sandstones where their matrix had been replaced by anhydrite.
7. Sapropelic Model Acting in the Lagoonal Environment. Kupferschiefer or its equivalent marl slate and, in lagoonal basins, lead- and zinc-bearing shales may be regarded as the prototype of euxinic metalliferous sediments. Indeed,

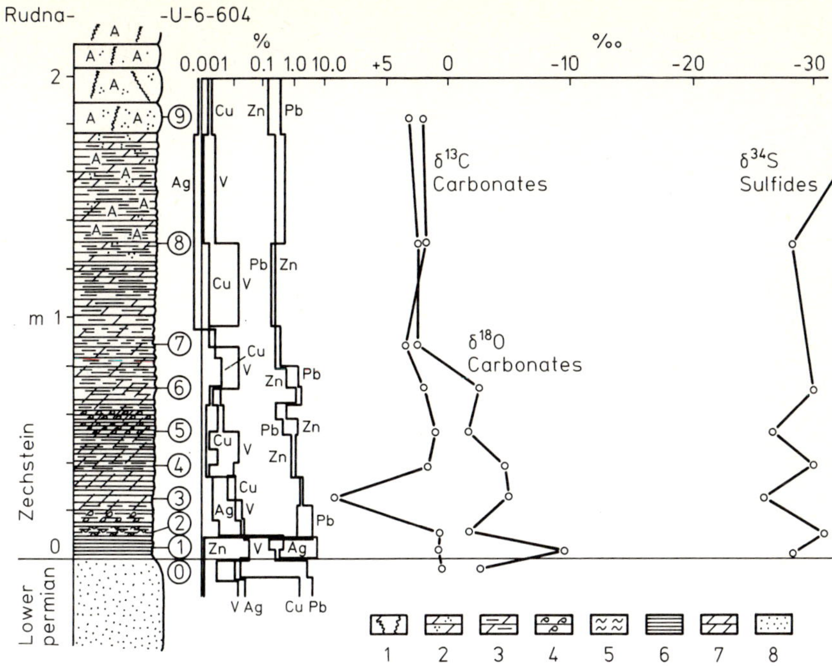

Fig. 8. Sediment sequence on the internal slope of Rudna lagoonal barrier. *1* Stylolitic dolomite; *2* dark weakly laminated dolomite with pearls of anhydrite; *3* argillaceous laminated dolomite; *4* lead-zinc-bearing dolomite with immature Brachiopods washed out from the barrier reef; *5* redeposited copper-bearing shales; *6* copper-bearing shales showing rhythmic structure; *7* basal dolomite; *8* Weissliegend sandstone

these are kerogen-rich sedimentary rhythms or saprovarvites composed of argillaceous-organic and dolomitic or calcitic laminae with a high or at least above-average content of metals accumulated by biological fixation (Harańczyk 1984b). This fixation is evidenced by the correlation between C_{org} and S^{2-} (Harańczyk 1961, 1972) and ipso facto with the metals saturating it. Consequently, it is not possible that much of the organic carbon could be oxidized during bacterial reduction of the sulphate ions, however, a considerable quantity of H_2 in organic compounds was used for that purpose (Harańczyk 1972, Rentzsch 1974). The organic substance of the shales is of sapropelic origin as indicated by the low C/N ratio of 20 to 50 (Rentzsch 1974). Copper-bearing shales are usually kerogen-rich low-bituminous rocks (Z. Sawłowicz, personal communication). There is strong evidence that biogenic-sulphide sulphur originated in the shales. This has been proven by a sulphur isotopic ratio $\delta^{34}S$ ranging from -26 to -32‰ (Marowsky 1969, Wedepohl 1971, Harańczyk 1984b). Lead isotopes indicate that different source material was involved in the genetic processes (Wedepohl et al. 1978).

The sapropelic model of biogenic fixation of metals is consistent with, and can be better understood, if the process took place in a sea during its regression stage. Marine regression stimulated stagnation and the seasonal bloom and fall

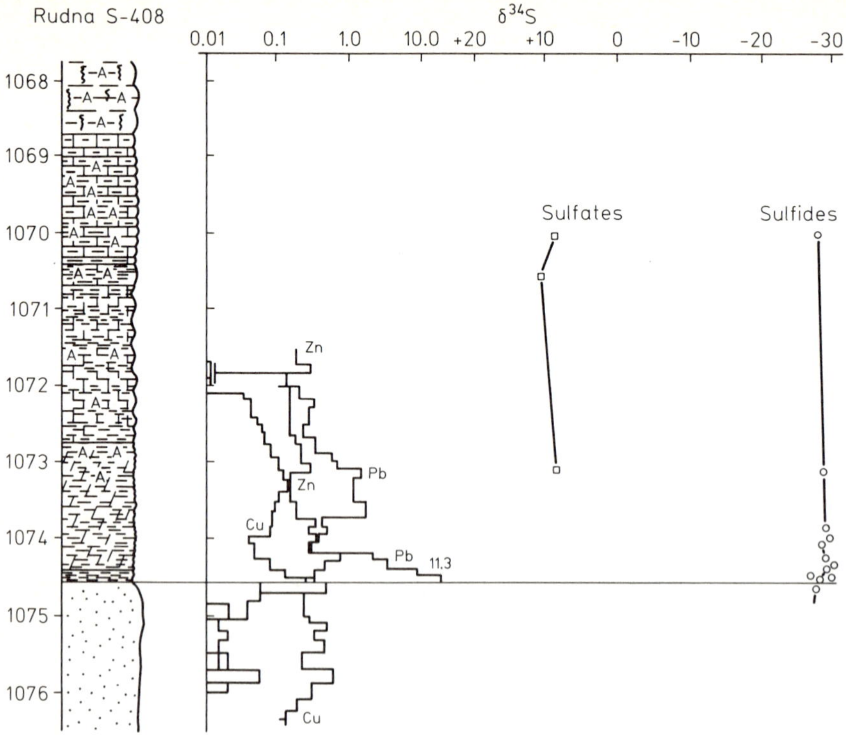

Fig. 9. Sediment sequence in the central part of the Rudna lagoon. For explanation of *symbols* see Fig. 8

of phytoplankton in euphotic zones, as suggested by Brongersma-Sanders (1965), thus developing a sea-bottom reduction zone in which the metalliferous anomaly in seawater was effectively eliminated. Separation and growing isolation of the Rudna lagoon produced brines with concentrated metals. When copper was exhausted in the isolated lagoonal basin, lead, zinc and finally iron saturated the biogenic-sulphide sulphur. Phytoplankton as well was blooming in euphotic surficial zones, however, with increasing salt concentration its growth was hindered, and consequently, beds of dolomarls, slightly laminated due to scarce organic matter, were deposited, and finally organic growth almost ceased. It is likely that the supply of heavy metals was exhausted earlier.

Discussion and Conclusions

The supergene processes involved during the first Zechstein subcycle produced lenses of redeposited shales, introduced sulphide mineralization to the underlying sandstones and contributed to the formation of the present size, shape and metal concentration of the largest European copper deposits in Lower Silesia in Poland. The process of supergene enrichment has so far been largely overlooked

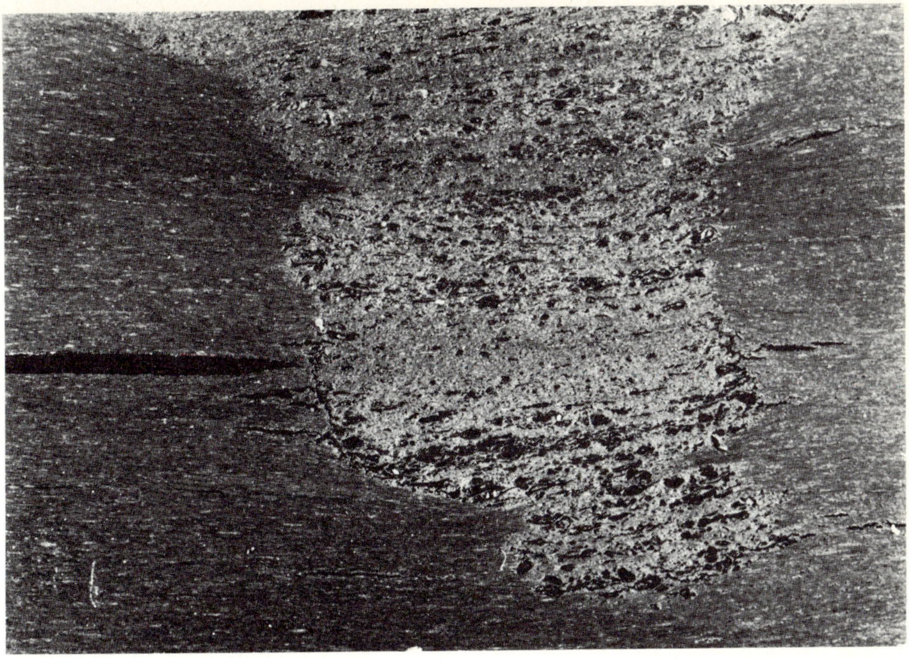

Fig. 10. Macrophotograph of a burrow perforated by scavenger in the copper-bearing shales from tidal flats. The canal was filled with foraminifera-rich sediments and subsequently replaced with bornite similar to that forming adjacent ore flats

in descriptions of copper deposits in sedimentary rocks (Tourtelot and Vine 1976), although the close relation between oxidation zones of so-called Rote Fäule and metal-rich Kupferschiefer beds has often been emphasized since Freiesleben's time. Our knowledge of subaerial processes is still far from being completely understood. Especially the cementation processes of the unweathered, deeper Kupferschiefer layers need additional sulphur isotopic data and microscopic studies. It seems very probable that the parageneses including native silver, digenite and covellite distinguished by Rentzsch and Knitzschke (1968) as well the occurrences of rare cobalt, silver, mercury, palladium and platinum minerals determined by Harańczyk (1972). Kucha (1982) and others may be regarded as parageneses of cementation minerals superposed on the primary chalcopyrite, bornite and chalcocite mineralization of the Kupferschiefer during its subaerial exposure.

The provenance of metals included in the Kupferschiefer is another intricate unsolved problem. It is clear that the sapropelic model for the fixation of metals dissolved in the early Zechstein Sea elucidates processes of efficient elimination of the metal anomaly in seawater. Admittedly, different opinions on this question are also held. Brongersma-Sanders (1965) proposed that the metals of Kupferschiefer were supplied by normal water. This hypothesis does not explain why the situation was never repeated. With the exception of the Kupferschiefer itself, there are no increased metal contents in other cyclothemic basal Kupfer-

Fig. 11. Macrophotograph of the basal chalcocite layer formed by cementation enrichment contacting with bioturbidite sandstone with admixture of reef talus. Red stains (Rote Fäule, II type) occur in the Weissliegend sandstone. Red Stains developed when sandstone matrix was replaced with anhydrite and quartz grains were regenerated and produced and a porous water interface

schiefer equivalents, although the Stinkschiefer at the base of the second cycle (Strassfurt-cycle) is the host rock of the Zechstein oil and gas deposits (Rentzsch 1974). Pertinent to the discussion is Wedepohl's suggestion (1971) that the high manganesian content of carbonate rocks overlying the Kupferschiefer may indicate hydrothermal submarine activity.

Areas of economic grade Kupferschiefer are not larger than 1% of the total area of Zechstein sediments (Wedepohl 1971), therefore, the question that now arises is what has determined the localization of rich primary Kupferschiefer? It must be added that a favourable euphotic zone and suitable environmental conditions existed in much of the Zechstein Sea area. However, these factors alone were insufficient. The author – due to the results of his studies and the foregoing considerations, has come to regard the presence of the Rotliegend rift basins in the basement, explained by Lorenz and Nicholls (1976) and Jowett (1984) in plate tectonic models, as the essential condition sine qua non for the area of sedimentation of a primary copper-rich shale bed. The genetic relation might be associated either with metal-remobilization action in a sabkha or in caliche style, from the red beds filling these sedimentation basins, however, this is difficult to reconcile with field evidence, or associated with reopening of rift dislocations, which serving as feeding channels, enabled an ascension of Cu, Ag, Pb, Zn and Co emanations (cf. Rentzsch 1974) from the resorbed oceanic lithospheric plate in a partly frozen Benioff zone. This oceanic plate was earlier involved in Hercynian collision and subsequent formation of the Rotliegend molasse and rift basins. However, other facts, e.g., with regards to Rücken Veins, must be obtained before it can be said with certainty whether or not biogenically deposited metals forming the Kupferschiefer, were of magmatic provenance associated with waning Benioff zone activity. The sapropelic model of elimination of the metal anomaly in the early Zechstein Sea, followed by enrichment during subsequent subaerial exposure best explains the so far ascertained regularities of ore distribution in old and new ore districts exploiting Zechstein black shales.

References

Bereś B, Kienig P, Kijewski P, Tomaszewski JB (1973) Lithostratigraphical division of carbonate rocks of the first Zechstein cyclothem i Fore-Sudetic Monocline. Przegl Geol 21:17 – 20 (in Polish)

Błaszczyk JK (1982) Facies dichotomy in lowest part of Zechstein carbonate sequence in Lubin copper district in Poland. Rudy Metale Nieżelazne 27:151 – 156 (in Polish)

Brongersma-Sanders M (1965) Metals of Kupferschiefer supplied by normal seawater. Geol Rundsch 55:365 – 375

Brongersma-Sanders M (1971) Origin of major cyclicity of evaporites and bituminous rocks: an actualistic model. Merin Geol 11:123 – 144

Eisenhuth K, Kautzsch E (1954) Handbuch für den Kupferschieferbergbau. Leipzig

Eisentraut O (1939) Der niederschlesische Zechstein und seine Kupferlagerstätten. Arch Lagersch Forsch, Berlin, No 71

Freiesleben JG (1815) Geognostischer Beytrag zur Kenntniss des Kupferschiefergebirges, vol III, IV. Freiberg

Harańczyk Cz (1961) Correlation between organic carbon, copper, and silver content in Zechstein copper-bearing shales from the Lubin-Sieroszowice region Lower Silesia. Bull Acad Pol Sci Ser Sci Geol 9:4 – 10

Harańczyk Cz (1964a) Investigation of copper-bearing Zechstein shales from the Wrocław Monocline Lower Silesia. Bull Acad Pol Sci Ser Sci Geol 12:13 – 18

Harańczyk Cz (1964b) Petrographic classification of Zechstein copper-bearing shales from Lower Silesia. Bull Acad Pol Sci Ser Sci Geol 12:19 – 24

Harańczyk Cz (1970) Zechstein lead-bearing shales in the Fore-Sudetian Monocline in Poland. Econ Geol 65:481 – 495

Harańczyk Cz (1972) Ore mineralization in the Lower Zechstein euxinic sediments in the Fore-Sudetic Monocline. Arch Miner 30:13 – 173 (in Polish)

Harańczyk Cz (1984a) Ore flats in the copper-bearing shales of the Fore-Sudetic Monocline Lower Silesia Poland. Amstutz Band (in press)

Harańczyk Cz (1984b) Lagoonal relief controlling distribution and isotopes content of the Zechstein metalliferous sediments and hosted sulphides. Inst Geol Warszawa Pap (in Polish) (in press)

Jerzykiewicz T, Kijewski P, Mroczkowski J, Teisseyre AK (1976) Genesis of the Weissliegendes sediments of the Fore-Sudetic Monocline. Geol Sudetica 11:57 – 97

Jowett EC (1984) The Rotliegendes in Central Europe: basin development controlled by plate tectonics. Przegl Geol 4:196 – 201 (in Polish)

Kłapciński J (1964) Palaeogeographical characteristics of the Zechstein of the Fore-Sudetic Monocline. Ann PTG 34:551 – 573 (in Polish)

Konstantynowicz E (1965) Mineralization of Zechstein sediments in Northern Sudetic Trough. Prace Geol PAN, Kraków, 28 (in Polish)

Krasoń J, Grodzicki J (1965) Sedimentationsentwicklung und Mineralisation des Weissliegenden Niederschlesiens. Coll Pap, Eisleben, DDR

Kucha H (1982) Platinum-Group Metals in the Zechstein Copper Deposits, Poland. Econ Geol 77:1578 – 1591

Lagny P (1980) Les gisements stratiformes associés aux évaporites. Position dans le temps et place dans l'espace des basins sedimentaires évaporatiques. Bull Centr Rech Explor Prod Elf-Aquit 4:44 – 478

Lorenz V, Nicholls A (1976) Permocarboniferous Basin and Range Province of Europe: an application of plate tectonics. In: Falks H (ed) The continental permian in central, west and south Europe. Riedel, Dortrecht

Marowsky G (1969) Schwefel-, Kohlenstoff- und Sauerstoff-Isotopenuntersuchungen am Kupferschiefer als Beitrag zur genetischen Deutung. Contr Miner Petrol 22:290 – 334

Metzler M (1975) Quelques donnés sur les dépots cuprifères de Basse Silesie Pologne. Bull Bur Rech Geol Min 2 II 5:407 – 422

Neuhaus A (1940) Über die Erzführung des Kupfermergels der Gröditzer Mulde in Schlesien. Z Angew Miner 2, 3

Oszczepalski S (1980) Palaeogeography, sedimentation and mineralization of the Z 1 carbonate series (Zechstein) in the western part of the Fore-Sudetic Monocline Western Poland. Contr Sedim 9:307 – 323

Renfro AR (1974) Genesis of Evaporite-Associated Stratiform Metalliferous Deposits – A Sabkha Process. Econ Geol 69:33 – 45

Rentzsch J (1974) The Kupferschiefer in comparison with the deposits of the Zambian Copperbelt. In: Cenenaire de la Société Géologique de Belgique. Gisements Stratiformes et Provinces Cuprifères. Liège, pp 395 – 418

Rentzsch J, Knitzschke G (1968) Die Erzmineralparagenesen des Kupferschiefers und ihre regionale Verbreitung. Freiberger Forschungsh C231:189 – 211

Rydzewski A (1969) Petrography of the copper-bearing shale from Fore-Sudetic Monocline. Inst Geol Warszawa Bull 217:113 – 170

Schneiderhöhn H (1921) Chalkographische Untersuchung des Mansfelder Kupferschiefers. Jahrb Min Beil 47:1 – 38

Smith DB (1980) The evolution of the English Zechstein basin. Contrib Sedimentol 9:7 – 34

Spangenberg C (1572) Mannsfeldische Chronica. Eisleben

Tourtelot B, Vine JD (1976) Copper Deposits in Sedimentary and Volcanogenic Rocks. Geol US Surv Prof Pap 907 C

Wedepohl KH (1971) Kupferschiefer as a Prototype of Syngenetic Sedimentary Ore Deposits. Soc Min Geol Jpn Spec Iss Proc IMA-IAGOD Meetings '70 IAGOD, vol III, pp 268 – 273

Wedepohl KH, Delevaux MH, Doe BR (1978) The potential source of lead in the Permian Kupferschiefer bed of Europe and some selected Palaeozoic mineral deposits in the Federal Republic of Germany. Contrib Miner Petrol 65:273 – 285

Formation Conditions of Copper-Sandstone and Copper-Shale Deposits

A. M. LUR'YE[1]

Abstract

The concept introduced here suggests a common provenance of copper origin for all cupriferous sandstones and shale-type deposits, the source of copper being terrigenous red-bed formations deposited under arid conditions. Copper was leached from these rock deposits by subsurface waters. Copper precipitation occurred at hydrogen sulphide barriers, among which two types are recognized: syngenetic (in unconsolidated sediments) and epigenetic (in lithified sediments). The latter are subdivided, according to the origin of their organic matter, into autochthonous and allochthonous groups. The decisive factor controlling the formation of sediment-hosted copper deposits with large reservers is the specific nature of the paleohydrogeological and geochemical environment.

Copper-sandstone and copper-shale deposits represent a highly specific group of deposits, distinguished by the following characteristic features: (1) invariable spatial association with red beds formed in an arid environment; (2) ore localization in grey sedimentary rocks in close proximity to the red beds; (3) consistent mineral composition and associations of major metallic minerals; (4) a zonal distribution pattern of the sulphides in the ore bodies.

Three hypotheses have been discussed by researchers for the origin of the metallic substances in this group of deposits: hydrothermal (endogenetic solutions), sedimentary (provenance area), and exogenetic-epigenetic (subsurface pore waters of arid red-bed formations). The validity of the above-cited hypotheses can readily be weighed: if it is true that they should provide a lucid understanding for the known spatial association of copper mineralization with arid red beds. This relationship, repeatedly verified by field observations, can be regarded as a well-founded empirical law; nevertheless, it is commonly ignored by the proponents of the hydrothermal hypothesis. A similarly adequate solution of the source problem has been offered by the supporters of the sedimentary hypothesis. According to the latter, advocated by Strakhov (1963), copper was leached from ore deposits of humid provenance by surface waters, and redeposited in arid regions in the form of carbonates that subsequently underwent conversion to sulphides during early diagenesis. However, the original copper carbonates have never been encountered. Consequently, it must be assumed that

[1] Institute of Lithosphere of the USSR, Academy of Sciences, Moscow, 109180, USSR

Geology and Metallogeny of Copper Deposits
(ed. by G.H. Friedrich et al.)

these carbonates were invariably precipitated only in those areas where the oxidizing environment in the sediments underwent a subsequent change to a reducing, hydrogen sulphide environment.

The general spatial association of copper mineralization with arid red-bed formations is feasibly explained by the epigenetic hypothesis, but from this viewpoint the characteristics of certain deposits, undoubtedly of sedimentary origin (for example, Mansfeld), has in past years remained incompatible. According to the epigenetic hypothesis, copper was leached from red beds by subsurface waters, and therefore its accumulation in marine sediments was regarded as inconsistent with this concept. In fact, the contribution of copper from bottom sediments to sea basin waters is quite natural, considering that influx to the seas is derived not only from the run-off of surface waters, but also from inflow of subsurface waters. If the waters of red-bed formations drained into the paleosea, then these waters could have been responsible for the copper supply. Hence, the most warranted hypothesis is that copper and associated metals were leached by subsurface waters from red-bed formations. In this respect the deposits under study are a group of closely related deposits, having in common the origin of their metal content.

Sulphide mineralization is usually restricted to grey sediments, occurring immediately adjacent to the red beds. This association indicates that the concentration of copper occurs at hydrogen sulphide barriers that are formed at the interface between sediments of markedly different characteristics. The distinctions between deposits are largely due to the conditions under which the hydrogen sulphide barriers are formed. According to the time of formation of the host rocks relative to mineralization, two types of hydrogen sulphide barriers can be distinguished: syngenetic and epigenetic types. The first occur in unconsolidated sediments, the second in lithified rock deposits.

Syngenetic Barriers

Syngenetic barriers are most frequently formed as a result of the transgression of the sea over an area of arid red molasse sediments. As a consequence of changes in the drainage pattern caused by the advance of the sea, the discharge area of subsurface waters in the red beds becomes the sea bottom, where interactions occur between these waters and the hydrogen sulphide-bearing bottom silts (Fig. 1).

All marine sediments that have formed under reducing conditions are potential hydrogen sulphide barriers. The real barriers are restricted to local areas where hydraulic communications exist between the subsurface waters of the red molasse sediments and the water bodies draining them. This hydraulic communication is a most essential factor in the ultimate control on the formation of copper deposits.

Confirmation of this is found in the following geological features of the deposits:

1. Copper mineralization is restricted exclusively to those basins of marine sediments where the basal section is comprised of red-bed formations. In the

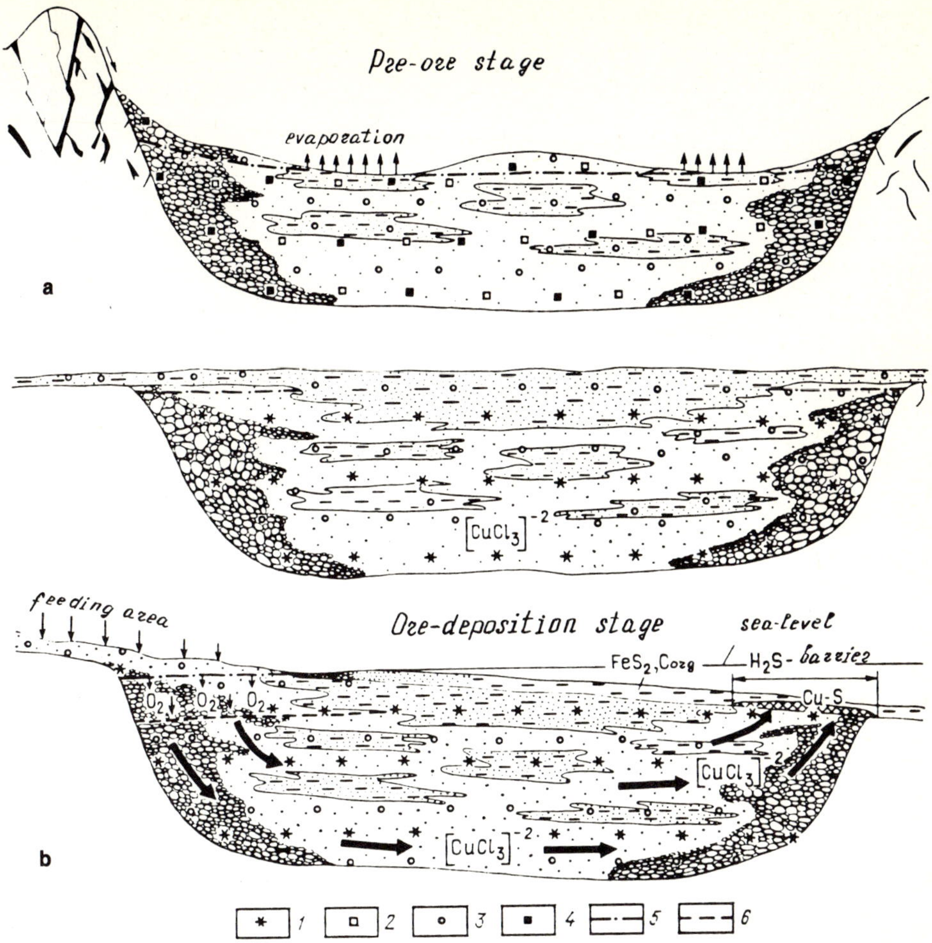

Fig. 1a, b. Formation of copper deposits in marine sediments. Pre-ore stage: **a** period of accumulation to terrestrial red molasses in intermontane troughs (active water interchange); **b** period of peneplanation (low hydrodynamical activity). Ore-deposition stage: period of transgression and submarine discharge. *1* Cupreous chloride complexes (copper in water solutions); *2–4* dispersed copper in sediments (*2* adsorbed copper; *3* intercrystalline copper; *4* oxides and dissolved carbonates). *5* Water table; *6* boundary between infiltration and connate waters

Western Ural region, the Kazanian marine sediments, containing copper sulphide accumulations in the basal section, are directly underlain by the red beds of the Ufimian (Lur'ye 1972). The ore-bearing deposits in the Lower Zechstein of Central Europe lie on the Rotliegend red molasse. The Nonesuch shales in the White Pine deposit, U.S.A., are superposed on the red beds of the Copper Harbor Formation (Ensign et al. 1972). The marine sediments of the Roan basin within the Zambian Copper Belt overlie terrestrial sediments that accumulated under arid climatic conditions; due to metamorphic alterations, the rocks lost their original coloration (Mendelsohn 1963).

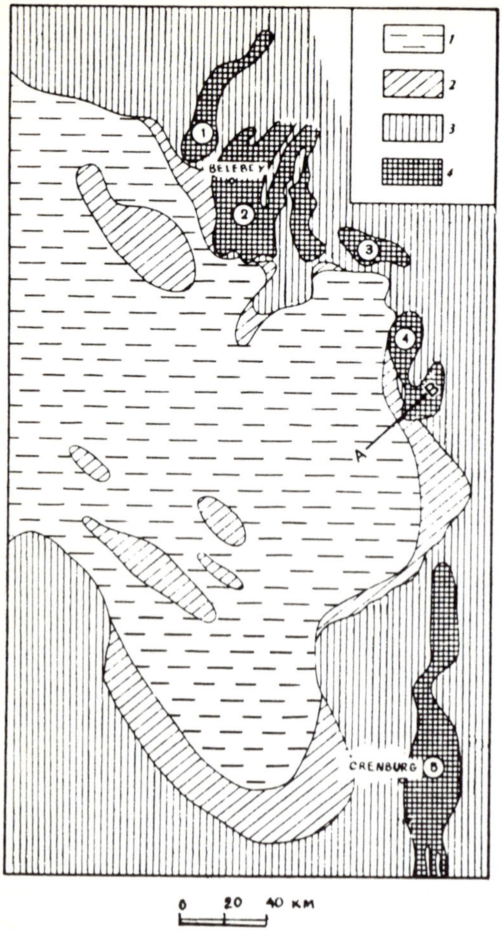

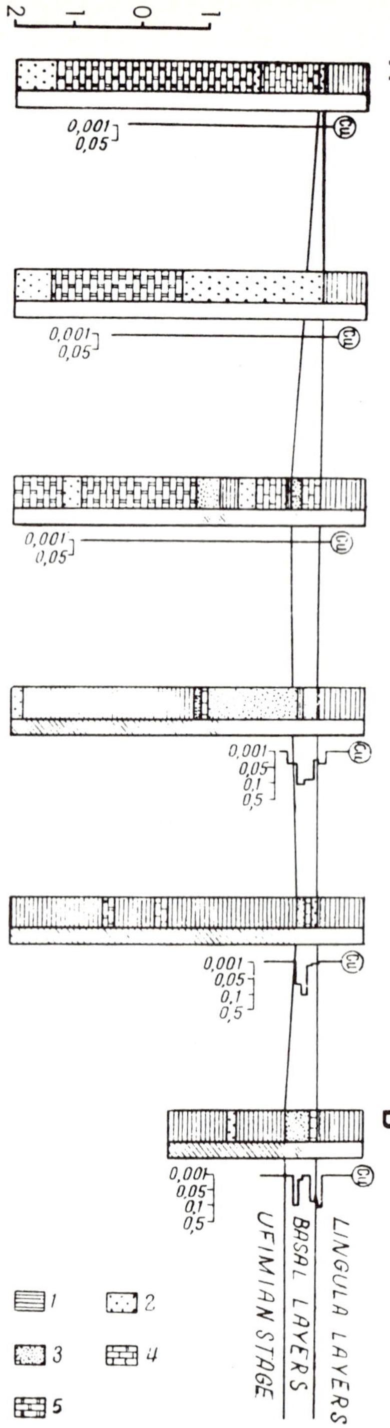

Fig. 2. Projection of ore zones in the basal section of the Kazanian marine sediments on the color distribution plan in the top of Ufimian sediments. For the *plan*, color of rocks at the top of the Ufimian stage: *1* grey sediments; *2* alternating red and grey sediments; *3* red sediments; *4* ore zones in the basal sections of the Kazanian marine sediments (*circled numbers 1* Kandrikulsk; *2 Belebe; 3* Urshak; *4* Fedorov-Sterlibash; *5* Salmish). – For the profile *A – B: 1* clay; *2* sandstone; *3* argillite; *4* limestone; *5* dolomite. For the *columns* at the *right*, the rock colors: red (*diagonal pattern*), grey (*no pattern*). Location of drill holes on the profile is without scale

2. Copper mineralization is typically restricted only to the lower section of the marine sediments. The remaining portion of the marine section, frequently reaching hundreds and thousands of metres in thickness, is barren. Copper-bearing rocks occur in immediate proximity to the underlying red beds. If the red beds thin out or undergo facies changes to grey sediments, then the mineralization is not seen in the overlying marine sediments. This is reflected very well in the Bashkirian ore zones where the red beds occur at the top of the Ufimian grade laterally into grey sediments (Fig. 2). The copper-bearing zones in the basal members of the Kazanian invariably overlie red beds, and have not been encountered outside the red-bed areas. The same observation applies to the Zechstein of Central Europe: copper mineralization occurs only directly above the red molasse. On uplifts of basement complexes, where red beds are lacking, copper mineralization has not been recorded.

Sections composed of alternating terrestrial red beds and grey marine deposits reveal the presence of several ore horizons e.g. the Kartamyshskaya Formation of Donbass (Lur'ye 1970), Upper Kazanian of the West Fore-Urals (Lur'ye 1973). Those depositional cycles which lack red beds are barren of copper ores. This relationship is exemplified in the Zechstein cycles, where deposits of non-ferrous metals are known to occur only at the base of the Werra series overlying the Rotliegend. The basal layers of all higher depositional cycles, which directly overlie evaporites, are barren.

3. The position of copper mineralization in the marine basin reveals a distinct dependence upon the structural features of the underlying red-bed deposits. The ore bodies at the base of the Kazanian marine section are located above "shoestring" sandstones at the top of the Ufimian (Fig. 3). The top of the Ufimian seems to be comprised chiefly of silty-argillaceous deposits. The sandstones are evidently ancient channel deposits. In those cases, where the sandstones do not extend along the entire channel of the Pre-Kazanian surface, but instead form isolated "windows", a particularly sharp increase of the copper content can be observed above these "windows".

From the paleogeographical sketch map of the Early Permian, showing the projections of the ore bodies in the Zechstein marine sediments, it is apparent that copper has concentrated chiefly above the red molasse sediment of the Rotliegend that fills the Variscian troughs (Fig. 4). Such superposed copper deposits are known in the Saale, Saar, North-Sudeten and Oder troughs. The distribution and configuration of the ore bodies depend to a large degree upon the structural features of the red-beds filling the troughs. The longitudinal axis of the ore-runs in the Mansfeld, Sangerhausen and Edderitz deposits (the Gellitzer line) coincides with the boundary where the porphyritic conglomerates at the top of the Rotliegend undergo lateral facies changes to arenaceous argillites (Hoyningen-Hüne 1963). Above this interface, the copper mineralization is characteristically at its maximum vertical extent, involving not only the cupriferous shale beds, but also the lower Zechstein limestone section. With increasing distance from this interface, the roof of the ore body drops further downwards into the lower beds of the cupriferous shales (Fig. 5).

The Richelsdorf deposits are located in the northeastern part of the Saar trough (Fig. 6). The closure of the trough resulted here from its conjugation with

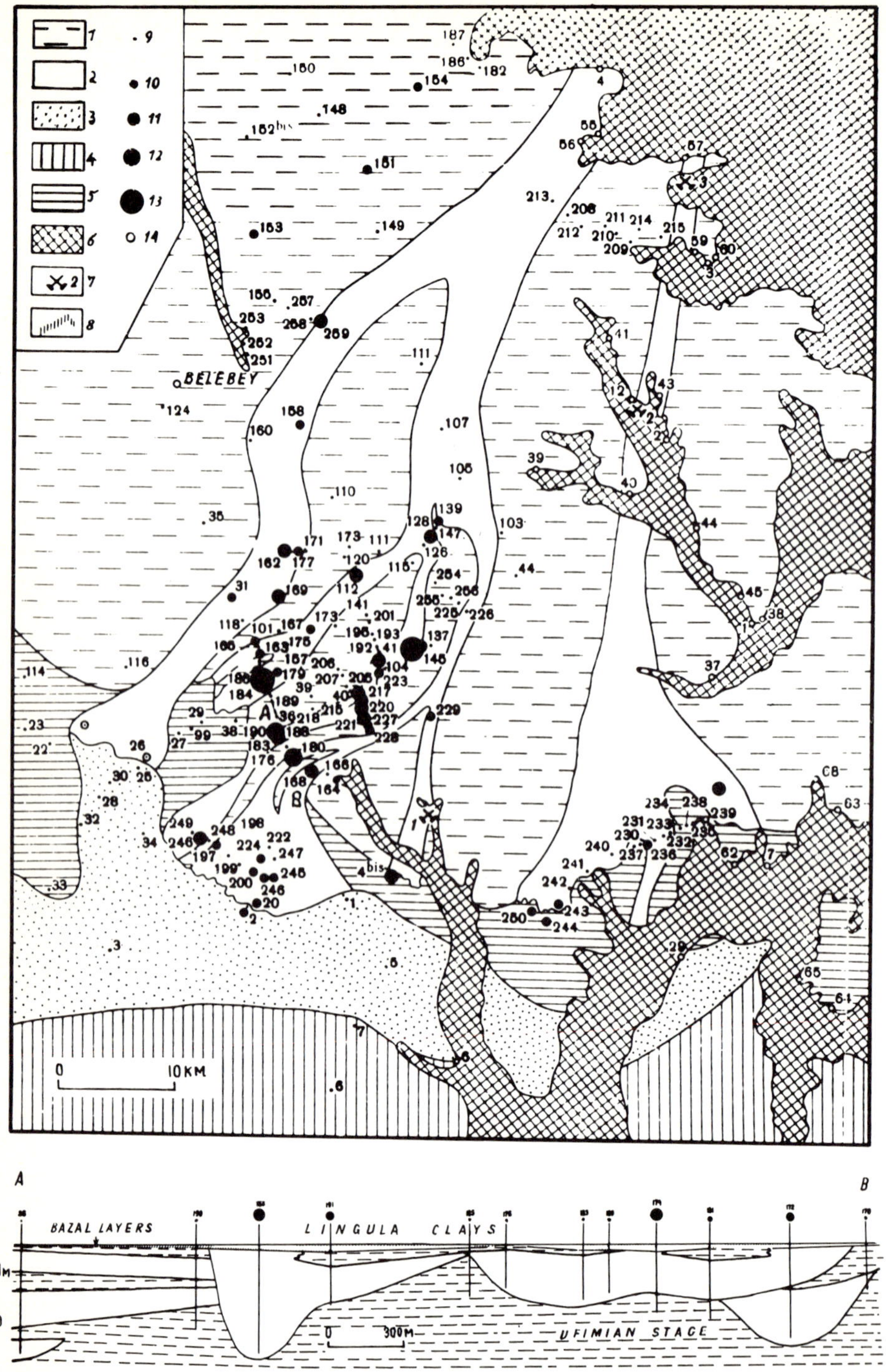
BELEBEY
0 10KM
A
B
BAZAL LAYERS
LINGULA CLAYS
UFIMIAN STAGE
4M
0
0 300M

Fig. 3

the Buchenau Transversal Uplift. The copper mineralization is restricted exclusively to the area overlying the conglomerates at the top of the Rotliegend. The contours of the ore bodies in the cupriferous shales match very closely the interface where the conglomerates grade laterally into sandstones (Richter 1941).

In the Zambian Copper Belt, the pre-Katangan erosional relief provided the control on the distribution of the deposits in the Roan basin. Topographically low areas in the basin are filled with terrestrial rock deposits that are similar to those comprising the pre-ore horizon. High areas reaching to the base of marine deposits are overlain by barren rocks, which are directly contiguous to the richest ores (Mendelsohn 1963).

In the White Pine deposit, the copper mineralization extends over a ridge buried beneath the red beds of the Copper Harbor Formation. This ridge had no influence on the facies in the Nonesuch Basin, but, nevertheless, an ore zone occurs above it (Ensign et al. 1972).

4. No changes in the position of the ore zone are apparent even when the environment of sedimentation underwent abrupt changes. The transition from basal sandstones to limestones and Lingula clays in the Kazanian section did not effect the distribution pattern of the copper mineralization. All the units of the section are ore-bearing in some places, and barren in others (Lur'ye 1972). The same observation applies to the White Pine deposits, where the ore section, 15 m thick, exhibits marked changes in its sedimentary environment, involving the uplift of the seafloor, partial draining of the basin, and subsequently more widespread transgression of the sea. These variations had no effect on ore emplacement: mineralization does not show any relationship to the regression and transgression of the sea (Ensign et al. 1972).

Notwithstanding the fact that the facies type does not control the position of ore localization in marine basins, this factor does determine the morphology of the ore body. Copper concentrates in those beds that are rich in carbon and reduced sulphur, and do not accumulate in facies of oxidizing environments, due to sections composed of several interbedded lithofacies, exhibit corresponding alterations of beds with different copper contents. In cases where lateral changes from reducing to oxidizing environments occur within the deposits, these facies changes provide the ultimate control on the distribution of the mineralization and so delimit the configuration of ore runs.

5. The orientation of the sulphide zones in ore bodies depends upon the position of the red-bed deposits. At the geochemical barrier, the redox potential undergoes a gradual change from high values in the red beds to low values in the grey sediments. The Eh – pH diagrams (Garrels 1962) show that decreasing Eh

←

Fig. 3. The Belebei area. Paleogeographical map of the end of the Ufimian and copper content in basal layers of the Kazanian marine sediments. Terrestrial condition of sedimentation (*1* – *2*): *1* coastel plain (red clays and silts with a trace of soil forming); *2* "shoe strings" and alluvial fans (red sands). Conditions of basin sedimentation (*3* – *5*): *3* bars (grey sands); *4* parts of the basin with high salinity (dolomite, gypsum, silts and clays with gypsum); *5* lagoons of low salinity (alternations of red clays, silts and more rarely, sands). *6* Areas where the top of the Ufimian stage was eroded; *7* dumps of former mine workings; *8* ore zones in the profile (copper content more than 0.1%). Copper content in basal layers of the Kazanian marine sediments (according to drill hole data) in kg m^{-2}: *9* less than 2.5; *10* more than 2.5; *11* more than 5; *12* more than 10; *13* more than 20; *14* outcrop

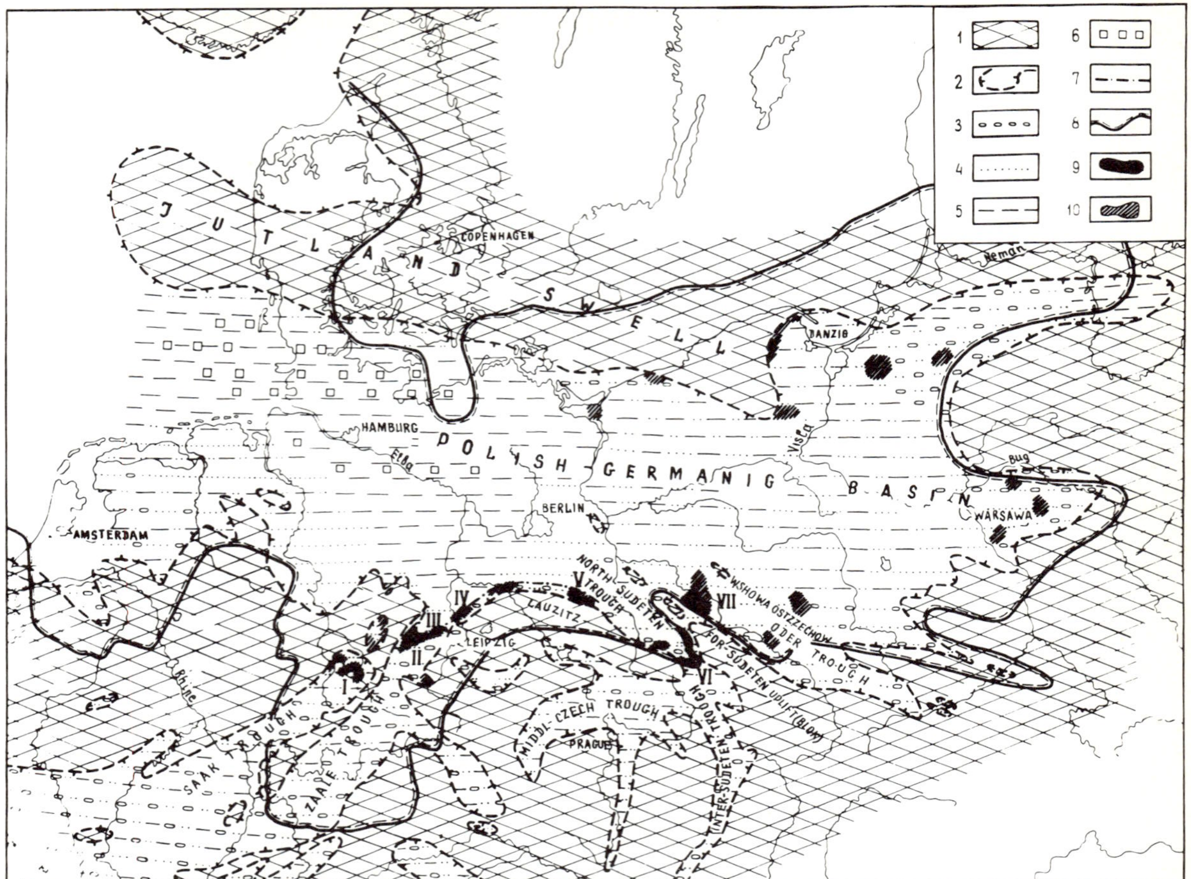

Fig. 4. Beds of Zechstein basin and position of the ore zones. *1* Source area in upper Rotliegend; *2* limits of sedimentation for regions of Upper Rotliegend; *3* – *6* rock composition of the Upper Rotliegend (*3* conglomerates; *4* sandstones; *5* argillites; *6* salts); *7* boundary of replacement of psephitic rocks of Rotliegend by argillaceous rocks; *8* boundaries of the Zechstein Sea; *9* copper de-

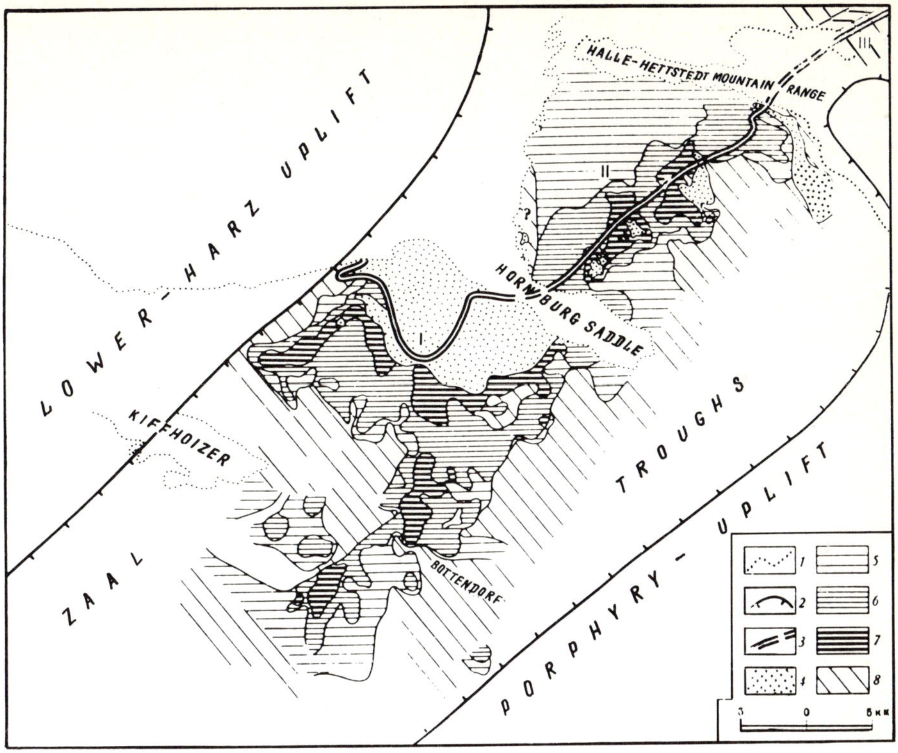

Fig. 5. Distribution of non-ferrous metals in the copper-shale horizon southeast of the Harz mountains, with elements of the paleography of the Upper Rotliegend. Compiled from papers of C. Freese asnd W. Jung (1965), E. Hoyningen-Hüne (1963), G. Knitzschke (1965), Autorenkollektiv (1968). *1* Outlines of present-day rock masses; *2* limits of sedimentation regions of Upper Rotliegend; *3* boundary of replacement of psephitic rocks of Rotliegend (in the southeast) by argillaceous rocks (in the northwest); *4* hematitic zones (Rote Fäule). Areas of copper mineralization (more than 5.5 kg $tonne^{-1}$) in beds of the copper-shale horizon (*5*–*7*): *5* in beds *1* and *2*; *6* in beds *1* to *4*; *7* in beds *1* to *6* (beds: *1* fine clay; *2* coarse clay; *3* Kammschale; *4* Schiefer Kopf; *5* Schwarze Berge; *6* Dachklotz); *8* areas with lead mineralization. Deposits: *I* Sangerhausen; *II* Mansfeld; *III* Edderitz

values in the Cu–Fe–S–O–H system are coincident with a change in the stability fields from chalcocite to bornite and then chalcopyrite. The changes in the sulphide composition of all ore bodies reveal similar sequential order of sulphide zones. The "chalcocite end" in the zonal sulphide series invariably faces the red beds. The copper content in the ores decreases with the distance away from the latter (commonly the most enriched are the chalcocite-bornite ores).

The widely held consensus which postulates that the supply of copper is derived directly from the continents in the process of ore genesis, does not provide a satisfactory explanation for the regularities outlined above. They can be comprehensibly interpreted only through the concept of leaching of copper and associated metals from subjacent red beds by subsurface waters. The inflow of subsurface waters into the Kazanian Sea could have occurred along the ancient channels as the top of the Ufimian (Fig. 3). The outflow of subsurface waters

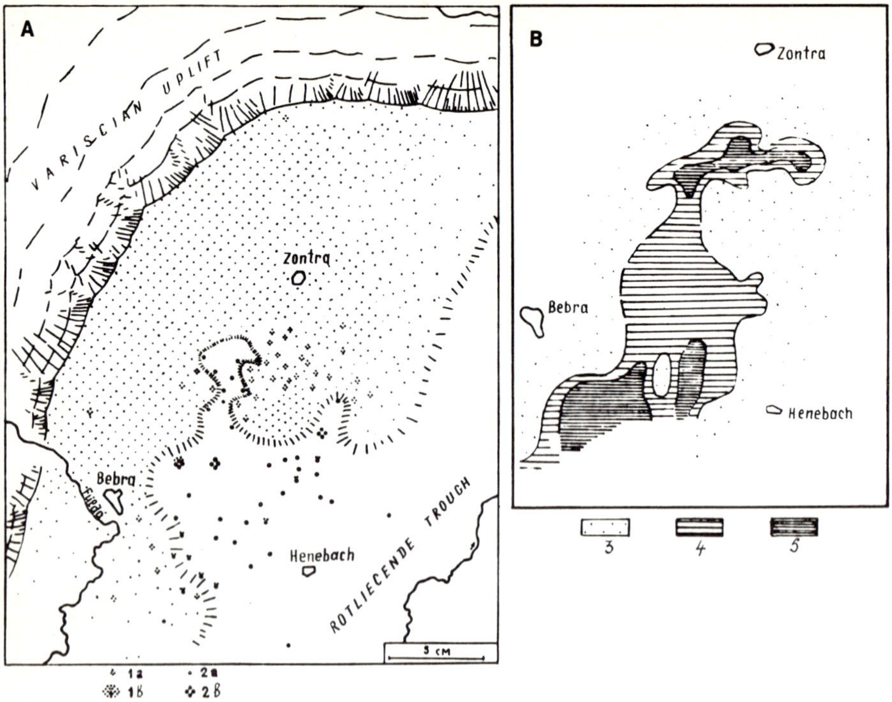

Fig. 6 A, B. Plan of the top of Rotliegend in the Richelsdorf deposits area (**A**), and copper content in basal layers of the Zechstein (**B**). (From Richter 1941). *1* Sandstones (*a* from drill hole data; *b* from outcrop data); *2* conglomerates (*a, b* the same as above). Copper content, in kg m^{-2}: *3* less than 10; *4* more than 10; *5* more than 20

from the Saar through into the Zechstein Sea would have been evoked by the impounding effect of the Buchenau Uplift (Fig. 6). The ore-bearing solutions would emerge as seeps within the conglomerate area at the top of the Rotliegend. This concept affords a feasible explanation for the matching of the contours of ore bodies in the Richelsdorf deposits with the boundaries of facies changes from conglomerates to more fine-grained rock types. In the Saale trough, the zones of facies changes from conglomerates to arenaceous-argillaceous deposits occurring at the top of the Rotliegend where the discharge areas for the subsurface waters flow out from the troughs into the sea (Hoyningen-Hüne 1963). The argillaceous sediments, as well as the basement complexes, served as obstructions impeding the flow of the subsurface waters along the seaward-inclined Saale trough (Figs. 4 and 5).

In the Roan basin, the subsurface waters at first oozed out within vast areas, forming extensive ore deposits in the basal marine sections. Then, as the permeability of the bottom sediments decreased under compaction, the discharge areas became increasingly more restricted and localized around the highs in the pre-Katangan relief where reservoir rocks of high permeability were preserved in the overlying sediments (disintegration products of the uplands). This resulted in the formation of concordant ore bodies in combination with flat-dipping crosscutting ore bodies. The formation of the copper ore zones at the bottom of the Nonesuch series can be explained by the presence of a burial ridge of rhyolites beneath the sediments of the Copper Harbor Conglomerate. Above the ridge the subsurface waters reached the sediments of the Nonesuch basin, where copper sulphides were precipitated (Ensign et al. 1972, Brown 1971).

It is common knowledge that the origin of hydrogen sulphide in the sediments is associated with biogenic sulphate reduction and also that copper solutions can exert a toxic effect on the barrier microflora. The inhibiting effect of the copper is not apparent as long as the amount of copper in the solution does not exceed the equivalent amount of hydrogen sulphide (Temple and Le Roux 1964). Consequently, it is conceivable that for the normal function of the hydrogen sulphide barrier, a definite sequence of events is required: first, the development of sulphate reduction processes in the bottom silts, and then the influx of copper-bearing solutions. If the influx of ore-bearing solutions into the sediments occurs prior to change of the oxidizing environment to a reducing environment, then no sulphide barriers can be formed. This is confirmed by interrelations of ore minerals: if there is an intergrowth of iron sulphides and copper sulphides in the ores, the latter consistently replace the first. If copper was originally present in the interstitial waters, then this copper would prevent the initial precipitation of iron sulphide until the almost complete extraction of copper from the solution.

The sulphate reserve is limited, insofar as the supply from bottom waters to the sediments ceases at depths between 4–6 m from the surface (Shiskina 1972). However, the high sulphur content in the ores, as compared to that in rocks of similar depositional conditions, can be explained by the presence of sulphates in the ore-bearing solutions, which is consistent with the concept of the formation of copper-bearing solutions under arid conditions and supports the sulphur isotopic data: the latter excludes the possibility that the entire supply of hydrogen sulphide could have been contributed by ocean waters. From the sulphur isotopic ratios, according to Vinogradov (1973), the inference can be drawn that the subsurface waters were the source of sulphate sulphur.

Taking into account the restrictions imposed by the toxicity of copper on the development of biogenic sulphate reduction, a rough approximation can be made of the rate of filtration of the ore-bearing solutions through the barrier. According to Kuznetsov (1976) the production of hydrogen sulphide by bacteria in silts can achieve several $mg\,l^{-1}\,day^{-1}$. Let us accept an average amount equal to $0.1\,mg\,l^{-1}\,day^{-1}$. This amount of H_2S can precipitate 0.4 mg of copper in the form of Cu_2S. Precipitation of copper from an ore-bearing solution with the concentration of $10\,mg\,l^{-1}$ requires the work of bacteria during $10:0.4 = 25$ days. With 10% water in the sediments (1 litre of solution for $10\,dm^3$ of sediment) copper toxicity will not be apparent if within 25 days the ore solution advances no more than $\sqrt[3]{10} = 2.15$ dm, i.e. the filtration rate should not exceed $21.5:25 = 1\,cm\,day^{-1}$.[2]

As a speculation it can be assumed that the formation of barren areas occurs where the influx of copper exceeds the allowable values. No matter how

[2] In all these calculations, the significance is in the order of magnitude of parameters but not in their absolute values.

paradoxical this may seem, it is possible that the more enhanced copper influx was the cause of the barren areas. In particular, this phenomenon can explain the fact that no correlation can be established between copper-bearing zones in marine silts and tectonic fault zones in the underlying red-bed formations. While the deposits are being formed, the discharge of copper-bearing subsurface waters occurs not in the form of submarine springs, but is distributed over the total area of the geochemical barrier by means of slow filtration through marine silts.

Epigenetic Barriers

Epigenetic barriers can be subdivided into two groups: allochthonous and autochthonous barriers. The first group involves those barriers where organic matter, which is the sulphate-reducing agent, was brought into the red-bed formation from nearby oil and gas source beds; the second group, where organic matter accumulated together with the sediments.

An obvious example of deposits formed at an allochthonous geochemical barrier is the Central ore zone of the well-studied Dzhezkazgan copper deposits (Fig. 7). The ore rocks of the minable series in these deposits accumulated in oxidizing environments. The grey color of the rocks is secondary in nature (Germanov 1962, Lur'ye et al. 1978, Feoktistov et al. 1971). The change in the

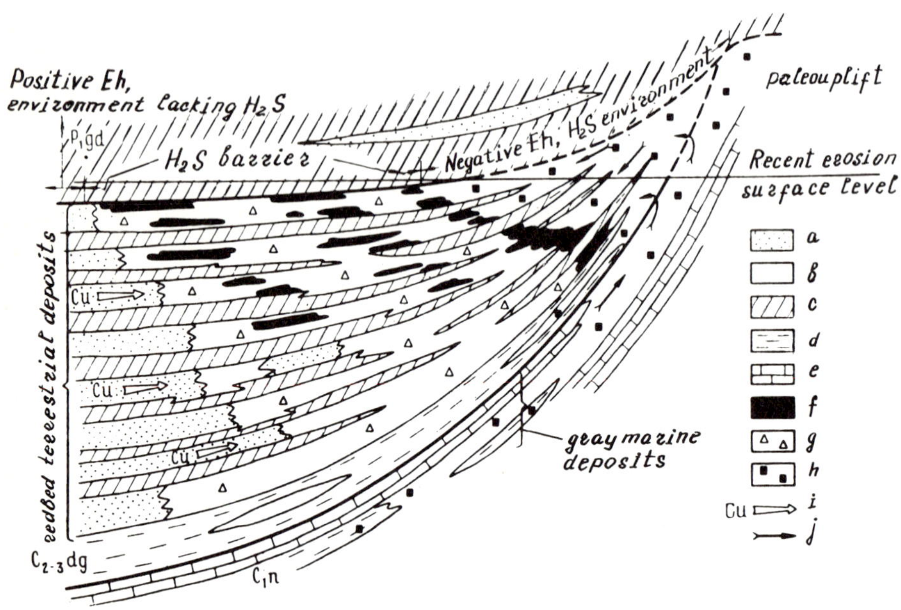

Fig. 7. Formation of copper deposits in lithified rocks as exemplified in the Dzhezkazgan ore deposits. Allochthonous hydrogen sulphide barrier after Lur'ye and Gablina (1978). *a* Red sandstones, conglomerates; *b* grey sandstones, conglomerates; *c* red siltstones, argillites; *d* grey siltstones, argillites; *e* limestones; *f* ore deposits; *g* disseminated copper sulphide; *h* disseminated pyrite; *i* flow of waters containing copper; *j* flow of waters containing organic matter

Dzhezkazgan deposits from oxidizing to reducing conditions and the formation of the hydrogen sulphide barrier, according to Germanov (1962), is associated with the penetration of hydrocarbons into the ore rocks from underlying Lower Carboniferous marine deposits. The geological and geochemical data available to date gives supporting evidence for this concept. In the Dzhezkazgan deposits a lithological window was identified through which the hydrocarbons could have penetrated into the ore rocks. This window is located near the paleo-uplift, where the minable series was not deposited. Approaching the paleo-uplift, all the red-colored rock types are replaced by grey sandstones. The intensity of reduction in the host rocks tends to increase towards this lithological window: the relict red-stained rocks completely disappear, and the sulphide composition of the ores undergoes a gradual change towards minerals that require lower redox values for their formation. Hydrocarbon compounds of the type associated with petroleum have been encountered in the reduced rocks (Poplavko et al. 1977, Susura 1980).

Numerous examples can be offered of epigenetic deposits with autochthonous hydrogen sulphide barriers. Among these are the Zhilandinskaya group of deposits in the Dzhezkazgan area (Esenov et al. 1975), and the Kargala deposits in the West Fore-Ural region (Lur'ye 1973). The first group of deposits is restricted to rock formations, which, according to their position in the section, are transitional from marine to terrestrial deposits. The copper mineralization is confined to sandstone layers containing coal detritus. The grey marls and argillites enclosing these beds, contain copper only immediately adjacent to the contact with the ore body.

The ore bodies in the Kargala deposits are confined to channels infilled with clastic material. The channels are incised in the "marly bed". The higher copper contents tend to occur near the accumulations of plant debris in the channel deposits. Copper-bearing marls occur only around the fringes of the ore bodies.

The origin of hydrogen sulphide at epigenetic barriers is more intricate, due to the fact that it involves the introduction of an important new factor – temperature. Pecularities of the mineral composition of the copper sulphides[3] infer that the deposits were formed at temperatures below 75 °C and therefore hydrogen sulphide may be of biogenic origin. From this approach to the problem, the resemblance of pecularities of the copper sulphides, accumulating in the sediments to those of lithified rocks, is a quite expectable feature, since it reflects similarities in the conditions of ore genesis.

As Rose (1976) showed, the most probable form of copper transport in the ore solutions is the form of cupreous-chloride complexes: $[CuCl_2]^-$ and $[CuCl_3]^{2-}$. The copper content in the form of chloride complexes in the solution can reach 100 mg l^{-1} at 0.5 m Cl^- and pH 7. Let us take a copper concentration in the ore solution of 10 mg l^{-1}. If rocks of a specific gravity equal to 2 have a water content of 10%, then for the conversion of this water into a mineralized solution, the amount of copper that should be extracted from 1 dm^3 (2 kg) of rocks would equal 1 mg, on 5×10^{-5}% of its weight. This figure is lower than the

[3] Monomineral inclusions of djurleite, bornite with dispersed phases of chalcopyrite and chalcocite, as well as *x*-bornite, are peculiarities of sulphides of copper in these deposits (Lur'ye 1978). *X*-bornite is not stable and decomposes at 75 °C.

abundance of copper in the red-bed formations by a factor of $10-10^2$. During multiphased water interchange, the amount of copper extracted from the source beds could be considerably greater.

Taking the rate of flow of subsurface waters equal to 1 cm day^{-1}, a total porosity of 10% and an effective porosity of 6%, then the calculated amount of ore solution flowing through an area of 1 m^2 of the hydrogen sulphide barrier would average 200 l yr^{-1}, from which 2 g of copper could be precipitated in the form of sulphides. The maximum mineralization of the ore bed in the Mansfeld deposits (250 kg m^{-2}) could be attained in a mineralization period lasting 125000 yrs.

As shown above, for normal functioning of the hydrogen sulphide barrier, the amount of copper supplied per time unit (the product of copper concentration in the ore-bearing solution and its flow rate) cannot significantly exceed the production of equivalent amounts of hydrogen sulphide. As there is no reason to suggest considerable increases in the sulphide reduction process, time remains the controlling factor, defining the growth of ore zones within the epigenetic barrier. Since the linear dimension of epigenetic barriers in the direction of solution flow is greater by a factor of 10^2, compared to that of syngenetic barriers, the mineralization period will be millions of years in duration rather than tens of thousands of years. The possibility of hydrogen sulphide influx from outside should not be neglected for epigenetic barriers. In this case limitation in the amount of copper supplied to the barrier, and thus the duration of formation of deposits, will be defined by other factors.

From the foregoing calculations it becomes immediately apparent that the source of copper is not related to the content of copper in red beds; it is only essential that copper should occur in a form that can be released from the rocks into the pore solution, and also that from such an immense reservoir as red-bed formations, the subsurface waters should flow towards a restricted area where a hydrogen sulphide barrier functions over a long period of time. From this, it is conceivable that the formation of large deposits requires a specific paleohydrogeological and geochemical environment. These genetic considerations should furnish the basic concepts for prospecting for copper-sandstone and copper-shale deposits.

Acknowledgements. I am indebted to Prof. A. W. Rose from Pennsylvania State University (USA) for his comprehensive comments on the first version of this paper.

References

Ackermann E, Forster A (1960) Grundzüge der Stratigraphie und Struktur des Irumiden Troges. In: Int Geol Congr. Rep 21 Session. Norden, part 18. Copenhagen, pp 182–192

Brown AC (1971) Zoning in the White Pine copper deposit, Ontonagon Country, Michigan. Econ Geol 66, 3, 4:540–572

Ensign KO Jr, White WS, Wright JC et al. (1972) Copper deposits in the Nonesuch Shales of Michigan. In: Ore Deposits of the United States. Izd Mir 660 pp

Esenov Sh E, Zaizev YA (eds) (1975) Geology and mineral resources of the Dzhezkasgan ore district. Nedra, 284 pp (in Russian)

Feoktistov BP, Kramarenko LE (1971) Bearing on the gray color of the rocks in the Dzhezkazgan deposits. Litologia Polez Iskopaem 3:119–125

Freese G, Jung W (1965) Über die Rotfärbung der Basalschichten des Zechsteins (Rote Fäule) und ihre Beziehung zum Nebengestein im südöstlichen Harzvorland. Freiberg Forschungsh, Leipzig, 193:9–23

Garrels R (1962) Mineral equilibria. Mineralniye ravnovesia. M Izd-vo Inostr Lit, 306 pp

Germanov AI (1962) Hydrodynamical and hydrochemical formation conditions of certain hydrothermal deposits. Gidrodinamicheskiye i hidrokhimicheskiye uslovia obrazovania nekotorikh gidrotermalnikh mestorozhdenii. Izv AN SSSR Ser Geol 7:79–98

Hoyningen-Hüne E (1963) Zur Paläohydrologie des Oberrotliegenden und des Zechsteins im Harzvorland. Ber Geol Ges DDR 1:201–220

Knitzschke G (1975) Die wichtigsten Erzminerale des Kupferschiefers sowie seines unmittelbaren Liegenden und Hangenden im Südöstlichen Harzvorland. Z angew Geol 12:625–637

Kuznesov SI (1976) The results and outlook of development of geological microbiology. In: Ecology and geochemical activity of microorganisms. Izd-vo ONTI Nauchnogo Centra Biologicheskikh Issledovanii. Puschino, 179 pp (in Russian)

Lur'ye AM (1970) Regularities in location of copper mineralization in Lower Permian variegated formation of Donbass. In: State and problems of Soviet lithology. V kn.: Sostoyanie i zadachi Sovetskoi lithologii. Vol II Izd Nauka, 248 pp (in Russian)

Lur'ye AM (1972) Mansfeld-type mineralization in Upper Permion Sediments of Western Urals Foreland. Academii Nauk SSSR vol 203, 3:658–661 (in Russian)

Lur'ye AM (1973) The types of copper deposits in redbed formation of the Russian platform. T. XV, 6:79–88 (in Russian)

Lur'ye AM (1978) Alteration of ores during epigenesis and metamorphism in the copper deposits of the redbed associations. Intern Geol Rev 20, 6:627–636

Lur'ye AM, Gablina IF (1978) Principle scheme for the formation of Exogenetic copper deposits. Printsipal'naya skhema obrazovania ekzogennikh mestorozhdenii medi. Dokladi AN SSSR. Vol 241, 6:1402–1405

Mendelsohn F (ed) (1963) The copper belt of Northern Rhodesia. Mednii poyas Severnoy Rodezii (pod redaktsiyey F. Mendelsona). M Izd-vo Inostr Lit, 472 pp

Poplavko EM, Ivanov VV, Serkis YuT (1977) On the geochemistry and formation conditions of cupriferous sandstones and shales. O geokhimicheskikh osobennostyakh i uslovia obrazovania medistikh peschanikov i slantsev. Geokhimia 88:1217–1233

Richter G (1941) Geologische Gesetzmäßigkeiten in der Metallführung des Kupferschiefers. Arch Lagerstättenforsch 73:61

Ridge TD (ed) (1972) Copper deposits in the Nonesuch Shales of Michigan. In: Ore deposits of the United States. AIME, New York, pp 602–627. Medniye mestorozhdenia v slantsakh Nonesuch, shat Michigan. V kn.: Rudniye mestorozhdenia Soedinyennikh Shtatov. Izd Mir, pp 602–627

Rose AW (1976) The effect of cupreous chloride complexes in the origin of red-beds copper and related deposits. Econ Geol 71, 6:1036

Strakhov NM (1963) Types of lithogenesis and their evolution in the history of the earth. M. Gosgeoltekhizdat, 535 pp (in Russian)

Shishkina OV (1972) Geochemistry of interstitial waters in bottom sediments of seas and oceans. Geokhimia morskikh i okeanicheskikh ilovikh vod. Izd Nauka, Moscow, 238 pp

Srakhov NM (1963) Types of lithogenesis and their evolution in the history of the earth. Tipi litogeneza i ikh evolyutsia v istorii zemli. M Gosgeoltekhizdat, 535 pp

Susura BB (1980) Geochemistry of epigenetic ore-deposition in the red-bed formations of the Chu-Sarysui depression in Kazakhstan (copper-sandstone type). Geokhimia epigeneticheskovo rudoobrazovania v krasnotsvetnikh formatsiakh Chu-Sarysuiskoy deressii Kazakhstana (tip medistikh peschanikov). Avtoref Diss Na Soiskaniye Uchen. St Kand Geol Miner Nauk M:27 pp

Temple KL, Le Roux NW (1964) Syngenesis of sulfide ores: Sulfate-reducing bacteria and copper toxicity. Econ Geol 59, 2:271–278

Vinogradov VI (1973) The source of sulphur in ore-deposits, as reflected by sulphur-isotope data. In: Int Geochem Congr 2:45–53. Istochnik seri rudnikh mestorozhdenii po izotopninm dannim. V kn.: Mezhdunarodnii geokhimicheskii kongress. M T 2, pp 45–53

Major Types of Copper-Bearing Zones in the Soviet Union

Y. V. BOGDANOV[1]

Abstract

The territory of the Soviet Union is characterized by orogenic, platform and geosynclinal, copper-bearing zones. Major epochs of sedimentary copper accumulation are: Early and Late Proterozoic and Late Paleozoic, favouring generation of orogenic copper sandstone, and Middle Mesozoic, favouring the emplacement of copper – pyrrhotite mineralization in early geosynclinal black shale trough zones.

Different ways of evaluating areas with good potential are shown, based on the cyclic recurrence and facies features of copper-bearing sediments, as well as structure of productive strata in deposits.

The cyclic pattern of copper-bearing molasse sediments determines the morphology of productive strata, their simple or multicyclic structure. Flyschoid sediments contain copper sandstone with a simple structure of productive strata, having relatively small size.

Ore-bearing deposits of arid red beds are divisible into the following zones: (1) copper and polymetallic mineralization in red and variegated sediments of greywacke composition and (2) copper mineralization in variegated and red sediments of arkose composition and displaying thick cycles.

Introduction

The extensive territory of the Soviet Union is one of the few regions of the earth characterized by widespread, copper deposits of sedimentary and sedimentary-volcanogenic origin of different age. Well known are such copper-bearing provinces in our country as Central Kazakhstan (Dzhezkazgan, Sary-Oba, etc.) and the North Transbaikal area (Udokan, Unkur, etc.), which are similar in many characteristics to such foreign ore-bearing areas as the Copperbelt in Central Africa, the Fore-Sudetic Trough in Lower Silesia (Lubin-Sieroszowice), Poland, and Central Afghanistan (Ainak, etc.).

Also well known is the great economic significance of copper sandstone deposits. Economically, ores of this type rank second among copper deposits

[1] VSEGEI, Leningrad, USSR

Geology and Metallogeny of Copper Deposits
(ed. by G. H. Friedrich et al.)

after ore bodies of the porphyry copper type. Copper sandstone is economically valuable due to the morphology of its ore bodies, which is favourable for exploitation (large sheet-form deposits), and a simple concentration process for most ores, which have a rather high (>1.5% Cu) copper content and contain admixtures of easily extractable precious and rare metals.

The potential for discovering new copper sandstone deposits is far from being exhausted. This potential is confirmed by the fact that during recent decades a number of intersting ore bodies were discovered in the USSR (the Igarka area, Sette-Daban, Western Darvaza) and abroad (Ainak in Afghanistan). Thus, it seems appropriate to characterize the copper-bearing zones and to discuss a more detailed classification in order to solve urgent problems of discovering additional deposits and evaluating areas with high copper potential.

Major Types of Copper-Bearing Zones

In the Soviet Union, several major copper-bearing zones are classified on the basis of their geotectonic position into orogenic, platform and geosynclinal formations. The stratiform copper deposits occurring within these zones differ greatly in many parameters (Fig. 1).

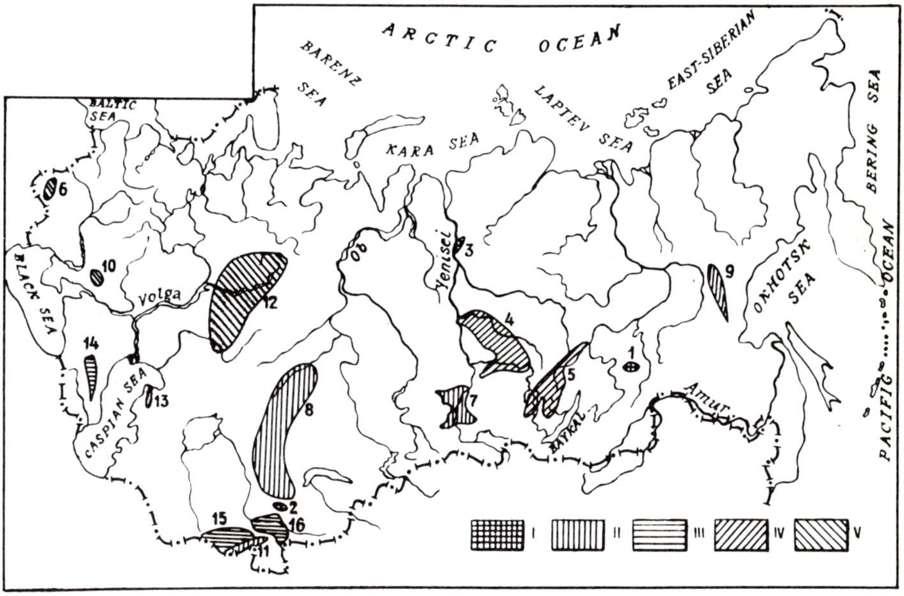

Fig. 1. Distribution of major copper-bearing zones in the territory of the Soviet Union. *I–III* Copper-bearing zones of fold belts: *I* Proterozoic; *II* Paleozoic; *III* Mesozoic-Cenozoic; *IV–V* copper-bearing zones of platform mantle. *IV* Early Paleozoic; *V* Middle and Late Paleozoic. Zones: *1* Kodar-Udokan (PR_1); *2* Talas (PR_2); *3* Igarka (PR_2); *4* Yenisei (C_{2-3}); *5* Lena (C_3-O_1); *6* Dnestr (D_1); *7* Minusa ($D-C_1$); *8* Tengiz-Sarysu ($D_{2-3}-P$); *9* Sette-Daban (D_3-C_1); *10* Donbas (P_1); *11* West Darwaza; *12* Fore-Urals (P_2); *13* Mangyshlak ($P-T$); *14* Near-divide ($T-J_2$); *15* Tadjic (K_1), *16* Fergana (Ng)

The specific character of copper mineralization in orogenic and platform zones which are underlain by red and variegated carbonate-terrigenous deposits, is determined mainly by the composition of the clastic material and its cement. Copper-bearing carbonate-cemented red beds of arkosic composition are characterized by a limited suite of accessory components (minor Ag, Mo, Bi). Carbonate-cemented red beds of greywacke composition concentrate, along with copper, admixtures of Ag, Co, Pb, Mo, Re and other elements. Quite different ore elements are accumulated in humid carbonate-free red beds. These are primarily, iron, bauxites and phosphorites.

Orogenic Copper-Bearing Zones

This type of extensive ore-bearing area is characterized by the largest number of stratified copper sandstone ore deposits. Copper-bearing red and variegated carbonate-clastic and volcanogenic-clastic rocks occur in early and late orogenic troughs of varied shape and size. Depending on the tectonic regime and the features of orogenesis, these rocks are characterized by a cyclic recurrence on different scale. The lithologic composition of these deposits is mainly determined by two factors, viz., the structure of the eroded basement blocks and features of preceding or synchronous volcanism, as well as by specific paleogeographic and paleotectonic conditions in the source and sedimentation areas. The above factors, to a large extent, predetermine the special metallogeny of the red and variegated formations.

In the North Transbaikal area, within the Olekma-Vitim mountain land, numerous troughs of Karelids are developed, filled in by metamorphosed Early Proterozoic red beds. The most interesting structure within this ore-bearing province is the Kodar-Udokan zone, containing the Udokan and several minor copper sandstone deposits (Fig. 2). The regional distribution of copper-bearing sediments displays a distinct zonal pattern relative to source areas. The stratigraphic and formational control of mineralization have been recorded. This evidence, along with the great areal extent of ore mineralization, provides the main basis for assigning the mineralization in question to a primary sedimentary origin (Bogdanov 1966). It should be noted that rich ores display many features of epigenetic concentration, viz., (1) an association with the most coarse-grained sandstone; (2) the presence of strongly recrystallized carbonate or carbonate-mica-quartz cement; (3) the development of various stringer – impregnated structures and texture in ores; (4) a crosscutting character of mineral zoning relative to the rock bedding (Bogdanov and Feoktistov 1982).

In the rather complicated accumulation history of copper-bearing sediments in the Kodar-Udokan trough, two major stages are distinguished; (1) early orogenic – accumulation of flyschoid greywacke deposits of the China Series (3000 – 4000 m) and (2) late orogenic-sedimentation of arkosic molasse sediments of the Kemen Series (4000 – 5000 m).

Ore bearing deposits of these series from two belts, extending up to 150 km and 20 – 25 to 40 km wide, which surround denudation areas or paleouplifts in the orogenic trough. In each belt, there is a gradation up the section from eche-

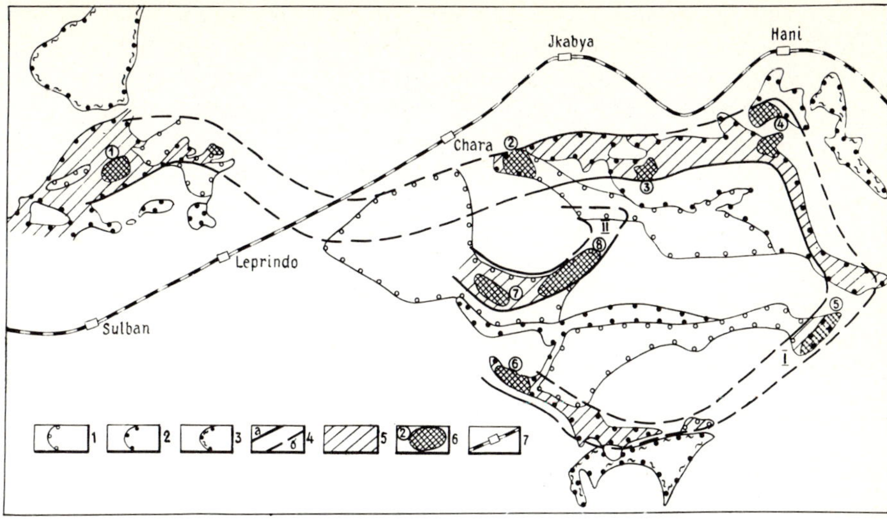

Fig. 2. Distribution of copper sediments of the Kodar-Udokan zone. *1* Sediments of the molasse siltstone-sandstone series; *2* sediments of the flyshoid sandstone-mudstone-siltstone series; *3* shale strata, enclosing schistose rock block of the flyshoid series formation; *4* copper-bearing belts, confirmed (*a*) and assumed (*b*) (*I* Aleksandrovsky-Chitkanda; *II* Naminga-Sakukan); *5* coastal-marine sediments with some copper-bearing horizons; *6* lagoonal-deltaic copper sediments. Areas *circled numbers*: Sulban (*1*); Unkur (*2*); Ikabya (*3*); Hani-Sakukan (*4*); Burpala (*5*); Pravo-Ingamakit (*6*); Udokan (*7*); Saku (*8*); *7* Baikal-Amur magistral

lonlike lenses of copper sandstone to more continuous copper slate layers (Fig. 3).

The outer belt is most uniform and corresponds to an older Aleksandrovsky-Chitkanda group of the China Series in the Udokan carbonate-terrigenous deposits. The ore-bearing part of the section is characterized by flyshoid cyclic recurrence and a greywacke composition of the clastic rocks. Copper mineralization is confined to coastal-marine sediments comprising lenses of silty-sandy lagoonal-deltaic deposits (at Krasnoe, Pravyj Ingamakit, etc.). The thickness of productive strata in these mineralized areas amount to tens of metres. Lenticular ore bodies occur in an echelon. Ores are dominated by chalcopyrite, bornite and, less frequently, chalcocite, occurring together with pyrite and pyrrhotite. Increased content of silver and molybdenum, and less frequently, of lead, nickel, cobalt, arsenic, and bismuth, are recorded. Ores are characterized by a comparatively narrow range of isotopic composition of sulphide sulphur (from $-15‰$ to $+10‰\ \delta S^{34}$).

The well-known Udokan deposit and a number of cupreous sandstone ore occurrences are confined to the inner belt, corresponding to the younger Naminga-Sakukan group. The productive strata at Udokan, about 400 m thick, are composed of metamorphosed clastic sediments of an arkose composition, having thick cycles. Lenticular ore bodies display an echelonlike occurrence in the productive sequence. Stringer-impregnated bornite-chalcocite and bornite-chalcopyrite ores dominate, pyrite-chalcopyrite ores being found much more

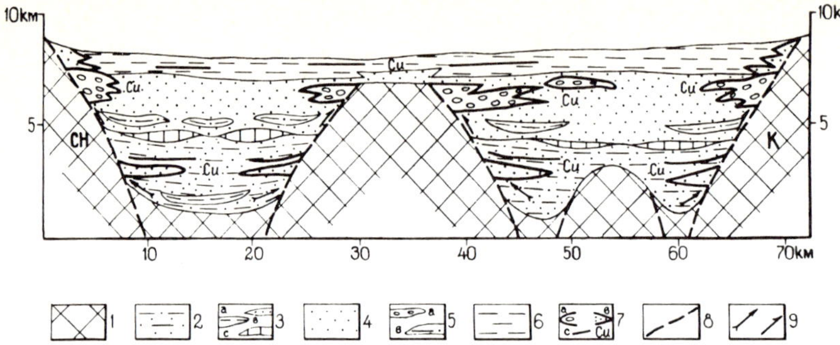

Fig. 3. Genetic model of copper sandstone deposits of the Kodar-Udokan zone. *1* Area of denudation – blocks of Archean rocks: Chara (*CH*) and Kalar (*K*); *2 – 3* China Series: *2* flyshoid sandstone-mudstone-siltstone sediments; *3* lenses of siltstone (*a*), sandstone (*b*), and carbonate rocks (*c*); *4 – 6* Kemen Series: molasse siltstone sandstone sediments; *5* lenses of sandstone (*a*) and siltstone (*b*); *6* mudstone-siltstone sediments of Naminga formation; *7* copper-bearing horizons, enclosing copper sandstones of Dzhezkazgan (*a*) and Dzhilanda (*b*) types and copper slates (*c*); *8* faults; *9* directions of migration of elision waters

rarely. These ores are characterized by low contents of accessory elements (silver and bismuth). The isotopic composition of sulphide sulphur is highly variable (a range of 50‰ δS^{34}), with light isotopes dominating, which points to the presence of biogenic sulphur in these cupreous deposits.

The small areal extent of deposits of the inner copper-bearing belt testifies to the fact that the prospects for discovering cupreous sandstone deposits of the molasse type, similar to the Udokan deposit, are limited. The potentials of the outer belt, which has a great areal extent, are much higher (Bogdanov et al. 1973).

In Central Kazakhstan, Upper Paleozoic cupreous deposits are localized within the extensive Teniz-Sarysu zone which has a north-south strike. The most interesting copper sandstone deposits have been discovered in the Dzhezkazgan area. There, two long ore-bearing belts of a north-eastern strike have been outlined, viz., the northern one – Taskuduk, and the southern one – Dzhezkazgan. These belts are composed of variegated lagoonal-deltaic silty-sandy deposits. Within the above belts, several copper sandstone deposits have been discovered, differing in size, and morphology and in the structure of the productive strata and certain ore bodies. The Taskuduk flyschoid belt includes the Sary-Oba, Itauz and a number of smaller deposits and ore shows, which are characterized by a simple structure.

In the Dzhezkazgan molasse belt, apart from the Dzhezkazgan deposit, only a few ore shows of copper sandstone are recorded. The productive strata of Dzhezkazgan, about 650 m thick, have a multicyclic structure with an echelonlike occurrence of lenticular ore bodies. Up the section of the productive strata, the morphology of ore bodies and mineral zones becomes more complicated (Bogdanov et al. 1973, Narkelyun et al. 1983).

The mineral composition of the ores is relatively simple. Impregnated copper ores are dominated by chalcocite, bornite and chalcopyrite. Pyrite, arsenopyrite, tennantite and other more rare minerals are also found. Ores are dominated by light sulphur isotopes, probably of biogenic origin (δS^{34} from $+1.00$ to -4.02‰). The economic potential of the more uniform Taskudak belt seems much higher. Since this belt was outlined, several minor cupreous sandstone deposits have been discovered within it. South of Dzhezkazgan, in the Chu-Sarysu Depression, Carboniferous variegated rocks on the slopes of various uplifts should be studied.

Among orogenic copper-bearing areas of a simpler structure, the Igarka area in the Lower Yenisei River should be mentioned. Not long ago, Vendian cupreous sandstone was discovered there. The copper-bearing red-bed formation is confined to the junction area of the Igarka protrusion of Baikalids and a piedmont trough. Stratified copper mineralization, localized on several horizons, is represented by stringer-impregnated bornite-chalcocite and chalcopyrite mineralization in grey clastic sediments (Narkelyun et al. 1983).

The economic potential of copper in the Igarka area has not been adequately studied. First, the Vendian variagated deposits, occurring near paleouplifts should be studied in detail and prospected.

With a certain degree of conventionality, the ore areas of Western Darvaza in Central Asia (P_2), Sette-Daban in East Siberia (C_1) and the Minusink Basin in the southern Krasnoyarsk area ($D-C_1$) can also be assigned to these types of copper-bearing areas. These copper-bearing areas display an increased content of pyroclastic rocks in the section and include sheetlike bodies of basaltic or andesitic volcanics. The evaluation of the economic potential of the above copper-bearing areas has not yet been completed.

Copper-Bearing Areas in Platform Regions

Within the sediments mantling the platform, there are practically no economic copper-bearing areas. Numerous ore shows of cupreous sandstone are recorded near large arched paleouplifts on the eastern margin of the Russian Platform, in the Fore-Uralian area and in the southern Siberian Platform in the Upper Lena River (Bogdanov et al. 1973, Narkelyun et al. 1983).

In the Fore-Urals, malachite and chalcocite mineralization is confined to Lower Permian alluvian and lacustrine continental deposits, displaying highly variable facies. There are small bodies of rich epigenetic copper ores, confined to fluvial facies. There is also a progressive decrease in age of cupreous sandstone from north to south.

Lower Permian sandstone is also reported from the southern Russian Platform, the slopes of the Bakhmut and the Kalmius-Torets basins. In this latter area, a rhythmic alternation of variegated mudstone, marl, siltstone and sandstone of the Kartamysh Formation, assigned to shallow marine sediments, is recorded. Ore-bearing horizons of small thickness are rather uniform along the strike and dip. Cupreous rocks contain low-grade impregnations of chalcocite, and less frequently of chalcopyrite, bornite and pyrite.

The economic potential of copper presence in red formations of marginal, tectonically activated parts of the Russian Platform have not been adequately studied. As a first priority, the slopes of synsedimentary paleouplifts (domes, arches, swells, etc.) should be studied, for they frequently control the distribution of copper sandstone deposits (Fore-Sudetic Block in Lower Silesia, the Igarka protrusion in the Krasnoyarsk area, etc.). Of certain interest in this respect is the Fore-Urals area near Orenburg, where lagoonal-deltaic variegated rocks underlie slopes of such paleostructures.

In the southern Siberian Platform, cupreous deposits occur within the lower Paleozoic of the Lena and Yenisei areas. These are thin horizons of cupreous siltstone, sandstone and marl of shallow, marine facies.

The most interesting, the Lena copper-bearing zone, 600 by 200 km, extends northeastwards and includes areas, underlain by the Verkholensk and Ilga formations (C_{2-3}), as well as by lead-zinc-bearing biogenic carbonate rocks of the Ust-Kut Formation (O_1). The thickness of copper-bearing horizons ranges from 1 to 10 m, rarely more. Towards the central parts of synsedimentary troughs, the thickness of the Ilga Formation increases from 60 – 70 to 300 m, and the number of copper-bearing horizons grows from one to three. Cupreous rocks are dominated by malachite. In places, chalcopyrite, bornite and chalcocite, as well as pyrite, galena and tennantite, occur.

The economic potentials of the Lena and Yenisei copper-bearing areas are not quite clear. Further studies on the structural control of cupreous sandstone should be carried out. Thus, of great significance is the study of the distribution pattern of cupreous sandstone, particularly of lagoonal-deltaic facies, on the slopes of various paleouplifts (the Nepa Arch, etc.). Of certain interest with regards to the copper presence are variegated deposits of the so-called podsolevoj[2] complex ($V - C_1$).

Copper-Bearing Zones of Geosynclinal Areas

In geosynclinal areas, stratiform copper deposits of a sedimentary-volcanogenic origin are mainly found among black shale formations of the initial evolution stages of troughs. Pyrite-polymetallic deposits, found among sedimentary-volcanogenic formations in such provinces as the Urals and the Ore Altai, are commonly assigned to the family of pyrite deposits (Savin 1978). Along with these deposits, an isolated group is frequently formed by copper-pyrrhotite deposits, localized in black shale formations. The Near-divide metallogenic zone of the Great Caucasus can be named among the areas with extensive showings of copper-pyrrhotite mineralization. Within this zone, a very thick (3.5 to 12 km) sequence of Lower-Middle Jurassic flyschoid clastic rocks occurs and encloses comprising copper-pyrrhotite deposits (Kizil-Dere, Katsdag, etc.). The exhalative-sedimentary copper-pyrrhotite mineralization is commonly localized in supravolcanogenic parts of the section. Ore bodies at the Kizil-Dere deposit are lenticular, less frequently vein-shaped, with characteristic pinches and swells. Bodies of massive copper-pyrrhotite ores are accompanied by aureoles of stringer-impregnated mineralization of the same composition. Rocks of the footwall of ore deposits are strongly quartzified, sericitized and carbonatized. The

[2] Podsolevoj complex – rocks underlying salt strata.

thickness of these zones of wallrocks alteration is 20 to 40 m. The ores, along with pyrrhotite, pyrite and chalcopyrite, also contain sphalerite, galena, and less frequently, cobaltite and magnetite (Savin 1978).

The presence of copper-pyrrhotite mineralization in shale structural-formational zones points indirectly to a possible presence of pyrite-polymetallic mineralization. Thus, many early geosynclinal black shale formations should also be studied.

Major Epochs of Copper Accumulation

The copper-bearing areas discussed are distinguished mainly by an extensive development of the following three types of structural-formational complexes, viz., orogenic and platform red and variegated carbonate-clastic and geosynclinal black shale deposits. Accumulation of the above deposits of copper-bearing complexes took place during certain evolution stages of geosynclinal-folded, orogenic and platform areas of different age. Thus, major ore epochs are distinguished, strongly differing in scale and character of mineralization. These are, primarily, the Early Proterozoic, Late Proterozoic and Late Paleozoic copper-bearing epochs.

The Early Proterozoic epoch of sedimentary copper accumulation was manifested during the stage of Karelian orogeny in the western Stanovoj Mobile Belt. There, the oldest copper sandstone and slate deposits were emplaced. In this region, cupreous red beds were accumulated in a system of north-south and east-west orogenic troughs. Cupreous sediments are commonly confined to the coastal-marine belt of an extensive ancient arid region. A multistage generation of cupreous sandstone took place during orogenic movements due to erosion of Archean blocks of the Aldan Shield.

Copper sediments are characterized by certain regular changes in time: gradations of greywacke into arkoses, flyschoids into molasses, increase of the thickness of productive strata of cupreous sandstone deposits, and a simpler composition of ores. The initial red colouring of rocks has almost disappeared due to regional metamorphism, corresponding to te greenschist facies or metagenesis stage.

The Late Proterozoic epoch of copper accumulation was manifested most intensively in red beds of foredeeps on the Siberian Platform. This is, primarily, the Igarka Trough, where in Riphean time (youngest Precambrian) several accumulation stages of cupreous sediments are distinguished, covering the marginal parts of arid areas and manifested most intensively near paleouplifts. Copper accumulation is much weaker in the Western Baikal area, where in Vendian, the emplacement conditions of cupreous sediments were characterized by a less arid climate.

The Late Paleozoic ore-bearing epoch was manifested in the development of copper-bearing arid carbonate-clastic and volcanogenic-clastic red beds in Caledonian and Hercynian orogenic troughs of the Sarysu-Teniz zone in Central Kazakhstan (C_{2-3}), Sette-Daban in East Siberian (C_1), Mangyshlak (P_1),

Western Darvaza (P_2), and Minusinsk depressions ($D-C_1$), as well as platform red beds of the Fore-Uralian area (P_1). This epoch is characterized by an absence of economic concentrations of cupreous sandstone among Mesozoic and Cenozoic red beds.

Of major interest are copper sandstone concentrations in the Dzhezkazgan area. They are characterized by: (1) association with coastal-marine (lagoonal-deltaic, etc.) greywacke deposits and (2) gradation of flyschoids into molasse sediments in time and increased size and grade of copper-bearing sequences.

Geosynclinal copper accumulation was restricted to occurrences of flyschoid black shale formations with copper-pyrrhotite mineralization (Kizil-Dere etc.) in the Mesozoic.

A Prognostic Evaluation of Copper-Bearing Areas

Regional prognostication is a process of successive differentiation of the regions evaluated, based on their potentials, with a view to establishing ore-bearing areas which are potential ore fields of deposits of the leading economic types. The lithological-petrological composition and the inner structure (rhythmic recurrence, etc.) of formational units predetermine their ore content. It is this very principle that determines the only correct order of regional prognostication: from studying the environment to the discovery of ore. A specific feature of regional prognostication is its areal character. This suggests that the standard objects chosen for prognostic evaluation should not be deposits, but ore-bearing areas (zones, areas, etc.).

In the course of evaluating metallogenic provinces for deposits of the cupreous sandstone and copper-slate type, primarily the geotectonic, formational and stratigraphic ore-controlling factors should be taken into account. Within large regions under consideration, of particular importance are orogenic troughs, depressions, and troughs with abundant mafic volcanics in the basement and flanks uplifts. Copper-bearing carbonate-terrigenous sediments are characterized by widespread red and variegated formations of a carbonate composition. They display a thick molassic bedding, or a medium, or less frequently, thin flyschoid bedding, the deposits of an arkose or greywacke composition being very thick. Copper mineralization is commonly confined to the upper part of megacycles.

Epochs of sedimentary copper accumulation are different for provinces of different age. In the North Transbaikal area, at least two copper-bearing time intervals are distinguished in Early Proterozoic; in Central Kazakhstan, two periods in Late Paleozoic.

In the course of evaluating metallogenic zones, the taxon prognosticated is an ore area, and only in certain instances, a potential ore field, characterized by a favourable combination of ore-controlling and ore-localizing factors, as well as the development of ore-bearing rocks. For the evaluation of zones, stratigraphic, paleotectonic and lithological-formational criteria are used (Bogdanov et al. 1973, Kutyrev 1984).

In the section of structural-formational complexes, several (two to three) copper-bearing intervals are distinguished, forming in plan extensive belts of copper-bearing coastal-marine deposits (Kodar-Udokan zone in the North Transbaikal area, Teniz-Dzhezkazgan zone in Central Kazakhstan, Copperbelt in Central Africa, etc.). Cupreous sandstone is commonly confined to lagoonal-deltaic deposits, accumulated on the slopes of paleouplifts, defined on the basis of thickness variations, changes in lithological composition, gradation from red beds to grey beds and other facies transitions. In the Kodar-Udokan zone, six ore areas are distinguished within two belts.

The main prognostication criteria for large-scale exploration are stratigraphic, paleotectonic, lithological, and mineralogical-geochemical ore-controlling and ore-localizing factors. A more detailed study of copper-bearing belts allows the exploration team to distinguish and trace the ore-bearing horizon and to give their preliminary evaluation, as well as to outline sites where productive strata of deposits, i.e. potential ore fields, are formed. Particularly favourable for copper accumulation are lenses of lagoonal-deltaic deposits. They are characterized by a rhythmic recurrency, expressed in alternation of cross-bedded sandstone and conglomerate-breccias with siltstone and mudstone. Relative to paleouplifts a progressive decrease in age of cupreous sandstone and a mineral zoning – chalcocite-bornite-chalcopyrite-pyrite – are observed. The following expected morphological and petrological types of copper sandstone deposits are established: Dzhezkazgan, Lubin-Sieroszowice, Dzhilanda and Cis-Uralian.

Copper sandstone of an arkose composition is characterized by monometallic ores. Besides copper, they also contain minor admixtures of silver, molybdenum, bismuth and other metals (Udokan, Ainak in Afghanistan, etc.). In cupreous sandstone of a greywacke composition, ores frequently have a polymetallic composition. Along with copper, they often contain appreciable concentrations of cobalt, lead, molybdenum, rhenium, silver and other metals (Dzhezkazgan, Unkur in the North Transbaikal area). Copper mineralization is commonly accompanied by authigenic mineralization in the form of hydromica, carbonate, zeolite and pyrite.

The cyclic pattern of molasse cupreous sediments determines the morphology of productive strata, their simple or multicyclic structure. Flyschoid sediments contain copper sandstone with a simple structure of productive strata, having relatively small size. The latter are determined by the dimensions of lenses of lagoonal-deltaic or other coastal-marine deposits.

Conclusions

1. The territory of the Soviet Union is characterized by copper-bearing zones, differing greatly in their geotectonic position. Orogenic, platform and geosynclinal zones are distinguished. Copper-bearing zones in orogenic and platform areas are characterized by variegated and red deposits, in arid environment sediments; those in geosynclines, by black shale formations.

2. Major epochs of sedimentary copper accumulation are: Early and Late Proterozoic and Late Paleozoic which favour the emplacement of orogenic cupreous sandstone, and Middle Mesozoic, which favours the generation of copper-pyrrhotite mineralization in early geosynclinal black shale trough zones.

3. Copper-bearing zones of arid red beds are divisible into the following types and subtypes based on predominance of deposits with different lithological compositions and cyclic recurrence.

Copper and Polymetallic Mineralization in Red and Variegated Sediments of a Greywacke Compostion

Based on cyclic recurrence:

1. Thin and medium cycles (flyschoids) of Sary-Oba, etc.; Dzhilanda group, Unkur, etc.; Kodar-Udokan zone, etc.
2. Thick cycles (molasses).

Based on facies:

1. Lagoonal-deltaic (Dzhezkazgan, etc.).
2. Lacustrine-alluvial (Fore-Urals, etc.).

Based on the structure of productive strata of deposits:

1. Multicycle (Dzhezkazgan, etc.).
2. Simple structure (Western Darwaza, etc.).

Copper Mineralization in Variegated and Red Sediments (molasses) of an Arkose Composition, Having Thick Cycles

Based on facies:

1. Lagoonal-deltaic (Udokan, etc.).
2. Lacustrine-alluvial (Central Asia, Fore-Urals, etc.).

Based on the structure of productive strata:

1. Multicycle (Udokan, Ainak, etc.).
2. With a simple structure (Igarka area, Copperbelt in Central Africa, etc.).

Types of copper-bearing zones with a two-stage structure are established. Up the section, flyschoid sediments of a greywacke composition grade into molasse sediments of a greywacke (Dzhezkazgan area) or arkose composition (Kodar-Udokan zone).

4. A prognostic evaluation of copper-bearing areas is based on formational analysis accounting both for specific features of lithological composition and rhythmic recurrence of formational units. The areal character of regional prognostication determines the following procedure: from establishing and studying favourable geological settings and formational combinations to discovering potential or real ore-bearing areas.

References

Bogdanov YuV (ed) (1966) Copper deposits of the Olekma-Vitim mountain land. Geology and the regularities of their distribution. Nedra, Leningrad, 385 pp (in Russian)

Bogdanov YuV, Feoktistov VP (1982) A genetic model of copper sandstone deposits of the Udokan type. Trans USSR Acad Sci 263, 4:949 – 952 (in Russian)

Bogdanov YuV, Bur'yanova EZ, Kutyrev EI, Feoktistov VP, Trifonov NP (1973) Stratified copper deposits in the USSR. Nedra, Leningrad, 312 pp (in Russian)

Kutyrev EI (1984) Geology and prognostication of conformable copper, lead and zinc deposits. Nedra, Leningrad, 248 pp (in Russian)

Mendelsohn F (ed) (1961) The geology of the Northern Rhodesian Copperbelt. Mc Donald, London

Narkelyun LF, Salikhov VS, Trubachev AI (1983) Copper sandstones and slates of the world. Nedra, Moscow, 414 pp (in Russian)

Savin SV (1978) Geological structure of the Near-divide metallogenic zone of the Great Caucasus. Geol Ore Dep 20, 6:13 – 23 (in Russian)

Volnov BA, Leleshus VL, Provotorov NG (1978) Copper presence in the south-western Darvaza. Trans Acad Sci Tadj SSR 3:37 – 40 (in Russian)

Red-Colored Terrigenous Sediments – Specific Copper-Forming Systems

B. B. SUSURA, V. O. GLYBOVSKY, and A. V. KISLITSIN[1]

Abstract

Remobilization of the geochemical background of the ore- and rock-forming elements of red-colored copper-bearing sediments in the Dzhezkazgan ore region has been studied. Potentialities of these deposits as specialized ore-forming systems are shown along with the paragenesis of copper and oil formation, with the biofactor actively participating. The systems elements are clearly distinguished in the geochemical and geophysical fields. The metallogenic classification of the copper-bearing, red-colored formations and the stratiform copper ores of the copper sandstones and slates type connected with them has been worked out. It served as the basis for the formation-metallogenic analysis of the ore content of the red-colored formations and stratolevels of various age (from Upper Proterozoic to Neogene) in Kazakhstan. Areas of occurrence of the commerically perspective formations of the Dzhezkazgan type have been distinguished.

The stratiform copper mineralization of the sandstone and slate type reveals a distinct spatial connection with a group of red-colored (variegated) terrigenous sediments and their detritus in volcanogenic and carbonate deposits of different age (from Early Proterozoic to Cenozoic), forming in practically all geotectonic zones of arid terrestrial sedimentation. The red-colored molasses retains, mainly, a metallogenic copper enrichment, irrespective of subsequent transformations up to the stages of deep metamorphism.

The stability of red-colored, cupreous-sedimentary rocks identifies them as specific metal-specialized ore-forming systems and ores derived from the remobilization of the geochemical background concentrations in the sediments. As for the hydrogenic-epigenetic concept of the genesis of copper sandstones and slates, it is explained by oxidizing conditions favorable for the migration of copper and other chalcophile elements (Gablina 1983, Lur'ye 1979, Perelman and Borisenko 1962, Rose 1976, Susura 1980). The mobilizing agents are fluid systems of various natures which are characteristic of the ore-forming process. The ore-concentrating factors in minable strata are geochemical hydrogen sulphide barriers. Such barriers occur rather commonly in the red-colored deposits. They are the syngenetic barriers, argillite and limestone intercalations

[1] The Kazakh Research Institute of Mineral Products, Alma-Ata, USSR

Geology and Metallogeny of Copper Deposits
(ed. by G. H. Friedrich et al.)

and lenses, enriched in organic matter, coal seams and scattered organic remains. Accumulations of hydrogen sulphide, hydrocarbons, bitumens and oils generated within the sediments themselves or introduced from the adjacent sediments serve as the epigenetic barriers.

Most detailed investigations of the hydrogenic-epigenetic concepts are exemplified by the Permian-Carboniferous red-colored strata in the Dzhezkazgan Region of Kazakhstan. Together with the underlying and overriding grey-colored terrigenous-carbonaceous deposits, the latter fills the Dzhezkazgan Sarysu depression. It is laterally adjacent to evaporitic sediments.

The stratiform deposits of the region occur in the lenticular zones of the grey-colored polymictic sandstones and conglomerates alternating with the red-colored aleuroargillites. Several metallogenic types of mineralization are distinguished. They differ in the position in the minable formation, likewise in scope, in their ore elements, as well as in mineralogical and element zoning.

In the lower part of the red-colored sequence, minor mineralization of the Jilandin types has a widespread areal extent. Some medium- and small-sized deposits and dozens of copper showings are traced in line for more than 100 km along the outcrop of this ore horizon at the edges of the Dzhezkazgan-Sarysu depression. The ores of this type are essentially cupriferous. The mineralogical zoning is represented by the replacement down-dip of chalcopyrite-bornite mineralization by chalcocite.

An ore bed of the Mansfeld type is established where the zone of the red-colored deposits contacts the overlying marl deposit of Late Permian age. The monometallic deposits of copper sandstones and limestones of some small-sized deposits and subeconomic ore occurrences are confined to this stratigraphic level.

In the sequence of ore zones, a replacement of chalcopyrite ores by bornite-chalcocite is observed stratigraphically upwards.

A multistage copper-polymetallic mineralization of the Dzhezkazgan type involves, as a rule, almost the entire sedimentary section. Archlike deposits with the proper outer and dissected inner outlines of boundaries encircle the consedimentation brachy-anticlines. The element and mineralogical zoning is expressed by a replacement of lead and copper-polymetallic ores by purely copper ones down the section and down-dip. A wide scope of admixed elements is typical of the ores. These regularities have been established for the Dzhekazgan Region and some of its new deposits.

The background contents of the ore-forming elements (Cu, Pb, Zn, Fe, S) in the minable red-colored deposits of the Dzhezkazgan region are close to Clark concentrations in sedimentary rocks. The transformation of the geochemical background is evident in the conjugate regions of the elements leaching and accumulation (Susura 1980). The lenselike leaching zones evident in the red-colored deposits of the consedimentation troughs encircle the ore accumulations within the paleo-anticline structures. The leached rocks are characterized by the low background (20% of clark) contents of copper, lead, zinc and are almost sterile with regard to sulphur. The reverse pattern exists for iron, implying that leaching from the cement of ore-bearing grey sandstones was followed by the formation of the grey border along their contact with the enclosing red-colored aleuroargillites (Perelman et al. 1962).

Partitioning of organic matter is exhibited by an increase of its content from 0.05% to 0.3% in rocks of ore zones (Arustamov et al. 1969) due to an increase of the mobile bitumen component (the bituminosity ratio increases from 15% to 90%) (Susura 1980). Different compositions and spectrophotometric characteristics of bitumens in the ore-bearing horizons at their irregular change in the deposits ore zone are indicative of the stratified mobilization and paragenesis of copper- and oil-forming components in the minable deposits.

Differentiation of rock-forming elements is established in the form of various deuteric changes of the cement of granular intercalations, i.e. silicification, albitization, carbonatization, chloritization and others. They are evident mostly in the ore zones and are not typical for unmineralized red-colored sections. Silicification followed by the deposition of essentially copper ores prevailed in the ore zone under acidic conditions. Albitization and carbonatization prevail under neutral and alkaline conditions. Lenticular zones of an intensive epigenetic carbonatization or anhydritization of the cap-rock type of the oil- and gas-bearing salt domes are clearly observed in the cement of deposits overlying the ores zones.

The zoning revealed in the remobilization of geochemical background concentrations identifies the red-colored deposits of the Dzhezkazgan region as an epigenetic stratified copper-forming system. The elements of this system (mobilization, transport and concentration) are well recognized in the field (Smelov et al. 1983). The mass transfer balance of the ore elements leached from the source rocks is comparable to the ore reserves.

A comparison of the paleo-hydrodynamic conditions with the isopleth maps of the elements and with deuteric changes of the rocks of the minable complexes indicates a major role in the ore genesis of the eliasite stage in the development of artesian structures (Susura 1980). The post-lithification folding and dislocations with a break in continuity intersect the ore zones without influencing the intensity of mineralization on the mineralogical and element zoning.

A thermobarogeochemical study of the fluid inclusions of the mineral parageneses of ores and rocks of the Dzhezkazgan region indicates the unique gas regime of different formation areas during ore deposition. The aqueous inclusion phase is geochemically analogous to the current formation waters of the red-colored exogenetic deposits ($\delta^{18}O$: -11 to -15.5 per mil: D: -88 to -120 per mil). From the temperatures of homogenization, a trend is established in the stratigraphic section in good agreement with the geothermal gradient of the Earth. At the depth of maximal burial of the minable deposits (about 2 to 3 km), the geothermal gradient probably determined the temperature of ore formation (from $+100°$ to $+140°C$). The hypergenic nature of mineralization is confirmed by the argon isotope (prevalence of the gaseous component) of the fluid inclusions of the ore sulphides, as well as by the biogenic nature of the sulphur ($\delta^{34}S$ -4.0 ± 52.0 per mil).

Specific conditions for the development of particular types of biocoenoses form in different zones of the red-colored formation. Heterotrophic microorganisms participate most commonly in the ore formation (Susura 1980). Under the oxidizing conditions of the red-colored zones, their aerobic biocoenoses produce organic acids and promote the leaching and dispersion of the ore elements in the form of organometallic compounds and complexes. Under reducing conditions of the ore depositional zones, the sulphate-reducing and mineral-degenerating heterotrophic microorganisms are most common. Their activity rise to the formation of hydrosulphuric barriers and mineralization of the organometallic compounds.

Figure 1 gives a probabilistic model of the ore strata in the Dzhezkazgan region. During the lithification of the sediments and under geostatic loads, the gas and liquid component segregates into the granular headers and migration occurs from the consedimentation troughs to the uplifts. The organic compounds of the oil series, responsible for the change from oxidizing conditions to reducing conditions, accumulated stage by stage in the relaxation centres. The organic matter resulted in iron reduction (grey color appeared) and an active biogenic reduction of sulphate. A sequential sedimentation of chalcophile elements from the metal-bearing formation waters was due to the concentration growth of H_2S according to the solubility products of the sulphide. The latter caused the element and mineralogical zoning of the ore zones. The mineralization of the Jilandin ore horizon is affected markedly by the underlying carbonate deposits which acted as a probable source of organic matter and/or hydrogen sulphide. The localization of ore of the Mansfeld type was determined by the syngenetic barrier in the zone of facies transition of the red-colored deposits into the overlying carbonate beds.

It may be assumed from the general metallogeny of the red-colored formations of the Dzhezkazgan type that the genetic concepts are generally similar for other specific stratified accumulations of economic minerals, i.e. manganese, strontium, native sulphur, asbestos, oil and gas. Most of these deposits are derivatives of the selective remobilization of the geochemical background concentrations of elements in the enclosing sediments.

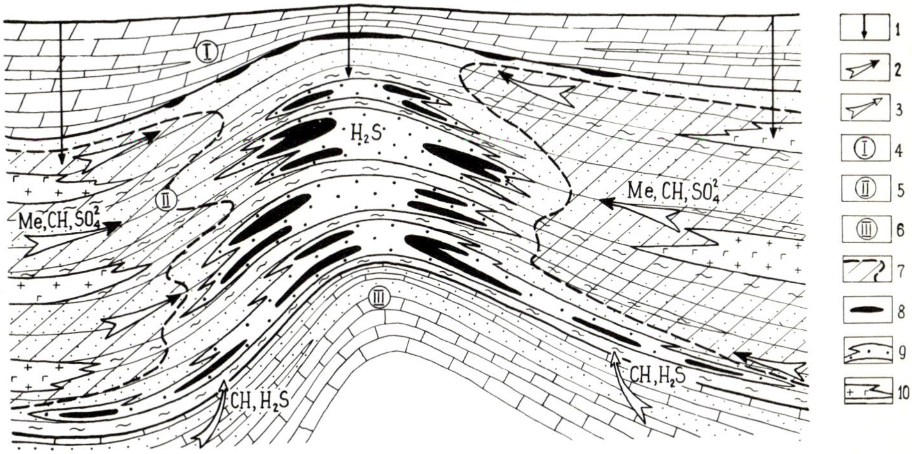

Fig. 1. Hydrogenic-epigenetic copper ore in the Dzhezkazgan region. *1* Directions of geostatic pressure; *2* directions of movement of gas-liquid ore-forming solutions; *3* directions of inflow of bitumens and hydrogenic sulphide; *4* clay-carbonate deposits; *5* red-colored terrigenous deposit; *6* terrigenous-carbonate bituminous deposit; *7* provenance area; *8* ore zones; *9* region of epigenetic rocks growing grey; *10* evaporites

In contrast to the terrigenous deposits of the Dzhezkazgan type, there are a large number of small, copper depositional centres in the red-colored carbonate-terrigenous sediments. The concentrating components are carbonified organic remains (the Urals type of the copper-bearing sediments and ores) scattered in the minable deposits, as well as intercalations and lenses of limestones, coaly argillites and coals (Donbas type of formations and ores). These accumulations fall short of an industrial scale. Such sediments do not attain the copper contents of the Jilandin and Mansfeld ore beds. Evidently, the ore horizons of a series of copper showings of the above-mentioned types are attributed to continental sabkha processes (Remfro 1974) and accumulation of copper-bearing peat bogs.

The copper-bearing character of the red-colored volcanogenic-sedimentary molasse is mainly defined by the metallogenic type of their terrigenous component. The influence of the effusive strata is shown in native copper mineralization.

At a later stage in the development of artesian structures, many copper-bearing beds (e.g. weathering crusts, karsts, eolian and vein deposits, oxidized and secondary sulphide enrichment zones) are related to infiltrations and reactions in the red-colored deposits. In most cases, they are indicators of the primary ore, but some may be of independent commercial importance (weathering crusts, karsts).

The copper-bearing metallogenic strata occupy a very definite position in the stratigraphic sequence of the red-colored deposits. The geotectonic series reflects a sedimentogenetic change from geosynclinal regions to the platform areas. The position in the facies change defines the formation structure, i.e. a set of sediment-forming rocks in a typical spatial relationship and, consequently, forming stratified ore runs. Asymmetrical and symmetrical facies changes within the sediments are distinguished. The former are formed in depressions within closed terminal basins, the latter occur at the transition into the sea basin. A climatic series reflects climatic conditions relating to the erosional and sedimentation regions. A mutual intersection of geotectonic and facies conditions forms a metallogenic type of deposit and defines specific features of its ore content (Fig. 2).

Such classifications of red-colored, copper-bearing sediments and ores connected with them is necessary for predictive metallogenic analysis, as only some of the mineralization is of commercial importance. The formations of the Dzhezkazgan type, including the major commercial types (namely, the Dzhezkazgan, Mansfeld and Jilandin deposits), are most prospective. Weathering zones and volcanogenic-sedimentary formations are of special concern.

From the metallogenic point of view, Kazakhstan is a unique region of stratiform copper deposits. Red-colored, copper-bearing formations of various ages (from Upper Proterozoic to Neogene), and of variable size (some large and a great number of stratified ore showings) are widely spread in the region. Time and again, the red-colored sedimentary areas ranged for hundreds of thousands square kilometres to cover different geotectonic zones. A complete metallogenic variety of sediments and related ores is known in the region.

Generally, Upper Proterozoic red-colored deposits form asymmetrical series. South of the volcanogenic complexes of the Kokchetav anticlinorium and via the region of variegated volcanogenic-sedimentary and terrigenous-carbonate depos-

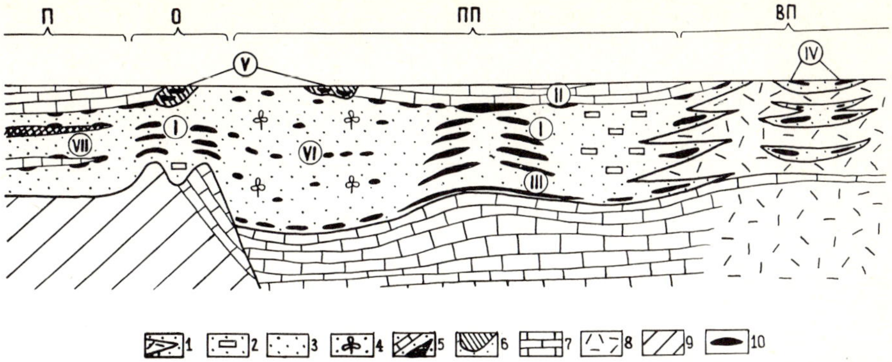

Fig. 2. Metallogenic series of red-colored copper-bearing strata. *1–6* Metallogenic types of formations: *1* volcanogenic-sedimentary; *2* evaporite; *3* Dzhezkazgan; *4* the Urals region; *5* Donbas; *6* weathering zones and karsts; *7* terrigenous-carbonate deposits; *8* volcanogenic strata; *9* basement rocks; *10* copper mineralization. Types of mineralization: *I* Dzhezkazgan; *II* Mansfeld; *III* Jilandin; *IV* volcanogenic-sedimentary; *V* weathering zones; *VI* the Urals region; *VII* Donbas. Geotectonic position of sedimentary zones: *BП* volcanic belt; *ПП* foredeep; *O* orogene; *П* platform

its, a change occurs in the thick terrigenous red-colored deposits of the Karatau-Talas zone and Chinghiz region. The latter resembles the Dzhezkazgan type formation.

The major areas of the red-colored sedimentation in the Silurian include appreciable zones of the Earth's crust, stabilized at the late geosyncline stage of Caledonides development.

In Early Silurian, they delineated the Balkhash massif which retained a marine regime. In Late Silurian, this epicratonal sea closed and the marine sedimentation changed to a continental type. As a whole, it may be stated that the areal constriction of the Silurian red-colored sediments coincided in time with the final epicentre in the Balkhash region. The lithofacies of rocks and the concentrating character for copper mineralization suggests that these deposits belong to the adjacent Donbas-Urals region – volcanogenic-metallogenic type.

The occurrence of the red-colored Devonian molasse was defined by the development of orogenic-volcanic belts of the Urals and Central Kazakhstan. These deposits form symmetric series of sediments in the orogenic intermountain area of the Urals-Mongolian folded belt and an asymmetrical series in the Urals trough and in the Western European platform (Fig. 3a). In the central part of the orogenic area, in the terminal basin (epicontinental marinelake), terrigenous-evaporite strata were deposited in the form of a giant lense occupying a central part of the Chu-Sarysu depression. At the periphery, they grade into the terrigenous strata of the Dzhezkazgan and then of the Urals region and Donbas types. These regions are overlain by the red-colored volcanogenic-sedimentary complexes of the Devonian volcanic belt of Central Kazakhstan and Mugodjary. In the Urals-Caspian region, a change occurs from the thick greywacke series of the Urals mountain structures to the fine-grained terrigenous-carbonate deposits of the Caspian region depression.

Several hundreds of stratiform copper showings of various types are found in the Devonian strata. In the upper part of the strata which contact the overlying

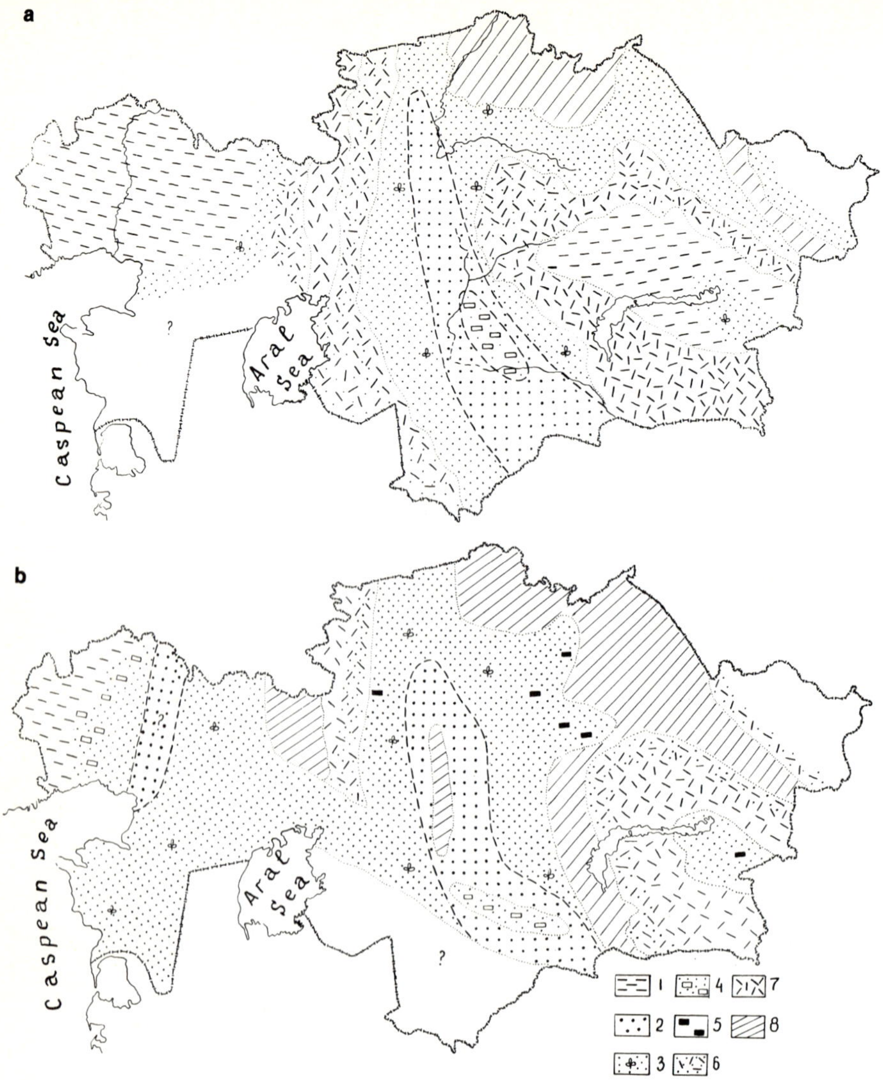

Fig. 3a, b. Areas of (**a**) Devonian and (**b**) Carboniferous-Permian red-colored copper-bearing deposits of Kazakhstan. *1* Grey-colored sea deposits; types of red-colored terrigenous sediments: *2* Dzhezkazgan; *3* the Urals region; *4* evaporite; *5* coal-bearing; *6* volcanogenic-sedimentary; *7* volcanogenic deposits; *8* regions of no sedimentation

carbonate beds of Late Devonian age, a copper-bearing horizon of the Mansfeld type is traced everywhere. Favorable conditions for the localization of commercial mineralization are observed in strata of the Dzhezkazgan and volcanogenic-sedimentary types in Ulutau, in Karatau, in the Chu anticlinoria and in the Shyderti-Chinghiz Tarbagatei region.

In the Famennian-Tournaisian period, red-colored sedimentation diminished sharply. This beds of the Donbas and the Urals type with insignificant copper showings are known only in the Chu depression and its margins.

The area of red-colored Carboniferous-Lower Permian sediments includes vast areas of Kazakhstan. Generally, these areas form a symmetric series of the intermountain sediments between the Valerianovsk and Balkhash-Late Paleozoic volcanogenic belts (Fig. 3b). In the central part of the area, evaporites of the inner zones of Dzhezkazgan-Sarysu and Chu depressions are isolated. They are bordered by the sediments of the Dzhezkazgan metallogenic type. In the west and southeast the latter grade into volcanogenic-sedimentary units, and in the northeast, they grade via the Urals region-Donbas strata into the deposits of the Karaganda and Kuzbas coal-bearing basins. In the north and in the west, the red-colored deposits are displaced by the terrigenous-carbonate facies of the Caspian region and the West Siberian platform.

The Dzhezkazgan type of sediment of the area mentioned encloses the Dzhezkazgan deposits and a number of medium-sized deposits of the Jilandin and Mansfeld types. The major commerical prospects are attributed to this ore horizon, particularly in the Chu-Sarysu depression and in the Turgai trough.

In the Permo-Triassic, the main areas of the red-colored sedimentation have shifted westward to the Kyzyl Kum, Caspian and West Urals regions. Red-colored sedimentation has been observed in the Teniz, Chu and South-Balkhash depressions and in the Bakanas synclinorium. These deposits resemble the Donbas and the Urals type including a great number of insignificant copper showings. Only Lower Triassic deposits in Mangyshlak and Permian red-colored deposits in the intersalt dome troughs of the Caspian region depression are close to the Dzhezkazgan type of the copper-bearing sediments.

For the red-colored Meso-Cenozoic deposits of Kazakhstan, one may outline a metallogenic analogy with the coeval formations of the Tajik zone and Fergana depression known for some copper showings of the Donbas the Urals type.

The Pre-Cretaceous Period is indicative of a widespread development of weathering zones. The processes were most intensive in the Turgai region which is a unique province characterized by the development of bauxite, nickel, copper-bearing and polymetallic weathering zones and karsts. A number of karst cavities in the Arkalyk region have formed a copper-bearing stratolevel of the Mansfeld type. In the products of re-deposited weathering crusts filling these funnels, copper ore accumulations have been revealed.

The metallogenic aspects considered of the red-colored formations and of various types of specialized copper-forming systems of the typification-metallogenic analysis of the red-colored deposit areas may be used in the study and evaluation of the stratiform copper content of other regions.

References

Arustamov AA, Esenov ShE, Parshin GB, Shtifanov VI (1969) Copper sandstones of Northern Dzhezkazgan. Nauka, Alma-Ata, 137 pp (in Russian)

Bogdanov YuV, Burianova EZ, Kutyrev EI, Feoktistov VP, Tifonov NP (1973) Stratified copper deposits in the USSR. Nedra, Leningrad, 311 pp (in Russian)

Gablina IF (1983) Conditions of copper accumulation in red-colored continental formations. Nauka, Moscow, 110 pp (in Russian)

Lur'ye AM (1979) Exogenetic copper deposits in sea deposits. Author's abstract of a doctor's of geology and mineralogy. Thesis, Moscow, 40 pp (in Russian)
Perelman AI, Borisenko EN (1962) Essays on copper geochemistry in hypergenesis zone. Proc IGEM AN SSSR 70:30–99 (in Russian)
Remfro AR (1974) Genesis of evaporite associated stratiform metalliferous deposits – a sabkha process. Econ Geol 69, 1:33–45
Rose AW (1976) The effect of cupreous chloride complexes in the origin of red-bed copper and related deposits. Econ Geol 71, 6:1036–1048
Smelov AA, Kolmogorov VYu, Junusov TD (1983) Petrophysical model of the ore-forming system of the Dzhezkazgan region and its prediction- and search importance. In collection: Metallogeny problems in Kazakhstan. Nauka, Alma-Ata, pp 194–207 (in Russian)
Susura BB (1980) Geochemistry of the epigenetic ore formation in the red-colored formation of the Chu-Sarysu depression in Kazakhstan (type of copper sandstones). Author's abstract of a candidate's of geology and mineralogy thesis. Moscow, 26 pp (in Russian)

Cupriferous Sandstones and Shales of the Siberian Platform

F. P. KRENDELEV[1], L. F. NARKELYUN[2], A. I. TRUBACHEV[2], V. S. SALIKHOV[2], P. N. VOLODIN[3], V. V. KUNITSIN[4], V. S. CHECHETKIN[4], and N. N. BAKUN[5]

Abstract

The Siberian platform is host to a variety of mineral deposits, including stratiform copper deposits in sandstones and shales. Six major copper provinces are recognized (Prieniseyskaya, Prisajanskaya, Pribaikalskaya, Prialdanskaya, Priverkhojanskaya, Prianabarskaya), and within these at least 15 major zones of deposits and occurrences, including the Udokan district. The cupriferous sediments occur in Proterozoic and Paleozoic detrital sediments ranging from sandstone to shale, usually either varicoloured or black. Some of the sediments contain carbonate (cement) and some contain volcanic material. The cupriferous sediments are localized around the border of the platform in basins filled with sediment shed from adjacent uplifts. More central parts of the platform contain lead-zinc, barite, fluorite, gypsum-anhydrite, and other evaporite minerals, whereas gold occurs in conglomeratic zones closer to the uplifts. The size and shape of ore zones depend closely on sedimentary facies. On a basin scale, the cupriferous sandstones and shales transgress into younger sediments toward the center of the basin, and are mineralogically zones from underlying pyrite and chalcopyrite into overlying bornite-chalcocite zones. Similar circumferential and beltlike zones of cupriferous sediments are recognized around and adjacent to other platforms on all continents. The deposits were formed during sedimentation and diagenesis, but have since been subjected to various combinations of epigenesis, metamorphism and weathering. The sources of copper are considered to be in the uplifted blocks that supplied the sediment, and in volcanic effusions.

The Siberian Platform and its folded border zones is one of the most important metallogenic regions in the world. It is formed almost entirely of sedimentary formations ranging in age from Archean to Cenozoic and including various mineral deposits (Bogdanov et al. 1973, Narkelyun et al. 1977). Cupriferous sandstones and shales are confined to Proterozoic and Paleozoic formations of the platform where the Prieniseyskaya, Prisajanskaya, Pribaikalskaya and Prialdanskaya copper provinces have been established (Fig. 1).

[1] Institute of Natural Resources, Butin Street 26, Chita, USSR
[2] Chita Polytechnic, Chita, USSR
[3] Central Research Institute of Geological Prospecting, Moscow, USSR
[4] Chitageology, Mingeo RSFSR, Chita, USSR
[5] All Union Institute of Research in Petroleum, Leningrad, USSR

Geology and Metallogeny of Copper Deposits
(ed. by G. H. Friedrich et al.)

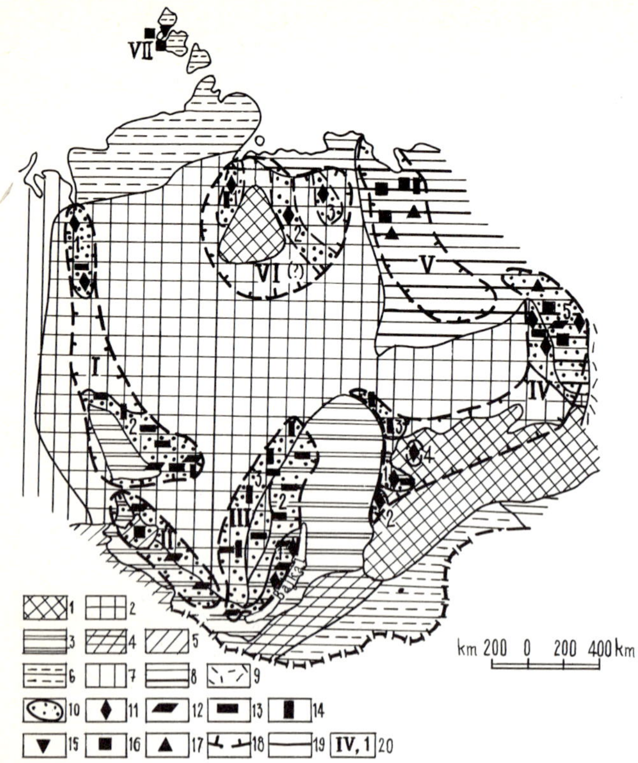

Fig. 1. Location of cupriferous sandstones and shales on the Siberian platform. The tectonic base is simplified after Yanshin (1966). *1* Basement uplifts; *2* cover of ancient platform; *3* Baikalides; *4* Caledonides; *5* Paleozoic basins; *6* Hercynides; *7* cover of epipaleozoic platform; *8* Mesozoides; *9* volcanogenic sedimentary complex; *10* zones of cupriferous sandstones and shales. *11* – *17* Deposits and occurrences of copper mineralization according to different metallogenic epochs: *11* Proterozoic; *12* Vendian; *13* Cambrian; *14* Ordovician; *15* Silurian; *16* Devonian; *17* Carboniferous. *18* Boundaries of copper provinces; *19* copper provinces (*Roman numerals*) and zones (*Arabic numerals*) within the provinces. *I* Prieniseyskaya: *1* Igarskaya, *2* Priangarskaya. *II* Prisajanskaya. *III* Pribaikalskaya: *1* Zapadno-Pribaikalskaya, *2* Verkhne-Lenskaya, *3* Angaro-Tungusskaya. *IV* Prialdanskaya: *1* Kodaro-Udokanskaya, *2* Kondo-Karengskaya, *3* Nuyjsko-Berezovskaya, *4* Olekmo-Tokkinskaya, *5* Vostochno-Aldanskaya. *V* Priverkhojanskaya. *VI* Prianabarskaya: *1* Prianabarskaya proper, *2* Kenelikanskaya, *3* Olenekskaya. *VII* Severo-Zemelskaya zone

Early Proterozoic copper deposits are the most ancient on the platform (the Kondo-Karengskaya, Olekmo-Tokkinskaya, Kodaro-Udokanskaya and Vostochno-Aldanskaya zones). Copper-bearing minerals are widely distributed in varicoloured detrital and volcano-detrital sediments in Riphean and Vendian strata of all the provinces. Early Cambrian cupriferous sediments are known throughout the platform. Middle and Late Cambrian copper-bearing sedimentary rocks occupy considerable areas. During Ordovician time copper accumulated in the Priangarskaya, Nyujsko-Berezovskaya and Angaro-Tungusskaya provinces. In the Priverkhojanskaya province copper is confined to the Silurian effusive-detrital suite. Devonian carbonate-detrital and effusive-detrital sediments have been established in the Priverkhojanskaya province as well as in the Vostochno-Aldanskaya zone and in the Ribinskaya depression of the Prisajanskaya province. The copper-bearing section terminates with Early Carboniferous effusive-detrital formations.

Thus, cupriferous sandstones and shales of the Siberian platform are to be found in sedimentary formations ranging in age from Lower Proterozoic to Carboniferous, their vertical stratigraphic thickness being 50 km. The mineralization is localized within narrow age intervals and is characterized by distinct stratigraphic control.

Accumulation of the cupriferous sedimentary formations took place within the platform depressions such as those of Kodaro-Udokanskaya and Ugujskaya, as well as in orogenic structures such as the Prieniseyskaya, Prisajanskaya, Priverkhojanskaya and Angaro-Lenskaya foredeeps and in marginal parts of the platform structures. All ore-bearing structures are located mainly in transition zones from geosynclinal to platform formations. This localization in transition zones is the essence of geotectonic control in the distribution of cupriferous sediments.

The copper deposits are closely connected with the host sediments. Various cupriferous sedimentary formations can be distinguished by their lithological associations. Varicoloured, low-carbonate sandstone, varicoloured calcareous shale, grey low-carbonate sandstone, and varicoloured volcanogenic-detrital cupriferous sedimentary formations are the most important lithological types. That the copper mineralization is confined to certain formations is the most important control on the location of copper deposits in the sedimentary suites of the Siberian platform.

Cupriferous sedimentary formations have vertically and laterally a genetic connection with other sedimentary formations of the platform containing metals and minerals such as lead, zinc, gold, rare metals, phosphorus, gypsum, anhydrite, barite, fluorite, celestite and rock and potash salts. The coarse detrital sediments are confined to the peripheral parts of the platform, adjacent to mountain structures that were the regions of erosion. Away from the mountains towards the internal parts of the platform, there are in sequence: detrital, carbonate-detrital, argillaceous-carbonate, carbonate-sulfate and saline formations. Gold mineralization is mainly restricted to the coarse detrital sediments; copper mineralization is confined to low-carbonate and carbonate-argillaceous sandstones; lead-zinc mineralization is restricted to argillaceous carbonate and carbonate; and mineral salts to sulfate-saline formations (Fig. 2).

On the Siberian platform copper has been recognized in conglomerates[6], gravelites, sandstones, aleurolites, argillites, argillaceous limestones, dolomites and tuffaceous rocks. According to lithological characteristics we differentiate three main types of copper deposits: (1) cupriferous conglomerates and sandstones – subcontinental deposits containing carbonate material up to 15%, including allochthonous organic substance (Early Cambrian formations); (2) cupriferous sandstones and siltstones of subcontinental, lagoonal, deltaic and littoral-marine types containing autochthonous and allochthonous organic matter

---

[6] Probably pebble conglomerates.

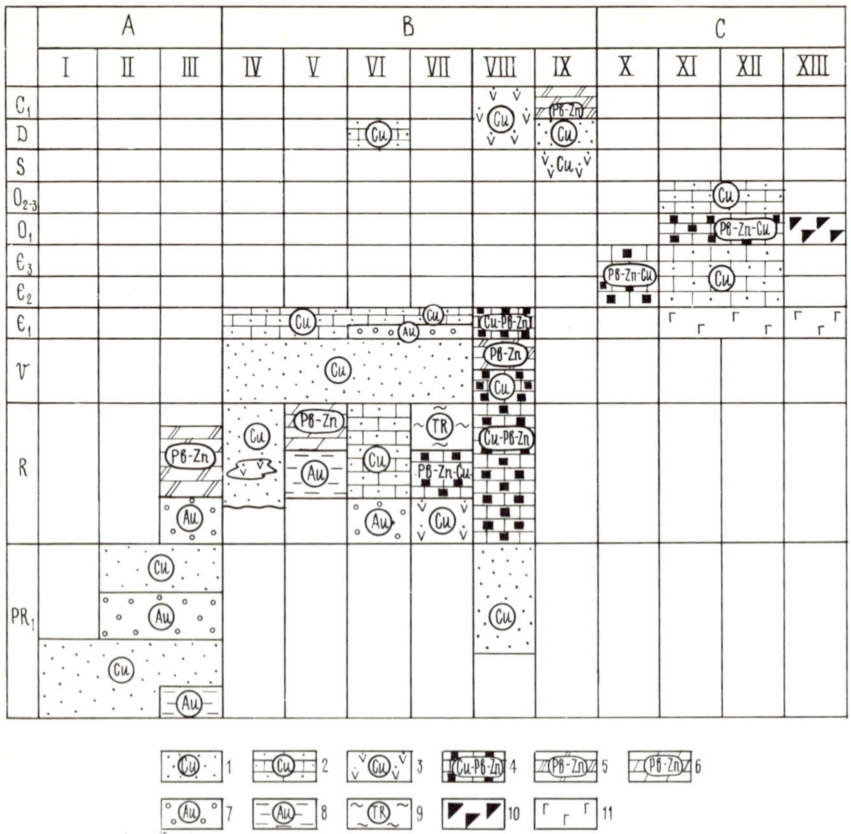

Fig. 2A – C. Stratigraphic-tectonic location of copper-sedimentary formations of the Siberian platform and their paragenetic associations with other mineral deposits. Sedimentary formations, cupriferous formations: *1* varicoloured-detrital; *2* varicoloured, carbonate-detrital; *3* volcanogenic detrital; *4* detrital-carbonate (with Pb and Zn), Lead-zinc; *5* carbonate; *6* argillaceous-carbonate. Gold-bearing: *7* conglomeratic-sandstone; *8* carbonaceous-argillaceous (black shale); *9* rare metallic-flyschoid; *10* sulphate-carbonate (with fluorite, celestite, barite); *11* sulphate-saline. **A** Interplatform-fold basins of the Olekmo-Vitimskaya mountain country. Zones: *I* Kondo-Karengskaya; *II* Olekmo-Tokkinskaya; *III* Kodaro-Udokanskaya. **B** Orogenic structures framing the platform: *IV* Igarskaya zone; *V* Priangarskaya zone; *VI* Prisajanskaya province; *VII* Pribaikalskaya province; *VIII* Vostochno-Aldanskaya (Eastern Aldan) zone; *IX* Priverkhojanskaya province. **C** Platform structures (edge zones of the platform): *X* Vostochno-Aldanskaya (Eastern Aldan) zone; *XI* Verkhnelenskaya (Upper Lena River) and Angaro-Tungusskaya zones; *XII* Priangarskaya zone; *XIII* interior parts of the platform

(Kodaro-Udokanskaya zone); (3) cupriferous shales, marls, argillaceous limestones and dolomites of lagoonal and littoral-marine origin containing about 40% carbonate with autochthonous organic matter in the form of algae, plankton, bitumen and microorganisms (Late Cambrian and Ordovician formations).

Sedimentary facies have a decisive influence on the location, morphology, parameters and composition of ore bodies in cupriferous sandstones and shales. For example, the thick ore bodies of the Udokan copper deposit are associated with the submarine-deltaic facies, whereas the smaller ore bodies, containing pyrite and chalcopyrite, occur in a bay-lagoonal facies.

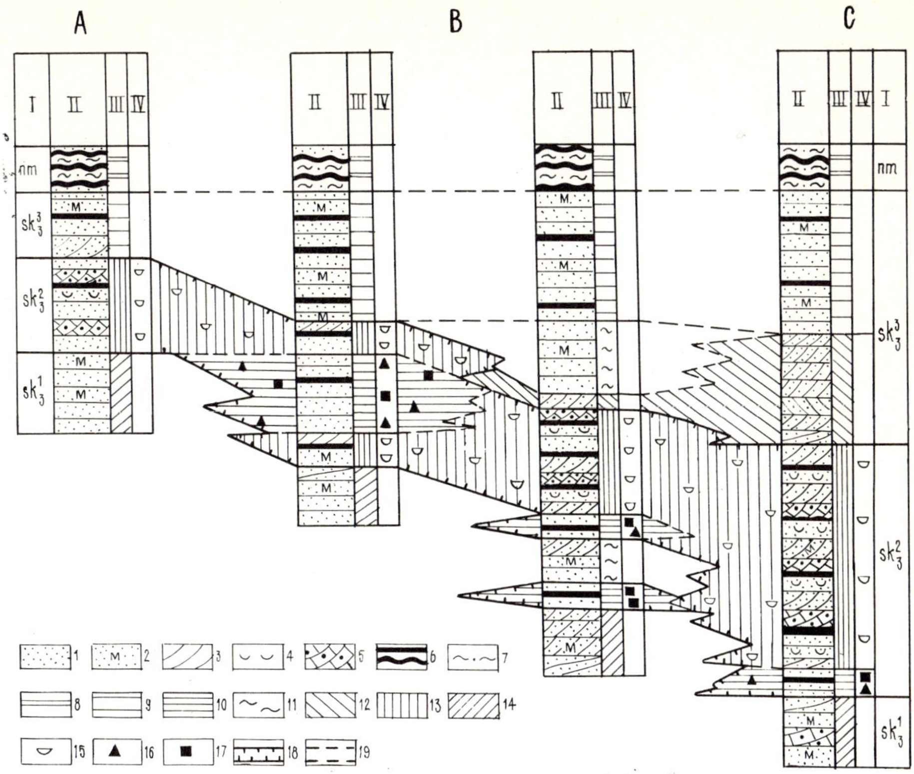

Fig. 3A – C. Location of lithological-facial types of deposition in the ore-bearing section of Udokan copper deposits (cross-section). Ore-synclinal: **A** south flank; **B** axial zone; **C** North flank. *I* Stratigraphic units: sk_3 Upper Sakukanskaya. Horizons: sk_3^1 under ores; sk_3^2 ore-bearing; sk_3^3 over ores; *nm* Namingskaya. *II* Lithological types of rocks: *1* quartzitic sandstones, vaguely layered and massive; *2* sandstones with magnetite; *3* cross-bedded quartzitic and calcareous sandstones; *4* sandstones with thin cross-bedding, or wavy or lenticular-wavy layering; *5* coarsely cross-beded calcareous sandstones with fragments of argillaceous rocks; *6* horizontal and cross-wave argillites; *7* siltstones. *III* Depositional environments of sediments: *8* periodically dried bays and lagoons; *9* shallow littoral zone of sea with relatively calm sedimentation; *10* lagoons; *11* bays; *12* subaerial part of delta; *13* submarine part of delta; *14* shallow seawaters. *IV* Mineralogical types of ores: *15* bornite-chalcocite; *16* chalcopyrite; *17* pyrite mineralization; *18* boundaries of ore-bearing horizon; *19* boundaries of depositional environments

Together with the transgression of favourable facies from regions of erosion towards the basin of sedimentation, there is a migration of mineralization upwards in the section (Fig. 3). A close connection of copper mineralization with certain facies and the age transgression of the copper mineralization has been determined for many stratigraphic levels and zones of the Siberian platform and other regions of the world.

Mineralogical and geochemical features of the copper deposits are quite specific, and that is why they differ distinctly from other genetic types of copper deposits. More than 50 ore minerals have been identified in cupriferous sandstones and shales. Most widely developed are chalcocite, bornite, and more rarely chalcopyrite, which generally forms zonal patterns, both vertically and laterally.

Table 1. Mineralogical and geochemical zoning of copper sandstones and copper shales deposits

Redox potential of ore-forming facies	Ore zones	Chemical forms of iron		Content of residual organic carbon (%)	Ore minerals	Impurity – elements and their mineral forms	Lithological and facial types of ore-bearing sedimentary rocks
		In nonmetamorphic rocks	In metamorphic rocks				
Oxidizing	Barren zone	Limonite is widely distributed	Magnetite and hematite are quite common	From 0.0 n to 0.n	Chalcocite (in small amounts); native copper (sparse)	Absent	Red and red-grey argillaceous continental lagoon sandstones, often indicating traces of salinity
Slight reduction	Chalcocite	Limonite; hematite (rare)	Magnetite, hematite, chalcocite	From 0.n to n	Chalcocite; bornite (rare)	Native elements: Ag, Au, Hg, Bi, platinum group elements	Grey, dark-grey and reddish grey sandstones, dark-grey siltstones, argillites and carbonaceous rocks (rare), deposited in delta or bay-lagoon environments in active hydrodynamic conditions
Reduction	Bornite	Limonite and hematite are rare	Magnetite, hematite; bornite (sporadic)		Bornite, chalcocite; betechtinite and Cu-sulphosalts are rare	Sulphides of Pb, Mo, Re, Bi (in wittichenite) As, Sb	

	Chalcopyrite	Protoxides of Fe (carbonaceous, biotite, glauconite) and sulphides of Fe are widely spread; limonite is practically absent	Sulphides of Fe predominate; magnetite and hematite are in small amounts		Chalcopyrite; bornite and pyrite (more rarely); galena and sphalerite (rare)	Sulphides of Zn, Co, Hg. Isomorphic impurities of Cd, Ge, Zn	Light-grey, grey and dark-grey fine-grained sandstones, siltstones, argillites and carbonaceous rocks deposited in deltas, bays and lagoons in quiet hydrodynamic conditions
Strong reduction	Pyrite	Sulphides of Fe predominate; red rocks with limonite are absent	Sulphides of Fe are widely spread; magnetite and hematite associated with pyrite are not present	n.	Pyrite; chalcopyrite and pyrrhotite (more rarely); sphalerite (rare)	Isomorphic impurities of Ni, Co, Se, Fe, As in arsenopyrite, Au	
Very strong reduction	Pyrrhotite	Sulphide iron predominates		From n to n0	Pyrrhotite associated with pyrite	Co, Ni: sulphide and isomorphic	Dark-grey, often almost black graphite-bearing siltstones, argillites, sandstones, and quartzites, deposited in littoral zones of the sea under quiet hydrodynamic conditions

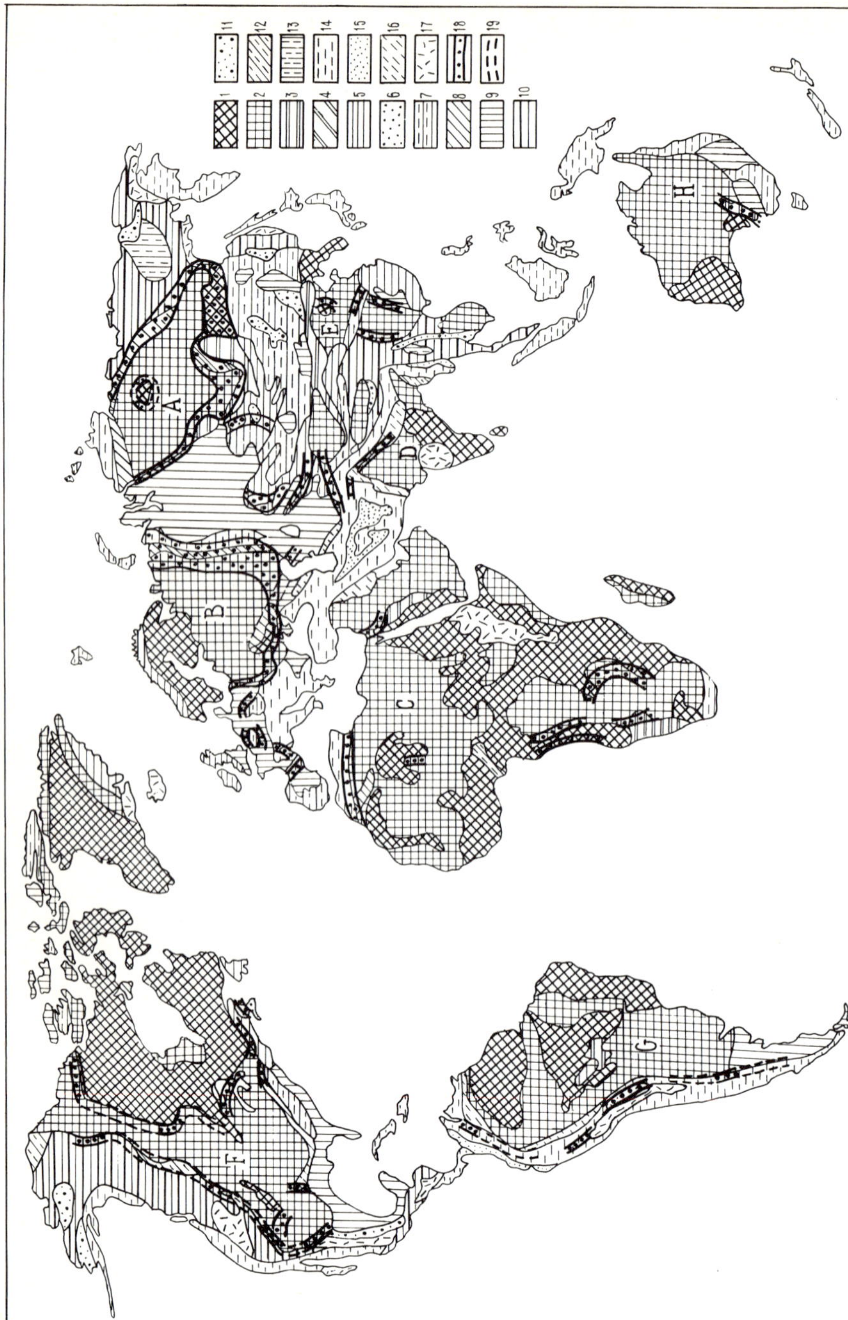

Fig. 4.

Copper ores also contain various minor elements such as Pb, Zn, Mo, Ag, Cd, Co, Ni, Bi and Au, which are forming separate minerals or occur as isomorphic substitutions within the principal ore mineral zones (see Table 1).

Thus, the localization of cupriferous sandstones and shales on the Siberian platform is controlled by stratigraphic, geotectonic, formational, lithologic-facial, paleoclimatic and paleogeographic factors. A favourable combination of the above-mentioned ore controlling factors has led to the formation of ore-bearing provinces stretching along almost the entire perimeter of the Siberian platform and making up one circumferential East-Siberian metallogenic belt of copper sandstones and shales (Narkelyun et al. 1983).

The analysis of published data on copper sandstones and shales of the world shows that they form circumferential and regional belts. We offer the following classification of the distribution of the copper belts concerned (Fig. 4): (A) Ancient platforms. (a) Closed belts on the circumference of ancient platforms [East-Siberian, East-European, Gondwana (South America and Africa), North American]; (b) Closed and semi-closed regional copper belts near shields (Aldansky, Anabarsky, Ukrainian, Canadian shields); (c) Linear belts of foredeeps and the marginal parts of the platform (foredeeps of the Siberian platform, Pricarpatsky, Appalachian and other foredeeps; marginal parts of the Siberian, East-European and North American platforms). (B) Young platforms: regional copper belts embracing geosynclinal massifs and prominences on the platforms (Czechoslovakian, French and others). (C) Geosynclinal folded regions: regional copper belts of internal troughs and basins (Minusinskaya, Tuvinskaya, Djezkazganskaya, Middle Asian and other basins).

With regards to the genesis of copper sandstones and shales of the Siberian platform, the following hypotheses exist: original-sedimentary, volcanogenic-sedimentary, hydrothermal-postmagmatic, diagenetic-epigenetic and connate-hydrothermal. The above-mentioned factors controlling the location of mineralization as well as its genetic features such as the concordant nature of ore bodies in the layered sedimentary suites, the multi-layer structures of mineralization caused by the cyclicity of the rocks, the close connection of mineralization with textural and structural characteristics of host rocks (cross-bedding, parallel-bedding and lenticular bedding), concretionary and socket-shaped segregations of ore minerals, sedimentary-diagenetic deformations of weakly consolidated ore sediments in landslide processes and the presence of asymmetric ore zoning (Bezrodnikh et al. 1977), testify to the original sedimentary genesis of copper sandstones and shales of the Siberian platform. However, later epigenetic and meta-

Fig. 4. Location of cupriferous sandstones and shales in the principal geotectonic structures of the world (on the basis of tectonic maps of the Physical-Geographical Atlas of the World, 1964, simplified). Ancient platforms: *A* Siberian; *B* East European; *C* African; *D* Indian; *E* Chinese; *F* North American; *G* South American; *H* Australian. *1* Exposure of folded basement of ancient platforms (shields); *2* sedimentary cover of ancient platforms. Baikalides: *3* exposure of folded structures; *4* foredeeps and aulacogens. Caledonides: *5* exposure of folded structures; *6* depressions. Hercynides: *7* exposure of folded structures; *8* marginal depressions; *9* sedimentary cover on Caledonian and Hercynian folded basement. Mesozoides: *10* exposure of folded structures; *11* intermountain depressions; *12* foredeeps; *13* central massifs. Alpides: *14* regions of folding; *15* intermountain depressions; *16* marginal depressions; *17* zone of intensive Upper Cretaceous and Tertiary volcanicity; *18* copper ore belts (sandstones and shales). *18* Established; *19* assumed

morphic processes led to the rock transformation, sulphide recrystallization and redeposition, the formation of Alpine-type veins and other changes (Bogdanov et al. 1973, Krendelev et al. 1983, Narkelyun et al. 1977).

Popov (1964) has stated that the process of mineralization and its later deformation and metamorphism comprised the following stages: sedimentogenesis, diagenesis, epigenesis, metamorphism and weathering. The principal stages that led to the formation of mineralization are sedimentogenesis and diagenesis; mineralogical-geochemical peculiarities of their development especially relating to the deposits which distinctly marked zoning in ore minerals have been considered in Table 1. Thus, the formation of sulphides was controlled mainly by redox reactions, quantity and quality of the buried organic matter and lithological-facial conditions.

The geotectonic position of cupriferous sedimentary suites close to the ancient massifs, which have above normal abundance of both copper mineralization and copper deposits of various genetic types testifies to the fact that these massifs may have been the source regions from which copper was transported in detrital remains or dissolved into the basins of sedimentation where the copper-bearing sediments accumulated.

We have come to certain conclusions concerning the sources of copper for the specific districts of copper sandstones and shales on the Siberian platform. It has been shown that the principal sources of copper for the Kodaro-Udokanskaya zone were the Archean suites of the Aldansky shield and Charsky block (Bogdanov et al. 1973, Krendelev et al. 1983, Narkelyun and Yurgenson 1968). Convincing evidence of a close geochemical connection of the regions of erosion and the basins of sedimentation in Pribaikalskaya, Prisajanskaya and Prienisejskaya provinces has been given in a number of publications (Bezrodnikh et al. 1977, Bogdanov et al. 1973, Narkelyun and Yurgenson 1968).

In all probability the source of copper for the formation of Devonian cupriferous sandstones in the Priverkhojanskaya zone and Sette-Davan mountains, may have been underwater volcanic exhalations, and for overlying copper-bearing carboniferous deposits the source of copper may have been eroding effusive-sedimentary rocks.

Thus, it can be stated that the principal sources of copper and accompanying metals in cupriferous sandstones and shales of the Siberian platform were the adjacent regions of erosion. This fact can be convincingly proven by a close metallogenic connection of the regions of erosion and the basins of sedimentation. Underwater volcanic effusions may be considered the sources of metals in certain zones of the platform.

References

Bezrodnikh YP, Narkelyun LF, Salikhov VS, Trubachev AI (1977) Conditions of formation of cupriferous sandstones in the south of the Siberian Platform. In: The basic problems of sedimentary ore formation. Ilim, Frunze, pp 112–136 (in Russian)

Bodganov GV, Burjanova EZ, Kutirev EI (1973) Stratified copper deposits in the USSR. Nedra, Leningrad, p 312 (in Russian)

Krendelev FP, Bakun NN, Volodin RN (1983) Cupriferous sandstones of the Udokan Range. Nauka, Moscow, 248 pp (in Russian)
Narkelyun LF, Yurgenson GA (1968) On sources of ore material in forming thc deposits of copper sandstone type. Lithol Ore Mineral 6:114–123 (in Russian)
Narkelyun LF, Bezrodnikh YP, Trubachev AI, Salikhov VS (1977) Cupriferous sandstones and shales of the southern part of the Siberian Platform. Nedra, Moscow, 223 pp (in Russian)
Narkelyun LF, Salikhov VS, Trubachev AI (1983) Cupriferous sandstones and shales of the world. Nedra, Moscow, 414 pp (in Russian)
Popov VM (1964) Stratified deposits of nonferrous metals and the problems connected with their genesis. In: Problems of ore genesis. Nedra, Moscow, pp 350–368 (in Russian)
Yanshin AL (ed) (1966) Geotectonic map of Eura-Asia. Moscow

Genetic Types of Copper Mineralization in the Igarka Area, West of the Siberian Platform

I. F. GABLINA[1]

Abstract

This paper discusses the general regularities of the distribution of the copper mineralization associated with the Vendian red-bed formation. Lean stratified mineralization is restricted to the zones transitional from the red-beds to the underlying and overlying grey-colored rock deposits. Rich ore deposits are spatially associated with epigenetic carbonaceous matter in the highly-permeable coarse-clastic red-beds near the paleo-uplifts.

The epigenetic host rock alterations associated with ore deposition, and the copper mineral zonation are discussed. A genetic classification of the copper mineralization in the area is suggested.

Geologic Setting

The Igarka copper-ore area is located on the western margin of the Siberian Platform (see Fig. 1 in Krendelev paper). According to its tectonic setting, it is positioned at the intersection of two large Precambrian structures: the Igarka Uplift of the Baikalides, and the foredeep, surrounding it at the east. The Igarka Uplift is a large block structure (up to 90 – 100 km in length) composed of highly deformed and metamorphosed terrigenous-carbonate rocks of the Riphean series. The foredeep was developing at the end of the Late Riphean. It is filled with a red-colored carbonate-terrigenous series of the Vendian, 1300 m thick, the Izluchinskaya Formation. The latter is underlain by marine terrigenous-calcareous, carbonaceous rocks of the Riphean Chernorechenskaya Formation, which is considered to be the major complex of oil and gas source-beds in the Precambrian of the Siberian Platform. At the end of the Vendian and beginning of the Cambrian, as a result of the transgression of the sea, the foredeep was overlapped by the Sukharikhinskaya Formation. From this period onward the area evolved under platform conditions (Rzhevskii et al. 1980).

The Igarka Uplift is divided from the downfold by a marginal suture, along which the Riphean basement rocks are gently overthrust into the Vendian sedi-

[1] Institute of the Lithosphere, USSR Academy of Sciences, 109180 Moscow, USSR

Geology and Metallogeny of Copper Deposits
(ed. by G.H. Friedrich et al.)

ments that fill the foredeep (Fig. 1). The Igarka Uplift represents a folded block structure, dissected by narrow basins into several blocks. One of these blocks constitutes the Chernorechenskaya anticline, which is divided from the Igarka Uplift by a trench-shaped intermontane trough (Fig. 1).

The *Izluchinskaya* red-bed formation is characterized by considerable facies changes. In a northeastern-southwestern direction, from the center of the foredeep toward the Igarka Uplift, the facies undergo a change from calcereous-argillaceous flood-plain-lagoonal sediments to alluvial-lacustrine-silty sandstones and then – to coarse deposits to river channels and piedmont detrital fan sediments (conglomerates, gritstones). In the same direction, the thickness of the formation decreases from 1200 – 1300 m to 200 – 300 m and less.

The Izluchinskaya Formation consists typically of red-colored rocks with a high ferric oxide content ($Fe_2O_3/FeO > 1$) and low organic-C values (commonly $< 0.1\%$, seldom reaching 0.28%), which tend to suggest that oxidizing conditions prevailed in the environment of their deposition. Only the flood-plain-lagoonal sediments, occurring northeast of the mineral estate, contain layers of grey dolomites, limestones, and argillites that in contrast to the former, are of high organic-C content (0.44 – 1.80%).

The marine sediments of the *Sukharikhinskaya* Formation (450 – 500 m) which lie conformably on the red-beds, are composed of limestones and dolomites, often silty and argillaceous, with sandstones and argillites intercalated in the bottom and middle sections. At the base is a quartz-sandstone member. The Sukharikhinskaya Formation consists characteristically of grey rocks with a high organic-C content (0.13 to 0.75%) and a predominance of ferrous oxides ($Fe_2O_3/FeO < 1$).

This formation is of uniform composition. The only exception is within the Graviika ore locus near the eastern flank of the Igarka Uplift, where all the carbonate facies of the lower section are replaced by sandstones of shallow water-coastal facies. The formation is of reduced thickness in this area.

Copper Mineralization

Copper mineralization is restricted to three stratigraphic levels: the base and top of the Izluchinskaya red-bed formation (horizons A and B) and grey rocks enclosed in the red-beds.

Horizon A is associated with the transitional zone from the Izluchinskaya red-beds to the underlying Riphean marine sediments. The contours on its outcrops beneath the Quaternary encircle the Chernorechenskaya anticline (Fig. 1). The copper-bearing horizon is of relatively uniform composition along its entire extension, and is composed of interbedded argillites, argillaceous limestones, marls, and black argillaceous-calcareous shales. Horizon A typically contains uniform mineralization along the strike and low metal content (although several times in excess of the normal Cu abundance). High copper concentrations are extremely scarce.

The mineralization is represented by finely disseminated impregnations of digenite ($Cu_{1.8}S$), bornite (Cu_5FeS_4), and chalcopyrite ($CuFeS_2$) which are zoned

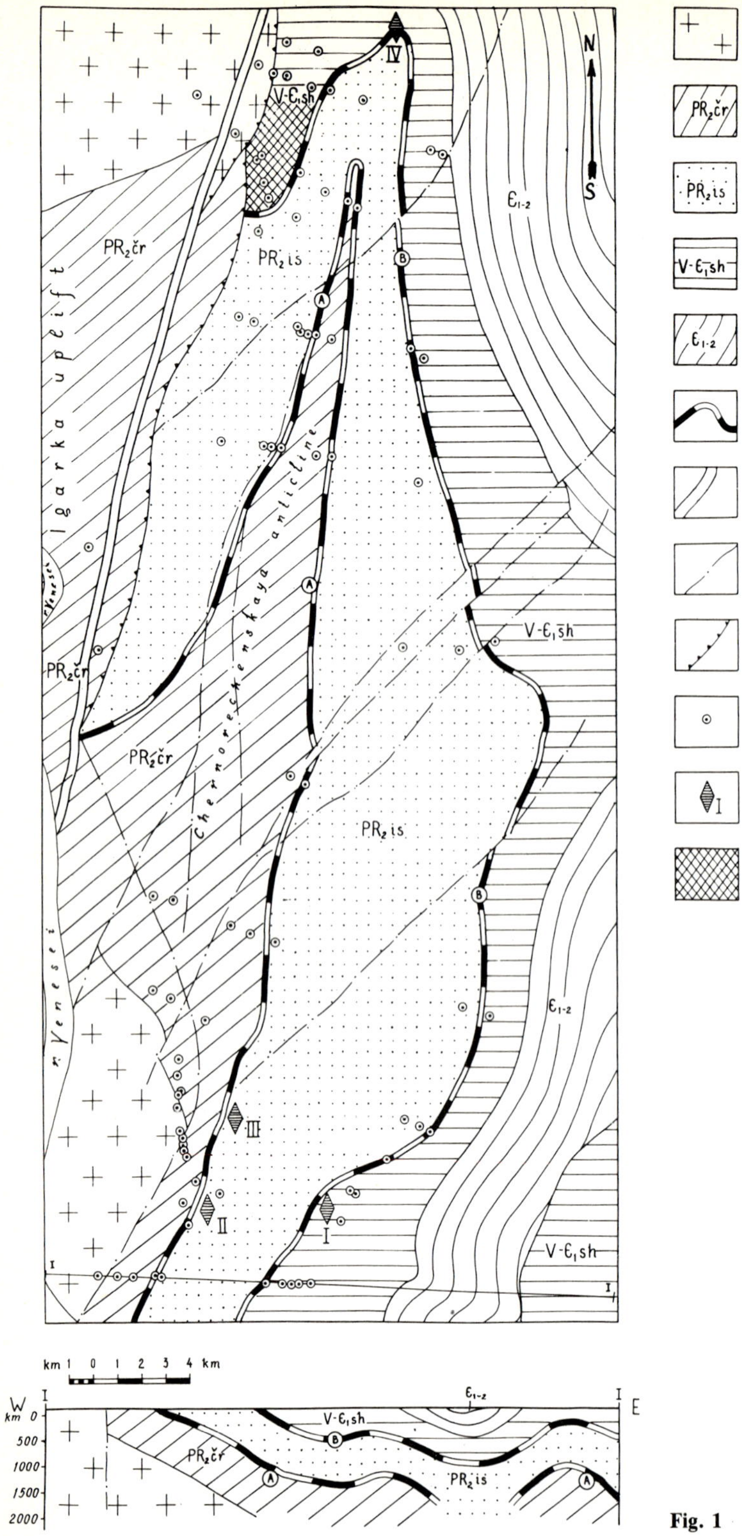
N
S
PR_2čr
$PR_2$is
V-$Є_1$sh
$Є_{1-2}$
Igarka uplift
r. Yenesei
Chernoreckenskaya anticline
IV
III
II
I
A
B
1
PR_2čr 2
$PR_2$is 3
V-$Є_1$sh 4
$Є_{1-2}$ 5
6
7
8
9
10
11
12
km 1 0 1 2 3 4 km
W
E
km 0
500
1000
1500
2000

Fig. 1

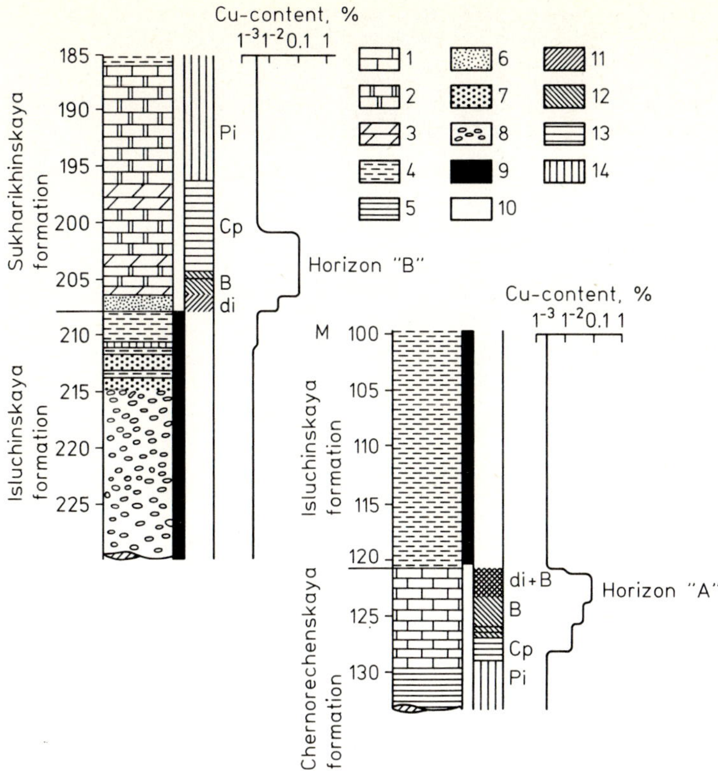

Fig. 2. Vertical mineral zonation in horizons A and B. *1 – 8* Lithologies: *1* limestones; *2* dolomites; *3* marls; *4* siltstones, argillites; *5* carbon-bearing shales; *6* quartz sandstones; *7* gritstones; *8* conglomerates. *9 – 10* Rock colour: *9* red, brown; *10* grey, black, green. *11 – 14* Metallic minerals: *11* digenite (*di*); *12* bornite (*B*); *13* chalcopyrite (*Cp*); *14* pyrite (*pi*)

vertically in the sequence away from the contact with the red-beds (downward in the section). The copper content of the ore rocks decreases in the same direction (Fig. 2). The vertical range of copper mineralization does not extend more than 5 – 15 m from the base of the red-beds. The metallic minerals occur in the form of small (up to 1 mm) isometric concretionary aggregations of metasomatic type, frequently surrounded by diffusion aureoles of the same sulfides (Fig. 3). The observed relationships between the metallic mineral aggregations and the host rocks serve to indicate that the mineral clusters formed within consolidated sediments.

Copper-bearing horizon B is restricted to the base of the grey marine sediments of the Sukharikhinskaya Formation that rest on the Izluchinskaya red-

Fig. 1. Map and cross-section showing geologic structure and distribution of copper mineralization in the Igarka area (adapted from V. F. Rzhevskii). *1* Rocks of the basement fold structure; *2* Chernorechenskaya Formation (PR_2 cr); *3* Izluchinskaya Formation (PR_2 is); *4* Sukharikhinskaya Formation ($V - \varepsilon_I$ sh); *5* Cambrian (ε_{I-2}); *6* copper-bearing horizons A and B; *7* deep-seated faults of the marginal suture; *8* other faults; *9* overthrust zone; *10* exploration wells; *11* ore occurrences; *I* Sukharikha; *II* Izluchinsk; *III* Rudnii; *IV* Chernorechensk; *12* Graviika ore deposits

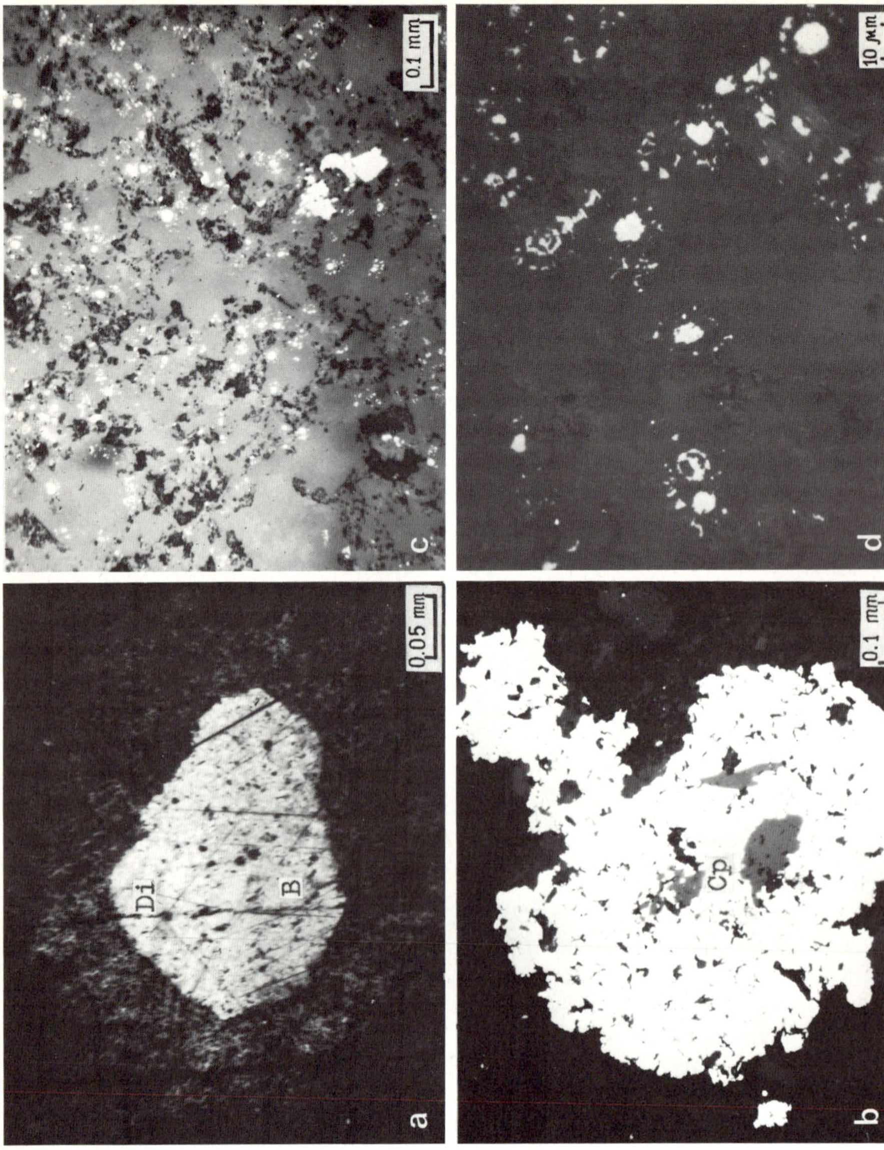

Fig. 3a – d. The detailed features of sulfides in horizon A (**a, b**) and horizon B (**c, d**, white-digenite and bornite). Photomicrograph of the polished section. Photo by N. Inyashkin. Abbreviations: *Di* digenite; *B* bornite; *Cp* chalcopyrite

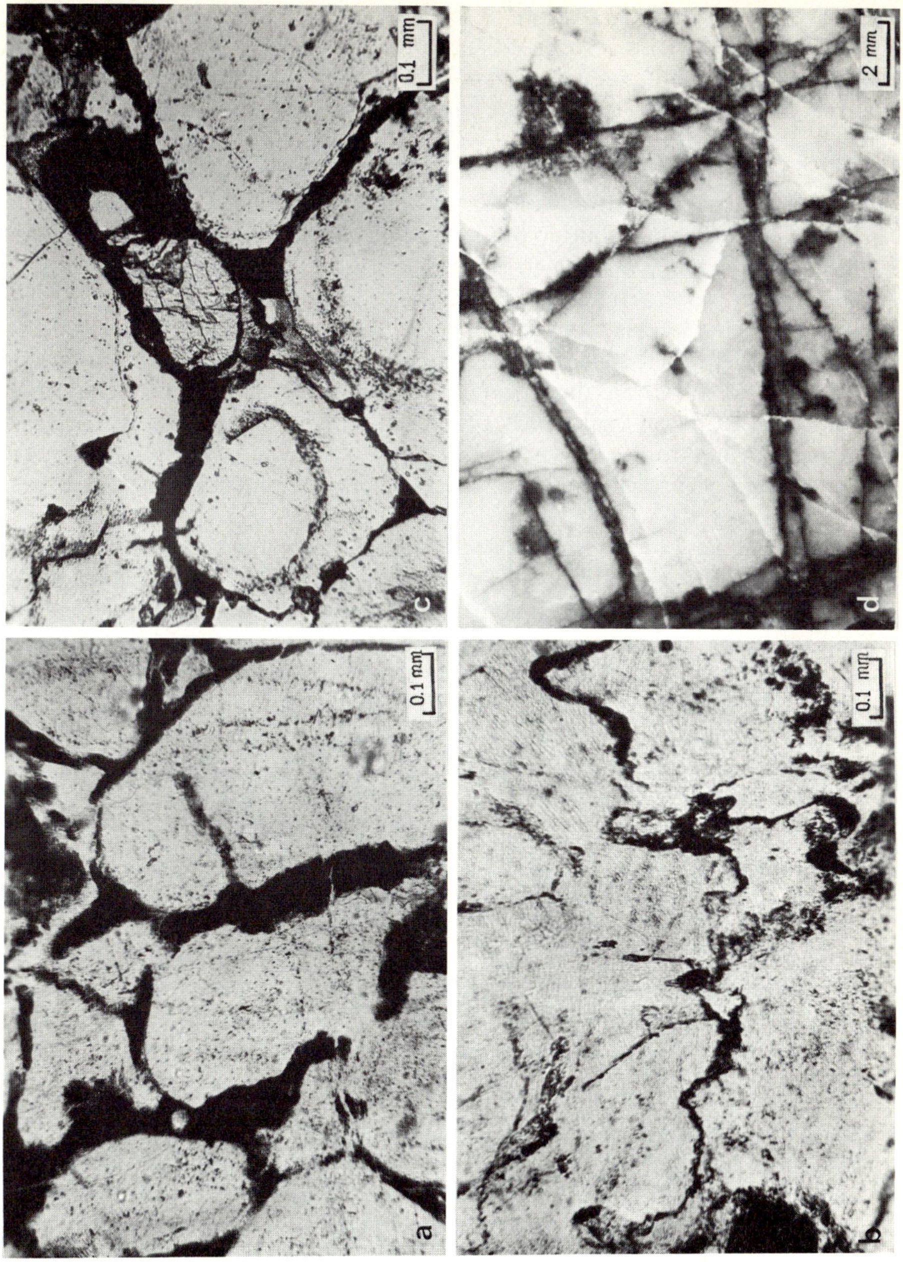

Fig. 4a – d. Photomicrographs of the specific features of the epigenetic organic matter (*black*) in sandstones (**a, b, c**) and dolomites (**d**). Photo by E. Zenkova

beds (Fig. 1). A distinctive feature of this horizon is the meager copper content, persistent along the entire extent of the horizon, and local enrichment only within discrete areas. The horizon is 10 – 30 m thick. The sulfide mineralization exhibits an irregular distribution pattern: the concentrations tend to increase towards intercalated sandstones or limestones with a high arenaceous, silty content. The metallic minerals are represented by the same paragenetic sulfide associations as in horizon A, but, in contrast, occur in the section in reversed order of succession (Fig. 2). The regular distribution of the metallic minerals in the vertical section in both horizons exhibits a symmetrical zonation in relation to the red-beds. The metallic minerals occur in horizon B in the form of finely disseminated impregnations in the ore rocks. Most typical are tiny (not exceeding 10 μm) globular forms or spherolites of radial-concentric structures. The rock texture is the controlling factor in their distribution (they form pore fillings in sandstones and inclusions in the authigenic cement), from which it can be inferred that their time of formation was closely related to the host rocks (Fig. 3).

Several ore occurrences in the Graviika copper deposits are associated with horizon B.

In the *Graviika ore deposits* horizon B shows greatly increased copper content in the ore rocks immediately above the coarse-clastic alluvial-deltaic facies of the underlying the red-beds. The ore body here is dome-shaped and, in general, configuration matches the contours of the sandstones of the coastal, shallow water facies. Its thickness varies from 50 – 100 m at the crest of the dome to a few centimeters on the flank.

The rocks of the Sukharikhinskaya Formation in the area of the Graviika ore deposits exhibit intense alteration. These secondary alterations are reflected in silicification, dolomitization, and calcitization of the rocks, as well as occurrences of epigenetic organic matter (EOM). The EOM is impregnated in the sandstones, occurring in the form of pellicular-pore cement and discrete oval-shaped aggregates, and it also penetrates into the quartz sandstones and carbonate rocks, pervading the numerous minor fissures, crush zones, and stylolitic sutures (Fig. 4). The EOM is of greatest vertical range in the shoaly sandstone deposits. As a consequence the latter become dark grey and in some places black in color. The pigmentation intensity, caused by the impregnated EOM, shows a conspicuous decrease upward in the section. When the sandstones undergo facies changes to dolomites the depth of penetration of EOM into the rocks of the Sukharikhinskaya Formation shows a sharp decrease in different directions away from the "sand bank", being limited, as a rule, to the basal sandstone layer (1 m thick).

Fig. 5a, b. Stratified section of the Graviika ore deposits, contoured on the basal sandstones of the Sukharikhinskaya Formation. **a** Epigenetic zonality in the wall rocks near ore. *1 – 3* Zones of epigenetic alterations: *1* quartz zone; *2* dolomitic; *3* calcitic. *4* Relative content of major oxides (SiO_2; MgO; CaO) in different alteration zones; *5* contour lines showing distribution of shallow-sandstone facies. *6 – 9* Wells with data on organic-C content in basal sandstones of the Sukharikhinskaya Formation: *6* <0.1%; *7* 0.1 – 0.5%; *8* 0.5 – 1%; *9* >1%. *10* Occurrences of epigenetic organic matter in rocks underlying the Izluchinskaya Formation; *11* limit of altered, lighter-colored rocks in the Izluchinskaya Formation; *12* limits of epigenetic alteration zones; *13* faults; *14* overthrust zone; *15* wells (with no data on organic-C). **b** Metallic mineral distribution pattern. *1 – 5* Metallic minerals: *1* sulfides of the chalcocite series; *2* bornite; *3* chalcopyrite; *4* galena; *5* sphalerite

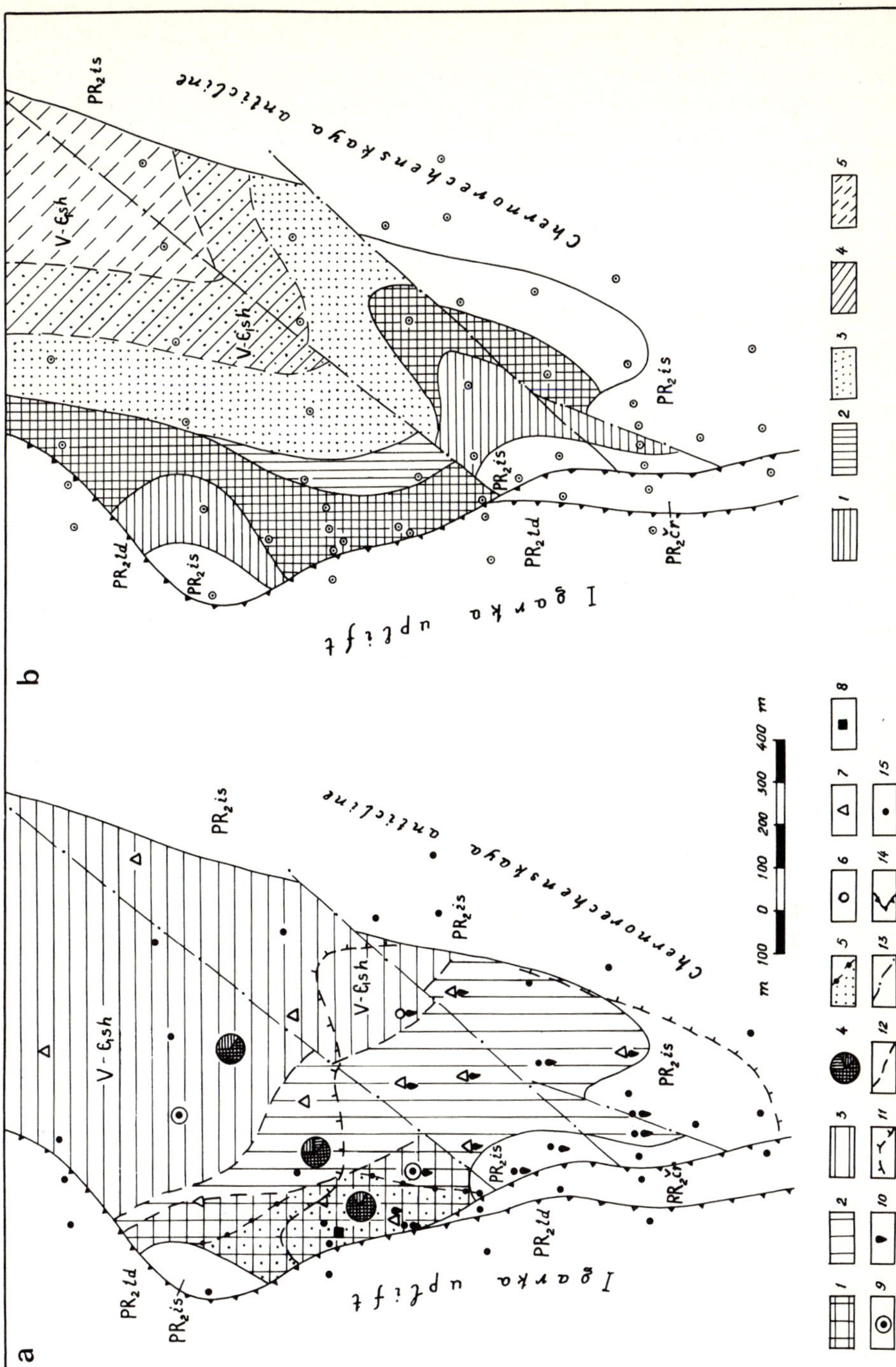
a
b
PR2is
PR2ld
PR2čr
V-Є1sh
Chernorechenskaya anticline
Igarka uplift
m 100 0 100 200 300 400 m

Fig. 5

EOM is rare in the overlying dolomites. In the rocks of the Sukharikhinskaya Formation, containing EOM, the organic-C content is several times higher than normal abundance, reaching in single samples to 1.09 – 1.71%.

A distinct zonation can be observed in the distribution pattern of other rock alterations, which are also controlled by lithofacies: the zone of silicification coincides in area with the shoaly sands facies: eastward it is replaced by a zone of predominant dolomitization, and further east of calcitization (Fig. 5A).

Commercial ore deposits are typically represented by the following highly-cupriferous association of djurleite ($Cu_{1.96}S$) + pink bornite ($Cu_{5+x}FeS_{4-x}$), which is superimposed on earlier paragenetic associations. Djurleite and pink bornite form lenticular veinlet deposits that crosscut the rocks and the disseminated digenite, bornite, and chalcopyrite globules (Fig. 6A). Of wide occurrence are copper sulfide pseudomorphs as replacements of pyrite.

The rich djurleite and djurleite-bornite ores occur in a band extending along the eastern margin of the Igarka Uplift. This ore zone changes eastward to a lean ore band of low-grade chalcopyrite and chalcopyrite-galena mineralization (Fig. 5B). A similar change in the mineral zones can be observed vertically upward from the top of the red-beds. The described distribution of the sulfides is evidence for a decrease in the copper concentration and Eh values of the environment, occurring upward in the vertical section from the top of the red-beds and in an east-west direction away from the Igarka Uplift.

The *copper mineralization* in the *Izluchinskaya Formation* of the Graviika mineral estate is restricted to highly permeable rocks – river channel conglomerates, consisting of poorly rounded coarse-grained pebbles of carbonate rock types (dolomites, limestones), and practically lacking sedimentary in-fillings. In the zone where the Chernorechenskaya anticline abuts the Igarka Uplift, the calcareous conglomerates and some of the wedge-shaped polymictic argillaceous gritstones are of lighter colours and intense carbonatization. These bleached conglomerates, in contrast to their red-colored analogs that replace them toward the north and east, are characterized by the presence of solid carbonaceous substances, similar in type to the EOM (Fig. 5A) that is impregnated in the rocks of the Sukharikhinskaya Formation. In the Izluchinskaya conglomerates the carbonaceous matter occurs in the form of large (up to 1 – 1.5 cm) oval- and drop-shaped inclusions in the carbonate cement, or in the form of a black mass densely pervading the entire rock section. According to data available from X-ray examinations, the carbonaceous matter of the inclusions can be assigned to the medium anthraxolite type. Its elemental composition, as determined in the VIMS laboratory, is as follows: C = 88.96; H = 3.64; (O + N + S) = 7.41 wt%, all on a dry, ash-free basis. Its carbon isotopic composition ($\delta^{13}C = -2.98‰$) is closely similar to that of petroleum of the Siberian platform. The specific forms of EOM occurrence, its restriction to highly permeable rock types in the Vendian deposits, and localization near the marginal suture in the zone of conjunction of two uplifted blocks, and its crosscutting position in relation to the lithofacies and stratigraphic boundaries – all these factors taken together give evidence for the migrational nature of the carbonaceous matter. These accumulations of carbonaceous substances in the permeable rocks of the Vendian can probably be regarded as traces of destroyed oil deposits.

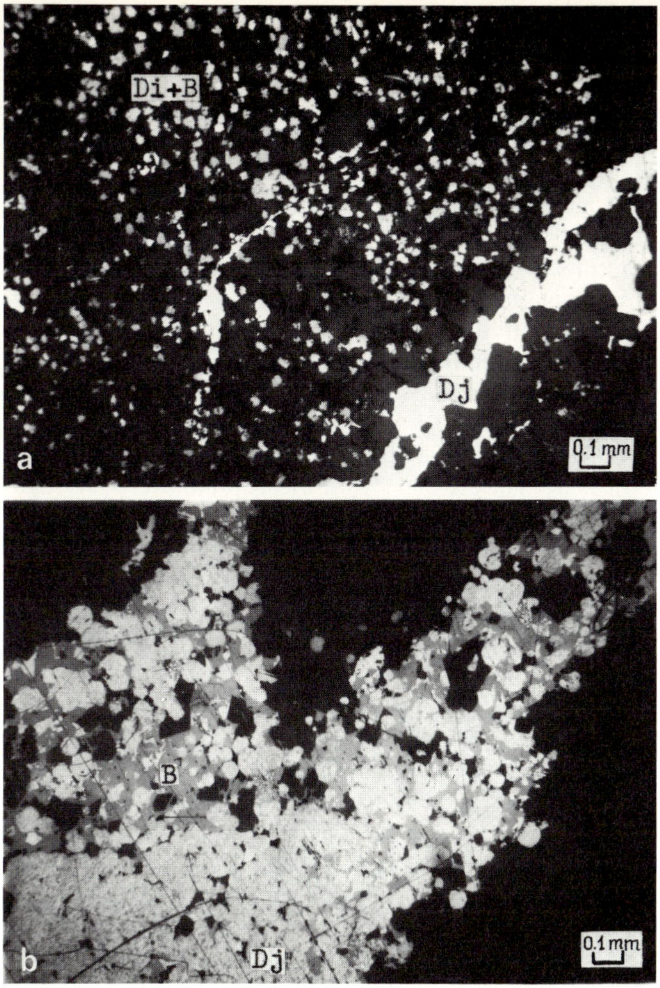

Fig. 6a, b. The detailed features of sulfides on the Graviika ore deposits: **a** in sandstones of the Sukharikhinskaya Formation; **b** in conglomerates of the Izluchinskaya Formation. Photo by N. Inyashkin. Abbreviations: *Dj* djurleite; *Di* digenite; *B* bornite

The lighter-coloured conglomerates, containing anthraxolite, typically reveal a pronounced increase in their organic-C content (up to 1.5 – 3.5%) and the predominant occurrence of ferrous oxides as compared to ferric oxides (Fe_2O_3/$FeO < 1$). They form a body, triangular in plan and lenticular in vertical section, that in the central part is comprised of a dolomite zone ("dolomite core"). The conglomerates resemble here dolomitic metasomes with ghost-relicts of primary, clastic structures and rare corroded quartz grains or silicate rock fragments. Calcite plays an insignificant role, occurring at times as pore-fillings in the dolomitic conglomerates. The calcite content tends to increase from the dolomite core toward the periphery of the bleached body. According to the data available from chemical analyses of their composition, the conglomerates of the dolomitic zone consist almost entirely of calcium and magnesium carbonates.

The calcitic zone surrounds the dolomite core, being transitional from the practically unaltered red-colored rocks to the dolomitized zone. This zone, on tracing the color boundary, curves around the bands of weakly permeable red-colored rocks, revealing an intricate flexuous configuration. The major epigenetic mineral in this zone is coarse crystalline calcite, occurring as metasomatic replacements of the cement and dolomitic pebbles. The maximal development of the calcitization process results in the conversion of the calcareous conglomerates into secondary limestones, containing ghost-relicts of original pebbles. In contrast to the metasomes of the dolomitic zone, they characteristically have a lower content of organic-C (0.09 – 0.27%, in occasional instances – up to 0.45%) and of MgO, the latter decreasing in amount at the transition from the dolomitic to the calcitic zone.

The sulfide mineralization coincides spatially with the calcitic zone and in part involves the dolomitic zone. The ore bed is of intricate multilayered configuration, matching the configuration of the interfaces of the transition from the red-colored rocks to secondary grey-colored types. The sulfides are represented principally by late paragenetic associations of highly-cupriferous sulfides (djurleite + pink bornite) and to a lesser degree – by chalcopyrite and galena. The sulfides occur as crusts and pore-cement in the conglomerates and frequently form corrosion borders in pebbles. In some cases three corroded-metasomatic types reveal relicts of the primary-globular structure (Fig. 6B). The above-cited characteristics suggest the long duration of the mineralization period, beginning in sediments with open pore-spaces and continuing during the process of lithification of the rocks.

The lateral zonation of the metallic minerals is reflected in the change of djurleite ores to bornites and then to chalcopyrites, occurring in the direction of the change in color of the host rocks from red to grey-colored types. In the same direction the zone of copper mineralization is replaced by a lead-bearing mineral zone. The mineral zones are repeated in the vertical section, revealing an orderly symmetrical arrangement relative to the red-beds.

Besides the previously described type of mineralization characterized by metallic mineral concentrations of commercial value, and associated with the grey-colored modifications of coarse-clastic facies of the Izluchinskaya Formation, metallic mineral occurrences are also associated with the calcareous-argillaceous, flood-plain, lagoonal facies (Chernorechenskaya ore occurrences). The latter are located northeast of the Graviika ore locus, within the zone transitional from the continental red-bed facies of the Izluchinskaya Formation to the grey lagoonal sediments. The ore mineralization is restricted to the layers of grey-colored dolomites, limestones, and argillites, which are typically of high organic-C content. The mineralization is of the stratified multilayered type (according to the number of grey-coloured rock layers), and shows similarity in its major features to the stratified ore deposits of horizon B.

General Regularities

The described types of copper mineralization in the Igarka area exhibit characteristics in their geologic position, mode of occurrence, configuration, and magnitude of the ore deposits and also in the time of ore deposition as related to the time of deposition of the host rocks. Notwithstanding, they are closely related, demonstrating affinities in their major genetic factors. The latter are as follows: (1) general association with red-bed formations; (2) localization of the ores in the area where the red-beds are directly marginal to the grey rock types (rich in organic carbon) which differ markedly from the former in their geochemical properties; (3) persistence of typical metallic mineral assemblages, reflecting the general similarity in the geochemical environment of ore deposition; (4) zonal distribution pattern of the metallic mineral species, the mineral zones exhibiting uniformity in their orderly arrangement in relation to the red-beds; (5) the joint occurrence in the ores of djurleite[2] together with non-equilibrium mineral associations [djurleite (or digenite) + pyrite; bornite + pyrite[3]] which imply low-temperature ($<93\,°C$) conditions of ore deposition, and the lack of subsequent high-temperature effects on these mineral assemblages.

Factors Controlling Copper-Ore Deposition

From the analytical treatment of the data outlined above, the inference can be drawn that the Izluchinskaya red-bed strata was the source of copper during the process of ore deposition. It is an acknowledged fact (Germanov 1962, Perel'man 1968, Lur'ye 1978) that confined connate waters of red-bed formations, due to their specific geochemical properties, are capable of leaching copper and other metallic minerals and can transport them in the form of dissolved substances. The metallic minerals precipitate in the form of sulfides upon encountering biogeochemical barriers, as a result of reactions between the metal-bearing waters with hydrogen sulfide or pyrite produced by sulfate-reducing bacteria. Such barriers can be marine sediments, rich in sedimentogenic OM, pyrite, and other reducing agents (syngenetic barriers), as well as epigenetic accumulations of OM, hydrocarbons, or H_2S (epigenetic barriers) (Lur'ye and Gablina 1982).

Formation of the copper-bearing horizons A and B could have resulted from the diffusion of copper from the pore waters of the Izluchinskaya Formation into the organic-rich rocks of the underlying Chernorechenskaya Formation and the bottom sediments of the Sukharikhinskaya Sea basin. The zonal distribution of the sulfides can be accounted for by the decrease in the Cu concentrations and in-

---

[2] Djurleite is a typomorphic mineral of copper sandstone and shale deposits and of secondary sulfide enrichment. In metamorphogenic veins of the Alpine type of deposits, chalcocite (Cu_2S) is present instead of this mineral. In magmatogenic ore deposits, copper sulfides are represented by chalcocite occurring in combination with high-temperature tetragonal forms of chalcocite (Gablina 1984).

[3] At 100 °C chalcocite and pyrite begin to react, resulting in the formation of bornite. At temperatures exceeding 228 °C reactions occur between pyrite and bornite that are conducive to the formation of chalcopyrite (Yund and Kullerud 1966).

crease in the H_2S concentrations with distance away from the Cu source, the red-beds. This affords a feasible explanation for the distribution pattern of the copper minerals and their localization in the contact zones of the marine host rocks and the Izluchinskaya red-bed strata. The specific features of the mineral sequences of sulfides in horizons A and B reflect the time of formation of the mineralization as related to that of the host rocks: syndiagenetic in horizon B and epigenetic in horizon A.

The formation of the Graviika ore deposits can be assigned to a late-diagenetic evolutionary stage of the Vendian, and seemingly, was a two-phased development. The first phase (creation of the epigenetic barrier) can be associated with the penetration of EOM into the reservoir rocks of the Izluchinskaya Formation, the EOM having been contributed, presumably, from the underlying oil and gas sources in the Riphean. Biochemical decay of the EOM probably caused the secondary alteration observed in the local area of the calcareous conglomerates: bleaching and dolomitization, concomitant with the leaching of SiO_2. The latter apparently was redeposited in the overlying sandstones of the Sukhatikhinskaya Formation, thus forming the zone of silicification.

The second phase (ore deposition) can be assigned to the period when, due to lithostatic compaction, the confined waters of the red-beds were squeezed out from the foredeep toward the Igarka Uplift, where the conditions favored their discharge. The inflow of these calcium sulfate waters of the red-bed formations was a favorable factor promoting the biochemical oxidation of OM that accumulated in the Izluchinskaya conglomerates. The decomposition products of biochemical OM decay consisted of H_2S and CO_2. Upon encountering the hydrogen sulfide or pyrite geochemical barrier, the nonferrous metals (copper and lead), which were brought in by the waters of the red-beds, precipitated. The direction of flow of the metal-bearing solutions is reflected in the zonal distribution pattern of copper sulfides: in the direction away from the red-beds toward the rocks of secondarily-derived lighter colors, the mineral sequences reveal the replacement in successive order of djurleite ($Cu_{1.96}S$) by bornite (Cu_5FeS_4) and further by chalcopyrite ($CuFeS_2$), which reflect the decrease in the copper concentration and concurrent increase in the sulfide-ion concentration in the same direction. The spatial separation of copper and lead minerals (shifting of the galena zone toward the barrier) is due to the difference in the geochemical mobility of these elements and also serves to indicate their migration from the red-beds.

It is known that the concentration of sulfate ions in the solution enhance the solubility of magnesium carbonates (reserve reaction of the Heidenberg reaction). Possibly the calcium-sulfate waters of the red-bed formations produced the dedolomitization and calcitization observed in the zone of sulfide formation.

Near the steep eastern flank of the Igarka Uplift, where the Izluchinskaya Formation consists of highly permeable rocks and the overlying marine sediments are of reduced thickness, the formation waters under pressure could have flowed vertically upwards. This unward migration could have been controlled by the permeable rocks of the coastal shallow facies, as well as the fractured zones of the marginal suture (Fig. 7).

The migration of hydrocarbons from the Izluchinskaya conglomerates into the overlying Sukharikhinskaya Formation was probably due to the upward flow

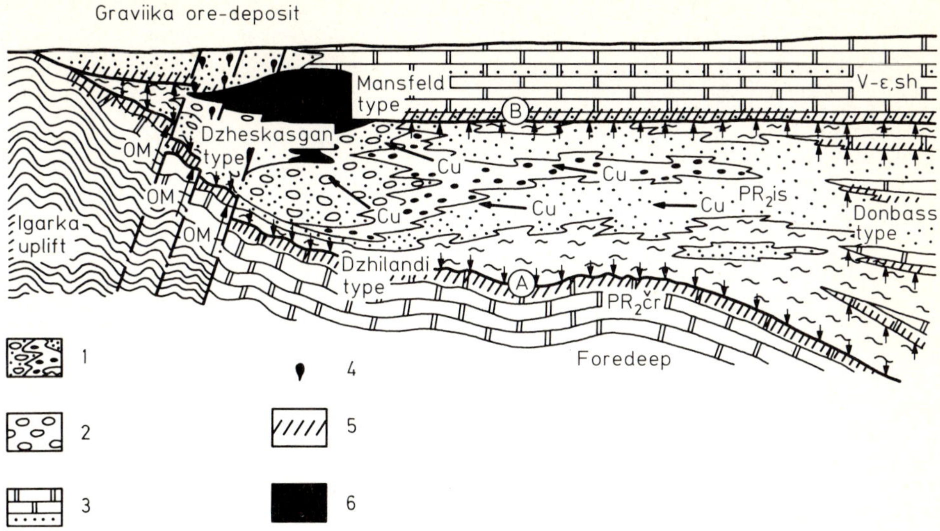

Fig. 7. The genetic model and types of copper mineralization in the Igarka area. *1* Red-bed formation; *2* bleached conglomerates of the red-bed formation; *3* grey marine and lagoonal deposits; *4* occurrences of epigenetic OM; *5* lean, stratified mineralization; *6* rich sulfide ores

of artesian waters. The rocks of the Sukharikhinskaya Formation, enriched in sedimentogenic and migratory forms of OM, served as a specific type of hydrogen sulfide "filter", upon which the metals, being carried upward by the upflowing waters, precipitated in the form of sulfide. With increased distance from the drainage zone, the vertical flow of the fluids changed to lateral migration within the permeable beds. Manifestations of this lateral migration are reflected in the zonal distribution pattern of the ore minerals (see Fig. 5B).

Conclusion

Several types of copper mineralization associated with the Vendian red-bed formation can be recognized in the Igarka area (Fig. 7, Table 1). According to the magnitude of the deposits they can be subdivided into two groups: (1) lean, stratified deposits, occurring at the top and base of the Izluchinskaya red-beds, and also in the enclosed grey layers of the flood plain and lagoonal facies that are transitional to shallow, marine facies; (2) rich accumulations of sulfide ores, spatially and genetically related to the permeable coarse-clastic rocks of the red-bed facies. The latter are represented by the Mansfeld and Dzhezkazgan types, the combination of both types being responsible for the Graviika ore deposits.

Table 1. Types of copper mineralization in the Igarka area

Type of mineralization	Subtype	Position of mineral zone with respect to that of the red-bed formation	Host rocks	Mode of ore deposition according to the time of formation of the host rocks	Type of biogeochemical barriers (or OM)	Distribution of mineralization	Magnitude of ore deposits
Mansfeld	Syngenetic	At the base of the grey marine series, overlying terrestrial red-bed formations	Basal members of the Sukharihkinskaya Formation (horizon B)	Syndiagenetic	Syngenetic (sedimentogenic OM)	Areal distribution	Noncommercial
	Epigenetic			Epigenetic	Mixed: sedimentogenic + epigenetic OM	Local restriction (above permeable rocks of coarse red-bed facies)	Commercial
Donbass		In the zone of lateral transition from terristrial red-beds to lagoonal-marine facies	Grey rock layers, enclosed in variegated lagoonal sediments of the Izluchinskaya Formation	Syndiagenetic	Syngenetic (sedimentogenic OM)	Areal distribution	Noncommercial
Dzheskazgan		Enclosed in terrestrial red-bed formations	Permeable coarse rock of secondary grey colors in the Izluchinskaya Formation	Epigenetic	Epigenetic (epigenetic OM)	Local restriction (to coarse-clastic facies)	Commercial
Dzhilandi		At the top of the grey marine sediments, underlying terrestrial red-bed formations	Rich in organic matter, grey marine sediments, transitional from the Chernorechenskaya to the Izluchinskaya Formation (horizon A)	Epigenetic	Syngenetic (sedimentogenic OM)	Areal distribution	Noncommercial

References

Gablina IF (1984) Mineral species of the chalcocite series in copper-sandstone and shale deposits. Minerali khalkozinovovo ryada iz mestorozhdenii medistikh peschanikov i slantsev. Zapiski Vsesoyuz Min ob-va 4:430 – 443

Germanov AI (1962) Hydrodynamical and hydrochemical factors controlling the formation of certain hydrothermal ore deposits. Gidrodinamicheskie i gidrokhimicheskie usloviya obrazovania nekotorikh gidrotermal'nykh mestorozhdenii. Izv An SSSR Ser Geol 7:79 – 98

Lur'ye AM (1978) Factors controlling copper migration in red-bed formations. Uslovia migratsii medi v krasnotsevnykh formatsiyakh. Geokhimia 6:926 – 932

Lur'ye AM, Gablina IF (1982) Controlling factors of copper concentration during the exogenetic process. In: Factors controlling the formation of rare and nonferrous metallic mineral deposits. Uslovia kontsetratsii medi v ekzogennom protsesse, pp 133 – 142. V kn: Uslovia obrazovania mestorozhdenii redkikh i tsevetnykh metallov. Nauka, Moscow, p 256

Perel'man AI (1968) Geochemistry of epigenetic processes. Geokhimia epigeneticheskikh protsessov. Nauka, Moscow, p 331

Rzhevskii VF, Miroshnikov AE, Dushatkin AB, Shklyarik GK (1980) Copper mineralization in the upper precambrian of the Igarka Area. In: Processes governing the accumulation of sedimentary and volcanogenic-sedimentary deposits of nonferrous metals (Siberia and the Far East). Medenosnost' verkhnedokembriiskikh otlozhenii Igarskogo raiona, pp 81 – 84. V kn: Protsessi osadochnova i vulkanogenno-osadochnovo nakoplenia tsvetnikh metallov (Sibir' i Dal'nii Vostok). Nauka, Novosibirsk, p 159

Yund R, Kullerud G (1966) Thermal stability of assemblages in the Cu-Fe system. J Petrol 7, 3:454 – 488

The Diverse Styles of Sediment-Hosted Copper Deposits in Australia

I. B. LAMBERT, J. KNUTSON, T. H. DONNELLY, and H. ETMINAN[1]

Abstract

The huge sediment-hosted copper deposits at Olympic Dam and Mount Isa are of unusual types. They formed epigenetically from hydrothermal fluids, as did several other important Australian copper deposits in sedimentary strata. Ore within a palaeoweathering zone at the top of a red-bed unit at Mount Gunson appears to have formed at relatively low temperatures over a long period and to have incorporated biogenic sulfur. Bedded-disseminated mineralization, which is widespread in South Australia, formed by reaction of cupriferous fluids with biogenic H_2S and early diagenetic iron sulfides. To date, stratiform copper mineralization has not been of major economic significance in Australia.

Introduction

Ores in sedimentary strata have provided roughly two-thirds of Australia's copper production and they contain the great bulk of the known reserves of this metal. The sizes, grades and ages of the most important of these copper deposits are given in Table 1. The regions in which they occur, shown in Fig. 1, are the Adelaide Geosyncline-Stuart Shelf (South Australia), the Mount Isa Inlier (Queensland), the Tennant Creek Inlier (Northern Territory), the Bangemall Basin (Western Australia) and the Cobar Trough (New South Wales).

This paper succinctly reviews the geology of these regions and discusses the general features and genesis of the main deposits. The emphasis is on the results of recent research in the two most important copper provinces, the Adelaide Geosyncline-Stuart Shelf and the Mount Isa Inlier.

[1] Baas Becking Geobiological Laboratory, GPO Box 378, Canberra ACT 2601, Australia

Geology and Metallogeny of Copper Deposits
(ed. by G.H. Friedrich et al.)

Table 1. Major sediment-hosted copper deposits in Australia

Region Deposit	Size[a,b]	Status	Age of host strata	Form of mineralization
Stuart Shelf Olympic Dam (Roxby Downs)	>2000 m.t. @ 1.6% Cu, 0.064% U_3O_8 0.6 g t^{-1} Au, (REE, F)	Preparations for mining	Middle Proterozoic	Hydrothermal mineralization in unmetamorphosed graben sequence dominated by sedimentary breccia units[c]
Stuart Shelf Cattle Grid (Mount Gunson)	5.9 m.t. @ 1.9% Cu, (Pb, Zn)	Ore nearly exhausted	Middle Proterozoic	Tabular deposit in breccia along unconformity surface at top of unmetamorphosed red-bed unit[d]
Adelaide Geosyncline Kanmantoo	12 m.t. @ 1% Cu	Closed 1976	Cambrian	Lenses of hydrothermal mineralization flattened parallel to axial plane schistosity in metasiltstone[e]
Mount Isa Inlier Mount Isa	>232 m.t. @ 2.9% Cu	Largest mine	Middle Proterozoic	Metamorphogenic hydrothermal mineralization in altered, brecciated dolomitic siltstone[f]
Tennant Creek Inlier				
Peko	3.1 m.t. @ 4% Cu, (Bi)	Closed 1976		Pipelike hydrothermal lodes with abundant quartz, magnetite, hematite, chlorite, in major shear zones[g]
Warrego	7.8 m.t. @ 2.6% Cu, 2 g t^{-1} Au, (Bi)	Mine	Early Proterozoic	
Gecko	5.2 m.t. @ 3.7% Cu 0.7 g t^{-1} Au, (Bi)	Care and maintenance		
Cobar Trough				
CSA	8.6 m.t. @ 1.7% Cu, 14 g t^{-1} Ag (Pb, Zn)	Mine Closed 1945	Siluro-Devonian	Hydrothermal stockwork to massive mineralization in silicified shear zones sub-parallel to slaty cleavage[h]
Great Cobar	4.2 m.t. @ 2.8% Cu, 2.2 g t^{-1} Au, 11.2 g t^{-1} Ag			
Bangemall Basin Nifty	1.3 km strike length secondary ore + primary Cu	Major prospect	Mid-late Proterozoic	Stratabound in sandstone-siltstone-shale sequence[i]

[a] Total size, i.e. ore mined plus proven and possible reverses as published in company reports and press releases. [b] m.t. = million tonnes; g t^{-1} = grams/tonne. [c] Roberts and Hudson (1983). [d] Knutson et al. 1983, Creelman 1983. [e] Verwoerd and Cleghorn 1975. [f] Perkins 1984. [g] Crohn 1975, Goulevitch 1975, Wright 1965. [h] Brooke 1975. [i] Western Mining Corporation, press release 1983; Goode and Hall 1981

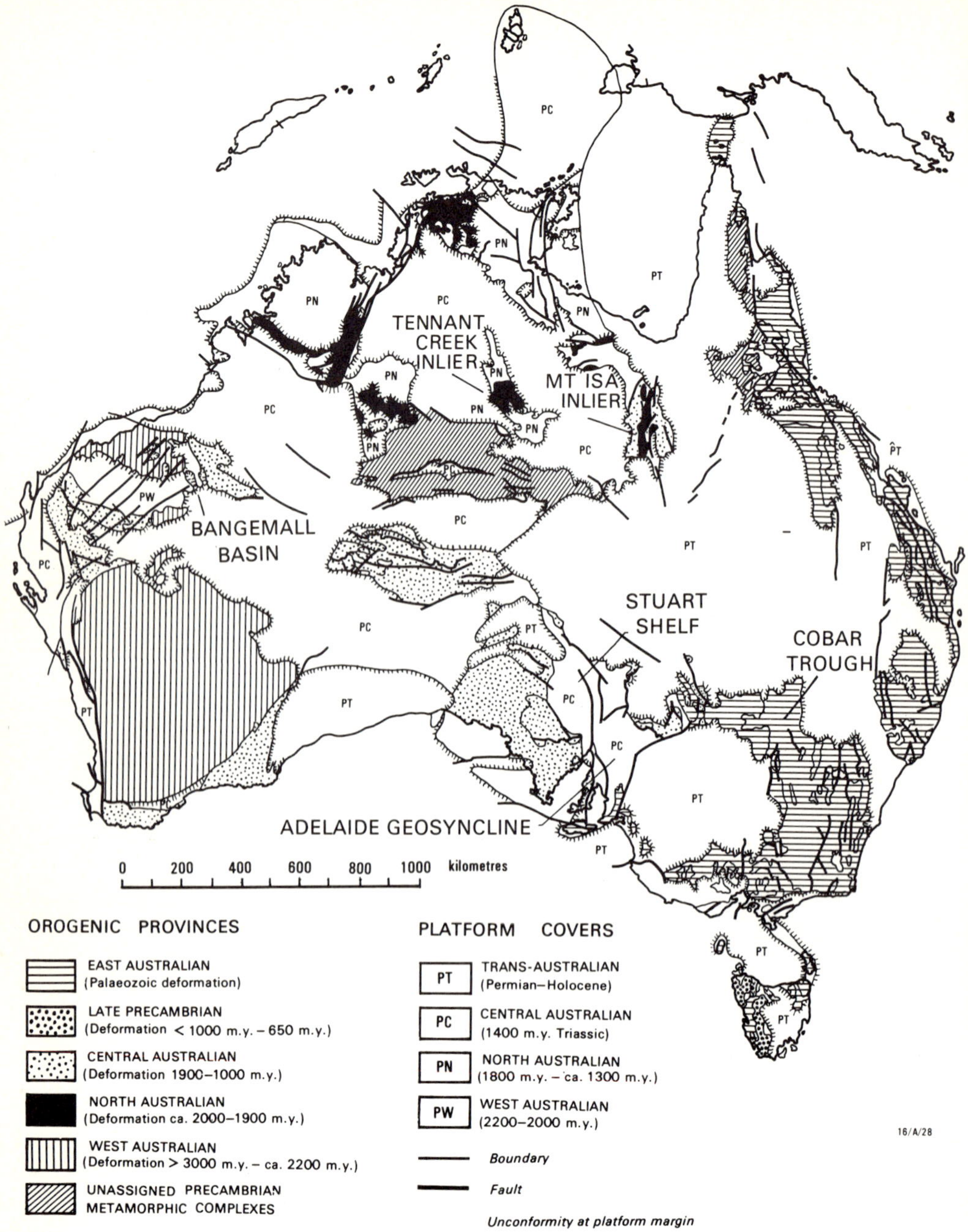

Fig. 1. Map summarizing major geological features of Australia (after Warren 1972) and localities of the main sediment-hosted copper provinces

Adelaide Geosyncline-Stuart Shelf

Regional Geology

The most recent review of the Precambrian geology of this region is by Rutland et al. (1981). Figures 2 and 3 summarize the general geology and stratigraphy.

The schists, gneisses and granites of the crystalline basement were metamorphosed and deformed approximately 1.8 to 1.65 b.y. B.P. In the west of the region, post-orogenic, middle Proterozoic volcanic and sedimentary strata accumulated in grabens. The youngest of the pre-late Proterozoic units is the Pandurra Formation, a thick continental red-bed sandstone.

Approximately 1.1 b.y. B.P., an aulacogen developed in the eastern part of the region, the site of accumulation of the thick, late Proterozoic (Adelaidean) clastic-carbonate sequence of the Adelaide Geosyncline. This was separated by

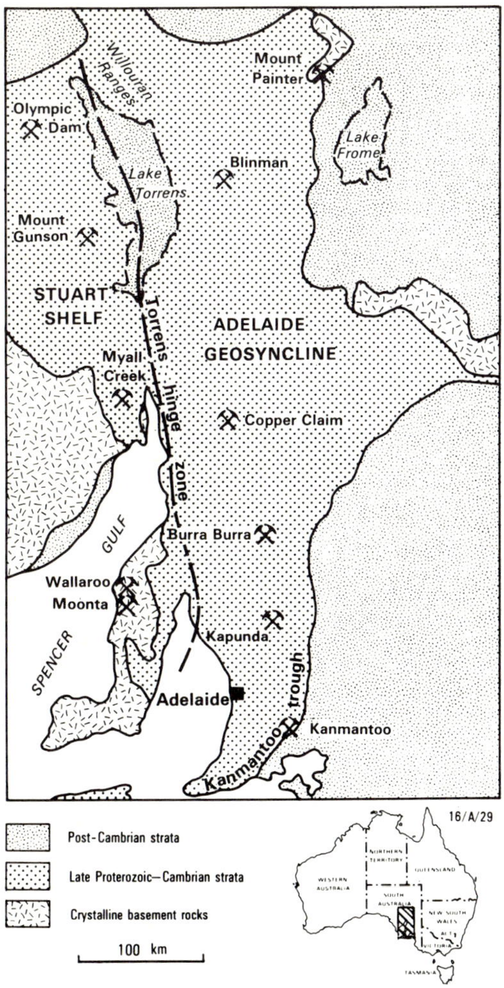

Fig. 2. General geological features and localities of main copper deposits, Adelaide Geosyncline-Stuart Shelf region, South Australia

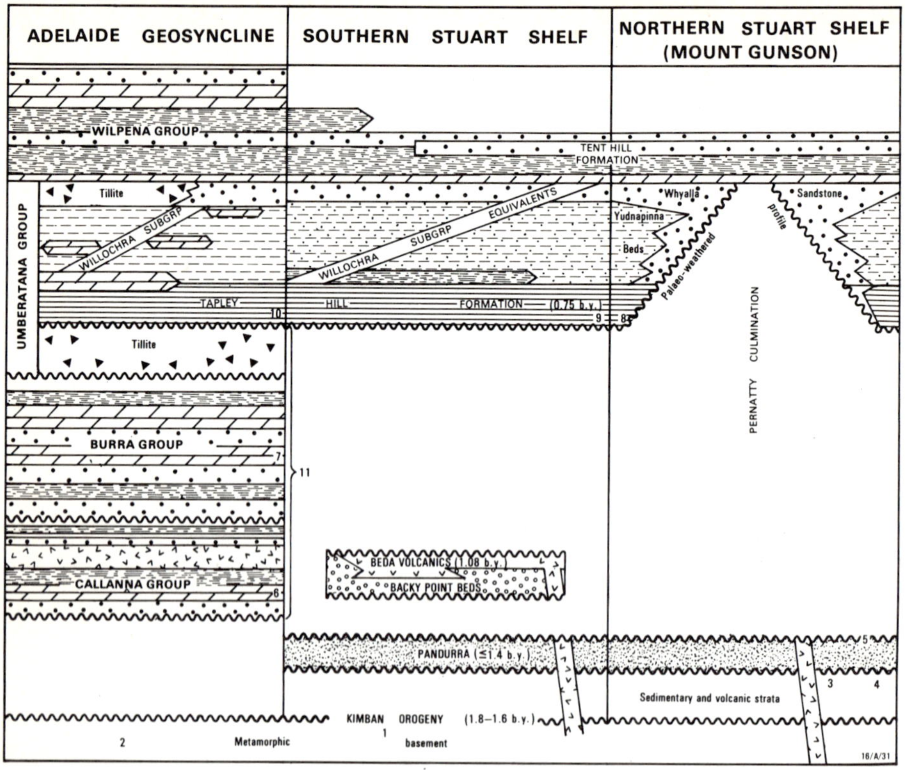

Fig. 3. Generalized stratigraphy, Adelaide Geosyncline-Stuart Shelf region, and stratigraphic positions of main copper deposits. *1* Wallaroo-Moonta; *2* Mount Painter; *3* Olympic Dam; *4, 5, 8* Mount Gunson; *6* Copper Claim, Blinman; *7* Burra Burra; *9* Myall Creek; *10* Kapunda; *11* Willouran Ranges

the Torrens Hinge Zone from the stable platform environment of the Stuart Shelf, where there is a relatively thin Adelaidean sequence. The late Proterozoic strata are characterized by a predominance of shallow water and emergent facies, subordinate mafic volcanism and two major tillite units. There is evidence of non-marine environments in some units.

A major erosional break preceded transgression and widespread deposition of post-rift Cambrian strata across the region. The Adelaide Geosyncline sequence was regionally metamorphosed and deformed during the late Cambrian-Ordovician, but that on the Stuart Shelf was not affected.

Mineralization

The locations of the main copper deposits are shown in Fig. 2; their stratigraphic positions in Fig. 3.

The features of Olympic Dam (Roberts and Hudson 1983), Moonta-Wallaroo (Johnson 1965) and Kanmantoo (Verwoerd and Cleghorn 1975) imply that they formed from hydrothermal fluids. Because of space limitations, only the largest of these, Olympic Dam, will be considered here.

The other two styles of copper mineralization to be discussed below, neither of which has obvious hydrothermal alteration zones, occur in the Mount Gunson-Pernatty Lagoon area (Fig. 2). The most important of these occurs in a breccia along the unconformity between middle Proterozoic red-bed and late Proterozoic pale sandstone. The second occurs as largely bedded sulfides in late Proterozoic dolosiltstone. Bedded deposits of this general style have been found at widespread localities in the late Proterozoic strata of this province; while none have continuity of high copper grades in primary mineralization, numerous, secondarily-enriched, near-surface ores have been mined in small-scale operations.

Olympic Dam

For a detailed description of this deposit, the interested reader is referred to the paper by Roberts and Hudson (1983). The post-orogenic middle Proterozoic graben sequence which hosts the Olympic Dam mineralization is over 1 km thick. It comprises mainly granitic and polymict breccia units overlain by iron formation and tuffaceous felsic volcanics (Fig. 4). The sequence is pervasively altered, and there is relatively intense hematite-chlorite and sericite-silica alteration associated with the mineralization.

The copper mineralization can be divided into stratabound and transgressive types (Figs. 4, 5) both of which contain significant uranium, rare earths, gold and silver. The ore is quite variable in appearance (Fig. 5). Sulfides are mainly disseminated through breccia matrices, but large clasts of sulfides are also common, and there is some massive mineralization. Textures indicate a complex paragenesis, involving microcavity and fracture filling, replacement and rimming of clasts. Reconnaissance fluid inclusion and sulfur isotopic analyses have been undertaken on Olympic Dam mineralization (D. Roberts, personal communication, 1984). Homogenization temperatures for inclusions in fluorite and quartz associated with sulfides range from 110° to 265 °C, and freezing point depressions indicate 0.5 to 14.3% NaCl equivalent. These indicate mixing of fluids of different salinities and temperatures. The $\delta^{34}S$ values for ore sulfides are concentrated in the range −5‰ to −11‰, and barite $\delta^{34}S$ values range from +4.6‰ to +13‰. In view of the evidence for hydrothermal alteration under relatively oxidizing conditions, these preliminary isotopic data are consistent with a magmatic sulfur component. The high contents of potassium, rare-earths, barium, fluorine and iron in the mineralization are consistent with hydrothermal activity associated with alkaline igneous activity.

Mount Gunson-Pandurra Formation

In the Mount Gunson-Pernatty Lagoon area of the Stuart Shelf, upfaulted blocks of Pandurra Formation red-bed sandstone formed a depositional high (Pernatty culmination; Figs. 3 and 6) during the Adelaidean. The relatively im-

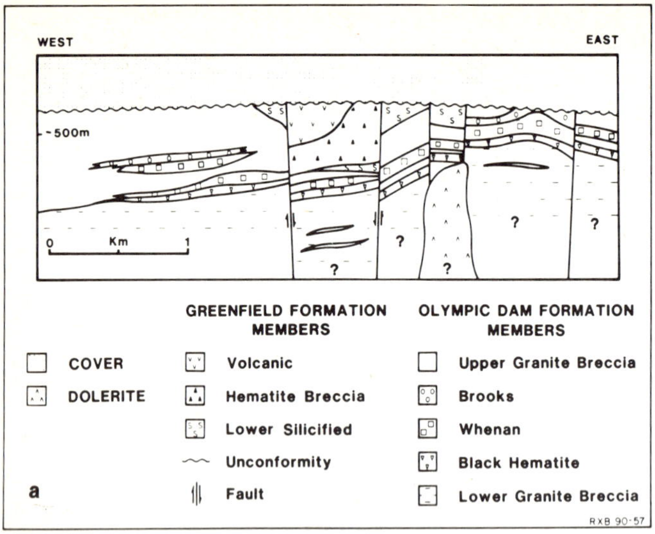

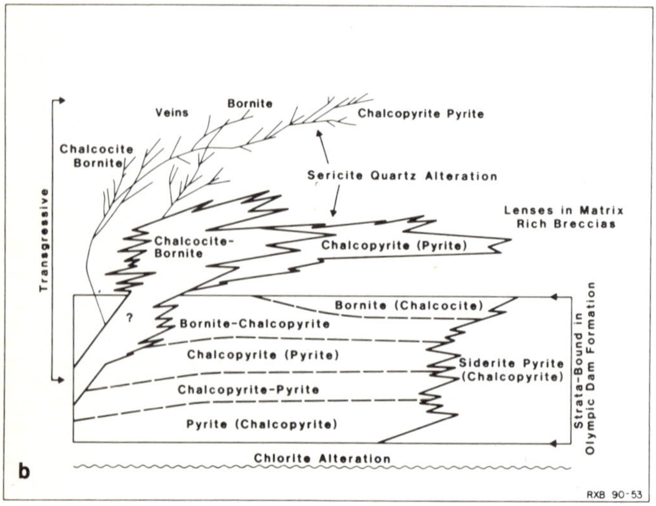

Fig. 4. a East-west cross-section, Olympic Dam area (from Roberts and Hudson 1983). **b** Representation of the two types of mineralization and mineral zoning, Olympic Dam deposit (from Roberts and Hudson 1983)

permeable Tapley Hill Formation is locally absent over this structure, the Pandurra Formation being overlain by Whyalla Sandstone. The only economically important mineralization known in the Pandurra Formation is in the brecciated palaeoweathering surface at the top of this culmination (Fig. 7). The brecciation appears to have formed partly as a result of periglacial weathering and it is locally intense around faults (Tonkin and Williams 1983). The main deposit is Cattle Grid, which has been described by Gersteling and Heape (1975), Creelman (1983) and Knutson et al. (1983).

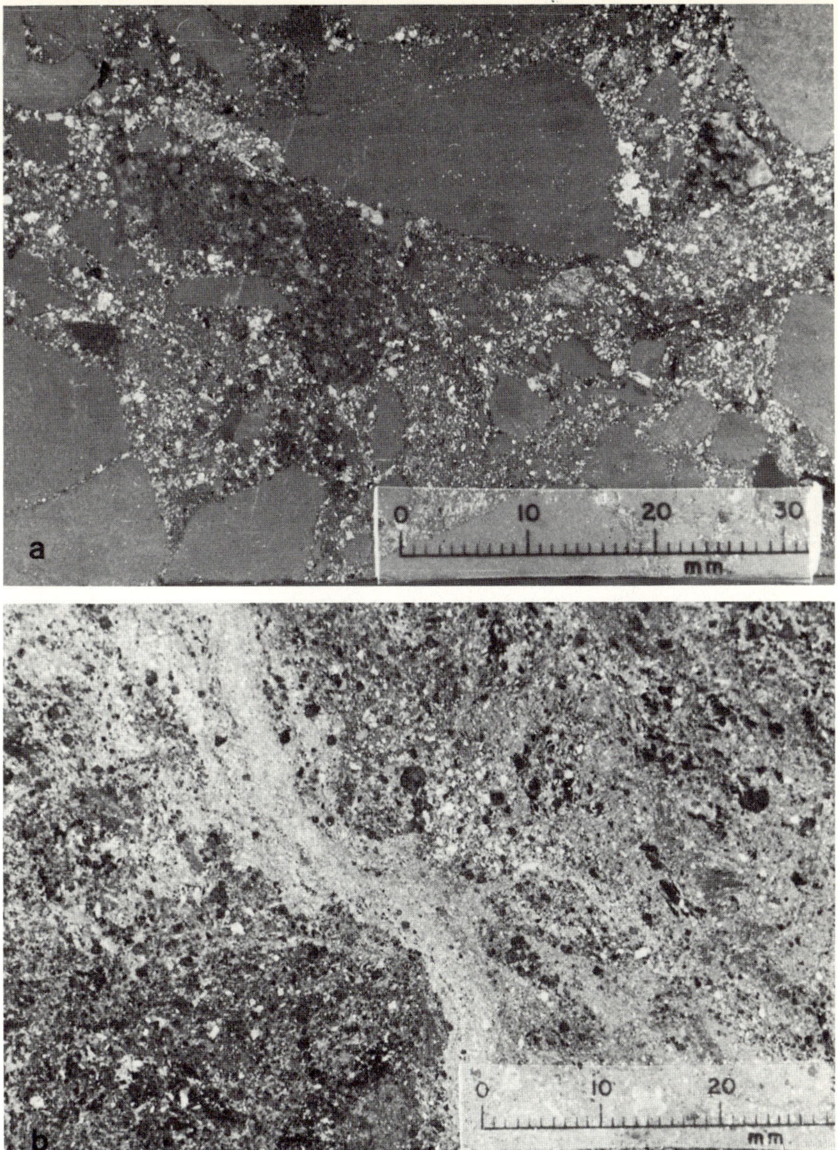

Fig. 5 a, b. Cut sections of Olympic Dam drill core. **a** Stratabound mineralization, polymict breccia clasts in chalcopyrite-rich (*white*) marix. **b** Transgressive chalcocite-bornite mineralization (Photographs kindly supplied by D. E. Roberts)

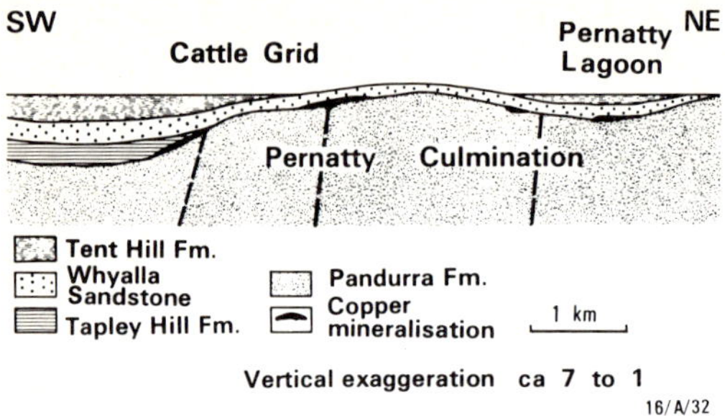

Fig. 6. Schematic section, Mount Gunson-Pernatty Lagoon area

Fig. 7. Chalcocite (*black*) and chalcopyrite (*light grey*) ore in brecciated Pandurra Formation, Cattle Grid mine, Mount Gunson

Cattle Grid is a tabular deposit, averaging roughly 4 m thick, in which the grade of copper falls off rapidly with decreasing brecciation. The mineralization mainly infills fractures and vugs, and it is dominated by chalcocite, chalcopyrite, bornite and pyrite; marcasite, sphalerite and galena are minor components.

The $\delta^{34}S$ values for the ore minerals are concentrated between −12‰ and −5‰ (Fig. 8). In the absence of evidence for hydrothermal mineralizing fluids, these are most readily interpreted in terms of bacterial sulfate reduction; the

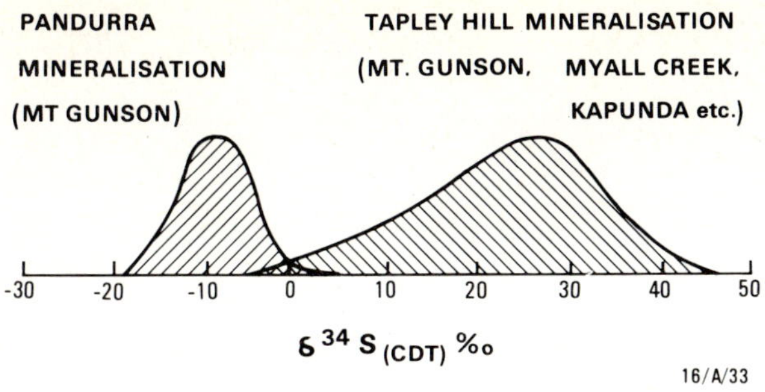

Fig. 8. Distributions of sulfur isotopic compositions of mineralization in Pandurra and Tapley Hill Formations

paucity of organic matter in the mineralized breccia, and the form and timing of the sulfide mineralization, imply that the H_2S was introduced in groundwaters.

The sulfide mineralization must have formed largely after burial of the palaeoweathering zone. It was the product of protracted movement of fluids through the permeable rocks of the culmination, resulting in a complex sequence of deposition, dissolution, redeposition and replacement. Metal-enriched fluids were probably generated as brines reacted with minerals in the Pandurra red-beds and in the older volcanic and sedimentary rocks. In addition, weathering of Beda Volcanics (Fig. 3) could have resulted in metal enrichments in waters in the underlying red-beds.

Mount Gunson-Tapley Hill Formation

The Tapley Hill Formation is the only mineralized late Proterozoic unit on the Stuart Shelf (Fig. 3). Copper, lead and zinc sulfides were widespread in minor amounts in the basal 5 m or so of this unit, but no economic deposits have been defined. The two main prospect areas have been at Mount Gunson and Myall Creek (Fig. 2). The following summary of the Mount Gunson mineralization draws information from the paper by Knutson et al. (1983).

The Tapley Hill Formation in the Mount Gunson area is up to 200 m thick and is dominantly dolostone adjacent to the Pernatty culmination, where it contains stromatolites, desiccation cracks, intraclast and edgewise conglomerates, and possible pseudomorphs after evaporites. It grades outwards into increasingly laminated strata with abundant carbonaceous dolosiltstone components, which were deposited beneath wave base. The mineralization comprises fine-grained disseminations of chalcopyrite, bornite, chalcocite-pyrite-sphalerite and galena. The copper minerals partly replace and rim pyrite (Fig. 9). In addition, there are discontinuous thin (millimetre-scale) sulfide veinlets and blebs which mainly fill dewatering channels and desiccation cracks, and probably formed during diagenesis. It is noteworthy that the mineralization is concentrated in the relatively permeable dolosiltstone laminae.

There is a general vertical zoning from copper at the base of the mineralization inwards to lead and zinc, and a similar zoning occurs laterally away from the palaeohigh. Relatively minor metal anomalies occur in places at the top of the Tapley Hill Formation, with decreasing Cu/Pb + Zn ratios downwards into the unit.

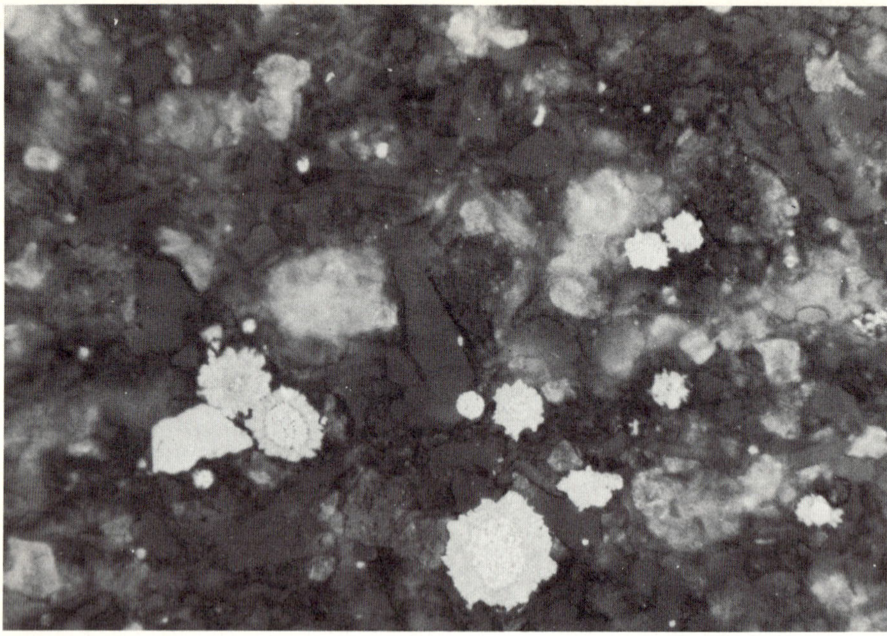

Fig. 9. Low-grade copper mineralization in unmetamorphosed Tapley Hill Formation, Stuart-Shelf. Framboidal pyrite (*white*) rimmed and partially replaced by bornite (*mid-grey*). Reflected light; approx. 200 μ across photomicrograph

Pyrite and ore sulfides in the Tapley Hill Formation at Mount Gunson have $\delta^{34}S$ values in a wide range from 0‰ to +45‰, indicating a different sulfide source from the mineralization in the Pandurra Formation (Fig. 8). Together with the absence of hydrothermal alteration, and the presence of carbonaceous matter, these isotopic compositions imply biological sulfate reduction in a restricted environment. Such ^{34}S-enriched sulfides, and the high $^{87}Sr/^{86}Sr$ ratios of dolomite from the Tapley Hill Formation (Lambert et al. 1984) are most readily interpreted in terms of predominantly continental waters in the Tapley Hill sea.

Dolomite from both mineralized and unmineralized Tapley Hill Formation has $\delta^{13}C$ values close to −3‰ and $\delta^{18}O$ values close to +26‰, indicating that the mineralizing fluids were isotopically similar to, or dominated by, the diagenetic fluids. However, alteration of detrital feldspar to clays is more advanced in the mineralized strata, implying that relatively high volumes of near-neutral fluids passed through these strata. This, and the concentration of copper in the dolosiltstone laminae, support protracted copper addition to the strata. The late copper appears to have been precipitated largely by reaction with iron sulfides.

The proximity and similar compositions of the deposits in the Tapley Hill and Pandurra Formations at Mount Gunson suggest their generation from the same, or related, metalliferous fluids.

Mount Isa Inlier

Regional Geology

The oldest rock exposed in this region comprise complexly deformed igneous and sedimentary rocks which have metamorphic ages around 1.87 b.y. B.P. (Page 1978, Blake et al. 1983). According to Plumb et al. (1980), these basement rocks are most abundant in a tectonic high separating the so-called eastern and western successions of unconformably overlying middle Proterozoic sedimentary and volcanic strata (Fig. 10), but recent studies imply that "basement" rocks are scattered throughout the region (Blake et al. 1984). The largest copper deposits in the region, Mount Isa and Mammoth, are in the western succession (Fig. 10).

The basal arenites of the western succession are overlain by up to 5 km of altered metabasalts and intercalated arenites of the Eastern Creek Volcanics. A sequence of quartz-rich arenites and conglomerates occurs unconformably above,

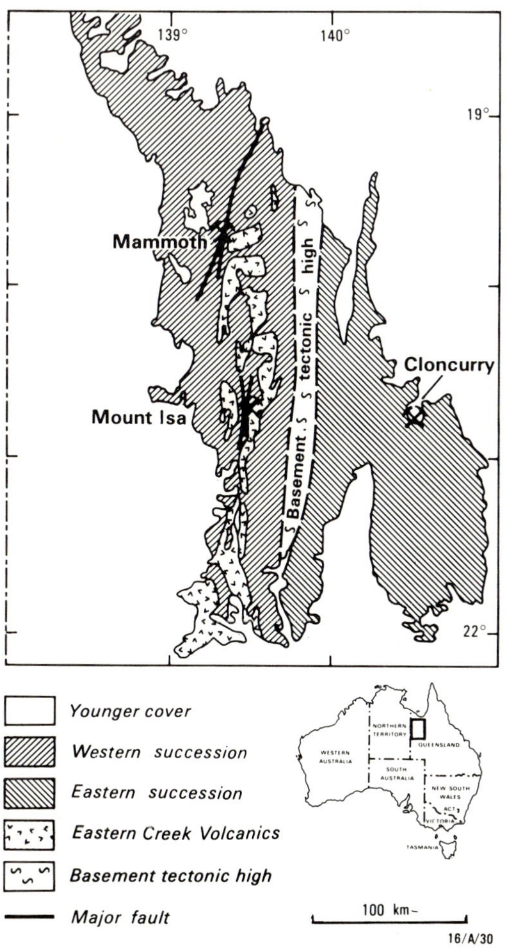

Fig. 10. The Mount Isa Block, showing gross geological features and localities of major copper deposits

and this contains local felsic volcanics at the top. This sequence is unconformably and disconformably overlain by Mount Isa Group strata, comprising generally shallow water to emergent, dolomitic and variably carbonaceous shales and siltstones (Neudert and Russell 1981), with adularia- and albite-rich tuffaceous beds. Tuffites from Mount Isa mine have been dated at 1.67 b.y. B.P. (Page 1981). Radiometric and structural studies imply periods of greenschist facies metamorphism at around 1.61 and 1.55 b.y. B.P., and a final deformation event at around 1.48 b.y. B.P. (Page 1978, Blake et al. 1983). Granites were emplaced before 1.8 b.y. and approximately 1.78, 1.67 and 1.5 b.y. B.P. (Blake et al. 1983).

Copper Mineralization

The great bulk of the copper deposits in the Mount Isa Inlier are small veins in faults and shears (Brooks et al. 1975). Roughly half of them are in metasediments, and may have a spatial association with mafic igneous rocks. Bedded deposits are rare.

Mount Isa is more than two orders of magnitude larger than Mammoth (approximately 1.5 million tonnes at 3% Cu), which is the second biggest ore body known in the region. Both of these deposits occur in brecciated, altered, pyritic strata adjacent to faults and altered mafic volcanics. Only the Mount Isa deposit is reviewed here; the features of Mammoth are available from the publications by Mitchell and Moore (1975) and Scott and Taylor (1982).

Mount Isa

Detailed information on this deposit can be obtained from the papers by Mathias and Clark (1975) and Perkins (1984).

There are closely adjacent but separate copper and lead-zinc ore bodies at Mount Isa (Fig. 11). The latter consist of fine-grained, laminar sulfides of syngenetic-diagenetic origin in the carbonaceous, dolomitic and tuffaceous Urquhart Shale of the Mount Isa Group.

In contrast, the copper ore comprises coarser-grained massive, disseminated and veinlet mineralization, mainly chalcopyrite, pyrite and pyrrhotite (Fig. 12). It occurs within so-called silica-dolomite (Fig. 11), a facies of the Urquhart Shale which extends upwards from the faulted contact with the underlying greenstones (probably Eastern Creek Volcanics). The silica-dolomite has an outer zone of recrystallized shale with pods of massive, recrystallized dolomite. Inside this is a zone of irregularly brecciated and recrystallized dolomitic shale, and a central zone of brecciated siliceous shale. The massive ore is developed within the central zone, against the basal greenstone contact.

The timing of the copper mineralization has been a source of considerable controversy. Several workers, most recently Finlow-Bates and Stumpfl (1979), have argued that the copper and lead-zinc ore bodies formed from the same mineralizing fluids, with the copper precipitating largely in near-surface feeder chan-

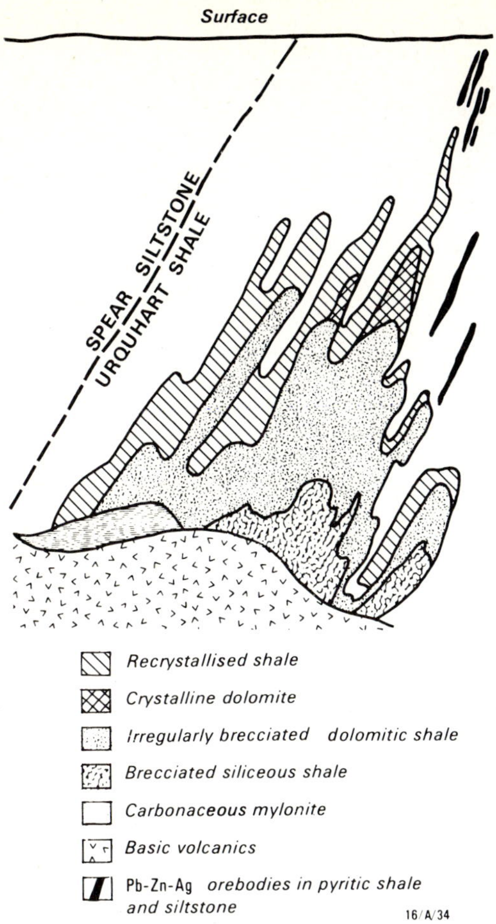

Fig. 11. Mount Isa mine (after Mathias and Clark 1975). Copper ore occurs throughout the "silica-dolomite", shown as *shaded zones* above the basic volcanics

nels. However, recent studies have produced evidence against co-genetic copper and lead-zinc mineralization. In summary, structural, textural and sulfur- and lead-isotopic data imply that chalcopyrite, and associated late-generation pyrite and pyrrhotite, postdate the early phases of regional metamorphism (Gulson et al. 1983, Perkins 1984).

The lead isotopic compositions of the various copper ore bodies at Mount Isa fall on several distinct trend lines (Gulson et al. 1983). These authors argued that the trend lines reflect the varying importance of the Eastern Creek Volcanics and basement rocks as metal sources, the latter providing a relatively high proportion of the metals in the largest ore bodies.

While details of ore genesis remain unresolved, the heated brines which formed the copper ore bodies appear to have been generated during metamorphism of the sequence (Perkins 1984). Sulfur-isotopic data imply that major proportions of sulfide were introduced with the copper (Gulson et al. 1983). This hydrothermal sulfide could have come from leaching of sulfide minerals from the underlying rocks, and/or thermochemical reduction of sulfate. The hydrother-

Fig. 12 A, B. Mount Isa copper ore. **A** High-grade chalcopyrite ore in silicified shale. **B** Chalcopyrite-pyrrhotite veinlet mineralization in brecciated dolomitic shale

mal mineralizing fluids, which probably ascended via faults and shear zones, caused metasomatism and hydraulic fracturing of the Urquhart Shale.

Other Sediment-Hosted Copper Provinces

Tennant Creek Inlier

In this region there is a low-grade metamorphosed, early Proterozoic greywacke-shale sequence with intercalated felsic tuffs and volcanics (Crohn 1975). Graded bedding and slumping are common in the sedimentary strata, which have been interpreted as accumulating in moderately deep water. The sequence has been intruded by massive and foliated granites, quartzo-feldspathic porphyries, diorites, dolerites and lamprophyres. Dating by Black (1983) indicates that the felsic volcanics, granites and porphyries formed about 1.87 – 1.85 b.y. B.P.

The main copper ores of the region, Peko and Warrego, have been described by Wright (1965) and Goulevitch (1975), respectively. These comprise chalcopyrite and pyrite, with minor bismuthinite and gold, as irregular dispersions in magnetite and quartz-magnetite ironstones. The ironstones are pipelike bodies in fold axes and shear zones cutting the greywacks-shale sequence. They are structurally-controlled hydrothermal deposits, which probably formed during the main and waning stages of deformation, around 1.81 b.y. B.P. (Black 1983).

Bangemall Basin

The discovery of several copper prospects in the Bangemall Basin, Western Australia (Fig. 1), was announced recently (Western Mining Corporation press release, September 1983). In this region, a thick, dominantly shallow-water sequence of sandstones and interbedded siltstones, shales and carbonates accumulated about 1.2 – 1.1 b.y. ago (Goode and Hall 1981). These authors consider that there is a volcanic ash component in the sequence. The mineralization is in a relatively highly deformed and weakly metamorphosed (to low greenschist facies) belt at the northeastern margin of the basin.

The 1983 press release stated that the largest of the prospects being assessed in this area, Nifty, contains secondary copper mineralization amenable to open-cut mining over a strike length of 1.3 km, and incompletely assessed primary (hydrothermal?) mineralization with grades of up to 15.8% copper. No detailed information is available on this deposit.

Cobar Trough

The geology of the Cobar region (Fig. 1) has been reviewed by Brooke (1975) and Pogson et al. (1976). In addition, synopses of recent information from a variety of studies in the region appear in the abstract volume of the 6th Australian Geological Convention (Geological Society of Australia, Canberra, 1983).

The oldest rocks known in the region are a sequence of Cambro-Ordovician turbidites, cherts and minor basic volcanics, which were metamorphosed and deformed under greenschist facies conditions in the early Silurian, and intruded by early to middle Silurian granites. The Cobar Trough formed in the latest Silurian and its subsidence appears to have been controlled by pre-existing faults. Relatively high relief generated at the eastern margin of the trough resulted in deposition of an early fringing sequence of conglomerates and sandstones. Thin-bedded turbidites were deposited under deeper water conditions to the west of, and overlying the coarse clastics, and were followed by shaley and silty beds. Finally, a sandy facies was deposited across the shallowing trough. Felsic volcanism occurred on the eastern shelf during the filling of the Cobar Trough. However, there are no known volcanic centres within several tens of kilometres of main mineral deposits, and the proportions of tuffaceous materials in the host strata are not generally considered to be high.

The copper-rich deposits of this field are in steeply dipping, silicified shear zones, sub-parallel to the regional cleavage. They comprise massive to stockwork, variably deformed mineralization, and occur in several stratigraphic units, mainly in fine-grained strata.

The sulfur- and lead-isotopic compositions of the Cobar ores fall in restricted ranges which are very similar to those for Silurian to early Devonian volcanogenic deposits in southeastern New South Wales (Sun 1983). While details of ore genesis are still unresolved, there is a consensus of opinion that the copper mineralization precipitated epigenetically from hydrothermal fluids generated by high thermal gradients in the trough, prior to or during regional metamorphism.

Summary and Conclusions

The most important Australian copper mine, Mount Isa, comprises stratabound, metamorphogenic mineralization in brecciated and metasomatized middle Proterozoic metasiltstone. The huge copper (gold-uranium-rare earths) deposit at Olympic Dam, which is not yet in production, formed from hydrothermal fluids permeating sedimentary breccias in tectonically active graben environments. Important structure-controlled hydrothermal deposits occur in metamorphosed trough sequences in the Tennant Creek (early Proterozoic), Kanmantoo (Cambrian) and Cobar (Siluro-Devonian) regions.

Relatively low-temperature epigenetic deposits have been mined from the palaeoweathering zone on a middle Proterozoic sandstone unit at Mount Gunson. Bedded-disseminated deposits are widespread in South Australia, and secondarily-enriched mineralization has been mined from these in numerous small-scale operations. The primary bedded sulfides formed as a result of long-term influx of low-temperature cupriferous fluids which reacted with biogenic sulfur and iron sulfides.

Acknowledgements. The Baas Becking Laboratory is sponsored by the Bureau of Mineral Resources, the Commonwealth Scientific and Industrial Research Organization and the Australian Minerals Industry Research Association Limited.

References

Black LP (1983) Tennant Creek geochronology. BMR 83. Yearb Bur Mineral Res, Aust, pp 109 – 110

Blake DH, Page RW, Wyborn LAI, Etheridge MA (1983) Mount Isa base metal province. BMR 83. Yearb Bur Mineral Res, Aust, p 50

Blake DH, Bultitude RJ, Donchack PJT, Wyborn LAI, Hone IG (1984) The geology of the Duchess Urandangi Region, Mount Isa Inlier, Queensland. Bur Mineral Res Bull 219

Brooke WJL (1975) Cobar mining field. In: Knight CL (ed) Economic geology of Australia and Papua New Guinea, vol I: Metals. Australas Inst Min Metal, Melbourne, pp 683 – 694

Brooks JH, Wilson IH, Saurers JD (1975) Minor copper deposits in the Mount Isa/Cloncurry District. In: Knight CL (ed) Economic geology of Australia and Papua New Guinea, vol I: Metals. Australas Inst Min Metal, Melbourne, pp 402 – 405

Creelman RA (1983) A petrological mineralogical and lead isotope study of copper in sandstones and related sediments. PhD Thesis, Macquarie Univ (unpublished)

Crohn PW (1975) Tennant Creek-Davenport Proterozoic Basins, regional geology and mineralization. In: Knight CL (ed) Economic geology of Australia and Papua New Guinea, vol I: Metals. Australas Inst Min Metal, Melbourne, pp 421 – 423

Finlow-Bates T, Stumpfl EF (1979) The copper and lead-zinc-silver orebodies of Mt. Isa mine, Queensland: Products of ore hydrothermal system. Soc Geol Belg Ann 102:497 – 517

Gersteling RW, Heape JM (1975) The Cattle Grid orebody, Mt. Gunson, South Australia. Aust Inst Min Metal Conf, S Aust, pp 103 – 112

Goode ADT, Hall WDM (1981) The middle Proterozoic eastern Bangemall Basin, Western Australia. Precambr Res 16:11 – 29

Goulevitch J (1975) Warrego copper-gold orebody. In: Knight CL (ed) Economic geology of Australia and Papua New Guinea, vol I: Metals. Australas Inst Min Metal, Melbourne, pp 430 – 437

Gulson BL, Perkins WG, Mizon KJ (1983) Lead isotope studies bearing on the genesis of copper orebodies at Mount Isa, Queensland. Econ Geol 78:1445 – 1466

Johnson W (1965) Copper and lead ore deposits of South Australia. In: McAndrew J (ed) Geology of Australia ore deposits. Australas Inst Min Metal, Melbourne, pp 285 – 297

Knutson J, Donnelly TH, Tonkin DG (1983) Geochemical constraints on the genesis of copper mineralization in the Mount Gunson area, South Australia. Econ Geol 78:250 – 274

Lambert IB, Knutson J, Donnelly TH, Etminan H, Mason MG (1984) Genesis of copper mineralization. Myall Creek Prospect, South Australia. Min Deposita 19:266 – 273

Mathias BV, Clark GJ (1975) Mount Isa copper and silver-lead zinc orebodies, Isa and Hilton mines. In: Knight CL (ed) Economic geology of Australia and Papua New Guinea, vol I: Metals. Australas Inst Min Metal, Melbourne, pp 351 – 372

Mitchell JW, Moore GP (1975) Mammoth copper deposit. In: Knight CL (ed) Economic geology of Australia and Papua New Guinea, vol I: Metals. Australas Inst Min Metal, Melbourne, pp 383 – 390

Neudert MK, Russell RE (1981) Shallow water and hypersaline features from the middle Proterozoic Mount Isa sequence. Nature (London) 293:284 – 286

Page RW (1978) Response of U-Pb zircon and Rb-Sr total-rock and mineral systems to low-grade regional metamorphism in Proterozoic igneous rocks, Mount Isa, Australia. J Geol Soc Aust 25:141 – 164

Page RW (1981) Depositional ages of the stratiform base metal deposits, at Mount Isa and McArthur River, Australia, based on U-Pb Zircon dating of concordant tuff horizons. Econ Geol 76:648 – 658

Perkins WG (1984) Mount Isa "silica-dolomite" and copper ore bodies: the result of a systematic hydrothermal alteration system. Econ Geol 79:601 – 638

Plumb KA, Derrick GM, Wilson IH (1980) Precambrian geology of the McArthur River Mount Isa Region, Northern Australia. In: Henderson RA, Stephenson PJ (eds) The geology and geophysics of northeastern Australia. Geol Soc Aust, Queensland Div, Brisbane, pp 71 – 88

Pogson DJ, Suppel DW, Gilligan LB, Scheibner E, Baker C, Sherwin L, Felton EA, Fail AP (1976) Recent studies of the tectonics, stratigraphy and mineralization of the Cobar-Mineral Hill region. Aust Soc Explor Geophys 7:31 – 34

Roberts DE, Hudson GRT (1983) The Olympic Dam copper-uranium-gold deposit, Roxby Downs, South Australia. Econ Geol 78:799 – 822

Rutland RWR, Parker AJ, Pitt GM, Preiss WV, Murrell B (1981) The Precambrian of South Australia. In: Hunter DR (ed) Precambrian of the southern hemisphere. Elsevier North Holland Biomedical Press, Amsterdam New York, pp 309–360

Scott KM, Taylor GF (1982) Eastern Creek volcanics as the source of copper at the Mammoth mine, north west Queensland. BMR J Aust Geol Geophys 7:93–98

Sun SS (1983) Implications of S and Pb isotope data to genesis of massive sulphide in the Cobar area. 6th Aust Geol Convent (Abstr), pp 307–308

Tonkin DG, Williams GE (1983) Fossil periglacial structures in the Cattle Grid mine, Mount Gunson. Adelaide Geosyncline Sedimentary Environments and Tectonic Setting Symposium. Geol Soc Aust, Adelaide, (Abstr), pp 59–60

Verwoerd PJ, Cleghorn JH (1975) Kanmantoo copper orebody. In: Knight CL (ed) Economic geology of Australia and Papua New Guinea, vol I: Metals. Australas Inst Min Metal, Melbourne, pp 560–567

Warren RG (1972) A commentary on the metallogenic map of Australia and Papua New Guinea. Bur Mineral Res Bull 145, p 85

Wright K (1965) Copper ore deposits of the Peko mine, Tennant Creek. In: McAndrew J (ed) Geology of Australian ore deposits. Australas Inst Min Metal, Melbourne, pp 183–185

Mineral Zoning in Sediment-Hosted Copper-Iron Sulfide Deposits – A Quantitative Kinetic Approach

E. Merino[1,2], C. Moore[1], P. Ortoleva[1,2,3], and E. Ripley[1,2]

Abstract

The zoning typical of many sediment-hosted copper-iron sulfide deposits cannot be well accounted for by equilibrium approaches, which lead to inconsistencies and which are inherently ill-suited to describe a spatial, nonequilibrium process such as zoning. Much better is a kinetic approach that consists of differential equations representing dissolution, nucleation, growth, flow, and diffusion, all of which are clearly involved in ore genesis. In this kinetic model, time and spatial coordinates appear explicitly, and the equations automatically incorporate feedbacks among mechanisms. Numerical solutions of the equations for many combinations of reasonable values of the parameters involved (such as velocity and composition of the mineralizing water, equilibrium and reaction-rate constants, initial amount of pyrite in the rock, diffusion and nucleation constants, and others) can yield mineral zoning of several types, including the typical chalcocite-bornite-chalcopyrite-pyrite found at the White Pine deposit.

Introduction

Many sediment-hosted copper sulfide deposits are zoned. Among them are the Kupferschiefer, the Zambian Copperbelt, White Pine, Seal Lake, Creta, Spar Lake, Nacimiento, and Corocoro. A summary of their typical zoning sequences with references is given in Table 1. Excellent review articles by Gustafson and Williams (1981) and Bjorlykke and Sangster (1981) outline the geologic and geochemical features of sediment-hosted stratiform deposits. Laznicka (1985) gives a comprehensive review of the zoning; see also Ripley et al. (1985). The host deposits are typically intracratonic, and contain both reduced and oxidized sediments. The copper mineralization occurs in the reduced sediments.

The purpose of this communication is to suggest that the observed mineral zones in sediment-hosted copper deposits are caused by the nonequilibrium, kinetic interaction of dissolution, flow, diffusion, nucleation, and growth.

[1] Geo-Chem Research Assoc., 400 E. First St., Bloomington, IN 47401, USA
[2] Department of Geology, Indiana University, Bloomington, IN 47405, USA
[3] Department of Chemistry, Indiana University, Bloomington, IN 47405, USA

Geology and Metallogeny of Copper Deposits
(ed. by G.H. Friedrich et al.)

Table 1. Summary of the most common zoning sequences in sediment-hosted copper deposits[a]

Deposit or district	Typical Cu – Fe sulfide zoning sequence	Reference
Zambia-Zaire	Pyrite-Chalcopyrite-Bornite-Chalcocite	van Eden (1974) Annels (1979) Fleischer et al. (1976) Bartholomé (1974)
Kupferschiefer	Pyrite-Chalcopyrite-Bornite-Chalcocite	Rentzsch (1974) Jung and Knitzschke (1976)
Redstone River	Pyrite-Chalcopyrite-Bornite-Chalcocite	Jefferson (1978) Lustwerk and Rose (1983)
White Pine	Pyrite-Chalcopyrite-Bornite-Chalcocite	Brown (1971, 1974, 1981)
Seal Lake	Pyrite-Chalcopyrite-Bornite-Chalcocite	Gandhi and Brown (1975)
Spar Lake	Pyrite-Chalcopyrite-Bornite-Chalcocite	Harrison (1974) Clark (1971)
Kansas	Pyrite-Chalcopyrite-Bornite-Chalcocite	Ripley et al. (1980)
Creta	Pyrite-Chalcocite	Dingess (1976) Hagni and Cann (1976)
Texas	Pyrite-Chalcocite	Smith (1976)
Kansas	Pyrite-Chalcocite	Ripley et al. (1980)
Nacimiento	Pyrite-Chalcocite	Woodward et al. (1974)
Corocoro	Pyrite-Chalcocite	Ljunggren and Meyer (1964)

[a] Note that especially for the pyrite-chalcocite deposits, the term "zoning sequence" is essentially a paragenetic sequence. In all of these deposits pyrite is inferred to be of early diagenetic origin, with chalcocite deposition at a later time. In the production of the pyrite-chalcopyrite-bornite-chalcocite sequence the distinction between zoning and paragenesis may be confusing in that different Cu or Cu – Fe sulfides may be forming in spatially distinct areas of the ore system at the same time.

Most of the deposits contain early diagenetic pyrite, which is generally recognized to have been partly replaced and rimmed by copper sulfides at temperatures of less than about 100 °C, under the action of basinal, meteoric brines with more than 0.1 molal Cl and pH 6 – 8. The typical zoning sequences are: barren rock-chalcocite-bornite-chalcopyrite-pyrite, and barren rock-chalcocite-pyrite. These zones can be in some way rationalized by equilibrium activity diagrams and by equilibrium log (oxygen fugacity) versus pH diagrams. As shown below, however, these diagrams lead to inconsistencies. Furthermore, it is unrealistic to expect equilibrium diagrams to portray accurately the mineralization of pyritic sediments, which is inherently a *non*equilibrium phenomenon. Also, zoning being inherently spatial (in that the grains of the minerals involved must be located in bands, and the bands must be placed in a definite sequence), it cannot be accounted for by equilibrium approaches, which characteristically contain no spatial coordinates.

Equilibrium Diagrams

Equilibrium activity diagrams pertinent to copper-iron sulfide sedimentary deposits are shown in Fig. 1. It can be seen that both the chalcocite-bornite-chal-

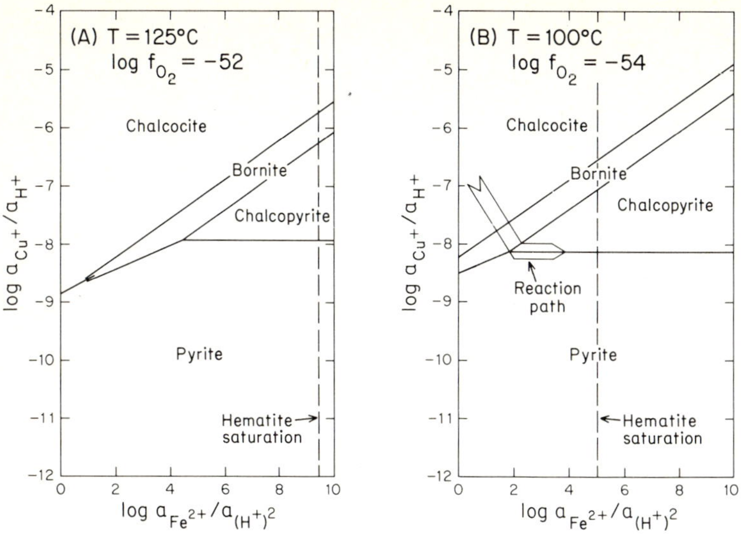

Fig. 1A, B. Stability relationships between Cu- and Cu – Fe sulfide minerals. Fluid path for reaction between a pyritic shale and a brine (Kettleman Dome, Merino 1975) is shown in **B**. Thermodynamic data from Helgeson et al. (1978)

copyrite-pyrite and the chalcocite-pyrite sequences could be produced by reaction between pyrite and a fluid whose composition evolved from top left to bottom right of the diagrams, either to the right of the bornite-chalcopyrite triple point or to the left of the chalcocite-bornite-pyrite triple point. (The digrams of Fig. 1 were constructed as shown by Helgeson (1968) and Brown (1970). Mineral-mineral reactions were balanced on sulfide, which is therefore assumed in the diagrams to be buffered by the system – in fact, the diagrams can be contoured for the activity product $a_{H^+}a_{HS^-}$. The temperature, oxygen fugacity, and water activity are fixed arbitrarily for the two diagrams; the values chosen for T and $\log f_{O_2}$ are intended to be roughly typical of the deposits in question.)

A $\log f_{O_2}$ versus pH diagram for the same system is shown in Fig. 2. The temperature chosen is 100 °C, total sulfide is 10^{-4} molal, and total chloride is 1 molal. The diagram suggests that chalcocite can form at fairly high oxygen fugacities (definitely above the pyrite-hematite boundary) and for realistic and significant amounts of dissolved copper (mostly complexed by chloride, see Rose 1976, Helgeson et al. 1978, and Crerar and Barnes 1976). The problem, however, is that the equilibrium crystallization of chalcocite falls in the field of predominance of SO_4^{2-} (aq), where the concentrations of S^{2-} and HS^- are insignificant. Geologic evidence in most deposits suggests that deposition of copper sulfides occurred under much lower f_{O_2} than indicated by the diagram. Indeed, the host sediment normally contains abundant organic matter and pyrite. A second difficulty in this diagram is that the crystallization of chalcopyrite is predicted to happen at very small concentrations of dissolved copper, of the order of 0.1 mg l^{-1}, thought insufficient to form ores (Anderson 1977). These two inconsistencies, together with the general considerations given at the end of the Introduc-

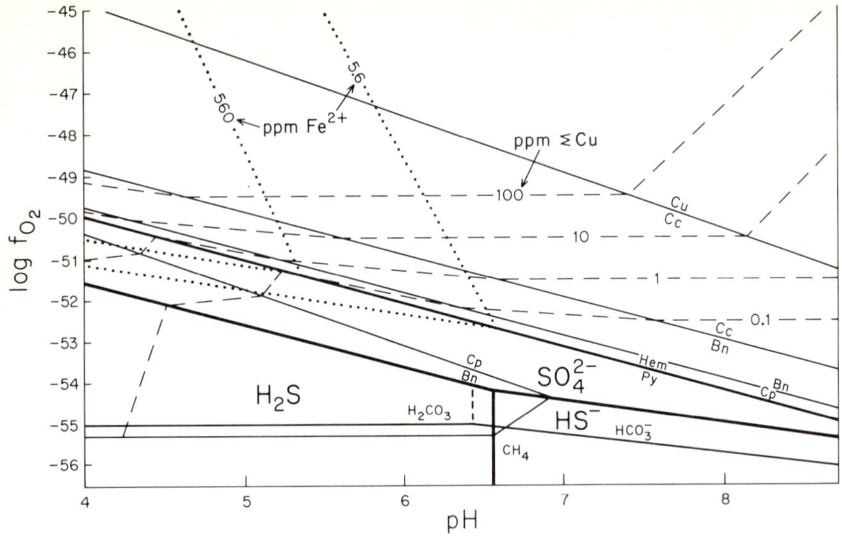

Fig. 2. Log fO_2-pH diagram showing stability fields of aqueous sulfur species, pyrite, hematite, and various copper minerals. Total ppm Cu is essentially that contained in $CuCl^0 + CuCl_2 + CuCl_3^{2-}$. Conditions of the diagram are as follows: $T = 100°C$, $\sum S = 10^{-4}\ m$, $Cl^- = 1\ m$. Abbreviations: *CU* native copper; *CC* Cu_2S; *BN* Cu_5FeS_4; *CP* $CuFeS_2$; *PY* FeS_2; *HEM* Fe_2O_3. Thermodynamic data from Helgeson et al. (1978)

tion, have compelled us to explore quantitatively the possibility that the crystallization and zoning of copper sulfides in sediments is a nonequilibrium process involving transport and reaction.

The Kinetic Approach

In the genesis of these sulfide deposits and their zoning aqueous solutions *both* flow through and react with permeable rock. The essential feature is that reaction, nucleation, and transport take place simultaneously, and that they interact.

There are very powerful techniques to model the reactions between waters and rocks (e.g., Helgeson 1968, 1979, Wolery 1979, Reed 1982), but because these approaches are blind to space and time they are fundamentally unable to describe transport, to predict spatial mineralization patterns, or to incorporate reaction-transport interactions. Similarly, there are sophisticated models to predict the flow of water through permeable rocks (e.g., Sharp 1976, Cathles 1981, Narasimhan and Witherspoon 1976, Kuyahorn and Pinder 1984), but these models take no account of chemical reactions, or even less of their interactions with transport. Attempts to juxtapose chemical models and flow models are bound to require prohibitively long computing times, even in the fastest computers, and, worse, such juxtaposition would still be unable to incorporate transport-reaction interactions.

In contrast, our approach (also attempted by Dobrovolsky and Lyalko 1979) is kinetic throughout. During the evolution of the system chemical equilibrium is nowhere assumed. We write differential equations representing mass conservation (or continuity), flow, diffusion, nucleation, and growth and dissolution rates. For simplicity, absorption and adsorption of aqueous species are ignored here, but they could be easily included in the continuity equations. The system of equations automatically incorporates feedbacks and interactions among the various mechanisms. Space and time appear explicitly. Any number of aqueous and solid species can be included; in our illustration below we have included the ten species (six aqueous, four solid) listed in Table 3. This general approach is common in chemical engineering and physical chemistry and has been applied to study the dynamics of redox fronts by Ortoleva et al. (1986) and Chadam et al. (1986a, b).

Mass Conservation. The continuity equation is simply

$$\frac{\partial c_i}{\partial t} = -\nabla \cdot J_i + R_i \tag{1}$$

and states that the time variation of the concentration (c_i) of species i (aqueous or solid) in a small volume of the system is due to the amount of i transported into that volume per unit time plus the total rate, R_i, of production or consumption of i through reactions within the volume of interest. ∇ is the spatial derivative vector ($\partial/\partial x$, $\partial/\partial y$, $\partial/\partial z$); and J is the total flux of species i. Neglecting hydrodynamic dispersion,

$$J_i = J_i^{\text{flow}} + J_i^{\text{diffusion}} = \boldsymbol{v} c_i - D_i \nabla c_i \,, \tag{2}$$

where $\boldsymbol{v}$ is the percolation flow velocity, D_i is the effective diffusion coefficient, and ∇c_i is the concentration gradient for species i. The rate of production or consumption, R_i, for species i is related to the rates of reactions, W_α, via the stoichiometric coefficients ν_α^i:

$$R_i = \sum_\alpha \nu_\alpha^i W_\alpha \,. \tag{3}$$

Reaction Rates and Nucleation. In the absence (to our knowledge) of well-established elementary reactions and kinetic data for the dissolution and growth of the iron-copper sulfides involved here, we have adopted the stoichiometric reactions of Table 2 as the basis for reaction-rate equations. For example, for the reaction of dissolution or growth of pyrite

$$FeS_{2(py)} + H_2O = Fe^{2+} + 2HS^- + \tfrac{1}{2}O_{2(aq)} \,, \tag{4}$$

and following our work on redox-front dynamics and deposition (Ortoleva et al. 1985, Merino et al. 1985, Chadam et al. 1985a, b), we construct the reaction-rate expression

$$W = k_{py}(P+p)^{2/3}(K_{py} - \{Fe^{2+}\}\{HS^-\}^2\{O_2\}^{1/2}) \,, \tag{5}$$

where K_{py} and k_{py} are the equilibrium and reaction-rate constants, the braces denote concentrations in moles per unit volume of bulk rock, P is the volume percent of pyrite present at a point at a given moment, p is a "switch" that

Table 2. Reactions used for the initial simulation

$FeS_2 + H_2O \rightleftarrows Fe^{2+} + 2HS^- + \frac{1}{2}O_2(aq)$
$Cu_2S + H^+ \rightleftarrows 2Cu^+ + HS^-$
$CuFeS_2 + H^+ + \frac{1}{2}H_2O \rightleftarrows Cu^+ + Fe^{2+} + 2HS^- + \frac{1}{4}O_2(aq)$
$Cu_5FeS_4 + 3H^+ + \frac{1}{2}H_2O \rightleftarrows 5Cu^+ + Fe^{2+} + 4HS^- + \frac{1}{4}O_2(aq)$
$Fe^{2+} + H^+ + \frac{1}{4}O_2 \rightleftarrows Fe^{3+} + \frac{1}{2}H_2O$

controls the nucleation of pyrite as explained below, the expression $(K_{py} - \{Fe^{++}\}\{HS^-\}^2\{O_2\}^{1/2})$ is the chemical affinity for the pyrite reaction, and W is the rate of growth ($W<0$) or dissolution ($W>0$) of pyrite. Let $K_{nucleation}(>K_{py})$ be the value of $\{Fe^{++}\}\{HS^-\}^2\{O_2\{^{1/2}$ above which pyrite nucleation starts. Equation (5) is reasonable because (a) it takes into account the surface area dependence of the rates through the 2/3 exponent, which converts volume into grain surface; (b) it yields the correct equilibrium condition when the rate W vanishes (assuming the solutions are dilute); and (c) it takes account of nucleation:

1. if $\{Fe^{++}\}\{HS^-\}^2\{O_2\}^{1/2} > K_{nucleation}$, then p is taken as a small positive number, W becomes negative even if there is no pyrite present, and pyrite nucleates and grows;
2. if $\{Fe^{++}\}\{HS^-\}^2\{O_2\}^{1/2} < K_{nucleation}$, then p is defined as zero; now if $P=0$ (i.e., there is no preexisting pyrite) $W=0$, which means that there is no pyrite growth, and
   if $P>0$ (i.e., there is some pyrite present already) then pyrite will grow if $\{Fe^{++}\}\{HS^-\}^2\{O_2\}^{1/2} > K_{py}$.

Thus, the formulation of Eq. (5) is able to "turn on" the precipitation of pyrite when the nucleation threshold is exceeded; allows for the dissolution of pyrite when the solution is undersaturated; allows for pyrite growth where there is already some pyrite present and the solution is supersaturated; and "turn off" growth where there is no pyrite present and the solution does not exceed the nucleation threshold. Rate expressions of the type exemplified by Eq. (5) may be inserted in the several expressions of the continuity equation (one for each species). These expressions then automatically incorporate feedbacks among reactions and between transport and reaction. More general expressions for non-dilute solutions can be obtained by replacing concentrations by activities.

To complete the theory we have to determine the percolation flow velocity $\boldsymbol{v}$. This is related to the texture-dependent permeability $\varkappa$ and the pressure p via Darcy's law

$$\boldsymbol{v} = -\varkappa \nabla p . \tag{6}$$

The conservation equation for water, assumed incompressible, yields

$$\nabla \cdot (\phi \varkappa \nabla p) = \frac{\partial \phi}{\partial \tau} \tag{7}$$

where ϕ is the porosity and τ is scaled time. The porosity is related to the grain size and number density of the various minerals making up the rock through the equation

$$\phi + \sum_i n_i L_i^3 = 1 \, , \tag{8}$$

where L_i is the local size of grains of i and n_i is the number of grains of i per unit volume of rock.

Modeling of the Cu – Fe Sulfide Zoning in Sediment-Hosted Deposits

In this attempt a modeling the zoning of sedimentary copper deposits we have used numerical values that allow quickness in computation and do not violate the quantitative relationships among imposed water compositions, initial rock composition, and equilibrium constants. We have chosen the ten-species system shown in Table 3. Boundary concentrations refer to the concentrations of all the elements in the system before the introduction of the inlet fluid (or fluid that carries out the alteration to copper sulfides). We have chosen that the Fe^{2+}, Fe^{3+}, O_2, and H^+ be in equilibrium with each other in the initial water and also that the initial water be in equilibrium with pyrite.

Scaling. It is necessary, again for computational ease, to scale concentrations and equilibrium constants as follows:

1. Concentrations are scaled by dividing the initial and boundary concentrations for each species by the larger of the two values. For example, if the initial

Table 3. Scaled parameters used for the discussed simulation (see text for details)

Species	Concentrations	
	Boundary	Initial
Cu^+	1.0	1×10^{-1}
Fe^{2+}	1.0	1.0
HS^-	1×10^{-1}	1.0
Fe^{3+}	3.2×10^{-2}	3.6×10^{-2}
O_2(aq)	1.0	6.4×10^{-1}
H^+	9.5×10^{-1}	1.0
Pyrite	–	1.0
Chalcocite	–	0.0
Chalcopyrite	–	0.0
Bornite	–	0.0
	Equilibrium constants	
K_{pyrite} =	6.5×10^{-1}	
$K_{chalcocite}$ =	9×10^{-2}	
$K_{chalcopyrite}$ =	9.5×10^{-2}	
$K_{bornite}$ =	8.0×10^{-4}	
$K_{ferrous\text{-}ferric}$ =	3.0×10^{-2}	

copper concentration is 10^{-18} and the boundary value is 10^{-15}, the concentrations scale to 10^{-2} and 1, respectively.

2. To scale equilibrium constants, the known equilibrium constant is divided by the corresponding ion product for the reactions, using in that ion product the maximum unscaled value for each species.

For example, for the values Fe^{2+} initial $= 10^{-12}$, Fe^{2+} boundary $= 10^{-14}$, HS^- initial $= 10^{-3}$, HS^- boundary $= 10^{-6}$, $O_2^{1/2}$ initial $= 10^{-30}$, and $O_2^{1/2}$ boundary $= 10^{-28}$; and given that the equilibrium constant for Eq. (4) is $K_{py} = 10^{-45}$, the scaled equilibrium constant turns out be $K_{py,sc} = K_{py}/(Fe^{22}$ initial)(HS^- initial)^2(O_2 boundary)$^{1/2} = 10^{-45}/10^{-12}(10^{-3})^2 10^{-28} = 10$.

Numerical Simulations. We have solved these equations numerically with a computer. The simulations require the following input of data: initial and inlet concentrations for each species (Table 3), diffusion coefficients for each mobile species, flow velocity of the fluid, and, for each reaction: equilibrium constant, rate constant, and supersaturation factor for nucleation. Also needed are certain ancillary parameters for controlling the computer program and describing the size of the system.

Of the above variables, only the equilibrium constants are well known. All the other parameters can only be estimated within a reasonable range. These are the parameters we have varied from computer run to computer run in order to observe what precipitation patterns, if any, develop as the system evolves. The lack of data, especially kinetic data, gives some freedom in choosing values for these parameters, but one must be careful, as the evolution of the system is highly dependent on some of them. A good example is the supersaturation factor, a small change which can drastically change the localization of a given phase or completely inhibit its nucleation and growth. Another example is the rate constant: when the rate constant is very high for a phase and that phase nucleates, the resulting rapid growth can effectively inhibit growth of other phases.

Because of these uncertainties we show here only that mineral zoning in sediment-hosted copper deposits *can* result. Much experimental work is needed to determine values for the kinetic variables, and very detailed field work describing the geometry and textures of the ore deposits is required. Assuming that this model is reasonable, it is interesting to note that one could work backwards to get order-of-magnitude figures for some of the parameters by attempting to duplicate known ore-body geometries.

Figure 3 indicates the results of one specific computer run, generated with the parameters of Table 3. We show here the concentrations of the various sulfide minerals on the ordinate versus space (distance from the inlet) on the abscissa. Water flows from left to right. Each graph represents a scaled time interval; the one for $t = 0$ represents the initial conditions (i.e., all pyrite), and the one for $t = 30$ represents the last calculated step of that particular simulation, which yields precisely the kind of zoning that is commonly observed.

Discussion. The only sulfide in the initial system is pyrite. As this initial pyrite is engulfed by the incoming water (with which it is out of equilibrium) it dissolves, yielding waters enriched in HS^-, Fe^{2+}, and O_2. These ions are transported with

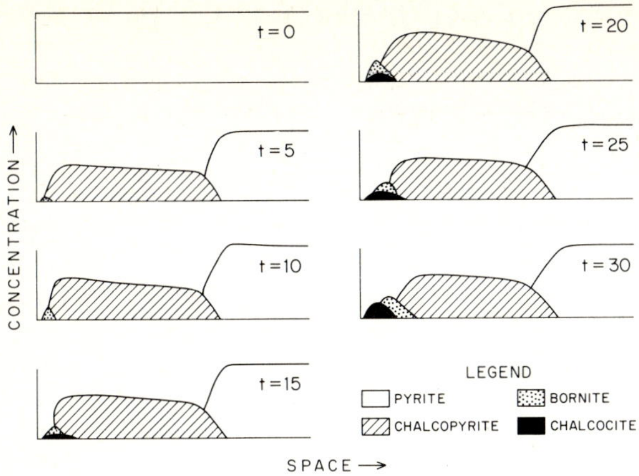

Fig. 3. Modeled evolution of a zoned sequence of copper-iron sulfides in a sediment-hosted deposit. *Symbols* are explained on the diagrams. Mineral concentrations were calculated by numerical solution of the model described in the text, and are here plotted vertically as functions of position and (scaled) time *t*. The parameter values adopted in the calculations are in Table 3

the fluid and, along with the Cu^+ in it, eventually exceed the nucleation concentration threshold, precipitating chalcopyrite at $t = 5$. The chalcopyrite grows as long as there is pyrite upflow or adjacent to it to provide the necessary ions. When all the upflow and adjacent pyrite is dissolved then the chalcopyrite too begins to dissolve. The concomitant release of HS^-, Fe^{2+}, O_2 and especially the previously fixed copper, along with the depletion of H^+, cause the solution to exceed the concentration product required for nucleation of bornite, and bornite precipitates. At this point an even larger amount of copper is fixed in the system. This situation is shown at time $t = 5$ in Fig. 3, and continues throughout. As time progresses all the chalcopyrite upflow of the bornite dissolves, cutting off the stabilizing supply of ions, and bornite starts to dissolve. This abundant release of Cu^+ and HS^- and decrease in H^+ cause chalcocite nucleation and growth, as shown in the plot for time $t = 15$ in Fig. 3. The concentrations of the copper-bearing species grow with time, but appear to approach a steady-state level (note chalcopyrite). Displacement of the mineral zones themselves is accomplished by dissolution at the upflow end and precipitation downflow.

The width and concentrations in each zone depend on the parameters chosen, most notably on the concentrations, degree of supersaturation needed for nucleation, and rate constant. High values of these parameters produce narrow, high-concentration zones. We have not determined whether total copper fixed in the system remains the same as the parameters are varied.

Preliminary investigations show that the chalcocite-bornite-chalcopyrite-pyrite zoning may be produced by a relatively wide range of values for input and initial water chemistries. Indeed, within a given range, it is insensitive to whether the incoming waters have a lower or higher pH than the initial waters. In addition, other sequences of precipitation can be produced, but the same zoning pattern eventually evolves. This may explain the occurrence at White Pine, where chalcocite apparently replaced pyrite, and thin bornite and chalcopyrite zones formed at the interface between chalcocite and unreplaced pyrite.

The results of three other computer runs are given in Fig. 4. Run *c* is the same as in Fig. 3. Run *a* is for the same parameters as *c* except that copper concentra-

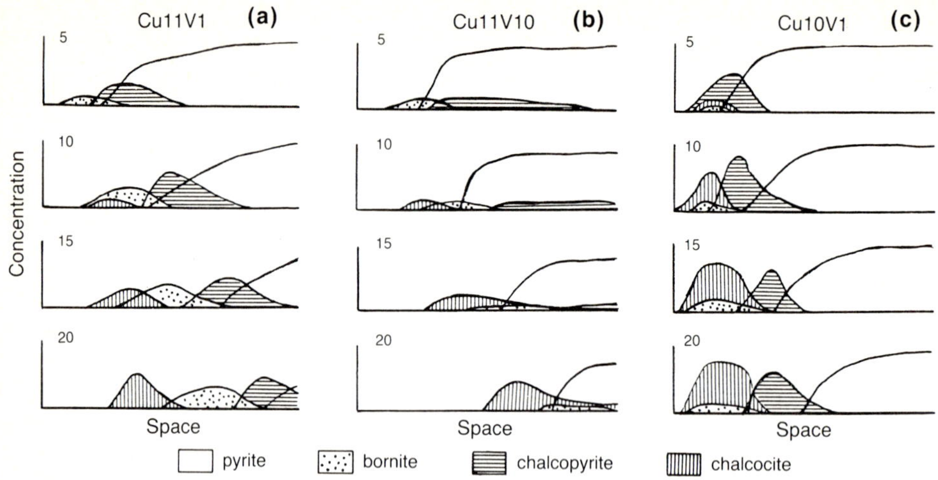

Fig. 4a–c. Other numerical solutions of the zoning model for iron-copper sulfides. Case (**c**) is essentially that of Fig. 3, and is repeated here for reference. Case (**a**) was run for a mineralizing solution ten times poorer in copper than that in (**c**). Case (**b**) was run for a solution ten times poorer in copper *and* ten times faster than in (**c**)

tion in the mineralizing water was ten times lower than in *c*. Run *b* is for a mineralizing solution ten times lower in copper and ten times faster than in *c*. Note that mineral zonings are produced in the two new cases, *a* and *b*; that the zones shift faster downflow in *b* than in *a* or *c*, and that a high degree of zoning is reached in case a earlier than in cases *b* and *c*.

The fact that these transport-reaction models succeed in duplicating well-documented features of zoned, sediment-hosted copper deposits as well as some aspects of roll-front uranium deposits (Moore et al. 1983, Chadam et al. 1986b) suggests that transport-reaction *coupling* is an important process in the formation of ore deposits and may play a role in other rocks showing chemical localization. It should be stressed that so far the model only shows the possibility that the zoning can arise in this way. Ongoing work is adding more support to these assertions, but the true test cannot be made until much more information is available on the kinetics of the reactions in question, the compositions of the boundary and initial fluids, and the geohydrology of the host sediments.

Conclusions

Understanding the genesis of the copper sulfide zoning in sedimentary ore deposits is not easy. Some of the difficulties one encounters in taking an equilibrium approach to account for the zoning are that the process of ore formation is inherently a nonequilibrium one that involves reaction and transport; that in the chalcocite field the predicted predominant sulfur species in sulfate, not sulfide; and that in the chalcopyrite field the largest dissolved copper concentration possible is 0.1 mg l^{-1}, which intuitively appears too low to give rise to an ore deposit. In contrast to the equilibrium approach, we present above a quantitative kinetic model of ore deposition that accounts for the zoning. The model is cast as a system of differential equations that include dissolution, nucleation, growth, flow, and diffusion. One strong advantage of the model is that it explicitly contains spatial coordinates, which are necessary to describe a *spatial* feature such as zoning. Time is also explicitly taken into account. The system of equations has been solved numerically for many combinations of reasonable values of the following parameters: velocity and composition of the mineralizing aqueous solution, composition of the formation waters before mineralization, equilibrium and rate constants for several reactions, diffusion coefficients of ions in aqueous solution, nucleation constants, and initial modal abundance of pyrite in the still-unmineralized sedimentary rock. Many of the numerical solutions yield, even for very low-copper waters, patterns of precipitation that are zoned and that evolve with time. Zonings obtained include the typical White Pine sequence of chalcocite-bornite-chalcopyrite-pyrite, as well as chalcocite-pyrite-chalcocite-bornite-pyrite, chalcocite-chalcopyrite-pyrite, and the striking bornite-chalcocite-chalcopyrite-pyrite.

Much work remains to be done, but the validity, viability, and power of this kinetic approach are, we think, clear. Interestingly, the approach also allows one to find out easily to which parameters the mineralization and zoning are most and least sensitive. Those most influencial parameters are the ones deserving the most careful experimental determination.

Acknowledgments. The authors thank the U.S. Department of Energy for financial support of this research under Grant DE-AC02-82ER12074 (Office of Basic Energy Sciences, Engineering Research Program).

References

Anderson GM (1977) Thermodynamics and sulfide solubilities. In: Greenwood HG (ed) Application of thermodynamics to petrology and ore deposits. Mineralog Soc Can Short Course, pp 136–150

Annels AE (1979) Mulfulira greywackes and their associated sulfides. Inst Min Metallurg Trans Sec B 88:15–23

Bartholomé P (1974) On the diagenetic formation of ores in sedimentary beds, with special reference to Kamoto, Shaba, Zaire. In: Bartholomé P (ed) Gisements stratiformes et provinces cuprifères. Liège, Soc Geol Belg, pp 203–213

Bjorlykke A, Sangster DF (1981) An overview of sandstone lead deposits and their relation to red-bed copper and carbonate-hosted lead-zinc deposits. Econ Geol 75th Anniv Vol, pp 179–213

Brown AC (1971) Zoning in the White Pine copper deposit, Ontonagon County, Michigan. Econ Geol 66:543–573

Brown AC (1974) An epigenetic origin for stratiform Cd–Pb–Zn sulphides in the Lower Nonesuch Shale, White Pine, Michigan. Econ Geol 69:271–274

Brown AC (1981) The timing of mineralization in stratiform copper deposits. In: Wolf KH (ed) Handbook of strata-bound and stratiform ore deposits, chap 1, vol IX. Elsevier, New York, pp 1–24

Brown TH (1970) Theoretical predictions of equilibria and mass transfer in the system $CaO-MgO-SiO_2-H_2O-CO_2-NaCl-HCl$. Ph D Thesis, Northwestern University, Evanston, Illinois (unpublished)

Cathles LM (1981) Fluid flow and genesis of hydrothermal ore deposits. Econ Geol 75th Anniv Vol, pp 424 – 457

Chadam J, Hoff D, Merino E, Ortoleva P, Sen A (1986a) Porosity instability at reactive percolation. J Inst Math Analysis (Oxford). Accepted

Chadam J, Merino E, Ortoleva P, Sen A (1986b) Geochemical self-organization, II: The reactive-infiltration instability. Am J Sci (submitted)

Clark AL (1971) Stratabound copper sulfides in the Precambrian Belt Supergroup, North Idaho and Northwestern Montana. Soc Min Geol Jpn Spec Iss 3:261 – 267

Crerar DA, Barnes HL (1976) Ore solution chemistry versus solubilities of chalcopyrite and chalcocite assemblages in hydrothermal solution at 200°C to 350°C. Econ Geol 71:772 – 794

Dingess P (1976) Geology and mining operations at the Creta Copper deposit of Eagle-Pitcher Industries, Inc. In: Johnson K, Croy R (eds) Stratiform copper deposits of the midcontinent region, a symposium. Oklahoma Geol Surv Circ 77:15 – 25

Dobrovolsky EV, Lyalko VI (1979) Dynamical model of fluorite mineralization of carbonate-bearing rocks. Am J Sci 279:1022 – 1032

Eden JG van (1974) Depositional and diagenetic environment related to the sulfide mineralization, Mulfulira, Zambia. Econ Geol 69:59 – 79

Fleischer VD, Garlick WG, Haldane R (1976) Geology of the Zambian copperbelt. In: Wolf KH (ed) Handbook of strata-bound and stratiform deposits, chap 6, vol VI. Elsevier, New York, pp 223 – 352

Gandhi SS, Brown AC (1975) Cupriferous shales of the Adeline Island Formation Seal Lake Group, Labrador. Econ Geol 70:145 – 163

Gustafson LB, Williams N (1981) Sediment-hosted stratiform deposits of copper, lead, and zinc. Econ Geol 75th Anniv Vol, pp 139 – 178

Hagni R, Gann D (1976) Microscopy of copper ore at the Creta mine, Southwestern Oklahoma. In: Johnson K, Croy R (eds) Stratiform copper deposits of the midcontinent region, a symposium. Oklahoma Geol Surv Circ 77:40 – 51

Harrison JE (1974) Copper mineralization in miogeosynclinal clastics of the Belt Supergroup, Northwestern United States. In: Bartholomé P (ed) Gisements stratiformes et provinces cuprifères. Liège, Soc Geol Belg, pp 353 – 366

Helgeson HC (1968) Evaluation of irreversible reactions involving minerals and aqueous solutions, I. Thermodynamic relations. Geochim Cosmochim Acta 32:853 – 877

Helgeson HC (1979) Mass transfer among minerals and hydrothermal solutions. In: Barnes HL (ed) Geochemistry of hydrothermal ore deposits. Wiley & Sons, New York, pp 568 – 610

Helgeson HC, Delany JM, Nesbitt HW, Bird DK (1978) Summary and critique of the thermodynamic properties of rock-forming minerals. Am J Sci 278A:1 – 229

Jefferson CW (1978) Stratigraphy and sedimentology, Upper Proterozoic Redstone Copper Belt, Mackenzie Mountains, N.W.T. – A preliminary report. In: Mineral industry report for 1975, Northwest Territories, India and Northern Affairs. Econ Geol Ser, pp 1 – 13

Jung W, Knitzschke G (1976) Kupferschiefer in the German Democratic Republic with special reference to the Kupferschiefer deposit in the southeastern Harz Foreland. In: Wolf KH (ed) Handbook of strata-bound and stratiform deposits, chap 7, vol VI. Elsevier, New York, pp 353 – 406

Kuyahorn PS, Pinder GF (1984) Computational methods in subsurface flow. Academic Press, London New York, 473 pp

Laznicka P (1985) Zoning and ores. In: Wolf KH (ed) Handbook of strata-bound and stratiform deposits, vol XI. Elsevier, New York (in press)

Ljunggren P, Meyer HC (1964) The copper mineralization in the Corocoro Basin, Bolivia. Econ Geol 59:110 – 125

Lustwerk RL, Rose AW (1983) Source and segregation of transition metals during diagenetic formation of the Redstone Stratiform Copper Deposit. MacKenzie Mountains, N.W.T. (abs). Geol Soc Am (Abstracts with Programs) 15:632

Merino E (1975) Diagenesis in Tertiary sandstones from Kettleman North Dome, California-II. Interstitial solutions: distribution of aqueous species at 100°C and chemical relation to the diagenetic mineralogy. Geochim Cosmochim Acta 39:1629 – 1645

Moore C, Auchmuty G, Chadam J, Hettmer J, Larter R, Merino E, Ortoleva P, Ripley E, Sen A (1983) Kinetic modeling of flow-reaction instabilities in porous rocks: Application to uranium roll fronts. 4th Int Symp Water-Rock Interaction, Misasa, Japan, pp 352 – 355

Narasimhan TN, Witherspoon PS (1976) An integrated finite-difference method for analyzing fluid flow in porous media. Water Resources Res 12:57 – 64

Ortoleva P, Auchmuty G, Chadam J, Hettmer J, Merino E, Moore C, Ripley E (1986) Redox front propagation and banding modalities. Physica D 18 (in press)

Reed MH (1982) Calculation of multicomponent chemical equilibria and reaction processes in systems involving minerals, gases, and an aqueous phase. Geochim Cosmochim Acta 46:513 – 528

Rentzsch J (1974) The Kupferschiefer in comparison with the deposits of the Zambian Copperbelts. In: Bartholomé P (ed) Gisements stratiformes et provinces cuprifères. Liège, Soc Geol Belg, pp 403 – 426

Ripley EM, Lambert MW, Berendsen P (1980) Mineralogy and paragenesis of redbed copper mineralization in the Lower Permian of South Central Kansas. Econ Geol 75:722 – 729

Ripley E, Merino E, Moore C, Ortoleva P (1985) Mineral zoning in sediment-hosted copper deposits. In: Wolf KH (ed) Handbook of strata-bound and stratiform ore deposits, vol XIII. Elsevier, New York, pp 237 – 260

Rose AW (1976) The effect of cuprous chloride complexes in the origin of red-bed copper and related deposits. Econ Geol 71:1036 – 1048

Sharp JM (1976) Momentum and energy balance equations for compacting sediments. J Int Assoc Math Geol 8:305 – 322

Smith GE (1976) Sabkha and tidal-flat facies control of stratiform copper deposits in north Texas. In: Wolf KH (ed) Handbook of strata-bound and stratiform ore deposits, chap 8, vol VI. Elsevier, New York, pp 407 – 447

Wolery TJ (1979) Calculation of chemical equilibrium between aqueous solutions and minerals: The EQ3/6 software package. Lawrence Livermore National Laboratory, Livermore, California. Publ UCRL-52658

Woodward LA, Kaufman WH, Schumacher OL, Talbott LW (1974) Stratabound copper deposits in Triassic sandstone of Sierra Nacimiento, New Mexico. Econ Geol 69:108 – 120

Results of Recent Exploration for Copper-Silver Deposits in the Kupferschiefer of West Germany

F.-P. SCHMIDT[1], C. SCHUMACHER[1], V. SPIETH[2], and G. FRIEDRICH[3]

Abstract

In two areas of the Werra-Fulda trough (Hessia, West Germany) exploration of Kupferschiefer-type copper-silver deposits was carried out by St. Joe Explorations GmbH Western Europe.

Anomalous high copper-silver contents were detected in the Spessart-Rhön area as well as in the Richelsdorf area. Especially in the Richelsdorf area all features connected with the occurrence of economic mineralization are similar to the deposits in Poland.

The relationship between grade of mineralization, spatial distribution of metals and minerals, paleogeography and the occurrence of "Rote Fäule" in two St. Joe drill holes and two holes drilled by the BGR (Bundesanstalt für Geowissenschaften und Rohstoffe, Hannover) is discussed.

Introduction

The Central European Kupferschiefer sea covered an area which today is located in West Germany, East Germany and Poland.

Economic copper-silver deposits occur in the Lower Zechstein strata of all three countries.

The largest ore deposits are in the Lubin district in Poland. At Lubin 1 000 million tons of ore are located in the Fore-Sudetic Monocline at the boundary of a Rotliegend basin to a Variscian basement high. The mineralization is concentrated in the Kupferschiefer itself as well as in the underlying Weissliegend and the overlying Zechsteinkalk.

Since 1978 St. Joe Explorations GmbH has conducted an exploration program which is focused on the discovery of a Polish-type, copper-silver deposit in the Lower Zechstein strata of West Germany. This work is done in joint venture with Preussag AG Metall, Goslar and is supported by the Government of the Federal Republic of Germany (BMWi).

[1] St. Joe Explorations GmbH, Georgstraße 50, 3000 Hannover 1, FRG
[2] Gewerkschaft Wilhelm, Hohenzollernstraße 2, 3000 Hannover 1, FRG
[3] Institut für Mineralogie und Lagerstättenlehre der RWTH, Wüllnerstraße 2, 5100 Aachen, FRG

Geology and Metallogeny of Copper Deposits
(ed. by G.H. Friedrich et al.)

Investigations of the parameters indicating a local 'epi(dia)genetic' copper-silver enrichment in the Lubin area of Poland and application of the results to previously untested areas in West Germany have led to the identification of several target districts.

In two target districts of Spessart-Rhön and Richelsdorf 60 drill holes have been sunk to depths between 150 m and 780 m. The investigations of the intersected strata have improved the knowledge of the paleogeography, the Variscian basement, the Rotliegend, and the mineralized Lower Zechstein sequence of the Weissliegend (S 1), Kupferschiefer (T 1) and Zechsteinkalk (Ca 1) in the Werra-Fulda trough.

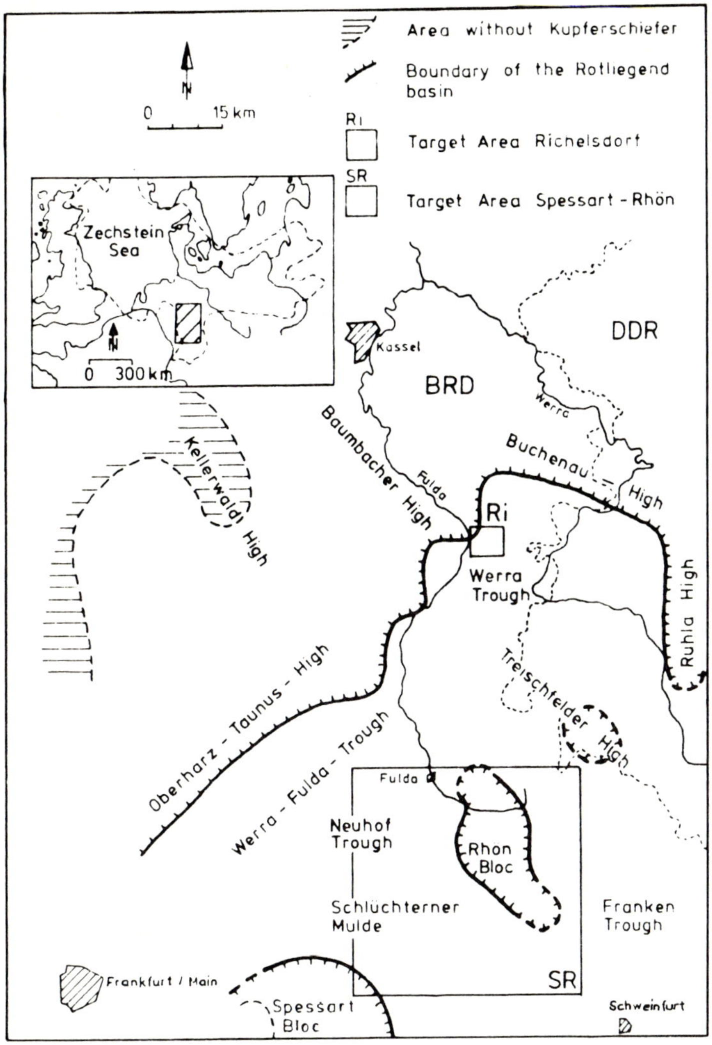

Fig. 1. Target ares for exploration on Kupferschiefer-type deposits in West Germany

Target Areas

Two main features are characteristic for Kupferschiefer-type deposits.

(1) The geological setting at the boundary between the Rotliegend troughs and the basement highs and (2) the occurrence of red-coloured, oxidized rocks, called "Rote Fäule" in the vicinity of the ore deposits.

The formation of copper-silver deposits related to Rote Fäule is interpreted as an 'epi(dia)genetic' enrichment according to Rentzsch (1974) and Rydzewski (1978).

In addition, the relationship to growth faults has been presumed by Ekiert (1960).

The Richelsdorf area and the Spessart-Rhön area which cover parts of the Werra-Fulda trough were selected as favourable sites for exploration by St. Joe Explorations GmbH (Fig. 1).

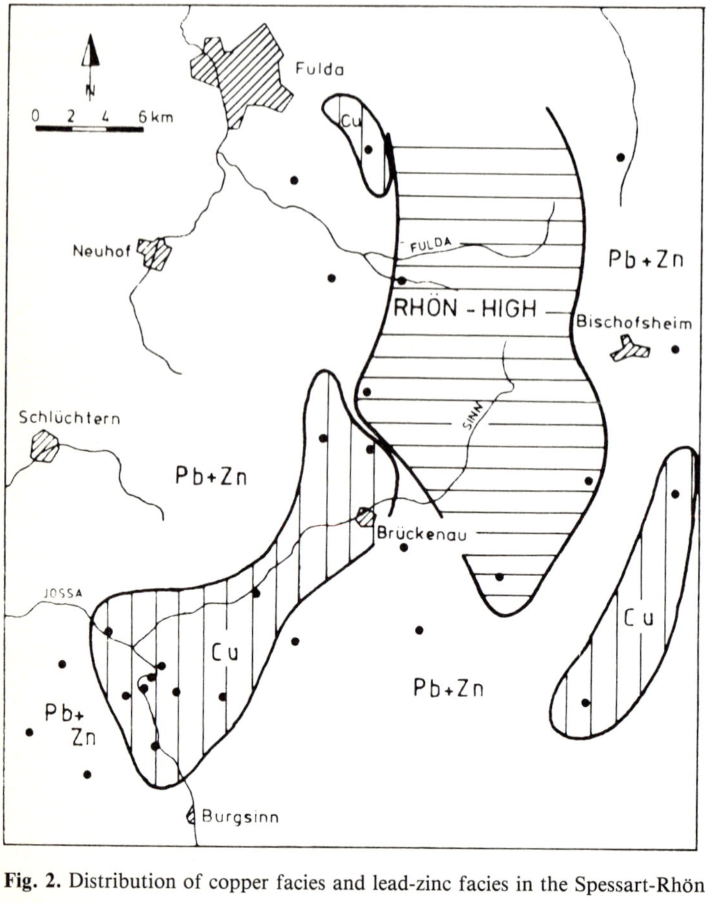

Fig. 2. Distribution of copper facies and lead-zinc facies in the Spessart-Rhön area

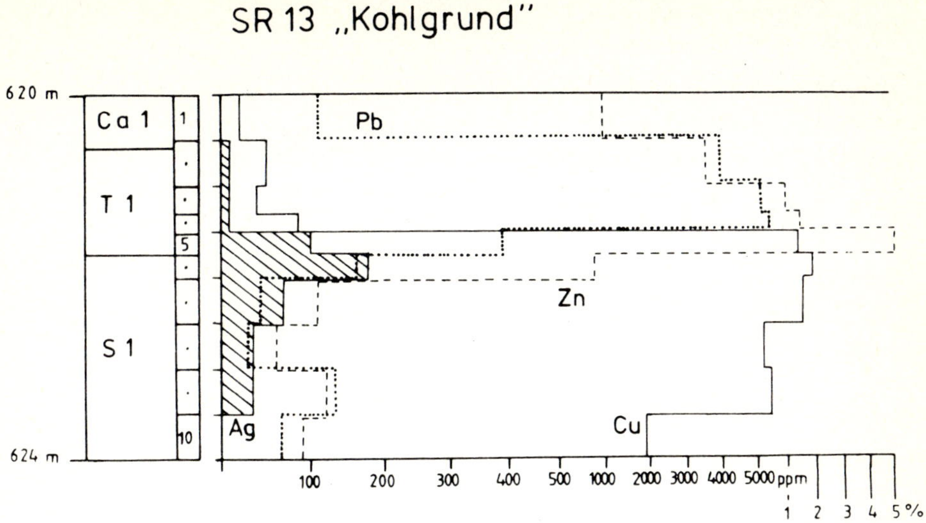

Fig. 3. Distribution of copper, lead, zinc and silver in the vertical section of drill hole SR 13

Spessart-Rhön Area

In the Spessart-Rhön area only few drill data were available. A recent publication by Schumacher et al. (1984) based on a large amount of new drill data gave rise to new ideas about the regional geological setting in this area and focused on the distribution of the boundary between the Rotliegend sediments and folded basement rocks.

The main structural element is the Rhön high, where an adjacent zone of copper predominance was discovered extending over an area of about 200 km^2 (Fig. 2).

The highest metal contents are 1.1% copper plus 70 ppm silver calculated over a suitable mining height of 2 m in the northern Rhön area.

Copper-arsenic sulphides and arsenides are abundant. In the economically interesting zone the minerals bornite, chalcocite and stromeyerite occur.

The Spessart-Rhön area is characterized by anomalous silver contents, caused by the presence of a tennantite phase rich in silver related to Saxon tectonics (Friedrich et al. 1984). In drill hole SR 13 the distribution of copper, lead, zinc and silver (Fig. 3) and the distribution of the ore minerals (Fig. 4) in the economical copper zone can be demonstrated.

Richelsdorf Area

In the Richelsdorf area metal enrichment is connected with the presence of Rote Fäule which represents the oxidized portion of a redox boundary. The Rote Fäule

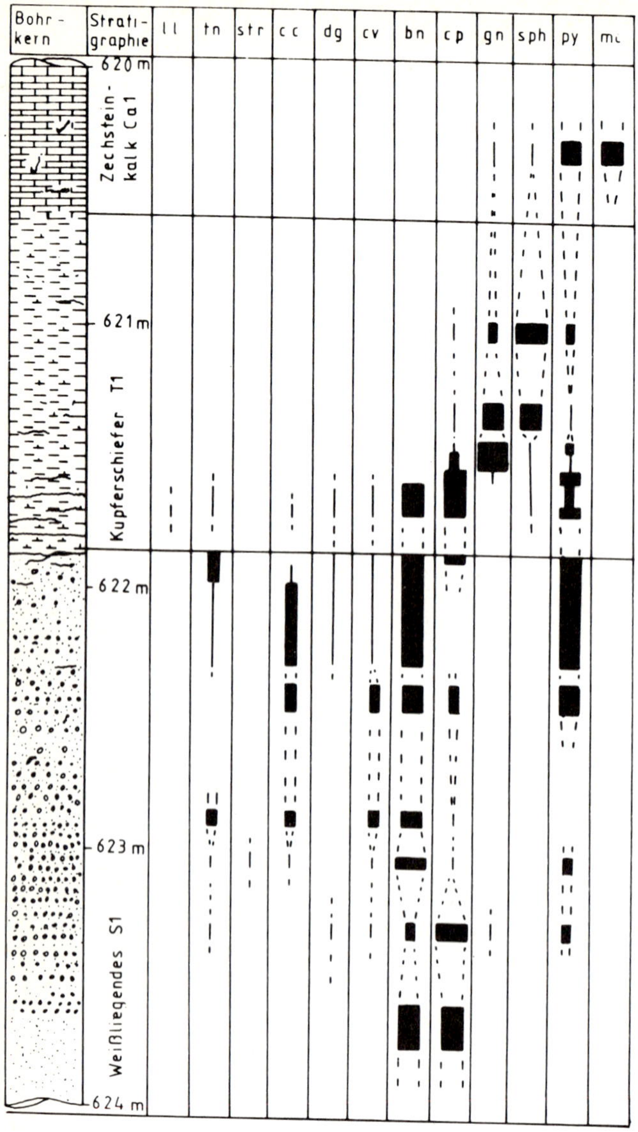

Fig. 4. Distribution of ore minerals in vertical section of drill hole SR 13. Abbreviations: *ll* löllingite; *tn* tennantite; *str* stromeyerite; *cc* chalcocite; *dg* digenite; *bn* bornite; *cp* chalcopyrite; *gn* galena; *sph* sphalerite; *py* pyrite; *mc* marcasite

contains hematite and iron-hydroxides. The main structural element is the Baumbacher high, where no Rotliegend sediments were deposited. The replacement of sulphides by hematite indicates an 'epi(dia)genetic' nature of the Rote Fäule, which extends into the reduced facies of the basal Zechstein strata formed under synsedimentary conditions.

Due to the changing redox conditions, a zonation of Rote Fäule facies, copper facies and lead-zinc facies occurs.

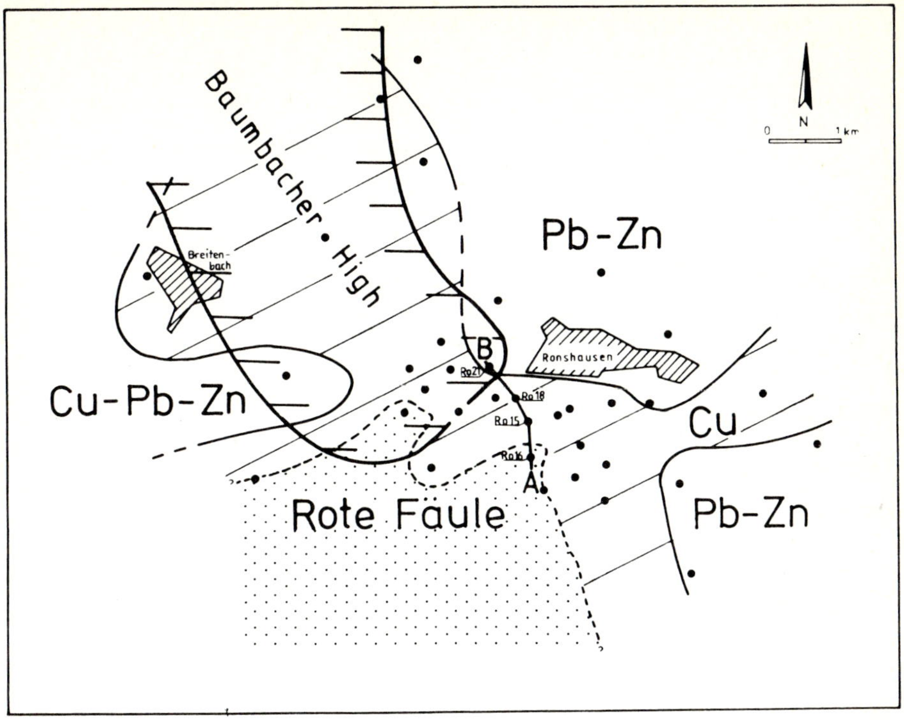

Fig. 5. Distribution of Rote Fäule facies, copper facies and lead-zinc facies in the Richelsdorf area

In the Richelsdorf area a Rote Fäule zone has been discovered in a number of widespread drill holes. At the northern margin of the Rote Fäule zone, copper and silver is concentrated. In several drill holes the mineralization was found to be of commercially interesting grade. The high-grade copper-silver ore zone follows the Rote Fäule margin in a sinous manner followed by a lead-zinc zone (Fig. 5).

Several copper minerals are formed depending on the abundance of copper. These are chalcocite, bornite, covellite and digenite, indicating a decreasing Cu/S ratio in a direction away from the Rote Fäule facies towards the lead-zinc facies (Fig. 6).

The economic ore zone is composed of chalcocite, bornite, covellite and digenite, whereas bornite and chalcopyrite are also found in the lead-zinc zone. The thickness of the mineralization is less than 2 m and crosscuts the strata at a low angle. This is the result of the downdipping of the redox boundary at a shallow angle to the north.

The relationship between paleoenvironment and metal distribution is demonstrated by the spatial distribution of the sandbars in the Kupferschiefer sea and the position of the Rote Fäule facies. Copper is concentrated in the slightly reduced environment of the interfingering facies zone between the sandbars and the open marine Kupferschiefer facies (Fig. 7).

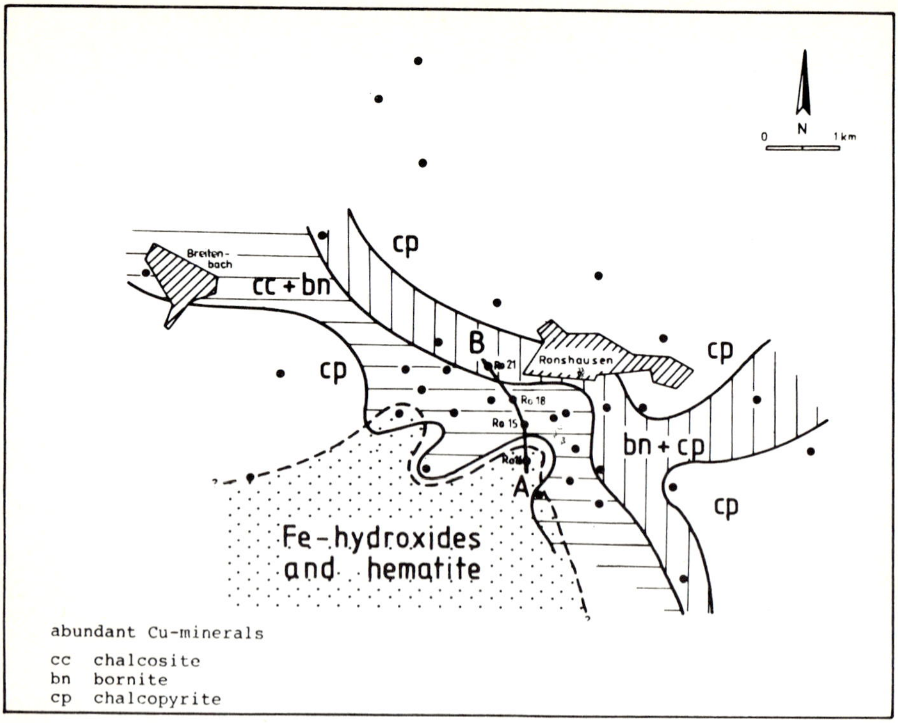

Fig. 6. Distribution of iron-hydroxides, hematite and copper minerals in the Richelsdorf area

The metal distribution is demonstrated in a south-north cross-section by drill holes Ro 16, Ro 15, Ro 18 and Ro 21 (Fig. 8).

The paleotopography is governed by prominent sandbars of the Weissliegend. On the top of the sandbars the Kupferschiefer is considerably reduced as shown by drill hole Ro 16. Here the thickness of the Kupferschiefer horizon is 0.06 m. The sandbar corresponds with the Rote Fäule zone and in the drill hole only 0.01% Cu with 5 g t^{-1} Ag over 2-m-mining height is present. Rote Fäule occurs in the Zechsteinkalk, the Kupferschiefer and the Weissliegend. In the copper facies close to the Rote Fäule zone, the copper content increases rapidly. In drill hole Ro 15 it is 1.3% Cu with 10 g t^{-1} Ag over 2-m-mining height.

The main copper mineral is chalcocite and the thickness of the Kupferschiefer is 0.62 m.

Drill hole Ro 15 represents the marginal position between the oxidized Rote Fäule facies and the reduced copper facies where drill hole Ro 16 is located.

In drill hole Ro 18 the copper content is of economic grade containing 2.8% Cu plus 26 g t^{-1} Ag over 2-m-mining height. The thickness of the Kupferschiefer is 1.9 m and contains a high portion of sand as a result of the vicinity to the sandbar. Abundant copper minerals are chalcocite, bornite, covellite and digenite which show various replacement features.

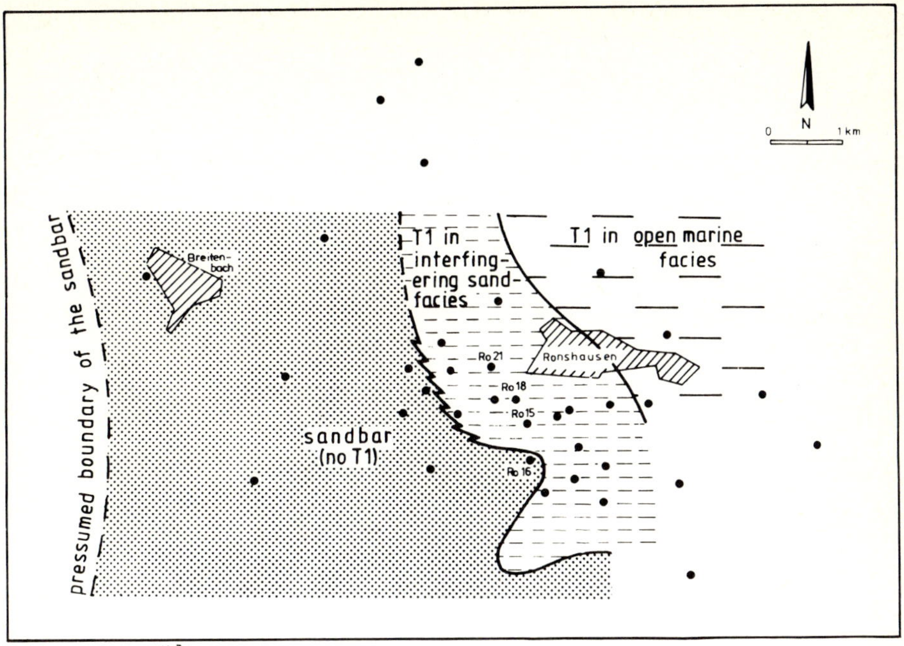

Fig. 7. The paleomicrorelief and facies types of the Kupferschiefer in the Richelsdorf area

The lead-zinc content increases slowly towards the hanging wall, where it is concentrated within the Zechsteinkalk. Drill hole Ro 21 represents the lead-zinc facies with a predominance of lead and zinc already within the Kupferschiefer. Thus, only 0.3% Cu with 1 g t^{-1} Ag over 2-m-mining height were found. The Kupferschiefer is 2 m in thickness, copper minerals are tennantite, bornite and chalcopyrite. Galena and sphalerite are the most abundant ore minerals. Generally, the lead-zinc zone is formed in a more strongly reduced environment than the copper zone.

The same geochemical transition from oxidized to strongly reduced environment is represented by the mineral zonation (Fig. 9) which starts with hematite and iron-hydroxides in the Rote Fäule zone, followed by the dominance of chalcocite and bornite in the slightly reduced copper zone and galena and sphalerite in the more strongly reduced lead-zinc zone.

Detailed investigations focussing on the relation between ore grade, metal distribution and mineral association with respect to the distribution of the Rote Fäule (Schmidt, 1985) showed that only through the interplay of primary synsedimentary features (e.g. organic carbon content and porosity of the host rock) and secondary 'epi(dia)genetic' features (e.g. influence of the Rote Fäule-forming solutions) economic mineralization was formed.

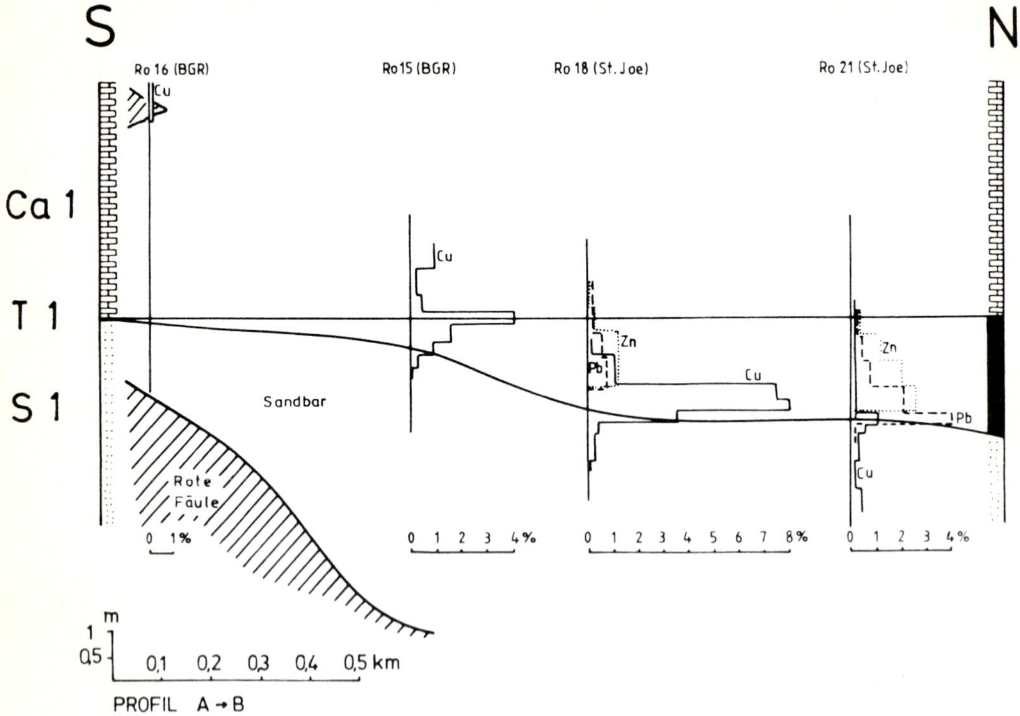

Fig. 8. Change of metal distribution in drill holes Ro 16, Ro 15, Ro 18 and Ro 21 in relation to the position of Rote Fäule

Conclusions

As mentioned before a post-sedimentary mobilization and deposition of metals in the host rocks is inferred which is in agreement with the genetic interpretations of the Kupferschiefer deposits in East Germany and in Poland.

The Rote Fäule facies is formed in the vicinity of sandbars which were shoals in the Kupferschiefer sea and consequently controlled the paleotopography. On the slope of the sandbars the Kupferschiefer contains a high amount of quartz-sand interlayers which cause a high thickness, whereas on the exposed centres of the sandbars the Kupferschiefer had not been deposited.

According to Rentzsch (1981) it must be emphasized that the Rote Fäule is not the common littoral facies of the Kupferschiefer. Where the Kupferschiefer lies directly over the folded basement, the Rote Fäule facies never occurs in coastal areas of the sea.

The Rotliegend red-bed sediments underlying the Kupferschiefer seem to be favourable source rocks of the metals concentrated in the Kupferschiefer (Lurje 1977). The metals were probably leached by chloride brines percolating through the Rotliegend sediments.

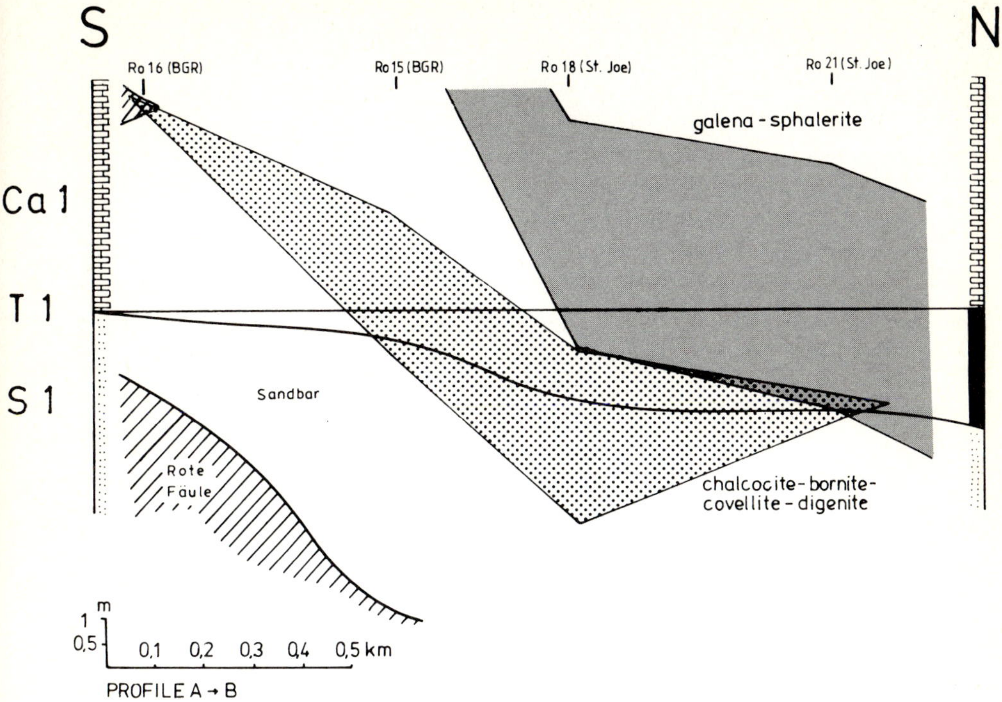

Fig. 9. Change of mineral association in relation to the position of Rote Fäule (see also Fig. 6)

As shown by Rose (1976) the metals could have been transported as chlorides like $CuCl_2^-$, $CuCl_3^{2-}$, $PbCl_3^-$ or $PbCl_4^{2-}$. The migration of the chloride brines was probably caused by increased subsidence in the centres of the sedimentary basins which gave rise to the accumulation of huge piles of sediments. The pressure of the sedimentary overburden moved the solutions enriched in metals towards the marginal parts of the basins. In the sandbar areas, where the Kupferschiefer is not present, or replaced by permeable sand-rich interlayers, the solutions spread out. On the outer flanks of the sandbars, where the Kupferschiefer was formed as an impermeable and reducing black shale, the metals were trapped and concentrated (Fig. 10).

The geological model used for exploration of Kupferschiefer-type copper-silver deposits was confirmed. Studies of controlling parameters, such as paleogeography, metal zonation, mineral zonation and the occurrence of Rote Fäule are important exploration tools on Kupferschiefer-hosted copper-silver mineralization.

Based on this tools, several holes drilled in the Spessart-Rhön and Richelsdorf area have encountered economic ore grades.

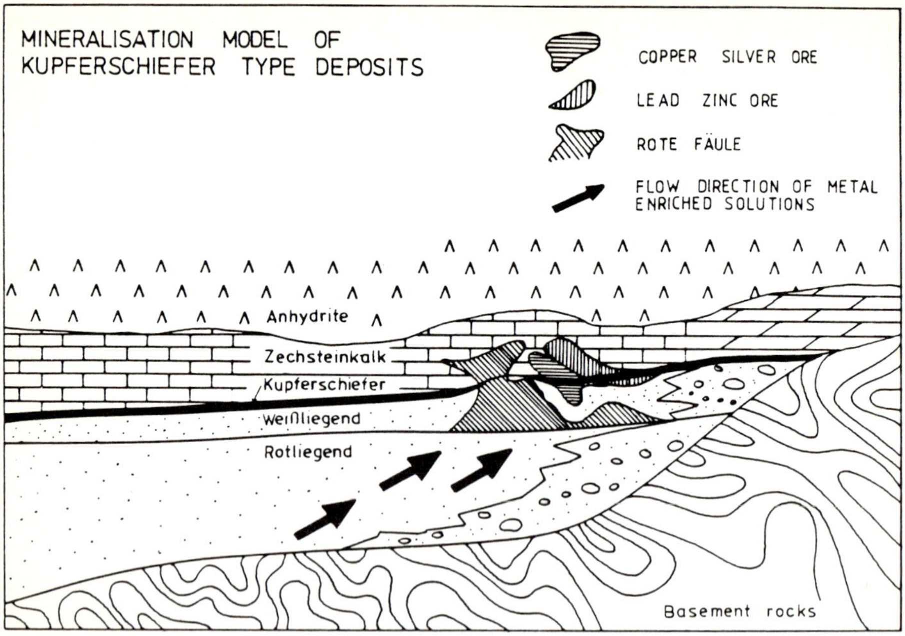

Fig. 10. Model of the formation of Rote Fäule controlled and epi(dia)genetically-formed Kupferschiefer-type deposits

Acknowledgements. The authors wish to thank Mr. T. N. Walthier, Vice President of Exploration, St. Joe Minerals Corp., USA, for his permission to publish the first results of the ongoing exploration campaign in West Germany. Our thanks are also due to Prof. F. Bender, Bundesanstalt für Geowissenschaften und Rohstoffe (BGR), Hannover, for making data available from drill holes Ro 15 and Ro 16.

References

Bundesanstalt für Geowissenschaften und Rohstoffe (BGR) (1983) Bundesbohrprogramm, Teilprojekt Kupferschiefer. Forschungsvorhaben NTS 30043 Abschlußbericht, Hannover

Ekiert F (1960) Neue Anschauungen über die Herkunft des in den Sedimenten des unteren Zechsteins auftretenden Kupfers. Freiberger Forschungsh C 79:120–201

Friedrich G, Diedel R, Schmidt F-P, Schumacher C (1984) Untersuchungen an Cu–As-Sulfiden und Arseniden des basalen Zechsteins der Gebiete Spessart/Rhön und Richelsdorf. Fortschr Miner 62, Beih 1:63–65

Lurje AM (1977) Zur Herkunft des Kupfers in den Basisschichten des Zechsteins und der Kasanstufe. Zeitschr Angew Geol 23:270–274

Rentzsch J (1974) The Kupferschiefer in comparison with the deposits of the Zambian copper belt. Centrenaire de la société géologique de Belgique gisements stratiformes et provinces de cuprifères. Liège, pp 395–418

Rentzsch J (1981) Mineralogical-geochemical prospection methods in the Central-European Copper Belt. Erzmetall 34, 9:492–495

Rose AW (1976) The effect of cuprous chloride complexes in the origin of red-bed and related deposits. Econ Geol 71:1036–1048

Rydzewski A (1978) Oxidated facies of the copper-bearing Zechstein shales in the Fore-Sudetic monocline. Prz Geol 26:102–108

Schmidt F-P (1985) „Rote Fäule“ kontrollierte Buntmetallvererzungen im Kupferschiefer Osthessens, BRD. Diss TH Aachen

Schumacher C, Kaidies E, Schmidt F-P (1984) Ergebnisse über den basalen Zechstein der Spessart/Rhön-Schwelle. Z d g G 135/2

Subject Index